The Forest Management Digest

ninth edition

Order Number: KAP22004

Editor—H. Allen Wickman, Department of Natural Resources, Forestry, Park Rapids, Minnesota
Consulting Editor—Brian I. Ingvalson, Ph.D., Ingvalson and Associates, Inc.

Ninth Edition, First Printing

HOBAR PUBLICATIONS
A DIVISION OF FINNEY COMPANY
3943 Meadowbrook Road
Minneapolis, MN 55426-4505
Phone: (800) 846-7027
www.finney-hobar.com

ISBN 9780913163368

About The Editor

Allen Wickman recently retired from a successful career as a Private Forest Management Specialist with the Minnesota Department of Natural Resources. He was last employed at the Park Rapids, Minnesota DNR Station.

Though he grew up in Minneapolis, he has had a life-long interest in forestry, wildlife, and nature in general. He also is an avid nature and wildlife photographer. Al earned a B. S. Degree in Forest Resource Management from the University of Minnesota in 1965.

Al not only worked in the forests, but works with its products as well. He and his wife, Renee, have a woodcraft business, specializing in hand-carved loons.

As a field forester for over 30 years, he always maintained a strong desire to share what he had learned about forestry, wildlife management and conservation.

For Al, editing was not a job - it was a hobby. Thank you Mr. Wickman from the thousands who have shared the products of your labor - THIS BOOK.

TABLE OF CONTENTS

CHAPTER 1 — HISTORY OF MINNESOTA'S FORESTS

Minn. 4-H Bulletin No. 87 and (or)
Forestry For Minnesota Schools — North Print Co.

The overall history of our forests in Minnesota is similar to the general history of forests throughout the United States. Our forests have been exploited, reconstructed, and now are managed in order to furnish our generation and future generations with products and services which only the forests can provide.

A study of Minnesota's forests must begin with an explanation of Minnesota's forest regions. Then we will discuss the past and present utilization of our forests and the various forest types. We conclude with a discussion of what the forests of Minnesota produce and what the future holds for the forests of our state.

MINNESOTA'S FOREST REGIONS

The forests of Minnesota originated after the glaciers retreated some ten to twelve thousand years ago. These great ice sheets left soils and topographic features which determined to a large extent the type of vegetation that later grew on them. With the warming of the climate after the glacial period, plants gradually built up on these soils until by the time that man arrived on the scene, there were three distinct natural vegetational regions: the evergreen forest in the northeast, the hardwood forest of the east central and southeast, and the prairies in the west and southwest of Minnesota (see Figure 3-1).

Evergreen Forest

The northeastern evergreen forest was the largest region of continuous forest. In the original forest grew the white, Norway, and jack pines; black and white spruce; balsam fir, tamarack, northern white cedar; and some broadleaf trees (notably the aspens and paper birch) which grow with conifers. These trees did not grow in one grand mixture but tended to occur in definite communities determined by soil

Photo Minn. D.N.R.

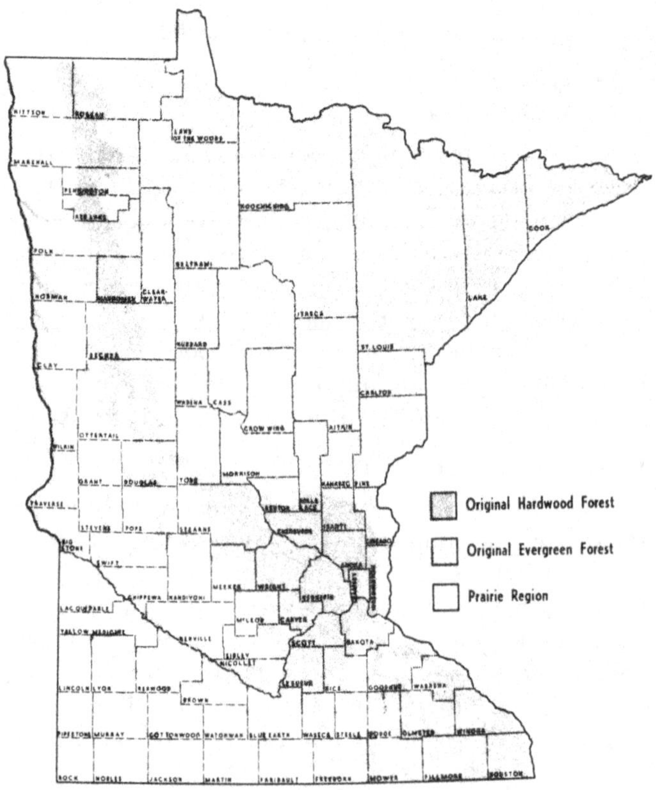

Figure 3-1. Minnesota's three natural vegetational regions

and moisture conditions and by the natural fire history of the area.

After the early cutting of these evergreen forests, particularly the pines, many forest fires swept this cutover region. These fires destroyed seed trees and young conifers and encouraged the reproduction of species such as aspen, birch, and oak which can reproduce by sprouting from the stump or suckering from the roots. Today most of the evergreen forest is a mixture of aspen and birch, and in places, oaks and other hardwoods, together with young stands of pine, spruce, and fir wherever fires and logging did not completely eliminate these trees or where new forests have been planted (see Figure 3-2).

Some conifers such as balsam fir are now gradually seeding in beneath aspen and birch forests. Balsam can withstand the shade and competition of these trees; and if fire is kept out, part of the second-growth aspen-birch forests will in time be replaced by fir and spruce. Unfortunately, the pines are not successful invaders of the less valuable aspen and birch stands because they do not grow well in shade.

Hardwood Forest

The original hardwood forest region of the southeast and east central portion of Minnesota consisted of stands of mixed hardwood species such as oak, elm, ash, black walnut, basswood, butternut, Kentucky coffeetree, maple, cottonwood, willow, aspen, and many others. These trees grew on soils which were capable of growing agricultural crops.

As the white man homesteaded into Minnesota, much of the hardwood forests of southeastern and east central Minnesota were cleared for agriculture. During the period from 1820 to about 1920, many of the trees of the original forests were used in building homestead and farm buildings. In many instances, the trees which were not used were simply felled, gathered into piles, and burned.

Today, remnants of this hardwood forest still exist along the rivers and in many small woodlands. The same types of trees are found today as originally covered the area, but the more valuable species such as black walnut and oak are not nearly as common as they once were.

Prairie Region

In the prairie region of western and southwestern Minnesota, the original forests occurred along the valleys and flood plains of rivers and streams. The species of trees found here consisted of willow, cottonwood, ash, boxelder, elm, and occasionally oak, maple, basswood, or other hardwoods. This area of Minnesota receives less moisture than the evergreen and hardwood forest regions, which is the main reason why all the land in western and southwestern Minnesota was not covered by forests. There was very little activity in cutting of trees in this area, and the present forest resembles the original forest for the most part.

PAST AND PRESENT UTILIZATION OF OUR FORESTS

The forests of Minnesota were utilized by the Indians long before the white man discovered Minnesota. The forest was a place for shelter; source of many forest products, such as birchbark, poles, firewood, foods (maple syrup, wild plums, berries, etc), and medicines; and a place to hunt game. However, the forest was also a hindrance to Indians and a record of fires set by them is found in our forests' history. These fires opened the forest, increased browse for game, and made better hunting grounds.

With the coming of the white man the forests were exploited for timber products and to develop agricultural lands. The first large-scale cutting of trees was probably by army troops stationed at Fort Snelling, which was the site of the state's first sawmill constructed in 1821. The first commercial sawmill was built in 1837 at Marine-on-the-St. Croix. The white pine lumber industry attained its full development in Minnesota from the period of 1890 to 1930,

1909 – Before Logging **1913 – Logged** **1960 – Same Area 47 Years Later**

Figure 3-2. A stand of Norway pine in northeastern Minnesota

and for several years Minneapolis-St. Paul was the largest sawmill center in the United States.

Wood was in great demand for all types of construction and for fuelwood so the cutting of trees was encouraged. Agriculturists expected that practically all of Minnesota would "go under the plow" and that farming would be possible in all areas of the state. Again, to clear the land for cultivation, logging the forest was encouraged. After logging, the stumps and logging debris were burned. This burning to clear lands for agriculture led to some very disastrous wild fires.

These fires are a very vital part of the history of forestry in Minnesota. In 1894, the Hinckley Fire, which covered a large area in Pine County, resulted in the loss of 418 lives. The Chisholm Fire in 1908 and the Baudette and Spooner Fires in 1910 swept vast acreages of northern Minnesota and caused much loss and suffering. The last and most notable forest fire in Minnesota was in the Cloquet-Moose Lake area in 1918. It caused the death of 438 people.

The logging of Minnesota's pine forests and the attempts to farm what was not very productive land continued until about 1930. By that time most of the original forests in northeastern Minnesota had been harvested and only a few remnants of the virgin pine forests were spared from the axe and land-clearing fires. One of these remnants became our first state park in 1891 when Itasca State Park was established. The retention of this beautiful area of old-growth virgin white and Norway pine was a culmination of cooperative efforts by state agencies, university professors, and the forest industries. This park which includes the headwaters of the Mississippi is today one of the finest parks for its size in our nation. It is also a reminder to us of the original forest of northern Minnesota.

Minnesota now has 13 million acres of forest land considered fit for raising commercial tree crops.

About 52 percent of the forests are managed by public agencies of the federal, state, and county governments; farmers own 24 percent; other small owners own 20 percent; and 4 percent is owned and managed by our forest industries.

Commercial forest land in Minnesota is being reduced at the rate of 114,000 acres per year.

CHAPTER 2 — HOW A TREE GROWS

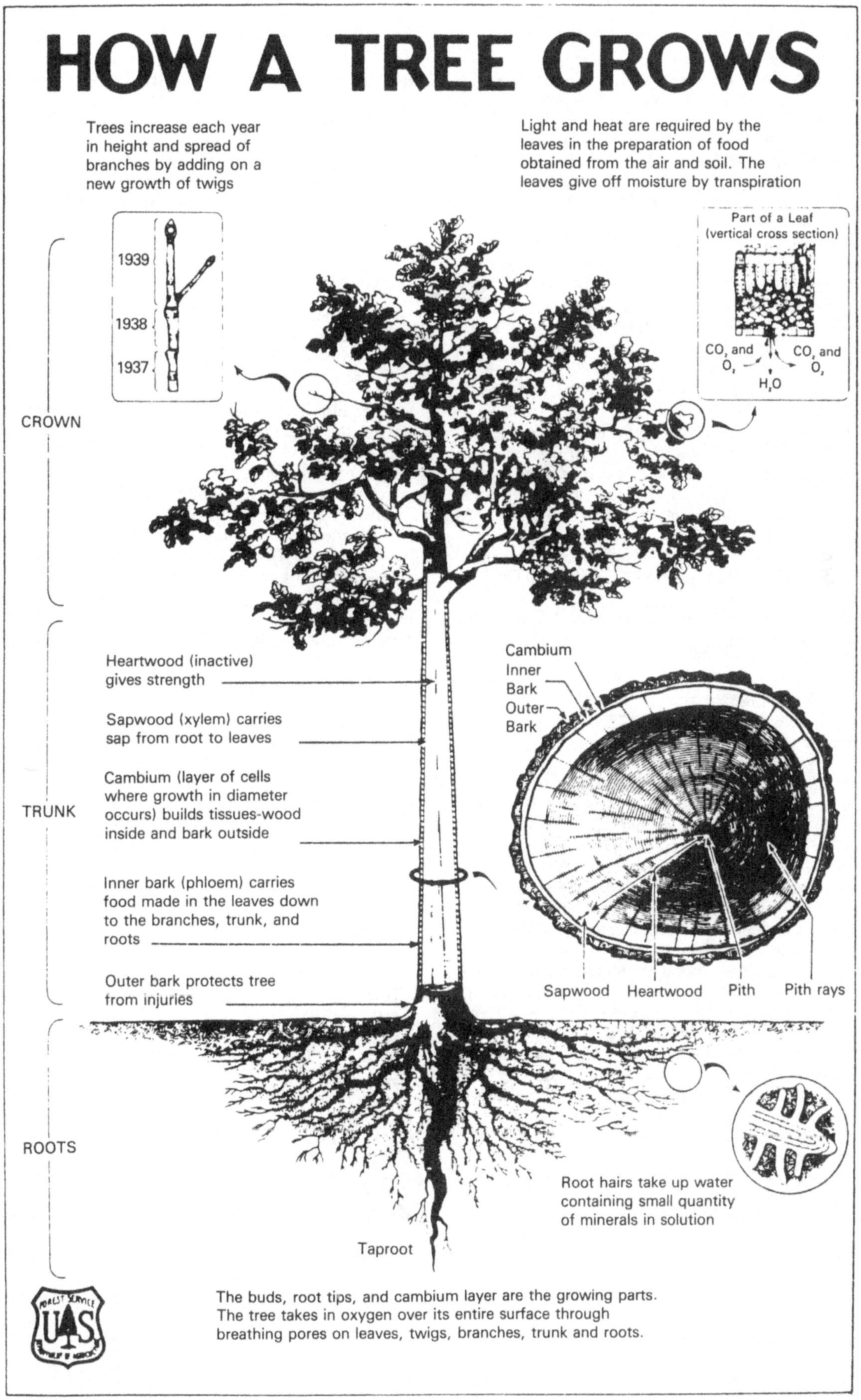

TREES AND THE FOREST MIRACLE

William R. Miles

Trees! Minnesota is blessed with an astonishing variety, as different as the parkland aspen in the Red River Valley, the towering pine along Lake Superior's North Shore, the precious black walnut on southeastern limestone bluffs, the green ash and spruce of western shelterbelts, and the majestic elm and red oak in towns like Winona and St. Cloud.

In Minnesota, trees are our constant companions. But few of us appreciate what a miracle of natural engineering a tree is, or what a starring role it plays in a forest community of other living and non-living things.

What Is A Tree?

A tree is a woody plant at least 20 feet tall when fully grown. It usually has only one stem—or trunk—with branches at the top. (Woody plants that have several stems and grow less than 20 feet tall are shrubs, not trees.) Trees come in different sizes and shapes. Some grow short and squatty; others tall and slim.

There are two main types of trees: conifers (softwood), and broadleafs (hardwoods).

Conifers bear seeds in cones. Spruce, pine, cedar, tamarack, and hemlock are common conifers. Conifers also have slender, pointed, needle-like leaves or overlapping scales. They are called evergreens because most keep their green leaves all year. In Minnesota, the tamarack is the only conifer that sheds its leaves in winter.

Hardwoods have broad leaves. Because most hardwoods lose their leaves in fall, they are called deciduous. They do not bear cones.

The terms "softwood" and "hardwood" are really misleading. Some conifers have harder wood than some hardwoods.

How Does A Tree Grow?

"Tall trees from little acorns grow." Every tree was once a seed—a tiny storehouse of energy—made by an older tree.

Nature programs the new seed for survival. To leave the parent tree and reach fertile soil, some seeds drift on the wind, others float on water, and still others attach themselves to the fur of animals who carry them long distances. Birds swallow some seeds, then release them in their droppings.

Once deposited in moist soil, the seed—fueled by the food stored in its hard shell—starts to grow. First, a small sprout—a tiny root—pierces the shell and begins to seek nourishment in the soil. Then another sprout, the stem, breaks out, seeking light. Sunlight, minerals, and water help the little tree manufacture its own food and grow taller.

The branches and leaves are called the crown. The leaves are tiny "factories." Using water absorbed by the roots and carbon dioxide from the air, they manufacture food. These leaf factories take their energy from the sun. The accompanying figure shows the process of photosynthesis—putting the sun's energy to work.

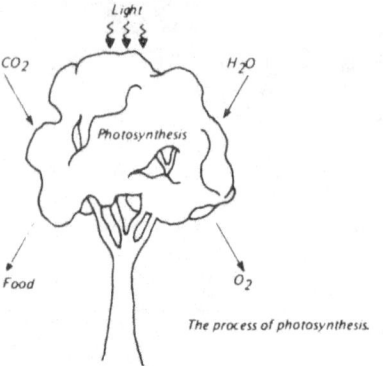

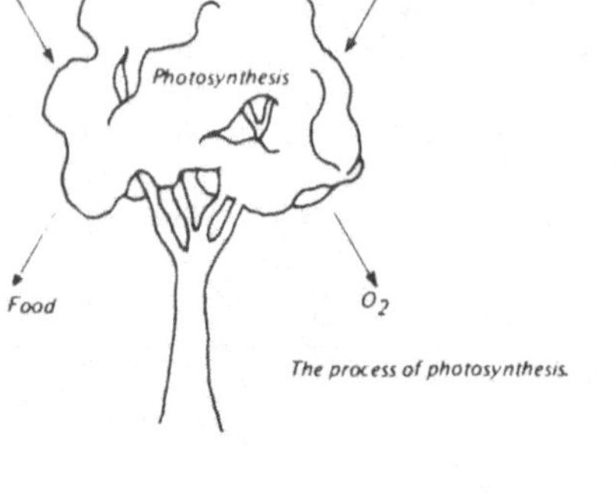

The process of photosynthesis.

After the leaves manufacture the food the tree needs for growth—primarily plant starches and sugars—it moves down the trunk to the roots to be used for growth or stored away for future use.

The trunk provides strength, supports the crown, and holds the pipelines for food-fluids to move up and down the tree and through the branches. These pipelines are made up of many small sections called cells. They fit end to end and form long tubes. These cells are so tiny they can be seen only through a magnifying glass. A toothpick contains millions of cells.

Roots anchor the tree to the ground so it can stand erect in strong winds. Roots absorb water and nutrients—dissolved minerals and nitrogen—which the tree used to make food.

A tree grows in three ways:

- Up, from the buds on the tips of the trunk and branches
- Down, from the root tips
- Around, by putting layers of wood around limbs, roots, and trunk.

The buds on the branches cause each limb to grow, making the tree taller. The limbs also spread to receive more sunlight.

The root tips cause roots to grow longer and spread in search of more water and nutrients.

The trunk of a tree never grows upward. If you drive a nail into the trunk of a young tree, it will always remain in the same place. To see how the trunk grows, examine the cross section of a stump or log.

The covering on the outside of the trunk is called the outer bark. It consists of dead cells and serves only to protect the vital interior. Just inside the outer bark is a thin, light-colored layer called the inner bark. The cells in the inner bark carry food, made in the leaves, to the branches, stem and roots.

Next comes a living, growing layer, the cambium, which is a thin, single layer of cells just inside the inner bark and invisible to the naked eye. During warm months, the cells in the cambium divide to form new cells. Cells produced toward the outside of the tree form bark; those produced toward the inside form wood. Cambium is also present in the limbs and roots and allows them to grow.

Now comes a wide band of light-colored wood called sapwood, a part of the pipeline system that carries water and food from the roots to the crown. The dark, inactive wood in the center of the trunk is heartwood. It provides

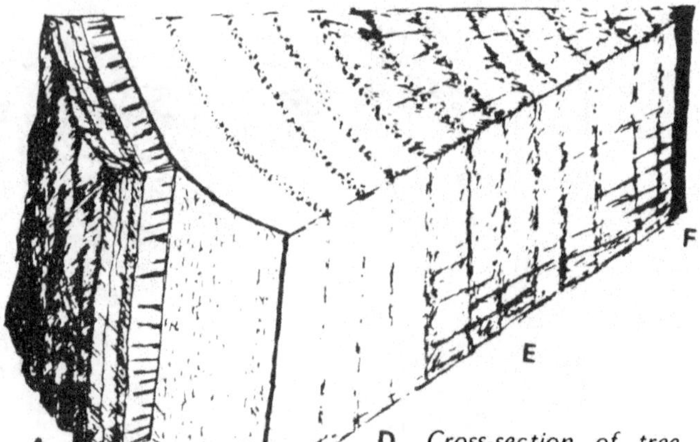

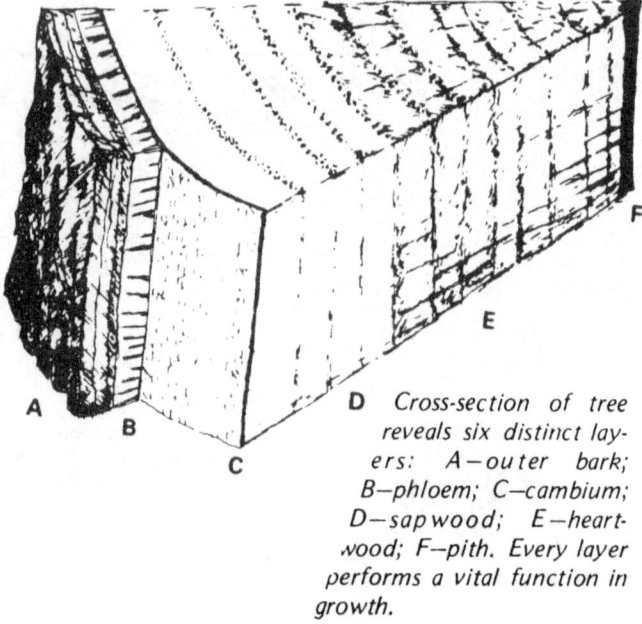

Cross-section of tree reveals six distinct layers: A—outer bark; B—phloem; C—cambium; D—sapwood; E—heartwood; F—pith. Every layer performs a vital function in growth.

support and strength to the tree. The trunk core is called the pith.

Each year, the cambium makes a new layer of wood for the trunk. Each layer is made up of two bands: light wood called spring wood and dark wood called summer wood. Usually the spring wood band is wider than the summer wood band. These rings, formed by the layers of wood, are called annual rings or growth rings.

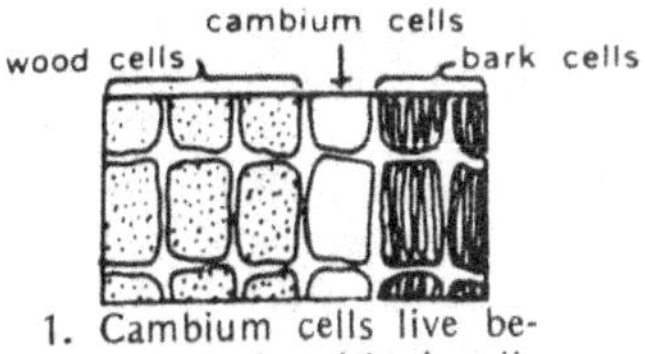

1. Cambium cells live between wood and bark cells and are visible only with a magnifying glass.

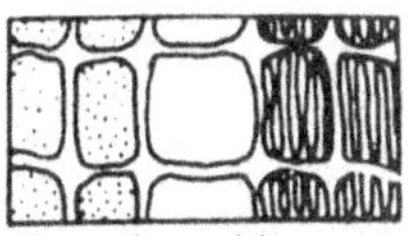

2. Growth widens tree trunks, limbs, and roots.

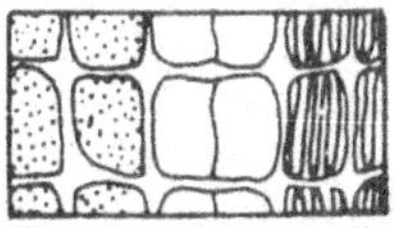

3. Then cambium cells divide, forming wood cells toward the center of the tree and bark cells toward the outside.

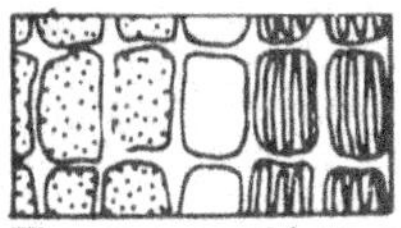

4. The new cambium cells begin the growth process again.

When A Tree Dies

Trees, like humans, are born, grow to maturity, and die. After they die, they decay, contribute minerals to the soil and create room for new plants to grow. This is part of the dynamic process called ecology.

A rotten log or decaying stump offers one of the best ways to study the ecology of a forest. What happens at the log—the competition and cooperation among plants, animals, and the influence of other factors in the environment like moisture, air, temperature, and light—reflects in a small way larger forces at work in the entire forest.

Shade-tolerant plants such as sugar maple, balsam fir, hazel, and dogwood, for example, sprout and grow on rotten logs and humus. These plants provide a special service as they grow: They help the log decay back into the soil.

By observing these same processes in the larger forest, we can understand how to manage and use a forest.

What Is A Forest?

A forest is a living association of plants and animals in which trees are the dominant species. From a distance, we see the upper layer of green crowns, called the canopy; the forest appears to consist only of big trees.

As we draw nearer, however, we see other plants—medium-sized trees and shrubs. The small trees growing beneath the forest canopy are called the understory. The ground cover in the forest consists of small plants, such as shrubs, vines, herbs, wildflowers, mosses. They grow in openings where sunlight reaches the ground. We call this beautiful storied effect the forest structure.

The base of the structure is the forest floor, the ground beneath the trees. If we could sweep the floor to remove dead leaves, we would expose the humus. A humus layer takes a long time to develop. It comes from the decay of forest litter—leaves, seeds, twigs, branches, dead plants and animals. This combination of decomposed plant and animal matter makes the soil granular and porous—easy for roots to penetrate.

Organic matter, the humus layer also helps seeds and young plants develop and grow. Experts estimate that, each year, two tons of such litter are converted to soil on each acre of forest floor. Bacteria and fungi help to break down the debris. Forest dwellers that find food and shelter on or underneath the forest floor—worms, insects, mice, moles, and shrews—also help to produce fertile soil.

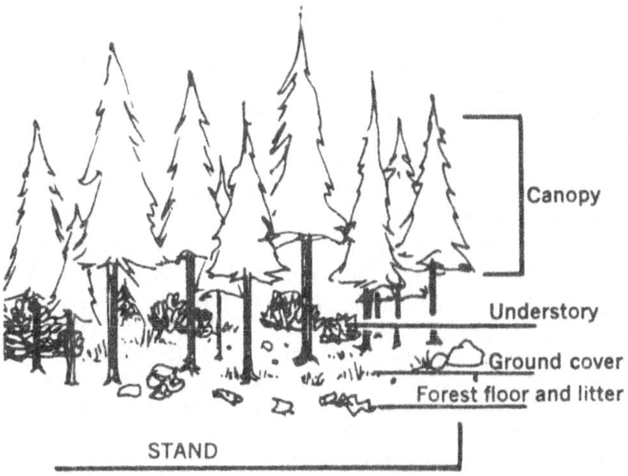

Species composition describes how a forest is "put together". A stand is an individual body of timber of similar age, species, and general appearance. A pure forest is one composed of a single species, such as a Norway pine plantation. A forest characterized by cone-bearing trees is a coniferous forest. One characterized by trees which lose their leaves is a deciduous forest. When a forest is composed of both conifers and deciduous trees, it is called a mixed forest. An aspen-fir-spruce forest is a good example.

Other life in a forest community includes birds chirping in treetops, frogs croaking on the ground, grouse drumming on a fallen log. Some life can be seen, but not easily heard—snakes slithering through litter, caterpillars devouring a leaf and spiders waiting for prey on webs that glisten in the sunshine. The joint activity of all living things in the forest—growing, reproducing, dying—improves the soil. And, for the livelihood of our forest community, fertile soil is essential.

Why Leaves Change Color

It requires no vivid imagination to picture Mother Nature going about on autumn days with a liberal supply of paint coloring the leaves of forest trees and other plants with brilliant shades of red, purple, orange, and yellow.

Many people suppose that Jack Frost is responsible for the color change, but he is not. Some of the leaves begin to turn before there is any frost. We know that color change is the result of chemical processes which take place in the trees as the season changes from summer to fall.

During the spring and summer months, leaves have served as factories where most of the foods necessary for the tree growth are made. This food-making process takes place in the leaf in numerous cells containing the pigment (coloring matter) chlorophyll which gives the leaf its green color. Besides the green pigment, leaves also contain yellow or orange carotenoids (pigments)—which, for example, give the carrot its color. Most of the year these yellowish colors are hidden by the greater amount of green coloring. But in the fall, partly because of changes in the length of daylight and temperature, the leaves stop their food-making process. The chlorophyll breaks down, green color disappears, and yellowish colors become visible, giving leaves part of their fall splendor.

At the same time other chemical changes may occur, causing formation of additional pigments that vary from yellow to red to blue. Some of them give rise to the reddish and purplish fall colors of dogwoods and sumacs. Others give the sugar maple its brilliant orange or fiery red and yellow. The autumn foliage of some trees, such as quaking aspen, birch, and hickory, shows only yellow colors. Many oaks and other species turn brown or red, while beech turns a golden bronze. These colors are due to the mixing of varying amounts of the chlorophyll and other pigments in the leaf during the fall season.

Color intensity may vary from tree to tree. For example, leaves directly exposed to the sun may turn red, while those on the shady side may be yellow. The foliage of some tree species just turns dull brown from death and decay and never shows bright colors.

Tree colors may vary from year to year, depending on weather conditions. When there is much warm, cloudy rainy weather in the fall, the leaves may have less red coloration. The smaller amount of sugar made in the reduced sunlight moves out of the leaves during the warm nights. Thus, no excess sugar remains in the leaves to form the pigments.

Only a few regions of the world are fortunate in having these showy displays. The eastern United States and southeastern Canada possess large areas of deciduous forests (broad-leaved trees) because weather conditions are favorable for bright fall colors. Some western areas of the United States, especially mountain regions, and eastern Asia and southwestern Europe also have bright coloration.

As fall colors appear, other changes take place. Where the base of the leafstalk is attached to the twig, a special layer of cells develops and gradually severs the tissues supporting the leaf. At the same time nature heals the break, so that after the leaf is finally blown off by the wind or has fallen from its own weight, a leaf scar marks the place where it grew on the twig.

Most conifers—pines, spruces, firs, hemlocks, cedars—are evergreen. The needlelike or scalelike leaves remain somewhat green the year round, though often becoming brownish green where winters are cold. Individual leaves may stay on the tree two or more years.

Through fallen leaves, nature has provided a fertile forest floor. Fallen leaves contain large amounts of valuable nutrients, particularly calcium and potassium, which were originally a part of the soil. Leaf decomposition enriches the top soil layers by returning part of the elements borrowed by the tree, and at the same time provides for more water-absorbing humus.

The Minnesota Volunteer

THE GROWTH PROCESS

The growth of new wood in any year forms a complete layer over the entire tree. That is why we can determine the age of a tree from the number of rings on the stump. Height growth occurs only in the new wood of any one year. If a nail is driven into the trunk of a ten-foot tree at five feet from the ground, it will still be five feet from the ground when the tree is 100 feet tall.

The form of trees depends on the growing space. We generally recommend a spacing of seven by seven feet when planting pine seedlings. We want the lower limbs to be shaded as soon as possible so that they will die and fall off, leaving tall, straight, clean stems, which are valuable for timber. Shade trees, which have room to spread out, have a large crown with spreading limbs and a short stem. Notice the trees at the edge of a forest; they have branches on the side toward the open, while the other side is free of branches.

All trees are composed of three parts: (1) the roots which anchor the plant and contain the rootlets necessary to absorb moisture and nutrients from the soil; (2) a stem or trunk which give rigidity to the structure; and (3) a crown bearing the leaves or needles which manufacture food used by the tree.

The growth process is complex. Moisture, containing mineral nutrients, is absorbed by osmosis from the soil through the very small roothairs. The nutrient-bearing moisture, travels up the fiberous roots and the trunk of the tree to the leaves essentially through the last annual growth ring. The leaves manufacture a simple sugar by the process of photosynthesis. The sugar solution travels down through the inner portion of the bark on the trunk and roots to supply the very thin growing cambium tissue located between the bark and the wood.

The tree grows in diameter by cell division (mitosis) of the cambium layer adding one growth ring annually. The larger portion of the annual ring is constructed during the spring; the smaller, denser and darker portion results from summer growth.

The tree grows in height and crown spread by the elongation of the buds on the branches.

Requirements For Growth And Survival

Each species of tree has specific requirements including: type of soil, mineral nutrients, amount of ground moisture, humidity, tolerance to shade, sunlight, temperature, frost free growing season, and insect and disease resistance. These factors, individually and collectively, will determine growth rate, maturity age, cutting practice, and what species will survive.

Effect of Sunlight

Shade tolerance, the ability of a tree to survive and grow in the shade of other trees, may be considered one of the most important factors which determines the basic forest management practices.

A tree may be defined as a woody plant having a well defined stem, a more or less definite crown, a height at maturity of at least eight to ten feet and a diameter at breast height of not less than two inches. In the United States, there are over 1,000 kinds of native trees. The Yearbook of Agriculture for 1949 entitled "Trees" lists 165 important forest tree species. Fifty of these are native to Wisconsin. Probably not more than 15 of these species occur naturally on any of our school forests.

Trees may be identified by their fruit, flowers, buds, bark, twigs and sometimes special characteristics such as thorns, method of branching and general shape and outline. When most of these means of identification are available, especially leaf identification, it is not too difficult to determine the species of a tree.

Since form and size, or color and character, of bark vary with size and growing conditions, too much importance should not be placed on them in identifying trees. The leaves from the lower branches of a tree may have a different outline than those from the tip of the tree, while leaves on the sprouts from a tree which was cut may be excessively large and of unusual shape. Notice the difference between the twigs and needles of balsam trees when one has had full sunlight and the others have grown in the shade.

CHAPTER 3
TREE IDENTIFICATION, RANGES & USES

How are trees identified? Actually several different characteristics can be used, depending on the knowledge of the individual and the season of the year. The person who is just beginning a study on trees will probably use one or two distinctive characteristics to determine correct identity. However, continued practice will result in including additional distinctive features to correctly determine a tree's identity.

To someone who has worked with trees for several years the identity of most trees is determined "on-the-spot" just because of the location and natural appearance of the tree. This type of identification determines that a given tree is a sugar maple "because it looks like one." Identifying trees in this manner is possible because the individual understands much about the natural habits of the tree and is able to rapidly separate out those distinctive and positive identifying features. The ability to identify a large variety of trees in this manner comes only after considerable practice.

One of the most helpful aspects of learning to identify trees is to know where to expect to find them. We have indicated that certain species prefer to grow on certain soils and with specific other kinds of trees. Let's look at an example, the northern hardwood forest which we referred to previously. We expect to find this forest type on upland areas with well-drained loam to clay loam soils. It usually will not be present on sandy soils or on poorly drained lowlands. We also know that within the northern hardwood type we can expect to find sugar maple, American basswood, yellow birch, American beech, northern red oak, eastern hemlock and occasionally white ash. With this information available we can eliminate many other species from consideration when attempting to identify trees in the northern hardwood type. Similar soils and species composition information is available about other forest types as well.

The form or overall appearance of trees can also be helpful in identification. The columnar appearance of the Lombardy Poplar is seldom mistaken for other trees even at a distance. Likewise, the upright, vaselike appearance of American elm is a useful aid in identification. Many trees, such as shagbark hickory and eastern white pine have distinctive appearances which are helpful in identification. The value of tree form as an aid in identification increases with greater knowledge and familiarity with the tree.

In addition to form and location there are several characteristics of trees which are used in identificiation. Some of the most common include:

Leaves

The leaves of trees are probably the most commonly used feature in identification. Initially they can be used to separate conifers or trees with needle-like or scale-like leaves from deciduous trees or those with broad leaves. Other characteristics of leaves which are used include:

1. **Leaf arrangement** - Different kinds of trees have varying patterns of leaf arrangement. This refers to the pattern of leaf attachment to the twig. Alternate leaves indicate an alternating pattern with only one leaf present at each location or node on the twig. Leaves oppositely arranged have two leaves present at each node, usually in a paired position. Some trees may have leaves which are attached in a whorled fashion with more than two leaves (usually three) at each node.

2. **Simple vs. Compound leaves** - Simple leaves are those in which the blade of the leaf is in one piece. Compound leaves have two or more leaflets which are attached to the main leaf stalk or rachis. The base of the rachis is called the petiole. Some trees such as honey locust may have doubly-compound leaves, in which smaller, sub-rachi which bear leaflets are attached to the main rachis.

3. **Leaf Shape and Margins** - Both of these features may be helpful in determining identification. In fact, in some trees the shape of the leaf is such that positive identification may be determined from this characteristic alone. The flat-topped leaf of the tulip tree provides a rapid means to correctly identify this species. Many other species such as the samples and many oaks have characteristic shapes which are helpful. The edge or margin of leaves are also quite variable. Some such as white oak are deeply lobed while others such as black and sugar maple, are lobed less deeply. On other types of leaves the margin may be smooth (entire) or cut into various irregular (toothed) patterns.

4. **Venation** - Leaves also vary in their patterns of venation (arrangement of veins within the leaves). Pinnately veined leaves like white oak have only one main vein with several smaller lateral or side veins. Palmately veined leaves have several main veins which develop from a central point at the base of each leaf. Maple leaves are good examples. Ginkgo trees have a third type of venation known as parallel, with many veins arising from the base of the leaf.

Bark

The bark of many trees is very distinctive and may provide a ready aid to identification. The scaly, shaggy bark of the shagbark hickory is seldom confused with other trees. Similarly, the mottled light and dark bark of sycamore trees is very distinctive. In some trees the bark does not remain the same throughout the life of the tree. On young trees of such species as sugar maple or white pine it may appear smooth, tight and grey in color. However, as these trees become older the bark becomes rougher, particularly due to the development of furrows and ridges. On very old trees the bark may appear very shaggy or blocky in appearance. Due to these variations in bark both among and within different species of trees, the inexperienced person should not form an opinion about the identity of most trees which is based only on bark characteristics. Rather, he should consider features of the bark along with other characteristics of the tree before a decision is made regarding its identity.

Twigs and Buds

The identification of many deciduous trees during the winter months is made easier by examining the leafless twigs and buds. In fact twig and bud characteristics are used to construct keys which can be used for identification purposes during the dormant season. When using twigs and buds for identification, emphasis is placed on the position, characteristics, and number of terminal (those on the ends of the twig) buds and lateral (those along side the twig) buds. Leaf scars (where last years leaf was attached) are also important identification aids.

Flowers

All trees when mature produce some type of flowers. The principal purpose is to produce fruits which will serve to reproduce the species. Flowers on some trees such as catalpa, and flowering crabs are very showy, but on most trees flowers are small and inconspicuous. Flowers on trees are also highly variable. Some trees may bear perfect flowers (both male and female parts present in the same flower) while others have imperfect flowers (only male or female flowers on a tree). To one who is familiar with flower parts these may be helpful in some instances.

Fruits

The value of tree fruits in identification is variable, depending on the kind of tree involved. Fruits when present on the tree or on the ground beneath the tree may be useful in making broad classifications, or may serve to assist in determining what the tree is not. For example it is difficult to confuse the large cones of white pine or the large, heavy acorns of burr oak with other pines or oaks. However, other species of pines and oaks have fruits which look alike. A detailed study is required to obtain positive identification.

These various features while distinctive must be examined carefully and in many instances several different features should be examined before the correct identification is determined. Quite often trees are not in their preferred natural location and identification becomes more difficult. Identification is also more difficult when only a few leaves, twigs, or other parts of the tree are available.

An additional aid in determining the identity of a specific tree is the use of dichotomous (two-choice) keys. These keys are constructed using the tree characteristics we discussed earlier. Once their use is understood the identity of a selected tree may be readily determined.

The use of dichotomous keys can be illustrated by looking at the following example. Suppose you had five different kinds of balls and did not know what kind each was. However, for purposes of illustration let's assume you had a "ball key" available. After examining each of the five balls you noted that while they were all alike in some respects, each had some characteristic which the other balls did not have. These characteristics included differences in size, hardness, shape, etc. Using these features the identity of each ball can be established using the following "ball key:"

1. Ball round or circular in shape	2
1a. Ball not round; pointed at both ends	Football
2. Ball 4-½ inches in diameter or larger	3
2a. Ball 4 inches in diameter or smaller	4
3. Ball approximately 10 inches in diameter, surface can be pushed in	Basketball
3a. Ball approximately 4-½ inches in diameter, hard surface	Softball
4. Ball approximately 2-½ inches in diameter, furry covering, soft surface	Tennis Ball
4a. Ball approximately 1-½ inches in diameter, hard, pebble-grained surface	Golf Ball

To use this key in determining the identification of one of the balls the following procedure is used. Let's assume you have picked up the golf ball (although you really don't know it was a golf ball) and want to determine its identity. You then look at the first entry (No. 1) in the key. You will note there are two ones and two statements following. You read both statements and note that the ball is not pointed but round, so you follow the first statement and note you are directed to No. 2. Again, after reading both statements you see that the second protion is correct for the ball in question and then proceed to No. 4. Reading both statements and comparing the size and hardness of the ball in your hand, you conclude according to this key, that it must be a golf ball. This procedure illustrates the use of a "two-choice" key. Trees are substituted for balls and the appropriate characteristics changed so the identity of an unknown specimen may be determined following the same procedures used above. A key for some common trees common to the Lake States follows.

Melvin Koelling, Extension Forester, Michigan State University

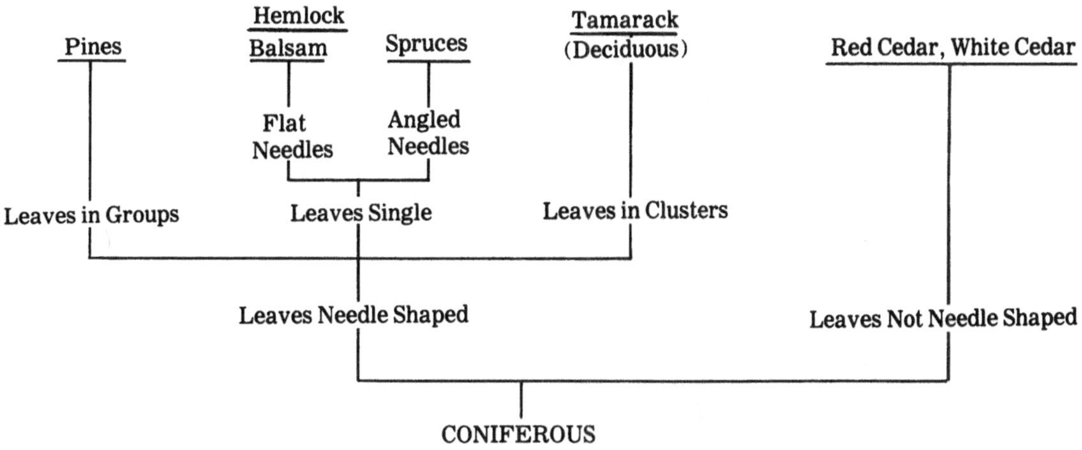

What kind is your tree?
TREE ROAD MAP
Read the signs, follow the arrows

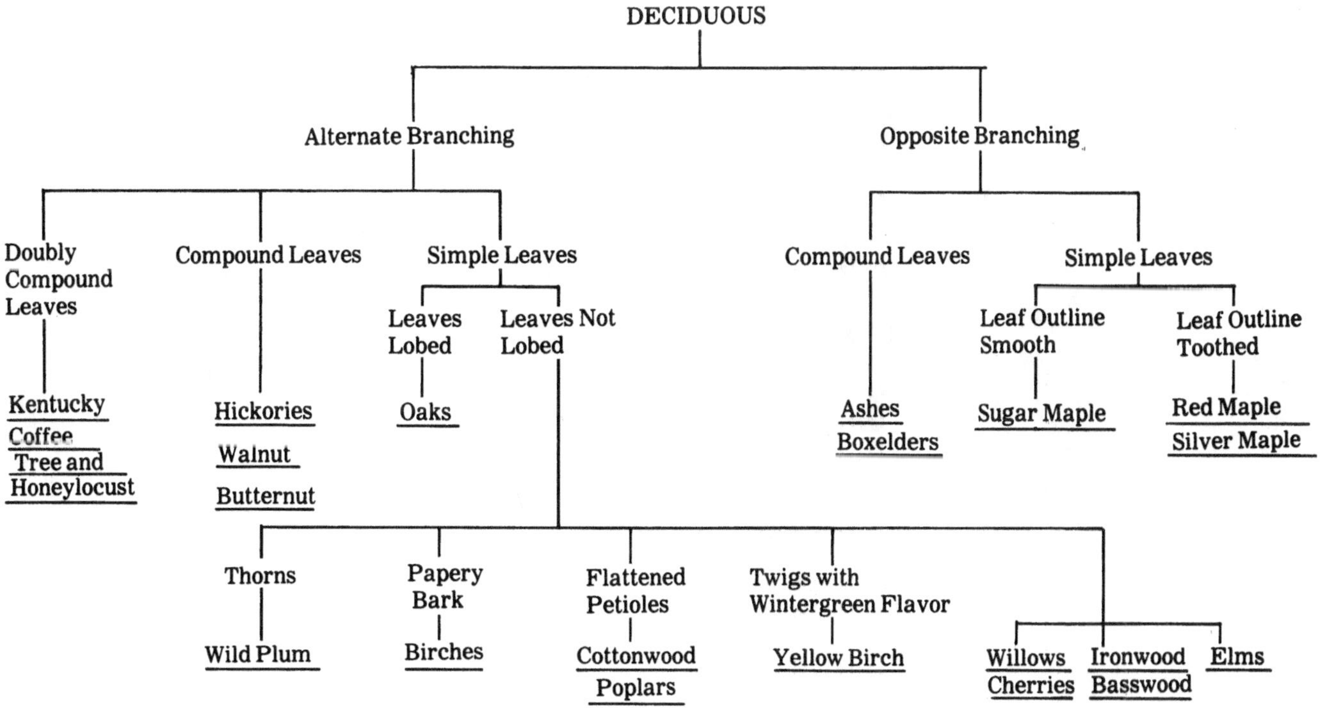

LEAVES may be deciduous (shed annually), or they may be evergreen or persistent (remaining on tree one to many years). Most cone-bearing trees and some broad-leaved trees are evergreen. Leaf arrangement may be obscure at growing tips, where leaves may not have reached full size. Leaves of some trees bear stipules (not shown), small leaflike appendages at base of petiole.

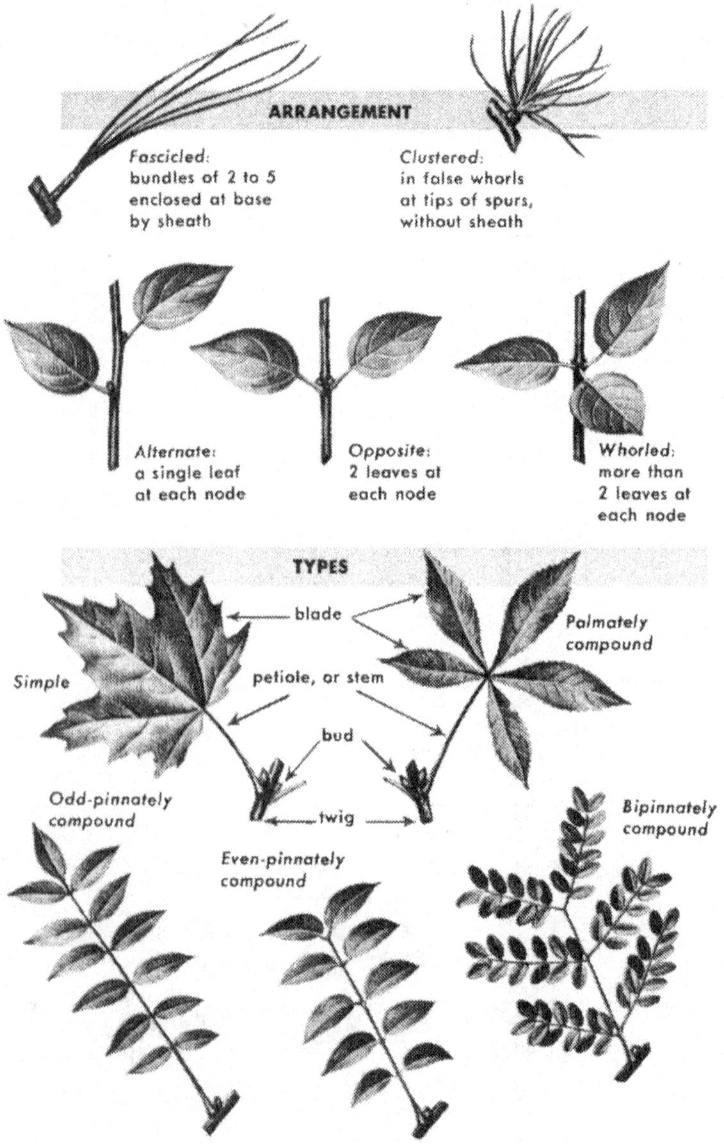

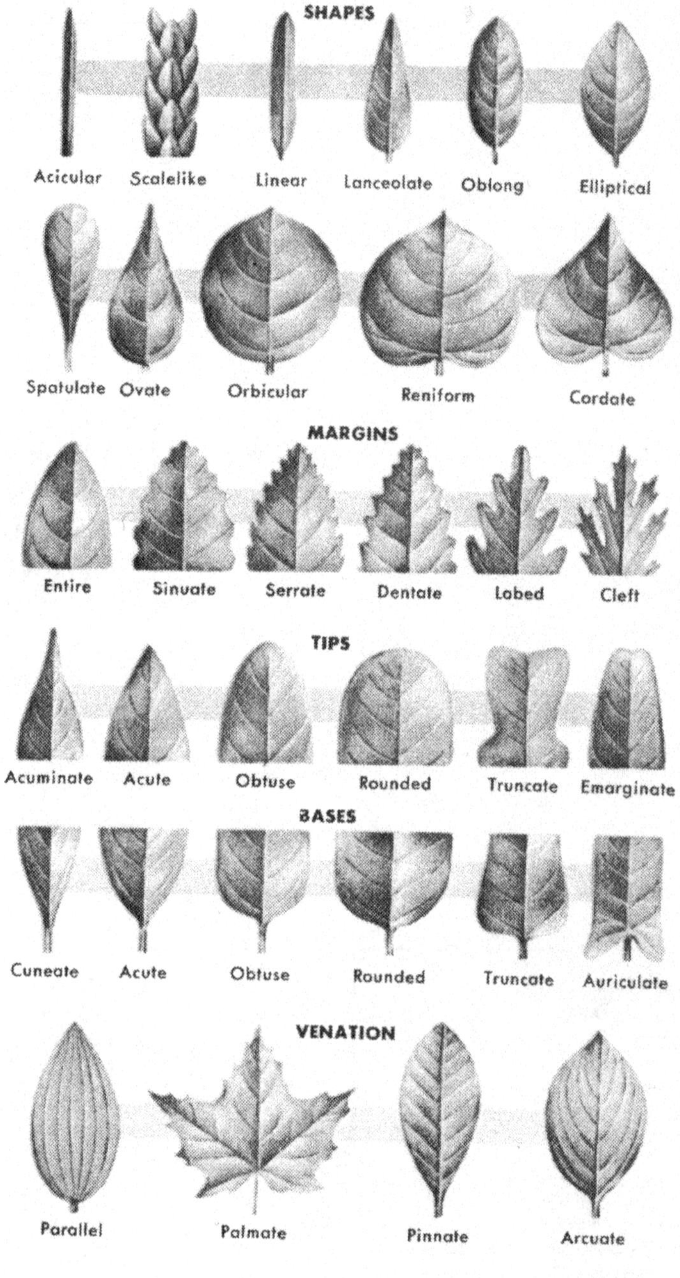

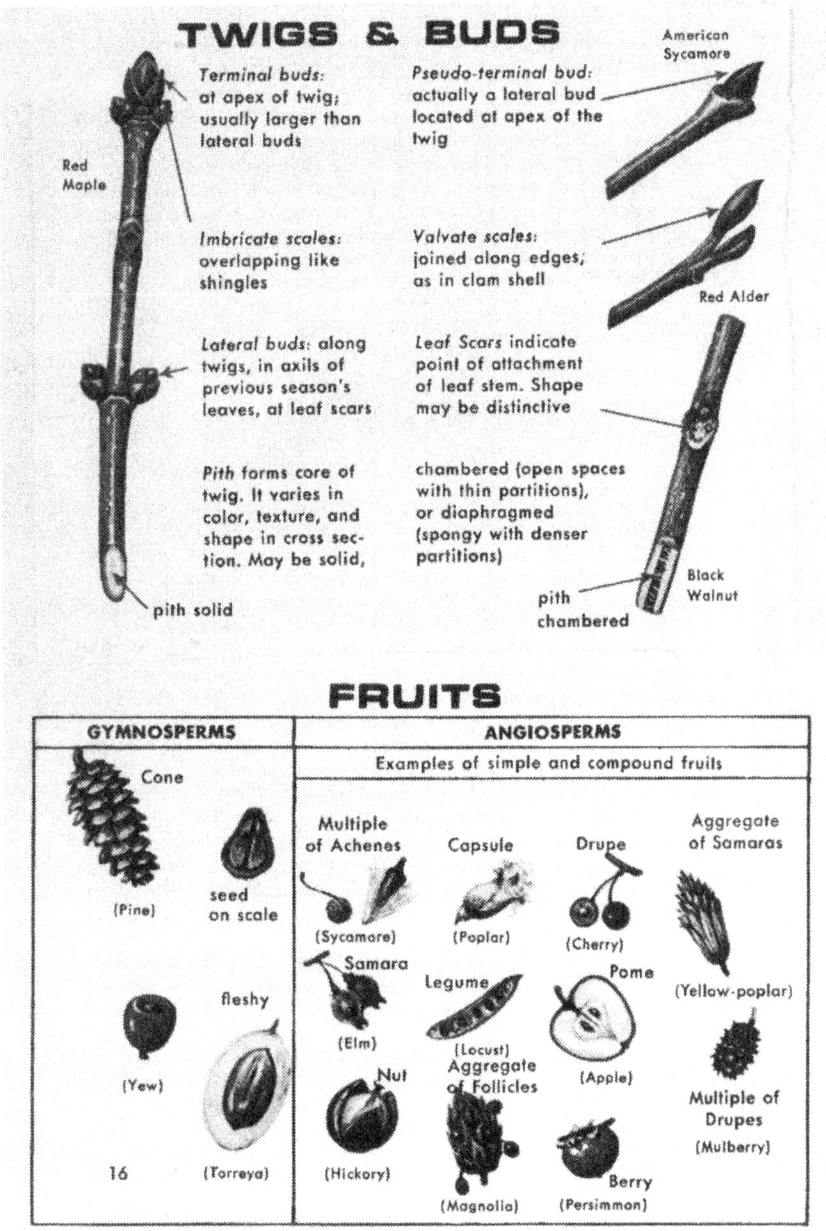

Important Trees of Eastern Forests—U.S.D.A.

Forestry Manual For Vo. Ag. Instructors
Wisconsin Department of Public Instruction

Tree Identification Key

1. Trees with needle-like or scale-like leaves 2
   Trees with broad leaves 10

2. Leaves in the form of needles 3
   Leaves in the form of scales 9

3. Needles in bundles or bunches 4
   Needles occurring singly 7

4. Needles in groups of two to five 5
   Needles in a cluster, deciduous in winter:
   Tamarack

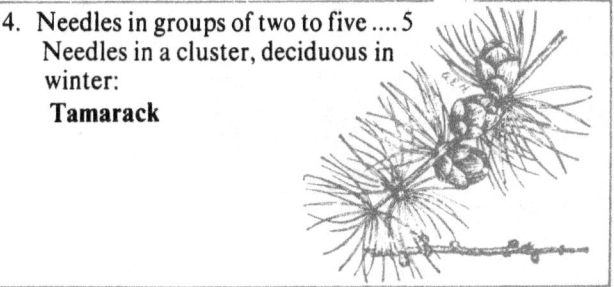

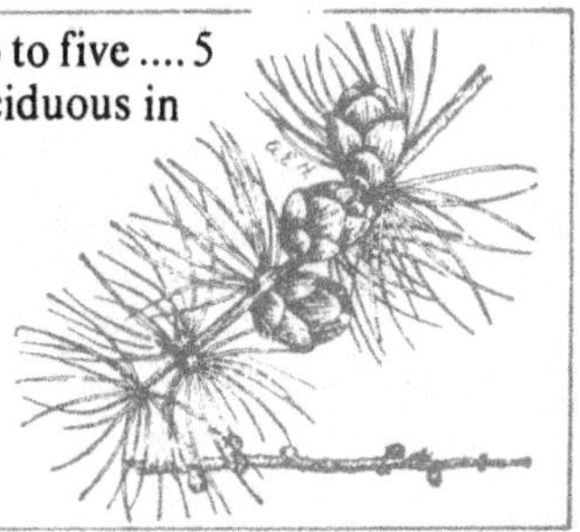

5. Needles in groups of two 6
Needles in groups of five:
Eastern White Pine

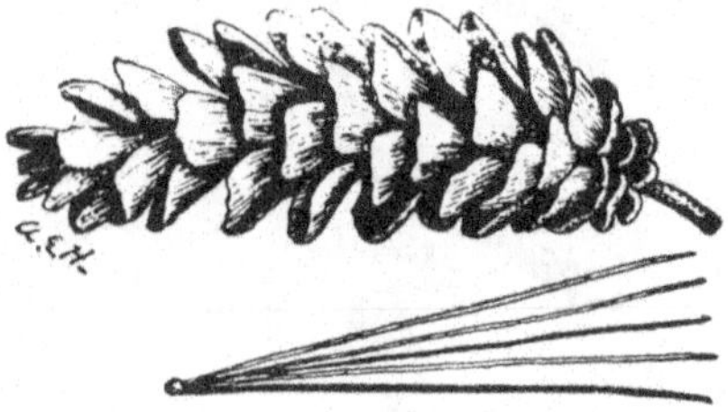

6. Needles 3/4 inch to 1-1/4 inches long:
Jack Pine

Needles 3 inches to 5 inches long:
Red Pine

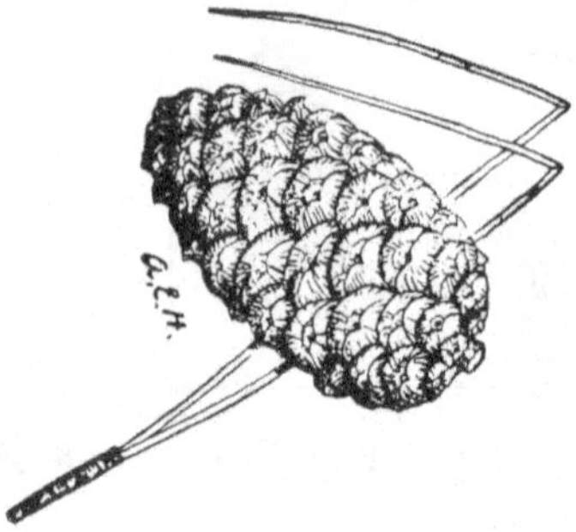

7. Needles 4 angled stiff, sharp pointed 8
Needles flat, rounded at tip
Twigs rough, cones hang down:
Hemlock

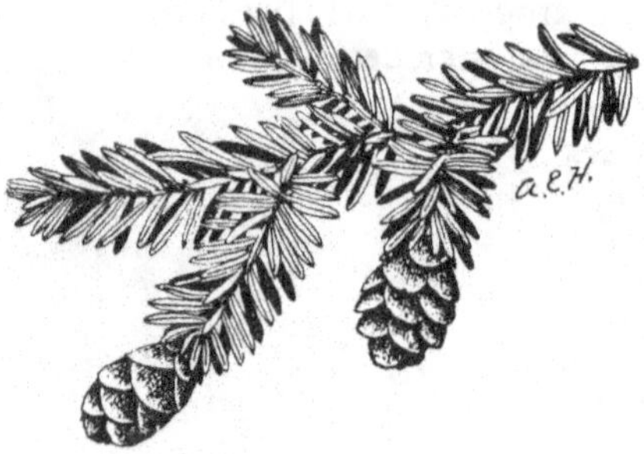

Twigs smooth, cones upright in top of tree:
Balsam Fir

8. Cones 1/2 - 1 inch, occasionally 1-1/2 inches long, cone scales with toothed margin, twigs hairy:
Black Spruce

Cones 1-1/2 to 2 inches long, cone scales with smooth margin, twigs not hairy, needles having a disagreeable odor when crushed:
White Spruce

9. Leafy twigs more or less flattened, cones woody:
Northern White Cedar

Leafy twigs not flattened, prickly, fruit is berry-like:
Eastern Red Cedar

10. Leaves and usually branches in pairs, opposite 11
Leaves and usually branches borne singly and alternate 17
11. Leaves not divided into leaflets, (single) 12
Leaves divided into 3 to 11 leaflets (compound) 14
12. Teeth of leaves many and sharp 13
Teeth of leaves few and blunt:
Sugar Maple

13. Leaves deeply 5-lobed, about 6 inches in diameter, margins of the middle lobe usually divergent:
Silver Maple

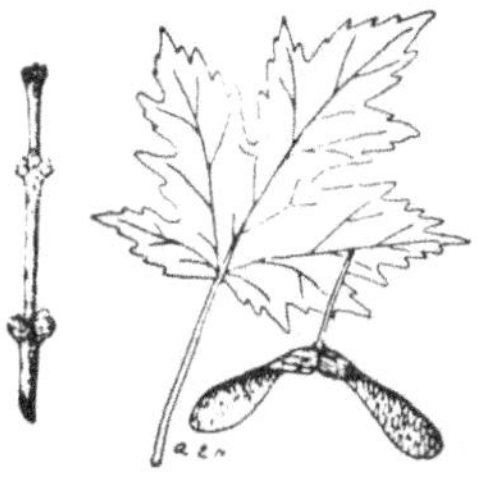

Leaves 3 to 5 lobed, about 4 inches in diameter, the sides of the middle lobe mostly converging to the apex:
Red Maple

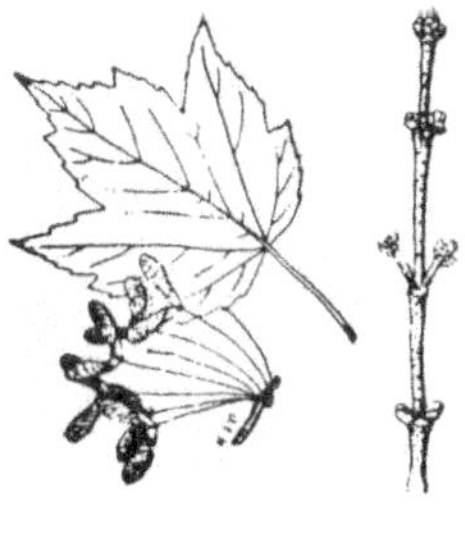

14. Leaflets 5 to 11 leaf margins entire or finely toothed 15
Leaflets 3 to 7 sharply toothed on margins:
Box Elder

15. Leaflets with stalks.. 16
Leaflets without stalks:
Black Ash

16. Leaves with 5-9 (mostly 7) leaflets, leaf scars notched at the top:
White Ash

Leaves with 7-9 leaflets, leaf scars nearly straight along upper edge:
Green Ash

17. Leaves divided into leaflets, compound 18
Leaves not divided into leaflets, simple 23
18. Leaflets rounded or bluntpointed, twigs spiny, fruit a flat bean-like pod 19
Leaflets long-pointed, twigs not spiny, fruit rounded or egg-shaped 20
19. Leaflets 7-19, round on top, short spines on each side of main leaf stalk, fruit a pea-shaped pod 2 to 4 inches long:
Black Locust

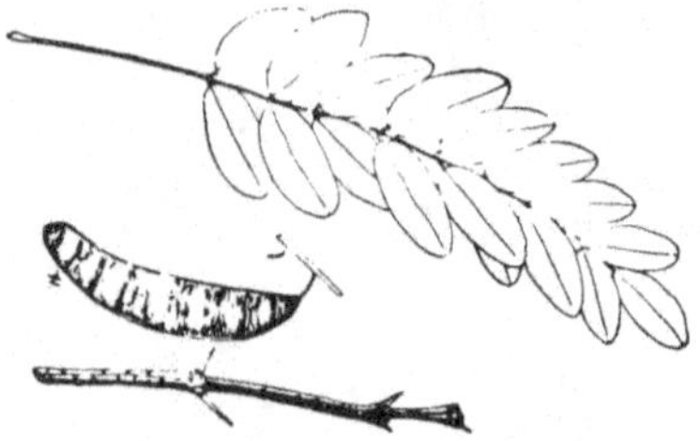

Leaves divided twice, 20-30 leaflets, fruit a pea-like pod 10 to 18 inches long:
Honey Locust

20. Leaflets 5 to 11, husk of nut splitting off 21
Leaflets 11 to 23, husk of nut not splitting off 22
21. Leaves 8 to 14 inches long, with 5 leaves, rarely 7 leaflets, bark in thin curly plates on large trees:
Shagbark Hickory

Leaves 6 to 10 inches long, with 7 to 11 leaflets, bark in old trees in shallow furrows:
Bitternut Hickory

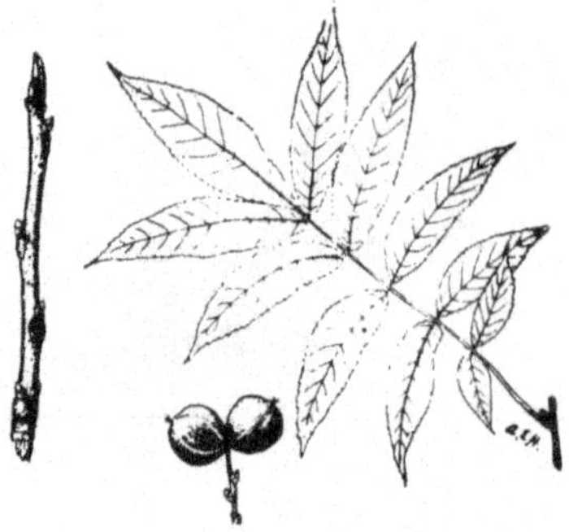

22. Leaflets 15 to 23, nuts or single or in pairs, with husk nearly round:
Black Walnut

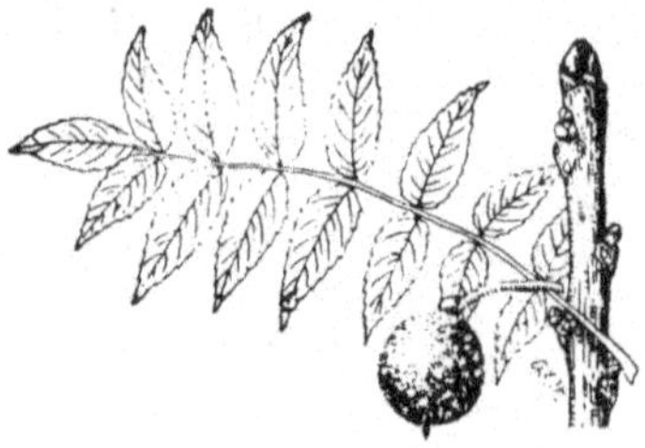

Leaflets 11 to 29, nuts generally in clusters of 2 to 5, sticky to the touch, eggshaped:
Butternut

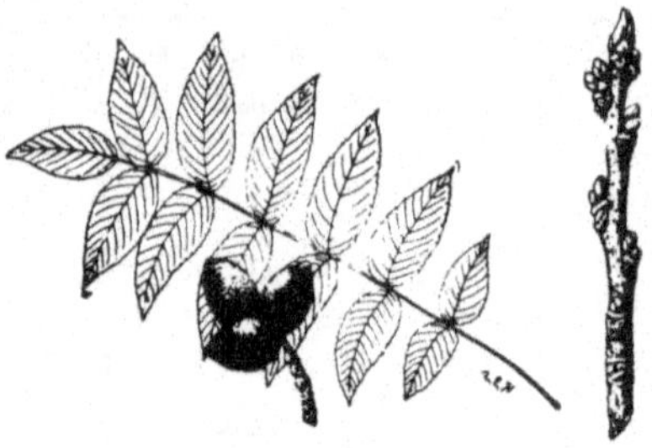

23. Fruit, not an acorn 24
Fruit, an acorn 36
24. Leaves with 2 sides unequal and 1 side larger at the base in 2 rows on twig 25
Leaves with both sides equal, spreading around twig 29
25. Leaves broad, heart-shaped, with leaf stalks more than 1-1/2 inches long:
Basswood

Leaves narrower, with leaf stalks less than 1/2 inch long 26

26. Leaves with 1 main vein and many parallel lateral veins 27

Leaves with 3 main veins from base:
Hackberry

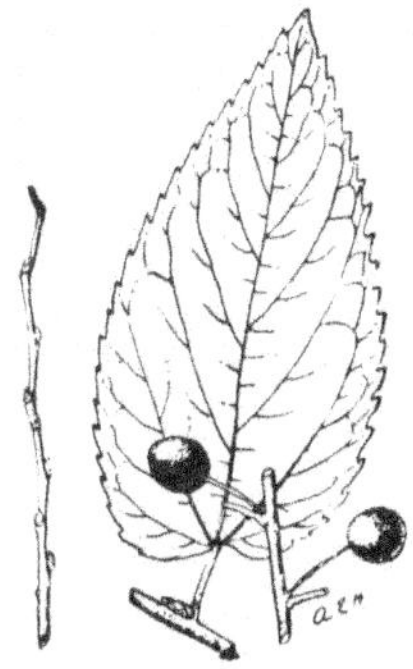

27. Twigs round, not corky winged 28

Twigs usually becoming corky winged, leaf margins doubly-toothed:
Rock Elm

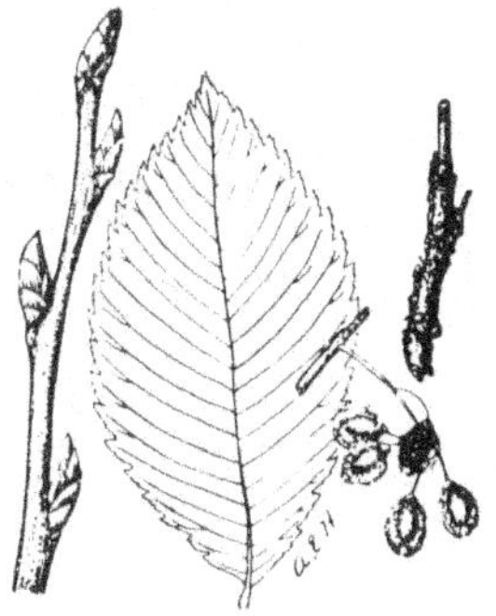

28. Outer bark layers of alternately buff and reddish brown color:
American Elm

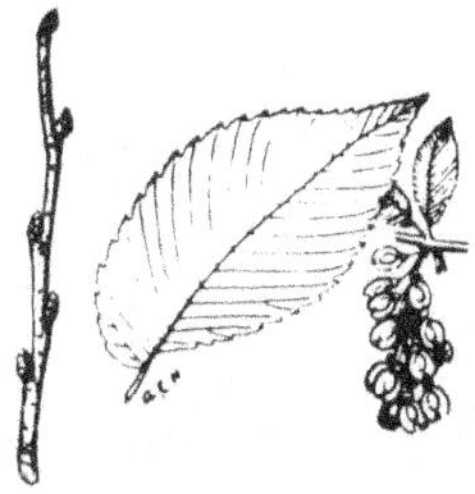

Outer bark layers dark reddish-brown, no distinct layers, inner bark gluey when chewed, leaves often creased along the midrib:
Slippery Elm

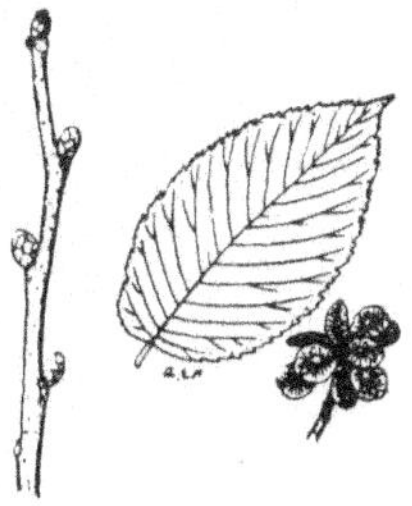

29. Leaf stalks more than 1-1/2 inches long, seeds cottony 33

30. Leafstalks flat 31

Leafstalks round:
Balsam Poplar

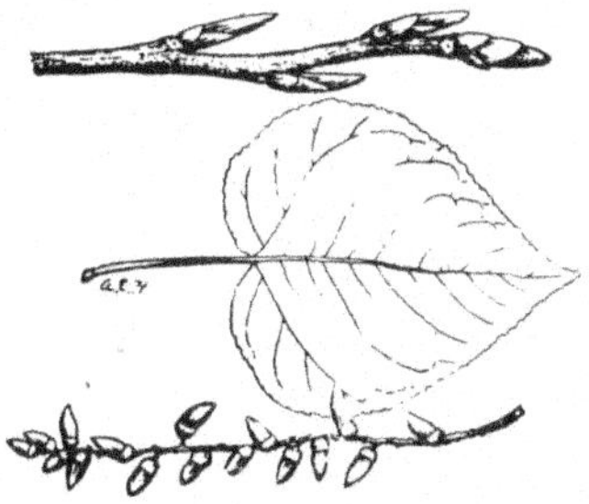

31. Leaves nearly round 32

Leaves triangular in shape:
Eastern Cottonwood

32. Leaves with coarse-toothed margins:
Largetooth Aspen

Leaves with fine-toothed margin:
Quaking Aspen

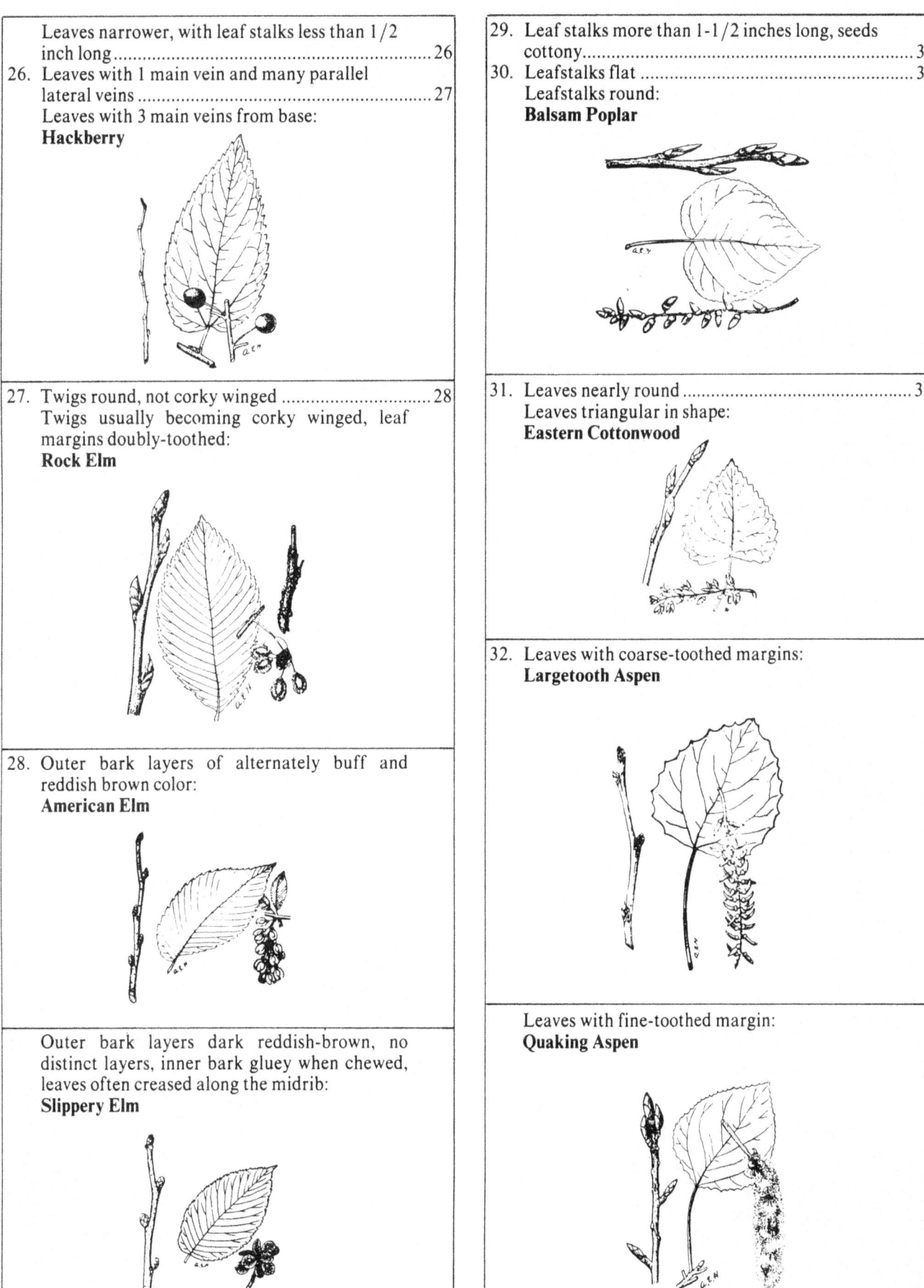

33. Leaf edges with teeth of 2 sizes and slightly irregular, fruit a cone 34
Leaf edges with uniform teeth, fruit not a cone....... 35

34. Bark yellowish or silver-gray, leaves with 9 to 11 main veins on each side, twigs when chewed give wintergreen smell:
Yellow Birch

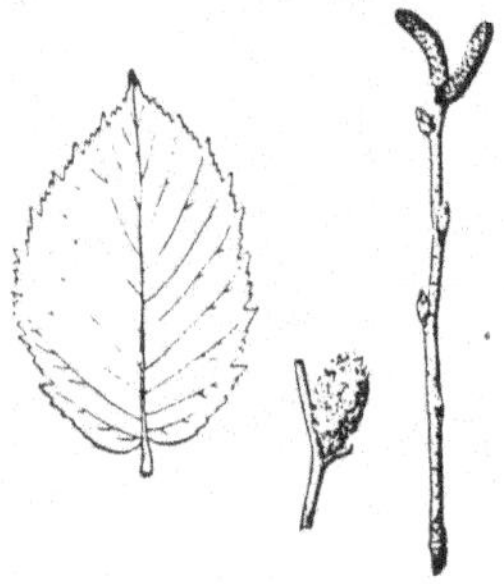

Bark white, leaves with 4 to 9 main veins on each side:
Paper Birch

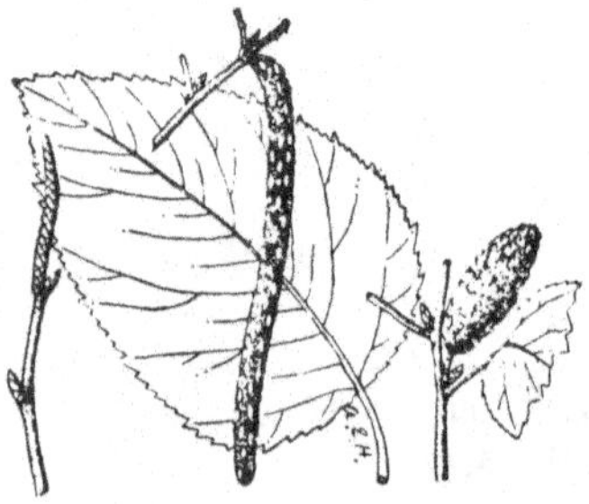

35. Leaves narrow, more than 3 times as long as wide:
Willow Species

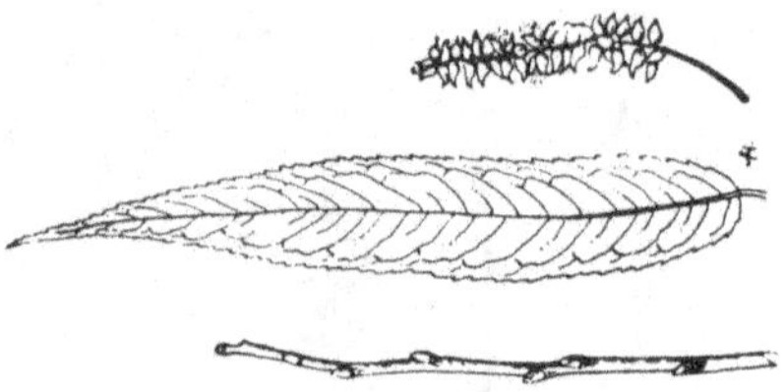

Leaves less than 3 times as long as wide, shiny on top, fruits edible cherries
Black Cherry

36. Leaf margin with rounded lobes............................ 37
Leaf margin without rounded lobes:
Northern Red Oak

37. Leaf edges deeply lobed... 38
Leaf edges many with uniform rounded teeth:
Swamp White Oak

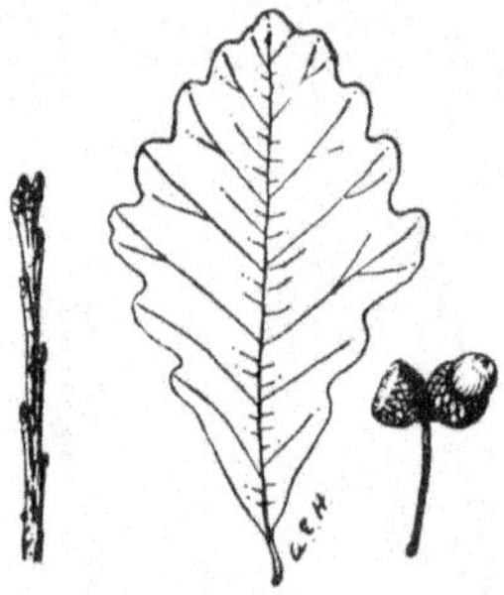

38. Leaves deeply and evenly lobed, acorns with shallow cup:
White Oak

Leaves deep-lobed only near the middle, shallow lobed above, acorn half-enclosed by the large cup with fringe-like border:
Bur Oak

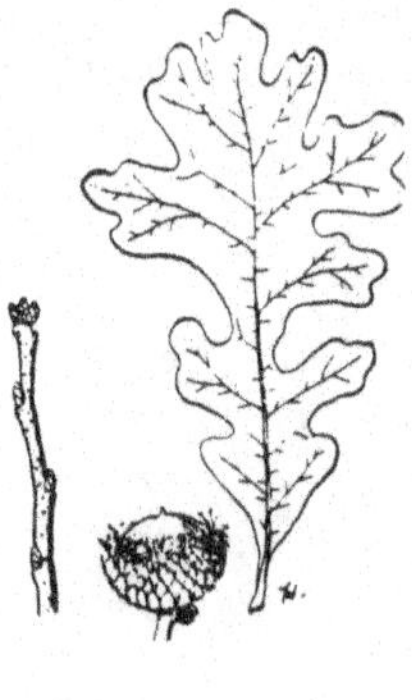

PINES *(Pinus)*

Pines are probably the world's most important conifers. They grow in most areas of the Northern Hemisphere and in a few areas south of the equator. About 100 species are recognized worldwide; 36 of these species are native in the United States.

Pine trees usually have deeply furrowed **bark**, resinous wood, and an upright terminal axis which produces the annual growth in the form of a long shoot with one or more whorls of spreading branches.

Pines are evergreen, with needlelike **leaves** usually borne in clusters or bundles of 2-5. Leaves remain on the tree from 2 to 12 years, depending upon the species.

Male **flowers** are clustered at the base of the current year's leafy growing shoots in lower portion of the crown. Female flowers occur on the same tree, more often on upper branches.

The **cones** are variously shaped, usually longer than wide, tapering to a point, maturing at the end of the second year (rarely the third), and either falling promptly or remaining attached to the tree for several years. Cones are composed of the hardened, woody scales of the ovulate (female) flowers. Seed are usually winged.

Pines are separated into two groups. The soft (or white) pines are identified by their soft, light-colored, close-grained **wood**, needles in clusters of 5 with the sheath of the needle-bundle being shed with the leaves, cones usually stalked, cone scales not armed with spiny tips. The hard (or pitch) pine characteristics are their coarse-grained, generally darker-colored **wood**, thickened cone scales mostly armed with spiny tips, needles in bundles of 2 or 3 with the sheaths of the bundles not shed, and cones short-stalked or (usually) not stalked.

The only soft pine of eastern forests is the eastern white pine *(Pinus strobus)*. All other species fall into the hard pine group.

EASTERN WHITE PINE *(Pinus strobus)* **needles** are soft bluish-green, flexible, 3-5 inches long, in bundles of 5, with 3-5 fine white lines (of stomata) on two surfaces of each needle. Needles remain on the tree for two years. Cones are fully grown in the summer of the second season, opening to discharge seed that autumn. **Cones** are 4-8 inches long, curved, stalked, with scales not spiny. The **bark** on young trunks and branches is thin, smooth, greenish (sometimes tinged with red), and lustrous. On old trunks, the bark is deeply divided into broad ridges that are covered with purplish scales. The **tree** commonly attains 100 feet in height and 4 feet in diameter, with a tall, straight stem and pyramidal crown. Branches are in definite whorls of long lateral branches sweeping upward in graceful curves. This is the largest conifer of eastern forests, is long-lived, and has been known to reach heights above 200 feet. The **wood** is light, straight-grained, easily worked, but not strong. It is used in cabinet work, interior finishes, woodenware, matches, and lumber. A large part of the winter diet of red squirrels (boomers) in the southern Appalachians is made up of the seed of eastern white pine.

RED PINE *(Pinus resinosa)* **needles** are soft, slender, flexible, lustrous, dark green, 5-6 inches long, and in 2-leaved clusters. **Cones** are symmetrical, ovoid, 2-2¼ inches long, light chestnut-brown and lustrous at maturity, and not armed with spines on the scales. The thick **bark** is divided shallowly into broad, flat ridges that are covered with thin reddish-brown scales. The **tree** usually grows to 70-80 feet in height with a trunk 2-3 feet in diameter. It may reach 100 feet in height; some trees have attained 150 feet in height and 5 feet in diameter. Red pine has a tall, straight trunk and a symmetrical crown. The somewhat pendulous branches almost reach to the ground on smaller trees. This species is an important timber tree. The **wood** is hard, very close-grained, light, and quite strong. It is used among other things for construction, pilings, and flooring.

JACK PINE *(Pinus banksiana)* **needles** are stout, flat, 1-1½ inches long, dark green, deciduous in their second and third years, and in 2-leaved bundles. The **cone** is usually erect, 1½-2 inches long, and generally curved in toward the branch. **Cones** are green or purplish when full-grown, turning light yellowish-brown as they ripen. They may remain on the tree for years. Cone scales are armed with small prickles that are often deciduous. The **tree** is frequently 70 feet tall with a straight trunk about 2 feet in diameter. The branches are long and spreading, forming an open crown that often has a ragged appearance. The wood is soft, light, and not strong. It is used mainly for pulpwood.

SPRUCES *(Picea)*

Spruces characteristically are pyramidal, with dense crowns and tall tapering trunks often buttressed at the base. **Bark** is thin and scaly. **Leaves** are linear (needles), evergreen, spiralling around the branches and extending out on all sides of the branches. Needles are 4-angled (flattened in a few species), stiff, and sharp-pointed. They are borne on peg-like stalks which remain on older twigs.

Spruce **cones** hang downward from the branches. The cones are egg-shaped, covered with thin woody scales, and often crowded on upper branches. Each scale bears at its base 2 long-winged seeds.

Spruces grow best in moist sites. They are widely distributed in the coniferous forests of cooler climates. Seven of the 30-45 species of spruce are native to the United States; 5 of the 7 are of commercial importance.

BLACK SPRUCE *(Picea mariana)* has blue-green **needles** that are rigid but blunt-tipped. Needles are ¼ to ¾-inch long, 4-sided but somewhat rounder in cross-section than the red spruce or white spruce. The **cones** are oval, pointed, ½-1 inch long, and persistent on the tree for sometimes as long as 20-30 years. Cone scales have rough margins. The **tree** is commonly 30-40 feet tall with a trunk diameter of 1-2 feet. Black spruce is one of the most abundant conifers of northern North America. It prefers wetland sites such as bogs, muck-filled seepages, and edges of streams and swamps. The **wood** is light, soft, and not strong. It is used extensively for pulpwood.

WHITE SPRUCE *(Picea glauca)* **leaves** are crowded on the upper side of the branches by the twisting upward of the needles on the lower side. Needles are ¾-1 inch long, and bluntly pointed. **Cones** are usually about 2 inches long, becoming pale brown and lustrous when mature. Cone scales are flexible and smooth-margined. Cones fall soon after they ripen in autumn, rarely persisting through the winter. The **tree** grows 60-75 feet tall with trunk diameter of 2 feet. Foliage of white spruce has a somewhat disagreeable odor when crushed. The **wood** is used for pulpwood, interior trim, and interior parts of furniture.

EASTERN REDCEDAR *(Juniperus virginiana)* is the most widely distributed conifer of tree size in eastern United States. It grows as a tall tree or a shrub and any size between, utilizing a wide variety of sites from dry hillsides to swamps. The **leaves** are usually opposite, being smooth, shiny, dark green, and glandular on older foliage. On young foliage, leaves are linear (somewhat needle-like), pointed, and prickly. The fleshy **cones** are round, 1/4-1/3 inch in diameter, and at maturity a bluish color with a grayish-white waxy covering. The **tree** is commonly 40-50 feet tall with a trunk diameter of 1-2 feet. The short, slender branches form a compact, pyramidal crown except on very old trees. The **bark** is light reddish brown, thin, and separating into long, peeling, fibrous strips.

SOUTHERN REDCEDAR *(Juniperus silicicola)* is very similar to eastern redcedar but has smaller cones and drooping branches. Southern redcedar prefers wet lowlands and swamps.

THE WOOD OF EASTERN REDCEDAR

Properties — Eastern redcedar is moderately heavy, hard, moderately weak in bending, and high in shock resistance. The wood is fine-textured and even-grained, with the well-known "pencil cedar" taste and odor. The wood works easily with tools, shrinks very little in drying, stays in place well after seasoning, and is very resistant to decay.

Uses — Eastern redcedar is usually available only in fairly small sizes, and generally is quite knotty. Principal use has changed from fence posts to novelty items. Lumber is used where its fragrance and reputed moth-repellent qualities are valued such as in storage chests, closets, and wardrobes. It is also used for millwork, pencils, woodenware, and containers such as buckets.

NORTHERN WHITE-CEDAR *(Thuja occidentalis)* is often called Arbor Vitae (Tree of Life) when used as an ornamental planting. The flattened **branchlets** have 4 rows of overlapping bright green scales, are much branched, and form horizontal fan-like sprays. Foliage has a spicy fragrance when crushed. Mature **cones** are light red-brown, 1/3-1/2 inch long, oblong, erect, and covered with 6-12 thin, blunt scales that open to the base. The **tree** grows to 40-50 feet in height with a short, often lobed and buttressed trunk 2-3 feet in diameter. The **bark** is gray to reddish-brown, often tinged with orange, and separated into flat connected ridges. The **wood** is light, soft, durable, brittle, coarse-grained, and fragrant. It is used for posts, rails, shingles, and pulpwood. This species is a favorite source of browse for deer and rabbits in northern swamps.

EASTERN HEMLOCK *(Tsuga canadensis)* is a graceful, lacy-foliaged tree of the cool moist forests. The flat **needles** are round-tipped, 1/3-2/3 inch long, and marked on the lower surface with two pale lines. Needles are narrowed to form short, slender stems that grow from rounded, dark orange-colored, persistent, woody pads on the twigs. **Cones** are roughly oval, ½-¾ inch long, with scales about as wide as long. The pendant cones are borne on short, slender stalks from the tips of branchlets, usually remaining on the tree until the following spring. Hemlock is remarkably tolerant to shading by overstory trees, and may remain in the understory in natural stands for 25-200 years. The **tree** is usually 60-70 feet high with a trunk diameter of 2-3 feet. It often has a broad-based pyramidal shape. Ends of branches are often drooping and "feathery." The **wood** is light, soft, brittle, and difficult to work. It is used occasionally for rough or construction lumber, and for pulpwood.

BALSAM FIR *(Abies balsamea)* is a small-to-medium-sized tree that grows in cold, moist climates. The dark green **needles** are ½ to 1 inch long, flattened, mostly 2-ranked (one row on each side of the branch) and marked on the undersides by 2 silvery-white bands. The cylindrical **cones** are erect, dark rich purple in color, 2-4 inches long, with scales longer than broad. **Bark** on old trees is thick, rich brown and separated into scaly plates. Bark on young trees often has resin blisters. Balsam fir has a symmetrical, slender, pyramidal crown that is quite dense. The **tree** grows up to 60 feet in height, with trunk diameter of 2 feet. The **wood** is light, weak, soft, and coarse-grained. Not much used for lumber, the balsam fir is utilized in the paper industry for pulp. It is planted as an ornamental, and prized for Christmas trees.

TAMARACK *(Larix laricina)* has one of the widest ranges of all American conifers. Its **needles** are deciduous, ¾-1¼ inches long, soft, flat, and roughly triangular in cross-section. They turn yellow and drop in September or October. Leaves are borne singly or alternate, but often form brushy clusters at the tips of short spur shoots. **Cones** are roughly rounded-oval, ½-¾ inch long, with about 20 scales that are longer than broad. The **tree** is 40-80 feet tall with trunk diameter of about 2 feet. It is most commonly found on moist organic soils such as those in swamps and muskegs. Tamarack often has a narrow pyramidal shape in forest stands, but develops a broad picturesque head in open areas. **Bark** is thin, bright reddish-brown to gray, and scaly. The **wood** is hard, heavy, durable and strong, but difficult to work with tools. It is used for posts, poles, ties, and as construction lumber.

MAPLES *(Acer)*

Five of the thirteen species of maple native to the United States are important timber trees. Worldwide, there are 115 species of maples *(Acer)*, some of which are shrubs.

Almost all maples are deciduous and have opposite, simple, palmately lobed, toothed, long-stemmed **leaves.** In a few species (including boxelder), leaves are pinnately compound.

The **fruit** occurs as paired, winged seeds that are often U-shaped or V-shaped, ripening in spring or fall. Each winged portion is a single 1-seeded fruit (samara). Fruit is borne in lateral or terminal clusters.

SUGAR MAPLE *(Acer saccharum)* is one of the largest and most important hardwoods of eastern forests. It is widely distributed, but not found in the South Atlantic or Gulf Coastal Plain states. The **leaves** have 5 lobes that are separated by rounded, shallow sinuses. Leaves are 3-5 inches across, rounded at the base, with sparse, large, pointed teeth on the margin. In autumn, sugar maple leaves are exceptionally colorful with brilliant reds and yellows. **Flowers** are yellow, polygamous, and occur on long, slender stems in clusters up to 3 inches long. They appear with the leaves in spring. The **fruit**, a U-shaped pair of winged seeds, ripens in the fall. The **tree** grows 75-100 feet in height with a trunk diameter of 3-4 feet. In crowded stands, sugar maple develops a tall, straight trunk and a narrow crown. Open-grown trees have a shorter stem and rounded crown. The **bark** of mature trees is thick, light gray to brown, and broken by vertical furrows into plate-like scales. The sap is used to make maple syrup and maple sugar. Sugar maple is a popular shade and ornamental tree.

THE WOOD OF THE HARD MAPLES

Properties — The wood of all hard maples is similar. It is heavy, strong, stiff, and has high resistance to shock. Shrinkage during seasoning is large. It has high resistance to nail-withdrawal, and is intermediate in ease of gluing. It takes stain satisfactorily and polishes well. Although usually straight-grained, maple occasionally has curly, wavy, or "bird's-eye" grain. The wood turns well on a lathe and is very resistant to abrasive wear. It has no characteristic taste or odor.

Uses — Maple wood is used principally for lumber, distilled products, veneer, crossties, and pulpwood. About 90 percent of the lumber is remanufactured into such products as flooring, furniture, boxes, crates, handles, interior finish, cabinets, woodenware, and novelties. It is especially suitable for bowling alleys and dance floors. It is distilled to make acetic acid and wood alcohol.

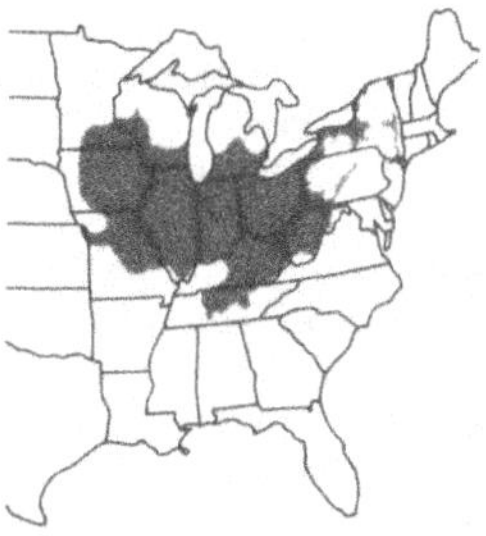

BLACK MAPLE *(Acer nigrum)* has 3-lobed **leaves** (occasionally 5-lobed), each lobe being abruptly sharp-pointed. Leaves are 5-6 inches long, and have downward-drooping sides. The **bark** on old trees is deeply furrowed and often almost black. Other characteristics of black maple that distinguish it from sugar maple are very slight. Black maple is also a source of sap for syrup and sugar. The **tree** commonly grows to 80 feet in height, with trunk diameters of 2-3 feet.

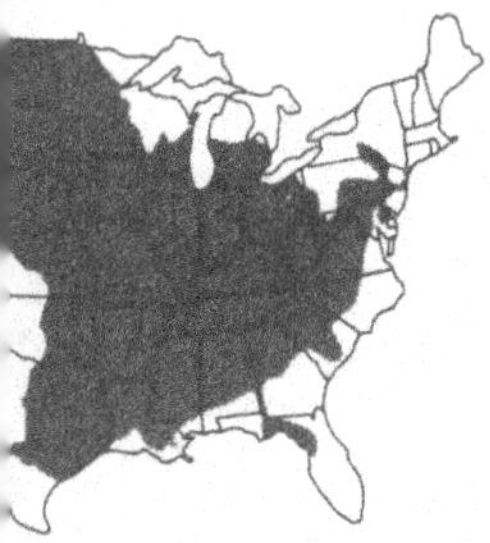

BOXELDER *(Acer negundo)* **leaves** are opposite (as are all maple leaves) but are odd-pinate, with 3-9 coarsely-toothed leaflets. The terminal leaflet is sometimes shallowly lobed. The **flowers** are small and yellowish-green, dioecious (male and female on different trees), and appear before the leaves do. The **fruit** occurs in V-shaped, winged pairs, forming long, drooping clusters that remain on the tree through the winter. The **tree** grows to 50-70 feet in height and 2-4 feet in diameter. It often divides near the ground into numerous branches that form a wide-spreading crown. Boxelder prefers moist sites, but adapts to a variety of conditions. The **bark** is pale gray or light brown, deeply divided, and with broad, scaly ridges.

RED MAPLE *(Acer rubrum)* **leaves** are 3-5 lobed, and have coarsely-toothed margins. Leaves are 2-6 inches long and somewhat broader than long. They turn brilliant scarlet, orange, or bright yellow in autumn. The polygamous **flowers** occur in short-stemmed clusters in springtime before the leaves appear. The paired, winged **fruit** is reddish, V-shaped, and ripens in late spring or early summer on drooping stems 3-4 inches long. On old trunks, the **bark** is thick, dark gray, and separated by vertical ridges into large plate-like scales. The **tree** may reach 120 feet in height and 5 feet in diameter under ideal conditions. Commonly it grows to 75-90 feet tall with diameters of 1½-2½ feet. Red maple is often planted as an ornamental. The buds and samaras are a primary food source for gray squirrels in late winter and early spring. Sprouts are a favorite deer browse.

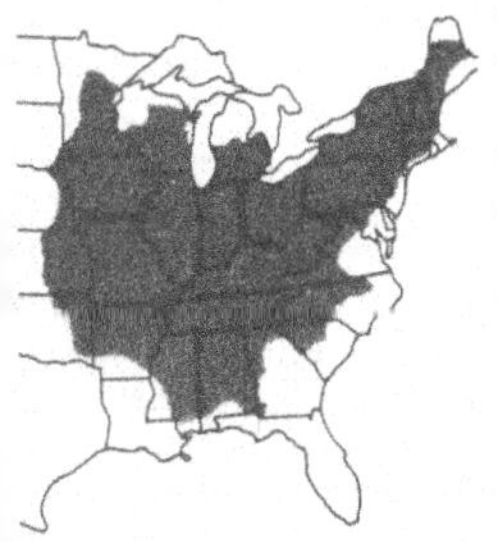

SILVER MAPLE *(Acer saccharinum)* **leaves** are deeply separated into 5 doubly-toothed lobes, the center lobe of the 5 sometimes 3-lobed. Leaves are 6-7 inches across, bright green on the upper surfaces, silvery white below. The bright red petioles are slender and drooping, up to 4-5 inches long. The greenish-yellow polygamous **flowers** are short-stemmed and occur in clusters in early spring long before the leaves appear. The paired, winged **fruit** (samara) is the shape of a broad V. The **tree** prefers moist sites, growing rapidly and attaining heights of 75-120 feet and trunk diameters of 2-4 feet. The **bark** of young trees and branches of older trees is smooth light gray. On old trunks, it is thick (½-¾ inch), reddish-brown, furrowed, and separating into large, thin scales attached in the center and pulling loose at the ends, giving the tree a ragged look. Silver maple is a favorite shade tree but its leaves are not colorful in autumn.

ASHES *(Fraxinus)*

White, black, and green ash are the most important of the 16 ashes native to the United States. All three species occur in eastern forests.

All species of ash have opposite, odd-pinnately compound **leaves** (with single terminal leaflets), each leaf having 3-11 leaflets that usually have toothed margins. Leaves are deciduous.

Some ashes are dioecious (male and female flowers on separate trees) while others are polygamous (with both sexes on the same tree). **Flowers** are borne in terminal or lateral branched clusters in spring before or at the time the trees leaf out.

The **fruit** of all ash species is a single-winged samara with a single seed (rarely 2 or 3 seeds). The fruit is usually borne in clusters on slender stems, maturing and falling in late summer and autumn.

Winter buds of ash are blunt, with 1-3 pairs of exposed scales. The terminal bud is larger than the others. Leaf scars are half-round and notched on the upper margin.

The **bark** is thick, furrowed or scaly, and usually ash-colored on branchlets. Some species of ash are classed as shrubs.

WHITE ASH *(Fraxinus americana)* is the largest and commonest of the ashes, and the most useful. The **leaves** are 8-12 inches long with 5-9 (usually 7) oblong leaflets 3-5 inches long that have smooth or finely-toothed margins. Upper surfaces of leaves are dark green and smooth; undersides are pale light green to whitish. The dioecious **flowers** open before the leaves in late spring, and occur in compact panicles. The **fruit** is 1-2½ inches long, and grows in crowded clusters 6-8 inches long. The **tree** may grow to 120 feet in height, but is commonly 70-80 feet tall with trunk diameters around 3 feet. The ashy-gray to brown **bark** is deeply divided by narrow ridges into net-like patterns.

THE WOOD OF WHITE ASH

Properties — White ash wood is heavy, hard, strong, stiff, has good shock resistance and excellent bending qualities. It holds screws and nails moderately well, but has a tendency to split. White ash lumber kiln-dries satisfactorily and holds its shape well after seasoning, but has low resistance to decay. The wood stays smooth under constant rubbing. White ash is sometimes confused with hickory, but the two woods are quite easily distinguished. The summerwood (darker) zone in ash has white dots or lines that can be seen with the naked eye; hickory also has these dots or lines in the summerwood, but they are not visible without magnification.

Uses — Much white ash wood is used for handles. It is the standard wood for D-handles for shovels and spades, and for the long handles of forks, hoes, rakes, and shovels. It is used in furniture, especially in the bent parts of chairs. It is used for slack cooperage and many types of sporting equipment including oars and bats.

BLACK ASH *(Fraxinus nigra)* **leaves** are 12-16 inches long with 7-11 pointed-oval, finely-toothed leaflets. Leaflets are smooth dark-green above, and paler below. **Flowers** are polygamous and appear before the leaves. The **fruit** (samaras) occurs in open panicles 8-10 inches long. Samaras are 1 to 1½ inches long, oblong, and with the thin wing extending below the center of the seed cavity. The seed cavity is indistinct in black ash. This species typically grows along streams, in bogs, and in poorly drained sites with high water tables. It grows most commonly in peat soils. However, it occurs on fine sands and loams underlain by clays where the surface drainage pattern causes high level of water table. The **trees** grow 60-70 feet in height (sometimes taller on the better sites) with trunk diameters of 1-2 feet; The slender upright branches form a narrow crown.

GREEN ASH *(Fraxinus pennsylvanica)* **leaves** are 6-9 inches long with 7-9 leaflets 3-4 inches long. The margins of leaflets are toothed, but only from the middle of the leaf to the pointed tip. **Flowers** are dioecious, appearing after the leaves. The **tree** commonly grows to 50-60 feet in height. Green ash is the most widely distributed of the ashes.

AMERICAN ELM *(Ulmus americana)* is a treasured ornamental and shade tree, lending its graceful, spreading form to many streets, parks, and campuses in eastern United States. Tragically, many of these stately trees have been wiped out by the Dutch elm disease in recent years. The **leaves** of American elm are oval, coming to a long slightly curved point. Leaves are rough-textured, rounded at the base on one side of the midrib, but shorter and almost straight on the other side. Leaf margins are sharply-toothed. The **fruit** is clustered on long stems. The small seed is encased in a flattened, papery, wing-like covering that is oblong in shape. In dense forest stands, the **tree** will have a fairly narrow crown and a long, clear bole. Open-grown trees usually fork near the ground and develop arching crowns. The characteristic "urn" shape of the elms permits easy identification even from a distance. This species attains heights in excess of 100 feet, and diameters of 6-11 feet. Old trees are sometimes enlarged at the base by large buttresses.

THE WOOD OF AMERICAN ELM

Properties — American elm is classed as moderate in heaviness, hardness, and stiffness. It has good shock resistance, but is moderately weak. Like all other commercial species of elm, it has excellent bending qualities. The wood is slightly below average in woodworking properties, but is among the best in ease of gluing. It is intermediate in nail-holding ability.

Uses — American elm lumber is used in the manufacture of containers, furniture, and dairy and poultry supplies. Because it bends so readily, it is much used for slack cooperage and kegs. Its veneer goes into fruit and vegetable boxes, crates, and baskets. It is used for crating of heavy articles, and in furniture for the bent parts of chairs and other items.

ROCK ELM *(Ulmus thomasii)* has a rather narrow, round-topped crown in contrast to the umbrella shape of the American and slippery elms. **Leaves** of the rock elm are similar to the other elms, but smaller. They are broadly-oval, pointed, rough-textured, deeply veined, 2-3 inches long, and have roughly-toothed margins. The oval, flattened **fruits** (samaras) are about ¾-inch long with a shallow notch at the end and an inconspicuous seed cavity. The **tree** grows to 60-80 feet in height on average sites, with diameters of 1½-2 feet. It often has a single stem instead of forking as other elms usually do. It often occurs on rocky ridges but grows best on moist, well-drained sandy loams.

THE WOOD OF ROCK ELM

Properties — The wood of rock elm is heavy and hard. It is stronger, harder, and stiffer than any of the other commercial species of elms. Rock elm also has higher shock resistance than any other American hardwood except hickory or dogwood. The wood of rock elm is difficult to work with hand or machine tools, but it has excellent bending qualities. The heartwood has only moderate resistance to decay.

Uses — Elm lumber is used principally for containers and furniture. The wood is used for veneer that is made into fruit and vegetable boxes and baskets. Large quantities go into crating for heavy articles such as furniture and appliances. The strength and toughness of this wood make it good for containers that must stand rough usage such as market baskets and bushel baskets. Considerable quantities are used in the manufacture of furniture, especially the bent parts of chairs, tables, etc.

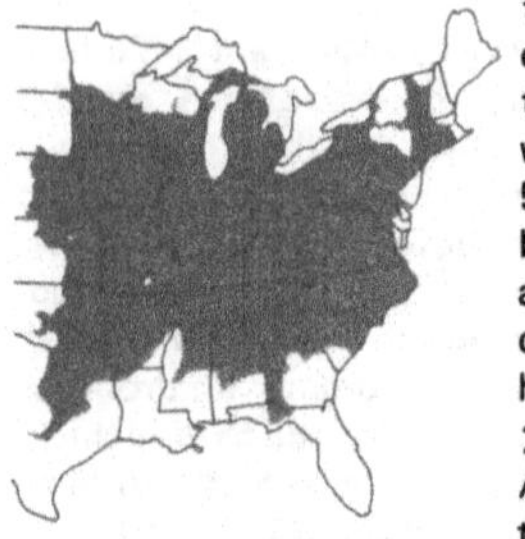

SLIPPERY ELM *(Ulmus rubra)* **Leaves** resembl those of American elm, being rough-textured, roughly oval, pointed with curving tips, unsymmetrical a the base, and sharply-toothed. The leaves are some what larger than those of American elm, being usuall 5-7 inches long. The **fruit** is also wafer-like, the see being enclosed in a flat leathery covering. The fruit i about ¾-inch in diameter and has a seed cavit covered with thick rusty-brown fuzz. The **tree** attain heights of 60-70 feet on average sites, and diameters o 2-3 feet. The form is similar to the "urn" shape o American elm. The **wood** of slippery elm is considere the best of the elms. It is tough, coarse-grained, heavy strong, and stains attractively. Its uses are mainly i furniture, crating, boxes, rough flooring, farm build ings, and gates. The common name of this specie comes from the mucilaginous coating of the fragran inner bark that is slippery to the touch.

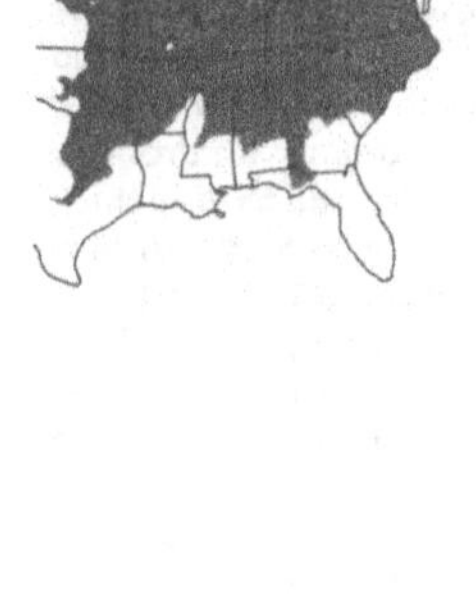

QUAKING ASPEN *(Populus tremuloides)* is the most widely distributed tree species in North America, although it is not a component of the southeastern forests below Virginia. The almost circular **leaves** give this species its name, each leaf trembling in the slightest breeze at the end of a long, flattened stem. Leaves are 1-4 inches in diameter, sharply pointed, roughly-toothed, and with a prominent, light-colored midrib. They are shiny green above, lighter-colored below. The **tree** may attain heights around 100 feet with a trunk diameter of about 3 feet, but usually is smaller (20-60 feet tall). The round-topped crown is usually narrow and symmetrical. The thin **bark** is light green and smooth except for warty bands or patches of dark-colored scaly protuberances.

THE WOOD OF QUAKING ASPEN

Properties — The wood of quaking aspen is classified as light, soft, weak, limber, and moderately low in shock resistance. Few hardwoods shrink as little in seasoning as aspen. It is low in decay resistance and nail-holding ability. It has little tendency to split, is easily worked with hand or power tools, and is fairly easy to finish to a smooth surface. It glues easily and ranks with the best of the hardwoods in paintability.

Uses — Higher grades of aspen are used primarily for lumber and wooden matches. Lower grade lumber is used largely for boxes and crates. Most aspen wood goes into pulp for the manufacture of book and magazine paper, and corrugated and insulating boards. It is preferred for high-grade excelsior. In recent years it has been approved for studs.

BALSAM POPLAR *(Populus balsamifera)* is one of the faster-growing trees of the northern forests, where it can attain heights of 100 feet or more, and diameters of 3-7 feet. **Leaves** are roughly oval, 5-7 inches long, 4-5 inches wide, pointed, and have conspicuous rounded teeth along the margins. Leaves are dark green on the upper surfaces and lighter green on the lower sides. The **fruit** is borne on stalks that hang from the branches in clusters. The buds are covered with a sticky, fragrant resinous material in the spring. The **wood** is used for veneer, furniture parts, and pulpwood.

BIGTOOTH ASPEN *(Populus grandidentata)* **leaves** give the species its common name, as the leaf margins have very conspicuous rounded teeth. Leaves are almost round, sharp-pointed, dark green on the upper surface but lighter green on the lower, and 2-3 inches long and wide. **Bark** is thin, smooth, and light-gray to green on younger trees and the upper portions of the stems of older trees. Near the base of older trees, the bark is dark brown tinged with red, and ¾ to 1 inch thick. This **tree** grows to 60-70 feet in height and about 2 feet in diameter, often with a narrow, round-topped crown. The soft, light-colored **wood** is used for interior parts of furniture, and for pulpwood.

EASTERN COTTONWOOD *(Populus deltoides)* is the fastest-growing commercial forest species in North America. It is also one of the tallest **trees** of eastern forests, attaining heights close to 200 feet and diameters of 4-6 feet. Carefully selected stock planted on favorable sites in the Mississippi Delta have attained 5-inch diameters and 30-foot heights in two growing seasons. The **leaves** are roughly triangular, toothed, pointed, 3-6 inches long, paler below than above, and with a flattened stem. The **fruit** consists of numerous green bud-shaped capsules that are clustered along short stems that hang from the branches in long, narrow clusters. Seeds are released from the capsules when ripe and drift on the breeze, each one suspended from a tuft of white "cottony" hairs. **Bark** on young trees and on upper stems of older trees is smooth and greenish. Bark of older trees is dark gray, heavily furrowed and ridged. Cottonwood does best on moist well-drained soils such as those found in many areas of the Mississippi Valley. It does not often develop into a well-shaped tree if it is more than 15-50 feet above the average level of the streams in the area.

THE WOOD OF COTTONWOOD

Properties — The woods of the several cottonwood species are similar, and are sold together for the same uses. Cottonwood is moderately light in weight, moderately soft, and moderately limber. It is moderately weak in bending and compression and moderately low in shock resistance. Shrinkage during drying is classed as moderately large, and the wood requires careful seasoning if warp is to be avoided. Cottonwood is low in resistance to nail-withdrawal, but does not split easily. It is difficult to work with tools without producing chipped or fuzzy grain, primarily because of the pockets of tension wood common to the species. It holds paint well.

Uses — Much cottonwood is cut into lumber and veneer that is remanufactured into containers and interior parts of furniture. The chief use of both the lumber and veneer in furniture manufacturing is as core stock that is overlaid with high-grade veneers. Cottonwood is much used for pulpwood, and is one of the few hardwood species that is planted and grown specifically for that purpose. The pulp is used in high-grade magazine and book paper.

PAPER BIRCH *(Betula papyrifera)* is a tree of the northern forests, where it is conspicuous for the white, papery **bark** on the trunks of mature trees. The bark often peels horizontally into long strips that are curled and ragged at the ends. This is the species utilized by Indians in making their legendary birch-bark canoes. The **leaves** of paper birch are pointed and doubly toothed, roughly oval, rounded at the base, and 2-3 inches long. The **fruit** is cylindrical, about 1½ inches long and ½ inch in diameter, and hangs from a slender stalk. This fruit (strobile) disintegrates when mature. The **tree** is usually slender and graceful, often growing in clumps of 2-4 individuals. It attains heights of 80 feet with trunk diameters seldom more than 2 feet. The crown is usually rounded and irregular. Paper birch grows in the South only in North Carolina, occurring there in the balsam-spruce forests at the very highest elevations. The **wood** of paper birch is hard, light, strong, and very close-grained. It is used for veneer, spools, bobbins, toothpicks, and pulpwood.

YELLOW BIRCH *(Betula alleghaniensis)* is named for its **bark** which has a yellowish-bronze color and peels into long, ragged, horizontal strips. **Leaves** are 3-5 inches long, pointed, sharply-toothed, and roughly oblong-oval. The **fruit** is an oval, erect strobile 1-1½ inches long and about ¾-inch thick. The **tree** may grow to 100 feet in height and 3 feet in diameter. It thrives in hilly terrain; in the southern Appalachians it is almost always found above 3,000 feet elevation. The tree furnishes browse for deer, and its buds and catkins are food for grouse and other wildlife.

THE WOOD OF YELLOW BIRCH

Properties — Yellow birch is heavy, hard, strong, stiff, and with very high shock resistance. Shrinkage is large during seasoning, and the wood has low decay resistance. The wood is difficult to work with handtools, but can be readily shaped by machine. It is high in nail-holding ability. It has pleasing grain pattern and takes a high polish.

Uses — Yellow birch is often mixed with hard maples, and used for lumber, veneer, distilled products, and crossties. The lumber and veneer go mostly into furniture, boxes, baskets, crates, woodenware, interior finish, and millwork. Yellow birch is one of the principal woods used for hardwood distillation for wood alcohol, acetate of lime, charcoal, tar, and oils. It is also used for pulpwood and slack cooperage.

OAKS *(Quercus)*

Taken together, the oaks represent one of the most important hardwood segments of eastern forests. They provide high-grade **wood** for furniture, flooring, veneer, and many other products useful and beautiful. Their **fruit** is the acorn, which is extremely valuable as wildlife food. It is the primary overwintering food source for most forest game species in the South. Acorns are eaten by deer, turkey, ruffed grouse, quail, bear, boar, and some songbirds. The **trees** are often decorative, and preferred by many homeowners as shade trees. In the fall, oaks add rich reds, browns, and russet colors to the autumn foliage display. Some species are deciduous, some are evergreen, and a few may be either depending on local conditions of site and weather.

Oak species are separated into two major groups: red oaks and white oaks. Generally the red oaks will have **leaves** with pointed lobes that are often tipped with bristles or spines. Red oak **acorns** are generally bitter, and require two years to mature. **Leaves** of white oak usually have rounded lobes. White oak **acorns** are not bitter, and mature in one season. The inside of the shell of the red oak acorn is hairy; white oak acorns are hairless inside.

All of the oaks have alternate leaves. Twigs usually have clusters of buds at the tips. Both sexes of **flowers** are borne on the same tree; the staminate flowers occur in rather showy pendants in early spring, clustered near the ends of the twigs.

RED OAK GROUP

NORTHERN RED OAK *(Quercus rubra)* **leaves** are 5-8 inches long and have 7-11 lobes. Each lobe is usually 3-toothed, sharply pointed, and with bristles on the points. Leaves are deciduous, turning red before they fall in autumn. The **tree** will reach 60-80 feet in height and 3 or more feet in diameter. The **acorns** are usually about an inch long with a flat, shallow cup at the base. The crown of the tree is usually rounded and comparatively narrow. The species is often planted as ornamental or shade tree.

THE WOOD OF THE RED OAKS

Properties — Wood of the oaks in the red oak group is similar to that of the white oaks. A major difference is that red oak (meaning the wood of all species in the red oak group) is extremely porous, and therefore not suitable for such uses as tight cooperage. The wood is heavy, hard, stiff, and has high shock resistance. It undergoes large shrinkage during seasoning. It is above average in all machining operations except shaping. The heartwood is low in decay resistance.

Uses — Wood of the red oaks is used for flooring, slack cooperage, furniture, millwork, boxes, crates, caskets, timbers, handles, coffins, pallets, agricultural implements, boats, and woodenware. The hardness and resistance to wear of red oak, plus its beauty, make it preferred for flooring for residences. When preservative-treated, red oak is used extensively for crossties, mine timbers, and fence posts.

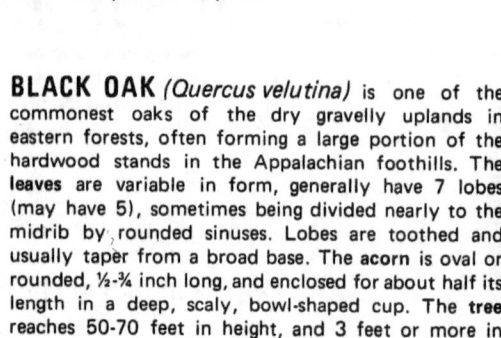

BLACK OAK *(Quercus velutina)* is one of the commonest oaks of the dry gravelly uplands in eastern forests, often forming a large portion of the hardwood stands in the Appalachian foothills. The **leaves** are variable in form, generally have 7 lobes (may have 5), sometimes being divided nearly to the midrib by rounded sinuses. Lobes are toothed and usually taper from a broad base. The **acorn** is oval or rounded, ½-¾ inch long, and enclosed for about half its length in a deep, scaly, bowl-shaped cup. The **tree** reaches 50-70 feet in height, and 3 feet or more in diameter. In early spring the unfolding leaves are a deep red, turning silvery within a few days. **Bark** on the trunks of old trees is usually thick, black, and deeply divided into broad, rounded ridges. The crown is usually rounded.

WHITE OAK GROUP

WHITE OAK *(Quercus alba)* has deciduous **leaves** 5-9 inches long with 7-9 rounded lobes. The depth of the sinuses separating the lobes varies, almost reaching the midrib in some cases. The base of the leaf narrows abruptly to become wedge-shaped at the stem. The **acorn** is about ¾-inch long, light chestnut-brown, enclosed about one-fourth of its length in a bowl-shaped cup that is covered with rough scales that are joined at their bases to form small knobs. The **tree** reaches 80-100 feet in height and 3-4 feet in trunk diameter. In the open, white oak develops a rounded spreading crown; in forest stands it has a tall, clear stem and smaller crown. Growth is good on all but the driest, shallow soils, but is best on deep, well-drained loamy soils.

THE WOOD OF THE WHITE OAKS

Properties — All of the white oaks are heavy, very hard, and strong. The wood is subject to large shrinkage during seasoning, and extra care must be taken to avoid checking and warping. Pores of the heartwood are impervious to liquids, making white oak the only successful wood for use as tight cooperage. Large amounts of higher grades are used for bourbon barrels. The heartwood is comparatively decay resistant, more so than that of red oaks. White oaks are above average in all machining operations except shaping. All oaks in the white oak group share just about the same properties and uses.

Uses — Most white oak is made into lumber for flooring, furniture, tight cooperage, millwork, timbers, handles, boxes, and crates. Perhaps the largest amounts go into high-quality flooring, barrels, kegs, and casks. It is prized for use in construction of ships and boats.

SWAMP WHITE OAK *(Quercus bicolor)* **leaves** are deciduous, 5-6 inches long, and with margins that have bluntly-pointed lobes of varying size and depth. Undersides of leaves are hairy. Leaves are roughly oval, broader near the apex and tapering to the base. **Acorns** grow usually in pairs on a fairly long stem (1-4 inches). Acorns are oval, about one inch long, with about a third of their length enclosed in the scaly cup. This species is commonly found on wet sites and on areas subject to flooding. On good sites, the **tree** grows quite rapidly, normally attaining heights of 60-70 feet and diameters of 2-3 feet. Some trees have been reported that were 100 feet tall and 7 feet in diameter.

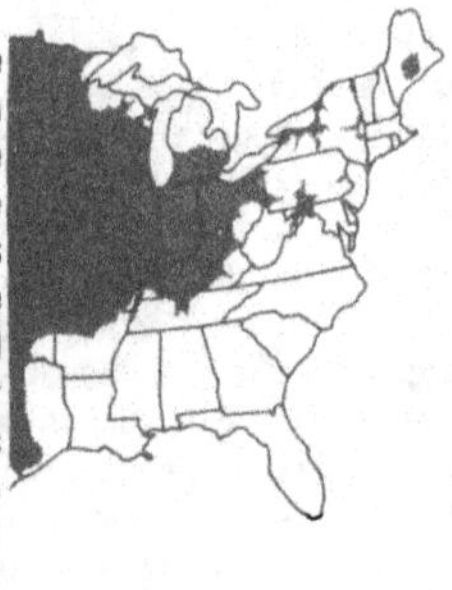

BUR OAK *(Quercus macrocarpa)* has deciduous **leaves** that are somewhat wedge-shaped, divided into 5-7 lobes by wide sinuses reaching sometimes almost to the midrib. The lobes are rounded, comprising a leaf that is quite large (6-12 inches long and 3-6 inches wide). The **acorn** is oval or broadly ovate, broad at the base and rounded at the depressed apex. The acorn is from ¾ to 1½ inches long, and half covered in the fringed cup that gives the species its common name. Twigs sometimes develop corky "wings" 1-1½ inches wide. The **tree** is a relatively slow grower; it is one of the most drought-resistant of the oaks, and does well on exposed sandy sites. On better sites, it reaches 100 feet in height and 3-4 feet in diameter. It is said to have reached a height of 170 feet and diameter of 7 feet in the Ohio Valley.

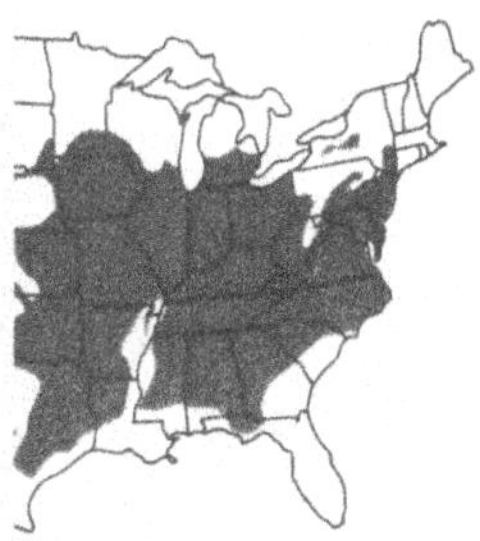

BLACK WALNUT *(Juglans nigra)* **leaves** are deciduous, alternate, pinnately compound, 12-24 inches long, with 15-23 sharply-oval, toothed, long-pointed leaflets 3-3½ inches long. Leaves turn bright, clear yellow in autumn. The **fruit**, borne singly or in pairs, is globular, pointed at the apex, 1½-2 inches in diameter with a thick yellow-green fibrous husk. The hard, woody nut is dark brown, oval to oblong, 1-1½ inches in diameter, and deeply divided on the outer surface into irregular ridges. The nutmeats are sweet and edible, and are a favorite food for red and gray squirrels. Black walnut develops best on deep, well-drained soils, and thrives in well-drained bottomlands and coves of the Appalachians. The **tree** reaches 100 feet in height and 2-3 feet in diameter.

THE WOOD OF BLACK WALNUT

Properties — Black walnut wood is heavy, hard, strong, stiff, and has good shock resistance. The heartwood is one of the most durable of any hardwood. Black walnut works easily with hand or machine tools, takes and holds paints and stains exceptionally well, is readily polished and easily glued. The wood finishes beautifully with an outstandingly handsome grain color and pattern. The heartwood is chocolate brown or darker, sometimes with purplish streaks.

Uses — Most black walnut goes into furniture of the highest quality. It is also prized for gunstocks and interior finishes. In furniture, it is used either as solid wood or as veneer. It is popular for any interior finish where its striking grain and color create special effects. The wood is particularly well suited to gunstocks because of its stability after seasoning, its fine machining qualities, its uniformity of texture, and its beauty. It became popular in Colonial days as an imitation of the more expensive imported mahogany.

AMERICAN BASSWOOD *(Tilia americana)* has large, alternate, roughly toothed **leaves** that are sharply pointed. Leaves are heart-shaped, conspicuously veined, 5-6 inches long and 3-4 inches wide. The **fruits** are hard and rounded, hanging suspended in clusters from a stalk that is attached to a papery-thin, strap-shaped bract. Each fruit contains two seeds and is covered with a thick, reddish-brown fuzz. American basswood is a tall and stately **tree**, sometimes reaching 120-140 feet in height and 4-4½ feet in diameter. The small slightly drooping branches form a broad, round-topped crown in open stands or a narrow pyramidal crown in dense stands. In closed forest stands the tree often has a straight bole that is clear of branches for half the tree's total height. Forest-grown basswood trees are easily damaged by fire; thus most large, older trees will be hollow or otherwise defective.

THE WOOD OF BASSWOOD

Properties — Basswood is light, weak, low in resistance to shock, and soft, but moderately stiff. It has large shrinkage, but stays in place well after seasoning. It has low nail-holding ability, but resists splitting. The wood of basswood is easy to work, holds paint well, and glues easily. It is prized by wood-carvers because it is easily worked and has an even grain.

Uses — Most basswood is cut into lumber for remanufacture into crates and boxes. Higher grades are used for sash, doors, and general millwork. Considerable amount goes into furniture, especially as core stock to be overlaid with high-grade veneers. Basswood is also used for pulpwood.

HACKBERRY *(Celtis occidentalis)* **leaves** are unsymmetrically oval, 2½-3½ inches long with sharply-toothed margins, and have curving, pointed tips. The globular **fruit** is borne singly on stems ½-¾-inch long, ripening in September but often remaining on the tree over the winter. The fruit is dark purple and about 1/3-inch in diameter. Hackberry is widely distributed in eastern United States and adapts to a variety of sites and climates. It is principally a bottomland tree, growing best on valley soils. In much of its range it is common on slopes and bluffs. The **tree** commonly grows to 30-40 feet in height, but on the best sites may reach a height of 130 feet and a diameter of 4 feet or more. The **bark** is smooth, dark brown to gray, and quite thickly covered with warty protuberances.

THE WOOD OF HACKBERRY

Properties — The woods of hackberry and sugarberry are quite similar, and share many of the same properties. The wood is moderately heavy, moderately hard, high in shock resistance, and with average resistance to decay. The grain is straight, but sometimes is interlocked. The wood is average in nail-holding ability, and works well with tools except that it shapes poorly. Shrinkage during drying is moderate. The wood holds its shape well after seasoning, and glues very well. It has no characteristic taste or odor.

Uses — Good grades of hackberry are used for furniture. Hackberry is also used for millwork and some athletic equipment. Poorer grades find their usage in crates and boxes. Quite often hackberry and sugarberry are mixed in with ash and elm at the sawmill, and these woods are sold together and find common usage.

WILLOWS *(Salix)*

There are more than 200 species of willows *(Salix)* distributed widely throughout the north-temperate zones of the world; about 70 are native to North America. Many of these are shrubs, and only 38 species reach tree size in the United States. Only black willow is important for lumber.

Willow **leaves** are deciduous, alternate, and typically long and narrow. They have finely toothed or smooth edges.

The **flowers** are borne in erect catkins in spring, usually before the leaves appear. The several species which bear the soft, downy catkins that are the familiar harbingers of spring are commonly called "pussy willows." A number of willows are planted as ornamentals.

The many species of willow are so similar that it takes a trained botanist to make precise identification. The **woods** of willow trees are also very similar, and often mixed together for pulpwood and other products.

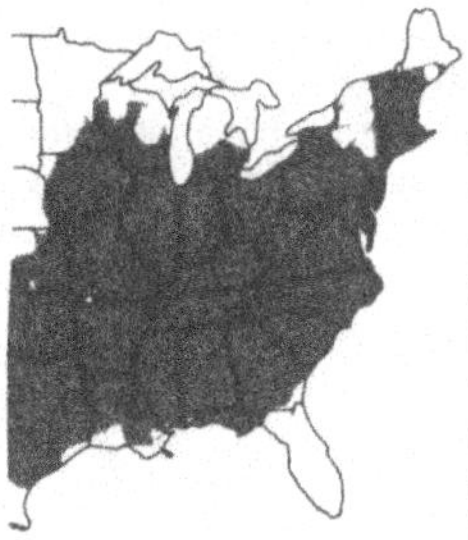

BLACK WILLOW *(Salix nigra)* is one of the largest of the native willows. Its **leaves** are long (4-6 inches) and narrow (½ to ¾-inch wide), pointed, and with finely-toothed margins. Black willow will grow on almost any soil but its shallow, wide-spreading roots need an abundant and continuous supply of moisture during the growing season. On the best sites in the Mississippi Valley, the **tree** will grow to 140 feet in height and 4 feet in diameter. Usually the tree will be 30-40 feet high, often with several stems clustered, and the thick spreading branches forming a broad, irregular crown. Trunks are often twisted, curved, or leaning.

THE WOOD OF THE WILLOWS

Properties — The various species of willow yield woods that share much the same appearance and properties. The wood is moderately light to light, moderately soft, very weak, high in shock resistance, low in nail-holding ability, but not easily split while nailing. Willow works well with tools, glues well, and takes and holds stains and finishes well. It has large shrinkage during drying, and is very low in durability.

Uses — Willow wood is used in boxes and crates, in furniture as core stock, for slack cooperage, and for woodenware and novelties. A specialty use is for artificial limbs. It finds common usages with cottonwood lumber, and the two are sometimes sold mixed at the mill.

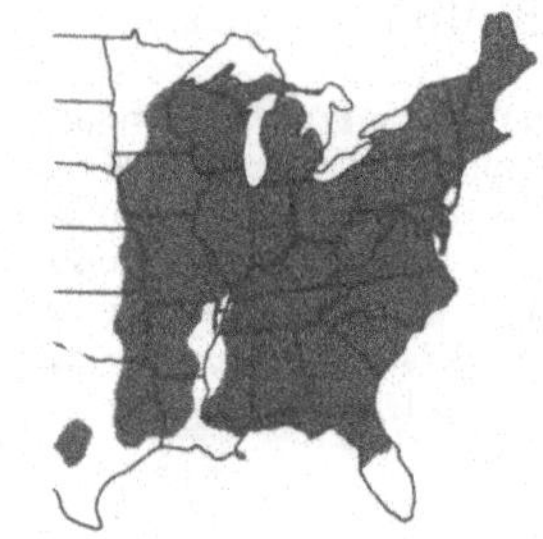

BLACK CHERRY *(Prunus serotina)* has deciduous **leaves** 2-6 inches long and ½ to ¾-inch wide, narrowly-oval or oblong, and pointed. Leaf edges are finely-toothed with incurving teeth. The white **flowers** bloom when the leaves are about half grown, occurring on racemes 4-6 inches long. As the **fruit** develops the racemes gradually droop as the berries reach full size (1/3 to 1/2-inch diameter). The cherries are dark red when they are fully developed, turning black with dark purple flesh as they ripen. They are food for wild animals and birds. Black cherry is the largest of the native cherries of the United States, and the only one of commercial value. The **tree** reaches 60 feet (sometimes more) in height and 3 feet or more in diameter. It grows on all kinds of sites except those that are very swampy or extremely dry. The **bark** on branches and on the trunks of young trees is thin, satiny, reddish brown, and with horizontal markings made up of patches or rows of lenticels. **Bark** on older trees is in small scaly plates with edges slightly upraised. The bark, twigs, and foliage of black cherry are bitter and astringent, as they are on all cherries. Black cherry is an intolerant species, which means that it will not survive for long when in competition with other trees for sunlight. It becomes established only in forest openings, and is typical of the many species of forest trees that can be regenerated only when all of the vegetation is cut on the area designated.

THE WOOD OF BLACK CHERRY

Properties — Black cherry is stiff and strong, with high resistance to shock, but is only moderately hard and heavy. Black cherry wood stays in place well after seasoning, and is comparatively free from checking and warping. It is difficult to work with handtools, but has excellent bending strength. It can be glued satisfactorily.

Uses — Nearly all black cherry harvested is sawed into lumber that is remanufactured into a variety of valuable products. Some of the highest-quality furniture on the market is made from solid cherry wood. It is also made into veneers that utilize and display the exceptionally attractive color and grain on less-expensive furniture. The wood has a distinctive light-to-dark reddish-brown color, and often has grain patterns that are quite striking. Because it is very stable after seasoning, it is preferred for use as backing blocks for electrotype plates and in pattern-making. It is also used for woodenware, novelties, and interior finishes for buildings. Lower grades may be used for the interior parts of furniture, and for molding and trim.

BUTTERNUT *(Juglans cinerea)* has deciduous, alternate, pinnately compound **leaves** 15-20 inches long with 11-17 pointed-oblong, finely-toothed leaflets 2-3 inches long. Leaflets are downy on the undersides. The terminal leaflet is stalked in butternut, but not in black walnut. The **fruit**, borne in drooping clusters of 3-5, is oval, 1½-2½ inches long, greenish-brown, and sticky on the surface of its thick husk. The **nut** is oval, light brown, long-pointed, and many-creased on the hard surface. The nutmeat is sweet. Butternut grows best on streambanks and on well-drained soils but occurs most frequently in coves, on slopes, and on terraces. It is not a common tree in any area. It is a fast-growing species on favorable sites. The **tree** rarely attains more than 100 feet in height and 3 feet in diameter. It often divides 20-30 feet from the ground into stout limbs that form the wide-spreading, round-topped crown typical of open-grown trees. The **wood** is light, soft, and not strong. It is used for interior finishes and furniture.

HICKORIES *(Carya)*

Hickories are an important segment of the hardwood forests of eastern United States, and are limited to those forests east of the Great Plains. Of the 15 species of hickories, eight are considered commercially important in the United States.

The **leaves** of all hickories are deciduous; they are compound, alternate, and odd-pinnate (with a terminal leaflet). Each leaf will have 3-17 roughly oval, pointed, toothed leaflets that are aromatic when crushed. Staminate **flowers** are 3-branched, slender, drooping catkins. The **fruit** is an edible nut, and an important source of food for squirrels. They are often preferred over acorns. In heavy mast years, they provide food through most of the winter. Sometimes they are eaten by wood ducks and mallards. The **trees** have smooth gray bark that becomes very rough and scaly on older trees, often separating into long strips. Pecans, both wild and cultivated, are members of this genus.

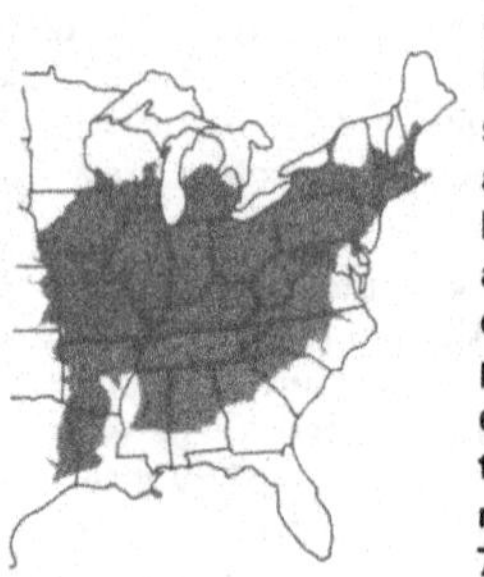

SHAGBARK HICKORY *(Carya ovata)* **leaves** are 8-14 inches long with 5 (rarely 7) tapered-oval, finely-toothed leaflets. The terminal leaflet is largest, being 5-7 inches long and 2-3 inches wide. The **nut**, borne singly or in pairs, is roughly oblong, 1-2½ inches long, and covered in a thick 4-part husk that splits to the base of the nut when ripe. The nutmeat is sweet with an excellent flavor. Shagbark hickory is named for the characteristic light-gray **bark** that separates into thick plates a foot or more long that curl outward at both ends. Older trees develop a rough, shaggy trunk. The **tree** grows on a variety of sites and soils, and may reach heights of 120 feet or more. It commonly grows 70-90 feet tall, and usually has a good form. Shagbark hickory is usually a minor component in the hardwood forests where it occurs. In the northern part of its range, it grows on upland slopes at elevations as high as 2,000 feet, where it is associated with oaks, other hickories, and various mixed hardwoods. In the South it is associated with a large number of hardwoods, especially those common to the bottomlands, and here prefers the deep, moist soils of alluvial origin.

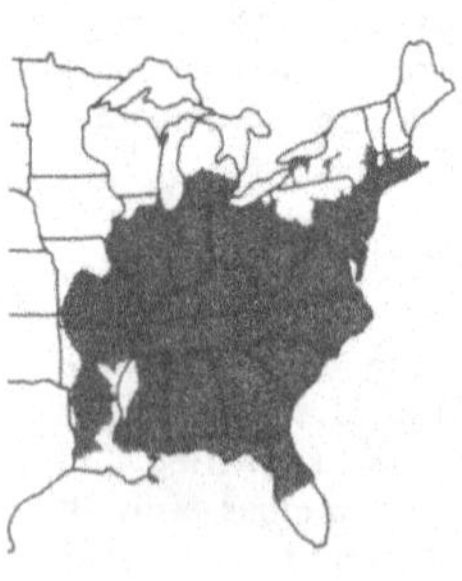

PIGNUT HICKORY *(Carya glabra)* **leaves** are 8-12 inches long with 5 (rarely 7) finely-toothed, sharp-pointed, lanceolate leaflets. The **fruit** is globular to pear-shaped, about 1½ inches long, and enclosed in a thin husk that remains closed or opens only part-way down the nut. The nutmeat is small and sweet. The **tree** is 50-75 feet in height, with trunk diameters of 3-4 feet. The spreading, often drooping branches form a tall, narrow head. This hickory inhabits dry ridges and hillsides with well-drained soils, but responds well to richer soils. It is also common on moist sites in the southern portion of the Appalachian region. The best development of this species occurs in the lower basin of the Ohio River. It is the hickory most commonly found in the Appalachian forest, but even there is a minor component of the total forest resource. As the tree matures, the dark gray **bark** becomes deeply fissured and separates into scaly ridges that form into rough diamond patterns.

BITTERNUT HICKORY *(Carya cordiformis)* **leaves** are 6-10 inches long with 7-9 long-oval, toothed leaflets that are dark yellow-green above and lighter below. The 4-ribbed **nut** is about an inch long, roughly spherical but often broader than long. The nutmeat is very bitter. The husk is 4-winged from the tip to about the middle, and covered with yellowish scales. The tree is probably the most abundant of the hickories, and found throughout eastern United States. It prefers rich moist sites such as bottomlands, there developing into a handsome tree 100 feet or more tall with a straight trunk 2-3 feet in diameter.

CHAPTER 4 — FOREST MANAGEMENT BY SPECIES

RED PINE

Description

A century ago, red pine made up about one-third of the 22 million acres of pine forest in Minnesota, Wisconsin and Michigan. Today it covers only a little more than one million acres — mostly planted since 1930.

Red pine is one of the most versatile of forest trees. It has high aesthetic value as well as being heavily used as a pulpwood and lumber tree. Because it is so versatile and useable, red pine is a highly recommended planting species.

Red pine requires direct sunlight and is quite intolerant to shade. Trees growing in medium to heavy shade usually wither and die in a matter of months or years.

Management

Red pine usually does best on coarse, well-drained sandy or sandy loam type soil. Red pine planted on sandy and relatively unfertile soil will do well with a minimum of attention, simply because there is very little vegetation competition. Good drainage is essential. Red pine planted in low areas seem to suffer from high mortality for a variety of reasons. Even a little dip in an otherwise flat parcel of land will tend to cause problems. Red pine does poorly on undrained or heavy clay loam soils.

Spacing

Recommended initial spacing is 600-800 trees per acre. This will result in merchantable trees at the first thinning. The most common mistake made in tree planting is to plant trees too close together. Be sure to give the trees needed growing space.

Planting

To achieve best survival, plant red pine seedlings as early in the spring as possible. Good soil moisture is essential at the time of planting to avoid high mortality. Containerized seedlings, however, may be planted throughout the summer with good survival.

Site Preparation

The control of competing vegetation in young red pine plantations is essential for good survival. Site preparation will vary, depending on the site. Your forester can tell you what kind, if any, is needed. Site preparation is discussed in more detail elsewhere.

Brush Control

Release from competing vegetation may be needed within five years of planting. The method of release should be either chemical or hand release. This is discussed in the under "Timber Stand Improvement".

Risk Of Loss From Damaging Agents

Although red pine is not the "problem free"

tree it was once thought to be, it still rates high in its ability to deal with its natural enemies. Like all trees, fire, insects and disease can affect the health and yield of red pine plantations. The best advice to be given here is to pay attention to the condition of your trees. If insects or disease become evident, be sure to contact your local forester immediately for further advice.

Yield

Red pine on better sites yield from 1.5 to 2.0 cords per acre per year of annual growth, and yield between 30-50 cords per acre at age 80.

Intermediate thinnings beginning at about age 25 will help to maintain good diameter growth in your plantation. Your forester can recommend the specific intervals of the thinnings.

Harvest

The final harvest of red pine usually occurs at a rotation age between 80-120, depending upon the site. Clearcutting, seed tree or shelterwood, are some of the harvesting methods used. Clearcutting is used most often because the seed tree and shelterwood methods depend upon natural seeding for reproduction which is not as dependable as hand planting.

Wildlife Considerations

Red pine stands are considered poor habitat for wildlife. If enough sunlight is allowed to reach the ground, however, small trees, shrubs, and other plants will offer diversity to the stand, increasing wildlife numbers and species. For approximately the first 30 years, red pine stands will offer winter cover for deer, but past 30 years the lowest live branches are too high above the ground to offer any thermal regulation. At the same time, the shading caused by a closed canopy will greatly reduce the understory layer of plant material, offering very little to wildlife.

Identification And Uses

Red Pine *(Pinus resinosa)* **needles** are soft, slender, flexible, lustrous, dark green, 5-6 inches long, and in 2-leaved clusters. **Cones** are symmetrical, ovoid, 2-2¼ inches long, light chestnut-brown and lustrous at maturity, and not armed with spines on the scales. The thick **bark** is divided shallowly into broad, flat ridges that are covered with thin reddish-brown scales. The **tree** usually grows to 70-80 feet in height with a trunk 2-3 feet in **diameter**. It may reach 100 feet in height; some trees have attained 150 feet in height and 5 feet in diameter. Red pine has a tall, straight trunk and a symmetrical crown. The somewhat pendulous branches almost reach to the ground on smaller trees. This species is an important timber tree. The **wood** is hard, very close-grained, light, and quite strong. It is used among other things for construction, pilings, and flooring.

Article — Minnesota DNR Forestry
Species Description — "Important Trees of Eastern Forests"
U.S.D.A. — Forest Service

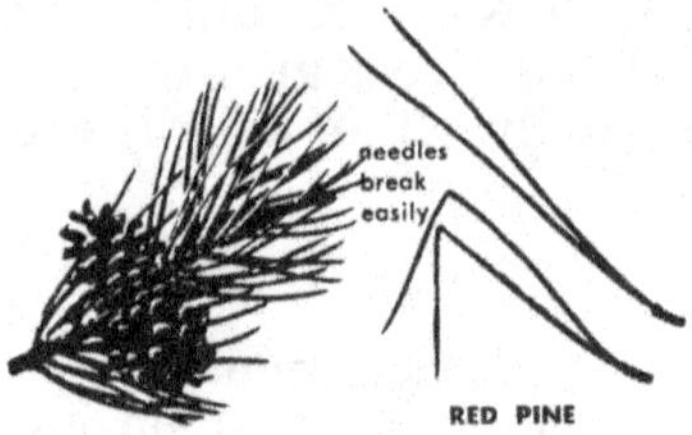

RED PINE

WHITE PINE

Description

Eastern white pine is the largest pine found in Minnesota. It was, and continues to be, highly prized for its valuable lumber. The large stands of tall white pine served to lure the timber barons to our state at the turn of the century. The tree is commonly 80-120 feet in height and has a diameter between 2-3 feet at maturity. While it is not unusual to have undisturbed trees live 200 years or more, rotation age is commonly 90-125 years. It will grow on a wide range of soils, but grows best on moderately moist, sandy loam soils. The most rapid growth rate occurs in the southeastern areas of the state, where 2-3 feet of height is not uncommon on good sites. Eastern white pine is an aggressive tree, and will readily invade openings such as old fields. Moderate shade tolerance allows the tree to become established in existing standings of pine or other species, provided the shading is not complete.

Management

Regeneration

A good white pine seed crop is produced every 3-5 years with the seed capable of traveling 200-700 feet or more, depending on wind currents. Soil scarification will help to prepare a moist mineral seedbed for natural regeneration. Hand planting approximately 600-800 trees per acre provides uniform stocking and spacing, and produces a high percentage of survival. In the northern two-thirds of the state, planting should be done under a light forest canopy to reduce damage by the white pine weevil and white pine blister rust. Eastern white pine can tolerate as much as 70 percent shading and maintain adequate seedling growth.

Culture

If trees have become established under a forest canopy, they should be released when the pine are approximately 20 feet in height. Examinations of the stand at this time will determine whether the first thinning should be accomplished simultaneously with the harvest of the overstory. The basal area should be reduced to 100-140 square feet per acre, and periodic thinnings of 10-15 square feet per acre should be practiced. Treatment in open grown plantations would be similar. Early pruning may be used to help prevent losses due to blister rust.

Harvest

Preparation for the final harvest should begin 15-20 years prior to the actual cutting. Reducing the basal area to 50-60 square feet per acre will open the stand sufficiently to reproduce white pine from seed. After the regeneration has reached approximately 20 feet in height, the older trees are harvested to release the juvenile stand to full sunlight.

Principle Enemies

Insect and disease are two serious enemies of white pine. The white pine weevil attacks the terminal leader of young pine, which cause serious deformities in the main stem. This insect can be discouraged by avoiding planting in full sunlight.

White pine blister rust is a fungus disease which attacks white pine. Although capable of infecting any age tree it is found most common in younger stands. Ribes plants (such as gooseberry), serve as the alternate host for the fungus; therefore, locations where ribes are prevalent should be avoided if possible. The state is broken into three zones, demonstrating the relative probability of a white pine becoming infected with blister rust. The highest hazard is the northeast one-third of the state; medium hazard in the northwest and central; and lowest hazard south of the Twin Cities. Selecting sites with favorable air movement will also lessen the incidence of the disease.

Insects and disease often enter into less vigorous, older, and fire damaged trees. These can be reduced by practicing proper management.

Uses

This tree produces a premium quality lumber for use in the construction and manufacturing industry.

Wildlife Considerations

Although wildlife receive virtually no food benefit from white pine, younger dense stands may be used for shelter during inclement weather. Deer and rabbits browse heavily on white pine seedlings, and this may become a problem in certain plantations.

Identification And Uses

Eastern White Pine
(Pinus strobus)

Needles are soft bluish-green, flexible, 3-5 inches long, bundles of 5, with 3-5 fine white lines (of stomata) on two surfaces of each needle. Needles remain on the tree for two years. Cones are fully grown in the summer of the second season, opening to discharge seed that autumn. **Cones** are 4-8 inches long, curved, stalked, with scales not spiny. The **bark** on young trunks and branches is thin, smooth, greenish (sometimes tinged with red), and lustrous. On old trunks, the bark is deeply divided into broad ridges that are covered with purplish scales. The **tree** commmonly attains 100 feet in height and 4 feet in diameter, with a tall, straight stem and pyramidal crown. Branches are definite whorls of long lateral branches sweeping upward in graceful curves. This is the largest conifer of eastern forests, is long-lived, and has been known to reach heights above 200 feet. The **wood** is light, straight-grained, easily worked, but not strong. It is used in cabinet work, interior finishes, woodenware, matches, and lumber. A large part of the winter diet of red squirrels (boomers) in the southern Applachians is made up of the seed of eastern white pine.

Article — Minnesota DNR Forestry
Species Description — "Important Trees of Eastern Forests" U.S.D.A. — Forest Service.

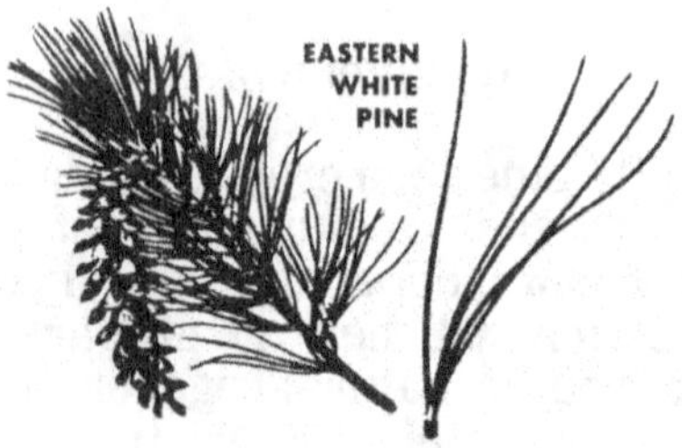

JACK PINE

Description

Jack pine is a medium-sized, short-lived tree. It grows further north than any other American pine and reaches some of its best development in northeastern Minnesota. Depending upon the site, jack pine matures between 40-70 years and begins to deteriorate between 50-90 years. Size at maturity can range from 40-80 feet in height and from 6-14 inches in diameter. It is considered a pioneer species and is very intolerant to shade. Due to its wide range, the species shows a great deal of genetic variation.

Jack pine is often found in pure even-aged stands, or mixed with aspen, white and red pine. It can maintain itself on very sandy or gravelly soils where other species scarcely survive; but it grows best in well-drained loamy soils.

Management

Regeneration

Reproduction of jack pine often occurs naturally in areas where logging or fire has occurred. This is due to the abundant release of seed from the serotinous cone which may remain sealed for 10-15 years, especially in northern Minnesota. Where the seed source is adequate, the stand of saplings may be so thick as to require weeding in the first ten years. Sometimes germination is less successful and interplanting is required to bring the stand to a desired stocking. Recommended stocking is 600-1000 trees/acre in order to maintain good form. Other methods of reforestation are planting and direct seeding. Preparation of the site to insure adequate light is necessary. On better sites, jack pine may be considered a temporary type that may be replaced by a more suitable species at the end of rotation.

Culture

Jack pine on better sites (site index 60), can be thinned to about 90 square feet of basal area at age 30-70 in order to maintain good diameter growth. Average annual growth ranges from .3 cords/acre/year on poor sites to .7 cords/acre/year on better sites.

Harvest

Clearcutting is the recommended method of harvesting mature trees where the next stand will be established by planting, direct seeding, or scattering serotinous cones. Rotation ages range from 40 years on poor sites (site index 40) to 70 years on better sites (site index 70).

Principle Enemies

Jack pine is very susceptible to fire at any age. Deer browsing and girdling by rodents can be a problem. Many insects also attack jack pine. Of special note are the jack pine budworm and bark beetles. Several varieties of stem rust also affect jack pine. Over-mature stands are often affected by heartrot fungi. Maintaining the stand in a healthy and vigorous condition is the best method of preventing or reducing damage.

Uses

Jack pine is used for pulp and sawtimber and is often

managed for these uses. Larger sawbolts and poles can be produced on the better sites. Thinnings can produce pulp and post materials.

Wildlife Considerations

Jack pine provides good winter cover for wildlife, especially for deer in areas where traditional deer wintering areas are absent or no longer adequate.

Identifications and Uses

Jack Pine *(Pinus banksiana)* **needles** are stout, flat, 1-1½ inches long, dark green deciduous in their second and third years, and in 2-leaved bundles. The **cone** is usually erect, 1½-2 inches long, and generally curved in toward the branch. **Cones** are green or purplish when full-grown, turning light yellowish-brown as they ripen. They may remain on the tree for years. Cone scales are armed with small prickles that are often deciduous. The **tree** is frequently 70 feet tall with a straight trunk about 2 feet in diameter. The branches are long and spreading, forming an open crown that often has a ragged appearance. The wood is soft, light, and not strong. It is used mainly for pulpwood, rough construction, boxes, crates, shipping containers and prefabricated buildings. Ornamental.

Line drawings for this series are by Mrs. Hoyle (1931) courtesy of the U.S. Forest Service.

Species Identification Information for this series is from "Important Trees of Eastern Forests," U.S.D.A. Forest Service

BLACK SPRUCE

Black Spruce is a very mundane tree, with none of the romance associated with the maples, birches, and pines. It has not found its way through song or verse into Canada's folklore, nor has it played a prominent role in the early history and settlement of the country. Even its Latin name, *Picea mariana*, meaning Maryland spruce, does not do justice to the tree. To Philip Miller, an English botanist who first named the tree in his Dictionary of Gardens published in 1731, "Maryland" epitomized the whole of North America, yet, in reality, Black Spruce does not even grow in Maryland.

Despite its lack of glamour, Black Spruce is a very hard-working tree. It is one of the most wide-spread and abundant trees in the whole of North America, and a major component of the northern coniferous forests that separate the densely populated parts of the continent from the snows of the Arctic. It is also a maintstay of Canada's pulp and paper industry and, for this reason, of vital economic importance to the country.

The Tree

Black Spruce is fairly small, slow growing, but long-lived tree. It can reach heights of 100 feet and diameters of 3 feet, but much more commonly reaches only 30 to 50 feet in height and 6 to 10 inches in diameter. In the far north, on mountain tops, and on cold, wet bogs it may be no more than a twisted, trunkless shrub, often forming dense, impenetrable thickets. It frequently lives to an age of more than 100 years, and in some cases even beyond 200.

In stands, Black Spruce trees have straight trunks with little taper. Stems are covered by thin, scaly, greyish-brown bark that is deep olive-green beneath. The crowns of young trees are narrow and symmetrical with short, slender, mostly horizontal branches; as the trees grow older the crowns become more open and irregular, with dropping branches that have upturned ends. Mature trees often have a compact top to their crowns, known as "club-top", above a very open and irregular lower crown consisting largely of dead branches.

The needles are stiff, blunt, four-sided, and about ½-inch long. When they emerge from the buds they are covered with a whitish bloom that later disappears, leaving them a dark lustreless bluish-green. The needles grow on the twigs from peg-like projections that are a characteristic feature of all species of spruce. The twigs range in color from light cinnamon through reddish-brown to dark brown, and are covered with fine, rust-colored hairs. The buds are covered with greyish scales that have long, slender points to projecting well beyond the tip of the bud.

Cones are usually about an inch long and egg-shaped, becoming almost spherical when they open. When first ripe they are purple in color, but later turn brown and eventually grey. The cones do not fall as soon as they are ripe but open at intervals, releasing seeds gradually over a long period. Some seeds may be retained in the cones for many years, still capable of germinating if they get the chance.

Distribution and Ecology

Black Spruce is found across the entire width of North America from the Atlantic shores of Newfoundland and Nova Scotia to the Bering Straight of Alaska, and from the borders of the tundra south to the Adirondack mountains of New York and Pennsylvania. It is a common tree in all the provinces and territories of Canada. It is one of the dominant trees of the north woods, Canada's vast boreal forest, and also found in the more mixed forests around the Great Lakes, in the Maritimes, and in New England.

The tree grows under a very wide variety of site conditions, from sea level to 6,000 feet elevation, and from dry sands and gravels to perpetually wet bogs and swamps. In the central part of its range it is usually displaced from the most productive sites by other tree species such as Birch, Aspen or Balsam-Fir, which grow more rapidly in early life. Near the southern limits of its range, Black Spruce is most common on peat bogs and for this reason is often referred to as bog spruce or swamp spruce; toward the north of its range it grows best on relatively dry mineral soils that warm up rapidly in spring.

Black Spruce has an advantage over many other tree species in being able to reproduce itself by two separate methods: by seed, or by "layering". The seeds are very small — about 400,000 to the pound — and as noted, are shed from the cones at intervals throughout the year. Heavy cone crops occur about every four years, but there is seldom a year without some cones; as a result most mature stands produce an almost continuous supply of seed. Regeneration from seed is most common on the more productive upland sites where seedbed conditions are usually favorable to germination and seedling establishment.

Layering is a method of reproduction in which living lower branches that touch moist ground or become covered by moss or litter, develop roots and eventually grow to become new trees. Sometimes, especially on the less productive wet sites, this is the only way Black Spruce is able to reproduce itself.

Many of Canada's most productive Black Spruce stands, including most of those in the Clay Belt of central Ontario and Quebec, have originated after fire. When fire sweeps through a forest, all trees and other plants, including Black Spruce, are killed — all, that is, except the cones of three tree species, Black Spruce, Jack Pine and Lodgepole Pine, which are usually above to withstand the fire's heat and flames. The intense heat of the fire causes surviving spruce cones to open, releasing large quantities of seed over the charred remains of the forest. If seedbed and weather conditions are suitable, large numbers of these seeds will germinate, eventually producing dense, regular stands of almost pure Black Spruce.

Enemies

Black Spruce has relatively few enemies when compared to other tree species. Browsing animals will eat its foliage only when nothing else is available, and then only sparingly. In addition, it is less susceptible to decay in the form of heart rot than most other trees, and even fire is, as we have seen, less of a catastrophe for Black Spruce than for most other tree species.

There are a number of insects, including several species of budworm, sawfly, and weevil, as well as several widely-distributed root rot diseases, that attack Black Spruce. However, it is not the preferred host of any of these pests, and consequently very rarely suffers serious damage. Perhaps the most serious pest of Black Spruce is the eastern dwarf mistletoe, which causes the development of so-called witches' brooms on trees in eastern and central parts of the continent. Infection by these parasitic plants results in reduced, stunted growth, and eventually death.

Commercial Uses

The wood of Black Spruce is moderately light and soft, resilient, straight-grained, and nonporous. It is very pale yellowish-white in color, with little contrast between sapwood and heartwood. Since the woods of Black, White, Red, Engelmann, and to a lesser extent Sitka Spruce, have very similar properties and are difficult to distinguish, they are generally treated as a group for commercial purposes and referred to as Canadian spruce.

By far the most important commercial use of Black Spruce is in the manufacturer of pulp and paper. Along with the other spruces and Balsam Fir, it is the mainstay of most of the Canadian pulp and paper industry. Its light color, low resin content, and long, strong fibers aid in the manufacture of paper and add strength to the finished product. Thus it is a favorite pulpwood for facial tissues as well as for newsprint and other paper products.

Black Spruce is used for lumber, though to a lesser extent than the other spruces because of the small size of most trees. However, it takes and holds paint well, and so finds use in siding for houses. Its fairly light weight and ability to hold nails have led to its use in boxes and containers; its lack of odor and taste and low resin content have made it a particular favorite for food containers. Spruce wood transmits sound vibrations better than any other material, and it is therefore very desirable for the sounding boards of pianos and other musical instruments.

Role In History

Black Spruce has been around for a long time. Cones, very similar to those of today, have been found beneath 100 feet of drift in northern Minnesota and dated at more than 38,000 years old by radio-carbon techniques. It is probable that Black Spruce was one of the first tree species to re-colonize the land following the retreat of the glaciers at the end of the last Ice Age; since that time it has played a very important role in protecting the northern environment. Yet it is only in the last 100 years — since the start of the pulp and paper industry in North America — that the tree has received any substantial amount of attention.

In the old days, loggers in the pulp and paper industry worked in isolated, rather primitive camps accessible only by water or narrow trails. Most logging was done in winter when the movement of logs by horse-drawn sleigh to "landings" at the water's edge or to the railroad was easier. Logs left at river or lakeside landings were carried downstream, sometimes hundreds of miles, to the pulp mills when spring break-up melted the ice and snow. Dams were often built to raise water levels and ensure that even small streams contained sufficient water to carry away the piles of logs. Logs often would become jammed or stranded, and river-driving crews were kept busy at the dangerous task of freeing these logs with long pike-poles. Today's pulpwood logger, who operates a huge largely automatic harvesting machine, and either commutes daily to work or lives in a large, modern camp, knows a very different life to that of his turn-of-the-century counterpart.

In the old logging camps, spruce gum was used by woodsmen as chewing-gum is today. In winter, "gummers" on snowshoes travelled around the forest knocking sticky chunks of exuded resin off spruce trees by means of chisels mounted on long poles.

Spruce beer is another traditional use of the tree that has never quite gone out of fashion. Joseph Banks recorded a traditional recipe in his dairy of travels in Newfoundland and Labrador in 1766:

"Take a copper that contains 12 gallons fill it as full of the Boughs of Black Spruce as it will hold Pressing them down pretty tight Fill it up with water Boil it till the Rind will strip off the Spruce Boughs which will waste it about one third take them out and add to the water one Gallon of Melasses let the whole Boil till the Melasses are disolved take a half hogshead and Put in nineteen Gallons of water and fill it up with the Essence. work it with Barm (yeast) or Beergrounds and in less than a week it is fit to Drink".

The refreshing taste of the drink was captured by Thoreau in *The Maine Woods*:

"Instead of water we got here a draught of beer, which it was allowed, would be better; clear and thin, but strong and stringent as the cedar-sap. It was as if we sucked at the very teats of Nature's pine-clad bosom in these parts — the sap of all Millinocket botany commingled, — the topmost, most fantastic, and spiciest sprays of the primitive wood, and whatever invigorating and stringent gum or essence it afforded steeped and dissolved in it — a lumberer's drink, which would acclimate and naturalize a man at once, — which would make him see green, and, if he slept, dream that he heard the wind sough among the pines."

Evidently Thoreau was plied with an excellent brew. Yet even in such a lyrical account as this, the final image conjured in Thoreau's mind by the spruce beer is of wind among the *pines* — maybe Black Spruce will never escape its lowly image!

Other Canadian Spruces

There are four other species of spruce native to Canada that should be distinguished from Black Spruce.

White Spruce *(Picea glauca)* occurs over much the same geographical range as Black Spruce, but can be distinguished by its needles which are strongly aromatic when crushed, by the lack of hairs on its whitish-grey to yellowish twigs, by its shorter bud scales which do not project beyond the tip of the bud, and its two-inch-long, slender cones which are shed from the tree each year. White Spruce is one of the most popular trees in Canada for pulpwood and lumber.

Red Spruce *(Picea rubens)* grows in the Maritimes, the lower St. Lawrence Valley, New England, and as far south as Carolina in the Appalachian Mountains. It has bright yellowish-green needles that are often curved, and cones that are narrow egg-shaped, pointed, and with stiff scales. It often grows in mixture with Black Spruce and the two species interbreed freely, producing various intermediate forms of hybrids. The wood is similar to that of White and Black Spruce and is put to similar uses.

Sitka Spruce *(Picea sitchensis)* is a native of the Pacific coastal forests from Alaska through British Columbia to northern California. It has flat, stiff, very sharp pointed needles that are yellowish-green above and whitened beneath; cones are broad cylindrical, two to four inches long, with loose-fitting, irregularly-toothed scales. The tree commonly attains heights of 125 to 175 feet and diameters of three to six feet in British Columbia where it is a very important timber-producing species. It is used for general construction, ship-building, plywood, pulp, and in earlier days, for aircraft construction.

Engelmann Spruce *(Picea engelmannii)* grows at elevations in excess of 2,500 feet in the subalpine forests from central British Columbia and Alberta south to Arizona and New Mexico. It has curved, bluish-green needles that emit a pungent, unpleasant odor when crushed, and one-inch to three-inch cones taped at both ends and with loose fitting scales. The crown is usually symmetrical, narrow and spire-like. Its wood is similar to that of White Spruce and is used for the same purposes.

Written by: Graham Page and James Richardson, Newfoundland Forest Research Centre, Canadian Forestry Service.

DEPARTMENT OF NATURAL RESOURCES

WHITE SPRUCE

Description

White spruce is a large tree native to the northeastern part of Minnesota; and where it is found to be predominant in the stand, there are only 79,000 acres of this forest type.

It reaches heights of 100 feet, and diameters over 20 inches on good sites. It is quite tolerant of shade, grows slowly in the understory, and responds well to release.

White spruce is generally found in mixed stands of balsam fir, aspen and paper birch. The best white spruce are found on heavy clay soils.

Management

Regeneration

White spruce will regenerate itself naturally if a seed source is present. However, this is not recommended, because good seed years are infrequent - generally one out of four years. Planting is the recommended method of regeneration. Stocking should be at least 600-1000 seedlings per acre for good form development.

Culture

After the first five years or so, the white spruce seedlings grow rapidly, rivaling the fastest growth among conifers. Once established, the stand should be thinned from 15-20 years, reducing the basal area to 90-100 feet^2/acre. Several intermediate cuts may be made periodically until the rotation age is reached.

Harvest

White spruce can either be clearcut and managed as an even-aged stand, or can be cut selectively as an uneven-aged stand. Selective cutting should be done by removing larger trees from the overstory, while maintaining a good stocking of young trees. While white spruce reaches maturity between 80-120 years, balsam fir and aspen reach their maturity some 20 years sooner. This may necessitate earlier harvesting. Large spruce left after a partial cutting are very susceptible to heavy loss by windthrow.

Principal Enemies

The yellow-headed spruce sawfly, spruce budworm, and wind are the major problems of white spruce.

The yellow-headed spruce sawfly, a defoliator, affects open grown ornamental and field grown plantation trees up to 15 feet in size. Normally, individual trees are attacked, making control a minor problem except on wet, organic sites which contain open grown trees.

Due to the heavy crown and shallow rooting, heavy winds can blow it down unless it is well stocked within the stand.

Uses

White spruce is managed for pulpwood and sawtimber. Because of the long fibers in the wood, it

makes excellent paper.

As sawtimber, it makes good dimension lumber for construction.

Identification and Uses

WHITE SPRUCE *(Picea glauca)* **leaves** are crowded on the upper side of the branches by the twisting upward of the needles on the lower side. Needles are ¾-1 inch long, and bluntly pointed. **Cones** are usually about 2 inches long, becoming pale brown and lustrous when mature. Cone scales are flexible and smooth-margined. Cones fall soon after they ripen in autumn, rarely persisting through the winter. The **tree** grows 60-75 feet tall with trunk diameter of 2 feet. Foliage of white spruce has a somewhat disagreeable odor when crushed. The **wood** is used for pulpwood, interior trim, and interior parts of furniture.

Line drawings for this series are by Mrs. Hoyle (1931) courtesy of the U.S. Forest Service.

Species Identification Information for this series is from "Important Trees of Eastern Forests," U.S.D.A. Forest Service

BALSAM FIR

Balsam fir is an important source of pulp and lumber, especially in eastern Canada, but most people probably know the tree best as a Christmas tree, popular because of its symmetrical cone-shaped crown with good needle retention and pleasant aroma. Fragrant mattresses of fir boughs have helped countless thousands of Indians, voyageurs, trappers, surveyors, prospectors, woodsmen, and travellers to sleep soundly in the forest. Other common names for this tree are fir, balsam, white fir, var, Canadian fir, and in Quebec, sapin.

Description

Balsam fir, *Abies balsamea* (L.) Mill., is an evergreen conifer that ranges from about 50 to 70 ft. in height and 10 to 20 inches in diameter when fully grown. In northern New Brunswick and the Gaspe where balsam fir probably grows best, taller trees up to 90 ft. can be found. Balsam fir is found in all Canadian provinces except British Columbia. It grows in pure stands, or mixed with other species such as spruce, white birch, aspen, and red maple.

The leaves, usually called needles because of their shape, are ¾ to one inch long. On the lower crown they are flat with blunt tips that are usually rounded but sometimes notched. Needles are attached to the twig singly and arranged spirally around it. On shaded branches, the needles are parted down the middle of the twig and extend on opposite sides. On branches in full light, needles stick out in all directions. These sun needles, usually have pointed tips, and are shorter and thicker than shade needles.

Balsam fir needles live for about 6 to 12 years. Because older needles die and drop unnoticed, a few at a time throughout the year, the trees themselves are truly evergreen. More needles drop in autumn than in other seasons.

The trunk, mainstem, or bole of young balsam fir is covered with smooth greyish bark dotted with resin blisters. On old trees, the blisters shrink and the bark becomes furrowed.

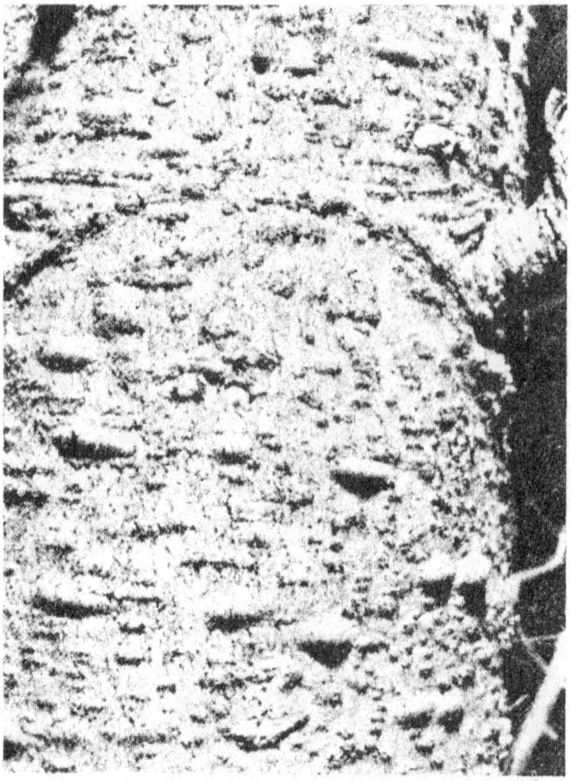

Bark is smooth or has blistered appearance.

The root system of balsam fir is shallow, usually no more than 12 inches deep, and spreading: some roots extend well beyond the perimeter of the crown. Nevertheless, fir trees are often uprooted by strong winds.

When balsam fir trees are about 20 years old, at which time they may be 13 to 27 feet tall, they start to produce flowers; female flowers near the top of the tree and male flowers lower down. This arrangement helps to ensure cross-fertilization, since the wind-borne pollen from the male flowers is usually carried to the receptive female flowers on other fir trees. Every 2 to 4 years when flowers are more plentiful than usual, a film of yellow pollen can be seen in spring floating on water puddles, lying on top of parked cars, and on window glass and sills. Some pollen may even sift into the house like dust to annoy the fastidious housewife.

In spring, both kinds of flowes are about 3/8 inch long. Male flowers disintegrate soon after the pollen is released but the female flowers develop into cones, 2-4 inches long, containing seed.

Cones are fully grown, ripe and dotted with cream-white pitch in late August and early September. At this time, squirrels begin to chew off the cones, drop them to the ground, and collect and store them for winter food.

Fir cones differ from those of most other conifers in that (a) they grow upright like candles, instead of hanging down in the usual manner, and (b) after ripening, the scales separate from the central axis and fall with the seed in October and November.

Foresters wanting seed for the tree nurseries have to collect the cones after the seeds ripen but before the cones fall apart and the seeds are dispersed. Each fir seed has a single wing that helps it float through the air for some distance. When seeds are collected and cleaned for use in a forest tree nursery, this wing is removed. Approximately 60,000 cleaned seeds weight one pound.

Growth Of Fir Stands

Balsam fir is a prolific producer of seed. Following good seed years, which usually occur at 2- to 4-year intervals, many thousands of small trees spring up on the forest floor. Most of the seedlings die young, but over 8,000 saplings (small trees with stems 1 to 4 inches in diameter) per acre have been counted and dense stands with trees less than one yard apart are common. Perhaps it was a voyageur, forced to carry his canoe through one of these thickets, who described it as "epais comme la barbe de Saint Michel" (thick as Saint Michael's beard). The name stuck and today many woodsmen, English speaking as well as French, refer to such thickets as "Saint Michel." At maturity 600 to 1000 trees per acre are considered sufficient for a good crop of wood.

Fir seedlings in the forest frequently grow slowly for the first five years or so, and may continue to grow slowly for many years if crowded or shaded by large trees. Fir trees are very tolerant of shade and some survive for years under heavy cover. In dense stands, lower branches become shaded and needles gradually die and drop, but dead branches persist for many years.

With wider spacing, say an average of 8 feet between trees, and lots of light, fir trees grow rapidly and on good sites can become 60 ft. tall and 12 inches in diameter in 40 years.

Pests

Balsam fir trees may live for 100 years or more but after about 60 years they become highly susceptible to damage by insects, disease, and strong winds. The most notorious insect of balsam fir is the spruce budworm which, despite its name, actually prefers to chew on fir needles. A few years of heavy feeding will kill a tree. When this happens over thousands of hectares, large amounts of wood are lost to forest industries. For this reason, much research has been concentrated on this pest and methods of controlling it.

Extensive areas of dead and dying fir helped fuel the famed Miramichi forest fire that swept approximately 1.5 million hectares of New Brunswick in 1825.

Uses - Past

Indians used the wood of balsam fir to start fires by friction, and the pitch as a dressing for wounds. Ernest Thomson Seton, the pioneer nature writer, depicts in "Two Little Savages" how Indians made fire with bow, drill, and fire block. Dry balsam fir was the favorite wood for drill and block.

A fascinating account of using pitch, from the resin blisters of young balsam fir trees, as a dressing for wounds is contained in Stuart Trueman's, "The Ordeal of John Gyles".

John Gyles, a Puritan boy from what is now Maine, was only 9 years old when captured by the Maliseet Indians in 1689 and carried away to their village on the Saint John River. During the 6 years Gyles spent with the Indians as their slave, he survived almost incredible hardships.

While with the Indians, Gyles observed Indian women making dishes from birch bark and sewing them so expertly, with fine fir roots, that they did not leak.

The species name, *balsamea,* assigned by the great Swedish botanist Linnaeus in the middle of the 1700's refers to the resinous blisters on the bark, a distinctive feature of this tree. The Latin word *balsamum* was used for any aromatic resinous substance obtained from plants. Fir pitch or Canada balsam obtained by puncturing and draining these bark blisters is still used in medicinal compounds, spirit varnish, and as a cement for mounting microscope specimens and parts of optical systems. Canada balsam has the same refractive index as glass.

Like the Indians, early settlers from Europe used balsam fir in minor ways only. It was not until the early 1900's when large pine and spruce became scarce and the pulpwood industry gained importance that fir also became important. Because the wood, unlike that of the pines and spruces, does not contain resin, it was preferred for food containers in the days when food was packed in wooded boxes.

Uses - Present

Wood from the balsam fir is important to the pulp and sawmilling industry, although statistical evidence of this importance is hard to find because fir and spruce pulpwood is usually piled together and fir lumber is mixed with and marketed as spruce. Also, western firs (alpine and amabilis) are combined with and usually called balsam fir in forestry statistics. Nevertheless, by making some assumptions, the harvest of balsam fir in the early 1970's is estimated as 14 million cubic metres per annum or about 12 percent of the volume of all wood harvested in Canada.

Besides providing much useful wood, millions of young fir, up to about 15 years old, become the traditional Christmas trees. Wild, uncultivated trees still supply a large part of this market, but more and more trees are being cultivated in natural stands or planted on fertilized soil, and protected from pests, weeded, and sheared to make them more attractive. Christmas wreaths are made from the tips of balsam fir branches bound together on a circular frame.

Balsam fir is rarely used for ornamental planting, although when young it has an attractive crown, dark green in colour and conical in shape, extending to the ground. In old age, the crown becomes ragged and widespreading. But for 20 or 30 years, planted singly or in small groups, fir are useful and pleasant ornamentals.

Similar Species

In eastern Canada, balsam fir is the only native fir but it is sometimes confused with spruce and hemlock. However, needles from the lower crown of balsam fir are flat with blunt tips, while those of spruce are four-sided and sharp-pointed. The bark of balsam fir is smooth except for resin blisters, while both spruce and hemlock have scaly bark.

In western Canada there are three native firs besides balsam fir. Two of these, amabilis fir and grand fir are confined to British Columbia, where there is no balsam fir. Only alpine fir overlaps the range of balsam fir in the lower foothills of Alberta. These closely related species are difficult to tell apart. Probably the best way is by the leaf stomata or pores that appear as a whitish line on each side of the midrib. On balsam fir, they are mostly on the lower surface with few or none on the upper surface, whereas on alpine fir they are numerous on both the upper and lower surfaces of the leaf. Also, alpine fir has a very narrow conical crown even when growing in the open.

Written by: B. C. Wile, Maritimes Forest Research Centre Canadian Forestry Service.

Cone Scale

TAMARACK

by B. C. Wile

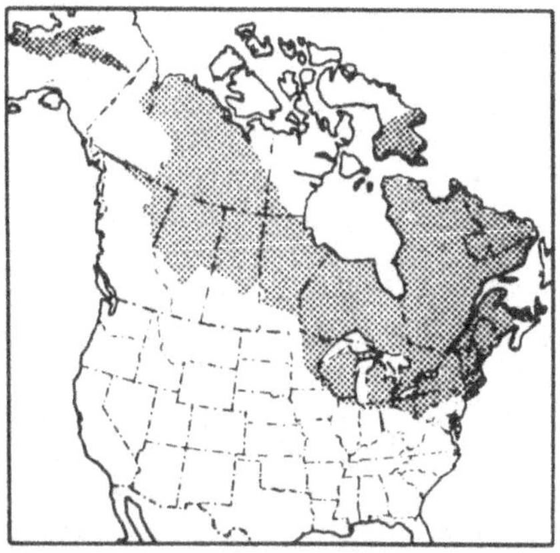

After most of the hardwood leaves have fallen, the green tamarack needles turn golden yellow extending the color of autumn and our appreciation of nature. Later the needles drop. Tamarack belongs to the group of trees called larches. There are about ten different larches in the world; three of these are in North America. They are the only native conifers that lose all their needles each autumn.

Tamarack [*Larix laricina* (Du Roi) K. Koch] is also called Hackmatack, Eastern Larch, American Larch, Alaska Larch, Black Larch, Juniper, meleze laricin, and meleze d'Amerique. The name tamarack may have been derived from the Algonquian languages but both persons give credit for the first botanical descriptions of the tree were Germans. Abbreviations of their names appear after the botanical name. Johann Philipp Du Roi, 1741-1785, made a special study of trees and shrubs, including some from North America. Kark Heinrich Emil Koch, 1809-1879, was a naturalist and traveller who published a book on tree identification ca. 1869.

Tamarack is a hardy pioneering tree that is found in every province and territory of Canada, and in the central and northeastern United States and Alaska. Its range extends north in the MacKenzie River Valley beyond the Arctic Circle to the northern limits of tree growth. It is one of the first trees to invade open bogs.

Tamarack commonly grows in bogs, swamps, and other open areas such as roadsides and cleared forest land because it is intolerant of shade and needs full light to survive. Given light and space, it grows best on moist, well-drained sites where it is one of the fastest growing conifers in northeastern North America.

Pure stands of Tamarack are usually small. More often it is mixed with other tree species such as Black Spruce on wet land, and Balsam Fir, White Birch, and Trembling Aspen on better-drained soils.

Description Of The Tree

Tamarack is a medium-size tree at maturity when growing on good sites. It ranges in height from about 15 to 22 m and in stem diameter from 30 to 60 cm. The bark is thin, smooth and bluish-grey when young, and reddish-brown and scaly when mature.

The leaves are about 20 to 30 mm long, needle-like, slender, flexible, light green in spring and summer and yellow in late autumn, prior to dropping. On new shoots, leaves are separate and arrange spirally. On older twigs, leaves are in clusters of about 10 to 20 on dwarf, spur-like shoots.

The small male and female flowers appear on the same tree in April to May after trees are about 15 years old. The male flower is yellow, the female reddish. The flowers occur over all the crown of widely separated trees but only at the top of trees growing close together. The female flowers develop into cones bearing seed.

Seed,
(natural size)

Cones mature within one year, are brown when ripe, and about 10 to 20 mm long. They open in late summer to release the seed.

The seed is small with a single wing that helps it to travel through the air. About 600,000 cleaned seeds (the wing removed) weigh one kilogram.

Tamarack wood is yellowish-brown, oily, the strongest, heaviest, and hardest of the eastern softwoods, and the best for heating. It is naturally resistant to decay, the most durable eastern conifer after Eastern White Cedar.

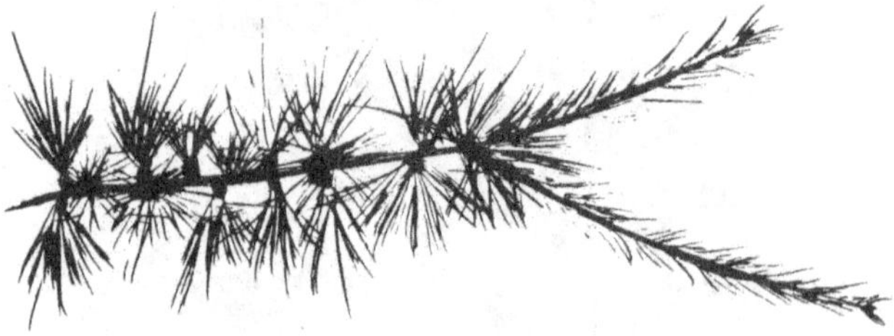

Leaves have two arrangements.

Past Uses

Indians used the fine roots to sew birch bark, the wood for arrow shafts, and the bark for medicine. Tamarack and cherry bark were steeped together to treat colds.

Early settlers from Europe were mainly interested in the wood for timbers, knees, and planking in ships. The knees, made with part of a large root left attached to the tree trunk, were

used to join ribs to deck timbers. In the 1840s and 1850s, the large export of ships' knees threatened to deplete stands of Tamarack in Nova Scotia. Tamarack was lighter than the oak usually used and gave Maritime vessels greater speed and buoyancy.

Present Uses

Tamarack wood is still used to a small extent for boat building, but the main uses are for piles, poles, mine props, railway ties, lumber, and pulpwood.

The economic importance of Tamarack is relatively minor. In national statistics, tamarack is combined with Western Larch, a more important species. Even then, the volume of larch wood in the national forest inventory is less than one percent of the total for coniferous species. Inventories of eastern provinces show similar quantities.

Closely Related Species

In all parts of Canada except southern British Columbia and Alberta, Tamarac is the only native larch, and there its natural range does not overlap those of Western and Alpine Larches. However, when planting there may be a need to separate tamarack from other larches, including exotic species. This may be done by measuring cone size. Mature Tamarack cones are usually 10 to 20 mm long: the two western larches and the three common exotic species (European, Japanese, and Siberian) all have larger cones, usually between 25 to 50 mm long.

Other distinguishing features are given in *Native Trees of Canada* (see reading list).

Improving Tamarack

The interest in planting tamarack is increasing. Tamarack grows rapidly, compared to most other northern conifers, and is adapted to a wide range of conditions, making it a useful tree for reforestation. From its wide geographic range the better strains can be selected, and the tree breeders can produce hybrids that grow faster than the parents, or have other desirable qualities. Tamarack was first crossed with Japanese Larch in 1956 at the Acadia Forest Experiment Station in central New Brunswick to produce hybrid trees that grew faster than the native tamarack. Since then it has been crossed with European, Western, and Siberian larches. Some of these new hybrids are of considerable interest. The work of selecting superior trees, of producing and testing hybrids, and developing methods for mass production has accelerated during the last decade.

The best known hybrid of larch, the Dunkeld Larch, does not involve Tamarack, but is a cross between Japanese and European Larch and was discovered at Dunkeld House in Perthshire, Scotland, in the early 1900s. Subsequently, it has been planted in many parts of the world. The first planting in North America was made at the Petawawa Forest Experiment Station in eastern Ontario in 1928. Growth has been impressive.

Major Pests

A small fly-like insect, about 10 mm long and black with a brown band around the body, called the Larch Sawfly is the most serious insect pest of larch trees in North America. It attacks all

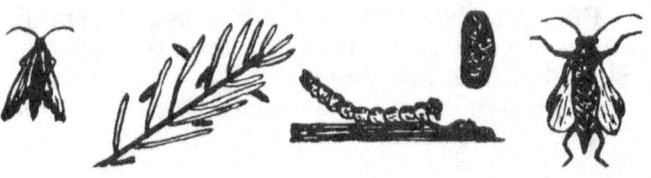

Larch Casebearer
Coleophora laricella (Hbn.)

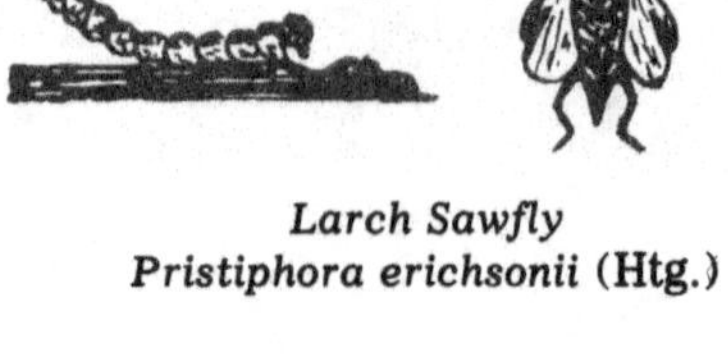

Larch Sawfly
Pristiphora erichsonii (Htg.)

larches, both native and exotic. The sawfly is so named because of its saw-toothed ovipositor used to cut slits along the underside of new shoots where the eggs are deposited. This causes the shoots to curl distinctively. After 7 to 10 days, the eggs hatch and the larvae begin feeding on the larch needles. Severe feeding will stop the tree from growing and if the infestation continues for about six years will kill the tree. Extensive and severe infestations have killed many Tamarack in Canada and the United States, and Western Larch in British Columbia. Part of this mortality, however, is caused by the eastern larch beetle which infests weakened trees during and following sawfly outbreaks.

Another less serious insect pest is a small (about 10 mm long), gray moth called the Larch Casebearer. The young larvae hollow out needles then cut off a portion and carry it to a twig or branch where it serves as a case or cocoon to protect them during winter. In spring, they crawl out of their cases and tunnel into new needles. When feeding is severe and sustained over several years, trees stop growing and die.

Still another pest is the Procupine. It is not a serious pest in natural stands of Tamarack but can be very serious in plantations. Plantations are expensive and any losses are important especially where selected trees are being tested. Porcupines feed on the inner bark exposing light patches of wood that can be seen from a distance. If the bark is removed all around the main stem or branch, turn upward to replace it, resulting in a deformed multistemmed tree of little commercial value.

Editor's Note:

The tamarack habitats are of critical importance to the Great Gray owl and its small mammal prey species. Some other species associated with tamarack are the White-winged Crossbill, Yellow-bellied Flycatcher, Northern Waterthrush, Palm Warbler, Connecticut Warbler, Black-backed Three-toed Woodpecker, Boreal Chickadee, and Hawk owl. It may be of interest to note that the Porcupine is the Fisher's main diet and that its role in the forest ecosystem is quite likely more valuable than currently realized.

—David H. Johnson

B. C. Wile is on the staff of the Maritimes Forest Research Centre, Canadian Forestry Service, Fredericton, New Brunswick.

As printed in National Woodlands.

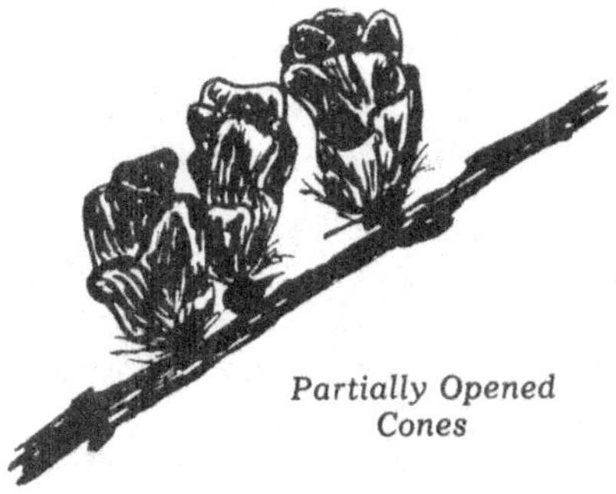

Partially Opened Cones

NA-02039
Rev. 1/82

DEPARTMENT OF NATURAL RESOURCES

WHITE CEDAR

Description

Northern white cedar occupies 498,600 acres - 3.6% of commercial forest land in Minnesota. White cedar may perpetuate itself in pure stands, whereas species of balsam fir, black spruce, tamarack and black ash will gradually replace it in mixed stands, particularly after disturbances. White cedar lives longer than associated trees, reaching 400 years of age on organic soil sites.

Management

Regeneration

Northern white cedar produces good seed crops every three to five years. Germination and early growth are best on moist seedbeds such as rotten wood, compacted moss in skid roads, and burned soils. Vegetation reproduction by layering can occur on organic soil sites. Northern white cedar can survive in shade for several years and responds well to release at nearly all ages.

Whenever possible, this type should be managed in fairly large, even-aged stands. This is best for timber production, deer yards, and would be well-suited for efficient cultural operations and mechanized harvesting. It is important to obtain satisfactory reproduction promptly after harvesting on brushy areas. If not stocked early with trees, these areas convert to lowland brush and become difficult and expensive to reforest.

Clearcutting is the best way to obtain even-aged stands of northern white cedar for timber and deer yards. Unless suitable residual stems are expected, clearcutting must be done in narrow strips or small patches because reproduction depends on natural seeding which has an effective range of only two to three times the height of the residual trees. Heavy slash left in some cutover areas is very detrimental to natural reproduction, therefore, it is advisable to log in winter and to utilize full tree skidding unless slash is needed to prepare a good seed bed by prescribed burning.

Rotation age varies from 70 to 90 years on excellent sites; from 100 to 140 on poorer sites.

Harvest

Clearcutting in strips or patches is the recommended method of harvest. In some stands, however, a shelterwood system can also be employed.

Uses

Northern white cedar products include posts, poles, and some sawtimber, with smaller amounts used for lumber, such as paneling.

Principal Enemies

Carpenter ants, both black and red, are the main insect enemies of northern white cedar. They frequently attack partially decayed heartwood in living trees. Butt-rot fungi that causes a white, stringy rot over a brown cubical rot are common in mature trees on the drier lowland sites.

The main damaging agents of northern white cedar are wind, deer, hare and impeded drainage. The relatively shallow root system of white cedar makes it susceptible to uprooting where trees are exposed to the wind. Small trees and reproduction are often over-browsed by deer and hare. Drainage impeded by roads and beaver has killed white cedar and associated trees on thousands of acres.

Wildlife Considerations

Northern white cedar stands have exceptionally high value for white-tailed deer. Mature cedar stands provide excellent winter deer cover and cedar is a preferred deer browse species.

Deer yard considerations should receive high priority in management techniques. Cutting should be done in small patches (½ acre up to 5 acres), to insure adequate growth of mature stands for desired browse and shelter; especially during deep snow periods in winter.

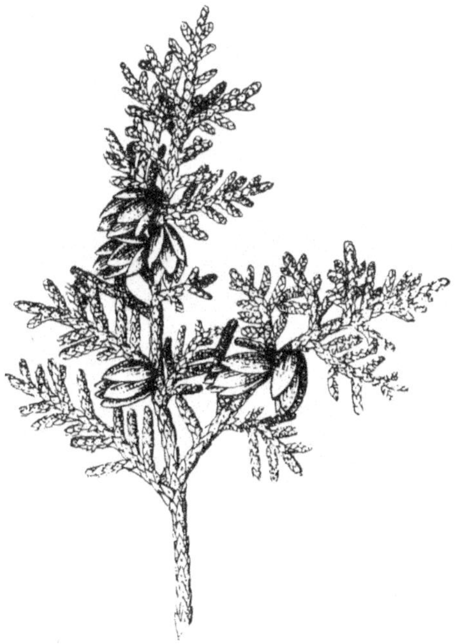

Identification and Uses

NORTHERN WHITE CEDAR *(Thuja occidentalis)* (white cedar, eastern white cedar, arborvitae, eastern arborvitae, swamp cedar). Medium-sized tree of northeastern United States, adjacent Canada, and Appalachian Mountain region. **Bark** reddish-brown, thin, fibrous, with narrow connecting ridges. **Twigs** flattened and branching in one plane. **Leaves** appearing flattened in 2 rows, scalelike, 1/16 to 1/8 inches (1.5-3mm) long, light yellow green, aromatic. **Cones** 3/8 to 1/2 inch (10-12 mm) long, pale brown.

Principal uses: Poles, railroad crossties, posts and ornamental fencing. Lumber for boats, paneling, boxes, millwork, tanks, and building construction. Cecar-oil, used in medicine. Ornamental.

Line drawings for this series are by Mrs. Hoyle (1931) courtesy of the U.S. Forest Service.

Species Identification Information for this series is from "Important Trees of Eastern Forests," U.S.D.A Forest Service

NA-02038
Rev. 1/82

DEPARTMENT OF
MINNESOTA
NATURAL RESOURCES

ASPEN

Mature Aspen - Back   Young Aspen - front

Description

The aspen type occupies 5,302,300 acres - 38.7% of the commercial forest land in Minnesota. Aspen itself comprises only 61% of this type, the remainder being other conifer and hardwood species. Quaking and big-toothed aspen occur on nearly every soil type, but grow best on deep, well-drained soils.

Aspens are highly sensitive to shade, soil compaction, fire and mechanical injury to the root system. They mature in 35 to 55 years and occasionally trees will survive 100 years or more. Without harvest, aspen stands will be replaced by more tolerant or longer-lived species.

Aspen typically sprouts many thousands of suckers per acre from the shallow parent root system after a stand has been opened by fire, windthrow or cutting.

Management

Regeneration

The recommended silvicultural system for growing and reproducing aspen is complete clearcutting at rotation age to regenerate pure, fully-stocked stands of suckers. By age two, when most suckering will have occurred, stocking should exceed 4,000 to 5,000 stems per acre. In young stands that were not established by complete clearcutting, the residual trees need to be removed as early as possible.

Only site index 55 or better sites should be considered for aspen timber management. Conifers are usually more productive than aspen on poorer sites, therefore, a forest manager may wish to convert to conifer species. On better sites, aspen grows rapidly and thins itself naturally through competition, insects and diseases.

Harvest

Clearcutting is the recommended method of harvest. The product leaves the woods as either 100 inch sticks, tree length or chips. Slash left on the ground following harvest does not hamper natural regeneration.

Uses

The major uses of aspen are for paper, particle board, and lumber. It is also used for fuelwood. Aspen promises to be an important economic tree in Minnesota as additional new industry and technology move into the state.

Insects and Diseases

Hypoxylon canker commonly infects stands of all ages and is the only disease causing significant mortality in immature stands. Big-tooth aspen is five times as resistant as quaking aspen. The only known control is to maintain full stocking throughout the rotation to incur at least partial replacement of growth loss. Shortened rotation ages can minimize losses.

Heart rot decays the heartwood of stands nearing maturity and this is the reason aspen should be harvested before it exceeds 55 years of age. Losses to this disease are not serious for shorter rotations.

Once past the juvenile stage, aspen is seldom killed directly by insects. However, insects such as the wood borers can damage and kill young suckers. Pole-sized and mature stands are affected by the poplar borer, however, they are more susceptible to the defoliating insects such as the forest tent caterpillar and the large aspen tortrix. Even repeated defoliation does not cause much direct mortality except on sites with high water tables - the main effect is temporary loss in growth. No direct insect control is now practiced in aspen, although dense sucker stands should be regenerated to minimize borer damage.

Wildlife Considerations

Of all the forest types in Minnesota, the aspen type provides the greatest wildlife habitat values. Not only is aspen important for varied wildlife food production, but aspen stands usually contain a mixed variety of other important wildlife species.

Young dense aspen suckers are important deer browse areas and provide good ruffed grouse brood cover. As aspen sucker stands grow into sapling and young pole stands, they remain as good ruffed grouse habitat for brood rearing, nesting and drumming sites. During winter and spring, mature male aspen buds are the preferred food for ruffed grouse.

For best wildlife values, aspen must be clearcut to provide for good aspen regeneration. In or near deer concentration areas, aspen should be cut in winter to provide needed deer browse. Aspen stands should be harvested in a planned checkerboard pattern of scattered clearcuts of 2-10 acres to provide age diversity among different stands.

Within larger cuts (over 25 acres), several scattered clones of mature male aspen should be left standing for a needed ruffed grouse food source.

Large cuts of 200+ acres in northeastern Minnesota are recommended in those areas designated as prime moose areas.

Identification and Uses

QUAKING ASPEN *(Populus tremuloides)* is the most widely distributed tree species in North America, although it is not a component of the southeastern forests below Virginia. The almost circular **leaves** give this species its name, each leaf trembling in the slightest breeze at the end of a long, flattened stem. Leaves are 1-4 inches in diameter, sharply pointed, roughly-toothed, and with a prominent, light-colored midrib. They are shiny green above, lighter-colored below. The **tree** may attain heights around 100 feet with a trunk diameter of about 3 feet, but usually is smaller (20-60 feet tall). The round topped crown is usally narrow and symmetrical. The thin **bark** is light green and smooth except for warty bands or patches of dark-colored scaly protuberances.

The **wood** of quaking aspen is classified as light, soft, weak, limber, and moderately low in shock resistance. Few hardwoods shrink as little in seasoning as aspen. It is low in decay resistance and nail-holding ability. It has little tendency to split, is easily worked with hand or power tools, and is fairly easy to finish to a smooth surface. It glues easily and ranks with the best of the hardwoods in paintability.

Principal uses: Higher grades of aspen are used primarily for lumber and wooden matches. Lower grade lumber is used largely for boxes and crates. Most aspen wood goes into pulp for the manufacture of book and magazine paper, and corrugated and insulating boards. It is preferred for high-grade excelsior. In recent years it has been approved for studs.

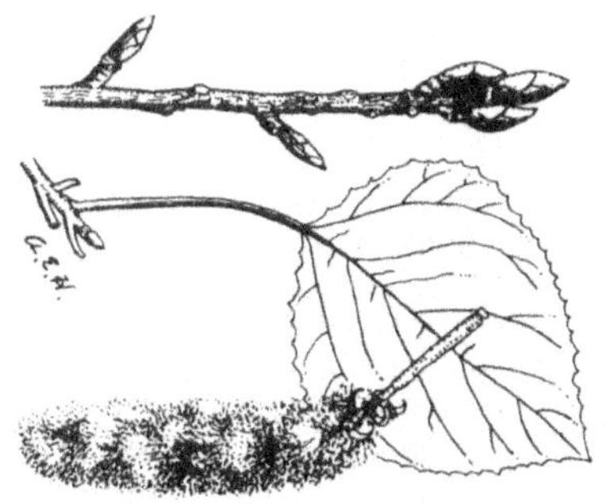

BIGTOOTH ASPEN *(Populus grandidentata)* **leaves** give the species its common name, as the leaf margins have very conspicuous rounded teeth. Leaves are almost round, sharp-pointed, dark green on the upper surface but lighter green on the lower, and 2-3 inches long and wide. **Bark** is thin, smooth, and light-gray to green on younger trees and the upper portions of the stems of older trees. Near the base of older trees, the bark is dark brown tinged with red, and ¾ to 1 inch thick. This **tree** grows to 60-70 feet in height and about 2 feet in diameter, often with a narrow, round-topped crown. The soft, light-colored **wood** is used for interior parts of furniture, and for pulpwood.

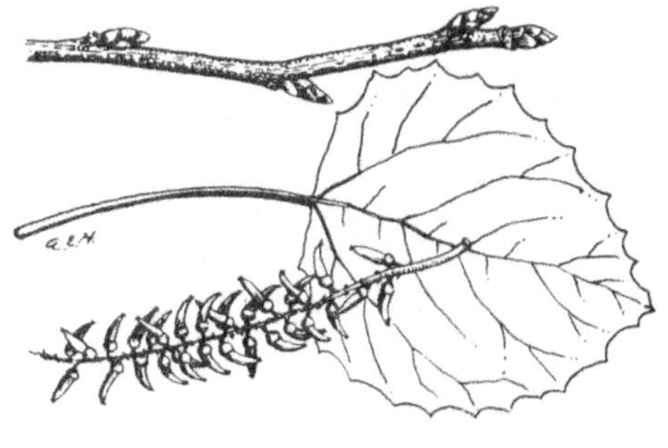

Line drawings for this series are by Mrs. Hoyle (1931) courtesy of the U.S. Forest Service.

Species Identification Information for this series is from "Important Trees of Eastern Forests," U.S.D.A. Forest Service

WILDLIFE AND THE WEED TREE

By Allen Wickman

Most people call it popple. Some call it a weed tree. But to foresters, wildlife managers and others involved in natural resource management, it is called aspen or poplar.

There are two common species of aspen in Minnesota; Quaking Aspen with leaves that tremble in the slightest breeze, and Bigtooth Aspen, noted for its leaves with fewer but larger "teeth".

For years, the importance of aspen both to wildlife and mankind has gone unnoticed, just as the look-alike monotony of an aspen stand goes unnoticed beneath the awesome sweep of a red or white pine. But let's set the record straight. In Minnesota, aspen is one of our most important trees economically, and of unmatched value to our state's principal game species, the whitetailed deer and the ruffed grouse.

This article is directed at answering two basic questions: 1) How much do grouse and deer depend on both the quality and quantity of aspen available to them? 2) How can logging influence the quality and quantity of aspen as it pertains to these species?

Let's examine the first question. How much do game populations in Minnesota utilize aspen? The answer is that their dependency on aspen is even greater than one might imagine. Of course, aspen is not the only wild plant important to grouse and deer. But without it, their populations would probably decrease.

A continuing decrease in young aspen stands in many areas of central and northern Minnesota has adversely affected whitetail populations. We can expect that our deer herd will continue to decline in forested areas which are not logged of aspen. This decline can only be partly off-set by mild winters and a reduction in hunting pressure.

A recent study by the Michigan Department of Natural Resources, Division of Game and Fish, showed a positive relationship between the amount of aspen in a township and the number of deer taken by licensed hunters in the study area. The more aspen the more deer. Sounds simple, but is it?

If we covered a township with aspen, would we have the maximum number of deer possible? Probably not. The Michigan study indicated that at least 15 percent of each township should be in forest openings. It also revealed that evergreens and the brush immediately surrounding an evergreen stand (especially younger white cedar stands), are excellent wintering areas for deer. So, we must provide deer with **both** food and cover during winter — especially during periods of deep snow and intense cold.

The Michigan study indicated that each deer-producing township should have a minimum of 35 percent aspen. Just as important, this aspen should be comprised of different age classes for maximum production of deer. About 25 percent of the aspen should be 1-10 years old; 25 percent should be 10-20 years; and another 25 percent in each of the two 10-year age classes up to 40 years. In other words, for maximum deer production, we should cut 25 percent of our aspen stands every ten years.

It has been discovered that aspen over 40 years of age loses its value for game production. From the forester's viewpoint, the aspen's pulpwood value may also deteriorate beyond that age. Unfortunately, because of local timber market conditions, many tracts of harvestable aspen have gone uncut. If we cannot sell the wood, who is going to cut it?

Aspen is even more important to grouse. Several studies have indicated that Minnesota grouse eat five to six times as many aspen buds as all other foods combined.

A grouse will consume about four ounces of aspen buds in a meal, usually lasting 15-20 minutes. This is equivalent to a 150-pound human eating 27 pounds of food in the same amount of time. Quite a meal!

Grouse usually eat only the flowering buds on male aspen trees. In fact, they seem to prefer those trees with the largest buds which, incidentally, contain the highest protein content. Trees with the largest buds are usually in the 25 to 35-year age class.

Young stands of aspen from 1-10 years of age provide excellent cover for young broods of grouse. The 10 to 25-year-old stands are used primarily for breeding and winter cover. As with deer, the ideal situation would be a blending of different age classes, which might be accomplished if the demand for Minnesota aspen was large enough.

We are paper users. The national average is 175 pounds of paper consumed annually for every man, woman and child. Of this, 19 percent is recycled paper. About half the paper consumed is newsprint. Yet, even with this tremendous paper consumption, Minnesota has considerably more pulpwood than we need for papermaking.

In recent years, aspen is being used more than ever. In 1930, only four percent of all pulpwood was aspen; today, that figure has increased to 50 percent. It is estimated that our demand for pulpwood products will increase four-fold in the next 30 years. By that time, we should be cutting aspen to the extent that game species will greatly benefit. However, until that time, we must apply multiple-use practices to manage our forests as best we can.

If we think we can have a problem, look to Canada! The provinces have ten times as much aspen, but are only cutting about six of every 100 harvestable trees. The implications this will have on Canadian game populations are frightening.

If we do not cut aspen, not only will we lose some wildlife habitat, we will lose our aspen stands as well!

Most existing aspen stands grew as a result of past fires and/or extensive logging of white pine in the early 1900's. Both factors resulted in a complete killing of overhead vegetation which is necessary for maximum regeneration of aspen.

To provide similar conditions today, we must completely cut all standing trees in an area that has at least some aspen intermixed. On most Minnesota soils, aspen will quickly regenerate and outgrow other tree species, so that eventually a "pure" stand will result.

What will happen if we do not completely clearcut (clean cut) an area? Why not log only part of a forest stand so things won't look so messy? The answer to these oft-asked questions is that we would lose most of our aspen and its game potential along with it. To understand how this would result, we must understand how aspen regenerate.

Each aspen tree produces over a million seeds, but unlike other tree species, very few of these seeds survive. Instead, aspen reproduce by means of lateral suckers. These root-like structures branch out from the main trunk within six inches of the ground surface. They vary in size from the thickness of a pencil to as much as two inches in diameter.

Young trees sprout up from these suckers and grow to exhibit the same genetic characteristics as the parent tree. This fall you may notice that one group of aspen will have colorful leaves, while a nearby cluster may still be green. Chances are that each group has developed from a different parent tree. These separate groups are called **clones**.

This is a simple explanation of how aspen rengerate. However, we have omitted one important factor. Suckers will not reproduce unless the sunlight can warm the soil sufficiently. This means a minimum of several days of 75 to 80 degrees during summer.

Whether it be conifers, hardwoods or brush, if a forest overstory is left standing, the soil will not warm sufficiently and suckers will lose their ability to sprout after three or four years.

An aspen stand must have abundant sunlight to survive. Complete clear-cutting opens the stand to sunlight, enabling the aspen to recover at a remarkable rate. Within a year, there may be over 40,000 young aspen stems per acre. These dense stands of aspen provide excellent protection for grouse from hawks, owls and other predators. Young aspen is also a favorite food of deer.

By the time the stand reaches 30 years of age, there will be about 1,000 to 1,500 stems per acre. Natural competition will eliminate many of the weaker trees, while disease, insects, storms and other natural factors will also take a toll.

A mature aspen stand will slowly change to either a hardwood or conifer type if not cut. The invading young trees are likely to be balsam, white pine, basswood, or other hardwoods that can grow in shade. This natural conversion of aspen to conifers or hardwoods or even to hazel brush, is called "breakup".

Mature conifer and hardwood stands may become so dense they prevent any sunlight from reaching the ground. Consequently there will be little or no brush to provide food and cover for wildlife. Younger conifer stands do provide winter cover for game species, but mature trees are of little value.

Our aspen stands have other problems such as **Hypoxylon canker** which appears as a black patch on the side of the trunk, often around a branch stub. This fungus affects about 13 percent of our aspen and usually causes a tree to die within three to seven years after infestation. It kills one to two percent of our aspen each year. It has been estimated that our supply of aspen could be doubled if this disease were eliminated.

Two other forest pests threaten the aspen. The fungus **Fomes ignarious** can affect as much as half of an aspen stand, rotting the wood and rendering it useless as sawtimber. The forest tent caterpillar, though not as damaging as Fomes or Hypoxylon, can completely defoliate trees in an area as large as 100,000 square miles. Fortunately, the caterpillar does not kill the tree with only one defoliation. However, growth is retarded and repeated defoliation will kill the tree.

In summary, I would mention the "forest edge" concept of game habitat improvement. Probably the best wildlife management technique, next to cutting and maintaining aspen, would be to convert our woodlands into a patchwork of conifers and hardwoods punctuated by clearings, trails and other openings. The greater the mixture, the more it will benefit wildlife.

Deer and grouse prefer edges, or wherever one type of forest blends into another. Brush and shrubbery grow in profusion along these sunlit edges to provide excellent cover and an abundance of wild berries and lush green plants for food.

Modern forest management encourages this patchwork design. Accordingly, an owner of a small woodland can incorporate this concept to attract more wildlife. You might cut at least part of that mature woodland that you now consider "too beautiful to cut". You might also cut out a few clearings, or plant conifers for winter shelter, But above all, **don't forget to cut that aspen!**

Your local foresters and wildlife managers will be happy to assist you in implementing good forestry and wildlife habitat practices. Try us!

DEPARTMENT OF
NATURAL RESOURCES

BIRCH

Description

The birch tree is well know for the famous Indian canoes and baskets that were made from its bark centuries ago.

Birch and aspen are commonly found mixed together in a forest stand. They are both "shade intolerant" species. They need full sunlight to grow and reproduce. This is the reason why birch and/or aspen are clearcut when harvested. These areas recover quickly after harvesting.

Management

Management techniques for birch depend on whether it is mixed with other hardwoods, if it is a pure stand, and the quality of the site. If it is mixed with other hardwoods, the birch should not be exclusively cut out of the stand. The shade of the remaining stand will prevent the birch from stump sprouting, its main means of reproduction.

In pure stands of birch, clearcutting or patch cutting is the recommended harvesting method. Birch requires no intermediate management between harvests and should not be thinned.

(A) Poorer sites should be clearcut for firewood every 40 years. These sites are good candidates for site conversion to species which are better suited to the site.

(B) On good sites, manage for sawlog and sawbolt material. Rotation age on good sites begins between 50-60 years. Harvesting should be done during the winter months, as spring and summer logging inhibits the tree's ability to sprout. A forester assisting the timber sale can minimize a large single cut "block" effect.

Uses

Paper birch is used for paper, lumber, dowels and novelty wood. It is also a popular fuelwood.

Principal Enemies

Birch trees are frequently attacked by the Bronze Birch Borer after a period of stress caused by fire or drought. Regular harvesting before the trees are overmature will lessen the chances of Bronze Birch Borer attack.

Birch Leaf Miner periodically attacks and defoliates birch. Control should be limited to ornamental yard trees, as control over an entire forest stand would not be economically feasible.

Wildlife Considerations

Some song birds eat the birch seeds, and dead birch trees are often used as feeding and nesting sites for cavity nesters. Some birds, such as the Black Capped Chickadee even nest on the top of broken off trees.

Identification and Uses

PAPER BIRCH *(Betula papyifera)* is a tree of the northern forests, where it is conspicuous for the white, papery **bark** on the trunks of mature trees. The bark often peels horizontally into long strips that are curled and ragged at the ends. This is the species utilized by Indians in making their legendary birch bark canoes. The **leaves** of paper birch are pointed and double toothed, roughly oval, rounded at the base, and 2-3 inches long. The **fruit** is cylindrical, about 1½ inches long and ½ inch in diameter, and hangs from a slender stalk. This fruit (strobile) disintegrates when mature. The **tree** is usually slender and graceful, often growing in clumps of 2-4 individuals. It attains heights of 80 feet with trunk diameters seldom more than 2 feet. The crown is usually rounded and irregular. Paper birch grows in the South only in North Carolina, occurring there in the balsam-spruce forests at the very highest elevations. The **wood** of paper birch is hard, light, strong, and very close-grained. It is used for veneer, spools, bobbins, toothpicks, and pulpwood.

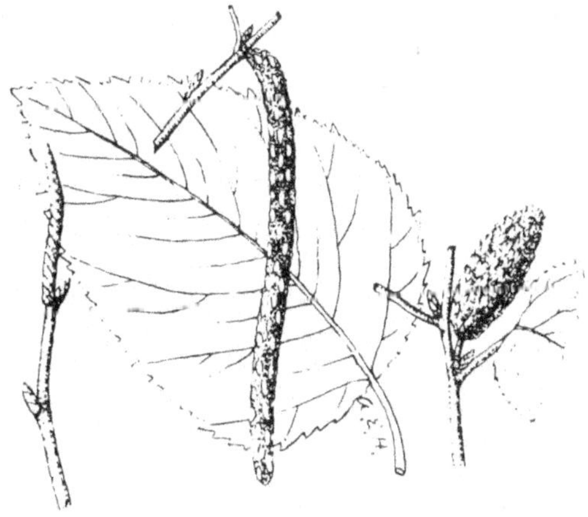

YELLOW BIRCH *(Betula alleghaniensis)* **BRITTON** (gray birch, silver birch, swamp birch; B. *lutea* Michx. f.). Large tree of northeastern United States and adjacent Canada and Appalachian Mountain region. **Bark** aromatic on young branches, yellowish or silvery gray, shiny, separating into papery, curly strips; on old trunks reddish brown. **Leaves** oval, 3-5 inches long, long- or short-pointed, sharply and double toothed, mostly with 9-11 main veins on each side, nearly hairless, dull dark green above, yellow green below. **Cones** ¾-1¼ inch (2-3 cm) long.

Principal uses: Lumber and veneer for furniture (one of the principal woods), cabinets, boxes, woodenware, handles, and millwork such as interior finish and flush doors; small turned products and barrel staves. One of the principal woods for distillation products, such as wood alcohol. Pulpwood, fuelwood, charcoal. Shade tree.

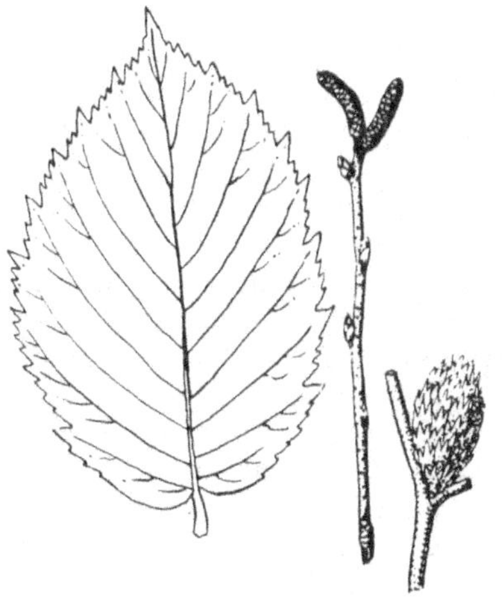

Line drawings for this series are by Mrs. Hoyle (1931) courtesy of the U.S. Forest Service.

Species Identification Information for this series is from "Important Trees of Eastern Forests," U.S.D.A Forest Service

NA-02042
Rev. 1/82

DEPARTMENT OF
NATURAL RESOURCES

NORTHERN HARDWOODS

Description

Sugar maple, basswood, and yellow birch are the primary species of the northern hardwood type. They will be found with a number of other species such as aspen, red oak, paper birch, ash, black walnut, butternut and elm. Pure stands are unusual. This stand diversity requires a management system which is very complex. The group is generally shallow-rooted, long-lived (sometimes 200-300 years), and very tolerant to shade. Less tolerant trees in a mixed stand will eventually be eliminated by the lack of sunlight. Due to its shade tolerance, the northern hardwoods become the climax species (continually perpetuating itself) on many sites, creating an uneven-aged stand.

The trees will grow best on moist, moderately well-drained soils. Quality of the trees vary considerably by location. Commercial yellow birch is limited to the northeastern portion of the state; sugar maple to the east, central and southeast; while basswood, which if found over the entire state, is ideally suited to the "Big Woods" area in the southeast.

Management

Regeneration

The northern hardwoods are prolific seeders and will maintain themselves with little soil disturbance. The exception is yellow birch, which requires some form of seed bed disturbance for good regeneration. Several varieties of hardwoods sprout from stumps.

Culture

Although pruning will add to the final value of the tree, it is more important to maintain the proper spacing and age distribution (selective harvest method) to produce a stand with clean, straight poles. Removal of cull trees allow more growing space for favored trees. If very high quality trees are desired and pruning is initiated, the trees should be pruned no higher than 17 feet. Thinning basswood to 2-3 sprouts per clump will allow sufficient diameter growth to produce good quality sawtimber. Thinning should be done in pole sized stands less than 45 years old to derive the most benefit from the resulting accelerated growth and form.

Open Grown Basswood Clump.

Harvest

This type is generally harvested by selectively marking trees to be removed every 8-15 years. Where selective harvesting is used, extreme care should be practiced to avoid damage to the remaining trees. In specific cases, clearcutting may be undertaken to alter the stand composition to a more desirable species. Selective harvesting will yield a regulated stand, with continuing income while upgrading the timber quality. The first two cuts in previously unmanaged stands are concentrated upon cull removal. Basal area should be maintained at 70-90 square feet per acre.

Management decisions should be made in consultation with a forester, especially if the timber is to be marked for a selective harvest.

Principal Enemies

Canker disease affect yellow birch and sugar maple. These infected stems should be removed to allow for reproduction. Frost cracking causes serious defects in sugar maple. Organisms causing rot and stain are usually introduced through injury and can be reduced by avoiding logging damage and keeping rotations to less than 100 years.

Uses

Northern hardwoods are used for lumber, railroad ties, veneer, plywood, excelsior, furniture, athletic equipment, maple syrup, etc.

Numerous sprouts and seedlings produce browse for deer and habitat for grouse. Mature trees have little wildlife value. Selected trees, such as hollow den trees may be reserved.

Whenever possible, cull trees should be girdled instead of cut down in order to provide nesting cavities, perches, singing sites, and territory marking posts for wildlife habitat.

Identification and Uses

AMERICAN BASSWOOD *(Tilia americana)* has large, alternate, roughly toothed **leaves** that are sharply pointed. Leaves are heart-shaped, conspicuously veined, 5-6 inches long and 3-4 inches wide. The **fruits** are hard and rounded, hanging suspended in clusters from a stalk that is attached to a papery-thin, strap-shaped bract. Each fruit contains two seeds and is covered with a thick, reddish-brown fuzz. American basswood is a tall and stately **tree**, sometimes reaching 120-140 feet in height and 4-4½ feet in diameter. The small slightly drooping branches form a broad, round-topped crown in open stands or a narrow pyramidal crown in dense stands. In closed forest stands the tree often has a straight bole that is clear of branches for half the tree's total height. Forest-grown basswood trees are easily damaged by fire; thus most large, older trees will be hollow or otherwise defective.

Line drawings for this series are by Mrs. Hoyle (1931) courtesy of the U.S. Forest Service.

Species Identification Information for this series is from "Important Trees of Eastern Forests," U.S.D.A Forest Service

Sugar Maple

SUGAR MAPLE *(Acer saccharum)* is one of the largest and most important hardwoods of eastern forests. It is widely distributed, but not found in the South Atlantic or Gulf Coastal Plain states. The **leaves** have 5 lobes that are separated by rounded, shallow sinuses. Leaves are 3-5 inches across, rounded at the base, with sparse, large, pointed teeth on the margins. In autumn, sugar maple leaves are exceptionally colorful with brilliant reds and yellows. **Flowers** are yellow, polygamous, and occur on long, slender stems in clusters up to 3 inches long. They appear with the leaves in spring. The **fruit**, a U-shaped pair of winged seeds, ripens in the fall. The **tree** grows 75-100 feet in height with a trunk diameter of 3-4 feet. In crowded stands, sugar maple develops a tall, straight trunk and a narrow crown. Open-grown trees have a shorter stem and rounded crown. The **bark** of mature trees is thick, light gray to brown, and broke by vertical furrows into plate-like scales. The sap is used to make maple syrup and maple sugar. Sugar maple is a popular shade and ornamental tree.

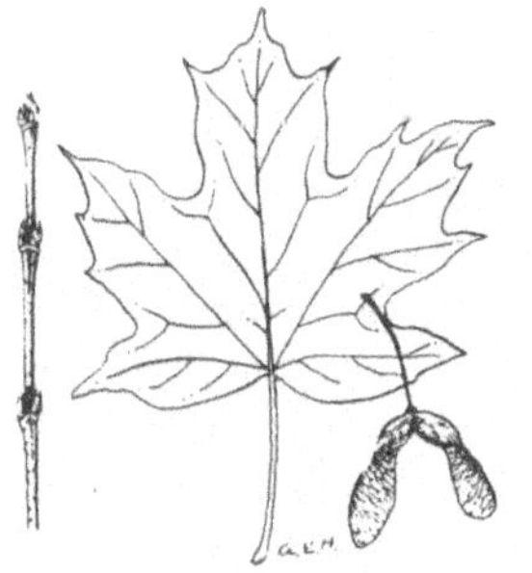

NA-02180
Rev. 1/82

DEPARTMENT OF
MINNESOTA
NATURAL RESOURCES

CENTRAL HARDWOODS

Description

This type consists of oak and hickory, and usually contains other scattered species such as black cherry, sugar maple, basswood, aspen, white ash, black walnut, butternut, and others. Red oak is the most abundant tree, with hickory and white oak usually being the second most abundant.

Southern Minnesota is on the northern most edge of the range for central hardwoods in the U.S. It is the most common timber type in southern Minnesota and can be found on most hillsides and woodlots. This type is found on upland soils and prefers moist, well-drained soils on north and east facing slopes for best growth.

Central hardwoods are generally moderate to intolerant to overhead shade. The origin of this type comes from site disturbances from fire or logging. Oak and hickory reproduce themselves by sprouting back from the stump which results in an even-aged stand.

The natural successional trend is for the shade intolerant central hardwood type to be replaced by shade tolerant northern types (maple, basswood). Management is necessary to help prolong the central hardwoods. The basic objective is to manage central hardwoods for the more valuable oaks and to disfavor other relatively low value species such as hickory.

Management

Regeneration

The basic tool in regeneration of central hardwoods is clearcutting the final product. This means cutting down all live trees larger than 3-4 inches in diameter so that oaks can obtain the full sunlight they need.

If the thinnings have done through the years, then some advanced oak regeneration should be present. These should be saved during clearcutting operations in order to fill in holes where sprouts from harvested stumps do not exist.

Seeding cannot be depended on as a total regeneration method. Yearly seed crops are sporadic and seedlings that do germinate often die quickly due to slow growth and brush and weed competition.

Culture

Timber stand improvement operations should be done over the life of a stand. Natural grown stands usually are too crowded to produce the fastest growth. Periodically thin out the stand as it becomes too crowded. This will reduce the rotation (harvest) age and bring a greater economic return earlier. How much to thin and to what spacing depends on the age, size, and composition of the stand. A forester should help you decide the best thinning procedures.

Bitternut Hickory

Harvest

Central hardwoods are grown until they reach maximum harvest size. This is when the majority of trees reach about 18 inches in diameter at 4½ feet above the ground. On good sites with proper management, this size can be reached in 60-80 years. In a natural stand where no management has been done, the rotation age will be from 80-120 years or more. If the stand is cut when it is "economically" mature instead of "physically" mature, the stump sprouts should be more abundant and vigourous.

Clearcutting is the best way to do the final harvest, and winter cutting is recommended. This will produce an even-aged stand which is much easier to manage and will have the most oak component. When the trees are being cut, the stumps should be kept as low as possible so they will decompose quickly and not damage new sprouts.

Since the composition of central hardwoods is so varied, it is best to have a 100% cruise done before the timber is sold. Contact a forester to help set up a sale.

Principal Enemies

Central hardwoods are subject to a number of damaging agents. The greatest losses are the result of disease organisms which discolor, decay, or deform standing timber. Practicing sound management is the best means of reducing these problems.

Oak wilt disease is perhaps the most well known. This can affect both red and white oaks, although white oaks are more resistant. Oak wilt disease usually results in small, dead patches of oaks within a stand. It can spread by both insect and root grafts. Partial control can be obtained by selectively cutting around the affected area to prevent the spread.

Butternut canker has resulted in severe losses of butternut trees in Minnesota. This disease has eliminated butternut as a viable tree species for forest managment.

There are various insects that can partially or totally defoliate central hardwoods. This will slow growth, but generally not damage the stand enough to warrant control measures. Live oak borers can cause significant lumber degradation, and their galleries may provide entry courts for decay fungi. Insects and disease often enter into less vigorous, older and fire damaged trees.

Uses

Central hardwoods provide a variety of products including fuelwood (hickory and oak are some of the best of any tree species), fence posts, railroad ties, veneers, cooperage, furniture, handles, lumber and flooring.

Wildlife Considerations

When clearcutting in larger than five acre patches, it is quite beneficial to wildlife to reserve clumps of oaks/hickories in the cutting area. The recommended amount is 12 to 18 trees/clump, with 2 to 3 clumps reserved in a 5-10 acre area; 4 to 5 clumps left in an 11-20 acre area, etc. Trees should be healthy, vigorous, and have full crowns. It takes approximately 20 years for seedling oaks to produce a most crop of acorns. Den, cull, and snag trees are important to wildlife (particularly hole nesters), and it is beneficial to leave at least 6-9 acres even though they have no timber value.

Identification and Uses

BITTERNUT HICKORY *(Carya cordiformis)* **leaves** are 6-10 inches long with 7-9 long-oval, toothed leaflets that are dark yellow-green above and lighter below. The 4-ribbed nut is about an inch long, roughly spherical but often broader than long. The nutmeat is very bitter. The husk is 4-winged from the tip to about the middle, and covered with yellowish scales. The tree is probably the most abundant of the hickories, and found throughout eastern United States. It prefers rich moist sites such as bottomlands, there developing into a handsome tree 100 feet or more tall with a straight trunk 2-3 feet in diameter.

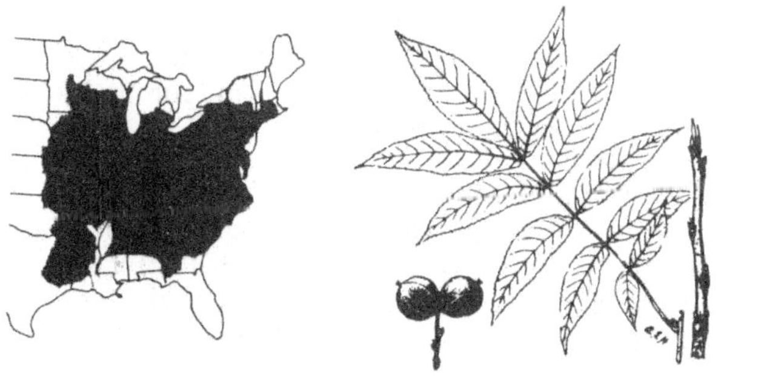

Line drawings for this series are by Mrs. Hoyle (1931) courtesy of the U.S. Forest Service. Species Identification Information for this series is from "Important Trees of Eastern Forests," U.S.D.A. Forest Service

SHAGBARK HICKORY *(Carya ovata)* **leaves** are 8-14 inches long with 5 (rarely 7) tapered-oval, finely-toothed leaflets. The terminal leaflet is largest, being 5-7 inches long and 2-3 inches wide. The **nut**, borne singly or in pairs, is roughly oblong, 1-2½ inches long, and covered in a thick 4-part husk that splits to the base of the nut when ripe. The nutmeat is sweet with an excellent flavor. Shagbark hickory is named for the characteristic light gray **bark** that separates into thick plates a foot or more long that curl outward at both ends. Older trees develop a rough, shaggy trunk. The **tree** grows on a variety of sites and soils, and may reach heights of 120 feet or more. It commonly grows 70-90 feet tall, and usually has a good form. Shagbark hickory is usually a minor component in the hardwood forests where it occurs. In the northern part of its range, it grows on upland slopes at elevations as high as 2,000 feet, where it is associated with oaks, other hickories, and various mixed hardwoods. In the South it is associated with a large number of hardwoods, especially those common to the bottomlands, and here prefers the deep, moist soils of alluvial origin.

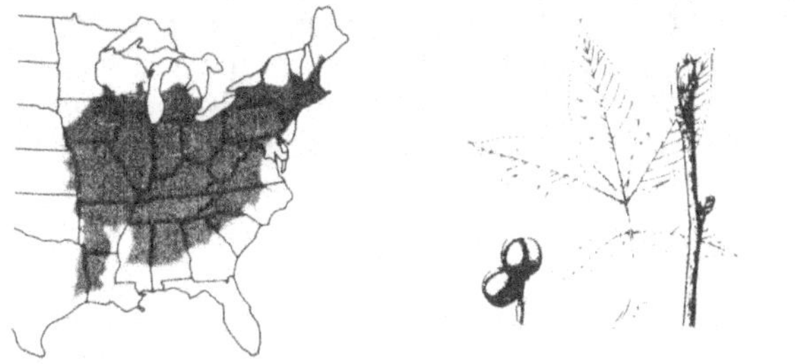

WHITE ASH *(Fraxinus americana)* is the largest and commonest of the ashes, and the most useful. The **leaves** are 8-12 inches long with 5-9 (usually 7) oblong leaflets 3-5 inches long that have smooth or finely-toothed margins. Upper surfaces of leaves are dark green and smooth; undersides are pale light green to whitish. The dioecious **flowers** open before the leaves in late spring, and occur in compact panicles. The **fruit** is 1-2½ inches long, and grows in crowded clusters 6-8 inches long. The **tree** may grow to 120 feet in height, but is commonly 70-80 feet tall with trunk diameters around 3 feet. The ashy-gray to brown **bark** is deeply divided by narrow ridges into net-like patterns.

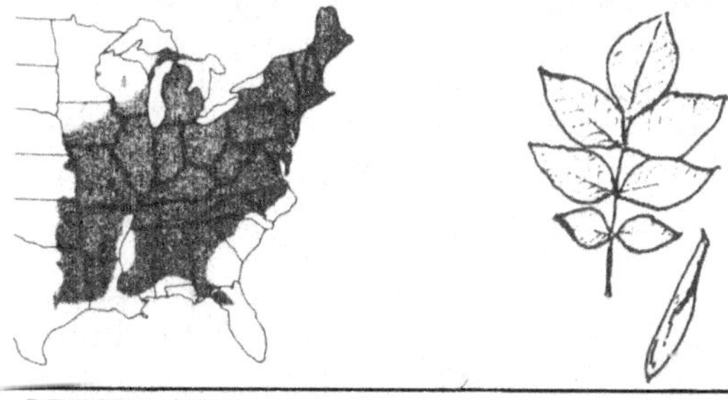

GREEN ASH *(Fraxinus pennsylvanica)* **leaves** are 6-9 inches long with 7-9 leaflets 3-4 inches long. The margins of the leaflets are toothed, but only from the middle of the leaf to the pointed tip. **Flowers** are dioecious, appearing after the leaves. The **tree** commonly grows to 50-60 feet in height. Green ash is the most widely distributed of the ashes.

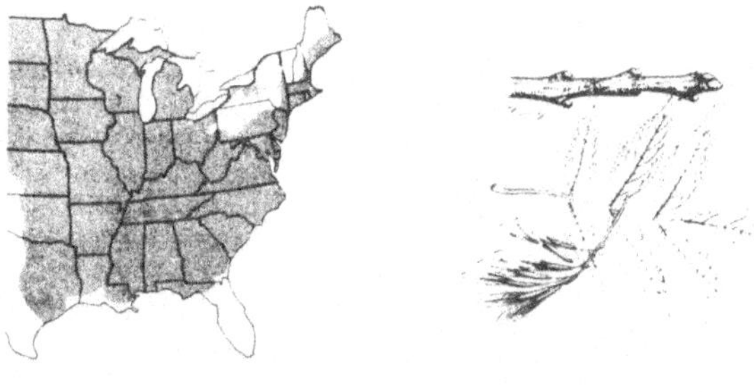

DEPARTMENT OF MINNESOTA NATURAL RESOURCES

BOTTOMLAND HARDWOODS

Description

The bottomland hardwood type consists of ash, elm, cottonwood, soft maple and willow. Black walnut may also be associated with this type in Southeastern Minnesota. Each of these species may be grown in pure or mixed stands on well-drained swamps and on river and stream banks.

On the best alluvial soils, the rate of growth is exceedingly fast. Young trees may grow from six feet to eight feet in height, and one inch or more in diameter per year. Cottonwood will reach sawlog and veneer size in 25-40 years. Most bottomland hardwoods are short-lived and will deteriorate rapidly after they have matured.

Black Ash

Management

Regeneration

The most important enemy to the bottomland hardwood is Dutch elm disease, which is transmitted by the bark beetle. Elm should be discriminated against when cutting for regeneration.

Normally this type is managed on an even-aged system. Openings are essential for the successful regeneration of certain bottomland hardwood species, because they require almost full sunlight.

When cutting lowland hardwoods, care must be taken to cut at the proper time to assure that sprouting will regenerate the stand. It is advisable to clearcut certain stands, and wintercuts will result in the most sprouting, thus leading to successful regeneration of those stands. Stands that contain a high percentage of ash or black walnut may be selectively cut so that maximum economic returns may be obtained.

Culture

Cottonwood and willow are intolerant to shade, which sometimes results in pure stands of both. Elm is intermediate in tolerance, while ash is moderate to tolerant.

Various types of stand management can be used to alter the overall stand composition. A professional forester should be consulted when making decisions pertaining to a bottomland hardwood stand.

Harvest

The objective of management should be to produce even-aged stands. This enables a clearcut to be used as a harvest method while insuring adequate reproduction. The stand can be improved by cutting cull trees. In addition, merchantable trees in need of harvest can also be removed.

Principal Enemies

Dutch elm disease is the most important enemy of the bottomland hardwood type. Other enemies include the ash sawfly, poplar borer, oystershell scale, Nectria canker, Cytosphora canker, and rot-causing fungi. These diseases and insects will not kill a tree. However, if extensive damage is seen, contact a forester.

Uses

Bottomland hardwoods are used for pulpwood, lumber, veneer, and plywood.

Wildlife Considerations

This forest type provides habitat for quite an array of wildlife. As the stand matures, an increasingly higher number of wildlife species, especially bird life, can be found in the lowland hardwood forest.

Mature and overmature stands provide cavities essential to many wildlife species, namely wood ducks, barred owls and raccoons. They also supply suitable nesting habitat for various woodpeckers and secondary cavity users.

Identification and Uses

BLACK ASH *(Fraxinus nigra)* **leaves** are 12-16 inches long with 7-11 pointed-oval, finely-toothed leaflets. Leaflets are smooth dark-green above, and paler below. **Flowers** are polygamous and appear before the leaves. The **fruit** (samaras) occurs in open panicles 8-10 inches long. Samaras are 1 to 1½ inches long, oblong, and with the thin wing extending below the center of the seed cavity. The seed cavity is indistinct in black ash. This species typically grows along streams, in bogs, and in poorly drained sites with high water tables. It grows most commonly in peat soils. However, it occurs on fine sands and loams underlain by clays where the surface drainage pattern causes high level of water table. The **trees** grow 60-70 feet in height (sometimes taller on the better sites) with trunk diameters of 1-2 feet. The slender upright branches form a narrow crown.

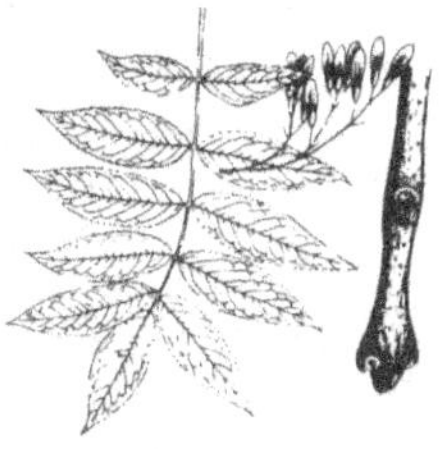

EASTERN COTTONWOOD *(Populus deltoides)* is the fastest-growing commercial forest species in North America. It is also one of the tallest **trees** of eastern forests, attaining heights close to 200 feet and diameters of 4-6 feet. The **leaves** are roughly triangular, toothed, pointed, 3-6 inches long, paler below than above, and with a flattened stem. The **fruit** consists of numerous green bud-shaped capsules that are clustered along short stems that hang from the branches in long, narrow clusters. Seeds are released from the capsules when ripe and drift on the breeze, each one suspended from a tuft of white "cottony" hairs. **Bark** on young trees and on upper stems of older trees is smooth and greenish. Bark of older trees is dark gray, heavily furrowed and ridged. Cottonwood does best on moist well-drained soils such as those found in many areas of the Mississippi Valley. It does not often develop into a well-shaped tree if it is more than 15-50 feet above the average level of the streams in the area.

BLACK WILLOW *(Salix nigra)* is one of the largest of the native willows. Its **leaves** are long (4-6 inches) and narrow (½ to ¼-inch wide), pointed and with finely-toothed margins. Black willow will grow on almost any soil but its shallow, wide-spreading roots need an abundant and continuous supply of moisture during the growing season. On the best sites in the Mississippi Valley, the **tree** will grow up to 140 feet in height and 4 feet in diameter. Usually the tree will be 30-40 feet high, often with several stems clustered, and the thick spreading branches forming a broad irregular crown. Trunks are often twisted, curved or leaning.

Line drawings for this series are by Mrs. Hoyle (1931) courtesy of the U.S. Forest Service.

Species Identification Information for this series is from "Important Trees of Eastern Forests," U.S.D.A. Forest Service

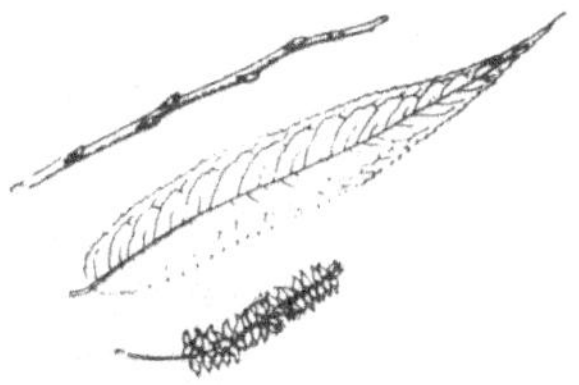

Black Willow

AMERICAN ELM *(Ulmus americana)* is a treasured ornamental and shade tree, lending its graceful, spreading form to many streets, parks, and campuses in eastern United States. Tragically, many of these stately trees have been wiped out by the Dutch elm disease in recent years. The **leaves** of American elm are oval, coming to a long slightly curved point. Leaves are rough-textured, rounded at the base on one side of the midrib, but shorter and almost straight on the other side. Leaf margins are sharply-toothed. The **fruit** is clustered on long stems. The small seed is encased in a flattened, papery, wing-like covering that is oblong in shape. In dense forest stands, the **tree** will have a fairly narrow crown and a long, clear bole. Open-grown trees usually fork near the ground and develop arching crowns. The characteristic "urn" shape of the elms permits easy identification even from a distance. This species attains heights in excess of 100 feet, and diameters of 6-11 feet. Old trees are sometimes enlarged at the base by large buttresses.

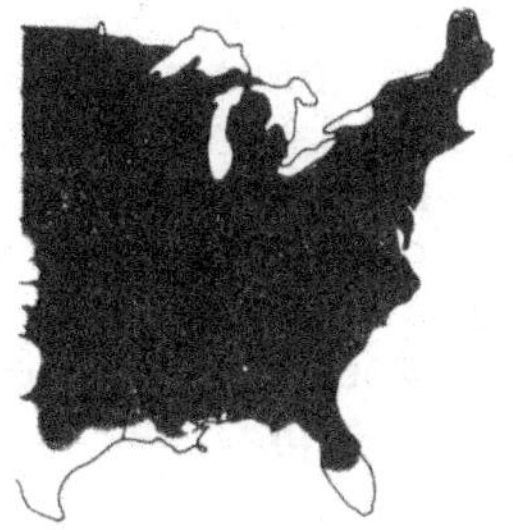

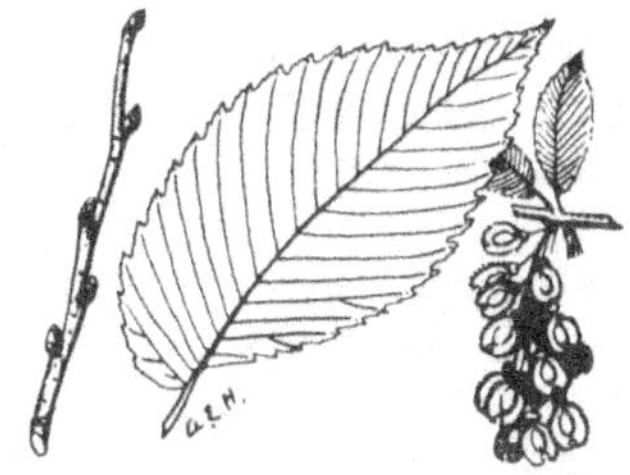

SLIPPERY ELM *(Ulmus rubra)* **leaves** resemble those of American elm, being rough-textured, roughly-oval, pointed with curving tips, asymmmetrical at the base, and sharply-toothed. The leaves are somewhat larger than those of American elm, being usually 5-7 inches long. The **fruit** is also wafer-like, the **seed** being enclosed in a flat leathery covering. The fruit is about ¼-inch in diameter and has a seed cavity covered with thick rusty-brown fuzz. The **tree** attains heights of 60-70 feet on average sites, and diameters of 2-3 feet. The form is similar to the "urn" shape of American elm. The **wood** of slippery elm is considered the best of the elms. It is tough, coarse-grained, heavy, strong, and stains attractively. Its uses are mainly in furniture, crating, boxes, rough flooring, farm buildings, and gates. The common name of this species comes from the mucilaginous coating of the fragrant inner bark that is slippery to the touch.

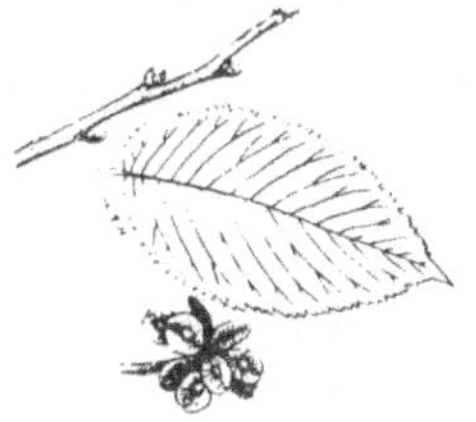

AMERICAN ELM

By G. D. Huntly

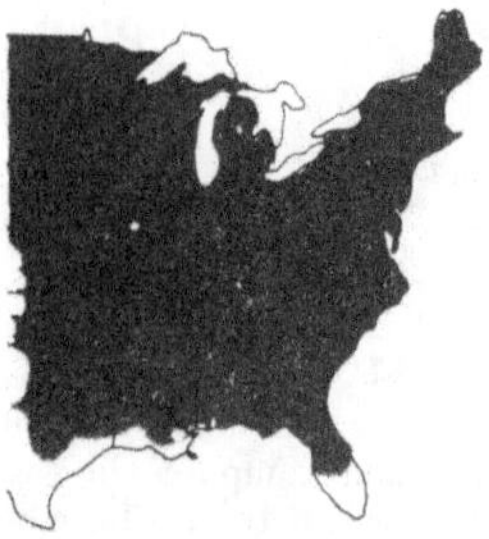

A Distinctive Profile

A mature American elm *(Ulmus americana)* is not hard to spot in the forest. It is one of the largest hardwood trees in Canada, and it rises straight and tall, forking out high above the ground into a graceful, arching crown shaped like a vase or umbrella. At close range it is recognizable by its dark green leaves with prominent saw-tooth edges that run from base to tip.

Habitat

Since it is found across the southern parts of central and eastern Canada from Saskatchewan to Nova Scotia, American elm obviously can survive in a variety of soils, but it does best on rich, moist, well drained sandy loams. It is found most often in wet flats and bottomlands mixed with red and silver maple, or balsam poplar, balsam fir and yellow birch.

A Closer Look

In the forest American elm can grow to 120 feet or more in height, and over 1 yard across the trunk. Open-grown specimens, i.e., those which stand alone, may attain a trunk width of 7 to 12 feet.

Open-grown elms in fields and along city streets may fork near the ground into several erect limbs, but forest-grown trees develop a long, straight main trunk before branching. Both types have the vase-shaped crown.

American elm is a relatively long-lived tree. Normally it reaches maturity in 150 years, and has an average maximum age of 175 to 200 years. A few specimens live longer than 300 years.

The grey or grey-brown bark of American elm has broad, intersecting, flat-topped ridges. As the tree ages, these ridges become more pominent and the bark turns to a mottled ash-grey. A cross-section of the outer bark shows alternating layers of off-white and reddish-brown cork and fibers.

The leaves have short stalks and are arranged alternately in two rows along slender, brown, usually hairless twigs. They are elliptical — 4 to 6 inches in length and 2 to 3 inches in width — with pointed tips and asymmetrical bases. The leaf edges are coarsely double-toothed, with the larger teeth at the ends of prominent straight veins and the small teeth between the veins. The upper leaf surface is dark green, varying from smooth to slightly rough, while the under surface is a paler green and covered with fine, soft, short hairs.

Before falling in the early autumn, the leaves turn a dull yellow. Once the leaves drop, the zigzag pattern of the twig — resulting from the alternate leaf arrangement — is more easily seen. The pointed winter buds, 2 inches long, are somewhat flattened and lie close to the twig. They have reddish-brown and slightly hairy scales.

In Canada, American elm flowers during April or May, some two to three weeks before the leaves unfold. The flowers are arranged in clusters on stalks that are about 4 inches long. Each flower contains both sexes. Soon after it has been pollinated by the wind, the flower turns from blue-violet to purplish-black.

Seed and Growth

The fruit of the American elm flower is an oval, light green, winged seed (called a samara) measuring 2 inches in width and 4 inches in length. The thin, papery wing entirely surrounds the seed, is deeply notched at its tip and has a fringe of short hairs.

A tree may begin to produce seed when it reaches 15 years of age. However, fruiting is rarely abundant before the age of 40, and then the tree normally has a good annual yield of seed until it is over 150 years old. Late spring frosts may kill the fruits in some years.

The winged seeds are so small that it takes 68,000 of them to make up a pound. They are ripe and ready for flight before the leaves are expanded fully; then they leave the tree in clouds. Most reach the ground within about 300 feet of the tree, although some may be carried further away by wind or water. Animals or birds such as squirrels, mice, opossums, ruffed grouse, bobwhite and Hungarian partridge eat part of the seed crop.

Although some seeds stay dormant until the following spring, most germinate soon after they fall and, within a couple of weeks, multitudes of elm seedlings spring up. One may see them in towns growing in flower beds, along curbs, or even in the spaces between concrete sidewalk slabs. Within a few days of germination, however, most of these die because they do not have enough room for growth. Nevertheless, a good number of seedlings become established where they have space.

After the first year, a white elm seedling grows best in full sunlight. Heavy logging that leaves elm trees as the main seed source in the residual stand will favor establishment of seedlings. Little damage is done to seedlings and young trees by mice, squirrels, rabbits and deer since white elm is not their preferred browse species.

Dutch Elm Disease

This wilt disease is by far the worst enemy of the elm. Caused by a fungus, the spores of which are spread chiefly by elm bark beetles, Dutch elm disease now stands from Nova Scotia to southeastern Manitoba. Once a tree is infected, it usually dies quite quickly. In rare cases, trees have been able to seal off the infection under layers of more resistant tissue and have survived.

Early signs of Dutch elm disease (mid-June to mid-July) are the wilting and curling of leaves on one or more branches. These leaves shrivel and turn brown, but usually stay on the tree. Later signs are yellow and drooping leaves, which drop prematurely. It is difficult to distinguish the late-season signs of the disease from the normal autumn coloring. In the spring, a diseased elm may display leaves that are smaller than usual, and some branches may be dead. These conditions may mean that the tree was infected the previous year but too late for the signs of the disease to have become apparent in that year.

Control of DED is difficult. The size of the elm bark beetle population must be reduced, primarily by means of sanitation. Sanitation involves the removal and destruction of dead and dying elms to prevent the buildup of beetle breeding material. Chemical insecticides may also be used to reduce beetle numbers and prevent their spread from diseased to healthy elms.

Chemical fungicides may be injected into the roots of an elm to protect it from DED, or to prevent the disease from killing an elm, if it is already present in a tree, but is detected in the early states. In addition, a preventive treatment may be applied to healthy elms, by root-flare injection of the fungicide.

The intensive treatment of elms for DED control must be conducted by trained, qualified aborists in both the public and the private sectors of the community. It is a costly operation, but the cost may be justified for individual elms of particularly high value.

Other Enemies

Elms may also be killed by phloem necrosis. Phloem necrosis is a vascular disease spread by an insect called the elm leaf hopper. Although phloem nercrosis is a serious problem in the middle and southern portions of the United States, the climate throughout the range of the American elm in Canada is such that the disease is unlikely to be established here.

Dothiorella wilt, sometimes called elm dieback, may also be a serious problem. Many of the symptoms of this fungal disease are similar to those of Dutch elm disease.

Elm is attacked by several common foliage-feeding insects that damage its leaves, such as canker worms, spiny elm caterpillars and forest tent caterpillars, as well as several scale (sucking) insects, but these rarely become a serious problem in the forest. They may, however, temporarily spoil the appearance of shade and street trees.

Open-grown elms, with their many branch forks and widespread crowns, may occasionally be injured by wet snow and ice storms, which break off branches and split large branch crotches. These wounds, as well as others, may be affected by wetwood, a bacterial infection. The sign of this disease is the presence of sap-like exudate oozing from unhealed wounds and flowing down the trunk. Not only is this unsightly, especially with street and shade trees, but any bark chronically affected by such sap will eventually die.

Uses of American Elm

The native people of Canada found many uses for white elm. Their elm-bark canoe was made either by tearing the bark from the tree in a long single strip, or by sewing shorter strips together. The bark was bent so that the tough outside layer would form the outside of the canoe. It was then fitted over a rigid pole frame. The ends of the canoe were shaped, brought together and then sewn with long, fibrous spruce roots passed through bone awl or needle. Strips of elm or other hardwood were used to strengthen the gunwale, as well as the bottom and sides of the canoe.

The natives also used elm bark in building their homes. The Ojibwa peaked lodge was made with a long ridgepole resting on a series of shorter poles, which sloped outward from the ridge to the ground on each of the long sides. This framework was then covered with sheets of elm bark.

The Iroquois longhouse was also covered with elm bark. The pole frames, posts, beams and rafters were first lashed together with strips of inner elm bark. Pieces of elm bark were then tied together and laced to the log framework, again with the inner bark strips.

The native people used the wood of American elm in making mortars and pestles for grinding their medicines. They also discovered that the tree had valuable medicinal properties. Elm bark mixed with red oak bark was a remedy for dysentery. A cough and cold medicine was made by steeping the bark of American elm in water. Elm root was boiled to produce a lotion for treating eye troubles. Wound poultices were made by wetting and pounding elm bark, and a decoction of inner elm bark served as a laxative, as well as a preservative for meat.

The early settlers found the hard, tough characteristics of American elm wood useful in the construction of farm implements, such as cultivators, hay rakes and reapers, for short-handled tools, and for pike-pole and cant-hook handles. Components of horse-drawn vehicles, such as wagon wheel hubs and spokes, were also made from elm wood.

Elm was not a preferred species for firewood because it was difficult to split and season. (It does, however, have a high heating value, comparable with that of oak, hickory and beech. A standard cord (approximately 3.6 cubic meters) gives the same amount of heat as 0.9 ton of coal.) Because the wood burns slowly, it was not suitable for domestic use among the early settlers. Nevertheless, they found that it made an excellent back log for fireplaces and campfires.

In the past, American elm was used extensively for kitchen table tops and kitchen cabinets, as well as for parts of wooden ice boxes and the early refrigerators. Vigorous scrubbing with soap and water easily removes the discoloration of weathering and dirt from elm wood surfaces. It is also free of stain, taste and odor, and has a considerable strength.

The excellent bending properties of American elm wood made it useful for such furniture parts as chair rockers, bent arms and curved pew seats. It bends without steaming and glues well. It is also easily stained to match oak pew ends. However, elm is difficult to work with tools, does not take a high polish, and is liable to splinter on the surface.

American elm was the main species used in the slack cooperage industry (nonwatertight barrels) as hoops and staves, since it is strong, steams well, and retains its form when bent. In addition, it does not taint or sour the substances with which it comes in contact. Elm slack barrels were used mainly to ship and store sugar and flour.

Because of its strength and good bending properties, American elm was also the principal wood used in rims for vegetable and fruit baskets. And, again because it imparts no odor or taste, it served for the sides and rims of cheese boxes.

In the early automobiles, American elm was considered indispensable for the production of back rails, wheel housings and other small bent parts, as well as for body-frame and vehicle-top construction.

American elm has been one of the most popular species for fencerow, street or shade trees. Its shallow, fibrous root system makes it relatively easy to transplant. It may be planted on a wide range of soil types and, once established, it grows quickly.

Today, American elm is used mainly for furniture and wall paneling veneer, and as pulp for papermaking. Other products are truck beds, pallets, and hockey sticks. Some of the traditional elm wood products still being made, but in smaller quantities, include dry-goods barrels, tool handles, fruit and vegetable baskets. In addition, although it is plagued by DED, the stately American elm, with its distinctive, arching, vase-shaped crown and its remarkable ability to withstand adverse environmental conditions, remains a superior shade tree in the urban landscape.

Written by: G. D. Huntley, Great Lakes Forest Research Centre, Canadian Forestry Service, Sault Ste. Marie, Ontario.

NA-02043
Rev. 1/82

STATE OF MINNESOTA
NATURAL RESOURCES

OAKS

Description

Oaks will predominate in pure stands, and are usually associated with other species of hickory, maple, basswood, and ash, with some cherry and walnut.

Oak growing on the best sites can produce 24 inch diameter trees in 70 years. They are long-lived trees, especially the bur and white oaks, which can live up to 300 years. Harvesting should occur at 60 years of age on the best sites, and 120+ years on poorer sites.

Oaks are considered intermediate in shade tolerance. The tolerance level of white and bur oak changes as they grow from seedlings to sawtimber. Seedlings and saplings are more shade tolerant than pole-sized stands. Red oak seedlings are less shade tolerant and make a poor competitor in the understory of a woodlot.

Oak forests occur in a wide range of soils and topography. Oaks survive best on moist, well-drained soils. In managing this type, the goals are to produce high quality timber in the shortest time.

Management

Regeneration

Regeneration is produced from natural seeding and stump sprouts. It is dependent upon cutting methods which bring more sunlight to the forest floor.

Due to lower shade tolerance, they will seldom reproduce without disturbance, and are frequently succeeded by a natural succession to the more tolerant hardwoods. Seldom will regenerated stands have a predominant percentage of oak, but multiple species can be managed to favor oak at final harvest.

Culture

Many woodlots may need timber stand improvement. Poor quality trees in the past have not been utilized by harvest crews. Left to stand, these trees take away valuable growing space from trees that could be the future crop. Similarly, low-value trees, such as hickory and boxelder may rob space from more important species. Some areas of a stand may also be overstocked in some size classes. The removal of these stems will provide for faster growth.

Harvest

Clearcutting provides full sunlight and necessary stump sprouts to regenerate the next stand. Clearcutting at maturity allows the sprouts from stumps to compete favorably with other regeneration.

Poor stump sprouting may occur in trees older than 45 years of age. Site determines the age when trees should be harvested. A tree diameter of 16-20 inches is an indicator to begin harvesting. In some stands, pruning up to 17 feet can increase veneer log quality. A forester should be consulted to help set up a harvest.

Principal Enemies

Oak wilt is a common disease that can kill in pockets. The red oak is most susceptible to this disease. The two-lined chestnut borer has caused considerable problems with several species of oak by boring tunnels into the inner bark and girdling the tree. Insects and disease often enter into less vigourous, usually older and fire damaged trees. Care should be taken to avoid wounding or pruning during the period of May through June in order to decrease the incidence of oak wilt disease.

Shoestring root rot is another serious mortality factor contributing with oak wilt and two-lined chestnut borer.

Uses

The high quality logs are peeled into veneer or sawn into lumber. These go to the furniture manufacturers throughout the nation. Markets for railroad ties are generally good, utilizing poorer quality wood. White and bur oak can be used for barrel staves.

Wildlife Considerations

The two most important wildlife foods in the U.S. are acorns and corn. In Minnesota, some 40 species of wildlife utlize oaks directly (acorns, buds, twigs, leaves).

To insure these species local survival, adequate supplies of acorn producing trees should be maintained. This can be done by reserving clumps of oaks in the cutting area. The recommended amount is 12 to 18 trees/clump with 2-3 clumps/5-12 acre cut; 4-5 clumps/10-20 acre cut, etc. Den, cull and snag trees are important to wildlife and between 8-15 trees per acre should be reserved whenever possible.

Identification and Uses

WHITE OAK *(Quercus alba)* has deciduous **leaves** 5-9 inches long with 7-9 rounded lobes. The depth of the sinuses separating the lobes varies, almost reaching the midrib in some cases. The base of the leaf narrows abruptly to become wedge-shaped at the stem. The **acorn** is about ¾-inch long, light chestnut-brown, enclosed about one-fourth of its length in a bowl-shaped cup that is covered with rough scales that are joined at their bases to form small knobs. The **tree** reaches 80-100 feet in height and 3-4 feet in trunk diameter. In the open, white oak develops a rounded spreading crown; in forest stands it has a tall, clear stem and smaller crown. Growth is good on all but the driest, shallow soils, but is best on deep, well-drained loamy soils.

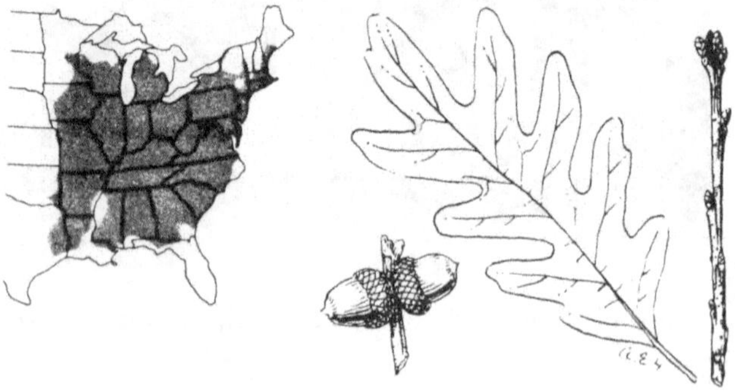

NORTHERN RED OAK *(Quercus rubra)* **leaves** are usually 5-8 inches long and have 7-11 lobes. Each lobe is usually 3-toothed, sharply pointed, and with bristles on the points. Leaves are deciduous, turning red before they fall in autumn. The **tree** will reach 60-80 feet in height and 3 or more feet in diameter. The **acorns** are usually about an inch long with a flat, shallow cup at the base. The crown of the tree is usually rounded and comparatively narrow. The species is often planted as ornamental or shade tree.

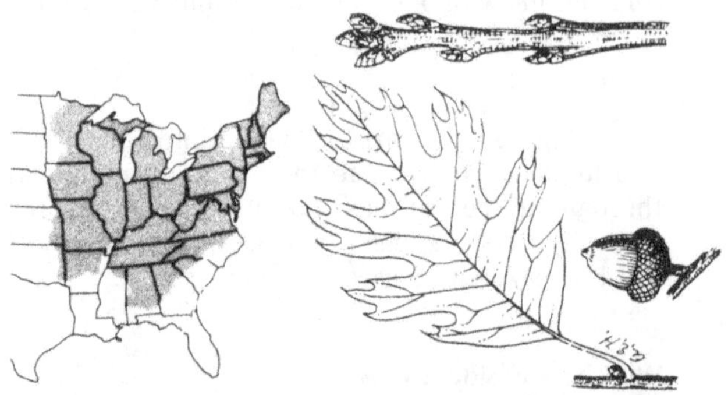

Line drawings for this series are by Mrs. Hoyle (1931) courtesy of the U.S. Forest Service.

Species Identification Information for this series is from "Important Trees of Eastern Forests," U.S.D.A. Forest Service

BLACK OAK *(Quercus velutina)* is one of the commonest oaks of the dry gravelly uplands in eastern forests, often forming a large portion of the hardwood stands in the Appalachian foothills. The **leaves** are variable in form, generally have 7 lobes (may have 5), sometimes being divided nearly to the midrib by rounded sinuses. Lobes are toothed and usually taper from a broad base. The **acorn** is oval or rounded ½-¾ inch long, and enclosed for about half its length in a deep, scaly, bowl-shaped cup. The **tree** reaches 50-70 feet in height, and 3 feet or more in diameter. in early spring the unfolding leaves are a deep red, turning silvery within a few days. **Bark** on the trunks of old trees is usually thick, black, and deeply divided into broad, rounded ridges, The crown i s usually rounded.

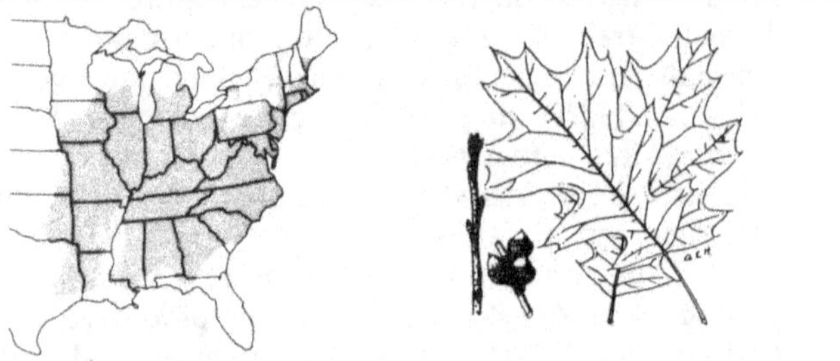

SCARLET OAK *(Quercus coccinea)* has **leaves** 5-8 inches long as does the northern red oak. However, the leaves of the scarlet oak have 5-9 pointed lobes deeply separated by wide sinuses that reach almost to the midrib. Lobes are toothed and pointed. In the fall leaves turn brilliant scarlet. The **acorn** is ½-1 inch long, oval, and enclosed one-half to one-third of its length in the deep, bowl-like cup. The exposed end of the acorn is often marked with circular lines. The **tree** grows 70-80 feet high and has comparatively small branches that spread to form a rather narrow, open, irregular crown.

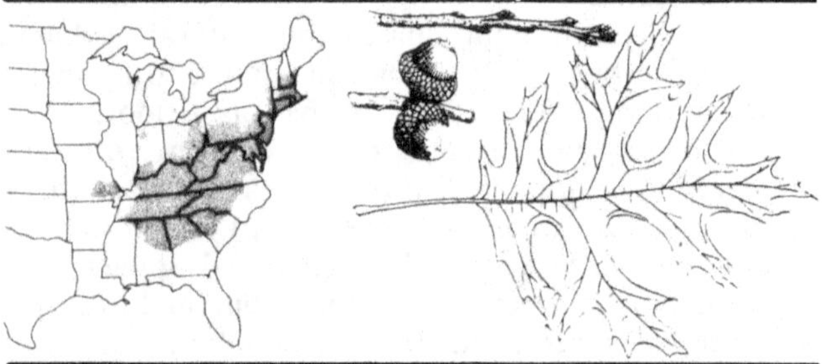

BUR OAK *(Quercus macrocarpa)* has deciduous **leaves** that are somewhat wedge-shaped, divided into 5-7 lobes by wide sinuses reaching sometimes almost to the midrib. The lobes are rounded, comprising a leaf that is quite large (6-12 inches long and 3-6 inches wide). The **acorn** is oval or broadly ovate, broad at the base and rounded at the depressed apex. The acorn is from ¾ to 1½ inches long, and half covered in the fringed cup that gives the species its common name. Twigs sometimes develop corky "wings" 1-1½ inches wide. The **tree** is a relatively slow grower; it is one of the most drought-resistant of the oaks, and does well on exposed sandy sites. On better sites, it reaches 100 feet in height and 3-4 feet in diameter. It is said to have reached a height of 170 feet and diameter of 7 feet in the Ohio Valley.

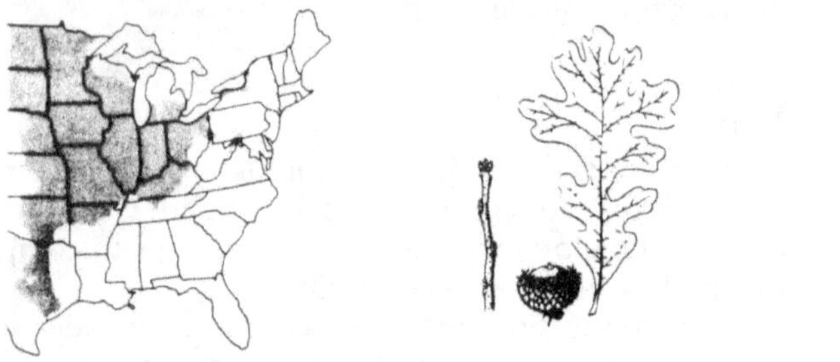

BLACK WALNUT
(Juglans nigra)
ITS GROWTH AND MANAGEMENT

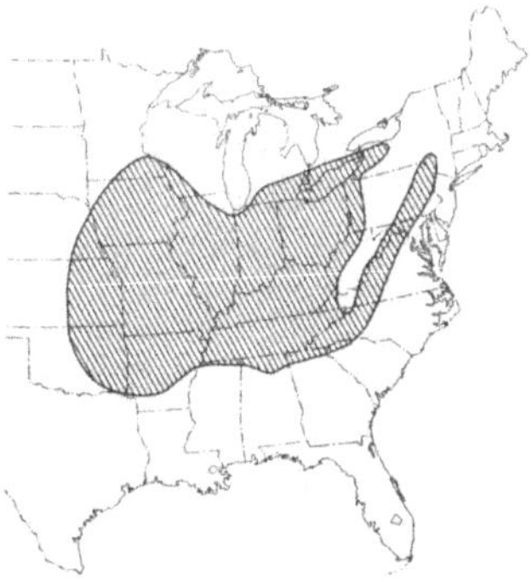

by Terry Helbig

Black walnut, or chocolate pine as it is affectionately referred to, is found throughout the eastern United States. It is one of the most highly valued of all North American hardwoods for both its wood and its nuts.

Although black walnut is usually a medium-sized tree of 70 to 90 feet in height and two to three feet in diameter, occasionally old growth trees are found which far exceed this.

Stories are often heard of individual walnut trees selling for $1,000.00 or more. In fact, a few individual trees have sold for over $30,000.00. Unfortunately, even trees of the thousand dollar variety are the exception, not the rule. With proper care and management, however a black walnut plantation can be one of the most satisfying and valuable investments an individual can make.

PLANTATION ESTABLISHMENT

Site Selection

A good soil is perhaps more important in a walnut planting than in that of any other species. Soils selected should be deep (15 feet or more to bedrock or water table), well-drained, moist (either flood plain or northerly or easterly slope aspect), fertile and nearly neutral in pH. The most important soil characteristics are texture, depth and drainage.

The best walnut sites are deep loams, loess soil and fertile, well drained alluvial deposits.

Spacing

The spacing of trees in a plantation depends on the landowner's objectives. Most plantations established solely for timber production are planted approximately 12x12. Trees put in for nut production are planted at about 20x20. A unique concept with walnut is intercropping. Here, trees are planted at spacings of 15x30 or 5x40. The 40-foot distance between rows allows the space to be used for growing corn, beans, hay or other forage crops.

Forest grown Black Walnut.

Site Preparation

The most critical factor in the early survival of walnut seedlings is complete control of all competition. On a cutover site, this means that all vegetation over ½ inch diameter should be killed. In open field plantings, all weed and grass competition needs to be controlled. Herbicides are more effective than cultivating to accomplish this. Mixtures or simazine and Roundup applied to a grassy field the year before planting have proven to be the best method of weed control.

Planting

Walnut may be machine or hand-planted. Root pruning is necessary in many cases to simplify the planting of the stock. Roots should not be pruned to less than 8 inches. In addition, all seedlings less than ¼ inch caliper at the root collar should be culled.

EARLY STAND TREATMENTS

Weed Control

A continuing program of weed control is imperative to the success of a walnut planting. This should be done for the first three years following planting.

Stem Form

During the early growth of a walnut plantation, corrective pruning often needs to be done to

Young Black Walnut that has been properly thinned.

maintain one terminal leader. A plantation should be inspected annually and trees with forks or multiple leaders should be corrected. This corrective pruning should not be confused with pruning of lateral branches. Lateral pruning should not be done until the trees are 15-20 years old.

Fertilization

Generally speaking, fertilization of young walnut plantings is not cost effective because the weeds get more advantages from the fertilizer than the trees do.

TIMBER STAND IMPROVEMENT

Pruning

As stated earlier, walnut trees should not be pruned to produce clear lumber until the seedlings are 15-20 years old. Branches on crop trees should be pruned before they are two inches in diameter and all pruning should be done during the dormant season. Not more than 100 stems per acre should be pruned and all pruning should be done before trees are 10 inches in diameter.

Thinning

Thinning is also needed to assure optimum growth. Frequent light thinnings are preferable to heavy thinnings in walnut plantations. Because of future high tree value, thinnings in walnut plantations are conducted on an individual tree selection basis rather than by removal of entire rows as is often done in pine plantations.

Single Tree Culture

The high value of walnut trees allows them to be managed on a single tree basis. Individual trees found in a woods can be identified and managed. The first step in this procedure is to release the selected trees. These should be trees that will be left to grow at least another 10 years. The rule of thumb is that the crown should have at least 12 feet of space on at least two sides. Pruning can be done as needed but branches removed should not exceed three inches in size.

HARVESTING

Walnut is a crop like any other tree and it should be treated as such. To be assured of receiving true market value for logs, landowners should seek professional assistance from foresters. Walnut grading rules may result in a higher value being placed on a limby tree if it is sold for veneer purposes than if it is sold strictly as a sawlog for lumber purposes. The forester can help the owner get a higher price for his trees in many instances.

OTHER CONSIDERATIONS

Insects and Disease

Walnut is a tree relatively free from serious pests. The major insect problems include the walnut caterpillar and casebearer. The casebearer is the more serious of the two pests because it can kill terminal buds and thus ruin good form. Anthracnose leaf spot is a major disease. This, however, is seldom a serious enough problem to warrant control.

Wildlife

The frequent nut crops of black walnut make it an excellent species to plant for wildlife, particularly squirrels. Deer also are frequently found in walnut plantations and damage from browsing and antler rubbing can often be found.

Nut Production

Black walnut is one of the few trees in which the fruits may be as valuable as the wood. Black walnuts are used for food and industrial abrasives. Once an area has proved it can produce large quantities of nuts, industrial hullers move in and do on the spot hulling of walnuts.

REFERENCES

Manager's Handbook for Black Walnut. General Technical Report NC-38. North Central Forest Experiment Station, USDA Forest Service, 1992 Folwell Ave., St. Paul, MN 55108.

Growing Walnut for Pleasure & Profit. Fine Hardwood Association, 5603 W. Raymond St., Suite "O", Indianapolis, IN 46241.

Black Walnut as a Crop. Proceedings of the Black Walnut Symposium, 1973. General Technical Report NC-4. North Central Forest Experiment Station, USDA Forest Service, 1992 Folwell Ave., St. Paul, MN 55108.

Growing Black Walnut in Minnesota. Extension Bulletin 396. Forestry Services 21, Agricultural Extension Service, U. of Minnesota, College of Forestry, Green Hall, St. Paul, MN 55108.

Black walnut.

CHAPTER 5
SPECIAL FOREST PRODUCTS

Wood Heat: Right For You?

Maybe, but some wood-stove buyers have found wood as costly to burn as fossil fuels.

Allen Wickman

To save on energy costs, many homeowers plan to install a stove or convert their furnace to wood. For some, this may be a mistake. Studies show that wood may actually be more expensive than other forms of energy. Too, a stove requires more tending than a furnace and can be dangerous because of faulty installation or lax maintenance. If you are considering the possibility of heating your home with wood, this article may help you decide what is right for you.

Cord Sizes

Any discussion of wood as a fuel begins with cords. A common measure of firewood, the cord has proved totally confusing for most fuelwood buyers—and no wonder! Wood is sold in three ways: (1) by the standard cord, (2) fireplace cord, (3) rick or face cord. It may be sold split or unsplit, neatly stacked or piled in a heap. Each way the measurement changes. How do you know how much wood you are actually buying?

Wood is usually sold by the cord, but I think a person is better off buying by the cubic foot. This cuts down on the confusion. The standard pulpwood cord is a stack eight feet long, four feet high and four feet wide. Multiply the length times the height times the width and you get 128 cubic feet.

Now let's cut this cord into 20-inch lengths so the logs will fit in the stove. When you pile and measure the 20-inch logs again, you assume there will still be 128 cubic feet. Not so. Cutting and restacking the logs has reduced the air space in the pile by about 18 cubic feet. The commonly accepted volume of a cord of wood cut into stove-length logs and neatly piled is 110 cubic feet.

What if you split the logs and measure the stack again? Would it shrink even more? No. The stack will actually gain in size after each log is split. A cord neatly cut into stove lengths (16 to 22 inches), split, and neatly stacked should equal 120 cubic feet.

Someone once decided to make fuelwood measurement even more confusing by introducing the "face cord," "rick," and "fireplace cord." The face cord is stove-length wood with a "face" of 32 square feet (a stack four feet high and eight feet long). The volume of a face cord depends upon its length. Rick and fireplace cords have an unsplit volume of 37 cubic feet and are also cut into stove lengths. The face cord, fuelcord, and rick are all about one-third the standard cord.

Six to ten **standard** cords of wood will keep the average home heated through our long Minnesota winter. But someone who orders eight fuelwood cords — or face cords — and expects to have enough wood may be surprised.

A comparison of heat values for different species of wood can be seen in Table #2, prepared by Lewis T. Hendricks, Agricultural Extension Service, University of Minnesota. At a 65 percent rate of efficiency, a furnace needs to burn 10.9 gallons of fuel oil to produce one million BTUs of heat. At 50 percent efficiency, a cord of red oak can produce 12 million BTUs, the equivalent of 130.8 gallons of fuel oil.

Consequently, to economize on energy costs, a cord of oak should not cost more than 130.8 gallons of fuel oil.

Other heat-value equivalents, in gallons of fuel oil per cord of wood, are as follows: white oak, 140 gallons; paper birch, 111 gallons; aspen, 80 gallons.

These equivalents are for cured wood at 20 percent moisture content. Wood not completely cured will yield fewer BTUs.

Finding Wood

After driving along miles of highway through stands of trees, you may have the feeling that fuelwood is easy to find. This isn't always so.

The first thing to consider is that about half of all wood is owned by private landowners. Most owners are not about to let strangers loose on their lands to cut timber. There are, however, a few landowners who realize the benefits of a properly thinned stand of hardwoods. The local forester is a good first step in finding fuelwood. If he can't show you where to find wood, chances are he will know someone who can.

Wood is often available on state, county, or federal lands, but a permit is necessary to take it. The number of permits may be limited, so check with the administrator of the land to see when permits become available. Many lumber mills and wood-using plants have wood residues which make good firewood. Prices are reasonable.

FUELWOOD MEASUREMENT UNITS

UNIT	Length	Split	Cubic Feet in Pile
Standard Cord	8'	No	128
Standard Cord	stove*	No	110
Standard Cord	stove	Yes	120
Fireplace Cord or Rick	stove	No	37
Fireplace Cord or Rick	stove	Yes	40
Face Cord	stove	No	About 37
Face Cord	stove	Yes	About 40

* Stove length is normally 16'' to 22''.

Species Selection

For fuelwood, broadleaf trees (hardwoods) are usually superior to needle-bearing trees (conifers) because the latter have more resin or pitch that cause creosote build-up in chimneys. Many people like to use conifers, such as jack or red pine, to start a fire and then switch to hardwoods. Conifers usually start burning easier.

The wood you select usually depends on what is available. People lucky enough to have a wide range of choices usually pick hardwoods. These woods have the most BTUs—heat-giving qualities.

A pound of wood will give off almost the same BTUs no matter what species. The main advantage of the heavier wood is that we handle it less. It takes nearly two truck-loads of aspen to give the same heat as one truck-load of oak. It is not practical to burn aspen if oak is available, unless aspen can be bought for about one-half the price of oak.

Gathering Wood

Imagine cutting fuelwood and you see the tree on the ground already bucked into stove lenghts. You may also see the truck waiting nearby to be loaded. However, you are likely to underestimate the distance between the two. Fuelwood seems to avoid growing within 50 feet of a road or clearing.

So how do you get the wood from here to there? Some people use a snowmobile in winter to haul it. Many just leg it. Most loggers solve the problem another way. They use large, rubber tired skidders capable of hauling several logs at the same time.

Another idea is to hire a logger to haul trees from point A where they were cut down to point B where they will be cut into logs. Hiring a skidder adds $5 to $15 to the cost of a cord of wood.

Heat Potential of Different Wood Species

Species	Pounds Per Cord Air Dried	Heat Units Per Cord* Air Dried	Percent of Short Ton of Coal
White Oak	4,300	23.9	92
Red Oak	3,900	21.7	83
Sugar Maple	3,900	21.8	84
Birch	3,800	18.2	70
Ash	3,800	20.5	79
Elm	3,100	17.7	68
Norway Pine	2,800	17.8	68
White Pine	2,700	14.2	55
Aspen	2,400	14.1	54
Balsam	2,200	13.5	52

*Available heat units per cord of 90 solid cubic feet (in millions of BTUs).

Injuries

The most efficient tool for cutting down a tree is a chain saw, but it also can be dangerous. Chain saw safety starts with wearing proper clothing and ends with proper tree felling and cutting techniques.

Hard hats are a must, unless you have an extremely hard head that can withstand the stunning shock of a falling branch. Eye and ear protectors are a matter of common sense, and leg protectors aren't such a bad idea either. After you have seen all the chain-saw nicks in boots that I have (don't ask whose boots, please), steel-toed boots make a lot of sense. Leather gloves are useful, too.

Needless to say, you must handle a chain saw with caution and respect. The problem is, people often start using a chain saw cautiously, then after nothing happens, they drop their guard. Then a painful, sometimes tragic, accident happens.

A few good rules to follow include:

* Don't be in a hurry.
* Keep both hands on the saw when it is running.
* Don't cut near the tip of the saw; this can cause kickback.
* Be sure to keep firmly balanced.
* Clear the cutting area of all debris that may interfere with your movement around the tree being cut.
* Be sure you know proper felling techniques; this helps take the guess work out of knowing where the tree will fall.
* Plan an escape path from the tree you are cutting **before** starting to cut.
* Never leave a cut tree 'hung up,' that is, caught in a nearby tree hoping that a strong wind will do the work of bringing it down for you.
* If a tree becomes hung up in a standing tree, never try to cut the standing tree. The mechanics of cutting two trees at once are too unpredictable.

Hauling

Living close to a fuelwood supply can help keep down expenses. Unfortunately, most of us don't. Even in forested areas, people commonly drive over 20 miles per load, round trip, to haul in their winter's supply of wood.

Three trips with an average one half-ton pickup truck will transport a cord of wood. It is possible to overload a truck and haul a cord in two trips, but overloading may cost more in the long run; repair costs these days are high.

Driving a pickup one mile costs about 35 cents, yet some people will drive a hundred miles one way to load their pickups with a third of a cord of wood. Ask why and they say, "I like the exercise," or "It's good recreation." But it would be cheaper and easier for them to exercise at home and have the finished product delivered to their doorstep. At current prices, driving a total of 60 miles to get a cord of wood home will add about $25 to its cost.

You may want to consider buying wood by the truck load in eight foot logs. You avoid hauling expenses since the wood is delivered to your door and the dangerous job of felling trees is left to the person who knows how to do it best, the logger.

Creosote build-up

If you manage to get your wood home without doing damage to yourself or to your machinery, remember that you have won the battle, but might still lose the war.

Burning those logs in your stove causes creosote build-up, a major cause of chimney and sometimes house fires. The condensation of unburned gases on the inner walls of your chimney causes the build-up. It usually develops when a fire burns 'low' for long periods or by a cool chimney, especially the part above the roof.

Checking the chimney for creosote once or twice every burning season is a must. Every home with a wood-burning stove should have a smoke detector in working order on every floor. Sleep with the doors closed.

When you switch to wood heat, seriously consider retaining your present source of heat as a back-up. One inconvenience of wood is that keeping a wood fire going requires attention every few hours. The back-up comes in handy when (A) you must heat your home while away on a trip, (B) the wood stove must be shut down for inspection and repair, or (C)

Table 1. Characteristics of woods for fireplace use[1]

Species	Ease of starting	Coaling qualities	Sparks	Fragrance	Heating class (1 best)
Apple	Poor	Excellent	Few	Excellent	2
Ash	Fair	Good	Few	Slight	2
Beech	Poor	Good	Few	Slight	1
Birch (white)	Good	Good	Moderate	Slight	2
Cherry	Poor	Excellent	Few	Excellent	2
Cedar	Excellent	Poor	Many	Good	3
Elm	Fair	Good	Very few	Fair	2
Hemlock	Good	Low	Many	Good	3
Hickory	Fair	Excellent	Moderate	Slight	1
Locusts (black)	Poor	Excellent	Very few	Slight	1
Maple (sugar)	Poor	Excellent	Few	Good	1
Oak (red)	Poor	Excellent	Few	Fair	1
Pine	Excellent	Poor	Moderate	Good	3

[1]Forestry Facts, December 1973. Quarterly publication of the Maine Cooperative Extension Service.

during periods when wood heat isn't adequate to keep the house warm.

Cure for Savings

Wood must be cured properly to give its maximum money-saving benefit. Uncured wood uses heat to evaporate excess water in its fibers. Uncured wood also adds to creosote build-up in chimneys, especially when it's burned at low temperatures.

Many people who cut fuelwood like to down their trees in late spring or early summer and leave the limbs intact. They claim that moisture evaporating through the leaves helps cure the wood faster. This may be true, but I prefer to buy my wood a year ahead and get it off the stump. That's when it cuts and splits easiest.

Curing wood one year before use is ideal. Six months of curing is the recommended minimum. Ash may be burned with fewer than six months curing and some local people burn paper birch that has been cured only a few weeks. But burning improperly seasoned wood requires a hotter fire and consumes a wood supply faster.

Wood should not only be cut into stove-box lengths—usually 16 to 22 inches—but should also be split to assure proper curing. Split wood dries much faster than unsplit. Birch often requires splitting. The bark traps moisture and causes the wood to rot.

Other Expenses

Heating your home with wood also involves other expenses, sometimes more than people bargain for. Besides the initial cost of a stove—less than $100 for a barrel stove and up to $2,000 for a wood-oil combination furnace—changes in the chimney and furnace ducts may be required. Ask a qualified furnace serviceman to look at your heating system and make recommendations.

Running a chain saw costs about $2 per hour. Used chain saws may sell for less than $50 while a good quality new one may cost $500 or more.

People looking for energy options, however, may find these expenses bearable. Wood will not solve the country's energy problem, but for a lucky few it may be the answer.

For more detailed information on heating with wood, read "Heating the Home with Wood", Minnesota Extension Bulletin #436, Agricultural Extension Service, University of Minnesota. Tables 2 and 3 in this article are from that bulletin.

Allen Wickman is a DNR private forest management specialist stationed in Park Rapids.

Table #2 courtesy of Lewis T. Hendricks, Agricultural Extension Service, University of Minnesota.

Table 2. Comparison of heating values for several species of wood

Species	Average specific gravity @ 20% moisture content	Average weight of 85 cu. ft. of wood at 20% moisture (pounds)	Possible recoverable heat units per cord of 85 solid cubic feet and assuming 100% efficiency[2] (in millions of BTU's) 20% moisture content	Available heat per cord at 50% heating efficiency (in millions of Btu's)	Units needed to give million Btu's of available heat (cords)
Hickory (average of several types)	.68	4,327	27.7	13.8	.072
Eastern hophornbeam	.67	4,267	27.3	13.7	.073
Apple	.65	4,140	26.5	13.2	.076
White oak	.63	4,012	25.7	12.8	.078
Sugar maple	.59	3,757	24.0	12.0	.083
Red oak	.59	3,757	24.0	12.0	.083
Beech	.59	3,757	24.0	12.0	.083
Yellow birch	.58	3,689	23.6	11.8	.085
White ash	.58	3,689	23.6	11.8	.085
Hackberry	.51	3,247	20.8	10.4	.096
Tamarack	.51	3,247	20.8	10.4	.096
Paper birch	.50	3,179	20.3	10.2	.098
Cherry	.49	3,120	20.0	10.0	.100
Elm (White or American)	.48	3,052	19.5	9.8	.102
Black Ash	.47	2,992	19.1	9.6	.104
Red Maple (soft maple)	.46	2,924	18.7	9.4	.106
Boxelder	.44	2,797	17.9	8.9	.112
Jack pine	.42	2,669	17.1	8.5	.118
Norway pine	.42	2,669	17.1	8.5	.118
Hemlock	.39	2,482	15.9	7.9	.127
Black spruce	.39	2,482	15.9	7.9	.127
Aspen	.36	2,295	14.7	7.3	.137
White pine	.35	2,236	14.3	7.2	.139
Balsam fir	.35	2,236	14.3	7.2	.139
Cottonwood	.33	2,108	13.5	6.7	.149
Basswood	.33	2,108	13.5	6.7	.149
N. white cedar	.30	1,913	12.2	6.1	.164

[1]Density-weight and volume at 20 percent moisture content.
[2]Conversion factors: at 20 percent moisture content, there are approximately 6,400 Btu's per pound of wood.

GROWING CHRISTMAS TREES

Marvin E. Smith

Growing Christmas trees can be a profitable and personally rewarding endeavor—if you like working outdoors (including the days when the birds are not singing); if you have a family that shares this desire; and if you consider some key points that I will briefly touch on in this paper.

Growing Christmas trees is a science and a business and must be treated as both. The science is knowing how to select, plant, protect, maintain, and harvest your trees. Knowing when, where, and how to sell trees to make a profit is the business aspect.

The grower must plan for the short term and the long term. What is the current Christmas tree industry like? What is the outlook in 5, 10, and 20 years? Good intentions alone are not enough. Incentive, know-how, and unflagging determination are required to produce a quality product. Quality is the hallmark of success in growing and marketing the plantation-grown Christmas tree. Where the grower is totally committed to producing top quality, he has little reason to fear an overproduction of trees.

A Christmas-tree enterprise has the special appeal of a short-term tree crop and is particularly well adapted to lands marginal for agricultural production. The full-time farmer, the part-time farmer, the city family with rural real estate—each owner group can learn to establish and manage a Christmas-tree enterprise. A danger, however, is that you may launch the business too rapidly and on too large a scale. It is easy enough to plant a 40-acre tract in one year. But 3 or 4 years later the task of shaping 50 to 60 thousand trees can be overwhelming. The point to remember is that the area in plantings should be correlated with what the owner either can do himself or can hire done.

In the instance of the 40-acre tract, a family seeking to combine a tree-farm business with outdoor recreation should plant no more than 4 to 6 acres each year. This procedure will spread the work load over time and result in smaller increments of work in any one year. When harvesting of trees begins, it can continue annually with about the same number of trees to harvest each year.

In other circumstances, a grower's scale of operation might embrace hundreds of acres planted to trees, hired field crews at peak work periods, and rail and truck hauling of trees to distant outstate markets. At the opposite end of the scale is the person whose immediate objective is Christmas trees, but who intends to harvest them as thinnings from a forest which will continue to grow and yield a variety of products such as posts, pulpwood, poles, and sawlogs. The income from the sale of Christmas trees can repay the accumulated costs of establishing the plantation.

Choosing a Location

The popular belief that trees will grow anywhere regardless of soil fertility and physical characteristics is questionable. They may grow, but that is not saying how well they will grow or how long it will take them to mature. Coarse-textured soils, sandy loam in nature, are preferred over the finer silt and clay soils. Most species will grow on these rather infertile soils. Another advantage of sandy soils

is the relative ease and economy of controlling weeds. On fine-textured soils, weed competition is severe and calls for frequent cultivation or the application of chemical weed killers.

Level and gently rolling land is an advantage to the grower from the standpoints of ease and economy of planting, harvesting operations, and some phases of cultural work. Where the topography is rolling, preference should be given to slopes where the tractor equipment can operate.

Other things being equal, the best choice of site is one having ready access from all-weather roads. Access is important when you are moving trees to market. Buyers may be discouraged by awkward shipping arrangements or excessive hauling distances to market. Remote locations also tend to complicate the task of recruiting seasonal labor.

What Trees To Plant

Choosing what types of trees to plant is a major decision, and it is not wise at this point for the novice grower to experiment or make broad guesses. The grower must decide what to plant so as to satisfy consumers' wants 7 or 8 years hence. He therefore must keep informed on trends in consumer tastes as they affect future demand for the various tree species.

Three species dominate the market for plantation-grown Christmas trees in the United States; Scotch pine, Douglas fir, and balsam fir account for 3 out of every 4 trees purchased. In the Midwest and the North Central region we are growing mainly Scotch pine, white spruce, red pine, white pine, and balsam fir. Other pines, Norway spruce, and some cedar make up the rest of the market.

Aside from the demands of the market, two factors that are limiting in your choice of species are climate and site. In respect to the influence of climate you should choose species native to the region or those that already have been successfully introduced. Site is the specific invironment in which a plant is growing. It is the sum of all the surrounding influences: soil, moisture, temperature, and light. Each species has a set of optimum conditions under which it will do best. For example, pines grow quite well on sandy and sandy-loam upland soils where moisture and nutrients are sometimes lacking. Spruces thrive best on loamy soils where moisture and nutrients are not limiting. Balsam fir will not grow well on impoverished soils and cannot withstand even the briefest drought periods.

First Plan—Then Plant

The successful Christmas-tree grower is generally one who has planned well. The initial layout is very important in regard to spacing of roads and lanes, row spacing and tree spacing within the row, and scheduling of production.

Plan to leave lanes or roadways at intervals of 200 feet or less to give access to every part of the plantation. For any kind of vehicle travel and when yarding trees to loading points, advance planning to provide easy movement of men and machines in the plantation will pay dividends later.

Row spacing and tree spacing within the rows determine the number of trees that can be grown per acre. What ultimately decides these questions is the growing space required for each tree. Trees should be planted at whatever spacing will provide ample room for them to develop without crowding. Some species need slightly more room than others. Based on the production of high-quality 5- to 8- foot trees, spacing recommendations and the number required per acre are as follows:

Species	Spacing	Number per acre
Scotch pine	6 x 6	1,210
Norway pine	6 x 6	1,210
Austrian pine	6 x 6	1,210
White pine	6 x 6	1,210
White spruce	5 x 5	1,740
Norway spruce	5 x 5	1,740
Balsam fir	5 x 5	1,740

It should be the aim of the Christmas-tree grower to regulate the annual planting of each species over a rotation period so that when harvesting begins the number of trees ready for market will be fairly constant from year to year. Under this plan, the plantation should be platted out in blocks of equal area. The necessary access lanes, roadways, and firebreaks can serve as block boundaries.

Choosing Planting Stock

As a general rule, buy the smallest stock that will suit your planting and growing conditions. Root pruning and other nursery practices account for the fact that transplants are usually sturdier than seedlings. They may be expected to have more fibrous roots and a thicker stem than seedlings. Also, the relation of weight of top to weight of roots is generally more favorable in the transplant than in the seedling. These advantages are the main reasons why transplant stock often fits the prescription for good Christmas-tree planting stock.

Many growers can point to a history of successful plantation etablishment using 2-0* and 3-0 seedlings. This experience only proves again that age alone is not an indicator of stock quality. On many sites and in many seasons 2-0 Scotch pine and 3-0 Norway pine can provide satisfactory survival and growth. However, the demonstrated superiority of transplant stock in withstanding weed and grass competition and in enduring drought periods is a strong reason for growers in less favorable environments to prefer the transplant.

**The first figure indicates years in the seedbed. The second figure represents years in the transplant bed.*

Quality planting stock can be purchased from either public or commercial nurseries. Many of the larger commercial nurseries produce seedlings and transplants especially for Christmas-tree production. Consequently, where the grower has specific needs that may include a variety of species, and perhaps certain seed sources for a species, the specialized commercial nursery is probably the best source of planting stock.

Shaping the Tree

Shaping is one of the keys to producing high-quality Christmas trees. Shaping is actually a combination of two processes—shearing and pruning. Shearing is the trimming or cutting back of the current growth of the terminal leader and side branches to improve the form and density of the tree. Pruning is the complete removal of competing terminal leaders or side branches that are protruding too far out from the tree. Shearing and pruning are usually done simultaneously, and the whole process is referred to as shaping. (Some people refer to the whole process as shearing or simply trimming.)

When the operation is repeated annually, it has the effect of developing a shorter, well-shaped, and compact tree from what otherwise would be a tall, rangy tree. Other purposes are:

1. To correct deformities
2. To remove multiple leaders
3. To prune lower branches to form a handle and complete base whorl.

Harvesting and Marketing

As the trees attain marketable size, the owner should make a detailed inventory of the trees he will have for sale that particular season. A good time to take this inventory is immediately after completion of the final shearing. Potential buyers may be expected to inquire about trees as early as July and August.

A good way for an owner to mark the trees he will have for sale is to tie a short piece of colored plastic flagging to each tree. Different colored flagging is used for several purposes, but it is most commonly used to denote grades. Some growers attach paper tags on the saleable trees. These tags are useful for pricing.

Marketing is a challenge, and the rewards will go to the grower who seeks early to learn how and where he can dispose of his trees at the best price. Essentially there are three ways a grower can market his product. He may sell his trees at wholesale either by his own efforts or through a broker, he may retail them himself, or he may combine retail and wholesale methods.

When wholesaling, the grower may sell his trees in the field (stumpage) or as cut trees piled on site or delivered. Retailing can take two forms—either as cut trees at a retail lot or as trees standing in the plantation. The latter method is particularly suited to locations near population centers. This so-called "choose and cut" retailing appeals to family groups who enjoy the adventure and tradition of finding and cutting a Christmas tree in a natural setting.

A Christmas-tree enterprise, especially as it approaches harvesting, represents a considerable investment of the owner's time and money. To protect his investment the owner should be prepared to take certain preventive or prompt control measures throughout the growing life of the trees. The hazards to growing trees are many, including fire, animals, rodents, diseases, insects, and trespassers. Protecting against hazards that can inflict crippling losses is a responsibility no grower should minimize or ignore.

Other Sources of Information

National Christmas Tree Association
611 East Wells Street
Milwaukee, Wisconsin 53202

SUMMARY OF PRUNING RECOMMENDATIONS

PINES

Start — When trees are 24" to 30" high.

Shear — Early June to mid-July before new growth has hardened.

Remove all but one terminal leader. Cut remaining terminal leader to desired length (10-12"). Make cut at 45° angle. Prune laterals of terminal whorl 3-5" shorter than terminal leader.

Sides

Prune all laterals around tree such that the tree of desired shape and density will be produced.

SPRUCE AND FIR

Start — When trees are 24-30" high or when annual growth exceeds 12".

Shear — During dormant season. (In West Virginia, between August 1 and April 1.)

Remove all but one terminal leader. Cut remaining terminal to desired length (10-14"). Make cut at an angle just above a single bud. Prune laterals of terminal whorl 3-5" shorter than leader.

Sides

Prune all laterals around the tree — at a bud such that the desired shape and density will be produced. Terminal buds on laterals can be pinched to reduce side growth.

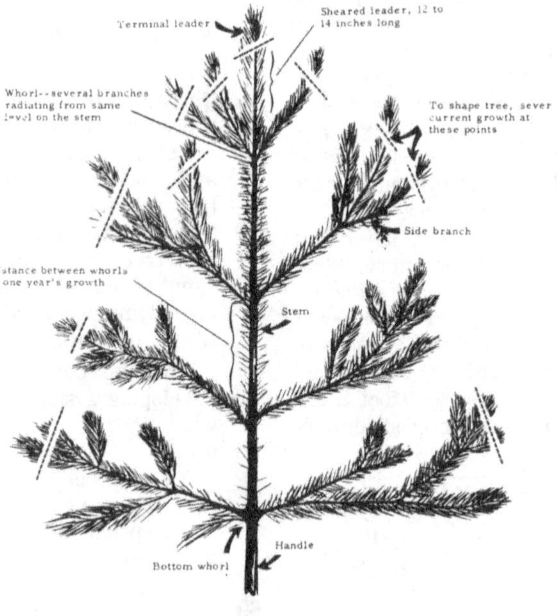

NOMENCLATURE OF A CHRISTMAS TREE

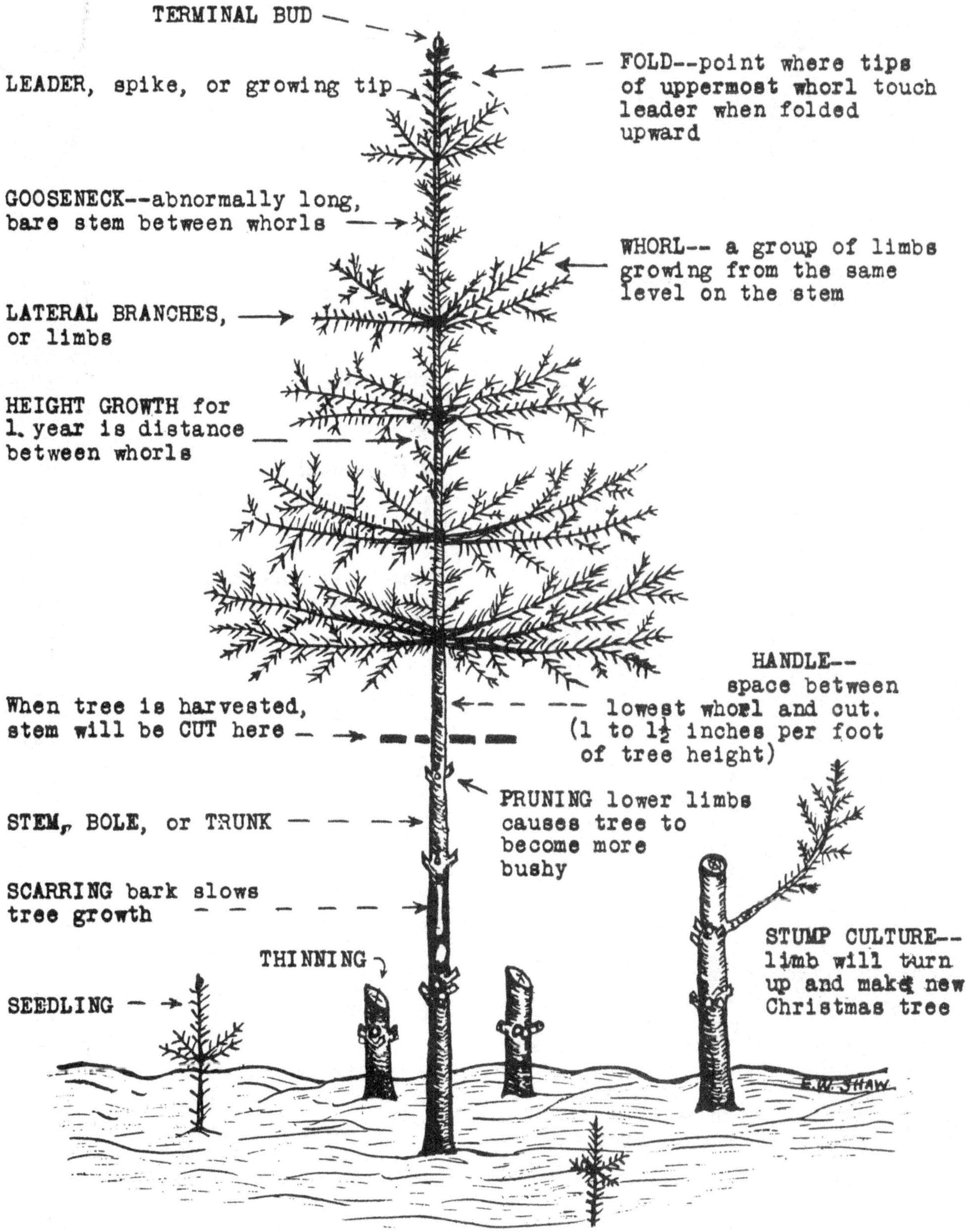

G3268

WISCONSIN WOODLANDS: Christmas Tree Shearing

Christy Hauge, Syd Hovde and Edward Steigerwaldt

The Wisconsin Christmas tree industry has come a long way from the days of wild balsam fir and spruce cut by rural landowners and farmers to be delivered unlabeled to city customers. Today, this sophisticated forestry business requires considerable knowledge in areas such as finances, land management, soils, seed sources, pest management, cultural practices and marketing. Most trees marketed come from established plantations producing quality trees in 6 to 15 years, depending on species. Pines, for example, grow to salable form twice as fast as do short-needled balsam and spruce.

Wisconsin has an estimated 38,000 acres of Christmas tree plantations, most located in the central counties. Christmas tree growers currently harvest about 3.5 million trees of all species in Wisconsin each year. The national harvest in 1982 was more than 28 million.

This fact sheet is prepared to assist growers in mastering the art of shearing to produce quality salable trees. Shearing increases the number of buds and therefore the future density and shapeliness of trees. Depending on location, most Wisconsin growers shear between June 15 and July 15. This is when new shoots complete growth and new buds begin to form. When sheared, pines will set a cluster of adventitious buds (see list of terms on p. 4) near the angle cut and the tree will produce more branches the next year. Spruce and fir do not set new buds. (see p. 2).

Shearing Tools

Hand Shears and Clippers

You need to use different tools to trim and shape trees of different sizes. Small hand clippers work well to cut out a double stem, dead branch, or extra long leader. However, when you shape trees, usually for the first time when they're knee-high or taller, use a hedge clipper operated with both hands. In the long run, it pays to buy quality tools that you can easily sharpen and clean of dirt and debris.

Knives

Hedge clippers work well for shaping first and second-year trees. For later shearing, many growers prefer to use knives. Knives can shape trees nicely when workers know how to use them properly and use good equipment, wear protective gear, and are carefully supervised.

You can buy knives in all sizes, but most operators prefer an 18 to 24-inch wooden-handled knife with an 18-inch blade. When using this shaping tool, wear leg guards. Some growers also require workers to wear a chain mail glove on the hand not holding the knife. Knives need periodically to be sharpened and cleaned of accumulating pitch.

Mechanical Trimmers

Larger operations usually need mechanized trimming. Growers now use several types of mechanical devices to shear trees.

Gas-Powered Trimmers. The first gas-powered trimmer for tree shearing was created by Howard Beneke, Sylvan Beach, N.Y. Beneke designed a long tubular "wand" which encloses a flexible shaft connected to a commercial mini-gas engine. The shaft turns a 10-inch circular saw with four teeth. The extended saw is balanced with an elastic cord and can cut on the up and down stroke when swung. Aluminum finger guards prevent the saw from gouging the soil.

Many older tree growers find the 10-pound unit easier to handle than swinging a knife or using hedge clippers all day. The long tubular arm lets growers easily shear 8 to 10-foot trees. Mini-engines, however, require constant care to maintain high production rates.

Battery-Powered Trimmers. The power trimmer Beneke developed has recently been modified to use a wet-cell, battery-powered motor. This trimmer operates quietly. You can run a fully-charged, motorcycle-type battery two to three hours before recharging it.

Other battery-powered units on the market are powered with nickle-cadmium batteries. The length of operation depends on battery charge and durability.

Electric-Powered Trimmers. Many growers have tried to bring a portable generator into the field. They have plugged clippers into long flexible outlet poles attached to the generator. Then the operators have moved down the rows parallel to the direction of the mounted generator.

Few large-scale producers favor using these trimmers because of high initial investment and because they can't be used on all terrain.

Automated Shearing Machines. A new development is a tractor-mounted shearing unit that uses a series of whirling knives to shear, after the operator selects the tree's angle of taper. It was designed for the Deep South where growers must shear twice a year and avoid snakes and alligators which endanger shearing crews.

It's not known whether any Wisconsin growers use this trimmer, but demonstrations using it on Scotch Pine have been successful. The trimmer's initial cost and maintenance is high.

Each tree grower must decide how many and what kind of shearing tools to use. A small tree plantation run by one or two people can use hand clippers, hedge shears and perhaps a power trimmer to shape taller, bushier trees. Operations run by five to 10 people can use hedge clippers on 3 to 5-foot trees and then switch to knives for trees over 5 feet tall. Some of the largest tree operators in Wisconsin use a combination of clippers for smaller trees and knives for the rest.

You can get advice on production of Christmas trees from the Wisconsin Christmas Tree Producers Association, Extension foresters or professional foresters.

Nicely shaped white spruce ready for tagging and marketing.

How to Shear the Short-Needled Species

Short-needled trees typically include white spruce, Black Hill spruce, Norway spruce, Colorado blue spruce, balsam fir, Fraser fir, and Douglas fir. Unlike pines, which must be sheared when new growth is succulent, you can shear short-needled trees any time of the year. These trees normally have buds all along the stem and don't produce extra buds at the point they are sheared.

The main objective in shearing short-needled trees is to develop a full, symmetrically shaped tree with a 40 to 70 percent taper.

Native Species

Trees native to Wisconsin or closely related trees include white spruce, Black Hill spruce, Norway spruce, and balsam fir. Here is a suggested shearing schedule:

Pre-shearing. Remove double leaders and correct deformities. Prune the base for the handle.

First Shearing. Trees should be 3 to 4 feet high.

1. Cut back terminals to 8 to 10 inches in length; cut slightly longer after the first frost.
2. Cut top laterals 4 to 5 inches shorter than leader.
3. Shape sides to a 40 to 70 percent taper. Try to shear new growth only.

Subsequent Annual Shearings.

1. Cut back leaders to 8 to 10 inches, depending on individual tree. Do not let leaders get too tall because spruce and fir will need more time than pines to develop lateral fullness.
2. Proportion laterals to desired taper.
3. Shear to maintain full symmetry.

Marketing Shearing.

1. Generally, shear balsam fir trees lightly but shear spruce fairly tightly.
2. Terminals can be 10 to 12 inches long.

Exotic Species

Use the same shearing guidelines for exotic species as for native species, but also consider these points:

Colorado blue spruce is especially slow growing, so take great care not to over-shear the laterals. Try to maintain 8 to 10-inch leaders, but shear laterals lightly to maintain desired taper.

Fraser fir lateral branches must be sheared lightly. Keep lateral shearing to a minimum until trees are 4 to 5 feet tall. Maintain terminals at 8 to 10-inch lengths.

Douglas fir is very susceptible to frost, in low-lying or frost-pocket areas. Terminals close to the ground may be left slightly longer to bring lower branches above frost danger.

For most growers, the short-needled trees can provide great flexibility in shearing schedules. Growers with both pine and combinations of spruce and balsam fir can shear pine first during the growing season and short-needled trees later. Although short-needled trees can be sheared any season, growth doesn't seem to be as vigorous after shearing in the dormant season or early spring.

Remember that spruce and balsam fir typically grow slower than pine, so keep the terminals within the 8 to 10-inch guideline to give laterals ample time to develop a dense foliage by mature Christmas tree height.

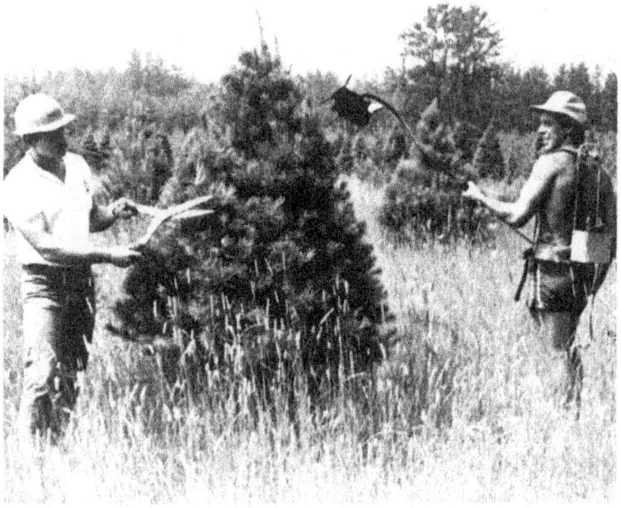

Two shearing tools: hand shears (large clippers) and a battery-powered trimmer.

Shearing the Long-Needled Species

The objective of shearing or shaping long-needled trees, such as Scotch pine, white pine and red pine, is to produce:

- One main stem or trunk and a single leader
- A symmetrical, well-balanced form
- Compact growth or full tree
- A handle, by pruning the base

Pre-shearing. Correct multiple tops and deformities. Leave only one leader.

First Shearing. In the third or fourth year after planting when trees are 3 to 4 feet high or when terminal leaders grow more than 10 inches:

1. Cut back central leaders to 10 inches at 45° angle.
2. Cut top laterals 4 inches shorter than leader or in proportion.
3. Cut side terminals to desired shape or taper. Cut only in new growth.

Subsequent Annual Shearings.

1. Cut back central leader to about 8 to 10 inches, depending on individual tree.
2. Cut top laterals in proportion to central leader—preferably 4 inches shorter than leader.
3. Cut side terminals and other branches as needed to get proper taper.

Marketing Shearing. Same as previous shearings, except leave terminals longer (about 12 to 14 inches), and cut side branches lightly to allow less vigorous laterals to fill in during the final growing season.

In general, cut the top leader at a 45° angle, and 10 inches long. Do not shear too close on young trees, because they should be allowed to grow. Remember, 3 years of live needles is all that can be expected at harvest with normal tree growth.

Slant the shears or knive angle to establish the proper taper. Remember to cut out double leaders, especially on Scotch pine. You don't need to shear branches growing toward the center of a tree. Red and Scotch pine can be sheared more tightly than other pines to establish proper form.

Terms

Adventitious Bud. A bud which develops at the base of a needle cluster, or on woody tissue on a branch or leader, when the end of the branch or leader is injured or cut off.
Axil. The point on a branch where two or more smaller branches form a fork.
Candle. The new bright green and tender growth all conifers grow in the spring.
Dominant. The tallest, fastest growing trees in a plantation or natural stand.

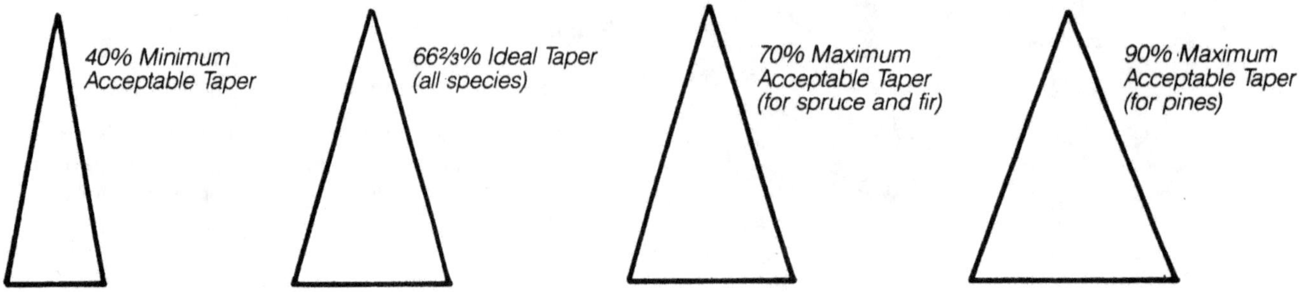

Handle. The base of the stem below the bottom whorl, cleared of branches for placement in tree stand.
Internodal. The stem between annual whorls of branches.
Internodal Buds. Single buds occuring irregularly on spruces and firs between annual whorls of branches. Internodal branches develop from these buds.
Laterals (Lateral Branches). All side branches of a tree growing from the stem.
Lateral Leaders. The terminals of a coniferous tree's side branches.
Leader Holder. A tool, usually made from a stick with a wire hook, that growers use to hold back the terminal leader when shearing laterals.
Node. The point where a terminal bud and some whorl buds form when new growth stops. You can figure an unsheared coniferous tree's age by counting the nodes on the main stem. (The number of whorls is commonly counted to determine age).
Succulent (also Succulence). The condition of new growth (candle) characterized in spring by high moisture content, light green color, and relative brittleness.
Symmetry. Shape or balance; how uniform the entire tree is.
Taper. How width is related to height. For grading, pines must have 40 to 90 percent taper; spruce and fir need 40 to 70 percent taper.
Terminal Bud. The bud which grows into an extension of the main stem of the tree or main stem of a branch.
Terminals (Terminal Leaders). The new growth produced by the terminal bud.
Whorl. Two to 10 or more branches growing in a ring at a node, surrounding the central leader or stem.
Whorl Buds. The rosette of buds forming a ring around the terminal bud at a node. (The lateral whorl branches form from these buds).

Christmas Tree Terms

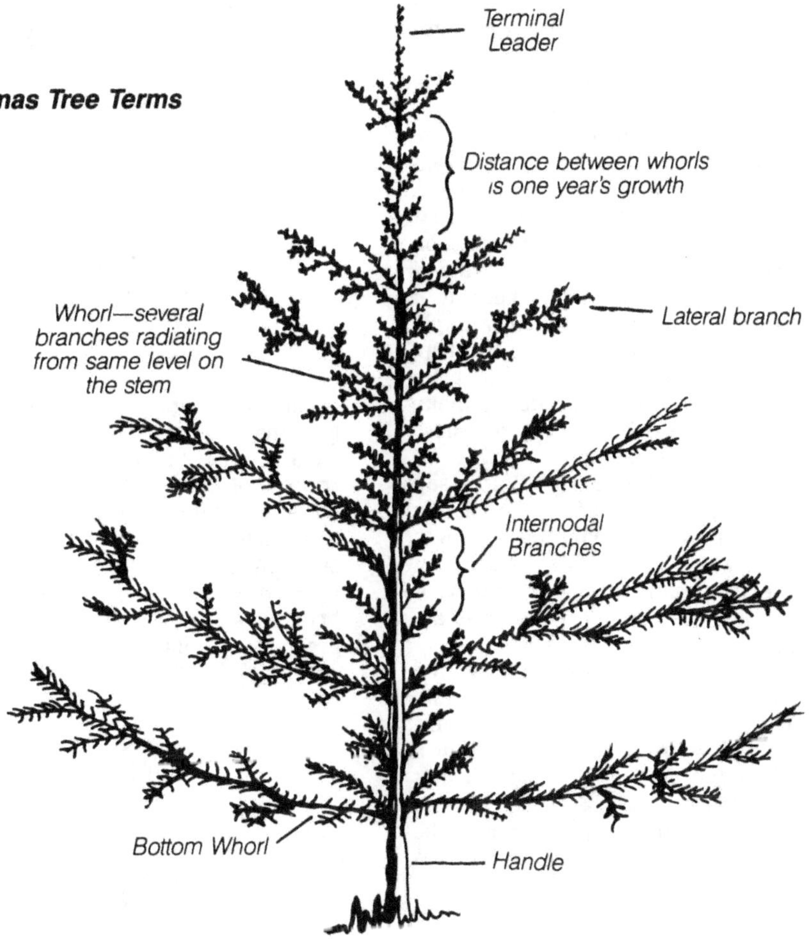

University of Wisconsin-Extension, Cooperative Extension Service, Charles F. Koval, director, in cooperation with the U.S. Department of Agriculture and Wisconsin counties, publishes this information to further the purpose of the May 8 and June 30, 1914 Acts of Congress; and provides equal opportunities in employment and programming including Title IX requirements.

Produced by the Department of Agricultural Journalism, University of Wisconsin-Madison.

G3268 WISCONSIN WOODLANDS: CHRISTMAS TREE SHEARING

20¢ SR-10-85-2.5M-E

This publication is available from your Wisconsin county Extension office or from:
Agricultural Bulletins, Rm. 245
30 N. Murray St.
Madison, Wisconsin 53715
Phone 608-262-3346

Editors, before publicizing, contact Agricultural Bulletins to determine availability.

Christy Hauge is Area Forestry Agent/Forestry Specialist, University of Wisconsin-Extension, Stevens Point. Syd B. Hovde is Area Forester, Department of Natural Resources, Wisconsin Rapids, and Edward Steigerwaldt, is a Consultant Forester, Tomahawk.

MAPLE SYRUP MAKING FOR BEGINNERS*

Gordon R. Cunningham

Maple syrup is a natural product of our northern woodlands. American Indians made syrup long before Europeans arrived. Maple syrup and sugar production is confined, however, to portions of the Northeast and upper Midwest, as the map in Figure 1 shows. In this restricted production area, less than 10 percent of the tappable trees are being tapped, creating a seller's market for maple syrup and by-products.

When springtime arrives, though, you need not be a commercial producer to enjoy maple syrup. By following these guidelines you can make you own.

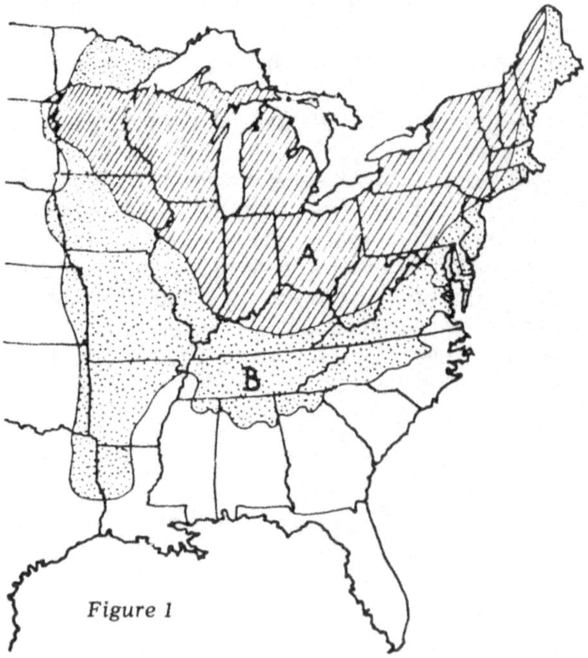

Figure 1

A and B show the ranges of hard maple trees. A indicates the range in which maple syrup is commercially produced.

Principles of Making Syrup

Maple sap begins to "run" in late winter when the temperature rises above freezing during the day and drops back below freezing at night. The ideal temperatures might be 20° to 25° F at night followed by 45° to 50° during the day. The season begins in the southern part of the states sometimes as early as January and spreads northward.

The principle of maple syrup production is one of concentration through evaporation. Evaporating (boiling or "cooking") the sap releases water as steam. The sap or sweetness concentrates until it becomes maple syrup. How much sap it takes to produce one gallon maple syrup varies depending on the sap's sugar concentration, but you can use the "rule of 86" to figure it. It takes 86 gallons of sap with 1 percent of sugar (plus other solids) content to make a gallon of maple syrup, 43 gallons of 2 percent sap to make a gallon of syrup, 28.7 gallons of 3 percent sap to make a gallon, and so forth. So, the sweeter the sap, the less boiling is needed to make syrup. Some maple syrup producers believe sap with less than 1.5 percent sugar is not worth collecting.

How can you test the sweetness of your sap? With a sap refractometer you need only a drop of sap. Unfortunately these instruments usually cost over $100. You can use a less expensive sap hydrometer, which measures the density of the sap and has a scale that reads in sugar content. A hydrometer usually costs less than $8.00 and requires approximately a cup of sap.

Before obtaining supplies for collecting, evaporating, and storing syrup, decide whether you are going to process into syrup all the sap you can collect or make only a certain amount of syrup. Predicting how many trees to tap to produce a certain amount of syrup is difficult. Although about ⅕ gallon of syrup per tap is produced in an "average" season, the actual sweetness and yield of sap in any one year can be from almost nothing to more than 20 gallons of 4+ percent sap, or a gallon of syrup per tap.

Maple Trees

Although 13 species of maple (Acer) are native to the United States, and all probably have been tapped, two are preferred for maple syrup production: sugar maple (Acer saccharum) and black maple (Acer nigrum). They are preferred because their sap is usually sweeter than that of other maple species. Red maple (Acer rubrum) and silver maple (Acer saccharinum) also produce sap, but it is less sweet and often contains excessive amounts of sugar sand (suspended solids of calcium and magnesium salts of malic acid).

The sugar maple (see Figure 2) is found in all states, usually on the better soils. It attains a height of 60 to 100 feet with a trunk diameter of 3 to 4 feet. In forests it develops a clean trunk to a good height, whereas open-grown trees form a dense, round-topped crown. The leaves are opposite, simple, 3 to 5 inches long and broad, and are usually 5 lobed, the lobes sparingly wavy-toothed. They are thin and firm, opaque, dark green above, paler beneath, in autumn turning to brilliant shades of yellow, orange, and red. The flowers are greenish and inconspicuous. The fruit differs from other maples in that it matures in the fall and germinates the next

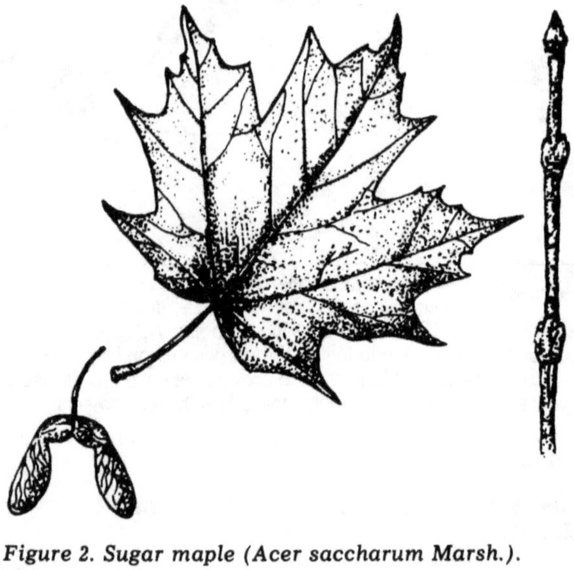

Figure 2. Sugar maple (Acer saccharum Marsh.).

spring. It is a two-winged samara or "key." The bark on young trees is light gray to brown and rather smooth, later breaking into long irregular plates of flakes that often loosen vertically along the side. The wood is heavy, hard, strong,

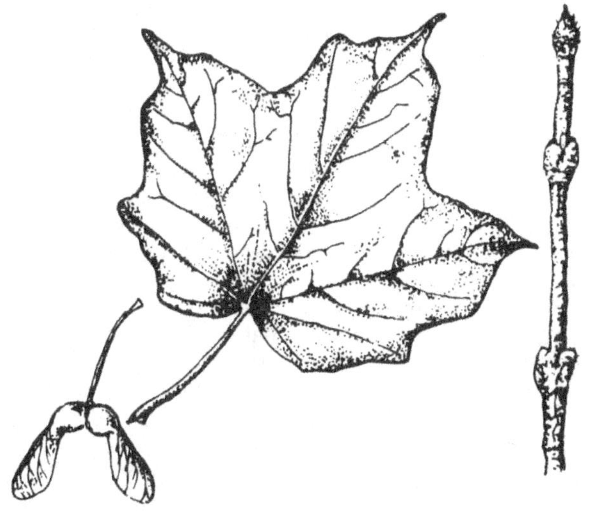

Figure 3. Black maple (Acer nigrum Michx. f.).

close-grained, and light brown to reddish, with lighter sapwood. It is used for flooring, furniture, shoe lasts, and a great variety of novelties. Low-grade material goes into railroad ties and fuel. Accidental forms with the grain curled, known as "bird's eye" and curly maple, are rather common and are prized in cabinet making. The sap is boiled down for maple sugar and syrup. The sugar maple can be planted in well-drained, moist soils where it is not exposed to salt and wind currents along heavily travelled streets and highways.

The black maple (see Figure 3) extends into southern Minnesota, Wisconsin, and Michigan, primarily on soils that have been glaciated. It grows well in river bottoms and can be found concentrated along streams. Black maple commonly grows to a height of 80 feet and has a 2- to 3-foot diameter trunk when mature. The leaves are opposite and simple and are usually 3 lobed. In addition, the sides of the leaves tend to bend downward, and the underside is hairy along the yellow veins. Its twigs are stouter, and the buds are hairier than the sugar maple's. The flowers are very similar to those for the

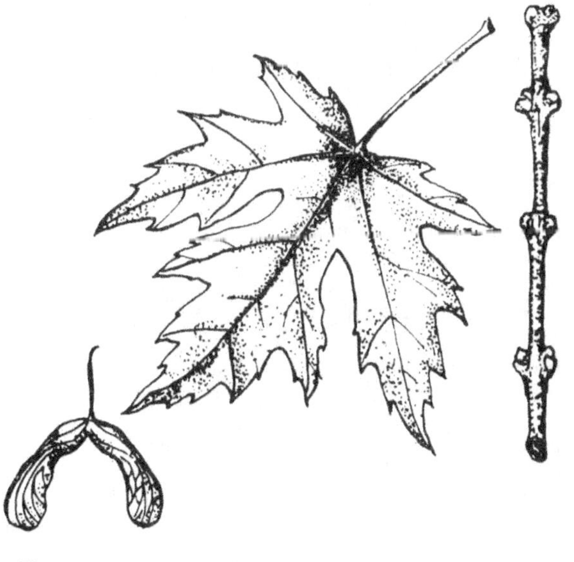

Figure 4. Silver maple (Acer saccharinum)

sugar maple. Its fruit, too, is similar to the sugar maple's; however it has a slightly larger seed cavity. The bark is more corrugated than that of the sugar maple. Like the sugar maple, its wood is commerically known as being hard. In addition, it is heavy, stiff, and very strong in bending. It is usually straight grained. The heartwood is uniformly light reddish brown whereas the narrow sapwood is white with a reddish tinge. Its uniform texture and hardness makes it a favorite for flooring.

The silver maple, or soft maple as it is often called (see Figure 4, is a common tree on river bottoms or other deep, moist soils in the northern half of the Lake States. It is usually 40 to 60 feet in height with a diameter of 3 to 4 feet, but it is often taller. A tree in Wyalusing State Park is over six feet in diameter. The leaves are opposite and simple with 3 to 5 lobes, ending in long points with toothed edges and separated by deep angular openings that extend underneath. The flowers appear in dense clusters in the spring before the leaves and are greenish yellow. The fruit ripens in late spring and consists of a pair of winged seeds or keys. The bark is smooth and gray, becoming darker and furrowed, separating into long flakes. The wood is hard, strong, close grained, rather brittle, and pale brown, with thick lighter-colored sapwood. It is occasionally cut for lumber. Sugar is sometimes made from the sap. The silver maple is not of value for forest planting. As a shade tree it is popular, but the branches are apt to be broken by storms, and it is often infested with the cottony maple scale. It is not being planted as much as formerly.

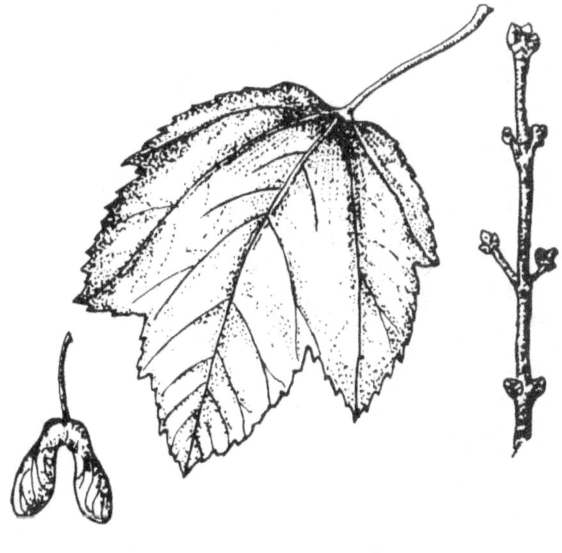

Figure 5. Red maple (Acer rubrum L.).

The red maple (see Figure 5) grows over most of the northeastern states and prefers moist soils, though it is common on drier sites in mixture with other trees. It is usually a medium-sized tree 40 to 60 feet high and 1 to 2 feet in diameter, but sometimes it is larger. The leaves are simple, opposite, 3 to 5 lobed, but not so deeply cut as the silver maple, and doubly serrate or toothed, dull green above, whitish beneath, and turning a brilliant red in autumn. The flowers appear in dense clusters in early spring before the leaves, and although they are small they are quite conspicous because of their red color. The fruit consists of a pair of winged seeds or keys that fall somewhat later than those of the silver maple and are only half as large though, like those of the silver maple they germinate at once. The bark is smooth and light gray on young stems and dark gray and rough on old trees. The wood, which is commercially known as soft maple, is heavy, not strong, close grained, and light brown in color and is used for furniture, woodenware, and fuel. The red maple is quick growing and makes a fair shade tree for light soils in the northern part of the state. As a forest tree it has little value. While widely distributed, it is a short-lived tree and forms only negligible part of commerical stands of timber.

Once you know which species are available for tapping, choose which trees you want to tap based on crown size and trunk size. Crown size is important because trees with large crowns usually produce large amounts of sweeter sap. A single taphole usually produces from 5 to 20 gallons of sap per season, usually yielding a pint to a quart of syrup when boiled down. Trunk size, traditionally measured at D.B.H. (diameter at breast height—4.5 feet), dictates the number of tapholes a tree can accommodate safely. The following is a rough guideline:

D.B.H. (inches)	Number of tapholes
less than 10	none
10 to 14	1
15 to 19	2
20 to 24	3
24+	4

In selecting a tree, species and crown size may be traded off: a red maple with a large crown might be favored over a sugar maple with a very small crown. Testing the sap sweetness, however, is the only way to identify the sweeter and more productive trees.

Equipment Needed

You can improvise, but you probably will be more satisfied with the results if you use designed equipment.

Collection Equipment:

— A brace and bit or drill for boring into the tree. The bit size should be $\frac{3}{8}$ or $\frac{7}{16}$ inch, depending on the kind or make of spile used.

— Spiles are hollow spouts inserted in tapholes. Obviously you need as many spiles as you have tapholes.

— A hammer to tap in the spiles and to pull them out after the season.

— A hypochlorite solution to flush out the tap holes, sterilize the spiles, and clean the equipment: 1 part household bleach (containing about 5 percent hypochlorite) to 9 parts water. Wear rubber gloves when using hypochlorite. It can be carried in a plastic detergent bottle or in a bucket or jar for the spiles and bit. Special pellets can be inserted into tapholes to keep them sterile throughout the season. These pellets, however, can cause some injury to the surrounding wood.

— A collection container. Commercially produced sap collecting buckets usually hold 13 to 16 quarts; they are galvanized, aluminum, or plastic. You can use clean number-10 cans instead. Plastic sap bags and plastic tubing are also used to collect sap.

— A large container to hold the sap. The average yield per taphole is from 5 to 20 gallons per season; however, a tap can produce 40 or more gallons in one season. This amount is usually received in 1 to 5 or 6 "runs." (A run is one period of freezing and thawing weather. It could be one day or several.) A new trash can, a clean 55-gallon drum (not previously used for oil or chemicals), or a new stock-watering tank can be used. How large a container you need depends on the number of taps you have and the flow. For example, if the average yield per tap was 10 gallons for a season, and if this yield was received during three equal runs, then a 30-gallon trash can could hold almost all the sap from a run for 10 taps ($10 \div 3 = 3\frac{1}{3} \times 10 = 33$ gallons), and a 330-gallon watering tank could hold the sap from a run of 100 taps.

—An outside area where you can build a fire under your kettle(s) or pan. If you use the old iron kettle(s) suspended from a crossbar, select an area free of combustible materials so that you will not start a grass or woods fire. And, of course, have the posts that support your crossbar far enough from the fire that they will not ignite. If you use a flat pan, you will need bricks or cinder blocks to support the pan over the fire, and you will need stovepipe or more bricks to build a chimney wide enough and tall enough to provide a good draft. If you use one or more specially built maple syrup evaporator pans, you will probably use the ready-made fire arch that is constructed for those pans.

—Heat. A wood fire is probably least expensive for evaporating small quantities of sap unless you are knowledgeable enough to install an oil burner or financially able to have one installed (many large evaporators are heated with oil burners). The best wood is dry, heavy hardwood such as beech, sugar maple, or oak. Wet woods give much less heat and soot up the bottom of the pan. A rule of thumb for the wood needed is that one standard cord of dry hardwood (a stack 4 feet high x 4 feet wide x 8 feet long) will cook about 25 gallons of syrup. This rule assumes about 10 gallons of approximately 2.5 percent (sugar in the) sap from about 85 trees. Sweeter sap will require less evaporating and thus less wood. Thinner sap will require more evaporting and more wood.

— Kettles or pans for evaporating the water from the sap by boiling. A pan should have 6-inch to 8-inch sides to confine the boiling sap (higher is fine). Flat pans in several sizes are available from dealers or can be made by sheetmetal shops. They can be made from stainless steel, tinned steel, or even heavy-gauge aluminum.

— A thermometer that reads up to 225°F, calibrated to 1 degree. Check candy thermometers.

— A container into which hot syrup can be filtered. A porcelain water bath used for canning, a clean milk can, a large surplus army cook pot, or any similar container can be used.

— A cloth to filter sugar sand and other solids out of the syrup as the sap is poured into the container: several layers or bags of flannel terry cloth, orlon, or wool. Filtering through a cotton filter then through an orlon filter will clear up the syrup even better.

— Clothespins (with springs) to hold the filter to the edge of the container described above.

Syrup Containers

Clean canning jars or bottles, regular maple-syrup cans, and bottles or jugs with tops are all suitable to store maple sugar in. Avoid untempered glass jars (the hot syrup could break them) and soft plastic bottles (the hot syrup could melt them).

Tapping

Once you select the trees and determine the number of tapholes per tree, use the drill and bit to bore the holes at a slight upward angle into the trunk. Before drilling each taphole, rinse the bit in the pail or jar of hypochlorite solution holding the spiles. The hole is bored 2 to 3 inches deep or before the heartwood is reached. If the tree has been tapped before, do not tap within 6 to 8 inches of last season's tap.

Flush the taphole with the hypochlorite solution in the detergent bottle. Take a sap spile from the hypochlorite solution and tap it into the hole until firm. Do not pound the spile too hard, or the wood around the taphole will split and the sap will leak out the split. Hang the sap bucket or bag on the spile or spile hook. If you use tubing, attach the tubing to its spile before tapping the spile into the hole.

Gathering Sap

Your outdoor thermometer will signal sap runs: below-freezing nights followed by above-freezing days. Empty the buckets or bags every day during a run if you can to prevent buckets or bags from overflowing. Also, fresh sap makes better syrup, and sap will spoil quickly during warmer weather. Up to 25 taps on tubing can be collected in one large (new) trash can. Evaporate the sap daily if at all possible because the longer it sits, the poorer the syrup's quality becomes. Filter the sap through flannel or cheesecloth to remove dirt and twigs.

Evaporating (Boiling)

Whether you have decided to make all the syrup you can or to take a chance that you will make about ⅛ gallon of syrup per tap in order to make a certain amount of syrup, the batch method, which will be described next, will work for you. Keep the rule of 86 in mind, as well as the fact that sap should be boiled as soon as collected to avoid spoilage. Also remember that you need much more sap than one panful when you begin boiling.

Unless you will be hanging kettles over an open fire, you will want to build an open-top "arch" firebox using bricks or cement blocks on which to set your evaporating pan. Add a sheet-metal or brick chimney to the firebox. Build the arch so that the fuel opening faces upwind and the smokestack is on the downwind end. Doing so will help prevent smoke from curling into the pan and giving the syrup a smoky taste. (During sap season, prevailing winds are usually from the northwest.) The design is flexible: build the arch to fit your evaporating pan. If it is located near the sap collection point, you will do less sap hauling.

Before boiling the sap you might want to try out the fireplace using the sap boiling container. Put some water in the container so that it does not burn up. This dry run will show you whether the chimney is drawing properly. Rearrange the blocks and chimney if necessary and try a second dry run.

Put the pan on the arch, pour in the sap until it is 2 to 3 inches deep, and start the fire. Evaporation really begins when the sap reaches the boiling point. Leave room for bubbling; rapid boiling yields higher-quality syrup. If foam develops, skim it off or reduce it with a little lard, butter, or several drops of commercial defoaming liquid. Try to keep about an inch of sap in the pan. As the boiling continues, add more sap. The closer the boiling sap gets to being syrup, the easier it is to burn your pan. Add sap as often as necessary until you have a fair amount of concentrated sap about to reach the finished syrup boiling point: 7° above the boiling point of water. Check the boiling temperature of water when you are boiling sap; keep a pan of water handy, bring the water to a boil, and check the temperature with your thermometer. Because of changes in atmospheric pressure, the boiling temperature can vary several degrees even in one day. If the boiling temperature of water happens to be 210° F when you are boiling the sap, you have maple syrup when the temperature of the sap reaches 217° F.

If you make more than a couple gallons of the syrup, you may want to check your finished syrup more accurately the way experienced producers do, with a syrup hydrometer. It measures the density of dissolved solids in the syrup. Since 98 percent of these solids are sugar, the hydrometer, for practical purposes, records the percentage of sugar in the syrup. Standard maple syrup is 66° Brix (or 66 pecent solids) at 68° F. Corrections for reading Brix above or below 68° F usually come with hydrometers.

When the temperature of the sap is approaching the finished syrup boiling point, control the boiling carefully to prevent burning and excess density. You can do this in two ways:

1. If your pan is deep enough that the sap can be boiled down to syrup density while still at least 1 inch deep, you can finish boiling in the pan to 7° F above the determined boiling point of water. When this temperature has been reached, quickly remove the pan from the fire and drain or pour off the finished syrup through your filter cloth or bag.

2. The second way to finish boiling is to drain or pour off most of the syrup when it is about 5° F above water's boiling point. If you leave the pan on the fire, add more sap immediately to the pan to prevent burning it; a hot, dry pan scorches quickly. Finish boiling the "drawn" syrup on the kitchen stove or a bottle-gas stove to 7° F above the boiling temperature of water. If the temperature goes above 7° F, sugar crystals will form in the syrup when it cools.

The syrup is now ready for filtering. Filtering removes sediment, such as sugar sand. Use several layers of cotton cloth on top of your wool or orlon flat filter or inside your filter bag. The hotter the syrup, the faster it will filter, so filter it as soon as it has cooled to 185° to 190°F. After filtering a batch of syrup, turn the filter material over or inside out and rinse with hot water to prevent clogging with sugar sand. When dry, the filter material can be used again. If it is used wet, the moisture will dilute the syrup.

Bottling or Canning

Bottling or canning must be done while the syrup is at least 180° F to ensure sterilization. Without wasting any time, remove the filter and pour the strained syrup into clean bottles or cans, filling them to the rim. Tighten the lids firmly and invert the containers to sterilize the lids. Return to the upright position and leave room between the containers so that air can circulate. If the containers are too close to one another, proper cooling cannot take place and the syrup will darken unnecessarily.

Making maple syrup is that easy. It involves some work, but you will have a unique natural food that has been made in our northern clime for centuries.

Cleaning Equipment

As the maple syrup season progresses, watch the buds on your maple trees. When they begin to swell and turn reddish, the sugar in the trees begin to turn to starch and the sap will acquire a bad (maple syrup producers call it "buddy") flavor. That ends the season. Throw away that sap, except for enough to fill your unheated pan. It will help to soften deposits inside your pan for easier cleaning.

When you have finished gathering the sap and making syrup, remove all spiles from your trees. The tap holes can be left open. They will heal over in several years. Wash all your gathering and syrup-making equipment in boiling or very hot water, and rinse thoroughly in hypochlorite solution. Use rubber gloves because the hypochlorite solution can "burn" your skin.

Reminders and Additional Hints

1. Have all your equipment before you begin: for tapping, gathering, evaporating, filtering, and bottling or canning.
2. Tap before the sap-flow season.
3. Use only clean equipment, and be sure to use covers if you have buckets.
4. A higher-quality syrup can generally be made earlier in the season because the maple season's early temperatures are usually lower and reduce the chances of sap fermentation.
5. When evaporating the sap, maintain a constant, intense heat source at all times until the finished syrup boiling point

is being approached, and check the boiling point of water at least twice daily. Changing weather affects the water's boiling temperature and hence the finished syrup boiling temperature.

6. Butter, cooking oil, or condensed milk may be used as a defoamer but may affect the maple flavor. A commercial defoamer may be purchased.

7. Between runs, clean all the equipment, including any sap storage tanks.

8. The shorter the boiling time, the lighter the syrup color.

9. Turn the filter inside out and rinse in hot water as needed to prevent clogging. Reuse the filter when it has dried.

10. Make sure the syrup is hot (180+°F) when it is packaged so that it will not spoil.

11. Once the syrup can or bottle has been opened, mold can grow on the syrup's surface. To avoid this, refrigerate and momentarily invert the container once a week.

12. There are strict processing regulations for syrup that is sold.

13. If you have only a few trees to tap, you might join a neighbor in making maple syrup.

14. For help with your maple syrup making, ask your Extension agent about nearby maple syrup producers. Any who are members of the Wisconsin Maple Syrup Council will be glad to help.

15. Enjoy your pure maple syrup!

* From UWEX Fact Sheet G3029, "Maple Syrup Making for Beginners," by Robson and Cunningham, Department of Forestry, University of Wisconsin, Madison, 1979.

Photos on this page by Allen Wickman

Sap collecting bags help keep sap "clean" but sometimes are chewed on by squirrels.

Sap is collected and placed in holding tank.

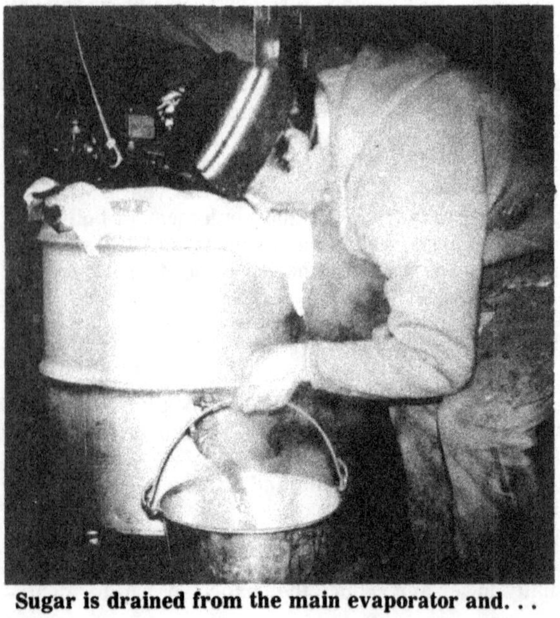

Sugar is drained from the main evaporator and. . .

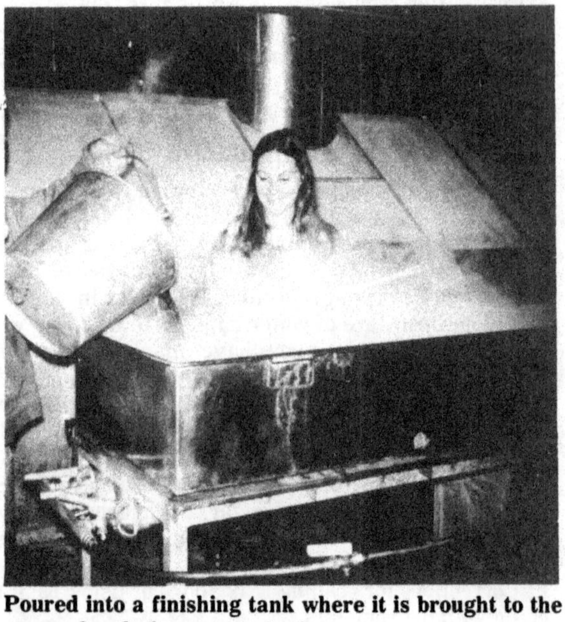

Poured into a finishing tank where it is brought to the proper level of sugar content.

PAPER AND PAPER MANUFACTURE

Paper's Place in Civilization

From the time man first appeared on earth he has sought to improve ways to exchange his thoughts with his fellow man and to record his ideas for future use.

Back many thousands of years he expressed himself by laboriously etching pictures and symbols on stones, walls of caves and bones. Later he employed other surfaces, like beeswaxed boards, palm leaves, bronze, silk and clay tablets. The ancient Greeks used a parchment made from animal skins. Four thousand years ago the Egyptians discovered a more efficient writing surface: papyrus, origin of the word "paper", a cross-woven mat of reeds which was pounded into a hard thin sheet.

Paper, as we know it today, was invented around the time of Christ by Ts'ai Lun, a Chinese court official. In all likelihood, Ts'ai mixed mulberry bark, hemp and rags with water, mashed it into a pulp, pressed out the liquid and hung the thin mat to dry in the sun. Unknown to him, Ts'ai's basically-wood mixture set off mankind's greatest revolution in communications. In his own homeland it carried the Chinese to a high cultural level. One emperor had a library of 50,000 books — at a time when most of the great leaders of Europe couldn't even write their names.

Nearly 1,000 years passed before the Chinese technique of paper making was brought into Europe by the Moors of North Africa. Ironically, along the route from the Orient, the method for making paper from wood was lost, and it took civilization nearly 2,000 years to find its way back to the most practical and most plentiful of all paper raw materials.

Until the early 1800's the Western world made paper from rags and cloth. Each sheet was individually turned out by dipping a screen into a vat of water-suspended fibers and filtering the water away from the fibers. A good worker could produce 750 sheets a day. It was an expensive, tedious process that couldn't answer the urgent demand for paper.

During the Revolutionary War the colonists starved for paper. Soldiers tore up old books to make wadding for their guns. Washington's generals sent him messages on scraps of paper. John Adams, in a letter to his wife, wrote: "I send you now and then a few sheets of paper; this article is as scarce here as with you." In desperation Washington ordered the discharge of some paper makers and sent them back to the mills.

In 1798 paper production went from the hand to the machine. Nicholas Louis Robert, a clerk at a paper-making mill in Essenay, France, devised a plan for a machine that replaced hand-dipping and produced paper in a continuous roll. The machine basically was a large endless wire screen, turned by hand to filter the pulp.

Robert couldn't get backing for his idea in France, so he sold the patent to the Fourdrinier brothers in England. The Fourdriniers built a practical machine, but it failed to provide inexpensive, plentiful paper. The embryonic industry was restricted by its raw materials. Rags were expensive and limited in quantity.

Nearly a hundred years earlier, however, in 1719, Rene de Reaumur, a French scientist, pointed to a solution. He noticed wasps using minute fibers of wood to make nests, the texture of which resembled paper. Reaumur wrote that the "wasp invites us to try to make fine and good paper from the use of certain woods."

His suggestion was ignored until 1850 when the German Friedrich Gottlob Keller read his treatise and developed a machine for grinding wood into fibers. In a few years Hugh Burgess, an Englishman, advanced mechanized paper making another step by inventing a chemical pulping process. Then, in 1865, C. B. Tilghman, an American scientist, solved a major chemical problem with his discovery of the sulphite process for dissolving unwanted resins in wood.

Two years later, when a pulp-grinding machine was imported from Germany to Stockbridge, Massachusetts, the age of economical, mass-produced paper was launched in America.

Newspapers multiplied. More, less-expensive magazines appeared on the stand, the school slate vanished, giving way to notebooks and lined paper. Paperback five-and-ten-cent novels gushed from the presses. Boston mills alone produced 75,000,000 paper shirt collars a year. Between 1889 and 1900 the production of paper doubled to about 2½ million tons a year.

Paper was first manufactured in America by William Rittenhouse at Roxborough, Pennsylvania (now a part of Philadelphia) in 1690. There paper was handmade from rags. It is from this kind of village craft that the paper industry has grown through initiative and free competition to be among the top ten industries in America today with annual sales of over $30-billion and paper and paperboard production, in 1975, close to 60 million tons.

Paper making changed from an individual art to an industrial art dependent on teamwork between managers, technicians and researchers. At the same time paper itself changed from a primarily cultural product used for newspapers, books and writing paper to a basic product, like steel, that could be adapted to innumerable purposes. Wrapping paper and bags flooded out to answer a crying need in the marketing of merchandise. Drugs and toilet articles were dressed in attractive wrappers and cartons. Signs and posters bloomed across the country. Corrugated and solid fiber boxes took a growing role in the economy and gradually replaced cumbersome wooden crates.

Without low-cost plentiful paper and paperboard, the industrial revolution in mass production, mass packaging and mass shipping could never have moved into high gear. Each year, literally each month, paper became a more important factor in the life of each person and the life of the nation.

How Paper Is Made

Paper making depends on the most basic manpower, the most advanced machinery and the highest technical skill. Although wood is the basis of paper making, only a little more than half of the fiber used comes from trees cut specifically for paper manufacturing. Approximately 22 per cent is from collected waste paper. The waste residues of lumbering operation and saw mills — wood chips, sawdust and the like — provide another 23 per cent of the basic raw material. An additional two per cent comes from non-wood based sources, such as discarded cotton fabric used to make cotton fiber papers.

Millions of tons of waste paper, including bond and writing paper, old magazines, old newspapers, used boxes and other paper and paperboard products, are collected each year and routed back to the mills. These reclaimed materials are repulped and used in the production of large quantities of paperboard and paper. The use of waste paper lessens the burden on the nation's solid waste disposal systems.

Although the paper industry is among the leaders in recycling, fiber reuse has some limitations. First, it is impossible to produce recycled paper of higher quality than the waste paper used, except by the addition of new wood fibers. Second, there is a limit to the number of times fibers can be recycled, because they eventually break into lengths that are too short for use. Thus, no matter how large the percentage of waste paper recycled, paper makers have to rely, basically, on wood.

Wood is made of billions of small cellulose fibers, bound together by a glue-like substance called lignin. Through these fibers flow the tree's sap and resins. When the sap, resins, lignin and other matter in the tree have been separated from the cellulose fibers by chemical means, the remaining fibers are called **chemical pulp**.

When the wood is reduced to small particles by rubbing against huge grindstones revolving at high speeds, the resulting product is called **groundwood pulp**. It lacks the strength of most chemical pulps but has qualities found useful in high speed printing. Most of the newsprint (paper on which newspapers are printed) throughout the world is made from mostly groundwood pulp.

The "mechanical" and the "chemical" methods of pulping are the two major processes used. However, chemists are continually trying to find better and more economical means of reproducing pulp. Some of the variations of pulp making are the chemiground-wood process, sulphate (or kraft), soda, semi-chemical, and cold soda processes.

In pulp making by chemistry, fiber separation is accomplished by cooking the wood in one of the several different methods, depending on the kind of pulp desired. The sulphate solution, for example, is especially used for making kraft (a Swedish word for "strength") paper for use as grocery bags and wrapping for packages.

First sliced into small chips, about an inch square and one-eighth thick, the pulpwoods are then fed into large vats, called "digesters" designed on the same principle as a kitchen pressure cooker. These digesters are 3-4 stories high. The chips and chemicals are steamed until the mixture is reduced to a wet, oatmeal-like mass. It is this cooking that dissolves the lignin and frees the fibers, suspending them in water.

The pulp is blown from the digesters under pressure to separate the fibers, washed to remove the chemicals and other materials from the fibers and then sent to the beaters. Pulp to be used for white papers is bleached white before entering the beaters. The most common form of beating consists of passing the suspended pulp between sets of metal bars or knives that complete the separation, reduce the fibers to proper length, and fray their edges so they will cling together when formed into a sheet. Added during the beating are color, if desired, size (which makes finished paper water resistant) and whatever other chemicals are required to make a specific type of paper.

After beating, the pulp flows to machines called refiners where the fibers are cut to uniform size and brushed to further improve their ability to cling together in the sheet.

Thus refined, the pulp — at this point 99 per cent water — is run onto an endless mesh screen at the "wet" end of the paper machines. This screen is called the Fourdrinier wire. Through a constant side-to-side vibration of the screen, the fibers are interlaced and much of the water extracted, leaving a sheet of paper which is quickly shunted through rollers for further drying.

Traveling at speeds sometimes faster than 3,000 feet per minute, it winds through a long series of steam-heated cylinders called "dryers," some wide enough to handle sheets 30 feet across. Here the last of the water is removed by heat, pressure and suction.

After drying, the procedure again differs according to the end product desired. Many papers go through a process called "calendering," which provides a glass-smooth finish by ironing the sheets between heavy, polished rollers. Others pass through tubs of chemicals that furnish additional coating. But however finished, the product that comes from the dry end is paper in large rolls which are then sent to converting plants to be made into thousands of different useful products.

Paper Today

Although paper has many uses, its most important contribution to our modern civilization is its use as a medium to record knowledge.

Engineering theories, technical manuals, blueprints, patents, mathematical equations, chemical formulas and, most important, the ideas and ambitions that drive mankind forward. In any area of the economy the more complex a project, like a new aircraft or an important change in federal law, the more technical know-how and therefore the more paper to tie in all details is required to insure its successful execution. The building of a rowboat involves less than a pound of paper in instructions, diagrams and bills for materials. The construction of an aircraft carrier, however, requires 200,000 pounds of paper, including a carload of blueprint paper.

Paper represents the perfect adjustment of a basic material to any use and purpose. It is constantly being substituted for other materials because it does the job more efficiently and economically.

Paper is versatile. Thick or thin, light or heavy, flexible or stiff, tough or fragile, plain or colored, large or small, its applications are limited only by man's ingenuity. It provides light-weight cartons for eggs as well as heavy-duty paperboard for shipping frozen poultry, meats and other foodstuffs. It is the ideal container for needles and nails, refrigerators and sofas, television sets and clothing, cement and candy. In the form of multiwall shipping sacks, it delivers each year more than 100 million tons of food, livestock feed, building materials, chemicals and drugs. Plentiful and low-cost, it provides more than 115 billion copies of newspapers and 230 million textbooks in the United States annually. From the viewpoint of protection, consider the breakage in electric bulbs that would occur without lightweight, shock-absorbing paperboard cartons.

With 6,000 plants in 49 states making and converting paper into many thousands of different useful products and with paper merchants in every important city, the paper industry is truly serving the whole of America.

Paper Tomorrow

While paper has been important to America's past growth and is essential to this country's present economy, research into new uses is bound to make paper an even more widely used commodity in the future.

Some of these new uses already in production are new kinds of filter papers; papers for printed circuits in the electronics industry and computer papers for electronic data processing. There are new transparent resin impregnated papers for many new kinds of construction uses and lamination for durable cabinet surfaces.

Paper is being married to other products such as wood, metal and plastics for hundreds of new uses requiring heat resistance and durability. Knitted paper and non-woven fabrics hold promise of a whole new field for paper with infinite new uses.

Homemakers across America are enjoying the use of special paper garbage disposal bags to replace noisy metal cans. Such disposable bags are now being used in many cities in different sections of the country.

Workers in atomic laboratories wear paper garments for protection against radio-activity. Hospitals are using disposable surgical gowns, bed sheets and pillowcases made of paper to lessen the danger of infection and to reduce costs.

But the paper industry has just begun to tap the vast potential of its product. Future uses of paper may exceed even the most imaginative dreams of the industry's scientists.

—American Paper Institute, 260 Madison Avenue, New York, N.Y. 10016.

HOW PAPER IS MADE

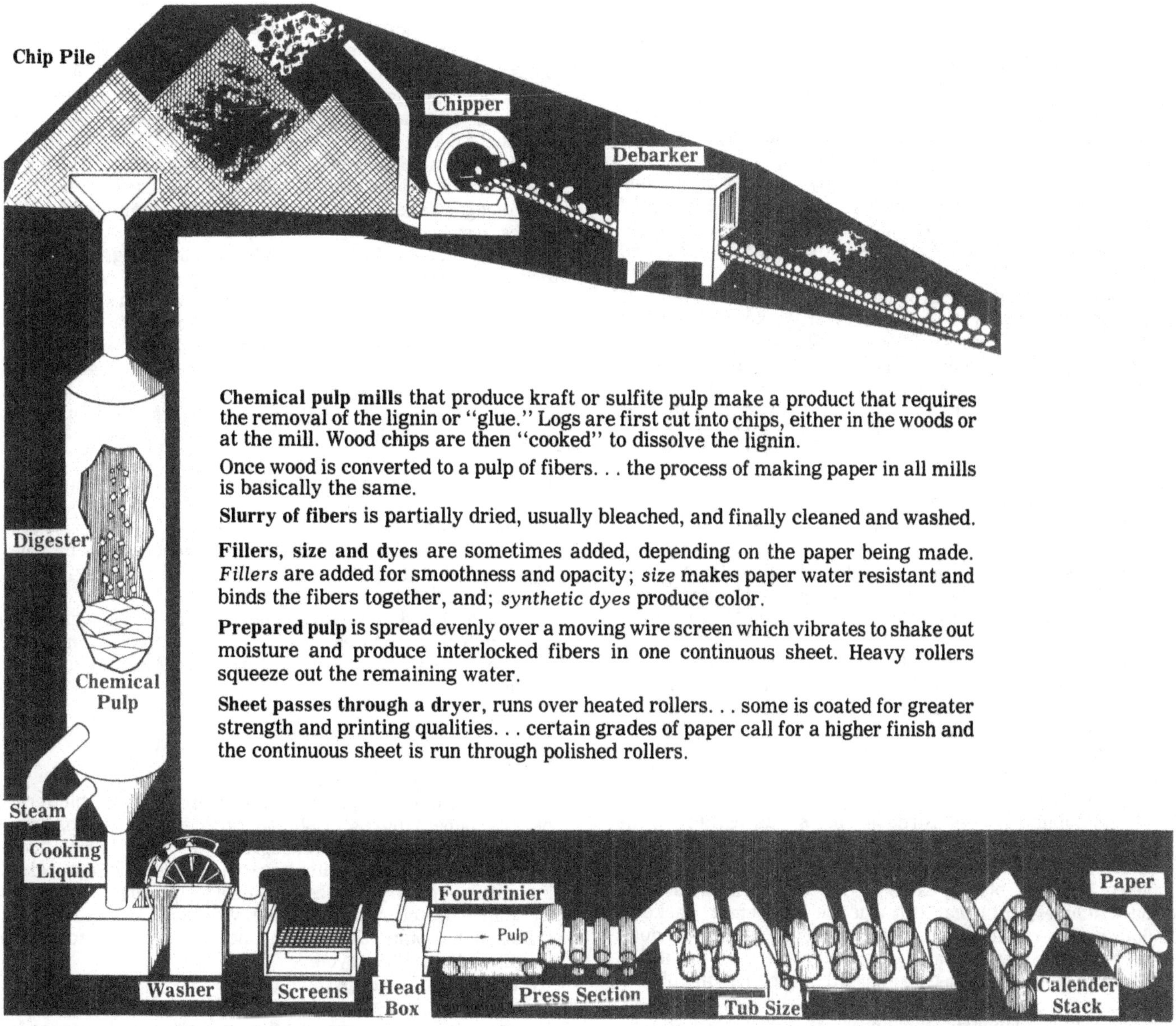

Chemical pulp mills that produce kraft or sulfite pulp make a product that requires the removal of the lignin or "glue." Logs are first cut into chips, either in the woods or at the mill. Wood chips are then "cooked" to dissolve the lignin.

Once wood is converted to a pulp of fibers. . . the process of making paper in all mills is basically the same.

Slurry of fibers is partially dried, usually bleached, and finally cleaned and washed.

Fillers, size and dyes are sometimes added, depending on the paper being made. *Fillers* are added for smoothness and opacity; *size* makes paper water resistant and binds the fibers together, and; *synthetic dyes* produce color.

Prepared pulp is spread evenly over a moving wire screen which vibrates to shake out moisture and produce interlocked fibers in one continuous sheet. Heavy rollers squeeze out the remaining water.

Sheet passes through a dryer, runs over heated rollers. . . some is coated for greater strength and printing qualities. . . certain grades of paper call for a higher finish and the continuous sheet is run through polished rollers.

Wood is the basic raw material in making paper. It has two main ingredients: cellulose fibers and a substance called lignin which helps hold the fiber in the trees together like a glue.

Minnesota mills use different methods to convert wood to paper. . . the groundwood and chemical processes.

Some **groundwood pulp mills** feed logs into grinders where they are pressed against huge rotating grindstones. Another groundwood process involves forcing chips through a refiner. In both methods the cellulose and lignin in the log is reduced to a fiber-laden slurry.

Minnesota Forest Industries • 200 Christie Building • Duluth, Minnesota 55802

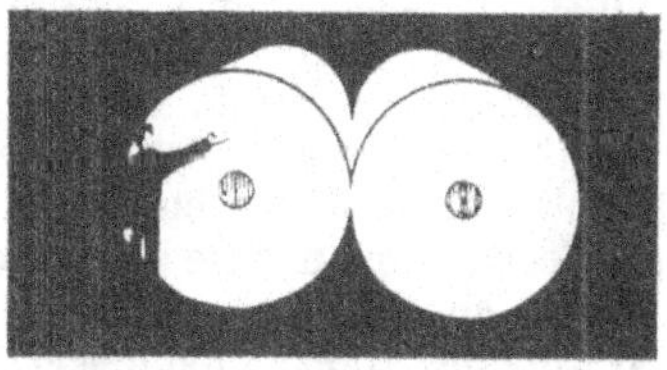

Paper can be shipped in huge rolls. . . in sheets (up to 38"x50") or packaged in sizes commonly used in businesses and schools.

PRODUCTS OF THE FOREST

By Lewis T. Hendricks and Tom Milton

Adapted from an American Forest Products Industries, Inc. poster entitled "Products of the Tree Farm," 1965; Extension Service Pub. No. 655, Mississippi State University; and Cooperative Extension Service Pub. No. 195, Virginia Polytechnic Institute.

Wood is considered the material of a thousand uses, perhaps a hundred thousand uses. Such a profound statement may be of little consolation to the wood products manufacturer caught in the midst of our present recession. When times get tough, most managers carefully examine their position, their products, and their competition, and begin working out a strategy. And they start thinking about new products, that niche in the marketplace, new ideas.

The following list categorizes many of the products and uses of wood. We hope it stirs your appreciation for woods' many uses. Maybe it will stimulate some ideas, but more importantly maybe it will help you gather your thoughts and regroup.

If an idea for a new product comes to mind, it may be helpful to review Ed Kallio's article on Marketing Research and the Small Forest Products Manufacturer in the September 1981 Marketing Bulletin. One other note — the Forestry Library on the University's St. Paul Campus contains a wealth of information on forest products and wood manufacturing. Many people are probably not aware of this fine resource.

Products of Sawmills and Specialty Mills

Products and their uses:

Timbers — Barges, bridges, building foundations, churches, dams, derricks, docks, factory and warehouse buildings, mine timbers, schools, ships, stringers, trailers, trucks, tugs.

Construction Lumber — Beams, boards, boat hulls and parts, dimension of all kinds, factory flooring, form lumber, heavy framing, joists, light framing, planks, posts, rafters, sheathing, sills, studs, subfloors, walls.

Finished Lumber — Baseboard, battens, casing, ceiling, flooring, lath, paneling, pickets, scaffolding, ship decking, siding, stepping.

Remanufactured — Airplane parts, agricultural implements, athletic equipment (baseball bats, skis, tennis raquets, etc.) balusters, bowling alleys and pins, bobbins, boxes, butchers' blocks, cabinets (for radio, television, sewing machines, etc.), car construction and repair, caskets, clothespins, conduits, crates, crossarms, displays, door jambs and frames, doors, dowels, floors, fixtures, furniture, furniture squares, glued laminated structural members, graindoors, gunstocks, gutters, handles, housetrailers, ladders, lattice, laundry appliances, machinery, matches, medical supplies, millwork, mouldings, musical instruments, novelties, pallets, panels, patterns, pencils, penholders, phonographs, playground equipment, plumbers' woodwork, professional instruments, printing material, pumps, radios, refrigerators, rollers for shades and maps, scientific instruments, ship and boat building including aircraft carrier flight decks, shiplap, shoe heels and lasts, shuttles, signs, skewers, spools, sporting equipment, stage scenery, surgical supplies, tanks, toothpicks, toys, trim, trunks, valises, vehicles, venetian blinds, wedges, window frames, wood pipe, wooden shoes, woodenware.

Ties — Railroad cross ties, mine ties, switch ties.

Cooperage (Staves) — Barrels, buckets, cooling towers, kegs, pipes, silos, tanks, tubs.

Miscellaneous — Acid washers, benches, corncribs, dunnage, elevators, fence pickets, grain bins, insulator pins, planks, reels, shingles, stakes, trestles, tunnel and mine props, wood chips for making wood pulp, wood turnings (for buttons, jewelry, etc.).

Residues — Fuel, planer shavings for compressed fuel logs and briquettes, poultry litter and raw material for hardboard, particle board and other bark, pulp and sawdust products (such as sawdust soil conditioner).

Products of Plywood Mills and Veneer Mills

Products and their uses:

Construction Plywood — Boxcar lining, boxes, cabinets, concrete forms, crates, door panels, finish, prefabricated houses, roofing, sheathing, siding, signboards, subflooring, truck floor and trailer panels, wainscoting, wall panels.

Marine Plywood — Canoes, motorboats, naval craft, racing shells, sailboats.

Compregnated Plywood — Airplane propellers, bearings, die stock, table tops, tubing, utensil handles, patterns.

Package Veneer — Baskets, crates, hampers, match boxes, wirebound boxes.

Face Veneer — Furniture, pullman car lining, show windows, store fixtures, wainscoting, wall paneling, wall paper.

Miscellaneous Veneers — Applicators, balloon sticks, book covers, candy and ice cream sticks, cigar boxes and wrappers, floral sticks, ice cream spoons, luggage, mustard paddles, novelties, square stick matches, surgical items, toothpicks, tongue depressors.

Residues — Fuel, raw material for other bark and pulp products, paper roll plugs, particle board.

Products of Hardwood Plants and Particle Board Plants

Products and their uses:

Hardboard — Battery separators, containers, decking, furniture and fixtures, paneling, sheathing, subflooring, signs, templates, toys.

Particle Board — Acoustical board, cabinet and wardrobe construction, door cores, moulded furniture parts, paneling, patterns, sheathing, store fixtures, subflooring, window displays.

Products of Pulp and Paper Mills

Products and their uses:

Sulphite Pulp — Paper and paperboard for bags, blotters, printing papers, boxes, bristol board, envelopes, folding boxboard, fruit wrappers, greaseproof packaging, insulation, labels, paper napkins, patent coated boards, photo processing paper, sanitary tissues, stationery, stencils, tag board, wallpaper, waterproof packaging, wrapping.

Dissolving pulps for cellophane, explosives, lacquers, plastics, photo film, rayon.

Sulphate Pulp — Paper and paperboard for bags, printing papers, bond paper, boxes, bristol board, chart paper, coating raw stock, condenser tissues, corrugated boxboard, envelopes, food containers, folding boxboard, insulation, ledger paper, liner board, offset paper, onionskin, parchment, sheathing paper, stationery, tag stock, towels, twisted cord and rope, waxed paper.

Soda Pulp — Paper and paperboard for blotters, printing papers, bristol board, corrugated paper, filters, insulating and wall boards, labels, liners for coated boards, stationery, testliners.

Semi-Chemical Pulp — Corrugated paper, egg cartons, insulating board, test liners, wall board, printing papers, glassine paper.

Groundwood Pulp — Absorbent papers, bags, boards, building and insulating papers, newsprint, printing papers, wall board, wood cement boards and blocks, wrapping paper, writing papers, roll roofing, shingles, hardboard, insulation board.

Residues (Liquor containing left-over cellulose and lignin not used in paper manufacture) — Sulphite liquors used in making adhesives, building briquettes, core binder, cymene, dyes, emulsifiers, ethyl alcohol, fatty acids, feeding yeast, fertilizers, fuel briquettes, linoleum cement, mordants, paint and varnish remover, plastics, road binder, tannins, vanillin.

Sulphate liquors used in making acetic acid, acetone, dimethyl-sulphide, fatty acids, furfural, methyl alcohol, oxalic acid, pine oil, rosin soap, rosin acids, tall oil, turpentine, ore flotation, pharmaceutical chemicals.

Soda liquors used in making acetic acid, acetone, calcium carbonate, methyl alcohol, oxalic acid, plastics.

Products of Wood Distillation Plants
Hardwood Distillation Products

Products and their uses:

Acetic Acid — Acetate solvents, cellulose acetate for rayon, photo film, lacquers, recording tape, and plastics; coagulant for latex, perfumes, and textile dyeing, manufacturing inorganic acetates, white lead pigments.

Acetone — Acetylene, explosives (cordite), solvent.

Charcoal — Activated carbon, black powder, explosives, chemical manufacture, fuel, livestock and poultry foods, manufacturing charcoal iron, medicines, metacase hardening compounds, producer gas, water purification.

Methanol — Antifreeze, dry-cleaning agents, formaldehyde, manufacturing chemical compounds, paints, pyroxylins, shellac, textile finishing agents, varnishes.

Pitch — Insulation in electric transformers, rubber filler.

Tar Oil — Flotation oils, gasoline (inhibitor oil), paints and stains, preservatives, solvent oils, wood creosote.

Softwood Distillation Products

Products and their uses:

Cedar Oils — Furniture polish.

Charcoal — Activated carbon, black powder explosives, chemical manufacture, fuel, livestock and poultry foods, manufacturing charcoal iron, medicines, metacase hardening compounds, water filtration.

Creosote Oils — Cattle and sheep dips, disinfectants, medicines.

Dipentene — Solvent for reclaiming old rubber.

Lacquer Solvent — Lacquers, paints, varnish.

Pine Oil — Disinfectants, fabric dyeing, flotation oil, paints.

Pine Tar — Coating and binding materials, disinfectants, manufacturing cordage, medicines, oakum, soaps.

Rosin — Paper sizing, varnish, soap, greases, waterproofing, linoleum.

Tar Oil Solvents — Disinfectants, flotation oils, paints, soaps, stains.

Wood Turpentine — Paint and varnish manufacture, synthetic camphor for celluloid manufacture.

Products of Wood Hydrolysis Plants

Processes — their products and uses:

Wood Hydrolysis — Acetic Acid: Textile manufacture, white lead pigment, cellulose acetate, perfume, cellophane.

Baking Yeast — Bakery products.

Butadiene — Synthetic rubber.

Carbon Acid — Industrial chemicals.

Ethyl Alcohol — Solvents.

Animal Food — Cattle feed, chicken feed.

Furfural — Resins, plastics.

Glycerine — Medicines, industrial chemicals.

Lignin Powder — Plastics and laminates.

Sugars — Stock feed, ethanol.

Wood Condensation — Furfural — Resins, plastics.

Soil Conditioner — To make soils more porous.

Alkaline Fusion — Oxalic Acid — Bleaching, industrial chemicals.

Pyrogallol — Stains.

Resins — Plastics.

Miscellaneous Products

Products and their uses:

Poles, Posts and Pilings — Antennae, arbors, bridges, channel markers, dams, docks, pole frame buildings, fence posts, flag poles, foundations, guard rails, jetties, levees, revetments, sign posts, tank traps, telephone poles, weirs, wharves.

Fuelwood — Home, commercial, industrial heating.

Sap and Gum — Balsam, birch beer, butternut syrup, gumthus, heptane, larch (Venetian turp.), maple sugar, mesquite gum, rosin, spruce gum, storax, turpentine, adhesives, ointments, chewing gum, perfumes, glass cement.

Bark — Adhesives, birch (flavoring) oil, cascara (drug), clothing (wood wool), drilling mud dispersants (oil industry), dye (osage orange and black oak), insulating wool, slippery elm (drug), soil building, tannins, (hemlock, chestnut, and tanbark oak).

Edible Fruits — Butternuts, chinquapins, hickory nuts, pawpaws, pecans, pinon nuts, serviceberries, walnuts, wild plums.

Needles — Pine and cedar needle oil.

Sawdust — Absorbent for explosives, artificial leather, artificial wood, body for paint, butcher shops, camouflage, clay products, composition flooring, curing concrete, filler for linoleum, filter for oil and gas, fireworks, glues, hand soaps, ice storage, insulating, insulating brick, livestock bedding, meat smoking, mild abrasives for cleaners, moth deterrent, nursery mulch, packing, plastics, soil conditioners and wood flour for billiard balls, bowling balls, explosives, moulded products.

Christmas Trees — Ornamentals.

Minnesota Forest Products Marketing Bulletin, Agricultural Extension Service, 240 Coffey Hall, University of Minnesota, St. Paul 55108.

Products Made From Trees

1 cord yields - 7,500,000 toothpicks
1 cord yields - 1,000-2,000 lbs. of paper (depending on the process)
1 cord yields - 942 1-lb. books
1 cord yields - 61,370 #10 envelopes (standard)
1 cord yields - 4,384,000 commemorative-size postage stamps
1 cord yields - 460,000 personal checks
1 cord yields - 89,870 sheets of letterhead bond paper (8½x11)
1 cord yields - 1,200 copies of the "National Geographic"
1 cord yields - 2,700 copies of the average daily paper (35 pages)
1 cord yields - 250 copies of the "Sunday New York Times"
1 cord yields - 30 Boston rockers
1 cord yields - 12 dining room tables (one table seats 8

Building an average 1800 square foot home uses 10,000 board feet of lumber - equivalent to 20 cords.

WOOD AND WATER

By Karl Ketter — Assistant Specialist, Forest Products

Hygroscopic is a word that applies to wood. It means that wood likes water. Wood will soak up water until the weight of the water in the piece of wood may even outweigh the wood substance itself! Wood products do not have to be put in a liquid bath to take up moisture. Wood can absorb water vapor from the atmosphere.

The amount of moisture contained in wood relates directly to the performance of products made from this remarkable material. Wood swells when it absorbs moisture and shrinks when it dries. No matter whether the product is furniture or houses, changes in wood dimensions can cause problems. Among other things, such fluctuations can cause loose joints in furniture, popping of nails on drywall, squeaks in floors and steps, openings and possible leaks around windows, and binding in doors and drawers.

Much research has been done on wood's relationship to water. The best advice in straight forward terms from the Forest Products Laboratory in Madison, Wisconsin, is: "To give best service, the wood should be installed at a moisture content close to the midpoint between high and low values it will usually attain in use."

Let us now take a look more closely at moisture content since this is how we measure the amount of water in wood. It is by controlling this moisture that we hope to overcome shrinkage and swelling problems. Moisture content is the ratio of water to wood expressed as a percentage. It can be calculated with the formula:

$$\text{Mc in \%} = \frac{\text{original weight} - \text{ovendry weight}}{\text{ovendry weight}} \times 100$$

This procedure would involve taking a number of pieces of the wood and weighing each of them as accurately as possible and placing them in an oven for a few days at 214 to 221°F. The pieces should be piled to allow the hot air to circulate around them. Higher temperatures should be avoided because resins may be driven off, thus affecting the weights. Immediately after the samples are dry, each should be reweighed. Knowledge of the original and the ovendry weight is all that is needed to calculate moisture content of a single sample. The MC of several pieces should be averaged.

There are also instruments to measure moisture content. Electric moisture meters work well between dry conditions and a 30 percent moisture content. One type of meter is used by inserting metal electrodes into the wood and the resistance is measured as an electrical current passes between the electrodes.

When the tree is first cut, it may have a moisture content of 100 percent or more and is called "green." When these higher moisture contents are reached, it would indicate that the tree contains more water than wood, by weight. The cut wood begins to dry and the moisture content drops. The water that was contained in the cell cavities of the wood is leaving the wood because the air is drier. It is the same when a kitchen sponge dries. The wood dires without any change in size until the moisture content reaches 25 to 30 percent, depending upon the species. **All the shrinkage takes place between moisture contents of 30 percent and 0 percent (no water condition). When this 25 to 30 percent moisure content range is reached in wood, it is called the fiber saturation point (FSP).** It is the condition at which the cell cavities are empty but water is still present in the cell walls. This portion of water is actually bonded to the wood. When this water is removed, a moisture content reduction is accompanied by shrinkage.

In an example developed by the Western Wood Products Association, they show how moisture content and size are related as a greenboard measuring 12¼" at 100 percent MC is dried to 0 percent MC.

Table 1* — Shrinkage across the width of a flatsawn board as related to moisure content.

MC	Actual Size	Total Shrinkage
100%	12.25"	0
75%	12.25"	0
50%	12.25"	0
24.4 (FSP)	12.25"	0
20%	12.08"	.17"
15%	11.91"	34"
10%	11.74"	.51"
5%	11.56"	.69"
0%	11.38"	.87"

From this table, we see that a 12¼" board shrank almost 9/10 of one inch as it dried! The advice to use wood that is at a moisture content close to what **it will be in service makes sense.**

*Source: Western Wood Products Association, Technical Guide, "Dimentional Stability"

The example above was one species. There are more than 100 commercially important trees used in this country. There is a certain amount of variability in moisure content of wood within a given tree as well as differences between species. Also, when lumber is cut from the log, the amount of shrinkage or swelling is different for the length, width, and thickness dimensions. These last differences are considerable. The greatest variation is a direction parallel to the growth ring (tangential). (See figure 1). This means that a flatsawn board will change more than a vertical grain board. The example above was a flatsawn board. This tangential variation is about 1½ to 2 times greater than shrinkage or swelling across the growth rings (radial). Dimension varia-

Figure 1 — Tangential and radial directions in a log.*

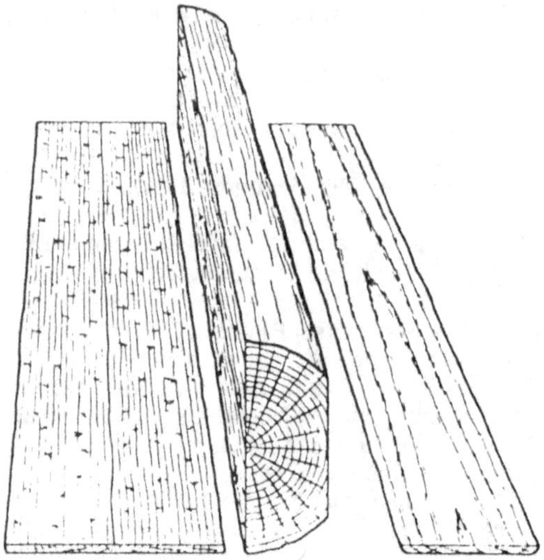

A — Radial or quartersawn

B — Tangential or flatsawn

* Source: U.S. Forest Service, Forest Products Laboratory, Madison, Wisconsin, Agriculture Handbook No. 188, Dry Kil Operators Manual.

Prepared by University of Minnesota Agricultural Extention Service and School of Forestry in cooperation with State Department of Conservation, Division of Lands and Forestry tions in the length is very small and less than 2/10 of 1 percent and can usually be disregarded in most applications.

The total amount of shrinkage when wood is dried from 30 percent to 0 percent MC also depends upon the species. As an example, red oak would shrink 4.0 percent radially (across growth rings) and twice as much, 8.2 percent tangetially (along growth rings). Old growth redwood would have about 2.6 percent total radial shrinkage and 4.4 percent total tangential shrinkage.

Generally, the softwoods would shrink or swell less than the hardwoods. Also, woods with high densities or weight tend to shrink and swell more than less dense or lighter weight boards.

The water leaving the wood must go into the atmosphere. How dry the wood gets depends upon the ability of the air to absorb moisture. Wood does not dry to 0 percent MC unless heat is added. The amount of water in the wood and in the air come into balance at the Equilibrium Moisture Content (EMC). This point of balance will fluctuate depending upon the temperature and the relative humidity of the air. As an example; if a room were kept at 70°F. for a year, the relative humidity would fluctuate between from 3 percent to 72 percent (from the dry heated air in winter to those muggy July days). Woodwork subjected to these conditions could fluctuate from .7 percent to 13.7 percent moisture content. The moisture content of wood is dependent upon the relative humidity of the air but the percentage figures will not be the same as each is a measure of a different phenomena. This range accounts for 50 percent of the total possible shrinkage for wood and would be a serious problem. Often a drastic change in the relative humidity of the air lasts only a short time while the wood requires more time to alter its moisture content. In fact, it is convenient to determine the average conditions of wood in use for various parts of the country. Table 2 summarizes the possible conditions.

*Figure 2—Recommended moisture content averages for interior-finishing woodwork for use in various parts of the U.S.

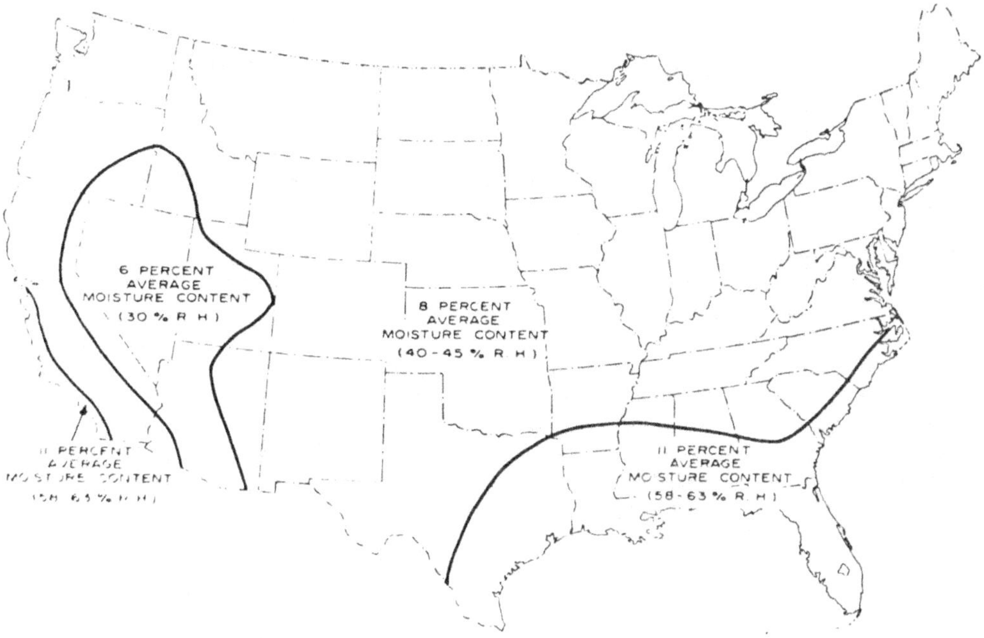

*Source: U.S. Forest Service, Forest Products Laboratory, Madison, Wisconsin Research Note FPL-0226, 1978, **Moisture Content of Wood in Use**

Table 2[*]--Recommended moisture content values for various wood items at time of installation.

Use of wood	Moisture content for-- Most areas of U.S. Average[2]	Individual pieces	Dry southwestern area[1] Average[2]	Individual pieces	Damp, warm coastal areas[1] Average[2]	Individual pieces
	Pct.	Pct.	Pct.	Pct.	Pct.	Pct.
Interior:						
Woodwork, flooring Furniture, wood trim Laminated timbers	8	6-10	6	4-9	11	8-13
Exterior:						
Siding, wood trim Framing, sheathing Laminated timbers	12	9-14	9	7-12	12	9-14

[1]For limiting range, see figure 2.

[2]To obtain a realistic average, test at least 10 percent of each item. If the amount of a given item is small, several tests should be made. For example, in an ordinary dwelling having about 60 floor joists, at least 10 tests should be made on joists selected at random.

[*]Source: U. S. Forest Service, Forest Products Laboratory, Madison, Wisconsin, Research Note FPL-0226, 1973, Moisture Content of Wood in Use.

"To give full and proper service, wood products should be installed at the moisture content close to the average it will maintain while in use." In most portions of the United States, including Minnesota, all interior wood should be kiln dried to an average moisture content of 5 to 8 percent. This is especially important if the wood will be remanufactured into large complex products such as cabinets, musical instruments, turning stock, or furniture. Wood that will be used outside such as siding and trim should be dried to an average of 12 percent for best performance.

Three Important Terms

Green Wood - Wood that has a moisture content above the fiber saturation point (25-30 percent). Wood of this moisture content is not suitable for any interior use.

Air Dried Wood - This is wood that is usually dried outside and due to climatic conditions in our area, seldom gets below 15 percent in moisture content.

Kiln Dried Wood - Wood that has been placed in a specially constructed enclosure to remove moisture is termed "kiln dried." In softwoods, wood dried in this manner to 19 percent moisture or less is frequently used in house construction. However, we feel that moisture content of these softwoods should be in the 12-15 percent range for best performance. More importantly hardwoods that are kiln dried should be reduced to a 6-10 percent moisture range, especially where interior use is a prime consideration.

HARDWOOD LOG LENGTHS

How accurately are hardwood logs cut for length in New York? A partial answer to this question can be had from the results of a recent lumber recovery study at 10 mills across the state during which 1030 hard maple logs were scaled and graded.

Only 25 percent of the logs had the ideal trim allowance of 3 to 4 inches. About 15 percent were cut shorter than this, but 60 percent were cut too long! Also, about 5 percent just missed the even foot by an inch or two.

So what! Well, reducing this excessive over-length can mean more money for a logger. For instance, calculations show that if he cuts about 1,000,000 board feet per year with overlengths such as these he could gain an extra 10 to 15,000 board feet of logs from the same quantity of timber if he cut all logs with only 3 to 4 inches of trim. Translated into money it could mean an extra $1,000 or more of income.

As for the mills--why not accept odd lengths? Quite a few already do and 20 percent of the logs measured were odd lengths anyway. It could encourage loggers to do a better bucking job by just knowing that every foot would be measured and paid for. Admittedly it's harder to "eyeball" lengths when scaling if they are a mixture of odd and even, but then it could probably be better to measure the lengths more carefully anyway since 15 percent of the logs may have insufficient or no trim allowance and 5 percent may be just short of the foot mark which is awfully hard to detect unless actually measured.

Finally--if all logs cut in New York each year were bucked accurately and well, there would be a potential savings of about 5,000,000 board feet of timber. Any part of this would be a welcome improvement.

* * * *

HANDY WOOD SELECTION GUIDE
It Pays to Use the Right Wood for Every Purpose

Wood can frequently replace or supplement expensive or scarce metals and plastics

Species	Wt.*	Hardness	Strength	Splits Easily	Dimen-sional Stability	Decay Resistant	Ideal Application
ASH, Black	39½	Medium	Medium	Moderate	Good	Moderate	Handles, dowels, blocks, balls, lamp parts.
ASH, White	41	Hard	High	Moderate	Good	Moderate	Rollers, handles, long dowels, athletic equipment.
ASPEN	25	Soft	Medium	No	Good	No	Light-weight use: dowels; plus for pipes, masking.
BASSWOOD	26	Soft	Low	No	Good	No	Light-weight use: nails easily, novelties, mouldings.
BEECH	45	Hard	High	Yes	Fair	No	Substitute for hard maple; distillation.
BIRCH, White	36	Medium	Medium	No	Good	Moderate	Fine turning wood; spools, dowels.
BIRCH, Yellow	44	Hard	High	No	Good	No	Finishesbeautifully: bobbins, mouldings, lab equipment.
CHERRY	35	Medium	High	Moderate	Excellent	Moderate	Furniture, food paddles, printing blocks, patterns.
CYPRESS	32	Soft	Medium	Yes	Good	High	Good for outdoor use: tanks, chemical plants.
ELM, Rock	44	Hard	High	Yes	Good	Moderate	Excellent for rollers, blocks, handles, farm implements.
ELM, Soft	37	Medium	Medium	No	Good	No	Bends, nails easily: aviary, poultry equipment; sporting goods.
FIR	34	Medium	High	Yes	Fair	Moderate	Good construction lumber; long softwood dowels.
GUM, Black	35	Soft	Medium	No	Fair	No	Bores nicely; rollers, handles, bushings, implements.
GUM, Red	35	Soft	Medium	No	Fair	No	Intricate turnings, furniture parts, wedges, bushings.
HICKORY	51	Very Hard	High	Moderate	Fair	No	Heavy duty tool handles, foundry blocks, wedges.
MAPLE, Hard	44	Very Hard	High	Yes	Good	Moderate	Resists abrasion: bearings, blocks, faucets, handles.
MAPLE, Soft	33	Hard	High	No	Fair	Moderate	Spool hubs, furniture frames, novelties.
OAK, Red	44	Very Hard	High	Moderate	Good	Moderate	Bungs, blocks, rollers, agriculture implements.
OAK, White	47	Very Hard	High	Moderate	Good	Moderate	Heavy rollers, plugs, tool handles, agricultural implements.
PINE, White	27	Soft	Low	Yes	Excellent	Moderate	Architectural designs, mouldings, models, patterns.
POPLAR, Yellow	28	Soft	Medium	Moderate	Fair	No	Display fixtures, lamp parts, musical instruments.
REDWOOD	28	Soft	Medium	No	Good	High	Good for outdoor use; tanks, mouldings.
SYCAMORE	34	Hard	Medium	No	Fair	Moderate	Butchers' blocks, rollers, scientific instruments.
WALNUT	38	Hard	High	Moderate	Excellent	Moderate	Trophy bases; display, lamp and furniture parts.

* Pounds Per Cubic Foot Dried to 12 Per Cent Moisture

Source: Chart compiled by Research Dept. of American Wood Working Co., Montello, Wisconsin.

GROWING SHIITAKE MUSHROOMS

By Martha Greenwald

Shiitake mushrooms are increasingly available in better grocery stores around the Midwest. Much of the market's new-found interest in the exotic shiitake (she-tah-key) has been generated by the research and marketing activities taking place at the Forest Resource Center, a nonprofit educational organization based in Lanesboro, Minnesota. Recent shiitake production at the Center has exceeded all expectations. Mushroom researcher Mitch Gilbert says "we are one of the few cultivators in the U.S. to have mastered the appropriate conditions for growth of shiitake."

Grants received by the Center in the last three years have allowed it to perfect its knowledge of growing this mushroom in the U.S., and to develop a comprehensive marketing program for wood-be growers. "We are doing a trial of 50 genetically distinct strains of mushrooms to see which are best suited for cultivation in a midwestern climate," says Gilbert. "We are also trying to see which strains are best suited for indoors forced fruiting."

In addition to their delicate, woodsy taste, shiitake are renowned in the Orient for their high nutritional value, providing protein, fiber, vitamins and minerals. Orientals have used shiitake as a food and folk medicine, and have also published research which suggests that shiitake can help the body's immune system.

Farmers with oak woodlots, unused outbuildings, and some spare time might also consider shiitake as a potential cash crop. The better grades of shiitake are selling in restaurants for from $7-$9 per pound, and in retail outlets for about $12 per pound. Gilbert says, "If you're contemplating commercial production of shiitake, the capital costs are modest. The big investment is labor." The Resource Center began its research on shiitake mushrooms 5 years ago because it was seen as a labor intensive cash crop that could also help improve forest management. Center director Joe Deden says that studies shows that "the market for shiitake mushrooms in the coming years could reach 5000 pounds per week in the Twin Cities and Rochester (Minnesota) alone."

One of the most difficult things about outdoor shiitake cultivation in the Midwest is managing the lengthy vegetative stage of growth, which lasts up to 18 months. Carefully selected healthy oak logs, with a diameter of 3" to 8" are harvested in the winter. Holes are drilled in the logs, which are then "seeded" with the mushroom spawn, which can be ordered in the form of inoculated sawdust or wooden dowels. The grower must take pains to make sure moisture content in the oak log remains at about 40% to 45% (roughly equal to its moisture at time of cutting) throughout the growing season. Because the Midwestern climate is normally drier than Japan, where shiitake grow naturally, growers typically need to irrigate logs to make sure moisture content remains at the required levels. Logs that are too wet or too dry could fall prey to weed fungus and the crop could be reduced or eliminated.

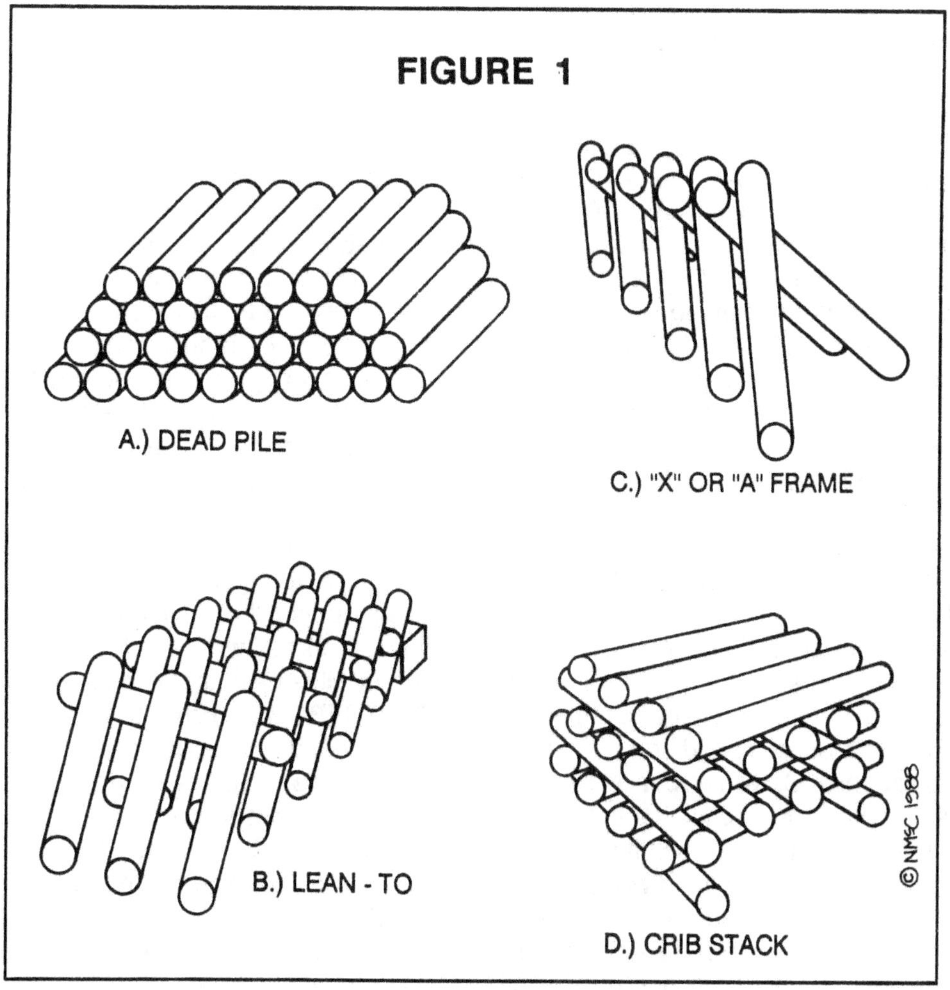

Logs should be stacked in a shady area, in an arrangement which allows each log to maintain equal exposure to moisture and ventilation. They should also be stored above grass or other vegetation which could expose them to weed fungus.

For a long period the spawn appear to slumber inside the log, gradually eating away at the wood until the mushrooms are ready to fruit. When the vegetative stage is over, the log will either naturally bear mushrooms, or can be forced to fruit. According to Gilbert, "this is done by soaking the logs in cool water for 24 hours. Then take them out of the soak and put them in an environment that gives them high relative humidity and low drying rates." The bulk of the logs at the Resource Center were innoculated in !987 and !988, and a large number are now ready to start producing. Gilbert thinks the center could now produce 200 pounds per week year-round, making it one of the biggest growers in Minnesota.

Early reports about shiitake were optimistic about the prospects for cultivation in the U.S. North American growers quickly learned that Japanese growing techniques do not work here. A sophisticated approach to log management and maintaining moisture content appears to be the key to success in shiitake cultivation.

For more information on shiitake mushrooms, contact the Forest Resource Center, Rural Route 2, Box 156A, Lanesboro, MN 55949. The Center publishes a newsletter called "Shiitake News," available four times a year for $25. The newsletter is a clearhouse for resource materials on shiitake mushroom cultivation.

(Martha Greenwald is a freelance writer based in Lanesboro, Minnesota, specializing in natural resource and environmental issues)

THINNING RED PINE

by Donald Johnson

THINNING - WHY AND WHEN

When trees are small, many are needed to establish a forest cover type and control the site. However, as the trees grow larger, fewer and fewer are needed. Some of the weaker trees die from crowding, and the more vigorous trees take advantage of the extra growing space. A natural stand of red pine may have several thousand seedlings per acre when young but fewer than a hundred when mature.

Plantations, of course, start with fewer trees per acre and require cultural practices to control competition and establish red pine trees as the forest cover type. Some older plantations began with more than a thousand trees per acre; but today, most red pine plantations start with fewer than a thousand and some begin with only a few hundred. The need for thinning is related to the number of trees per acre established and the size of trees desired.

Thinning is the way a forest manager controls the development of a red pine stand and produces some early economic returns. Growth is put on the best trees available by cutting the competing trees. The crop trees are then able to maintain a uniform growth rate, which is an important measure of wood quality. Periodic thinnings remove injured and diseased trees, thus helping reduce the risk to crop trees. Thinning also increases timber yield by harvesting some trees that might otherwise die. Further, rotation length can be controlled by the thinning methods and schedules chosen.

Red pine stands need thinning when the crop trees are crowded enough to slow diameter growth to about 10 or more rings per inch. According to an old definition, the object of thinning is to provide the remaining trees with "room to grow but none to waste." Some guidelines to help you detect when a red pine stand may be in need of a first thinning are:

If merchantable products can be cut, consider thinning when: 1) the live crown is less than one-third of the trees's total height, or 2) tree height is more than five times the average spacing, or 3) the average diameter at breast height (the DBH) in inches is equal to, or greater than, the average spacing in feet.

These guidelines take into account the general effects of age, site, and spacing. Plantations with more than a 6-foot average spacing between trees usually do not need thinning until merchantable products can be cut, regardless of site quality. Plantations with less than an average of 6 feet between trees may need thinning before merchantable products can be cut. (This is called a precommercial thinning.) However, trees planted on high-quality sites will grow faster than trees planted at the same spacing on low-quality sites, and they will need thinning at a younger age. While plantations on low-quality sites will need thinning under the same guidelines, the trees will be older than those on the high-quality sites because of the slower growth rates.

COMMERCIAL THINNING

The wood from red pine trees is suitable for a variety of products. Stands managed for sawtimber can produce pulpwood, posts, poles, cabin logs, and small sawlogs from the periodic thinnings. Management objectives and expected markets play major roles in the choice of thinning practices. Thinnings should be scheduled to produce more products for which there are good markets and fewer products for which there are poor markets.

In Northwest Lower Michigan, for example, we now

have markets for utility poles, sawlogs of all sizes, and large-sized pulpwood material. The market for small sized pulpwood material is somewhat limited. The utility pole market is probably the most lucrative. It is also the most demanding in terms of product quality and the management needed to produce an acceptable utility pole.

Two important aspects of all thinning prescriptions are the density the stand will be after thinning and the method used to select which trees are to be cut or left. In most instances in our area, since we are now thinning plantations for the first time, entire rows will be cut or left. This makes the selection of trees to cut somewhat less important.

MINIMUM RECOMMENDED STAND DENSITY

The minimum recommended stand density to leave after commercial thinning is related to the average diameter of the trees in the stand, and this varies over a fairly wide range. It assumes a uniform spacing that will provide each tree with all the room it can use. While stands should not be thinned below the minimum density, they can be left at higher densities. In fact, it will be necessary to leave most stands at higher than the minimum density because any one thinning should always cut less than half of the trees. Stands with a high risk of snow, sleet, or wind damage should be thinned even more lightly, removing one-third or less of the trees. Stands left at higher densities will require more frequent thinnings to keep them from exceeding the maximum recommended stand density.

THINNING METHOD

The first thinning in a stand should provide an access system that will be useful throughout the life of the stand. The typical row thinning methods used in our area also provide access to the stand.

First thinnings in northwestern Lower Michigan usually employ a row thinning method that removes entire rows of trees at a predetermined interval. Besides providing access to the stand, this type of thinning also makes room for the trees to be felled. Some common thinning intervals are: 1) remove every third row, 2) remove every other row, 3) remove two rows and leave two rows, and 4) remove every fourth row and thin selectively in the area between the cut rows.

Removing every third row is a fairly conservative method that removes one-third of the trees and provides a cut row on one side of each row that is left. This method is useful where higher densities after cutting are desired, such as with managing for utility poles.

Removing every other row reduces the stand by one-half and is favored by many loggers. In many instances, this is probably too severe and can result in damage to the stand from wind and snow. It may also result in a longer period of time before the stand is in need of another thinning.

Removing two rows and leaving two rows is a variation on the previous thinning method in that it removes one-half of the stand. It is used in plantations that were planted at a close spacing and where removing only one row does not provide enough room for harvesting equipment to be used effectively.

Removing every fourth row with selective thinning between is the most conservative method. It results in removal of between one-fourth and one-third of the stand. This method is used where higher stand densities are desired for production of utility poles.

The frequency of thinnings depends upon the stand density after thinning. If stand density is left near the maximum recommended level, thinnings will be required more frequently than if stand density is left near the minimum recommended level. Site quality and tree age also influence how fast stands grow toward the maximum recommended density, which relates to how frequently thinning will be required. The combination of density after cutting, thinning method, and frequency of thinning for the life of a plantation make up the thinning regime.

THINNING REGIMES

The various thinning regimes possible between the minimum and maximum recommended stand densities allow the forester to customize a thinning program for specific management objectives. For example, red pine plantations managed to produce large trees as soon as possible might have thinning regimes that cut the smaller trees and leave residual stand densities at or near the minimum recommended. Plantations managed for timber products with quality requirements based upon taper or knot size, such as utility poles, might have thinning regimes that cut both large and small trees and leave residual stand densities about midway between the minimum and maximum recommended. Plantations managed primarily for piece products and a long rotation might have thinning regimes that cut the larger trees and leave residual stand densities near the maximum recommended.

CONCLUSION

Through thinning regimes, foresters can control the composition and growth of a plantation. The initial thinning can develop the access trails useful for all future thinnings and many other management activities. By choosing the stand density when the plantation is to be thinned, the residual stand density to leave, and the method of selecting the trees to cut, the forester can have a major impact on the number, size, and quality of trees harvested.

Donald Johnson is a forester who is also Kalkaska County Extension Director, Michigan State University Cooperative Extension Service.

National Woodlands Magazine

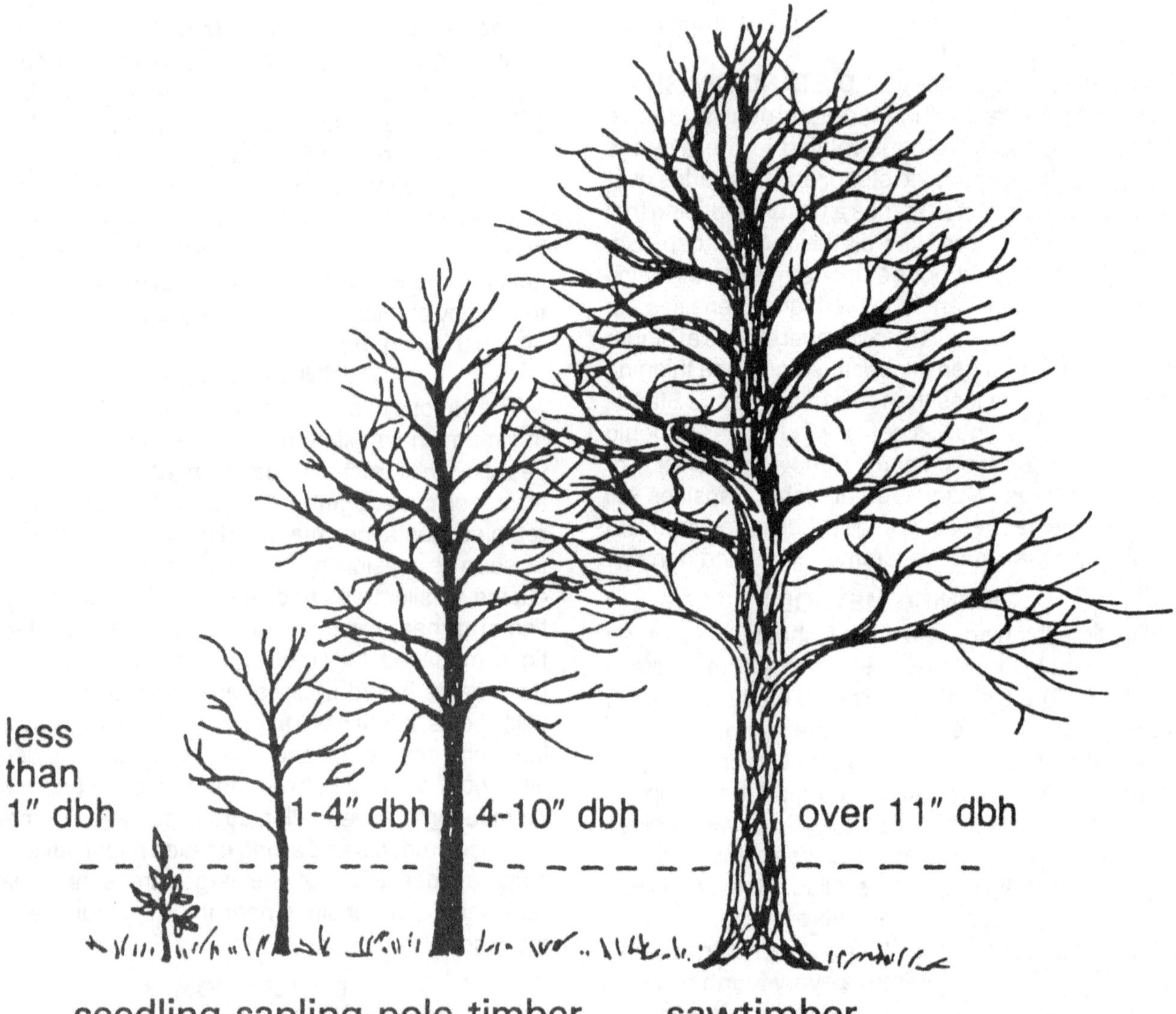
less
than
1″ dbh
1-4″ dbh
4-10″ dbh
over 11″ dbh
seedling-sapling-pole timber
sawtimber

CHAPTER 6 — HARVEST

Minn. 4-H Bulletin No. 87 and (or)
Forestry For Minnesota Schools — North Print Co.

Harvesting means removing trees from a forest and using those trees to benefit mankind. The harvesting process is called logging and the people who harvest the trees, loggers.

Logging is one of man's oldest professions. We have used trees for shelter, fuel, and food since earliest recorded history. You may wish to study early accounts of tree harvesting which are included in references in the back of this publication.

In the United States, our American Indian predecessors were using trees for shelter and food when the pilgrims landed at Plymouth Rock. Consider the birchbark canoe, bows and arrows, maple sugar, and the history of the Indians' use of fire to furnish fuel and to improve wildlife habitat.

Those pilgrims, landing in a boat constructed of wood, constructed shelters of wood as one of their first jobs. Throughout early United States history, trees were an item of demand and controversy. The reservation of white pine trees by the King of England for use as spars in ships of the British Navy was one of the issues which led to the Revolutionary War.

As America gained her independence and settlers moved westward, the forest hampered agricultural development. Wood was needed to build homes, factories, railroads, bridges, and to furnish fuel for U.S. citizens and industries. This led to the exploitation of forests and the unfortunate idea that "timber was a one-time crop." Now we recognize that wood is a renewable resource and plan for the future growth and harvest of many crops of forest products from the same land area.

Logging in Minnesota's forests started in the early 1800's and continues today. Perhaps the heyday of the logging industry occurred during the 1880-1920 period when white pine abounded and large sawmills were cutting logs from the St. Croix and Mississippi watersheds to International Falls and Duluth.

Today, harvesting forest crops is still a vital component of our economy and forestry ranks as Minnesota's third largest industry, employing more than 40,000 people in forest products and allied manufacturing units. Over $500 million is generated each year in Minnesota from manufacturing of wood and fiber products. All of this depends on the harvesting of trees from our forest lands. Consider the job of logging: the harvesting of trees

THE HARVESTING PROCESS

The harvesting process consists of the following steps:

(1) Selection of area and trees to be logged;
(2) Felling, limbing, and bucking the trees;
(3) Skidding the trees or logs;
(4) Loading and unloading the trees or logs;
(5) Hauling the trees or logs to the mill;
(6) Preparing the land for a new crop of trees.

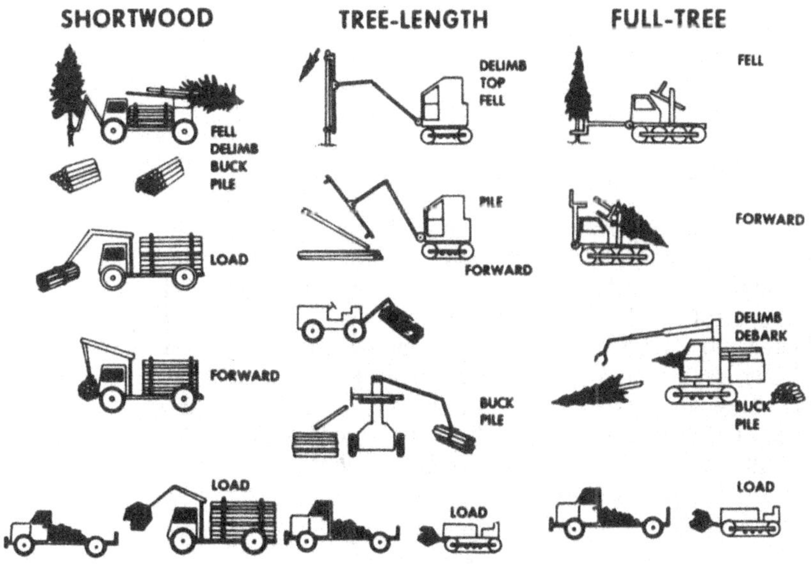

Selection of Area and Trees to be Harvested

The selection of the area and trees to be harvested is part of the management plan which has been developed by the forest landowner. The age of the mature forest stand which is ready for harvesting, called the rotation age, is determined years beforehand. For example, in Minnesota, aspen may be logged at 40 years, red and white pine at 120 years. It is also important to consider what products are desired from the forest in determining cutting. Once these decisions are made, certain areas will be selected for harvest. These areas are then marked on the ground (commonly by spraying the boundary trees with paint), the volume of trees to be cut determined, and a timber sale made.

Felling, Limbing, and Bucking

The logger purchases the right to harvest trees, then cuts them down: this is called felling. Years ago this was done with an ax and crosscut saw; now it is done with power saws. A new technique involves the use of hydraulically operated shears mounted on the front of a crawler tractor. The shears clip off trees 20 inches in diameter or smaller at the stump. However, the power saw is the most used tool in felling trees. Limbing means removing limbs from the bole (trunk) of the tree. Topping means cutting the top from the bole. Bucking means cutting the tree into certain lengths. For cordwood bolts in Minnesota, lengths are 100 inches. This is known as a shortwood system in pulpwood harvesting. Logs to be cut into lumber are generally longer, up to 40 feet in the Pacific Northwest.

In Minnesota pulpwood operations, logging the entire tree is becoming more common. In the "tree-length" method, the tree is felled and the top and limbs removed before dragging it to the landing. In the "full-tree" method, the entire tree including top and limbs are dragged to the landing where the top and limbs are removed. There are three distinct systems in Minnesota pulpwood harvesting: shortwood (100-inch bolts), tree-length, and full-tree.

Skidding or Yarding

The process of dragging the felled tree or logs to a central point (landing) for loading onto a truck is called skidding or yarding. Formerly, oxen or horses did this, now tractors or skidders are used. The rubber-tired skidder with chokers (cables attached to the logs) is commonly used in this operation. In the West, cable systems skid logs to the landing. Also, helicopter systems are used to reach more inaccessible forest stands, and balloon logging is experimental in Oregon.

Loading and Unloading

At the landing the logs or tree-length boles are loaded onto trucks. In the "glory days of logging," this was accomplished using crosshauls powered by horses or oxen. Now a front-end loader or a loading machine mounted on a truck does this. Unloading takes place at the mill or railroad yard. This, too, is accomplished with the same loading machinery.

Transportation

Trucks are the standard method of transporting the logs to the mill or railroad shipping point.
The log truck, which may be carrying cordwood in 100-inch sticks or bolts or sawlogs to the mill, has replaced horses and oxen which once pulled sleighs to the railroad or to lakes or rivers where the logs were moved by water during the spring breakup to the mills.

Skidding

Loading

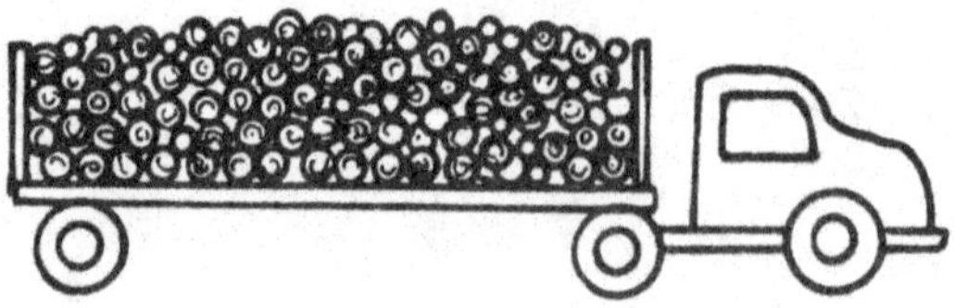

Hauling

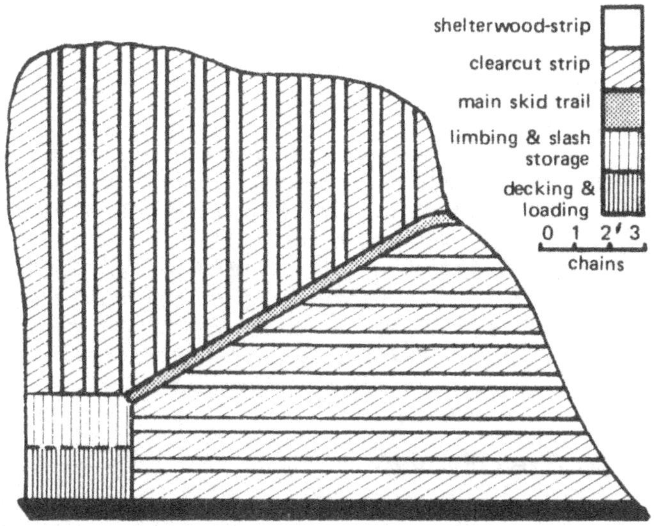

Preparation for a New Crop of Trees

In any forest management plan, the preparation of the land for regeneration of a new stand of trees will be considered in the method of harvesting. For trees which cannot tolerate shade, such as the pines and aspen, the slash must be removed either by piling it or by burning. Controlled burning is preferable since it converts the tops, limbs, and smaller vegetation to ashes, which provide a fine seed bed with nutrients immediately available to the seedlings. Piling and/or burning also clears the area for future planting.

Harvesting patterns are also important in determining future roads and landings for machine movement in the new forest. Consider where lanes should be created for firebreaks to allow thinning, pruning, and finally harvesting the present crop of mature trees.

SILVICULTURAL METHODS OF HARVESTING

Silviculture is the art of producing and tending a forest. One of the ingredients in tending the forest includes the harvesting of trees. There are four main methods of harvesting:

(1) Clearcutting
(2) Seed-tree
(3) Shelterwood
(4) Selection

Clearcutting and selection represent the two extremes.

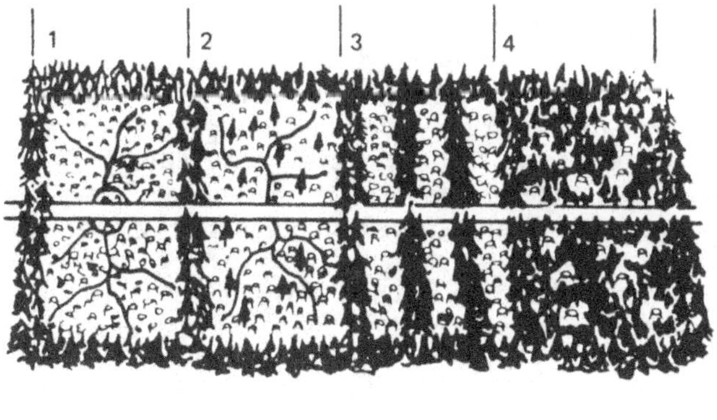

Clearcutting Method

An area of the forest is marked for harvesting and every tree, regardless of size and species, is removed. Clearcutting is used when the forest to be established consists of species which will not live under shade. It is also used where the present forest consists totally of trees which are mature or overmature. It may also be applied where the expense of other methods of cutting would prevent their use.

Clearcutting in Minnesota is used with stands of jack, Norway (red), and white pine; white and black spruce; and aspen. These are all trees which cannot reproduce or grow well under shade. They also are species which do well as even-aged stands (all about the same age or age class).

Clearcutting is also used with those species which can reproduce in nature after fire or some other disturbance. In Minnesota, these are the pines, spruces, and aspen-birch and in the West, Douglas-fir, the pines, spruces, and redwood.

In the new forest, we may expect the clearcut area to be naturally reseeded from standing trees in the surrounding forest. We may artificially spread seed on the harvested area, or we may plant the area with trees of desired species which were produced in a nursery. This latter method is becoming more popular as it insures the type and spacing of trees and to some degree eliminates the problems of early survival of the seedling, brush competition, and too much competition between trees of the same species. With aspen, we expect the suckering from stumps and roots of the cut trees to provide the future forest.

Clearcutting with proper regeneration is a necessary method of harvesting the forest. Again with aspen, it provides excellent habitat for wildlife such as deer, moose, and ruffed grouse.

There are some drawbacks to clearcutting. For several years the area may be unsightly. However, those forested areas of northern Minnesota which are traditionally pictured as representing the primeval forest were created by nature's methods of clearcutting: the wildfire or windstorm. Also, if the cutting area is on a steep hillside and improper techniques are used in the harvesting job, excessive soil erosion may occur. This is

Clearcutting method.

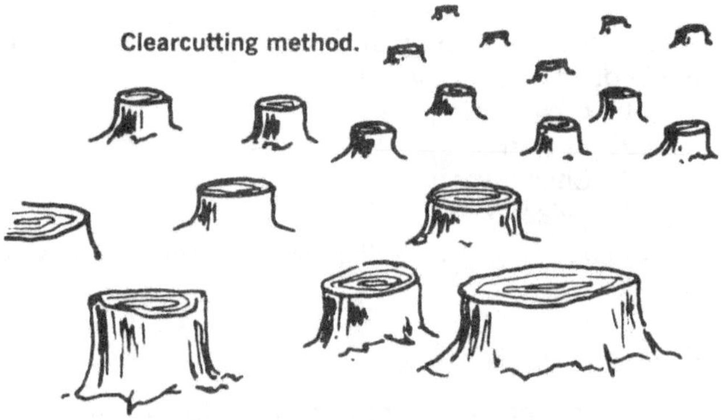

Seed-tree method.

Shelterwood method.

more a problem in mountainous country with the main erosion problems found in conjunction with road building.

Clearcutting is a very proper method of harvesting. It has occurred in nature over millions of years since trees first grew on the land. It has been used by man in Europe for centuries without loss of soil nutrients or degradation of the land. It was used by the Indians in this country, through the setting of wildfires, for thousands of years before the white man, farmer-logger, came to this continent. It will be used in the future if we are to provide those essentials of timber and fiber for human welfare.

Seed-Tree Method

In this method, certain trees are left in the harvested area, either singly or in groups, to furnish seed which will fall on the cut area, germinate, and grow to become a future forest. Once the new forest is established the seed trees may be harvested. This system is rarely used in Minnesota, but is more common in the southern and western forests of the United States.

The seed-tree method is generally applied with species which do not tolerate much shade. Since natural seed source is from the standing trees in the cut-area, it is not necessary to depend on adjacent trees to the cutover area for natural seeding, thus the area cut may be larger than in the clearcutting method.

One of the main disadvantages in this system is the possibility of loss of the seed trees by wind, fire, or insect attack before they produce seed. Because of the increasing dependence on tree planting, the seed-tree method has not been used as extensively in recent years. Aesthetically, there is little difference between the clearcutting and seed-tree methods.

Shelterwood Method

In this method, portions of the forest stand are removed gradually over the latter part of the rotation period. As the name of the method implies, the trees not cut serve to shelter the new crop of trees which starts to grow in the areas where trees were removed. So a new crop of trees is started before the mature trees have all been removed.

Obviously, the intensity of shelter or protection of the new trees can be varied by cutting many or few of the mature trees. Thus, the system provides for a considerable range of tree removal, similar to a heavy thinning: the best trees are left standing as a shelter while the poorest trees are removed.

This method can be used with even- or uneven-aged stands. It is generally done over the last one-fifth of the rotation age. For example, when growing white pine on a 100-year rotation age (that length of time from the germination of the seedlings to the final harvest), shelterwood cutting might begin at age 80 and the final cut would be at age 100.

Several factors must be considered in this method. First, the new crop must be shade tolerant enough to survive and grow normally. Second, the trees which are left should not be subject to windthrow.

Third, the standing trees must not be susceptible to damage in the removal of the partial cuttings; and fourth, the cost of logging may be more expensive than in the clearcutting or seed-tree methods.

There are many variations of the shelterwood system, generally classified under:

(1) Uniform method—harvesting throughout the entire stand,

(2) Group method—groups of trees are cut and removed, or

(3) Strip method—strips of trees are removed.

The latter method is sometimes used in cutting spruce-fir in northern Minnesota. Because of the economics of harvesting, the shelterwood method has not been utilized as much as may be desirable. This method is preferred from an aesthetic standpoint to the clearcut or seed-tree method. It is especially adaptable to white pine and northern hardwoods; it cannot be used with jack pine or aspen; and it may have some potential with Norway (red) pine.

Selection method.

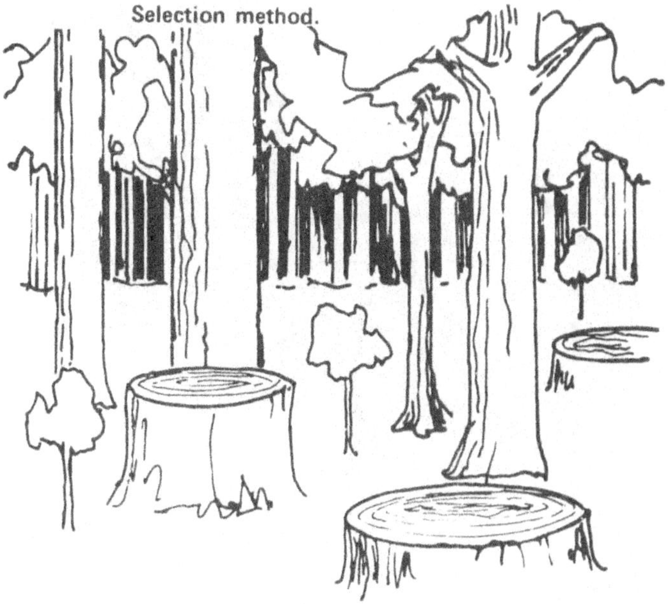

Selection Method

As the name implies, the individual tree is considered and selected or left in this method. It is primarily employed with northern hardwoods and in stands where single or groups of conifers, such as white pine, are found. The selection method is aimed at retaining an uneven-aged composition within the forest. Only mature trees or species which are highly desirable are removed. For example, in Minnesota, black walnut is a highly sought-after tree and loggers may enter a stand to remove only this species. Another example might be the removal of mature trees, with crowns occupying too much space.

Selection cutting can proceed throughout the life of the stand since there are older, middle-aged, and very young trees in the uneven-aged stand. This method depends on the ability of new trees to reproduce and grow without hindrance from shading. Generally, natural reproduction is expected but planting is possible, in fact desirable, when growing certain species such as black walnut.

There are a number of factors to consider in using the selection method, several of which were discussed with the shelterwood method. Tolerance of shade, ability to reproduce, wind-firmness, damage from logging, and economics are the most important. Also, the selection method affords the least impairment of stand beauty if the slash is removed or cut and scattered throughout the stand. This system is most widely used in Minnesota in the northern hardwood types found in central and southeastern Minnesota.

A well managed forest.

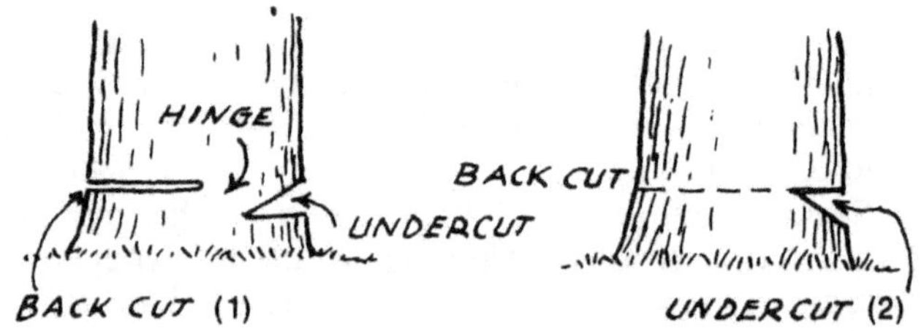

Two types of undercuts
(1) The strip of uncut wood acts as a hinge when tree falls toward undercut.
(2) This undercut results in a log of full length.

Harvesting Woodland Products

Forestry Manual For Vo. Ag. Instructors
Wisconsin Department of Public Instruction

Harvesting is important to the timber owner. The quality and quantity of the timber is mainly a result of forest management. Any type of harvesting should provide for eventual reproduction of the stand.

Proper Cutting of Trees

The following precaution and practices will prevent excessive logging damage to remaining trees.

1. Don't fell trees on young growth, Pile the slash in areas that have no reproduction
2. Avoid felling trees so that no damage occurs to remaining trees.

Broken branches and skinned stems are often the direct cause of later defects. When cutting the tree, determine the best direction to fall, considering lean, wind direction, possible breakage, and damage to saplings or reproduction. Make the undercut as close to the ground as possible, usually under 12 inches on a sawlog, or under 8 inches on a pulpwood tree.

Cutting the Logs

Producing logs for home use or the commercial market demands more skill than felling the tree and cutting the logs to length. The appearance and quality of logs produced determines what a buyer will pay. For home use it is important to produce logs which will give the greatest yield of usable material.

Bucking is sawing felled trees into shorter cuts. How a tree is bucked depends on several factors; the minimum log size that is marketable under existing specifications of available markets or log grading rules and the amount of defect in the tree.

Before any forest landowner cuts logs, there are several important points to consider:

1. Follow the buyer's specifications carefully. Failure to do so may result in rejected logs, or a loss in footage.
2. Keep your logs straight, avoiding crooks and sweep.
3. Cut the logs for highest quality and group defects when possible.
4. Leave a trim allowance on all logs in accordance with the buyer's specifications.

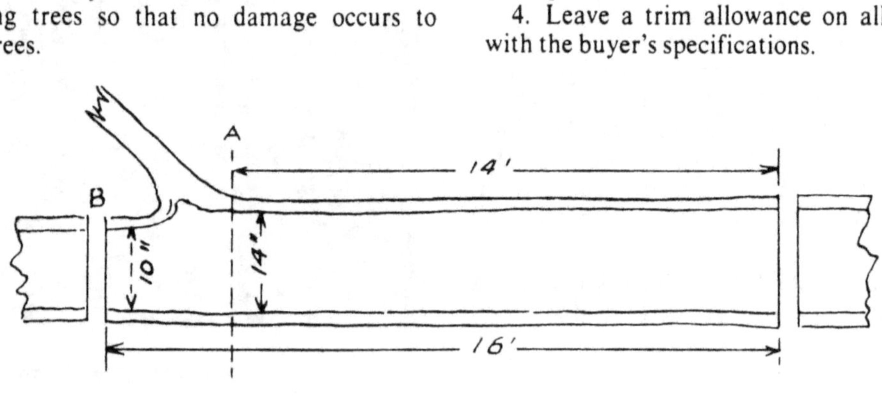

Cutting at A instead of B yielded an additional 40 feet.

Figure 40

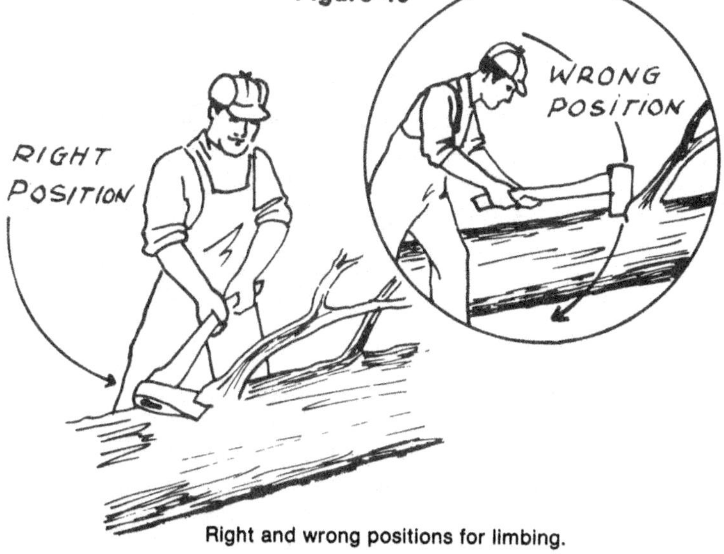

Right and wrong positions for limbing.

5. Leave logs of no value in the woods (cull logs), if they cannot be used for pulpwood, fuelwood, or chemical wood.

6. Prevent the log from splitting when cutting log lengths.

7. Trim all limbs close to each log.

8. Look out for metal in the logs. If you know it is there, cut it out or cut off the section containing metal.

9. If the base of the tree is hollow, taking a short log first may make the next log higher in quality. If a large hole is present at the base of the tree, cutting a long log results in loss of quality, footage and money.

10. After a few logs are cut, ask the buyer to check the logs and to show his method of scaling.

11. Standard hardwood log lengths are measured to the longest full foot. Softwood log lengths are measured in even lengths. Be sure to add trim allowance (4 inches) to these lengths.

12. Butt logs should be cut 14 to 16 feet long, if this can be done without sacrificing quality, and if you can follow buyer's specifications.

Logging Equipment

Hand Tools

Although power equipment has replaced the use of most of the hand tools in modern harvesting methods the use of the following tools are still quite popular as part of logging operations.

Axe—Axe patterns are of a great variety-largely to suit individual preference. The double-bit axe is favored by many woods workers. It has one keen blade for chopping; the other blade is less sharp for grubbing around roots or cutting hardwood. A single-bit axe is still in common use and is an efficient woods tool.

Crosscut Saws—Most crosscut saws have two types of teeth: cutting teeth which sever the fibers on each side of the cut and raker teeth which rake the shaving-like chips into the gullets between the cutting teeth and rakers and then out of the cut. The well-sharpened crosscut saw makes shavings-not sawdust.

Bow Saws—These are used both in felling trees and bucking logs up to 16 inches in diameter and commonly have blades 36 to 48 inches long. The blade is thin and this minimizes wedging. Adjustable frames maintain needed tension on the blade.

Wedges—Made of untempered steel or wood are useful for felling, bucking and splitting. They are usually 6 to 8 inches long and taper from the edge to the shank. Steel wedges should never be used behind a chain saw. Wooden wedges or wedges of some soft metal, such as aluminum, are used instead.

Cant Hook—Is helpful in rolling or prying logs and in bucking. A variation of the cant hook, called the peavey, has a steel spike in the end that can be stuck in the ground or into wood for use as a pry.

Power Tools

Power Chain Saw—Machines powered by gasoline, electricity or compressed air. Innovations and improvements are constantly being made in the saws and cutting chains. Current models have solid state ignition eliminating spark plug problems. Lightweight one-man saws for felling and bucking are commonplace among woods workers.

Figure 41

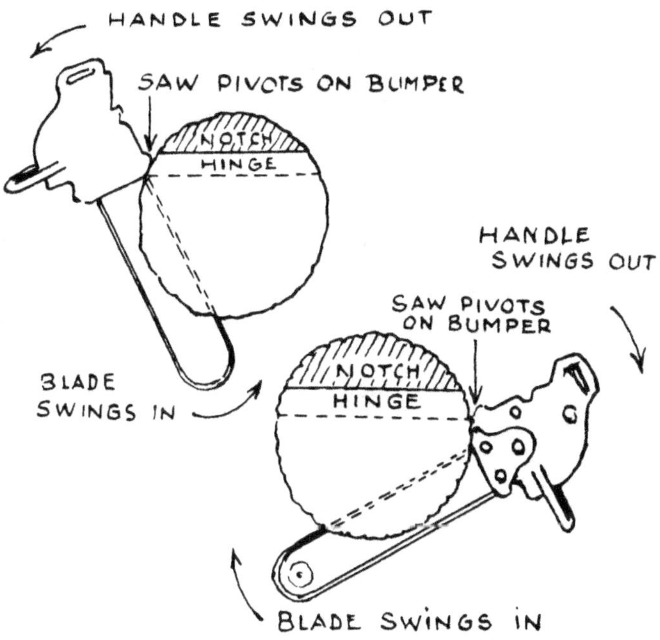

Use of a one-man chain saw in felling. Small tree can be cut in one sweep.

Skidding Equipment

The skidding equipment needed to move products from the stump to an assembly point depends on the size, length of the product, skidding distance, lay of the land, soil conditions, season of the year and how the logs were felled.

Animals or machines supply the power needed to move the products from the stump to skidding terminal - skidway, landing or the assembly point. Often a horse can handle small products economically over short distances of several hundred feet. Large material and longer hauls

Figure 42

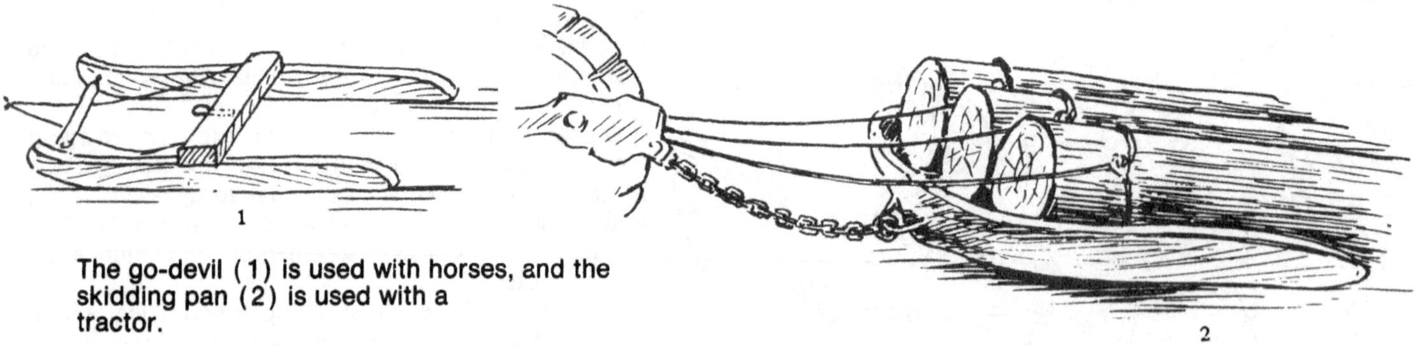

The go-devil (1) is used with horses, and the skidding pan (2) is used with a tractor.

require a team or small four wheel skidding or crawler tractors. By reason of their maneuverability, adaptability to a wide range of topographic conditions and overall economy, tractors have opened new possibilities in logging on every scale. In some agricultural areas of the state, farm-type wheel tractors have been used successfully in skidding small material such as pulpwood, poles and small logs.

For heavy logging, the crawler or four wheel skidding tractors are recommended. With suitable auxiliary equipment such as winches and arches, these machines can be adapted to any particular job.

For most farm woodlot skidding operations, the farm tractor used in conjunction with a logging chain or chain and skidding arch or tongs provides adequate mechanical equipment to do a successful job with a minimum financial investment.

Loading Equipment

The simplest loading possible is from a skidway located to permit gravity loading onto a truck or other conveyance. Two skid poles are set to permit rolling the round pieces onto a truck, wagon or sled.

Round pieces, if not too heavy, can be rolled up by hand on skids from the ground level but the job is easier with draft power and cross haul line.

The A-frame jammer also provides a relatively simple method of loading logs. It can be easily moved and requires the necessary cable, blocks and tackle. This method is worth using if there is much loading out to be done from the ground level.

Another loading device available on many farms is the front-end loader of a tractor. If logs are not excessively heavy, this tractor attachment works satisfactorily.

New wood-handling equipment, saving time, work and money, has started to serve the commercial logging industry. Equipment that can load logs or pulpwood sticks directly on self-contained pallets using hydraulics, means greater production, lower costs and less maintenance of equipment—all with a limited use of man power.

The cost of much hydraulic-powered equipment precludes use in farm woodlots and small private holdings. A hydraulic loader now used for both pulpwood and sawlogs can run into considerable capital outlay. In commercial operations, however, this cost is not excessive when reduction in man power, need for more efficient machinery and increased production are all considered.

Hauling

Wagons, trucks and sleds, depending upon the season of the year, are the usual types of conveyances for moving timber products. Trucks are generally used for long distances. The average farm wagon is not built to transport heavy logs. A tractor-trailer combination, such as might be available on some farms, is satisfactory. The load must be properly blocked, balanced and securely wrapped with chains to keep it intact during transit.

It is not unusual for an owner of a small woodland tract

Figure 43

Loading logs with a crosshaul.

to sell his forest products at the skidway or roadside and thus eliminate the loading and hauling. It hardly pays to buy special equipment and conveyances for this purpose and the usual vehicles found on the farm are often too light for hardwood sawlogs.

Processing Lumber

The operation of a farm or rural business is often made less expensive and much more convenient because of the availability of local wood products. Farmers are the country's biggest users of wood products. An individual farmer may use 100 or more fence posts and 2-3 thousand board feet of lumber annually.

A farm woodland that is large enough and managed properly can supply many of the home-needed wood products.

In order to efficiently use home-grown timber, the farmer should have a knowledge of how to convert it to useful products, its adaptability in construction, its durability and proper information on seasoning and preservation to extend its usefulness.

Custom Sawing

Scattered throughout the state are numerous custom sawmill operators. These businessmen have a self-contained portable sawmill that is mounted on rubber tires and able to come to a woodlot to saw logs. This service eliminates the need for the farmer to load and haul logs to a stationary sawmill that may be some distance away. Most custom operators saw logs all year. A minimum of two thousand feet is usually required. The farmer also must furnish several additional men to help. The farmer pays a fixed fee per thousand board feet of lumber produced. An advantage of this service is that the farmer can have his logs sawed into the dimension lumber he wants. Most custom sawyers can cut between 800-1,000 board feet per hour, depending upon the size lumber being produced.

Seasoning

Wood is seasoned to improve its suitability for the purpose it is to be used, and for weight reduction to lower shipping costs. Seasoning is accomplished by lowering the moisture content of the wood.

If lumber cannot be used within a few days after sawing, it must be properly piled to prevent warping and twisting. Much locally sawed lumber is unfit for use because of improper handling. Hardwoods are especially subject to warping and require care and time for proper piling for drying. But this work more than pays its way in quality of the product. Future timesaving and convenience also can be assured by making separate piles for different species and varying lengths.

The amount of time required to thoroughly air-dry lumber varies by species, thickness and local climatic conditions. When lumber has lost all but 15% to 20% of its moisture, it is said to be air-dry and is a stable product under average conditions in Wisconsin.

Under ideal drying conditions in late spring and summer, the minimum time required to season one-inch boards in properly built piles is as follows:

40 Days	50 Days	60 Days	70 Days	80 Days
Basswood	Aspen	Ash	Beech	Spruce
Cottonwood	Elm	Hemlock	Birch	White Oak
Pine	Maple	Hickory	Cherry	
			Red Oak	
			Black Oak	
			Walnut	

If lumber is piled in late fall or winter, the time required for air-drying may be two or three times the minimum shown above.

Piling and seasoning are unnecessary for many purposes of rough construction, provided the lumber can be used as soon as it is sawn and is not permitted to warp. Hog houses, shed, cribs and other outbuildings are commonly put up with green lumber.

Piling Lumber

The lumber should be piled immediately after sawing. There should be no delay or the wood will stain and deteriorate rapidly.

The site should be well drained and exposed to the wind. Avoid low, or damp ground, and remove all debris that would affect the ventilation.

When building the pile, the foundations must be solid and high enough to allow air to circulate freely under the pile. Allow 18-inches off the ground. The foundation should be sloped one-inch per linear foot from front to back to allow rain or snow to run off. Use blocks for the base and place heavy timbers on them to prevent the pile from sagging. Place the heaviest dimension stock on the bottom of the pile. Long boards should be on the outside, with shorter ones on the inside. No loose ends should be allowed to overhang without support and the boards should be at least one-inch apart to allow for air circulation. The pile should be built up, using only lumber of the same thickness in each layer, and pitching the front end of the pile outward or forward one-inch for each vertical foot to prevent rain and snow from entering between the boards.

The stickers used should be of equal thickness; for hardwoods, they should be one by one and one-fourth inches wide; for softwoods, they should be one by three inches wide. When piling, each sticker should be directly over the one below and each tier should be directly over a beam in the foundation. In most cases the sticker spacing required is from three to four feet. Checking of end boards can be reduced by extending the outer edge of stickers one inch beyond the ends of the boards at the front and rear of the pile. This practice will partially shade the ends and keep the rate of moisture loss nearer to that of the center of the pile. This means the outside stickers may have to be slightly wider. To reduce stain or decay, use only seasoned stickers or green stickers sawed from rot free heartwood, because stain occurs in sapwood.

Roofing is necessary to protect the pile from sun and rain, and to reduce checking and warping. Roofing can be done by laying boards of low-grade material about one-half inch apart, and placing another layer directly over the first so that the spaces between the lower boards are covered by the middle of the boards in the upper layer. Since good ventilation is necessary, the roof should be at least six inches above the top of the pile.

To insure good drainage, the roof should be sloped one-inch for each foot of length, and be projected at least two-feet out at the front and back of the pile. This will prevent water from draining down the pile.

Considerations for Safety While Managing and Harvesting Timber

Clothing

Wear clothing that will protect you from the needless scratches which might otherwise occur. Trousers should be cuffless and reasonably close-fitting around the ankles to help prevent tripping. Foot gear should be substantial, with a sole that will minimize the possibility of your slipping and falling.

Protective Equipment

A hard hat should always be worn while working in the woods. Safety glasses or ordinary glasses equipped with hardened lenses are desirable. Many serious eye injuries are caused by brush and twigs. When operating a power saw or other noisy equipment, use earplugs or ear muffs.

A ballistic nylon called B-64 nylon has been developed which can give the wearer a great deal of protection from accidental power saw cuts. This material is available in the form of knee pads which can be worn on the inside of the trouser leg. It also is available in some boots, where it is sewn into the instep, and gives the operator much

protection in this area from power saw and axe cuts. This material has also been found to be effective in protection of legs of men trimming Christmas trees with machetes.

Buoyant vests should be worn when working in water. This lifesaving equipment should be used while in boats or on docks or other areas where one might fall into the water.

Tools

Chain saws should be operated and maintained in accordance with the manufacturer's operating instructions. Keep the chain sharp and maintain correct tension on the chain. Adjust the idling screw so the chain will stop when the motor idles. Keep fuel in an approved can. Let the motor cool before refueling. Refuel only in a cleared area. Don't smoke while refueling. Wipe spills off motor. Keep a fire extingquisher handy. Carry a chain saw with the motor stopped and the guide bar pointing to the rear. If carried on your shoulder, protect your neck with leather or heavy canvas. Never start the saw when it is on your knee. Investigate the use of a safety chain on your saw to reduce the chance of kickbacks.

Figure 44

(A) Right start -- with the saw on the ground, not the knee.

(B) Right carry -- with the saw stopped, guide bar to the rear.

The axe can be an extremely handy tool, but can also be dangerous if not properly maintained and used. Keep your axe sharp. A dull axe is an extra hazard. The head should be tight on the handle. Cracked or slivered handles should be replaced. Keep the wedge sheathed when carrying the axe to or from the job or when transporting it.

Consider the use of wood, soft metal, or plastic wedges rather than steel because they will not produce the flying steel particles such as will develop on a mushroomed head of a steel wedge.

Injector tools should be sheathed or otherwise covered while being transported and not in use.

Girdling tools should be used in accordance with the manufacturer's instructions. Safety glasses and hard hats should be worn while operating this equipment. Emergency shutoffs should be provided.

U.S. Bureau of Mines approved respirators should be used when handling chemcials requirng this protection.

Always comply with the chemical manufacturer's recommendations.

Travel

Always be alert when walking through wooded terrain and try to avoid areas where rocks could roll or dead trees could fall. If you must walk on down timber, watch out for loose bark, and, during peeling season, watch out for slippery bark in areas where peeled timber has been produced.

Try to plan your work so that you need not travel alone in wooded country. If you must, leave a written itinerary so that others will know where to look for you should you be injured or otherwise need help. Always carry with you a compass, knife, and matches in a waterproof container.

When traveling in your car, carry equipment to keep you warm in winter should the car break down. Also carry an axe, shovel, chains, and if possible, flares.

If you must travel on ice, always assume that it will not carry your weight until it is proven otherwise.

First Aid

Have a first-aid kit available for emergency treatment. If at all possible, try to get first-aid training so that in the event of an emergency, you will know how to act. Remember not to move an injured person until you have determined the extent of the injuries, unless it is

imperative that he be moved to a safer area.

Treat all injuries, no matter how slight. Get tetanus shots to protect yourself from this disease.

Chemicals

All chemicals should be handled in accordance with the manufacturer's instructions. Always read carefully the instructions on the label of the container and follow them to the letter. Remember that most herbicides, insecticides, and rodenticides are very toxic substances.

Tree Felling

No logging operation calls for more skill, know-how and caution than the felling process! Studies and experience prove that felling and bucking are a *major* cause of frequent and severe injury!

Space yourself well away from other fallers to protect yourself and them. Don't work up-slope or down-slope from other fallers. Fell trees down-slope; never directly up-hill. Don't work during a high wind.

Size up the tree before felling. Check for lean, loose limbs, loose bark, broken tops, rot. Note the wind direction. Check the close by area for loose limbs and snags. Don't expect a nearby tree or snag to fall in the direction of a tree felled into it; the top may break off and come whizzing right at you. Scan the ground where the tree will fall, for poles or debris that might be thrown back at you.

1. Check for widowmakers. Clear brush and debris from around the tree base. Select and clear and escape path at least 25 feet to the rear of the tree and at a 45 degree angle from the direction of fall. Plan an alternate route, just in case. When working with a partner, plan in advance which direction each of you will go. Don't use chain saw to cut brush, as it will cause it to whip.

Figure 45

Check top for widow-makers; clear out underbrush.

2. Prepare retreat path. Make the undercut facing the direction of fall. Use good, deep undercuts, penetrating to about 1/3 the diameter of the tree. Use deeper undercuts or cornered notches for heavy-lean trees. Height of undercut opening should be at least 2 1/2 inches for each foot diameter of the tree. Clean out the back of the undercut.

3. Notch for safe fall. Sound warning before starting chain saw for back cut. Make sure everyone is in the clear. Make back cut at least 2 inches above the undercut to prevent kickback. Never cut through to the undercut. Set the saw dogs (fingers) securely so they will not clip. Stand against the saw to keep it from kicking. Watch your footing.

Figure 46

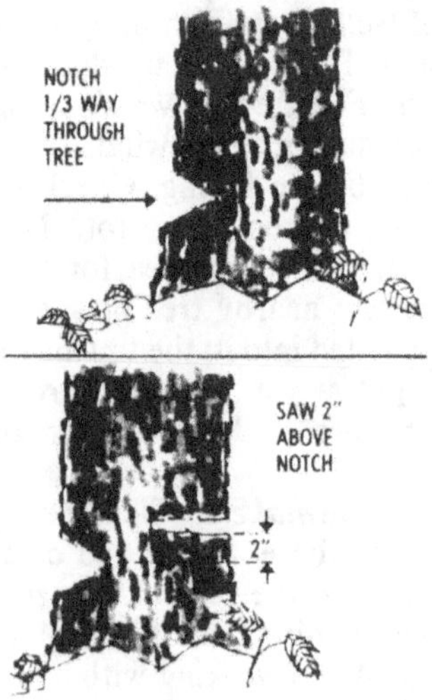

Notch deeply enough; saw above the notch, as detailed above.

Figure 48

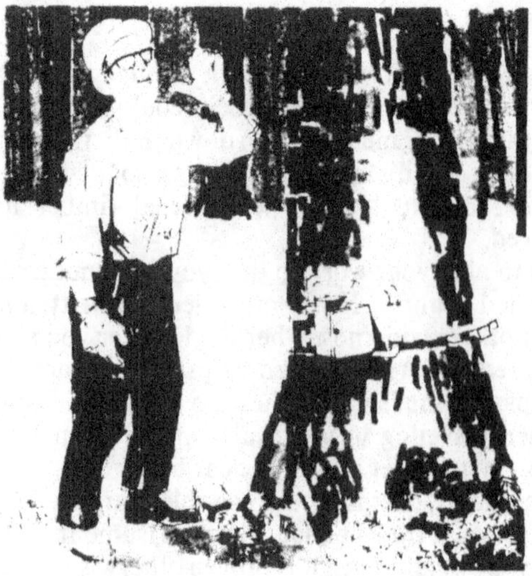

"Timber-r-r!" Give warning before starting felling cut . . . and again as tree starts to fall.

4. *Get in the clear.* Sound warning before completing back cut. When the cut starts to open, remove and shut off the saw, place it on the ground away from direction of fall. Retreat along escape path for at least 25 feet. Watch for flying limbs.

Trees don't always fall exactly as planned. If and when they are hung up, never cut down the tree supporting a hung tree, or otherwise try to saw it down. Never climb a lodged tree to shake it loose. Let the skidder pull it loose. NEVER leave a lodged tree.

Figure 47

Good notching makes for safe felling.

Figure 49

Get into the clear! Leave saw at base of tree. Retreat over cleared route at 45° angle from direction of fall, and at least 25 feet back from tree.

Figure 50

Hung tree -- and the only safe thing to do is pull it down with a tractor.

Figure 51

Limbing -- cut smaller branches with ax; stand on side opposite from where limbing is done.

Figure 52

Bucking -- work above the tree, in a cleared area with log in solid position.

Limbing and Bucking

Clear a chopping and sawing area around the felled tree. Make sure footing is good. Start limbing from butt end of tree. Keep on uphill side of tree. Stand with feet in the clear and limb from opposite side if possible. Use axe on smaller limbs. Watch the spring or jump of limbs in a bind. Cut spring poles from the underside. Cut any supporting limbs last and with extreme caution — as the log may roll. When cutting large limbs, be alert for the chain saw binding and kicking back.

Clear work area and be sure you have solid footing. If necessary, block the tree to prevent rolling. Work on the uphill side; if this is impossible, have tree moved. Start the undercut first. On large logs stand against saw to avoid kicking. Watch for bind or kickback. When walking on logs, watch for loose bark. Don't jump from a log; put your saw or axe down and slide off.

In bucking windfalls, watch for binds. Choose cuts so that bind is relieved most gradually, and work from the safe side. Visualize what may happen when a root wad is bucked off, and plan accordingly with an escape path cleared before starting. Watch out for rocks and dirt that may fall from the root wad. Watch for hidden roots.

Recent studies indicate that 24 per cent of the lost time injuries in the Wisconsin logging industry take place during the tree felling process. These same injuries account for 37 per cent of the industry's compensation costs.

Figure 53

Hard hats save lives . . . and eye protection saves sight.

Skidding

Every piece of equipment used in skidding operations can be a threat to the safety of the skidder; he must know everything about the equipment, including its limitations.

Know your signals and give them clearly. If chokers are hung up, let the operator pull them clear. Keep yourself out of reach in case they snap clear. Keep in the clear of logs that might roll, and watch out for unexpected movement of chain or choker cable. Place chokers near, but not less than 2 feet from, the log's end. Stay behind and to the side of a moving load; never ride it. Stay clear of cables under tension. Inspect cables frequently for broken strands; watch for jaggers. Wear gloves.

Know and obey signals. Use care in mounting a tractor; guard against slipping. Move tractor only when you know everyone is in the clear. Never pull a load around a sharp turn. Drop the load, make the turn, then winch the load in. Couple the load in close. Short hookups are less likely to catch on obstacles. Use caution in going up or down steep slopes, never go straight across. Don't trail the winch cable, wind it on the drum. Always operate the winch from the tractor seat. Don't raise logs when skidding upgrade, the tractor may upend. It is better to pull dead trees down with the cable, rather than push them down with the tractor. Allow no riders on the tractor or load.

Loading, Hauling and Unloading

Loader operators should be sure all crew members are clear before moving the loader. Build balanced loads. Make sure that logs are well saddled and stable before getting next log. Use special care on short peakers and crooked logs. Keep load built inside of stakes; never put logs on top of stakes. Never suspend or swing logs over crew members. If the driver or a crew member is alongside the truck, stop loading. Hold peakers in place until the first binder is in place and tightened. Lower hooks, tongs or grapple to ground when not in use.

Stay out of the space between loader and truck being loaded. Don't get between the truck cab and the log being loaded. Keep tag lines on end hooks from entagling your feet or body. Hold the end of the tag line, with slack under control. When setting hooks or tongs, keep hands from between them and the log. Keep clear of flying hooks if they cut out of the log. Stay out from between the swing of the loader and the log deck, or any other object.

After the truck is loaded, sound a warning before heaving binders over the load. Wear your hard hat. Don't back up unless you have someone to direct you. Test truck brakes before leaving the landing.

While hauling, be sure to know and obey the rules of the haul road and public highways. Watch for vehicles entering from side roads. Hold speed well within your ability to control the truck. Remember that some types of road surfaces — gravel, mud, snow — make for poor braking or stopping conditions. Check the condition of your load frequently. Check the truck regularly; especially brakes, lights, wipers, horn and emergency brakes. Remember — NO RIDERS!

Keep all binders on and tight until the unloading straps have been placed and tightened, or until a mechanical unloader is in position to prevent logs from rolling when the binders are removed. Make sure everyone is in the clear before releasing the binders, stakes or otherwise dumping the load. Use quick release binders with long trip line on unstable loads. Never go between the brow log and truck for any reason. Watch for logs swinging around or moving in the wrong direction when releasing stakes and binders. Have a plan of action in case a log rolls toward you.

Timber!

TIMBER HARVEST ON PRIVATE LANDS

Why Harvest?

Harvest For Income
Different species of trees mature over different time periods. For instance aspen matures in 40 years, while Norway pine matures at 100 years. As a stand of timber passes maturity, it starts to deteriorate. When the state of deterioration increases the stands value decreases. It makes good economic sense to harvest when the stands value is at its' peak - at maturity.

Harvest To Improve The Forest
Timber harvest is a forester's most important tool to increase the productivity of a forest. Harvesting not only yields a profit, but can be manipulated to increase the future productivity of the forest as well.

Harvest to Benefit Wildlife
Harvesting is often done in a manner that benefits wildlife. Snag trees may be reserved because they serve as homes for many species of birds and animals. Oak trees may be reserved for acorn production. Aspen harvest may be staggered over several years to create blocks of different age classes - benefiting grouse and other wildlife.

Do It Yourself Harvesting

Are you planning to do your own harvesting?
Some landowners decide to harvest their timber instead of selling it to a logger. They hire someone to cut the trees and skid the trees to a landing where they are loaded on a truck and hauled to a mill. Some landowners have their own logging equipment, but most don't. A landowner should be able to answer **all** the following questions **before** attempting to harvest:

What are the long term management goals for the forest?

What type of timber harvest will benefit my forest the most (clear cut, selective, etc.)?

What species and products should be harvested?

What are the exact costs of harvesting, skidding and hauling?

Is there a definite market for the wood?

What is the value of the wood?

If labor is hired, what is the landowner's responsibility for workmen's compensation?

What product specifications does the consumer-buyer have? Most mills have specific species, length, diameter and quality restrictions.

Many landowners that have attempted "do-it-yourself harvesting" have lost good sums of money and ruined their forests in the process. A word of advice - don't try it unless you know exactly what you are doing.

Obtaining A Forester's Assistance

Following are three reasons a landowner should locate a forester for timber harvest assistance:

Maximize Your Profits
A forester can advise on how to put timber up for bids. Many landowners are tempted to contact a logger down the road to harvest their timber. Sometimes this works out fine, but more often the landowner will end up with less money than he could have had if the timber were put out on bids. If a logger feels he is in competition with other loggers for the same patch of timber, he will normally bid a higher price than he would otherwise pay. There have been occasions when landowners have had an offer from a logger and then requested a forester's advice. With professional assistance, the same timber was advertised for bids and brought more than double the original offer!

Improve Your Woodland
Depending on how its done, harvesting can either harm or benefit a woodland. A forester can advise a landowner on the best techniques of harvest. Proper harvest practices will increase the future productivity of a forest. Improper harvesting may ruin it.

Harvesting techniques will vary depending on the composition of the forest, soil type, access, regeneration desires, local market conditions, desired wildlife benefits and other factors.

A forester is best qualified to provide a harvest plan that will consider all the variables that must be considered so that the harvest operation will be of most advantage to the landowner.

Minimize The Risk
Foresters are able to provide landowners with payment advice and contract forms that help to prevent problems as the harvest progresses.

There are several ways that a landowner may collect payment for his timber sale. Each payment plan must be tailored to match the situation. Contract forms should be used so that all people involved in the timber sale agreement are protected.

Locating A Forester
The easiest way to locate a forester is to: (A) contact your nearest DNR Forestry office; (B) contact your local County Agent; (C) check your phone book yellow pages for Consulting Foresters; (D) inquire at a local wood manufacturing company.

In many parts of our state, a landowner has a wide selection of foresters to choose from. Most landowners first choose to obtain a free forest management plan provided by their local DNR Forester. If the plan recommends a sizable harvest, the landowner often chooses to contact an industry or consulting forester to assist with the sale, since the DNR Forester is limited in the amount of service he can provide.

The Forester-Landowner Relationship
Most landowners know very little about forest management and harvest, so they put alot of faith and trust in the advice of their forester. It may be tempting for a landowner to let the forester "do it all". It is suggested that a landowner use the opportunity to broaden his or her knowledge of forestry in general and specifically the resource potential on the property in question. Ask questions! Some of the questions a landowner should have answered include:

What species and volumes of timber are on the property?

What products do they make?

What is the potential value of the forest? (Both short and long range.)

What are the management options? (And costs.)

What are the local market conditions?

What is ready to harvest?

How should it be harvested?

Costs
Most foresters charge a fee for timber sale assistance. Find out what fees are involved. The higher the value of timber on the property the more important it is for the landowner to get the proper management advice and assistance. A landowner with ten acres of aspen has less to lose than a landowner with ten acres of

Norway pine or black walnut. Many landowners have no idea of what they have on their property - hence the need for a forest management plan and answers to the questions above.

Setting Up The Sales

Boundaries

Once a landowner has determined what should be sold, it's time to mark off the sale boundaries. Boundary marking is not necessary if every tree to be harvested is individually marked - as long as the person doing the marking knows where the property boundaries are. In many cases, a forester may provide this service. As a matter of courtesy its good to contact any landowners adjoining the area intended for cutting to let them know what's going on. If there is any chance of a potential property boundary dispute, it is wise to have the neighbor involved in determination of the lines. Don't harvest until it is certain there will be no boundary disputes.

Bid Invitations

After the boundaries are marked, it's time to advertise the timber for bids - or is it? If market conditions are poor, it might be wise to hold off on sending out bid invitations until market conditions improve. This may help maximize profits. Regardless, the landowner should sit down with a forester and prepare a bid invitation.

There are many topics to address when setting up the bid invitation. Some important items include:

(A) How is payment to be made for the wood?

(B) Should the wood volume be measured? If so, how?

(C) How much time will the logger have to cut the wood?

(D) How is the wood to be cut and skidded?

(E) Where is the wood to be piled?

(F) Where are the access trails to be located?

(G) What trees are to be reserved from harvest?

(H) What must be done with the unused portions of harvested trees (slash)?

(I) Will there be periods when logging is prohibited?

(J) Will a performance bond be required of the logger?

A logger has the right to know any factors that will affect his rate of production **before** he bids on the wood. These factors must be spelled out in the bid invitation.

The bid must also include:

(A) Landowner's name and address.

(B) Legal description of the wood for sale plus a map.

(C) An estimate of species, volume and products.

(D) The time, date and place where the bids will be opened.

Payment

On some sales, a landowner will choose to require 100% payment of the appraised value of wood before harvesting starts. At other times, a landowner may require a down payment on the date of sale, with the balance to be paid before logging begins.

Another popular payment method is to require a $1/3$ down payment, with another $1/3$ to be paid before logging starts (which could be several months); the balance is required by the time the sale is $1/2$ cut.

The main idea is to collect the money faster than the wood is being cut - this will help prevent money collection problems once the sale has been cut. The logger must be made aware of the payment schedule before he bids.

Measurement

Trees must be measured after they are cut to determine their actual volume. This process is called scaling. The scaling process may be done on the property before the wood is moved, it may be done at the mill where the logger sells his wood, or at both of the above.

Some landowners prefer to sell their timber by the tract on a "lump sum" or "sold as appraised" basis. In this type of sale, the logger is asked to bid on a timber tract. The landowner collects payment based on the logger's bid, and no scaling is required to determine the logger's payment. This type of sale puts alot of pressure on the logger to know exactly what he is buying.

A logger must know the type of scaling procedure being employed before he bids, because this can affect his rate of production.

Timber Values

Foresters can supply landowners with information on the average market prices of various species of wood. The actual value of any tract of timber depends on many variable factors. All factors are related to the loggers costs of production and the price he can get for the products he will sell.

Landowners tend to underestimate a logger's overhead expenses - such as loading and hauling as seen here.

These factors include:

(A) Species. For example, pine is more valuable than aspen.

(B) Size. Big trees are normally worth more than small trees, if they can be sawn into lumber.

(C) Quality. Knots, rot, crookedness and other defects can all lower a tree's value.

(D) Total volume. A logger has expenses everytime he moves his equipment from one tract of timber to another. The longer he can keep his equipment on a single tract the more he can save on overhead costs. These savings can be passed on to a landowner in the form of a higher price for a larger volume sale.

(E) Distance to market. The distance from a market will affect a logger's hauling costs. The closer a tract is to the market, the lower the hauling costs.

(F) Ease of Removal. Is there good access to the timber for sale, or will logging trails have to be constructed? Is the topography level or hilly? These factors also affect logger's costs and value of stumpage.

(G) Market Conditions. Markets have a profound impact on the value of timber. Some markets fluctuate while others remain quite constant. It's best to sell timber on an up cycle of the market if possible.

It is also wise to include a clause saying: "The landowner reserves the right to reject any or all bids".

A Matter of Courtesy

It' a matter of courtesy to contact all bidders after the bids are opened. The bidders took the time to investigate a sale and to bid on it. The landowner should take the time to respond to all bids.

All bids should be opened at the same time, which is the time listed on the bid invitation.

High Bidder

The high bid is the winner? Usually, but not necessarily! Before accepting high bid for the timber being sold, a landowner should be sure to check out the high bidder. Is he honest? Will he pay for the wood? Will he follow instructions? Will he get the job done on schedule? Too many landowners make the mistake of accepting the bid first and checking out the logger second. Remember, you are in control.

The Contract

Is a timber harvest contract really necessary? Won't a handshake do? Let these examples help you decide.

Mrs. Doe sold timber to logger, Mr. Buck. Mr. Buck decides to sell tree tops and unused portions from his operation to Mr. Jones. Mr. Jones proceeded to enter on to Mrs. Doe's property to remove the wood he bought from Mr. Buck. Mrs. Doe did not like Mr. Jones and did not want him on her land. What could she do? Fortunately her timber sales contract stated, "This agreement shall not be assigned in whole or in part by either part hereto without consent of the other party". Mrs. Doe brought this to the attention of Mr. Buck and her problem was solved.

A logger, Mr. Cutter, hauled wood on Mr. Jones' property during the spring breakup causing several deep ruts. To put it simply, the road was a mess. How could Mr. Jones get Mr. Cutter to fix up the road? The landowner had leverage because part of the contract read: "The Purchaser agrees to repair immediately at his own expense, damage caused by him or his agents to roads, gates, fences, bridge or other improvment on the Seller's property".

How about the logger who harvests a landowner's wood but does not pay for it? One clause of the standard contract reads: "All timber included in this agreement shall remain the property of the Seller until paid for in full".

Loggers need protection too. There have been cases where timber has been put up for sale by people who had no right to sell it. Loggers must be certain that they have full right to the timber they purchase. A good contract will provide this assurance.

Administration

Once timber has been sold a landowner may have a tendency to feel the sale is over. It's not. Scaling may be necessary on a sale. Beyond that comes the important task of sale supervision. Is the logger living up to his contract? Is the slash being taken care of properly? Is the logger cutting everything he is supposed to cut, or is he creaming out only the best timber? Is the timber being utilized properly, or is good merchantable timber being cut and left to rot? Are reserved trees being protected from injury? A consulting forester is best able to inspect and enforce your sales contract.

Taxes

Landowners selling timber may be eligible for special tax breaks such as the capital gains treatment or the timber depletion allowance. Some consulting foresters specialize in assisting with these matters. Ignoring of these topics may cost a landowner hundreds - or even thousands of dollars.

Looking Ahead

The future of the forest should be considered before any timber is put up for sale. The type of harvest must be tailored to match the long term management goals of the forest.

Once a harvest has been completed, it's time to follow through with the management plans. In many cases a forest will regenerate itself with minimum intervention necessary by the landowner. This is especially true when forester's harvest recommendations have been properly followed by the logger.

On the other hand, there may be a need for timber stand improvement, slash utilization, site preparation and/or planting. Your DNR forester should be consulted on these matters. Inquire about any possible government financial assistance for forest improvement practices.

A properly organized and supervised timber harvest will benefit the forest, wildlife, the landowner, the logger and even the local economy.

Summary

Forest harvest and forest management go hand in hand. A landowner should take the time to do things right if he or she is concerned about the long-term productivity of the forest. Even the short-term financial return greatly depends on how a timber sale is handled. If in doubt, contact a professional forester for assistance. Be sure to address the topics mentioned in this article.

Timber Harvest Definitions

Board Foot - A piece of wood that is 1 ft. square and 1 inch thick. Sawtimber is usually measured in board feet. Sales price is based on the value per 1,000 board feet. 1,000 board feet may be expressed by the initials MBF.

Clear Cut - All wood in a designated area is cut regardless of size, species, etc.

Consulting Forester - A self-employed professional forester.

Consumer - The company or individual who purchases rough wood products with the intent of remanufacturing or reprocessing them into a usable form.

Consumer Scale - Wood is hauled off the land and is measured at the mill (where the logger sells his wood). The logger reports this mill scale to the landowner. Payment to the landowner is based on this scale at the mill.

Cord - A pile of wood that is 4 ft. wide, 4 ft. high, and 8 ft. long or any pile that is equal to 128 cubic feet.

Cruise - When a forester is making a field inspection to determine the amount of timber to be sold he is said to be making a "cruise". A field inventory of timber species, volume, products, quality and vigor.

Cull - (1) A log of merchantable size, but made unmerchantable by poor form, rot, or other defects; (2) The deduction made from the gross volume to adjust for defect; (3) To reject a log in scaling.

Cunit - A unit of measure in cube scaling equal to 100 cubic feet of wood.

Face Cord - Stove length wood with a "face" of 32 square feet. Actual volume depends on its length. About 1/3 of a standard cord.

Industry Forester - A professional forester working for a wood consuming industry.

Landing - A spot in the forest where wood is hauled, piled and loaded onto logging trucks. Wood is usually scaled at the landing before it is hauled away.

Lump Sum Sale - The logger bids one total price for all wood designated to be cut. No scaling is necessary. This is very similar to sold as appraised wood, except there may be no volume estimates given by the landowner. Wood is sold by the tract and not by a unit of measurement such as cords or board feet.

Peeler - A log from which veneer stock will be cut.

Piling - Round timbers to be driven into the ground to support other structures.

Pole - A round timber 16 feet in length or longer used to support power, telephone or other structures.

Post - A short timber up to 16 feet in length used in an upright position to support other structures for fencing.

Pulpwood - Measured in cords but sometimes weighed and converted to cords. Wood normally 96" to 100" in length. Small end diameter pulpwood 3", hardwood 4". This wood is normally too small or defective to be used for sawtimber.

Rick - One-third of a standard cord. 37 cubic feet unsplit, 40 cubic feet split.

Sawlog - Wood used for sawtimber measured in board feet. Minimum length 8'. Minimum top diameter: softwoods 6"; hardwoods 8".

Scaling - To determine timber volume in cords or board feet by measurement (usually with a scaling stick). Payment for the wood is usually based on this measurement.

Selective Cut - Only part of the timber is cut from a designated area. The wood to be cut is marked with paint or ribbon or is designated by a size diameter limit.

Slab - The exterior portion of a log removed during the sawing process.

Slash - The debris that is usually left after a logging operation. This debris consists mostly of tree tops and limbs but may also consist of cull trees that are cut and left.

Sold As Apppraised - Wood is sold "on the stump" and the sales price is based on the appraised volume determined by the forester. This volume is only an estimate. The actual volume cut will most likely be different. But the sale price is based on the estimate and not the actual volume cut.

Stumpage Price - The price a logger is willing to pay for wood as it is in the woodland or "on the stump".

Trim Allowance - Extra wood material allowed over and above the standard length of the log in order to assure lumber of standard lengths.

Veneer - High quality, straight, defect free logs or bolts used for veneer. Minimum diameter is 12", minimum length 8'6". Measure in board feet. Highest valued of all lumber.

Normal Wood Product Dimensions

Product	Length	Small End Diameter	Unit of Measurement
Veneer	8'6" +	12"	bd. ft.
Saw Log (hardwood)	8' +	8"	bd. ft.
Saw Log (softwood)	8' +	6"	bd. ft.
Pulpwood (hardwood)	100"	4"	cords
Pulpwood (softwood)	100"	3"	cords
Posts	5' to 7'	3" to 8"	cords
Poles	12' to 45'	5" to 7"	cords
Pilings	15' to 60'	8" +	cords
Bolts	8'6"	6" +	cords

This beautiful spruce plantation is the result of follow-up on a timber harvest that took place about ten years previously.

TIMBER HARVEST AESTHETICS

by

Roger Hoeksema and William Hoppe

A growing concern of many landowners is the condition in which their woodlot is left after a timber harvest. With some planning, the visual character of a woodlot can be protected. Three different problems associated with timber harvests should be considered: the tree tops, trails and skid roads, and landings.

Tree tops remaining after a timber harvest can give a "bombed-out" appearance to a woodlot for a few years. This problem can be remedied by either utilizing the tops for firewood or by "lopping".

With today's demand for firewood, if you don't want the tops, chances are somebody else does. In fact, additional income can be obtained by selling the tree tops. If you don't know of anyone who burns wood, a simple newspaper ad will usually bring "firewooders" pounding on your door. To get the highest price for your tops, sell them using competitive bids.

If you want the tops for yourself but have more wood than you can use, consider a "share" agreement. Let someone else have a share of the wood for free but require that a specified share be stacked by your house for your use.

In the unlikely event that you don't want the tops for firewood and can't find anyone who does, the appearance of a woodlot following a timber sale can be improved by lopping. Lopping is simply making a few cuts to get all parts of the tops to within a given distance of the ground, usually four feet. This allows you to see over the tops easily rather than looking into a wall of leaves and branches. Cutting the trees when the leaves are already off will also improve the visual quality of a woods. The leaves remain on tops much longer than they do on standing trees.

Where the woodlot adjoins agricultural fields or other areas, there can be a problem of tree tops being left out in the fields. This problem can be remedied by cutting up the tops for firewood, by hauling them back into the woods, or by leaving the edge trees uncut. The latter method is probably the best. Edge trees are often limby, containing little timber volume. However, they are usually good wildlife trees because they have more homes and food for wildlife than trees further in the woods. They also provide excellent wind protection for the woods and are effective as screens for wildlife.

The condition of trails and skid roads is another concern which can affect the visual appearance of a woods. When timber is harvested in wet weather, the trails can be rutted. One way to prevent this from happening is to give the logger ample time to

cut the trees. Otherwise, he may be forced to work in the woods when he would prefer not to and should not. In addition, be sure your timber sale contract spells out when cutting can be done in sensitive areas and how trails and skid roads are to be left. If ruts are made, the logger should be required to level them off. Small trees which were tipped over and bulldozed during road building should not be left leaning against other trees, nor should their roots be left sticking up. The logger should utilize these trees or lop them and tip the roots back flat to the ground. To further improve and protect woods trails after logging, seed them with one of the following mixtures:

Mixture	Rate per acre
a. Kentucky 31 Fescue	40 lbs.
Sericea Lespedeza	20 lbs.
b. Kentucky 31 Fescue	60 lbs.
Weeping Love Grass	2 lbs.
c. Crownvetch	15 lbs.
Red Top	3 lbs.

Landings, where logs are concentrated for cutting and loading, can be another problem. Logging crews often leave trash and logging debris on the site. Trash such as oil cans, pop and other bottles, broken cables, old tires and other equipment should be either buried or hauled to a dump. Cleanup of these areas should be another requirement in your timber sale contract.

Logging debris, such as cull logs, tops, butt ends and, in the case of a portable mill, slabs, edgings, and sawdust should be piled back into the woods, buried or removed. If the landowner intends to use this material for firewood, then it should be piled where he can get at it. The landing should then be smoothed and possibly seeded, depending upon the intended future use of the area.

Not all logging jobs have these problems but they should be taken into consideration. Some of the things can be handled by the landowner, others by the logging crew. Those that are to be handled by the logger should be part of the conditions of the sale and should be spelled out in the timber contract because they do cost time and, therefore, money.

HOW MUCH IS WOOD WORTH?

Allen Wickman

Landowners in the process of marketing their timber want to know its approximate worth. Often they call a local forester and ask for a quote on recent stumpage (price of wood on the stump) prices. Actual prices often vary from average timber prices because of the factors listed on this page.

Location

It's expensive for a logger to move his equipment. The distance he has to move will have a bearing on how much he is willing to pay for the wood. He will most likely be able to bid higher on wood if a short move (such as 1 mile) is involved than if a long move (such as 50 miles) is required.

The location of the wood in relation to good all weather roads is also important. The closer it is to good roads the more it should bring on the market.

Timber for sale located way out in the "boondocks" is going to be worth less than if it were located closer to good roads.

Size of Sale

Most loggers don't seem to be interested in small timber sales of 50 cords or less. That's understandable. Once they start cutting on an area they like to stay there a while. They earn their money cutting wood and frequent moving cuts into their production time. Many loggers could cut a 50 cord sale in 1-3 days. That's not much time to spend on a job.

There are exceptions. If a 50 cord sale is located close to other wood to be cut by the same logger, the landowner will have a much better chance of making a sale.

We once tried a group sale. Three landowners combined their wood into a single sale. All wood was to be sold to one logger. There was very little interest by the loggers because the sales were located several miles apart. If all sales were within a one mile radius, the interest of local loggers would have been much greater.

Trail Building

Loggers often construct trails to give them better access to the wood they will be harvesting. Since trail building is often time consuming and expensive, it is certainly going to have a bearing on the price they will pay for the wood. Wood that requires little or no trail building will be worth more than wood that will require extensive trail construction—everything else being equal, of course.

The loggers' willingness to build trails is also dependent on the quantity and value of the timber to be harvested. A logger may be willing to put a one mile trail into a high value Norway pine sale. He may not be interested in putting even a ¼ mile trail if it were low value aspen being sold instead of Norway pine.

Volume Per Acre

We once tried to sell 1,000 cords of jack pine. Loggers were not interested. Why? The wood was scattered over 300 acres. This averages out to 3⅓ cords per acre. From a logger's standpoint that is a pretty low volume per acre. Unless the wood for sale is of high value (such as sawtimber), loggers are usually not inclined to be interested unless a minimum of 10 cords per acre are to be harvested.

As timber density is reduced so is the logger's ability to harvest a given quantity of timber within a specified period of time. The amount of timber a logger is able to produce has a big effect on his profits (or losses) so timber density is very important to a logger. 15 to 20 cords of harvestable material per acre is considered good and anything over that is excellent.

Market Conditions

The market demand for many types of timber (especially pulpwood) seems to fluctuate. When demand is low stumpage prices often decrease. As market demand increases so do prices. Any landowner selling timber would do well to check local market conditions before deciding when to put his timber up for sale.

When market conditions are poor it may be difficult to sell timber at all. I am sure many landowners have put a lot of time and effort into selling wood only to have failed because the market improved to the point where the wood could sell, the landowner was too discouraged to attempt to sell it.

Distance From Market

Of course the distance from local mills will affect the price of wood. Wood located more than 100 miles from the nearest mill may not even be merchantable. The same wood within 10 miles of a mill (that can use it) should bring a prime price.

Summary

As you can see, several factors affect the value of wood. This is only a partial listing of some of the major ones. Attempting to consider all these factors may confuse some people. In this case, my recommendation is to ask several loggers to give estimates or bid on the wood you would like to sell. Foresters may also be able to provide a ball park figure on how much your wood is worth.

CHAPTER 7
Timber Stand Improvement [T.S.I.]

Forestry Manual For Vo. Ag. Instructors
Wisconsin Department of Public Instruction

Many unmanaged forests could be improved if certain treatments known as forest improvement were applied to the forest during development. This involves the application of treating individual trees through cultural practices. These practices are intended to make the forest better by increasing the quality of trees and at the same time speeding up the growth rate. These cultural practices are: (1) noncommercial thinnings; (2) commercial thinnings; (3) prunings; and (4) control of undesirable species.

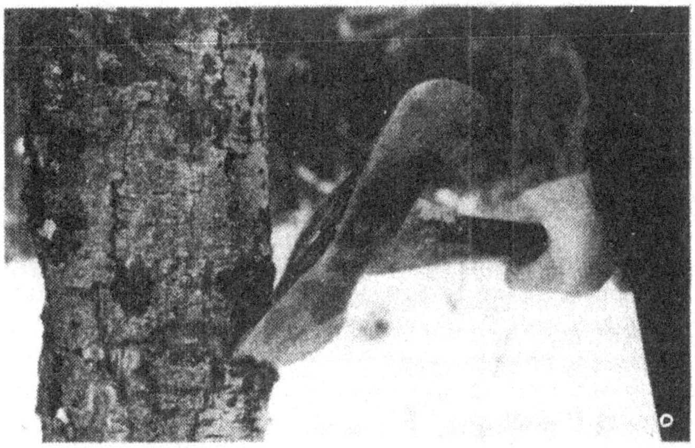

Figure 1. Killing the tree by girdling.

NONCOMMERCIAL THINNINGS

Thinning means cutting some of the trees in the forest stand. Noncommercial thinnings are those which improve the quality of the forest stand but are not intended to return an immediate profit. None of the trees felled is removed and used. Noncommercial thinnings include weeding, cleaning, liberation, and improvement cuttings (figure 1).

Weeding is the removal of all plants competing with the timber species. Rarely is there a market for any of the weed species of trees removed for they are too small. Usually they are cut and left where they fall. This very intensive practice is hardly ever used in forests, but is frequently practiced in forest nurseries where both woody and herbaceous weeds are removed.

A similar method, cleaning, is a treatment applied in stands of trees up to the sapling size (less than 4 inches in diameter). It is intended to release desirable trees from the competition of undesirable trees of about the same age. These trees overtop the desirable species or they are likely to. None of the undesirable species removed is used. This practice and liberation cutting are referred to as timber stand improvement (T.S.I.).

The removal of trees which overtop seedlings or saplings is often referred to as a liberation or release cutting. This treatment is made to release young stands, not past the sapling stage and usually between 1 and 20 years old. The older and overtopping trees may be a desirable species but poor form (called wolf trees), or species less desirable than the young stand. If the older or overtopped trees are marketable, they can be removed by a commercial sale. But often there is no market for such trees so they are killed or disposed of cheaply without endangering the stand's desirable trees. Methods of killing or disposing of the trees include girdling, felling, and poisoning.

The final noncommercial thinning operation, an improvement cutting, is a thinning made in a stand past the sapling stage (more than 4 inches in diameter at breast height—D.B.H.) to improve composition and character by removing trees of less desirable species, form, and condition (figure 2).

The undesirable trees may be removed by making a commercial timber sale if it is marketable size. However, if the trees are culls and cannot be sold, it is necessary to kill them and leave them in place.

COMMERCIAL THINNINGS

The primary purpose of commercial thinning is to improve the quality of the trees left standing. Secondarily, it is to return a profit from the removal of trees which are interfering with the quality of growth of the

Figure 2.

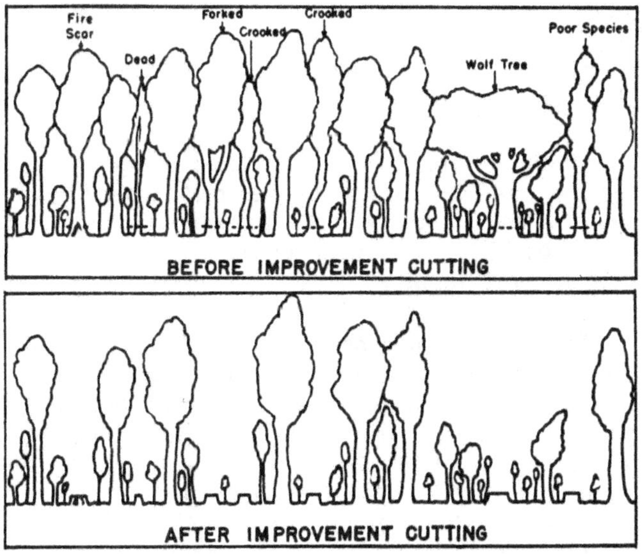

remaining trees and removal of trees which would not live until the next thinning or final harvest.

Commercial thinnings are most often applied to even-aged stands (all the trees are nearly the same age) such as plantations (figure 3). These stands may contain more trees than the land is capable of sustaining. When the trees become so crowded that the tops are competing for available sunlight, enough trees should be removed periodically to maintain normal growth of trees to be kept through to the final harvest. Crowding is evident when the live crown of the average tree in the stand becomes one-third or less of the total height of the tree. Another indication of crowding is when the smaller, individual trees begin to die.

The result of overcrowding in any forest stand is a keen competition for the available water, light, and the soil nutrients. This competition results in a forest stand in which there are suppressed or deformed trees; low value species crowding out valuable species; and slow growing trees. Figure 4 shows the effect of thinning on tree growth.

Some crowding is desirable in young stands so that the lower branches will be shaded out causing them to die. This develops long, clear stems that will later yield knot-free lumber. The thinning operation should be conducted when the trees to be removed have a marketable value for such products as fuelwood, fence posts, or pulpwood.

A simple rule-of-thumb that most foresters follow in spacing the thinned trees for maximum timber production is to add 6 to the d.b.h. in inches. The resulting figure in feet is a good spacing for trees of that diameter. For example, the recommended spacing for trees averaging 6 inches in diameter at breast height would be 6 plus 6, or 12. This means an average spacing of 12 feet between trees so that a well-stocked stand of 6-inch d.b.h. trees should have about 300 trees per acre.

Thinnings are usually conducted every 5 to 10 years or until the crowns of the trees come together again. If the stand has been properly thinned, there should be 150 to 200 trees per acre at the time of the final harvest (approximately 120 years for Norway pine).

These simple rules are not a substitute for common sense in the field. A clump of eight or ten good trees, such as a basswood clump, with sufficient room on the outside but crowded on the inside, might be thinned to two or three of the best trees; or cutting of perhaps only three trees might give the rest enough room to grow. Each tree should be assessed for its chances of growing into profitable timber.

Figure 3.

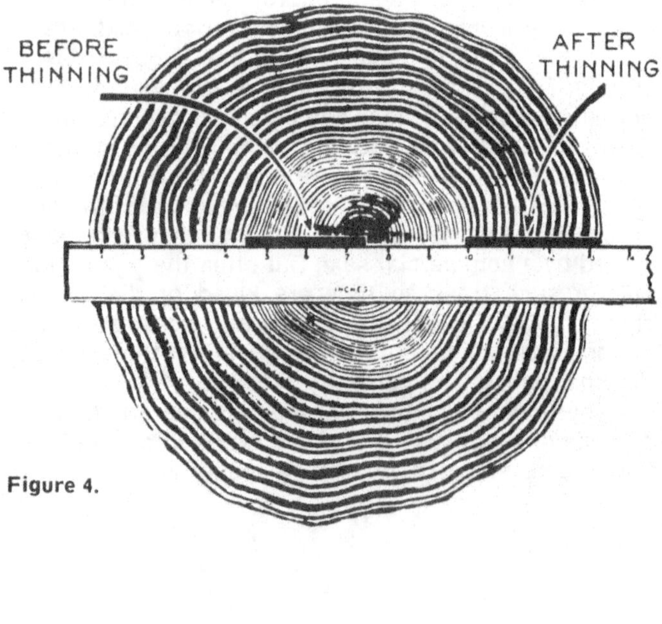

Figure 4.

If some of the trees to be removed in a commercial thinning have no marketable value the forester may recommend that they be killed and left. Three common methods are used to kill trees for this purpose. The first is to simply kill the trees by cutting with a power saw or other cutting tool. The second is to girdle the trees which will be removed at a later date. This involves cutting completely through the sapwood around the trunk of the tree with a hatchet or an ax. Once the cambium is severed, the tree will die. The third method is to poison the trees. Poisons are either injected with a special tool or they are applied in frills (cuts made through the bark and cambium) using a hatchet at intervals every 6 inches around the trunk. Lethal dosages of herbicides such as Ammate or 2,4-D are used to kill undesirable trees.

PRUNING

The removal of live or dead branches from the main stem of a standing tree is known as pruning (figure 5). **Forest owners can afford to prune the most promising trees in some open-growth young stands. Pruning increases the yield of knot-free, high-grade lumber that can be obtained from any one tree, particularly in the butt log (first 16 foot log) of selected crop trees.**

Figure 5.

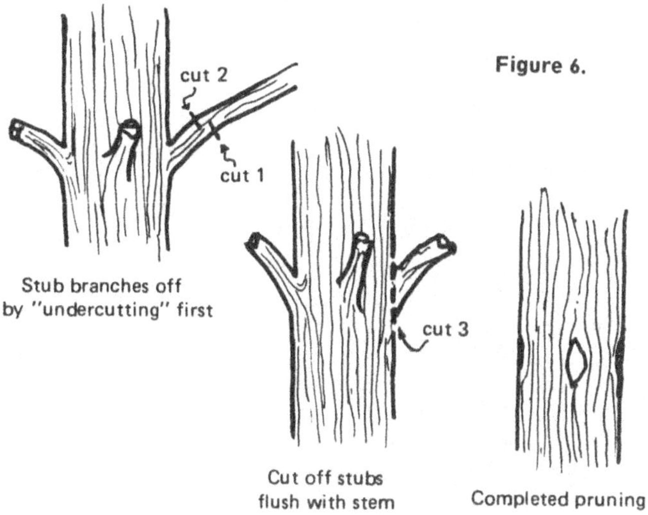

Figure 6.

If you look closely at a forest stand you will notice that trees around the perimeter of the forest have branches almost all the way to the ground. When you enter the interior of the stand you notice less branching on the lower portions of the trees. This is nature's way of pruning trees. The lower branches die because they are shaded out by the upper branches. However, if there is no natural pruning taking place in a forested stand, foresters will sometimes use a pruning saw.

Only about 125 to 150 trees per acre should be pruned. These should be trees that are sound and well-shaped and can be left for crop trees at the final harvest. The first pruning in any stand should be done when the trees are young and 3 or 4 inches in diameter. At this age the wounds heal rapidly leaving only small knots. The best time to prune is just before the growing season begins and before the buds begin to swell. All pruning cuts should be as close to the main stem **as possible without injuring the bark (figure 6).**

Individual trees should be pruned to leave the upper 30 to 40 percent of the tree in the live crown. Second prunings are usually needed in later years to get a knot-tree, 16-foot bole or butt log. This operation should be conducted after the trees reach 8 inches in diameter or when the limbs to be removed exceed $1\frac{1}{2}$ inches in diameter. The trees in the forest stand should remain growing for at least 25-50 years to obtain a profitable return from pruning.

Management Practices Which Influence Growth Rates

Forestry Manual For Vo. Ag. Instructors
Wisconsin Department of Public Instruction

Intermediate Cuttings

Intermediate cuttings are prescribed management operations in an immature forest aimed at the improvement of that forest. A farmer increases the yield of his corn crop by cultivation; the yield of a forest can be increased by judicious intermediate cuttings.

It must be emphasized that immature forests vary in size from small samplings to sawtimber and are in contrast to mature forests which are ready for a final harvest as we have discussed.

Intermediate cuttings are classified into six practices (1) Cleaning, (2) improvement cuttings, (3) liberation cutting, (4) releasing, thinning and pruning, (5) salvage and sanitation cuttings and (6) timber stand improvement.

Improvement cutting

An improvement cutting is a cutting made in a stand past the sapling stage to remove the poor form, defective, and inferior species which are competing for growing space. This cutting is similar to cleaning except that it is carried on in pole and larger size timber.

Improvement cutting is the most common intermediate cutting practice in Wisconsin forests today. Millions of acres of forest land, particularly the hardwood pole stands, can use an improvement cutting. Where this work is done, the residual forest is composed of fast growing, higher quality, higher valued species of good potential.

The products so removed may be saleable depending on distance to a market. Many improvement operations will pay for themselves, whereas in some areas an investment in labor is required.

Liberation cutting

A liberation cut is the removal of large spreading "wolf" trees and low valued overstory species which are hindering the development and growth of the young stand beneath. The term "release" cutting is synonymous and probably more descriptive.

Wolf trees are large spreading trees of any species occupying excessive crown space. These trees are limbly and of poor merchantable height. Removal of these trees is recommended because the space occupied can be devoted to growing a great deal more usable wood for the owner.

Liberation cuttings are common to release a high-valued conifer from under a low-valued species such as scrub oak or poor quality aspen. Overtopped plantations or natural conifers are slow growing and are liberated to accelerate their growth rate.

The release may be partial or complete depending upon the conifer involved. The complete removal of the overtopping trees in one cut may result in sunscald, in which case the overstory should be removed in two or more cuts. The complete release of sapling or smaller white pine is not recommended since open grown white pine is an invitation to insect infestation by the white pine weevil.

Plantation Management

Release, Thinning and Pruning

Release, thinning and pruning are a major part of good management practices. Production of quality useable timber in the woodlot is dependent to a great extent on the spacing of the trees in the stand with close spacing being the key factor in obtaining high volumes.

Close Spacing

One of the reasons for close spacing in the establishment of a plantation is to crowd out or suppress grass, weeds, and brush so that the trees may have the sole use of soil mositure and nutrients. Another important reason is to develop trees with clean stems and good form.

Where trees are grown at an excessive spacing interval, they are inclined to become heavily branched with live lower branches persisting for many years. The height growth, too, may be less than that of trees spaced closer. In the end then, one is likely to have trees with heavily branched stems and of poor form. With closer spacing the branching is finer, the lower limbs die sooner, and the height growth may be somewhat better. What one is actually after is a tree whose stem as nearly as possible resembles a cylinder. A tall slender stem with a form close to that of a cylinder will have much more volume than a short stem that more nearly resembles the shape of a cone.

Pine Plantation Management

Release, Thin, Prune

Forest plantations usually need one or all of the following three practices releasing, thinning and pruning.

1. Release—Removing or killing undesirable trees that are over-topping and competing with planted trees. Pine overtopped by undesirable hardwoods, for example, may be only four or five feet tall 25 years after planting. Free growing trees of the same age may exceed 35 feet.

2. Release planted red and jack pine from lower quality trees such as scrub oak, aspen, black locust and box elder that are over-topping the plantation wholly or in part. This practice is not recommended for white pine. This release may take the form of cutting, girdling, poisoning, or ground or aerial spraying with suitable herbicides. Ask your forester about the cheapest and most effective method to use.

3. Thinning—A cutting in a plantation which produces saleable wood products and increases the growth rate of remaining trees. Trees must be thinned when they become too crowed. As trees grow, they require more space. Thinning is generally applied to conifers. By proper thinning a landowner:

a. gains an early return on his investment.
b. speeds the growth of remaining trees.
c. reduces insect and disease outbreaks.
d. improves the health and beauty of the plantation.
e. improves wildlife habitat.

Pine forests, you may recall, are generally evenaged and require increasing crown space as the forest grows larger. Thinning is merely anticipating nature's method

COMPARISON OF THE FORMS OF OPEN-GROWN AND CLOSELY-GROWN TREE TRUNKS

of "shading out" some of the trees which fall behind, and to cut these trees before they prohibit crown expansion within the forest. Periodic thinnings, not to exceed 10 years, maintain a high growth rate on the residual timber and increase production through the harvest of what would normally be lost in natural mortality.

A thinned stand of pine maintains fewer trees of larger size than an unthinned stand of the same age. The larger dominant trees are left as "crop trees" in contrast to the removal of the large mature trees in the selection harvesting of mature hardwood.

Why are the larger pine left in thinning whereas the larger mature hardwoods are cut? The answer to this question depends upon its maturity. Thinning is carried on only in immature stands and favors the fastest growing, most productive timber. Selective cutting applies only to mature all-aged forests.

Technical advice should be obtained to determine the volume of wood to be removed in a thinning.

Thinning may or may not result in merchantable products depending upon the size of the trees and the available markets.

Thin all species of pine trees when they become too crowded. Table 1 shows the recommended number of trees per acre at various average diameters.

Table 1

Average diameter breast height	Trees per acre	Approximate Spacing Between Trees (Feet)
4″	1,000	6 to 7
6″	450	10
8″	200	15
10″	180	16
12″	130	18
14″	95	22

Determine average diameters by sampling 25 to 50 trees at random in the plantation. The number of trees per acre can be determined by counting the number of trees on a typical acre. An acre measures 208 feet square.

Selecting trees to be removed can be a complicated task. In even-ages plantations of pure red pine, row thinnings are sometimes prescribed by foresters for the first thinning if the arithmetic works out properly. For example, a 6-inch average diameter plantation may have 750 trees per acre. From the table we find that only 450 trees should be occupying the acre. Hence, a thinning removing every third row will benefit all trees that remain and reduce the number to 450 per acre.

Thinning, however is required in order to maintain or increase the growth rate. One can recognize a plantation in need of thinning by the following:

1. A possible slowdown in height growth.
2. Dead limbs on more than one-half of the stem.
3. A decrease in diameter growth as shown by increment borings.
4. Crowns of the taller trees interlaced and competing with each other for light.

Thinning objectives and results

When the above indications of a slowdown in growth rate are recognized, a thinning should be planned. The objective of the thinning is to remove a part of the trees and give the remaining trees a chance to resume normal growth. Otherwise a life and death struggle will exist between all trees and may result in stagnation of the stand.

If a tree is to put on the maximum of additional wood, the crown must have room for development. In any thinning operation the better trees should be left for future crops and a space of several feet opened up around their crowns by removing less desirable growing stock.

The first thinning, depending upon the initial spacing, may or may not pay for itself in usable or marketable wood products. Products from the first thinnings can be used for posts and pulpwood. Later thinnings at 5- to 8-year intervals will yield additional pulpwood, poles and eventually small sawlogs.

The diameter growth response to a thinning may not be immediately apparent at breast height; however, additional and increased diameter growth is taking place in the upper portion of the stems. It simply means that the trees are attempting to correct their form and develop stems that will closely resemble cylinders. When the better form has been attained, additional diameter growth will appear evident in the lower portion of the stems.

Thinnings principally remove trees which in the unthinned stand eventually die in the struggle for existence and may decay before the crop is harvested. Thus, thinnings save what otherwise may be a total loss. In addition, an actual increase in the volume growth will follow because of the more vigorous condition and better crown development of the individual trees in a properly thinned stand. Improved soil conditions and a greater amount of available nourishment also contribute to increased production. The systematic use of thinnings may increase the total amount of usable wood produced by 50 percent or more over that produced in an unthinned stand.

CHAIN SAW GIRDLING — HOW GOOD?

An evaluation of chain saw girdling to treat unwanted northern hardwoods was made by Jack B. Cody and Kenneth F. Burns of the Applied Forestry Research Institute, State University of New York, Syracuse, N.Y. After four full growing seasons following the treatment, they found that with treatments done in January, February, and March on trees 4" to 10" in DBH:

1.Properly made a single chain saw girdle without chemicals is as effective as other treatments. In order to be effective the girdles must completely sever the cambium and the ends meet. Overlapping the end cuts is not sufficient.

2. After four growing seasons all treatments approached 90 percent or more trees killed.

3. Use of the chain saw girdle will save both labor and chemical cost and is by far the cheapest treatment tried. Such girdling also avoids environmental concerns about the use of chemicals.

ILLUSTRATIONS OF TREE CROSS SECTIONS AND INCREMENT BORER
Cores Showing Changes in Rate of Growth of Plantation Trees.

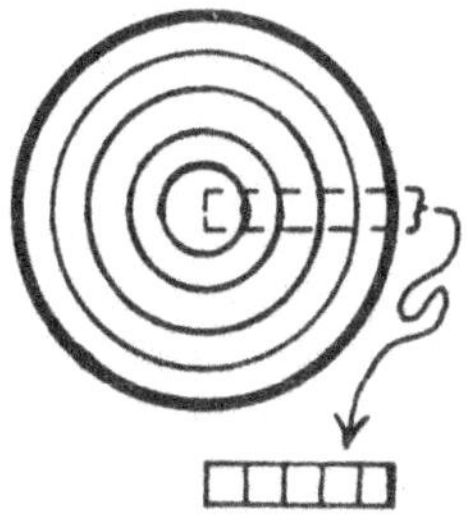

1. Normal growing plantation tree.

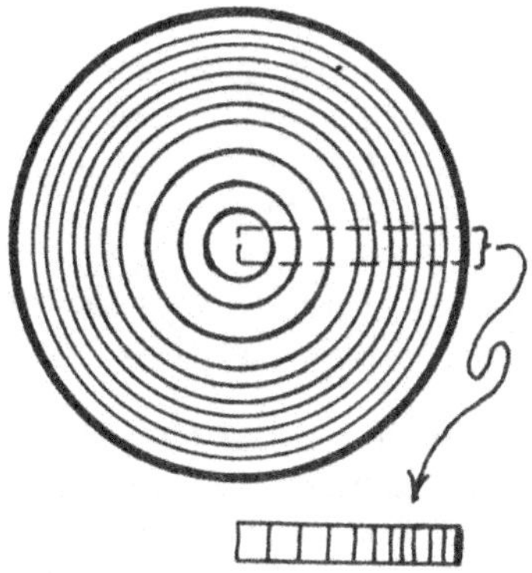

2. Same tree several years later in need of more growing space. Note slowdown in diameter growth.

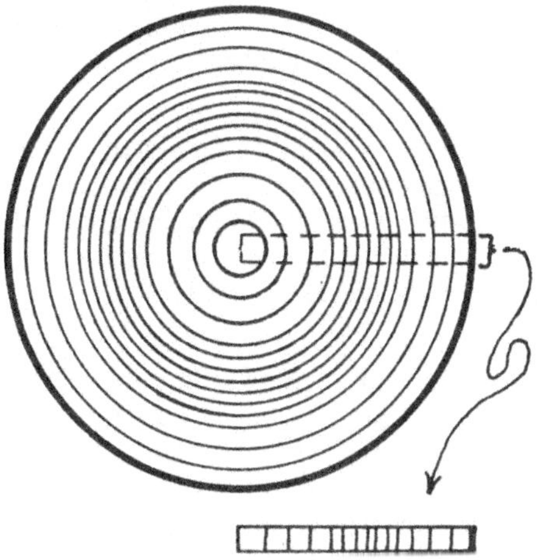

3. Same tree several years after a thinning. Note an increase in diameter growth.

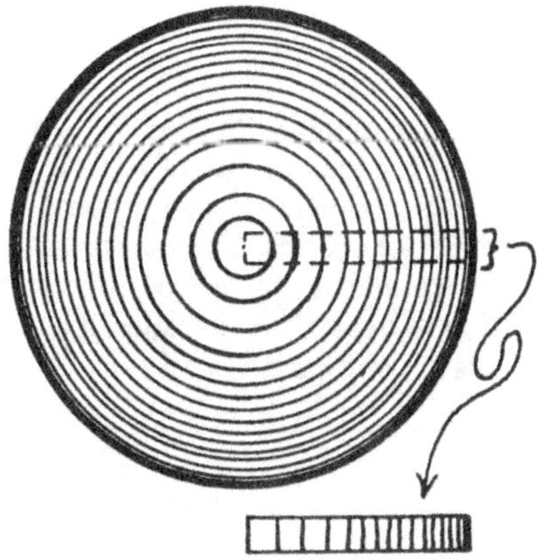

4. Tree of the same age in an unthinned plantation. Note exceptionally slow growth - plantation should have been thinned several years ago.

A small handsaw or pole mounted saw is used to remove the branches as close to the trunk of the tree as possible to hasten healing of the wound. Using an axe to remove the branches, or leaving a stub projected from the trunk, delays healing and may result in an infection of wood rotting fungi. Portable (power pack) saws are also being used extensively for pruning.

It is best to prune in fall or winter using a saw or pruning shears—never an axe. Prune as close to the trunk as possible; never leave a stub.

Red, White, and Jack Pine

Don't prune trees on the outside edge of a plantation if there are open areas adjacent to the plantation. Such a pruning allows undesirable drying winds to blow through the trees and over the forest floor. If there is danger that a surface fire might spread into low branches in the outer row, it may be desirable to prune lower branches up to about 4 feet.

Scotch Pine

Scotch pine has a large number of insect enemies, especially on sandy soils in Wisconsin. The tree rarely has value for any product other than Christmas trees because of insect damage.

Except under very unusual conditions, Scotch pine should be harvested as soon as possible. Clear-cutting is recommended. Release, thinnings and pruning are not recommended.

After the area is cleared it should be allowed to "lie fallow" for two years before replanting to reduce populations of insect enemies. Volunteer seedlings of Scotch pine should be eliminated and the area planted to a better species.

Mixed Pine Plantations

Many of the plantations in Wisconsin and lake states are not pure stands of one species. There are often mixtures of two or more species of pine and other conifers such as spruce, cedar and balsam.

Detailed recommendations for each possible mixture would require more space than is available. Mixtures frequently require special techniques that can best be prescribed by a professional forester.

Some rules of thumb on mixtures, however, follow:

1. Scotch pine with poor form should be removed.
2. Spruce grown in mixture with pine will probably be overtopped by the more rapid juvenile growth of the pine. Such plantations will require specialized thinning procedures to maintain spruce in the stand.

In individual tree selection thinnings, trees to be thinned should be marked before they are cut. Remove slow-growing and poorly formed trees before removing vigorous, high-quality trees. The temptation is usually just the reverse-taking the best trees first. It should be resisted because such a practice seriously damages the ability of the stand to remain healthy and productive.

Thinnings should be spaced at about 10-year intervals. Professional foresters use a measure known as "basal area" when determining the amount of thinning required. Their system of measurement is more accurate than the average diameter and number of trees per acre method described here. Seek out the advice of a forester before engaging in a large-scale thinning operation. He will be able to save you time and money . . . perhaps even your plantation.

Prune about 150 red and white pine trees per acre (17′ x 17″ spacing) which you have selected as final crop trees. These trees will be harvested for lumber and other high-quality products when the stand approaches maturity. The pruned trees will not be cut until they are at least 12 to 16 inches in diameter. Pruning is not recommended for jack pine.

The chosen crop trees should have the lower limbs pruned away so that knot-free lumber, poles, and piling will result. Pruning should proceed to at least 17 feet above the ground in one or more stages. Trees should not be pruned over one-half their total height at any one time, but may be pruned the full 17 feet by the time the trees are 34 feet tall.

The earlier the pruning, the better. If a tree is over 8 inches in diameter, pruning will have very little effect on ultimate value of the tree.

Only the straighter and more vigorous trees should be chosen for pruning as final crop trees.

Some landowners prefer to prune all the trees to at least 8 or 9 feet above the ground. Since many of these trees will be thinned for pulpwood, such pruning is not absolutely necessary because knots are not degrading in pulpwood.

But pruning all the trees does have other benefits:

1. Reduces fire hazard.
2. Simplifies the work of pulpwood cutters, in that they do not need to limb as many pulp sticks.
3. Makes the plantation more accessible for marking trees that should be thinned, and for locating crop trees.
4. Opens area for recreational use.
5. Salvage cutting

Salvage cutting is a cutting to utilize injured or dead timber damaged by fire, insect, disease or wind. All intermediate cuttings in immature stands are aimed at increasing the quality and quantity of growth with the exception of salvage cutting.

A salvage, as the name implies, is merely the utilization of timber to prevent financial loss and buildup of insect and disease problems.

Cleaning or Weeding

A cleaning cut is a cutting made in the sapling stage (under four-inch diameter) for the removal of poor form, defective, and inferior species which are overtopping or threatening to overtop superior species. The purpose of a cleaning is to improve the character and composition of a sapling stand. This cutting represents an investment since no merchantable products are obtained.

Figure 34

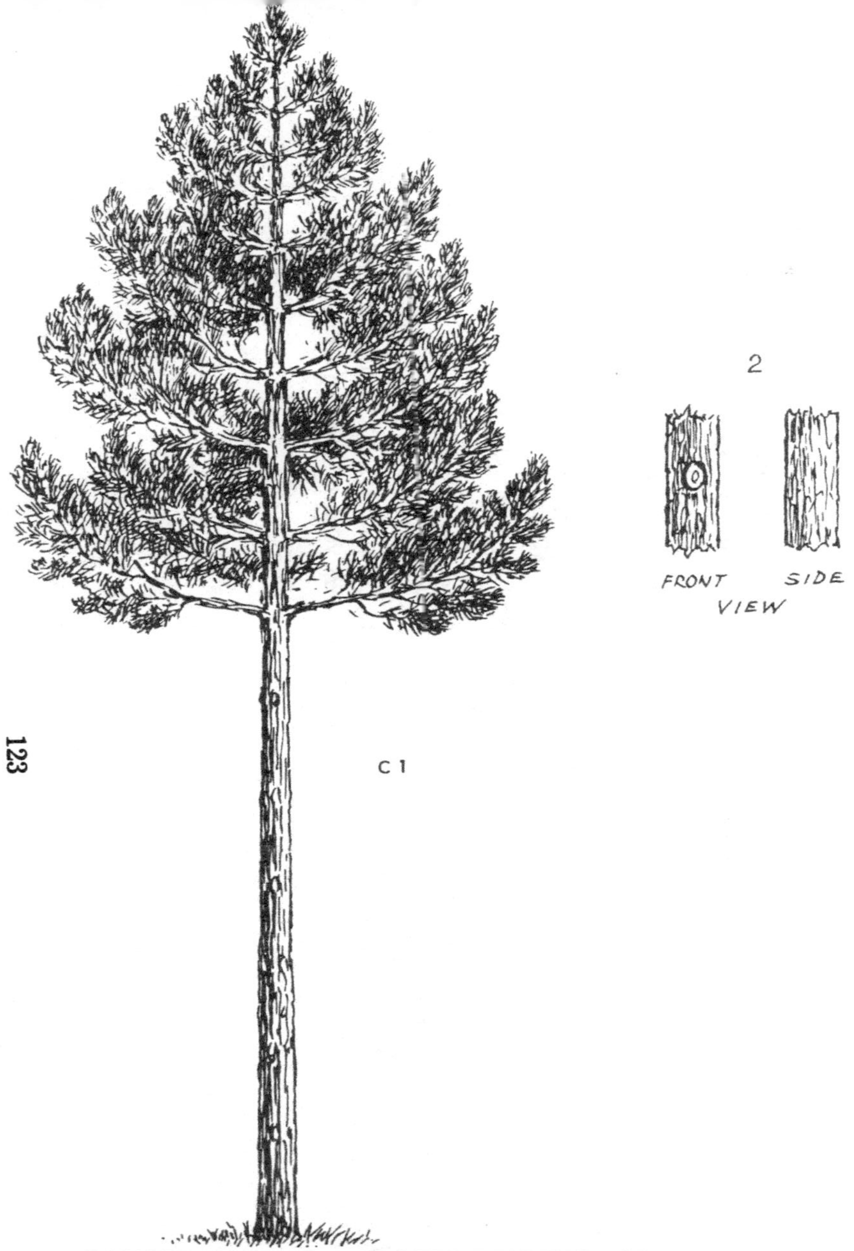

ILLUSTRATION OF A PROPER PRUNING JOB

1. Pruned to correct height.
2. Branches cut off flush with stem without wounding or injuring bark.

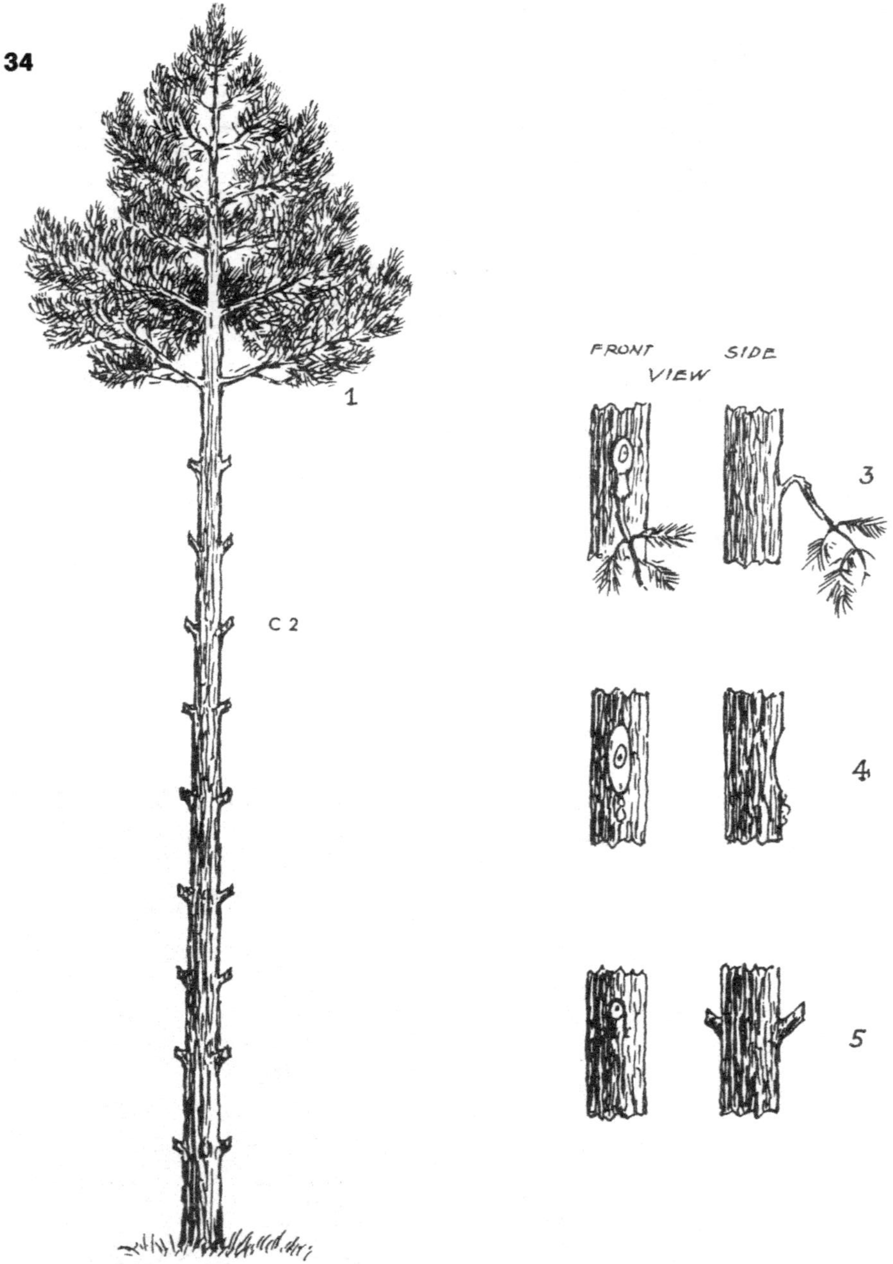

ILLUSTRATION OF A POOR PRUNING JOB

1. Tree pruned too high.
2. Branches not cut flush with stem.
3. Tearing and injury of bark by improper cut.
4. Too deep a wound.
5. Dead stubs lower portion of stem not removed.

NOTE: In both of these examples the live crown represented two-thirds of the total height of the tree before pruning.

July '59
Jens V. Sivers

Improve Your Woodlot by Cutting Firewood

By Ken Lancaster and Clyde Hunt

U.S.D.A. FOREST SERVICE
NORTHEASTERN AREA
STATE & PRIVATE FORESTRY
BROOMALL, PA. 19008

INTRODUCTION

Over the centuries, wood has been a major source of fuel and heat. Our forefathers revered wood and used it to fill every need from heating their homes to making tools. They made maximum use of the back woodlot.

The choice of wood for fuel was not only logical, but a most natural one, for trees are a renewable resource. They can be harvested periodically without permanent disturbance to the natural environment.

Only in the last fifty years or less has wood lost this important role. As a nation, we shifted to the convenient fuels: oil and natural gas. Ironically, it is our heavy dependence on these nonrenewable resources that created the shortages and exhorbitant prices of today's energy crisis. Consequently, wood has come into its own again, as more and more people look to the woods for the traditional source of heat.

The renewed use of wood for fuel creates an opportunity to correct some of our past mistakes. Our woodlots have been mismanaged, overcut, or often neglected. Past cutting practices left our woodlands with an overabundance of crooked, diseased, and otherwise unsaleable trees. These hamper the growth of the more desirable individuals, the straight, and healthy trees that are needed for lumber and veneer. To establish a good forestry program, the first step is to remove these less desirable trees for fuelwood, especially those trees that compete with the best crop trees.

THINNING HARDWOOD STANDS

Trees need room to grow at their maximum rates. If too close together, they compete for water, nutrients, and sunlight, and grow more slowly. A young stand of trees starts with 4000 to

6000 stems per acre. At maturity, when they measure about 20 inches in diameter, less than 100 trees per acre will survive. Most of the young trees die before they are large enough to harvest for sawlogs.

This is a natural selection process and it is slow, requiring 150 to 200 years or more to complete. Thinning hardwood stands when they are young hastens the process by permitting the more desirable trees to grow rapidly throughout their lives. Removing competing trees by frequent thinnings enables the stand to produce larger, higher quality trees. This promotes a greater volume of wood per acre in a reduced period of time—much less than 100 years.

Before thinning your hardwood stand, know the value of the trees you are thinning. As a rule sugar maple, ash, white and yellow birch, are more valuable than red maple, beech, or aspen (poplar); oaks are more valuable than the hickories. Check with your forester for advice on this point and for hints on tree identification.

You should begin thinning as early as possible to gain the benefits of repeated thinnings. The best time to start thinning a hardwood stand is when the trees average between 4 to 10 inches in diameter at breast height ($4\frac{1}{2}$ feet above ground). Trees of this size class, commonly referred to as poles, respond rapidly to thinning. At that point in life, intense competition from surrounding trees starts to slow their growth. (See figure 1.)

This does not mean that stands of larger size trees, averaging 10 to 12 inches, should not be thinned. Such hardwood stands are, however, approaching commercial sawtimber size. In most cases, the thinnings can be sold as sawlogs. Before doing anything in these stands of larger trees, you should get technical assistance from the local service forester or a consultant forester. These professionals will help you select the trees to be cut and those that should be left for future growth.

Figure 1.–A young stand of hardwood in need of thinning. In this stand a number of trees are unable to reach full sunlight. These "suppressed" trees, will naturally die and drop out of the stand.

SELECTING CROP TREES

The best way to thin a young polestand is the "crop tree selection method". This is a simple method for thinning stands to the greatest advantage of the best trees in the stand. Cut competing trees for firewood.

The trees selected as crop trees should be of valuable species. They will probably be the most valuable individuals in the stand. They should be straight and tall with relatively small branches, and should show signs of self-pruning: the lower 10 to 16 feet of the tree should have few or no branches. A few small dead branches in this section are a good indication that the tree will develop into a good quality tree, free of defects. (See figures 2 and 3.)

Figure 2.–A crop tree is straight and tall. Its relatively smooth bark is free of seams, breaks, and large wounds.

Look up into the crown. The crown of a crop tree needs three to four feet of open space on at least two sides. Those trees touching the crown of your crop tree are the competitors. They may be removed for fuelwood. (See figure 4.) In most cases, removing one or two side competitors will provide the crop tree with the space it needs, but don't hesitate to remove more if necessary.

The field procedure for selecting crop trees is this: start 10 to 20 feet into the stand or from the property line. Select a crop tree and identify it, either by tying a ribbon around it at breast height, or by using a spot of paint. Then, pace about 20 feet (eight steps) on a line parallel to the edge of

Figure 4.–The crown of the crop tree C needs room to expand for maximum diameter growth. Crowns that are narrow, compressed because of side competition, are not reaching full growth potential. Trees A and B are competitors of the crop tree C.

Figure 3.–Crop trees should not have: A. Swollen stems: B. Seams or breaks in the bark: C. Mechanical wounds caused by logging or other equipment: D. Poorly healed branch stubs. All of these defects indicate internal damage or disease. Such affected trees and crooked trees are best removed for firewood.

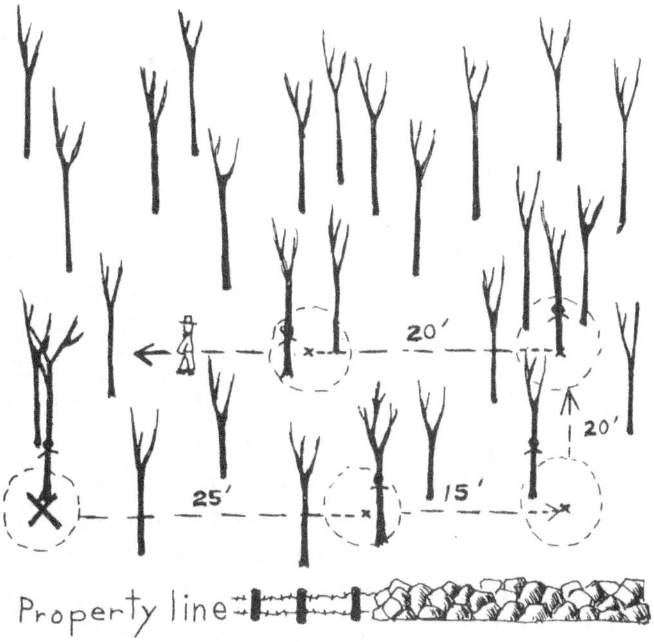

Figure 5.–Try to space crop trees 20 feet apart. You will end up with a sufficient number of trees per acre, spaced, in some instances, 15 feet apart and in others, 25 feet apart.

the stand or property line. Mark the closest crop tree within a 5-7 foot radius. If there are no trees that meet the crop tree specifications within this circle, pick the best of the lot and mark it. If there are no trees, pace out another 5 feet (two steps) and try again. If there aren't any trees within 7 feet of the second spot skip it and start over again at the next spot 20 feet away. (See figure 5.)

Follow this procedure until you reach the far edge of your stand or the distance you have previously decided to go into the stand.

When you reach the end of the first line, pace off 20 feet at a right angle to the first line. Pick a crop tree, mark it, then go back along a line parallel to the first line. Use the paint spots or ribbons on the crop trees of the first line to guide you as you proceed on the second line.

Ideally, you should pick a crop tree every 20 feet and release its crown. As nature doesn't space trees evenly, it is impossible to adhere rigidly to this distance, but by using it as a guide you will end up with a sufficient number of released trees (about 100 per acre). As you walk through your stand, don't hesitate to pick a good crop tree even though it's growing within 15 feet of the last one. If it is impossible to locate an ideal crop tree within 25 feet, pick the best one you have.

This is a slow procedure at first. But as you gain experience and confidence in yourself, the work will progress faster and be most enjoyable.

THE HARVESTING OPERATION

After you have selected the crop trees for release, you can begin harvesting your stand. First, remove trees that touch or are too close to the crowns of the crop trees. They are direct competitors. (See figure 6.)

Figure 6.–Remove the side competitors on at least two sides, allowing from 3 to 4 feet between crowns.

Figure 7.—These smaller understory trees exist without full sunlight. Such suppressed trees, are also a source of firewood. Removing them improves the looks of your woodlot.

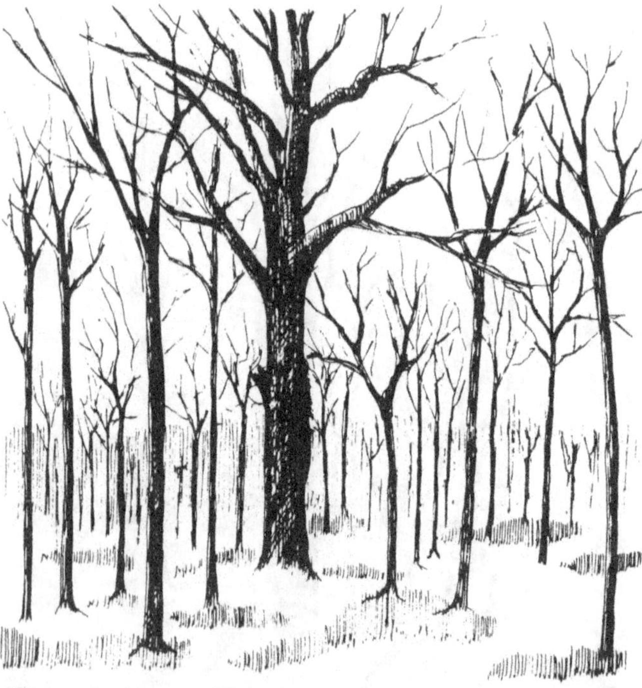

Figure 8.—Large, old and overmature "wolf trees" take up more than their share of growing space. You might consider giving their space to the next crop of seedlings.

In some high-quality stands, the trees to be removed are as high in quality as the crop tree. Although this may be disturbing, remember that most of the trees you are removing will not live to maturity. At some future time, they will be shaded out and die. Furthermore, the crop trees you release will grow faster so they will regain some of the growth you lose by removing the competition.

Sometimes a compromise is in order. If two high-quality trees are side by side, the best decision might be to accept both as crop trees and to release each on two sides.

In most polestands, there is an abundance of "understory" trees that are much smaller than the crop trees. Their crowns are seen below the crowns of the larger trees. In such a position, they are deprived of sunlight—nature's way of removing them from the stand. Harvest any understory trees big enough for firewood. Their removal will have little effect on the growth of the crop trees, but they will provide enough firewood to make the effort worthwhile (figure 7).

After releasing the crop trees, your next concern is the dead, dying, and deformed trees that hinder the development of the area. Any of these trees that have not been removed in thinning should also be harvested for firewood. (See figure 8.)

Work safely in your woodlot. Felling trees and falling branches—both live and dead—present potential hazards. Stay alert until they are safely burning in your fireplace.

A pamphlet of this length cannot cover every situation. The system though simplified is silviculturally sound. It will improve the quality and composition of the stand, with the opportunity to increase growth and volume. However, you should contact your local service forester before you start. He will help you to identify species and choose crop trees, and will clear up any problems you may have in thinning your woodlot.

CHAPTER 8
FORESTRY AND SOILS

By Michelle Lackey Olsen
Consulting Forester in the
Alexandria, MN area

Soils? That's something that concerns the agricultural farmer — not the tree farmer. Wrong! Consideration of soil and its properties should be as important to forestland management as it is to farmland management. After all, both are concerned with, among other things, producing a crop while maintaining the quality of the land. There are experts — soil scientists, foresters — who can help make decisions regarding proper forest management on your particular soils but it's good for an actively-involved landowner to have a little understanding of the subject. Potential problems can be spotted and the right questions can be asked. The following chapter is designed to give some basic information about soils and then to describe some of the relationships between soils and forestry practices. Use of this knowledge should help you, the landowner, make decisions that will be beneficial to both the soil resource and the forest resource.

WHAT IS SOIL?

Webster's Dictionary (Seventh New Collegiate edition) defines soil as "the upper layer of earth that may be dug or plowed" or, more specifically, "the loose surface material of the earth in which plants grow". Soil is composed of both minerals and organic matter. Parent material is the mineral or organic matter from which the soil is developed. Examples of parent material are limestone, granite, sphagnum moss (forms peat), and sandstone. It should be added that soil is always changing. These changes are caused by weathering, addition of organic material from the vegetation growing on it, and the land use practices of man. The following paragraphs describe some of the physical and chemical characteristics of soil and how these properties may be related to use of the soil for forestry purposes. It will become apparent that all the properties are closely linked and dependent upon each other. This makes it difficult to isolate one property from the others when considering the impacts of the soil on forestry practices and vice versa.

PHYSICAL PROPERTIES OF SOIL

Soil is made up of individual grains called **soil separates.** These grains vary widely in size. The following table tells the name associated with each grain size according to the system used by the U.S. Department of Agriculture (from Brady, 1974).

	Size (diameter)	
Separate Name	**Millimeters**	**Inches**
Very coarse sand	2.00-1.00	0.074-0.040
Coarse sand	1.00-0.50	0.040-0.020
Medium sand	0.50-0.25	0.020-0.010
Fine sand	0.25-0.10	0.010-0.004
Very fine sand	0.10-0.05	0.004-0.002
Silt	0.05-0.002	0.002-0.0007
Clay	less than 0.002	less than 0.0007

Sand and larger particles can be separated from silt and clay by sieving through mesh screens. Silt and clay can only be separated by dispersing the soil in water and measuring the amount of material that settles out in a given time with an instrument called a hydrometer. For example, sand settles out in 40 seconds, silt takes almost seven hours and anything left suspended in the water after that is clay.

Soil texture is determined by the relative proportion of the various separates in a soil. The textural triangle shows the percentage of each separate found in the different texture classifications. For example, a soil that is 60% sand, 25% clay, and 15% silt is called a sandy clay loam.

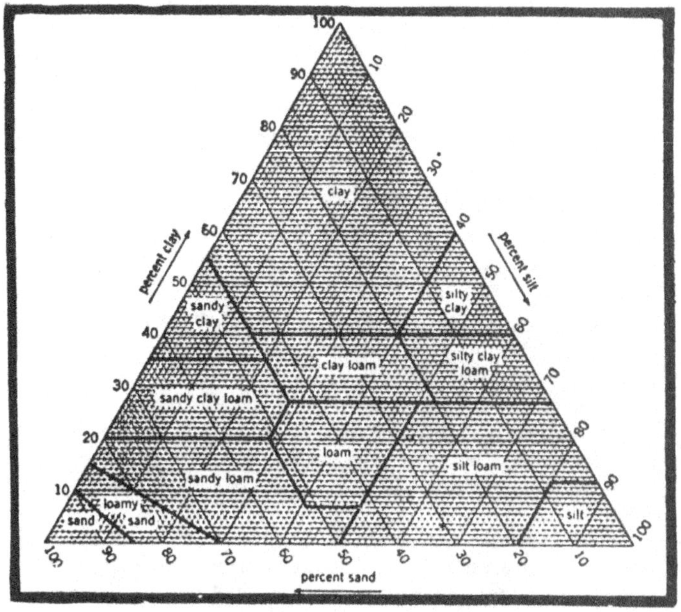

Figure 1. After determining the percent of various soil separates in a soil sample the texture is found from the above **Textural Triangle**. *(Brady, 1974; Soil Science Society of America, 1978).*

This method of classification, however reliable, is inconvenient for most landowners to use because it requires special equipment or a laboratory analysis of the soil to obtain the percentages of each soil separate. For most forestry purposes a field test can be performed that will give adequate information about the soil texture. With practice, landowners should be able to use the "feel test" in Table 1 to determine basic soil textures found on their property.

Table 1. Physical characteristics and field determination of soil textural classes.

SAND — Contains less than 15% silt and clay. Loose and single-grained. Individual grains can readily be seen and felt. Squeezed in the hand when moist, it will form a cast which will crumble when touched. Very fine sands have a certain amount of cohesion when moist. Leaves no coloring on fingers when squeezed.

LOAMY SAND — Contains 10-30% silt and clay. Loose and single-grained and most of the individual grains can be seen and felt. Feels gritty. When moist, will form a somewhat cohesive cast but cannot be molded into a ball. Leaves coloring on fingers when squeezed.

SANDY LOAM — Contains 15-57% silt and clay. Has enough sand to feel gritty but also enough silt and clay for considerable cohesion. Many individual grains can be seen and felt. Dry clods easily crushed. Squeezed when moist, forms a cast that will bear careful handling without breaking.

LOAM — Contains relatively even mixture of sand, silt, and clay. Dry clods crushed under moderate pressure. Gritty yet fairly smooth. Ball formed by squeezing moist soil is stickier and can be handled quite freely without breaking. May ribbon slightly when moist soil is rubbed between finger and thumb. Ribbon surface is rough.

SILT LOAM — More than half of the particles are silt. Dry clods broken readily and when pulverized feels foury or silky. When wet, runs together and puddles. Squeezed moist soil forms cast that can be freely handled without breaking. Ribbons slightly but has broken appearance.

CLAY LOAM—Contains 27-40% clay. Dry clods very hard and difficult to break. Moist soil sticky and forms cast that will bear much handling without breaking. Forms ribbon that breaks easily and has surface that feels slightly gritty.

CLAY — Contains 40% or more clay. Dry clods are dense, compact, and extremely resistant to crushing by hand. Moist soil is very sticky and smooth. Forms a long flexible ribbon with a smooth surface when rubbed between thumb and finger.

There are other textural classes (example: sandy clay loam) but the above are the ones most usually found in the surface soil.

It is common to hear references to "light" soils or "heavy" soils. In this context, "light" refers to sandy soils and "heavy" describes clayey soils.

The presence of particles larger than sand can be indicated by the adjectives gravelly (particles up to 3" in diameter), cobbly (3-10"), or stony (larger than 10"). The percentage of soil volume occupied by these large pieces should also be estimated.

Soil texture is significant to forestry practices in several ways. It has an effect on water- and nutrient-holding capacity which both help to determine the productivity of a soil. Accessibility and use of machinery will also be limited by the textures present as will root penetration of trees and shrubs. These factors will be discussed in more detail in the section on forestry considerations and the soil.

Soil structure is the arrangement of the particles in the soil. The particles may be loose with no bonding or cohesion between them (sand on a beach, for example) or there may be a certain amount of bonding or cementing that holds the particles together in various shapes and forms. Clay particles have an electrostatic attraction to each other so soils with significant amounts of clay are likely to be found in some type of chunk or aggregate. The different structural aggregate types such as granular, crumb, blocky, platy, etc. are hard for the landowner to recognize. It is more important to understand that various practices can have a negative effect on soil structure and should be avoided if possible. Probably the worst problem is compaction of the soil by heavy machinery, grazing, and recreational activities. Compaction reduces air space in the soil, water infiltration, and root penetration. It is more of a problem in soils that are sandy loams or heavier. Again these implications will be discussed in a later section.

Closely related to soil structure is **bulk density** which is the mass of dry soil per unit volume. This is a way of measuring how much space there is in a soil for air and water. The right proportion of air, water, organic matter, and soil particles is required for good root growth. It is usually not necessary for landowners to be able to determine bulk density but it is a

Table 2. Soil drainage classes.

DRAINAGE CLASSES	COLOR-MOTTLING	DEPTH TO PERMANENT WATER TABLE
1. Excessively and somewhat excessively drained. Water removed rapidly. Usually coarse-textured or sandy. May be steep and lose much water as runoff.	No mottling. Subsoil is bright red, brown, or yellow depending on parent material.	More than 60"
2. Well-drained. Water removed readily and available to plants throughout most of growing season. Roots not hurt by excessive wetness for any significant length of time. Soil usually medium-textured.	No mottling above 36". Indistinct mottling may be found below 36". Subsoil is as above.	More than 60"
3. Moderately well-drained. Water removed somewhat slowly during some periods. Usually wet for only a short time during growing season but may be long enough to affect some types of vegetation. Often have a less permeable soil layer or periodic heavy rainfalls or both.	Indistinct mottling within 36" of surface. Subsoil is as above.	Generally more than 60"
4. Somewhat poorly drained. Water removed slowly enough that soil is wet for significant periods of time during growing season. Restricts growth of most crops. Drainage problems caused by a less permeable soil layer, high water table, seepage from other area, nearly continuous rainfall or a combination of these.	Distinct mottling in upper 20 inches of soil. Soil at depths of 20-36" may be gray or grayish-blue (gleyed).	36-60" depending on time of year
5. Poorly drained. Soil saturated or remains wet for long periods during growing season. Water on or near surface means that soil must be artificially drained to grow most crops. Poor drainage caused by factors in 4. above.	Distinct mottling or gleying immediately below surface layer. Subsoil is gleyed.	18-36" may be closer to surface during certain times of year
6. Very poorly drained. Water remains on or near surface during most of growing season. Again artificial drainage required for most crops. Ponding is common, especially in depressions.	Distinct mottling or gleying immediately below surface layer. Subsoil is gleyed. Surface may be peaty or mucky.	Less than 18"

(Portions of this table are from **Forest Soil Management for Field Foresters; Part 1 — Forest Site** published by Minnesota Department of Natural Resources Division of Forestry in June 1983.)

helpful concept to understand. If bulk density is very high, as in a compacted soil, serious problems can result. Erosion potential is increased and productivity is decreased.

Soil drainage is just what the name implies: how fast water moves through the soil. Another way to think of it is as the frequency and duration of periods when soil is completely free from saturation. It refers only to natural drainage and not that altered by artificial means such as tiling. Color of the soil is the most frequently used indicator of drainage class. **Mottling** is the presence of a mixture of variation in soil colors in one area. These colors are in irregular spots or blotches. Mottling usually signifies a drainage problem. The drainage classes and a brief description of each are listed in Table 2.

Drainage is influenced by climate (mainly amount and timing of precipitation), topography, soil texture, and soil structure. It will also be affected by the presence of fragipans or hardpans. These are hardened natural soil layers that have excessive cementing which makes them much less penetrable by water. Stones and gravel will speed up movement of water through the soil. Drainage has an impact on moisture and oxygen supply to trees and vegetation and on accessibility to a site. Poorly drained soils result in a lack of oxygen and a buildup of carbon dioxide in the soil, both of which are detrimental to plant roots. Growth rates may be reduced by lower soil temperatures caused by poor drainage.

Effective rooting depth of a soil is that depth that can be used by plant roots. This depth is determined by the distance down to a layer of saturated soil or to a layer that is not penetrable by the roots. The latter category includes bedrock, dense gravel, soils with high bulk density (such as a compacted soil), and fragipans. Saturated soils inhibit root growth of most vegetation because they are lacking in oxygen. Water is occupying pore spaces that should be filled with oxygen so a deficiency results. Most species will not survive or do well on saturated soils but a few (tamarack, black spruce, white cedar) have lower oxygen demands and more tolerance for carbon dioxide buildup and can be found on these sites. The type of root system present on a species also determines the rooting depth required. Trees with a taproot will obviously need a much deeper soil than those with a shallow root system. Most trees must have at least 18" of rooting depth if they are to grow adequately and not be subject to windthrow.

As mentioned previously, soils consist of both minerals and organic matter. The mineral particles are sand, silt, and clay. The **organic matter** comes from vegetation on the surface of the soil. In a forested area there is a considerable amount of dead plant material that is an important part of the soil system. Organic matter in these forest areas can be separated into two types. The **litter** layer consists of leaves, needles, twigs, branches, bark, etc. that are still recognizable and show few if any signs of decomposition. Under the litter is **humus** which is the decomposing organic matter. It is usually not possible to identify the original material in humus because decomposition has progressed too far. Humus may be found as a distinct layer on the mineral soil surface or it may be mixed in with the top layer of mineral soil. The former is called a mor humus and usually occurs under conifer forests. The latter, a mull humus, is more common under hardwoods.

Organic matter in the soil contributes greatly to productivity of a site. It increases the water- and nutrient-holding capacity and helps to maintain a favorable structure. The organic material not only holds nutrients but it is in itself a source of nutrients as decomposition occurs. The forest floor or layers of organic matter on the soil surface absorbs much rainfall so more can eventually soak into the soil instead of running off. This decreases erosion and increases the amount of water available to plants growing in the soil. In the absence of a forest floor, bare soil tends to be clogged up by raindrops and water infiltration is reduced. Forestry practices which destroy or remove the organic layers or the topsoil which has humus incorporated into it should be avoided.

CHEMICAL PROPERTIES OF SOIL

By definition, **soil pH** is the negative log of the hydrogen ion concentration in the soil. In practical terms, it is a measure of the degree of acidity or alkalinity of a soil. It is expressed as a value from 1-14. The different pH values are described in the following table from the USDA-SCS Soil Survey Manual.

Term	pH	Term	pH
Extremely acid	1.0-4.4	Neutral	6.6-7.3
Very strongly acid	4.5-5.0	Mildly alkaline	7.4-7.8
Strongly acid	5.1-5.5	Moderately alkaline	7.9-8.4
Moderately acid	5.6-6.0	Strongly alkaline	8.5-9.0
Slightly acid	6.1-6.5	Very strongly alkaline	9.1-14.0

Each change of 1.0 in pH is a tenfold change in acidity or alkalinity. For example, pH 5 is ten times less acid than pH 4 and one hundred times less acid than pH 3.

Soil pH has a direct influence on the availability of nutrients for tree and plant growth. Due to complex chemical properties, each of the essential plant nutrients has a certain pH range within which it can be readily used by the plant. Outside of that range, the nutrient is released much more slowly or not at all. (See Table 3.) The most fertile soils for conifers have a pH of 4.8-6.8 and for hardwoods, 5.3-7.5. Very high (greater than 9.0) or very low pH (less than 4.0) can actually be toxic to plants. It is difficult and very expensive to try to change soil pH so landowners should manage for tree species that are best suited to the pH present on their property.

The **essential plant nutrients** mentioned above are necessary for the normal growth of plants. There are seventeen such nutrients. **Macronutrients** are those nine elements used in relatively large amounts. The eight **micro-nutrients** are used in very small quanitities but are just as important to plant growth. Three of the macronutrients — carbon (chemical symbol: C), hydrogen (H), and oxygen (O) — are derived mostly from the air and water. Carbon and oxygen are taken from the atmosphere and, through photosynthesis, are changed into forms usable by the plant. Hydrogen is from water that enters the soil. These three elements make up 94-99.5% of living plant tissue.

The remaining macronutrients must come from the soil itself. These nutrients are nitrogen (N), phosphorous (P), potassium (K), calcium (Ca), magnesium (Mg), and sulfur (S). The micronutrients — iron (Fe), manganese (Mn), boron (B), molybdenum (Mo), copper (Cu), zinc (Zn), chlorine (Cl), and cobalt (Co) — are all obtained from the soil.

There are several sources of nutrients in the soil. Organic matter becomes a major provider of nutrients as it decomposes. Weathering or breakdown of parent material and rock releases certain essential elements. Some nutrients are introduced into the soil from precipitation and dust particles. The application of fertilizers also increases the amount of nutrients in the soil. Generally soils with a higher content of humus and clay particles can hold and make available to plants larger quantities of nutrients.

Nutrients are lost from a soil system through erosion, leaching (especially on sandy sites), harvesting of vegetation and fire (both reduce organic matter). **It is easier and less expensive to preserve the nutrients already present than to try to replace lost nutrients.** This should be kept in mind when making forest and land management decisions.

Table 3. Relative availability of essential plant nutrients at various pH levels.

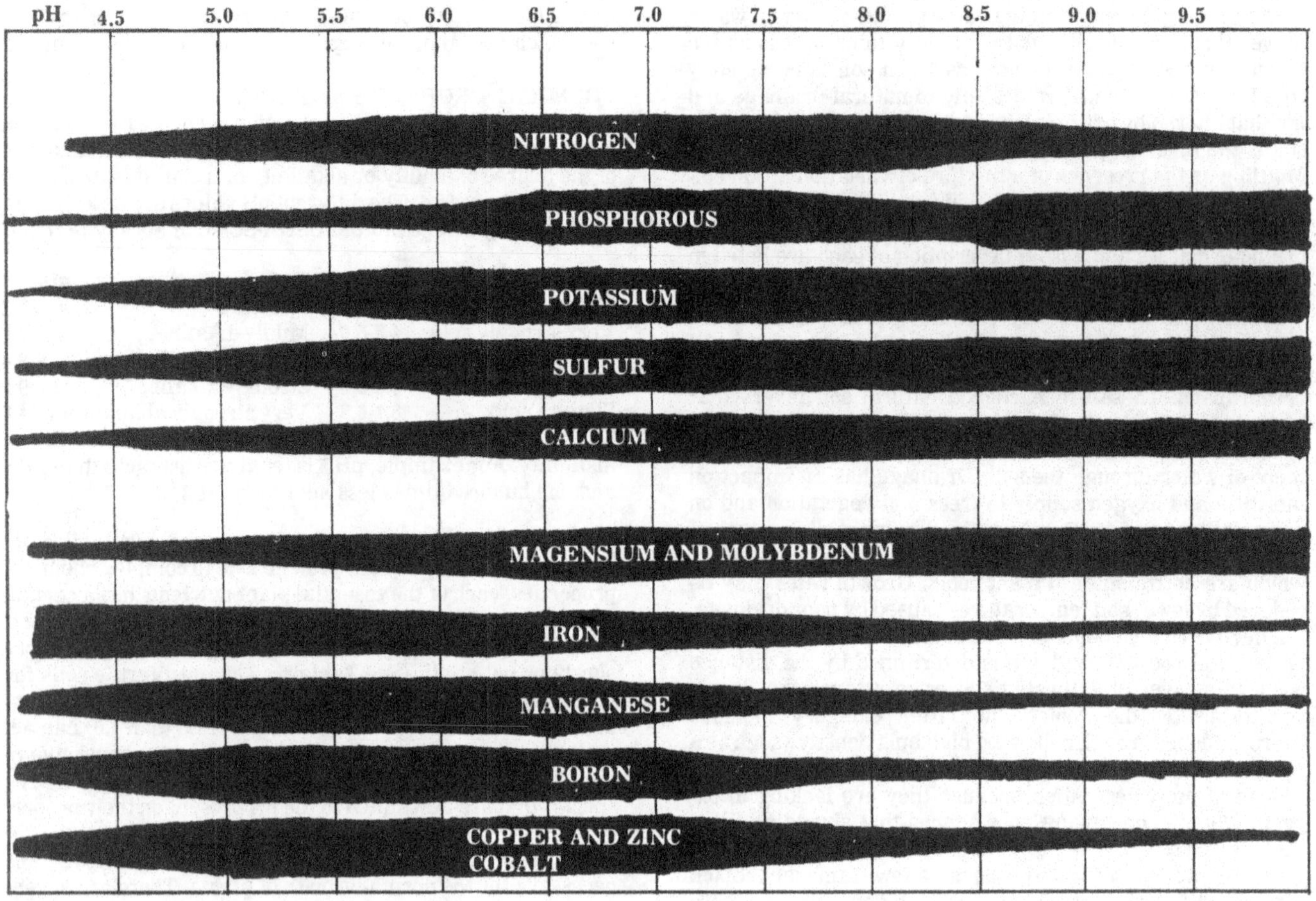

(Adapted from Brady, 1974, and other sources.)

SOIL PROFILE

A **soil profile** is a cross section of soil from the surface down to parent material or bedrock. The best way to see a soil profile is to dig a pit that will expose this vertical cross section. Look for it also at building excavation sites. The profile is made up of several layers called **horizons** which run roughly parallel to the soil surface. The horizons will vary in number and depth from soil to soil. They are not always clearly defined. Table 4 describes characteristics of the horizons most commonly found in a profile. The diagram shows the position of each horizon relative to the others. The soil profile and its horizons are used to help classify and name soil types.

The A horizons are often called the topsoil. Cultivation and erosion may change the original profile by distributing and removing the A horizons.

TABLE 4. Major horizons in a soil profile. (Adapted from Brady, 1974.)

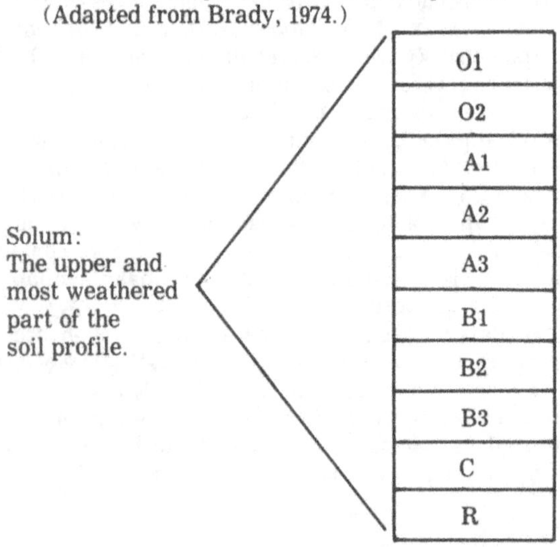

O1—Organic. Surface layer of litter consisting of leaves, needles, bark, fruit, twigs, stems, etc. Partially decomposed but the original forms are still recognizable.

O2 — Organic. Well decomposed humus whose original form is no longer recognizable.

A1 — Mineral soil mixed with decomposed organic matter (humus). Very dark in color.

A2 — Mineral soil horizon that has lost most of its clay, iron, and aluminum by leaching so it has a high concentration of sand and silt.

A3 — Transition zone between A and B horizons, more like A than B.

B1 — Transition zone between A and B horizons, more like B than A.

B2 — Horizon that accumulates most of substances leached out of A2 horizon.

B3 — Transition zone between B and C horizons, more like B than C.

C — Mineral soil horizon that is out of the zone of active soil formation and outside the zone of major biological activities.

R — Bedrock

EFFECTS OF SOIL PROPERTIES ON FOREST PRACTICES

The soils present on a property should be one of the major influences on the forest management of that site. When planning a regeneration project or deciding which trees to favor in an existing woodlot, it is helpful to know soil requirements of the different species. Each type of tree has specific needs concerning moisture, pH, rooting depth, and nutrients. Moisture and nutrient content are determined, to a great extent, by soil texture and organic matter. Sandier sites tend to be drier and less fertile than soils with a higher clay content. They also have less organic matter because they produce less vegetation to begin with. Rooting depth depends on drainage, texture, and structure. Compacted and very heavy soils and those with poor drainage severely restrict root growth.

Soils that are poorly drained may require direct seeding for regeneration instead of planting because access is not possible during the planting season. On the other hand, very sandy soils may also present regeneration problems in periods of drought. There isn't much that can be done to control the amount of precipitation that falls on an area so only the most drought-resistant species should be planted on these sites. Even then some mortality is to be expected depending on the severity of the drought. Trees that are stressed by an excess or deficiency of moisture are prone to more damage by insects and disease than healthy, vigorous trees. This is yet another reason to make sure that only the most appropriate species are chosen for each area. Table 5 summarizes some of the information necessary to successfully match trees species to a site. (See also Table 6).

Table 5. Site conditions and important Minnesota tree species that commonly occur together. Geographic location of site will also influence choice of species. See species range maps elsewhere in this book. (Some information for this table was taken from Fowells, 1965).

SPECIES	INDICATOR PLANT GROUP	MOISTURE CONDITION	SOIL TYPE	OTHER
Conifers				
jack pine red pine	A	dry to moderately dry, well-drained	sand	acid soil, tolerates very little competition from other plants
white pine	B	moderately dry to moist, well-drained	sand or loam	
white spruce balsam fir	C	moist	loam or clay	more demanding in nutrient requirements than other spruces & pines
black spruce tamarack	D	moist to wet	peat	occasional flooding okay
Hardwoods				
bur oak	A	dry to moderately dry	sand or loam	SW slopes
aspen birch	B	moderately dry to moist, well-drained	loam	very intolerant to competition from other plants
red oak	B	moderately dry to moist	loamy sand to clay	NE slopes
sugar maple basswood black walnut	C	moist, well-drained	loam, silt, or clay	deep soils
elm green ash cottonwood	C	moist	loam, silt	bottomlands, streambanks, alluvial sites
black ash	D	moist to wet	peat	occasional flooding okay

When selecting tree species for a site, it can be very useful to notice what other species of shrubs and plants are growing on that site. These **indicator species** have fairly specific requirements for moisture and nutrients. By taking note of which indicator species are doing well on a site, it is possible to predict which tree species are suitable for that site also. A landowner who is familiar with the vegetation on his/her property has probably already noticed some of these relationships. Blueberries seem to be found on the same sites that support jack pine. Maple-basswood forests abound with trillium and jack-in-the-pulpit. Labrador tea thrives in black spruce bogs. Table 6 lists some of the indicator species associated with four site types. This is one more tool that the landowner can use to choose appropriate trees for a site. While it does not involve direct observation of soil properties, the presence of indicator species certainly says something about soil conditions on a site.

TABLE 6. Indicator species for four site types. Actual plants found will also depend on the geographic range of the different species.

A. **Dry, nutrient-poor sites**
dwarf juniper
bearberry
wintergreen
blueberry
snowberry
woolly yarrow
everlasting
northern bedstraw
lousewort
pipsissewa
sweetfern
trailing arbutus
jack pine
Norway pine

B. **Dry to moist, medium nutrient sites**
round-leaved dogwood
hog pea-nut
yellow clintonia
bunchberry
liverleaf
twinflower
ground pine
black snakeroot
twisted-stalk
juneberry
meadowrue
bloodroot
highbush cranberry
baneberry
woodbine
white pine

C. **Moist, nutrient-rich sites**
alternate-leaved dogwood
leatherwood
spikenard
jack-in-the-pulpit
wild ginger
blue cohosh
ostrich fern
woolly sweet cicely
lopseed
large-flowered bellwort
trillium
wild black currant
red-berried elder
nannyberry
sugar maple

D. **Wet, nutrient-poor sites**
speckled alder
bog birch
bog rosemary
leatherleaf
creeping snowberry
Labrador tea
small cranberry
swamp cinquefoil
three-leaved false Solomon's seal
bog laurel
sphagnum moss
dwarf alder
black spruce
tamarack

Machinery for harvesting, timber stand improvement (TSI), and other work projects should not be used on heavy soils when they are wet. It is easy to get stuck and easy to seriously compact the soil by repeated trips under these conditions. It takes no more than 4-5 trips over one spot by a piece of heavy machinery to ruin the structure of the soil by compaction. The effects of compaction will be discussed in the next section. Trafficability on light soils is generally not much of a problem because water moves out of these much more quickly.

Sites with poor drainage usually require winter access because the soil water is frozen then. In Finland and Scandinavia, these sites, especially the peat bogs, are often drained to improve forest productivity. This is uncommon in the United States but low areas are drained here for agricultural purposes.

The natural fertility of a soil determines the amount of competition that desirable tree species will receive from other vegetation. In very fertile soils, not only will tree seedlings grow well but so will everything else. More site preparation and release work will be needed to give the seedlings an advantage over the competition. Release from competition, whether chemical or mechanical, may have to continue for up to five years depending on the tree species and amount of other vegetation present.

Fertilization of less productive forestry sites is being done in some parts of the country. Much of the work is still experimental. It is a costly practice and must be repeated several times to gain any appreciable benefits. At the present time, it is not of great importance in the Lake States region.

On coarse-textured soils, chemical pesticides or fertilizers may be leached from the upper horizons rather quickly. Some effectiveness will undoubtedly be lost by this rapid movement. An accumulation of the chemical in a lower horizon may be toxic to the desirable tree species as well as to the "weed" species for whom the herbicide was intended. Be sure to follow chemical label directions to minimize these problems.

EFFECTS OF FORESTRY PRACTICES ON SOIL QUALITY

Erosion

In the Lake States, erosion is usually not much of a problem on the average forest activity site. However it can be. Just about any activity that removes cover or the protective litter layer and exposes mineral soil or compacts the soil increases the potential for erosion of the topsoil. Timber harvesting, TSI, and site preparation are three of these practices that are common in many woodlots. In timber harvesting and TSI, it is not felling of the trees that increases erosion hazard, but rather the hauling of the logs out of the woods. Hot fires, whether planned or accidental, also destroy the litter layer and expose the soil to the eroding forces of wind and water. Through a chemical reaction, fire can also create a water-repellent crust on the soil surface. This crust inhibits water penetration and eventually results in runoff and erosion. Heavy grazing can remove enough vegetation and compact the soil enough to cause erosion.

Road building for access into woodlots is capable of causing very serious erosion problems. In fact a recent DNR study determined that it has the greatest potential for forestry-caused water pollution in the state of Minnesota. Road surfaces are compacted and usually have little vegetation on them. This means lower water infiltration rates and higher runoff rates. Improper drainage causes washouts and gullies, both of which carry away large quantities of soil.

Given mineral soil exposure or compaction, the risk of erosion varies from site to site and is dependent upon several factors. Table 7 describes some of the factors which influence the extent of the erosion hazard.

TABLE 7. Relative erosion hazard in relation to site factors in the Lake States. (Source: Dick Rossman, Minn. DNR Forest Soils Specialist.)

SITE FACTORS	HIGH EROSION HAZARD	MODERATE EROSION HAZARD	LOW EROSION HAZARD
Soil	Silt loam, loam, very fine sandy loam	clay loam, clay, sandy loam	loamy sands, sands
Slope	Steep Over 30%	Moderate 10-30%	Gentle 0-10%
Precipitation	Heavy snowmelt or prolonged intense summer storms	Moderate snowmelt, short heavy rains or moderate rains	Light snowmelt or light rains
Vegetation and other organic matter on and in the soil	None to very little	Moderate amounts	Large amounts

There are two major results of erosion in a forest ecosystem — loss of valuable topsoil and lower water quality. The topsoil is the most fertile layer of soil, especially if it has a considerable amount of humus incorporated into it. Loss of this soil means lower productivity for the site. Regeneration and growth will be decreased.

Unfortunately the eroded topsoil often ends up in a stream, river, or other body of water where it can significantly degrade quality of the water. Even though this section is supposed to focus on soil quality, it is a good time to describe the effects that soil moved by erosion has on water. Increased amounts of sediment may cover the gravelly spawning sites of fish or fill in small pools in a streambed. The clarity of the

FIGURE 2. Although this figure is from data from a very heavy rainstorm in the Utah mountains, it demonstrates the dramatic effect that ground cover can have on erosion potential. (Taken from **Forestry and Water Quality** by George W. Brown, School of Forestry, Oregon State University, Corvallis, Oregon, 1978.)

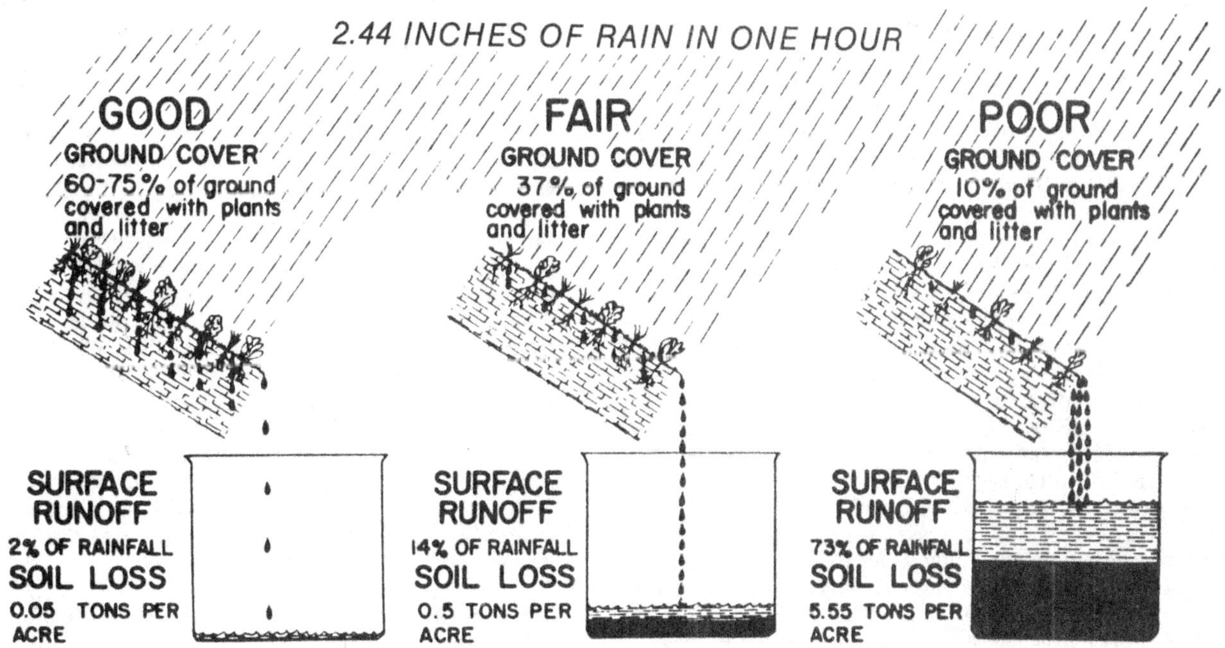

Figure II-4. The effect of watershed condition on rainstorm runoff and erosion, Subalpine Range Ephraim Watershed, Utah (after Noble, 1963).

water is decreased and it may look dirty. The nutrients in the soil increase growth of algae. The "algae blooms" cause an unpleasant taste, odor, and color in the water. Anyone who lives in a city that gets its water from a river or lake has probably experienced this distinct change in water quality. The increase in algae also changes oxygen levels in the water. This can drastically affect fish and other aquatic organisms that have very specific oxygen requirements. The whole character of a lake or stream can be changed by the introduction of eroded soil into it. One final way in which water is polluted by runoff is through chemical pesticides that are carried along with the soil.

The news isn't all bad though because there are many ways in which a conscientious landowner can decrease erosion potential. First of all maintain cover in the forms of vegetation and litter. Plant trees, shrubs, or ground cover where soil is exposed. Logging will often be less damaging if done in the winter when the ground is frozen. Avoid any type of potentially damaging practice in high risk areas such as on very steep slopes. This might mean that every acre cannot be intensively managed.

Plan all roadbuilding carefully before starting any construction. Run across slopes not up and down them. Reduce cut and fill operations. Proper drainage is especially important because roads usually change the normal movement of water. Keep ditches and culverts unclogged. If you must cross a stream, make sure that streamflow is not disrupted. A road should cross a stream at a 90° angle. Install drainage dips and water bars on roads with long slopes to slow the flow of water over the surface. Provide a buffer strip between road and stream. Minimize area used for roads whenever possible. In some areas ACP-FIP cost-share programs may be available to offset some of the costs of forest road building. Practices covered by these programs include road stabilization, culvert installation, improved road design, and the seeding of unused roads. Check with your local forester or ASCS office to see if this program is available in your area.

Compaction

Soil compaction is caused by grazing, recreational activities (trails, picnic grounds, off-road vehicles), and the use of heavy equipment for timber harvesting, TSI, and site preparation. Road surfaces are also compacted. The medium-textured soils (loams) suffer the most from compaction. Normally they have good structure with ample pore space for water and air. The mix of different-sized soil particles fills in these pore spaces under pressure. Clayey soils are also subject to compaction but since they have less pore space to start with they have less to lose. Compaction is minimal on sandy soils. Wheeled machinery puts more pressure on the ground than track vehicles and thus results in greater compaction.

The result of compacted soil is decreased productivity of a site and increased runoff and erosion. Bulk density of the soil is increased which means that there is less pore space available for air and water. Water infiltration is slowed considerably and it is much more difficult for roots to penetrate the soil. In extreme cases, tree roots may be exposed. All these factors combine to reduce growth and increase stress on tree species. Stressed trees, as mentioned before, are more vulnerable to damaging insect and disease attacks than healthy trees. Natural regeneration is also slower. Erosion problems have already been discussed.

There are several ways to lessen compaction. Use heavy machinery on heavy soils only when the ground is dry or frozen. Limit the number of trips over any one area if it isn't a road. Grazing should not be allowed in the woodlot. It makes the area look like a park but the price is damaged trees and no natural regeneration coming in to replace older trees. **An improved, non-wooded pasture gives better results on a much smaller area.**

There are also ways to rehabilitate places that are already compacted. Plan no activity in that spot for several years. This will allow the litter layer to build up, decompose, and eventually be incorporated into the soil by soil fauna. Some of the soil structure can be regained by this means. If more immediate results are required, plowing or cultivating may help but it will also injure any shallow-rooted species growing on the site.

Nutrient Loss

The loss of nutrients can be caused by the same things that cause erosion. The topsoil lost through the erosion process is the most fertile soil. In addition, the removal of the litter layer takes away the nutrients that would be released by its decomposition and incorporation into the soil. The litter layer is often removed by a hot fire, site preparation, logging, and TSI. One method of site preparation that can result in heavy topsoil loss is the formation of windrows. Unless care is exercised, the topsoil will be scraped into the rows along with the logging debris. Another practice that decreases the nutrients available on a site is whole tree logging where leaves, branches, etc. are cut off the tree at the landing instead of in the woods. This results in a concentration of organic material in one spot instead of having it distributed throughout the woodlot. Of course, with a clearcut this is more serious than if only selected trees are removed. Loss of nutrients is of greater significance on the more infertile soils. Since they produce less to start with, it will take longer for them to replace the lost material.

The end result of nutrient loss is decreased productivity on a site. Growth and regeneration will be slowed.

To avoid nutrient loss, protect the litter layer and topsoil. Do heavy site preparation involving dozers and windrows when the ground is frozen. Less topsoil is likely to be scraped away but the equipment operator must still be made aware of the potential problem. A quick, cool fire will skim over the surface of the litter layer without completely consuming it. This is preferable to a hot, intense fire. Any ersosion prevention measures will also protect the valuable nutrients on a site.

Other

Puddling is caused by removal of the litter layer or wet tilling of heavy soil. Impact of raindrops on the clay soil and the tilling change the structure of the soil. Pore space is reduced and a crust that is almost impenetrable is formed. The result is, among other things, higher seedling mortality and much slower seedling establishment. To avoid this problem, protect the litter layer and don't till fine-textured soils when they are wet.

Some of the effects of chemical use on soil, particularly sandy soil, have been previously discussed. Other problems associated with heavy chemical use involve the buildup of these chemicals in the soil. The danger is not so much that the soil itself will be damaged but that the pesticide will enter the food chain and affect animals at all levels. Fortunately many chemicals break down naturally within a couple days and some of the more dangerous cumulative ones have been removed from the market. Since forestry uses of chemicals are quite small compared to other agricultural uses, the potential problem is also less. Be sure though that all label directions are carefully followed. Also avoid any situations where chemicals are likely to end up in bodies of water through drift, runoff, direct spraying, etc.

SOURCES OF ASSISTANCE AND FURTHER INFORMATION

There are several sources that landowners may use to answer their soil-forestry questions. Most professional foresters are able to give this type of advice and offer recommendations after a field examination. Government, industrial, and consulting foresters are all available to the private landowner.

County Soil and Water Conservation Districts (SWCD) and the federal Soil Conservation Service (SCS) also provide field examinations and assist in carrying out recommended practices. They usually know the local area and its peculiar problems well and are a good source of detailed local soil maps. Cost-share programs for soil conservation practices are administered by these offices.

Agricultural and forestry county extension offices are affiliated with state universities. Extension agents are qualified to answer your questions and also have much written material in the form of fact sheets, pamphlets, etc., that are usually free to interested persons. Check your county seat or the university for these offices.

Soil labs will analyze your soil and make specific recommendations upon request. Agricultural universities often have a soil lab for this purpose. Private soil laboratories can be located by calling a local farm service center or cooperative.

RECOMMENDED READING AND REFERENCES

Armson, K. A. 1977. **Forest Soils Properties and Practices.** University of Toronto Press, Toronto. 390 p.

Brady, N.C. 1974. **The Nature and Properties of Soils, 8th Edition.** MacMillan Publishing Co., Inc., New York. 639 p.

Fowells, H. A. 1965. **Silvics of Forest Trees of the United States.** Agricultural Handbook No. 271, USDA-Forest Service, Washington, D.C. 762 p. (For more details soil requirements of tree species.)

Harlow, W. M. and E. S. Harrar. 1969. **Textbook of Dendrology.** McGraw-Hill Book Company, New York. 512 p.

Pritchett, W. L. 1979. **Properties and Management of Forest Soils.** John Wiley and Sons, Inc., New York. 500 p.

1978. **Glossary of Soil Science Terms.** Soil Science Society of America, Madison, Wis. 36 p.

Material for this chapter, its tables and figures, is taken from numerous sources including publications listed in the Recommended Reading and literature from forestry and soils courses at the University of Minnesota. Since many of the tables are a compilation of data from several sources, it was not always possible to cite specific sources.

Special thanks to Dick Rossman, Soils Specialist for the division of Forestry, Bemidji Region, Minnesota DNR, and Al Wickman, PFM Specialist for the Division of Forestry, Park Rapids Area, Minnesota DNR, for their assistance.

Soils and Windbreak Growth

All species of trees and shrubs do not grow at the same rate nor do they grow to the same mature height. Likewise, adapted species vary in their growth on different soils within a geographical area.

The amount of available soil moisture during the growing season and soil aeration are two important factors affecting tree growth. These are largely determined by soil texture and depth and by climate.

On deep, fertile, well-drained sandy or silt loam soils without hardpans and with adequate available water, all species grow well. These soils allow extensive root penetration and provide all the essentials for good tree growth.

Only a few species of trees can tolerate saline, alkali, or very acid soils. Some trees grow fairly well on shallow soils for a short time, then grow slowly or stop growing and die. Disease problems also multiply on these soils.

Trees make only fair to poor growth on poorly drained clayey soils. They are less porous, have a slower water-infiltration rate, and are subject to runoff. They can store considerable soil moisture but much of it is not available to plants. Aeration is poorer, root systems are shallower, and tillage is more difficult than in sandy or silt loam soils.

Recent studies by the author on windbreak growth in a central Plains county, following a 5-year drought, show that all adapted species were growing well on deep sandy loams and on soils that have a stable water table. In 20 years, cottonwood had grown to 60 feet; Siberian elm, 50 feet; ash and hackberry, 35 feet; bur oak and Osage-orange, 25 feet; and ponderosa pine and eastern redcedar, 20 to 25 feet. In adjacent windbreaks on clay soil, only Siberian elm, oak, Osage-orange, pine, and redcedar survived, but vigor was only fair and growth was about one-third less than that on deep sandy loams. On very sandy soils, only pine and redcedar survived and growth was poor.

Since 1951, teams of SCS soil scientists and foresters have studied soils and windbreaks to determine windbreak growth on different soils. They carefully described the soils, took measurements of tree height and age, and observed general vigor and condition of the trees. If available, this information has been included in published soil surveys.

The following table, abstracted from the published soil survey for Reno County, Kans., shows the kind of information that is available in published

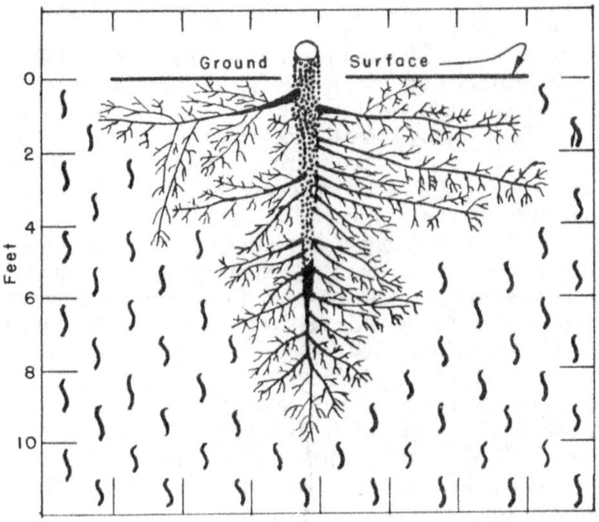

A good loam soil allows moisture penetration and deep, uniform distribution of tree roots.

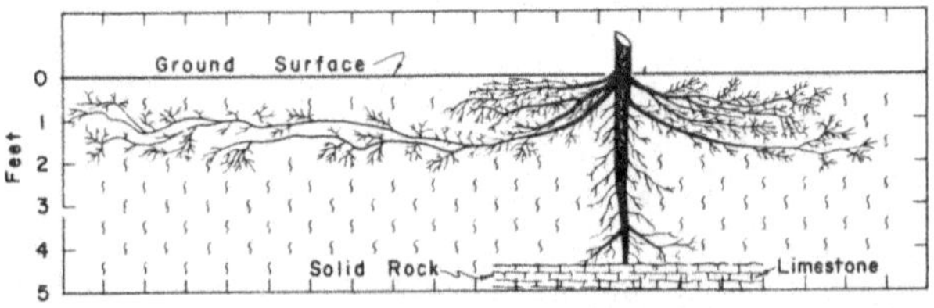

These tree roots could not grow any deeper because of the solid rock.

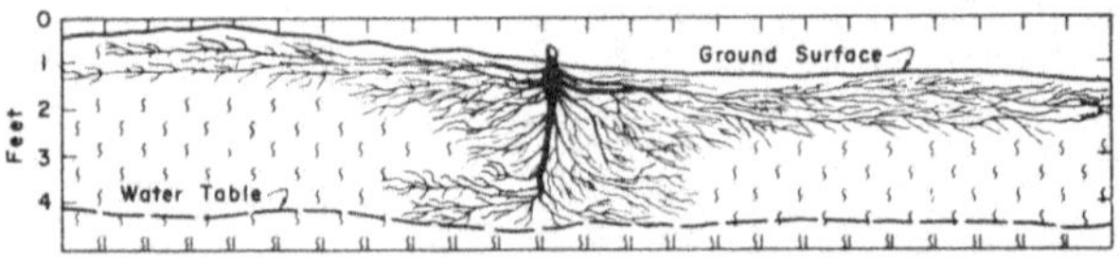

A high water table prevents deep root penetration. These roots stopped at 4 feet. Tree roots must have air to grow.

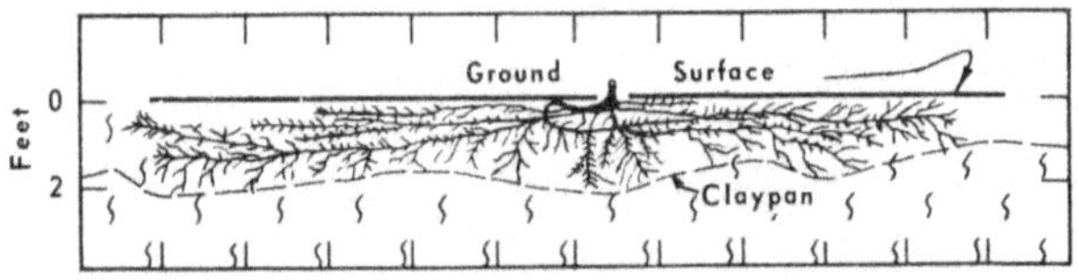

The claypan soil restricted the roots of this tree to 2 feet in depth.

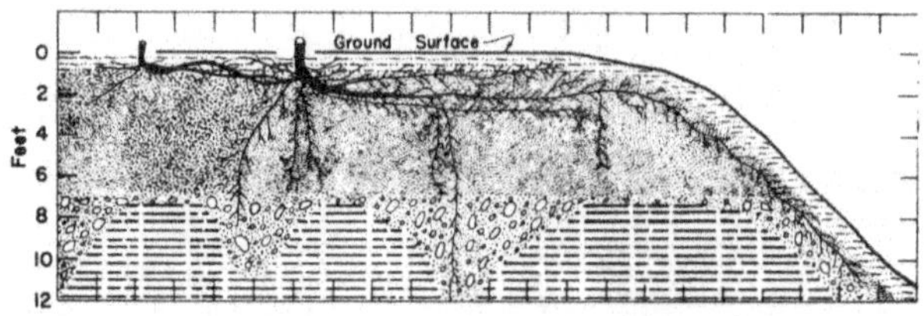

Tree roots follow moist sand and gravel pockets.

GROWING TREES FROM SEED

Growing trees from seed can be both educational and fun. It isn't a difficult project, but it does take time and patience. You'll find that this experience is most meaningful and rewarding when you undertake the whole process, from collecting the seed to the final transplanting of the seedling.

By growing trees from seed, you'll learn many fascinating things about trees and how they reproduce. Each species has individual seeding characteristics. Some trees, such as pines, elms, and maples, have seeds with wings, so they can be carried by the wind. Some seeds have barbs or hooks that cling to the fur of animals and are carried to another location, and some are surrounded by a tasty pulp that animals or birds eat and deposit in distant places. Still other seeds are in a fruit that explodes and shoots them out, and others are in the form of a nut or acorn that squirrels bury and forget.

Some seeds will grow on moss, rotten wood, or almost any type of forest floor: others must have specific types of soil.

After the seeds start to grow (germinate), you'll discover that each kind of seedling also has its own characteristics. Some have needles, some are broadleaved, and some are susceptible to disease or special insect pests that must be guarded against. Some grow fast, others slowly. Some need lots of sunlight, others need shade.

As you proceed in this project, the wonderful workings of our largest plants will become familiar to you and you'll develop an appreciation for these partners in our environment.

COLLECTING AND STORING SEEDS

Always collect seed from healthy trees that bear large quantities of seed.

Conifers

Conifers (evergreens) have seeds in cones, the fruit of the coniferous trees. The easiest to collect in Minnesota are the cones of white pine, red (Norway) pine, and white and black spruce.

It is the seed inside the cones of evergreen trees, not the cones, that must be planted. Normally, two seeds lie at the base of each cone scale. When the seed has ripened, the cone scales usually spread and the seeds fall out. So you must collect the cones from the trees before they open and release the seed, but not before the seed is mature.

Most coniferous trees have cones with seed that ripens in the fall, usually in September and lasting into October. Immature cones are green and gradually turn brown as they ripen.

There are two good methods for determining whether or not the cones are mature enough to collect. First, you can inspect the seed. If the seeds are not milky, but well filled and solid, they are mature. Secondly, you can observe when squirrels begin cutting and hoarding the cones. When they do, the cones are mature enough to collect.

CHAPTER 9 — REGENERATION

Remember that the seed source is of utmost importance in forest planning. Follow these simple rules for selecting tree seeds:

- Secure seed from as local a source as possible.
- Gather cones from well-shaped, thrifty trees, preferably those in mature stands.
- Do not gather cones from deformed, suppressed, and diseased trees.
- If you purchase seed, buy it from a reliable dealer who knows his seed source.
- Remember that it usually is safer to use seed from a colder region than seed from a warmer one.

Keep your cones separate if you collect more than one kind. Norway and jack pine cones sometimes can be picked or cut from the lower limbs of standing trees. Do not pick up cones from the ground unless you're sure they were cut by a squirrel and are good. The easiest way to collect cones is from the tops of trees cut down during a logging operation. But be sure to get permission to do so.

Store the cones in a warm, dry room and use either artificial or natural heat to open them. A small number of cones will open readily when spread on canvas in full sunlight. This method is recommended, since there is no danger of damaging the seed during extraction. You can use the oven of a kitchen stove, but be careful not to heat the cones too much or too long or you'll reduce the vitality of the seed. Excessively high temperatures may even kill it. A temperature of 120 F. for 4 hours should open the cones without seriously injuring the seed.

After the cones are open, the seeds can be shaken from them. You can rake the cones back and forth, shake them in a bag, or tumble them in a churn. To remove the wing attached to the seed, gently rub the seeds between your hands or over a screen. Pouring the seeds from one container to another when the wind is blowing or with a fan blowing on them will remove the wings and other debris.

Whether you purchase or collect the seed yourself, store it in an airtight jar or bottle until spring. Place the bottle where rapid temperature changes do not occur, such as in a fruit or vegetable room in the cellar. A cool, dry place is best.

Hardwoods

Elms, silver maple, cottonwood, and the willows have seeds that mature in the spring (May-June). Their seed is wind-blown, so look for drifts of seed in sheltered corners and alongside street curbs. Seeds that ripen in the spring lose their ability to grow after 3-4 weeks, so plant the seeds as soon as possible after gathering. You'll have particularly good luck with elm and silver maple seeds.

Some of the trees whose seed ripens in the late summer or early fall include: sugar maple, boxelder, the oaks, basswood, honey locust, hackberry, the ashes, black walnut, butternut, Kentucky coffee tree, wild plum, lilac, honeysuckle, and Russian olive. Some of these seeds fall to the ground; others can be picked from the trees.

Almost all seed can be attacked by insects while it is still in the fruit, nut, acorn, etc. To check for this possibility, open a few seeds from each tree you collect from. If the seed is damaged, you may wish to try another tree. Healthy seed is firm, greenish-white, dry, and has no indication of insect damage.

Time of flowering and seed ripening for common Minnesota trees

Kind of tree	Flowering period	When seed is ripe
Ash	April-May	August-September
Aspens	April-May	May-June
Balsam Fir	May	August-September
Basswood	June-July	September-October
Paper Birch	April-May	August-September
Yellow Birch	April-May	September-October
Butternut	April-May	September-October
Red Cedar	April-May	September-November
White Cedar	April-May	August-September
Cherries	April-June	July-September
Cottonwood	April-May	May-June
Boxelder	April-May	September-October
Elms	March-April	May-June
Hackberry	April-May	September-October
Hemlock	May-June	September-October
Hickories	April-June	September-October
Ironwood	May-June	August-September
Red and Silver Maples	March-May	April-June
Sugar Maple	April-May	September-October
Oaks	April-May	September-October
Pines	April-June	August-September
Spruces	May-June	August-September
Tamarack	May	August-September
Walnut	May-June	September-October
Willow	March-May	May-June

Storing methods for hardwood seeds collected in the fall vary with the species. Seeds that can be planted in the fall are: sugar maple, boxelder, ash, basswood, lilac, honeysuckle, and Russian olive.

Some seeds should be stratified, which means they should be placed between layers of moist sand or peat moss and kept moist and cool (40° F.) for 2-4 months, depending on the species. Seeds that should be stratified include: black walnut, butternut, oak, hackberry, and wild plum. For complete information on treatment of seeds, see "Woody Plant Seed Manual," U.S. Department of Agriculture Miscellaneous Publication 654.

PURCHASING OR SELLING SEEDS

You may wish to purchase seeds that you can't collect. It is best to order such seeds in the fall to avoid the rush during spring planting time. For a list of seed dealers, write to: Extension Forester, Green Hall, University of Minnesota, St. Paul, Minnesota 55101.

Some nurseries will purchase seeds or cones. You might check with your local forester for potential markets. Selling seeds could be a club money-raising activity.

SEEDBED

Location of the Seedbed

Although a seedbed takes little space, there are important requirements that affect its location. Some of these requirements are:

- An adequate source of water.
- A well-drained site that is exposed to the sun and protected from drying winds, but has some ventilation.
- A light sandy loam soil, although it is possible to raise seedlings in almost any type of soil.
- Protection from animals and livestock.

A part of the family garden usually fits these conditions well.

Preparation of the Seedbed

Dig and spade the soil thoroughly, breaking up all lumps. Remove all debris and stones. There should be little organic material present. Under no circumstances should you add barnyard manure of any kind to the seedbed soil.

Now make a wooden frame of 1 inch x 4 inch or 1 inch x 6 inch lumber that will border a seedbed 4 feet wide and 12 feet long. You also can use cement or cinder blocks. Set this frame on edge about 1 inch into the ground. From the surrounding area, add enough soil to raise the surface of the bed about 1 inch. If the soil is loam, round it upwards toward the center of the bed. If it is sand, you can leave it flat. Firm, but do not pack, the soil with a board.

6

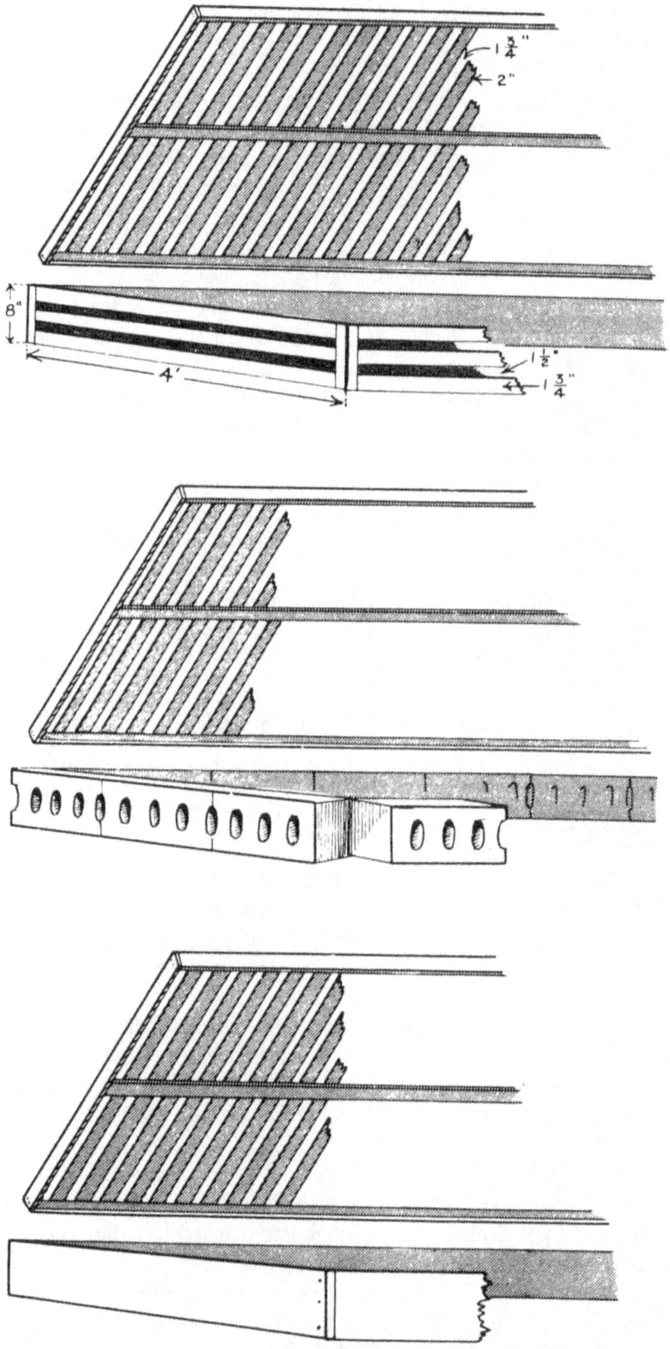

A lath cover over the seedbed provides partial shade and protection from birds. The seedbed frame can be made from lath (top drawing), cinder blocks (center drawing), or solid boards (bottom drawing).

A completed seedbed with a lath covering.

Planting the Seeds

Spruce and pine seeds are so small that thousands of them are required to make a pound.

Species	Number of seeds per pound	Average germinating capacity, percent
White pine	27,000	64
Red pine	52,000	75
Jack pine	131,000	69
Black spruce	404,000	61
White spruce	240,000	49

You will want to grow about 50-75 seedlings per square foot. Even if your seed is better than 75 percent fertile, you should plant about 200 per square foot. If the seedlings come up too thick, you can thin them later.

You can sow the seed broadcast or in drills 3 inches apart. If you broadcast seed, distribute it as uniformly as possible. You can punch holes in the lid of a pint jar and use it as a shaker to apply the seed.

After sowing, cover the seed with ¼ inch of clean, dry sand sifted through a sieve with ¼ inch mesh. A cover of more than ¼ inch sand may keep the small seeds from germinating, while one less than ¼ inch permits too rapid drying out. Cover the bed with a layer of burlap, and fasten it securely to keep the wind from blowing it off.

Treatment for Damping-off Fungus

Get ferbam (ferric dimethyl dithiocarbonate) or a similar fungicide. Such fungicides are available from most garden or drugstores.

8

Mix 2 tablespoons of the fungicide with 1 gallon of water. Apply this solution at the rate of 1 pint to each square foot of seedbed as soon as the seeds are planted and covered with sand and burlap. Use a fine sprinkling can to apply the fungicide, but be careful not to wash out the newly planted seeds. If you don't have a sprinkling can, puncture tiny holes in the bottom of a tin can. Spread the solution north and south and then east and west to assure complete coverage of the bed.

When 50 percent of the seeds have germinated, apply another ferbam treatment to control damping-off disease. Use the same strength and volume as for the first treatment. Then, every 3 weeks until September 1, apply the same mixture and be sure to cover the seedlings thoroughly.

CARE OF SEEDLINGS

Shade Requirements

Seedlings injure easily from too much heat when they are young and tender. Fifty percent shade seems to be about right. You can use a snow fence as a seedbed cover or build a frame covered with wooden laths. Do not space the laths over 1½ inches apart. The frame also will afford protection from birds that pull out tiny seedlings.

Weeding the Seedbed

Weeds will smother the trees if you don't remove them promptly. Also, if you let the weeds get as much as 1 inch tall, it is hard to pull them up without injuring the root systems of the tiny trees. You can cut off large weeds, but it is best to weed often so large weeds don't develop.

Thinning

After the first 6 weeks, you can thin the rows to allow better spacing. Leave at least 50 seedlings per-square foot.

Watering

You will get better growth by watering seedlings if rainfall is low or comes at irregular intervals. Prolonged drought will kill seedlings.

During the first 2 months, apply water about twice each week if it doesn't rain. Later, one heavy soaking to a depth of 6 inches will be satisfactory. Be very careful not to wash the tiny seedlings out of the soil.

Fall Mulching

After the first heavy frost, place a 3- or 4-inch mulch of leaves or straw over 1st year seedlings. You also can apply sawdust to a depth of ½-1 inch, depending on the size of the seedlings. The winter mulch will help prevent frost heaving. Leave it on the seedbed until all the frost is out of the ground in the spring. You can mulch the seedlings the second winter also if heavy frost heaving occurs in your area. It's best to be safe.

Fertilizing

At the beginning of the second growing season, about April 1, apply a topdressing of 6 tablespoons of 10-6-4 or 10-10-10 fertilizer to each 16 square feet. Repeat this application about June 15. **Don't burn your trees.** To avoid burning them with fertilizer, apply it on a dry day when seedlings are dry. After you apply the fertilizer, immediately sprinkle the seedlings with water to remove all the fertilizer from the leaves or needles.

Transplanting

As a general rule, pine seedlings can be planted directly from the seedbed after 2-3 years of growth. However, spruce seedlings should be transplanted at the beginning of the third growing season. Take great care when handling evergreens: an exposure of roots for even 30 seconds to the direct rays of the sun and wind often is enough to kill the seedlings. After digging up seedlings, immediately place them in a pail with enough water to cover the roots. A garden fork is a good tool for digging seedlings.

You should root-prune the seedlings when you transplant them. The long side roots as well as the tap root should be cut clean so that a small compact system is left to develop numerous fine roots that can be handled without loss at planting time.

Extension Folder 249 — 1970

William R. Miles and James H. Smith

AGRICULTURAL EXTENSION SERVICE
UNIVERSITY OF MINNESOTA

Site Preparation For Upland Conifer Planting

By Allen Wickman

Rock Rake - used for summer site preparation.

Introduction

Would you plant a tree if you knew it was going to die before the age of two? Of course not! Yet every year thousands of trees are doomed to an early death because they are planted in adverse conditions.

There are several ways a young tree seedling may die. Some of the more common problems include: (1) improper handling and care before planting; (2) not packing soil firmly around tree roots; and (3) planting trees in dry soil. We have another problem that I think kills more trees than a combination of all of the above (although the other certainly take their toll). This problem is called inadequate site preparation.

What is Site Preparation?

In simple terms, site preparation means to prepare the ground so that a tree may be planted in it. Site preparation can be good or poor depending on how it's done. To help illustrate, I will compare it to planting a garden.

Point #1 - Trees Need Soil Preparation

It's a rare individual who will plant garden seeds in soil that has not been prepared first. Types of soil preparation are varied but the objective is the same - to reduce competition from unwanted vegetation so the desired plants have an increased chance of survival. Trees need soil preparation too!

Point #2 - Weeding Benefits Trees

If weeds are left to grow at the base of a tomato plant, they are certainly going to cause problems for that plant. The same is true for a newly planted tree. Grass and weeds and woody vegetation (brush) have caused high mortality in many pine plantations. Site preparation will help reduce grass and weeds if it is done correctly.

Point #3 - Trees Need Sunlight

We all know that tomatoes and corn require a lot of sunlight. Nobody in his right mind is going to plant a tomato plant and then put a bushel basket over it. It would die from lack of sunlight. But many trees are planted in this very manner! A bushel basket doesn't blot out the sun, but competing vegetation, such as brush and weeds do. Results are the same. The plant dies.

Some trees like white pine, balsam fir, and maple can tolerate a fair amount of shade. But most species require large doses of sunlight if they are to be strong and vigorous. The lower the amount of sunlight the higher the mortality. This lack of sunlight won't kill a tree immediately. Sometimes the tree will be sickly for several years before it will succumb. Even shade tolerant trees will usually grow faster as the amount of sunlight is increased.

Rule of Thumb

All this information can be summed up fairly simply; Treat a tree as you would any garden plant. What's good for a plant will be good for a tree. After all, a tree is a plant. This sounds ultra-simple but it's surprising how many expert gardeners throw all their plant knowledge to the winds when they plant trees.

Large Scale Planting

It's not too difficult to provide a few yard trees with the care they need for growth and survival. But what do you do if you want to plant an acre? Ten acres? A hundred? Will you have time to weed, water and watch? Once most trees are planted they are "on their own". Because of this it is very important to get a young tree off to as good a start as possible. This is why proper site preparation is so critical.

Furrowing

Furrows are grooves (or trenches) cut into the ground from 1" to 10" deep. About 4" to 6" deep is ideal. Furrows are usually made with a tractor and a one-bottom plow. Furrows should only be deep enough to break through the upper layer of soil.

Furrowing is most common in open fields and semi-brush areas. Its main benefit is to eliminate or reduce competition from grass and other plants. The furrows also tend to provide trees with more moisture. Furrows are excellent for tree growing. They normally give us our highest percentage of seedling survival but are not so effective in heavy brush, recently cut-over aspen, or standing timber. All grassy areas should be furrowed before planting. If a planting machine is used, it should have scalpers capable of making a furrow. Trees should never be planted directly in grass covered sod.

Grubbing

Grubbing is done with a large caterpillar tractor with a flat "dozer" blade or rock rake on the front. Trees and brush are rooted out of the ground and the debris is piled or windrowed. The more roots taken out, the better the job. Roots left in the soil will send up new sprouts that will compete with seedlings.

Timing is very important. The best time to grub is in July, August or September. Most of the tree nutrients are above ground at this time so the roots are in their weakest and most "vulnerable" stage. There will be some roots left with all but the best of grubbing jobs. Roots will have much less vigor (for resprouting) if the timing for the job is right.

Grubbing works well in all situations where trees and brush will compete with seedlings for survival. There is a large amount of top soil disturbance. Some dirt will end up in the brush piles or windrows which will make them more difficult to burn. Piles usually are easier to burn than windrows, so I recommend grubbing and piling in July or August for the most effective site preparation.

Shearing

Shearing is done with a large caterpillar much like grubbing. A special blade called a K-G blade is used. This blade is sharp at the base so that it will cut off trees and brush at ground level. Most roots are left in the ground. Debris is usually piled or windrowed.

Shearing can work well in areas where little resprouting is expected. A rather "pure" stand of pine that has been clearcut may provide the ideal situation. Pine rarely reseeds or resprouts after logging. If they did, we would not have to replant.

Shearing is mostly a wintertime operation. I don't recommend it on aspen, or hardwood areas that are to be planted to pine *unless* the area is treated with chemicals to control brush *before* the pine are planted. Advantages of shearing include (a) less top soil disturbance, (b) less dirt in windrows, (c)windrows burn easier.

Discing

Discing is often employed after land clearing to chop up any roots left in the ground. Farmers clearing land for agriculture will often (1) grub, (2) pile, and (3) disc. There is no doubt that this is one of the most expensive methods but is also one of the best.

Grubbing and discing may not be as expensive as it may seem. Many types of site preparation (expecially shearing) may require 1 to 3 herbicide applications to hold down broad leaf vegetation. Grubbing and discing may reduce or eliminate the need for herbicides. Herbicide treatment can cost $20 - $25 per acre per "shot". Reducing the need for herbicide can save money in the long run.

Discing may also be used alone. Double discing (discing the area twice) is more effective than single discing. Discing as site preparation should be limited to areas of brush and small diameter aspen and other hardwoods.

We have had good luck grubbing in June and discing in August. The discing will destroy any unwanted brush and tree sprouts that regenerate after clearing.

Roller Chop

The roller chop is a large barrel with blades sticking out at right angles. These blades can do a fairly effective job of chopping up roots and killing brush and yound hardwoods (or pine?). Timing is important again. Roller chopping done during the winter or spring is just about worthless. Green stumps will also retard its effectiveness since the blades will ride up over the top of the stumps.

Controlled Burning

Areas of cutover conifers (pine, spruce, etc.) are sometimes prepared for planting by controlled burning. This is very dangerous work and should only be attempted under the direction of a forester or a person thoroughly trained in this field.

Burning is usually quite economical. The drawback is its limited use and its extreme danger. Trees (except possibly jack pine) should not be planted directly in the ashes. The ground may heat up in the summer sun and the seedlings could "bake". Furrowing after burning will often work quite well.

Herbicides Role in Site Preparation

. Foresters seem to be giving increasing attention to the use of herbicides for site preparation. Brush and grass control in forest plantations has always been a problem. It is not uncommon to see a plantation fail because seedlings have been "drowned out" by much faster growing brush and grasses. During our severe drought of 1976 many pine plantations bit the dust because grass used up all the available soil moisture.

We promote good quality site preparation to reduce the threat of unwanted vegetation taking over a planting site. Some types of site preparation create a great deal of top soil disturbance. Some landowners are quite negative about this disturbance. Herbicides may help solve the problem.

When do we need herbicides?

The need for herbicides is usually in direct relation to soil fertility. The higher the soil fertility the greater the need for herbicide brush control.

Dozing, discing and furrowing are all widely used methods of site preparation. However, these methods are no match for highly fertile soil. Herbicides used alone—or in combination with mechanical site preparation methods—may be the answer.

How are herbicides used?

Some chemicals such as simazine are called pre-emergent herbicides. These are applied to the ground and kill grass and brush seeds before they have a chance to sprout. These herbicides don't work on grass and brush that has already sprouted.

Established grass can be controlled with Roundup or Kerb. However, Roundup may kill all vegetation sprayed while Kerb kills most grasses but few broad leaved weeds. Tordon has also been found to be quite effective against most broad leaf plants (and conifers) but not grasses. Landowners must pick the chemical that matches their needs.

Sometimes we have used site preparation that has been quite effective against broad leaved plants—and then the grasses and annual weeds took over! This possibility must always be considered when choosing a chemical.

A typical procedure many landowners have used is (a) clearing land, (b) planting and (c) relying on herbicides to control the brush. This has only been partially effective. A better idea might be to (a) clear land, (b) treat with herbicide and (c) plant. Herbicides applied before seedlings are planted have fewer restrictions on how and when they can be used—and are usually more effective if applied properly.

If used wisely, herbicides can help us be much more effective in our reforestration attempts. They may reduce the need for site preparation techniques which may cause a great deal of top soil disturbance.

Summary

There are other types of site preparation I have not discussed. Their merits can be judged by the criteria in this article.

No matter what method is used, timing is very important. July to mid-September is the best time for most ''site prep'' jobs including grubbing, discing and roller chopping. Shearing is mostly a winter operation and herbicides are often most effective if applied in May or June. If pine seedlings are on the area herbicides cannot be applied until late July or August. The pine may be harmed if herbicide is used before they ''harden off''.

An old timer once told me he could kill an oak by cutting it after the dark of the moon in August. It will send up sprouts but the sprouts will eventually die. I won't guarantee this is going to work but it ties in very nicely with the period we recommend for site preparation.

When one considers the followup work necessary and the potential plantation survival, we may end up with a paradox. The cheapest site preparation methods may cost us the most and vice versa. Of course, the amount of money we spend doesn't make the big difference. If we want good site preparation we have to get to the ''root'' of the problem.

Editor's Note:

The current site preparation trend in Minnesota is away from summer time rock raking. Winter shearing followed by July or August discing is becoming more popular since the top soil is better preserved. Herbicide treatment is sometimes used instead of discing. In some cases herbicides may be used alone for site preparation. Contact your local forester to see what the preferred methods are in your area.

PLANTING GUIDE
FOR TREES AND SHRUBS

(1) IF SOD, WEEDS AND BRUSH ARE HEAVY, PREPARE GROUND BEFORE PLANTING BY ONE OF THE FOLLOWING METHODS:

(A)

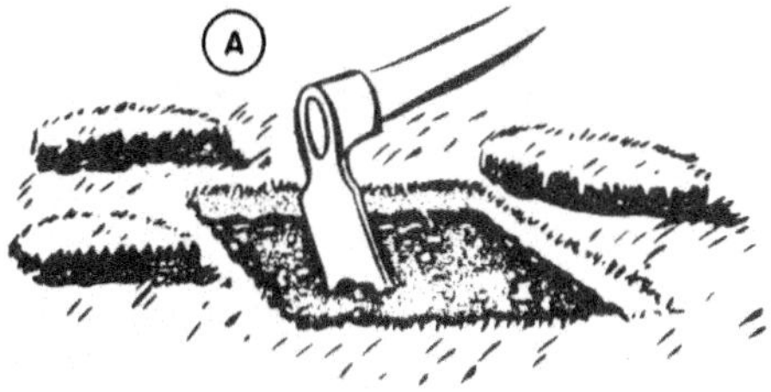

STRIP OFF SOD - 12 TO 18 INCHES SQUARE.

(B)

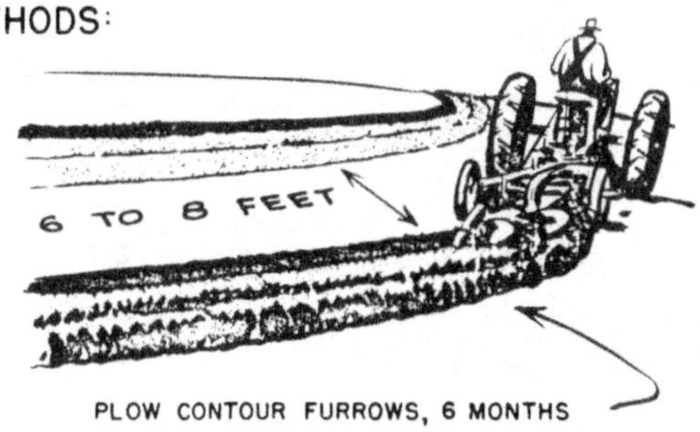

PLOW CONTOUR FURROWS, 6 MONTHS TO YEAR BEFORE PLANTING AND SET PLANTS IN FURROWS

(2) IF TREES CANNOT BE SET OUT WITHIN A DAY OR TWO, UNPACK AND COVER THEIR ROOTS IN A TRENCH.

A. DIG TRENCH IN SHADE AND OUT OF THE WIND.
B. KEEP ROOTS MOIST AT ALL TIMES.
C. WHEN READY TO PLANT, CARRY TREES IN CONTAINER WITH ROOTS COVERED WITH WET MATERIAL SUCH AS MOSS OR FINE SHAVINGS OR PAIL OF WATER.

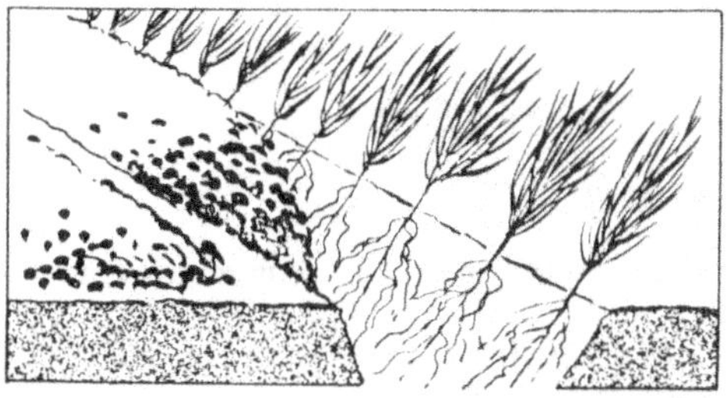

(3) SET TREES BY ONE OF TWO METHODS SHOWN BELOW:
(USE 6 TO 8 FEET SPACING FOR TREES, LESS FOR SHRUBS)

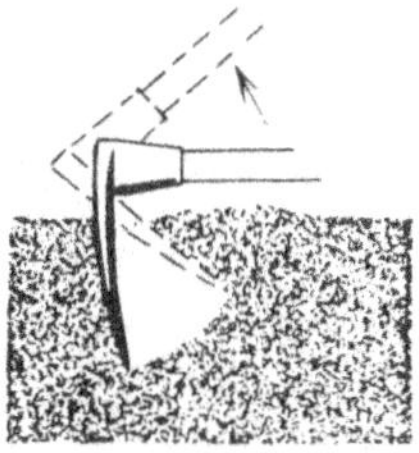

(A) DIG HOLE LARGE ENOUGH TO HOLD ROOTS.

(B) SET IN HOLE AS DEEP AS TREES WERE IN THE NURSERY.

(C) PACK SOIL FIRMLY.

(D) PRESS AND LEVEL SOIL WITH FOOT.

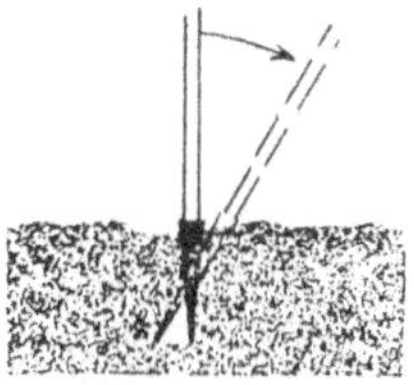

(A) DRIVE MATTOCK OR PLANTING BAR INTO SOIL WITH BLADE PERPENDICULAR. PUSH HANDLE FORWARD.

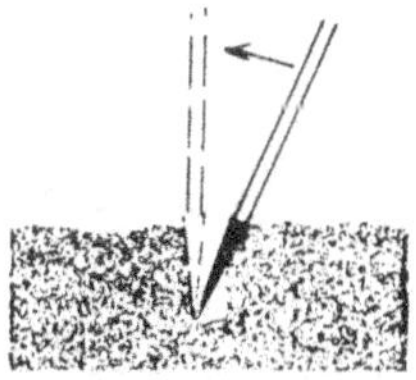

(B) RAISE BLADE AND DRIVE BACK INTO SOIL AT SAME ANGLE TO GET NEW HOLD PULL BACK TO ENLARGE HOLE.

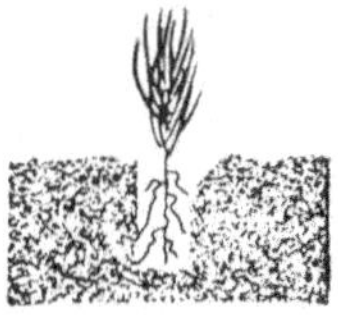

(C) SET AS DEEP AS TREES WERE IN THE NURSERY.

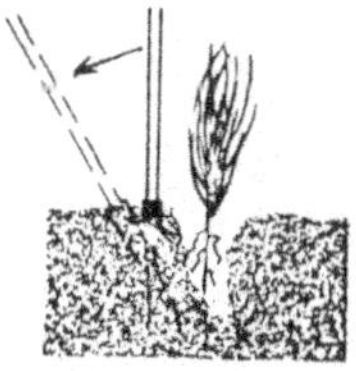

(D) CLOSE BOTTOM OF HOLE WITH PLANTING BAR OR MATTOCK.

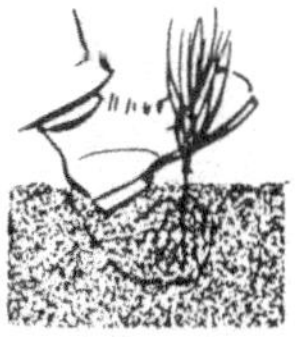

(E) CLOSE TOP OF HOLE WITH HEEL:

NOTE: THE SECOND METHOD WORKS BETTER ON LIGHT TEXTURED THAN HEAVY SOILS.

COMMON PLANTING MISTAKES
THAT WILL KILL TREE SEEDLINGS

By Allen Wickman

Care of Trees

- **Failure to protect tree bundles from drying effects of the wind.**
- **Failure to keep bundles covered or in shaded areas to protect from over heating by the sun.**
- **Failpre to keep partially used bundles properly covered.** Wind will often blow the cover away from trees while planters are working.
- **Failure to keep roots moist.** It is best to pack roots with damp moss. Do not submerge roots in water since this will wash soil particles off the roots. Rapid root drying often occurs when a planting bucket is near empty since air circulation is increased.
- **Failure to keep tree roots at the bottom of the bucket** - where they are protected from effects of sun and wind.

Planting Errors

- **Soil not packed tight around tree roots.**
- **Planting too deep.**
- **Planting too shallow.**
- **Allowing roots to curl back toward top of hole.**
- **More than one tree in a hole.**
- **Improper spacing.**
- **"Second hole" not closed** - a common problem when planting bars are used.
- **Not planting in mineral soil** - trees are sometimes planted in duff and ground litter where soil cannot make proper contact with tree roots.
- **Not allowing proper root spread.**

Other Common Mistakes

- **Planting in grass without making a sufficient trench,** (furrow).
- **Planting in areas of poor drainage** (low spots).
- **Planting in dry soil.**
- **Planting a species not suited to the locality or conditions.**
- **Planting at wrong time of year.**
- **Planting in areas where brush competition will crowd out seedlings.**
- **Planting trees where they don't receive sufficient sunlight.**

SPACING

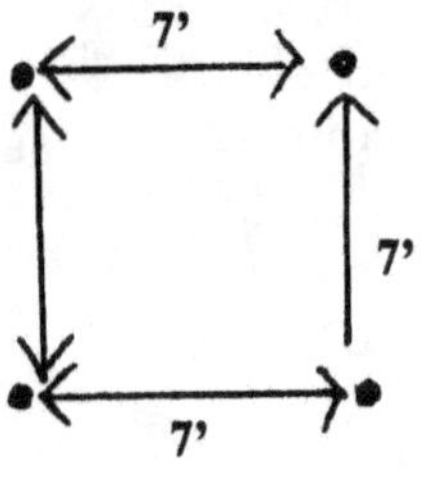

• **Tree**

TREE DEPTH

Excerpts From: **"Planting Trees For Farmstead Shelter"**

Marvin Smith and Harold Scholten
Agriculture Extension Service
University Of Minnesota

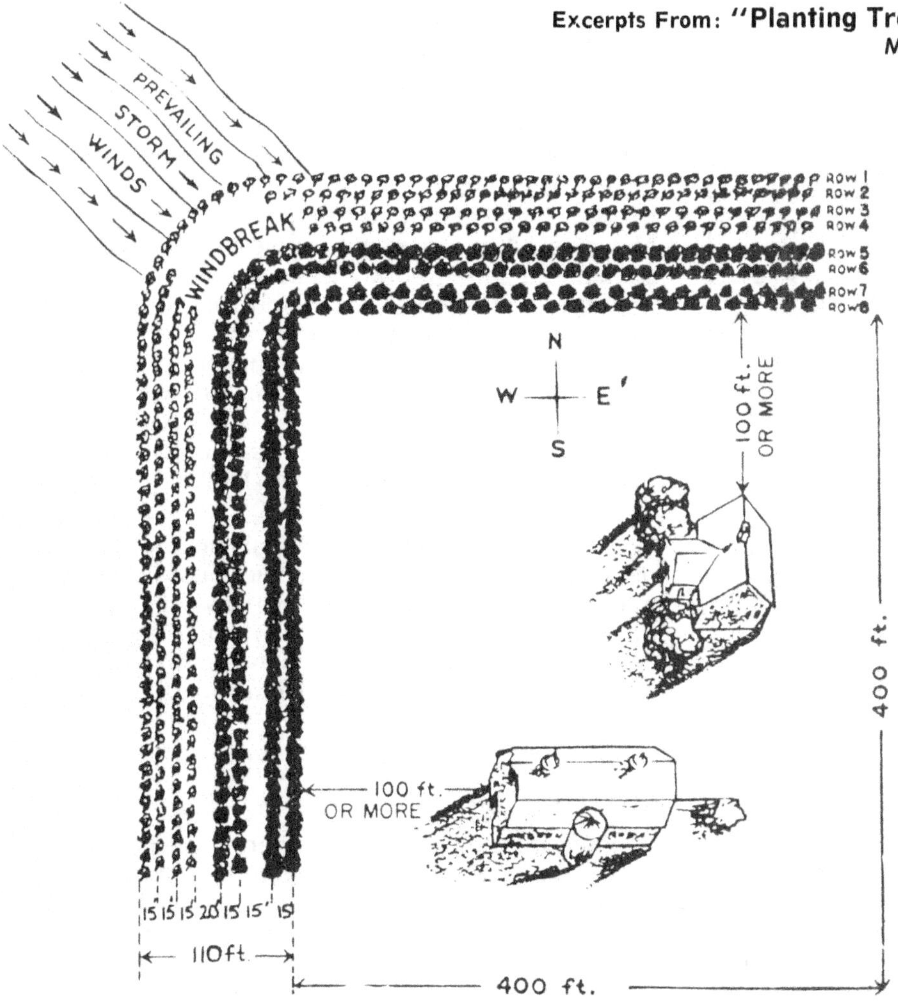

Figure 5. Standard plan. The indicated spacing between rows is only a guide and shows minimum spacings. Make spacing at least 4 to 5 feet wider than the width of your cultivating equipment.

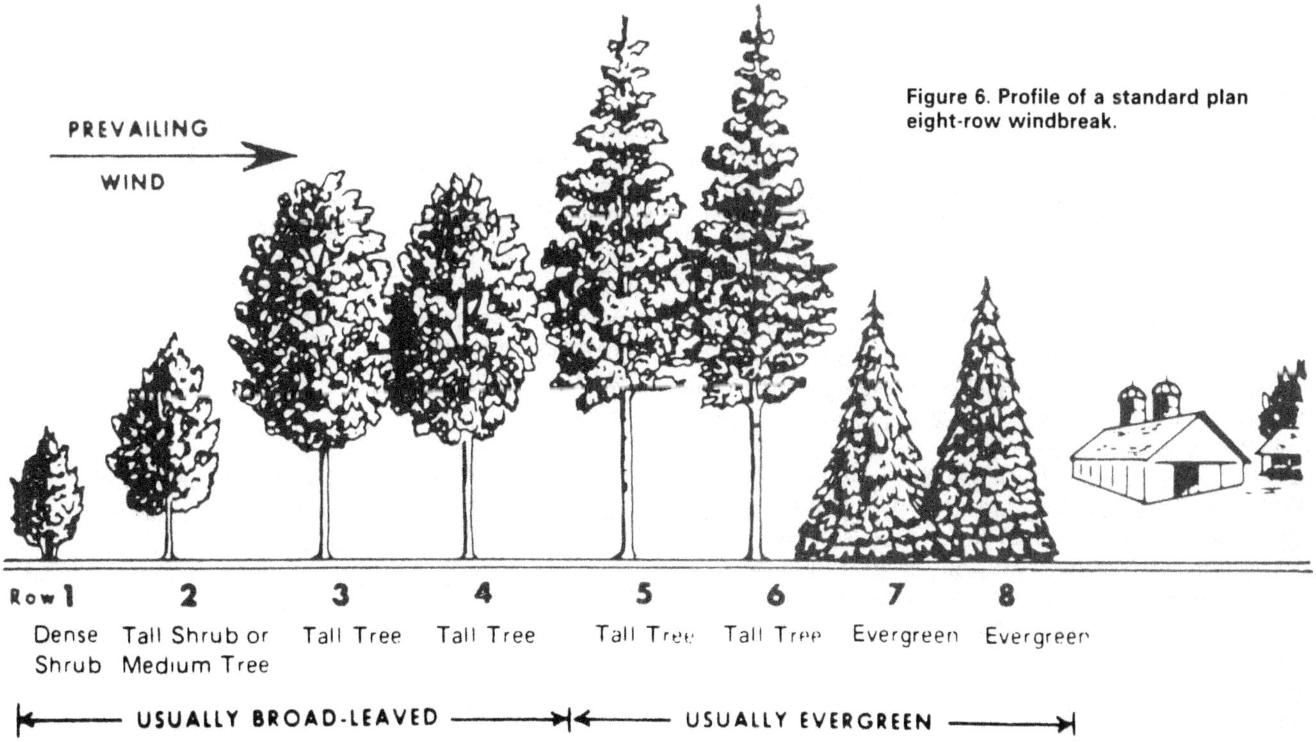

Figure 6. Profile of a standard plan eight-row windbreak.

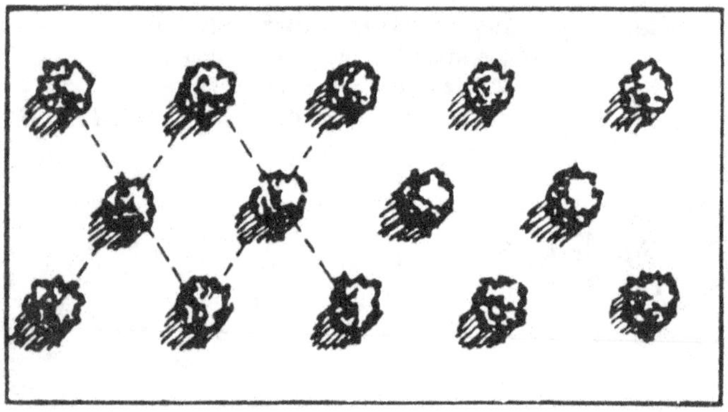

Figure 7. Proper alignment of trees—alternated or staggered.

A windbreak that includes these features provides: (1) protection to ground level, (2) maximum ultimate height, (3) year-round protection, (4) rapid growth, and (5) reasonably long life.

What to Plant

Choose trees of proven hardiness in your locality. To check on how well certain trees will do, compare them with others of the same kind that have grown successfully on similar soils in your neighborhood. For specific recommendations, see your county extension director, Soil and Water Conservation District (SWCD), or local forester. Some suggestions follow.

SHRUBS

Plant shrubs about 4 feet apart within the row. Choose from: Tatarian honeysuckle, White Belle honeysuckle, purple-osier willow, lilac, Caragana, Amur honeysuckle, Zabels honeysuckle, and redosier dogwood.

TALL SHRUBS OR MEDIUM HEIGHT DECIDUOUS TREES

Plant these trees 5 to 8 feet apart in the rows. Choose from: golden willow, laurel-leaf willow, Russian olive, American wild plum, Amur maple, boxelder, chokecherry, Siberian crab, or one of the hardy strains of flowering crab.

TALL DECIDUOUS OR CONIFER TREES

Spacing of trees within the row (as well as between rows) is very critical. The closer the spacing the sooner the lower branches of adjacent trees begin to touch, the earlier you get protection from wind and snow. However, when lower branches begin to crowd they will die out from shading and this dying of branches progresses up the tree over time. Also, the closer the spacing, the greater the competition for soil moisture and the more tree vigor and growth will be reduced. Therefore, the eventual or final spacing of trees in well-established shelterbelts should be a minimum of 16 feet—preferably 18 to 20 feet. If you wish to plant closer than 16 feet you must plan on thinning later so that the permanent trees are 16 to 20 feet apart. Thin before lower branches begin to die from crowding. To end up with 16 to 20 foot spacing after thinning, you could plant at 5-, 6-, 8-, 9-, or 10-foot intervals.

You should consider the following species for your shelterbelt: Deciduous trees—green ash, hackberry, cottonwood, Norway poplar, white willow, and silver maple. Silver maple and white willow grow rapidly and give early protection, however, they should be considered as temporary trees and removed later as they take up too much room and will crowd adjacent rows of trees even when rows are 20 feet apart. In southeastern and southern Minnesota, consider also walnut, butternut, and hard maple. Conifers—red (Norway) pine, ponderosa (western yellow) pine, Scotch pine, white pine, Norway spruce, Douglas fir (of hardy seed source), and Siberian larch (sheds needles in fall).

The standard 8-row windbreak includes four rows of tall trees. These may be all deciduous species or you may choose to include one or two rows of pine. Be sure to locate the pines next to the inside spruce rows. Leave a 16- to 20-foot space between the last row of deciduous trees and the adjacent row of pine.

CONIFERS FOR INNER ROW

Plant trees 8 feet or more apart in the row. Choose from: Black Hills spruce, Colorado spruce, white spruce, Norway spruce, and white cedar. Eastern redcedar should also be considered. However, since it (as well as other junipers) is the alternate host for the Cedar Apple Rust Fungus, it should not be planted if there are apple trees within the farmstead.

Spruce are recommended for the inner two rows because they have dense foliage and maintain their limbs close to the ground for most of their life. This growth characteristic helps prevent dust and snowstorms from filtering through the treebelt.

Other alternatives to two rows of spruce are (1) spruce and white cedar, and (2) spruce and flowering shrub or flowering crab, with the cedar, shrub, or crab used in the innermost row.

If limited space prevents your planting the standard eight-row windbreak, plan to use fewer rows rather than crowding trees. For example, five rows of trees with room to grow give better results than eight overcrowded rows. If you must plant fewer than eight rows, use this guide:

If you only have room for:	Use one of these combinations from figure 5:
7 rows	1,2,3,4,6,7,8 or 1,2,3,5,6,7,8
6 rows	1,2,3,4,7,8 or 1,2,3,5,7,8
5 rows	1,3,4,7,8 or 1,2,4,7,8
4 rows	1,2,7,8 or 1,4,7,8
3 rows	1,7,8 or 1,3,7

TREE CHARACTERISTICS
DECIDUOUS TREES

	Height	Width	Growth Rate	Hardiness Zone	Moisture Tolerance	Drought Tolerance	High pH Tolerance	Shade Tolerance	Salt Tolerance	Fall Color	Fruits/ Flower	Form	Remarks
Green Ash	50′	30′	Fast	2	Yes	Yes	Yes	Yes		Yellow		Round	Very adaptable
Marshall's Green Ash	50′	30′	Fast	2			Yes			Yellow		Round	Shiny green foilage
Summit Green Ash	50′	30′	Fast	2			Yes	Yes				Oval	Narrower than Marshall's
White Ash	60′	30′	Med-Fast	3b			Yes	Yes	Yes	Yellow-Red		Oval	Native stock recommended
River Birch	45′	25′	Fast	4a	Yes			Yes	No			Oval	Reddish bark; prefers moist, sandy soil
Ohio Buckeye	40′	25′	Slow-Med	2						Yellow	Yes	Oval	Good for small yards
Amur Corktree	35′	30′	Med	4b		Yes	Yes			Yellow	Yes	Vase	Few pests; corky bark; hardy on protected sites in zone 4a
Flame Flowering Crabapple	25′	15′	Med	2							White	Round	Disease resistant; holds fruit
Radiant Flowering Crabapple	25′	15′	Med	2							Red	Round	Bright red fruit; susceptible to scab
Red Splendor Flowering Crabapple	25′	15′	Med	2							Red	Oval	Holds fruit; disease resistant
Vanguard Flowering Crabapple	15′	10′	Med	2							Pink	Vase	Protect all crabs from sunscald
Ginkgo	45′	30′	Slow	[illegible]		Yes	Yes		Yes	Yellow		Pyr	Few pests; use male only; may be hardy on protected sites in zone 4a
Hackberry	70′	40′	Med	3b	Yes	Yes	Yes			Yellow	Yes	Round	Pebbled bark; very adaptable; may be hardy on protected sites in zone 3a
Imperial Honey Locust	50′	30′	Fast	4b		Yes	Yes		Yes	Yellow		Round	Compact form; lacy foilage; may be hardy in southern part of zone 4a
Skyline Honey Locust	60′	45′	Fast	4b		Yes	Yes		Yes	Yellow		Pyr	Casts light shade; easy on turf; may be hardy in southern part of zone 4a
Ironwood	45′	25′	Med	2		Yes		Yes	No		Yes	Round	Few pests; may be hard to find good "clump" tree
Kentucky Coffeetree	60′	35′	Slow	4a		Yes		Yes			Yes	Round	Attractive bark, sturdy branches; slow to leaf out in spring
Japanese Tree Lilac	20′	15′	Slow	3a							White	Oval	Flowers late June/early July
American Linden (Basswood)	60′	40′	Med	2	Yes			Yes	No		Yes	Oval	Fragrant flowers/late June
Littleleaf Linden	45′	45′	Slow	4a					No		Yes	Round	Pest resistant; fragrant flowers; may be hardy in southern part of zone 3b
Greenspire Linden	45′	25′	Slow	4a					No		Yes	Pyr	Narrower than Littleleaf; may be hardy in southern part of zone 3b
Redmond Linden	45′	25′	Slow	4a								Pyr	Reddish winter branches; may be hardy in southern part of zone 3b
Amur Maple	20′	20′	Med-Slow	2						Red-Orange	Yes	Oval	Excellent fall color
Norway Maple	40′	30′	Med	4b				Yes	Yes	Yellow	Yes	Round	Protect all maples from sunscald; also hardy in southern part of zone 4a
Cleveland Norway Maple	40′	30′	Med	4b				Yes		Yellow		Oval	One of the best Norways; also hardy in southern part of zone 4a
Emerald Queen Norway Maple	40′	30′	Med	4b				Yes		Yellow		Round	Dark glossy foilage; also hardy in southern part of zone 4a
Schwedler Norway Maple	40′	30′	Med	4b								Round	Leaves red in spring, turning green in summer; also hardy so. part of zone 4a
Red Maple	50′	30′	Med	2	Yes			Yes	No	Red-Yellow	Yes	Oval	Best on acid soil; native stock recommended
Sugar Maple	60′	45′	Med-Slow	2				Yes	No	Red-Orange	Yes	Oval	Best on heavy moist soil; excellent fall color
Mountain Ash	35′	20′	Med	2							Red-White	Oval	Red fruit attractive to birds; protect from sunscald
Pin Oak	60′	45′	Med	4a	Yes				No	Red	Yes	Pyr	Requires acid soil; graceful sweeping branches; trial in zone 3b
Russian Olive	20′	15′	Med	3b	Yes	Yes	Yes		Yes		Yes	Round	Silvery foilage; fragrant flowers; may be hardy on protected sites in zone 3a
Black Walnut	60′	45′	Med	4a							Yes	Round	Hard on turf

EVERGREEN TREES

	Height	Width	Growth Rate	Hardiness Zone	Moisture Tolerance	Drought Tolerance	High pH Tolerance	Shade Tolerance					Remarks
Eastern Arborvitae	40′	15′	Slow-Med	2	Yes		Yes	Yes					Often sheared for use as hedge; susceptible to winter browing
Balsam Fir	40′	25′	Slow	2	Yes			Yes					Aromatic; protect from NW winds; requires cool, moist atmosphere
Douglas Fir	40′	30′	Med	4a	Yes								Plant trees from Northern Rocky Mountains only
White Fir	35′	25′	Slow-Med	4a									Avoid dry, windswept areas; pyramidal form
Eastern Hemlock	50′	35′	Med	3a				Yes					Must be protected from wind and winter sun
Eastern Larch	60′	35′	Slow-Med	2	Yes								Needles brilliant gold in fall; needles drop in fall
European Larch	25′	45′	Med-Fast	4a	Yes								Also turns gold in fall and drops needles; may be hardy in southern 3b
Austrian Pine	40′	30′	Med	4a	Yes	Yes							Symmetrical; stiff dark green needles; good windbreak
Ponderosa Pine	40′	35′	Fast	3a		Yes	Yes						One of the best for windbreaks; needs large space – ornamental
Red Pine	60′	35′	Med	2		Yes							Does well in exposed sites; not recommended for small sites
Scotch Pine	50′	35′	Fast	3a									Orange bark; not recommended for Western Minnesota
Colorado Spruce	40′	25′	Slow-Med	3a	Yes		Yes						Often needs to be replaced after 20 years
Norway Spruce	50′	35′	Med-Fast	3a	Yes								Pendulous appearance when old
White Spruce	40′	25′	Med-Fast	2	Yes								Pyramidal form

HARDINESS

Our winters dictate selecting plant species that grow well in Minnesota. Trees that have grown in northern nurseries are better acclimated to withstand our winters. The map below can be used as a guide to selecting hardy species for your area. The hardiness zones, however, are not absolute rules. For instance, a Zone 4 tree planted on an exposed, windward site in southern Minnesota may winterkill; whereas the same tree, if planted in a protected, windless Zone 2 location, may survive.

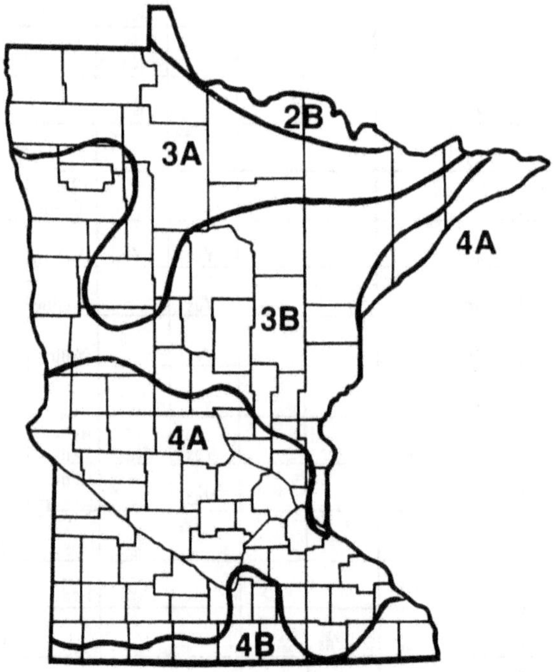

The "Tree Characteristic Chart" is from a booklet titled the "Planting Handbook".

This handbook was published by the Minnesota Department of Agriculture Shade Tree Program with the cooperation of the Minnesota Department of Natural Resources and the University of Minnesota Agricultural Extension Service, April 1980.

It is designed to help individuals plant replacement trees for those lost to Dutch elm disease or simply a new one where there's a need.

The handbook is not intended as a guide to landscaping. Those wishing to landscape their yards should consult their local nursery.

VARIATIONS IN SIZE, GROWTH RATE, AND LONGEVITY OF NORTH AMERICAN CONIFERS

From: "How Forest Trees Grow" Theodore T. Kozlowski

Common Name	Scientific Name	Maximum height	Maximum diameter	Growth Rate	Longevity
		(feet)	(feet)		(years)
Arbovitae (see Whitecedar)					
Douglas fir	*Pseudotsuga menziesii*	270	15	Rapid	500-800
Balsam fir	*Abies balsamea*	85	3	Rapid	100-150
Fraser fir	*A. fraseri*	65	2.5	Moderate	200-300
Grand fir	*A. grandis*	250	6	Moderate	200-400
White fir	*A. concolor*	200	6	Moderate	100-400
Hemlock (eastern)	*Tsuga canadensis*	160	6	Slow	300-600
Juniper (see Redcedar)					
Larch (see Tamarack)					
Jack pine	*Pinus banksiana*	90	2	Rapid	80-150
Jeffrey pine	*P. jeffreyii*	130	9	Moderate	300-500
Loblolly pine	*P. taeda*	190	5	Rapid	150-250
Lodgepole pine	*P. contorta*	150	3	Slow	120-300
Longleaf pine	*P. palustris*	150	4	Rapid	300-400
Pinon pine	*P. edulis*	50	3	Very slow	150-400
Pitch pine	*P. rigida*	100	3	Rapid	100-200
Ponderosa pine	*P. ponderosa*	235	9	Moderate	300-500
Red pine (Norway Pine)	*Pinus resinosa*	150	5	Moderate	200-400
Shortleaf pine	*P. echinata*	150	4	Rapid	200-300
Slash pine	*P. elliottii*	130	3	Rapid	150-250
Sugar pine	*P. lambertiana*	250	10	Rapid	300-600
Virginia pine	*P. virginiana*	100	3	Moderate	100-200
White pine (eastern)	*P. strobus*	220	6	Rapid	300-500
White pine (western)	*P. monticola*	120	8	Rapid	200-500
Redcedar (eastern)	*Juniperus virginiana*	100	4	Slow	150-300
Redwood	*Sequoia sempervirens*	365	20	Rapid	800-1500
Giant Sequoia	*S. gigantea*	350	38	Rapid	2000-3000
Black spruce	*Picea mariana*	100	3	Slow	150-250
Red spruce	*P. rubens*	120	4	Slow	200-300
Sitka spruce	*P. sitchensis*	300	16	Rapid	400-750
White spruce	*P. glauca*	120	4	Slow	150-350
Tamarack	*Larix laricina*	100	3	Moderate	100-200
Whitecedar (northern)	*Thuja occidentalis*	125	6	Slow	300-400

VARIATIONS IN SIZE, GROWTH RATE, AND LONGEVITY OF NORTH AMERICAN BROADLEAVED TREES

Common Name	Scientific Name	Maximum height	Maximum diameter	Growth Rate	Longevity
		(feet)	(feet)		(years)
Black ash	*Fraxinus nigra*	90	5	Slow	200
Green ash	*F. pennsylvanica*	85	2.5	Rapid	200
White ash	*F. americana*	125	6	Rapid	260-300
Bigtooth aspen	*Populus grandidentata*	80	3	Rapid	70-100
Trembling aspen	*P. tremuloides*	120	4.5	Very rapid	70-100
Balsam poplar	*P. balsamifera*	100	5	Rapid	100-150
American basswood	*Tilia americana*	125	5	Rapid	100-140
American beech	*Fagus grandifolia*	120	4	Slow	300-400
Grey birch	*Betula populifolia*	60	1.5	Rapid	50
River birch	*B. nigra*	100	5	Rapid	—
White birch	*B. papyrifera*	120	5	Rapid	80-100
Yellow birch	*B. alleghaniensis*	100	4	Rapid	150-300
Blackgum	*Nyssa sylvatica*	100	4	Rapid	—
Yellow buckeye	*Aesculus octandra*	100	4	Rapid	60-80
Butternut	*Juglans cinerea*	110	3	Rapid	80
Catalpa	*Catalpa speciosa*	120	5	Rapid	100
Black cherry	*Prunus serotina*	100	5	Rapid	100-200
Cottonwood (eastern)	*Populus deltoides*	175	11	Very rapid	60-100
Black cottonwood	*Populus trichocarpa*	225	8	Rapid	150-200
Flowering dogwood	*Cornus florida*	50	1.5	Slow	125
American elm	*Ulmus americana*	120	11	Rapid	150-300
Red elm	*U. rubra*	90	4	Rapid	300
Hackberry	*Celtis occidentalis*	130	5	Rapid	75-150
Bitternut hickory	*Carya cordiformis*	85	4	Slow	175
Mockernut hickory	*C. tomentosa*	100	3.5	Slow	200-300
Pecan (hickory)	*C. illinoensis*	180	6	Moderate	300
Pignut hickory	*C. glabra*	120	4	Slow	200-300
Shagbark hickory	*C. ovata*	120	4	Slow	250-300
American holly	*Ilex opaca*	140	4	Slow	100-150
Honeylocust	*Gleditsia triacanthos*	140	6	Rapid	120
Ironwood or Hophornbean	*Ostrya virginiana*	55	1.5	Slow	—
Black locust	*Robinia pseudoacacia*	100	5	Rapid	60-100
Red maple	*Acer rubrum*	120	5	Rapid	80-250
Silver maple	*A. saccharinum*	120	7	Rapid	50-125
Sugar maple	*A. saccharum*	135	5	Slow	200-300
Red mulberry	*Morus rubra*	50	1.5	Moderate	125
Black oak	*Quercus velutina*	55	7	Moderate	150-200
Blackjack oak	*Q. marilandica*	55	2	Slow	100
Bur oak	*Q. macrocarpa*	170	7	Slow	200-400
Northern red oak	*Q. rubra*	150	11	Rapid	200-400
Pin oak	*Q. palustris*	120	5	Rapid	125-150
Post oak	*Q. stellata*	100	4	Slow	250
Scarlet oak	*Q. coccinea*	110	4	Moderate	150
Southern red oak	*Q. falcata*	110	7	Moderate	200-275
Swamp white oak	*Q. bicolor*	100	7	Slow	300
Water oak	*Q. nigra*	125	5	Rapid	175
White oak	*Q. alba*	150	8	Slow	300-600
Persimmon	*Diospyros virginiana*	130	7	Slow	60-80
Sweetgum	*Liquidambar styraciflua*	200	6	Rapid	200-300
Sycamore	*Platanus occidentalis*	175	14	Rapid	250-300
Black walnut	*Juglans nigra*	150	7	Rapid	150-250
Black willow	*Populus trichocarpa*	225	8	Rapid	150-200
Yellow popular	*Liriodendron tulipifera*	200	12	Rapid	200-250

TIMBER PRODUCTION

Based on Site Index Yields in:

(a) Cords per acre per year (b) Cubic ft. per acre per year

	40 S.I.		50 S.I.		60 S.I.	
SPECIES	CDS	CU. FT.	CDS	CU. FT.	CDS	CU. FT.
Norway Pine	.40	31	.60	47	.90	70
White Spruce	.25	20	.50	39	.90	70
Jack Pine	.40	31	.50	39	.60	47
White Pine	.25	20	.50	39	1.0	78
Black Spruce	.45	35	.75	59	1.0	78
Balsam Fir	.35	27	.50	39	.70	55
White Cedar	.45	35	.65	51	.75	59
Tamarack	.30	23	.45	35	.65	51
Oaks	.20	16	.40	31	.50	39
Aspen	.10	8	.30	23	.50	39
Birch	.10	8	.30	23	.50	39

(1 cord)= 78 Cu. ft.)

	65 S.I.		70 S.I.		80 S.I.	
SPECIES	CDS	CU. FT.	CDS	CU. FT.	CDS	CU. FT.
Norway Pine	1.1	86	1.3	101	1.80	140
White Spruce	1.0	78	1.1	86	1.40	109
Jack Pine	.72	56	.75	59	.90	70
White Pine	1.2	94	1.45	113	1.75	137
Black Spruce	1.1	86	1.25	98		
Balsam Fir	.75	59	.85	66	1.05	82
White Cedar	.86	66	.95	74		
Tamarack	.67	53	.70	55		
Oaks	.70	55	.90	70		
Aspen	.60	47	.70	55	.85	66
Birch	.60	47	.70	55	.80	66

(1 cord = 78 cu. ft.)

SITE INDEX, ROTATION LENGTH, AND FINANCIAL YIELD
Of Common Minnesota Tree Species
(for use in FY 83)

Species	Site Index Adjective	Site Index Range	Rotation Age	Dollar Yield 1/ (per acre) All Regions
Norway Pine	Excellent	63+	80	$ 1107.00
	Good	54-63	80	999.00
	Medium	45-54	100	891.00
White Pine	Excellent	74+	80	$ 1134.00
	Good	64-74	80	1026.00
	Medium	54-64	100	918.00
Jack Pine	Excellent	71+	50	$ 364.00
	Good	58-71+	60	325.00
	Medium	45-58	60	286.00
Balsam Fir	Excellent	63+	50	$ 100.00
	Good	52-63	50	90.00
	Medium	41-52	60	75.00
White Spruce	Excellent	68+	60	$ 595.00
	Good	57-68	60	544.00
	Medium	46-57	70	476.00
Black Spruce	Excellent	42+	80	$ 400.00
	Good	35-42	90	272.00
	Medium	28-35	90	224.00
Tamarack	Excellent	57+	90	$ 138.00
	Good	47-57	100	102.00
	Medium	37-47	100	60.00
Cedar	Excellent	38+	80	$ 341.00
	Good	30-38	80	242.00
	Medium	22-30	90	143.00
Oak-Hickory 2/	Excellent	75+	80	$ 962.00 ($468.00)
	Good	65-75	80	851.00 (414.00)
	Medium	55-65	80	555.00 (270.00)
Elm-Ash-Cottonwood 2/	Excellent	75+	70	$ 704.00 ($320.00)
	Good	65-75	80	638.00 (290.00)
	Medium	55-65	80	572.00 (260.00)
Maple-Basswood 2/	Excellent	75+	70	$ 520.00 ($182.00)
	Good	65-75	80	480.00 (168.00)
	Medium	55-65	80	420.00 (147.00)
Walnut	Excellent	75+	80	$12760.00
	Good	65-75	80	11165.00
	Medium	55-65	90	7018.00
Aspen (Bi & B/G)	Excellent	80+	40	$ 168.00
	Good	70-80	40	108.00
	Medium	60-70	40	56.00

1/ Dollar Yield is based on predictions of volume yield (from yield tables, growth models, and Phase II Inventory Data), on a given site index for the specified rotation age, times the average selling price for that species (assuming an appropriate mix of products, as found on DNR timber sales).

2/ Dollar yields are for the southern three regions, the northern three regions' yields are shown in parenthesis.

VOLUME OF TIMBER SOLD AND AVERAGE PRICES RECEIVED BY PUBLIC AGENCIES – 1988

* Note: The Actual range of prices varied by more than 50%.

Product and Species	MBF Sold	Avg. Price
Sawtimber		
Aspen	49981.6	$15.89
Balm of Gilead	174.2	$12.12
Birch Species	4372.8	$18.71
Ash Species	960.5	$38.76
Elm Species	302.1	$34.24
Oak Species	1884.8	$60.50
Other Hardwoods	3837.7	$30.12
Balsam Fir	2457.4	$11.83
White Spruce	3283.2	$35.65
Black Spruce	818.9	$29.25
Tamarack	333.8	$17.46
White Cedar	349.0	$25.53
Jack Pine	12030.0	$43.49
Red/White Pine	30960.1	$66.85
All Species	111746.2	$35.19
PULPWOOD	**Cords Sold**	
Aspen	971126.9	$3.55
Balm of Gilead	9562.8	$2.31
Birch Species	40397.4	$1.96
Ash Species	2364.9	$3.05
Elm Species	153.0	$1.89
Oak Species	7284.9	$3.88
Other Hardwoods	16696.8	$2.37
Balsam Fir	129522.2	$3.49
White Spruce	39059.9	$9.29
Black Spruce	67970.3	$9.39
Tamarack	9819.6	$4.67
White Cedar	3331.8	$6.83
Jack Pine	75803.0	$9.69
Red/White Pine	17842.0	$7.48
All Species	1390936.0	$4.32
FUELWOOD	**Cords Sold**	
All Species	86454.6	$2.50

TREE SEEDLINGS GROWN IN CONTAINERS

CATCH ON ALMOST EVERYWHERE

by J.D. Schultz

Adoption of container rearing technology by tree nurseries has become widespread in several major areas of the United States, notably the Pacific Northwest and South. And these operations are common all across Canada. The technique has been slow to catch on in the Midwest and Northeast, however.

A few nurseries in these latter areas have been raising trees in containers for several years, and there is promise that more may do so in the future, but there is a natural reluctance to gear up for such operations in these regions long-accustomed to growing millions of tree seedlings outdoors in extensive nursery beds. The costs of establishing entirely new facilities, and of incurring higher electrical and heating energy requirements when the energy crunch tightens even more, have been responsible for much of the hesitancy to plunge into container operations. Purchasers of the seedlings have also questioned whether or not the trees would survive and grow any better than the traditionally grown nursery trees.

In the Midwest, two major forward thrusts in the container rearing business have been taking place. The Forestry Department at Michigan State University, under the direction of Dr. James Hanover, has been conducting tree improvement studies for several years. Dr. Hanover is a forest geneticist and the research by him and his students has been aimed at finding species of trees that grow more rapidly, become established more easily, produce more plant biomass, and possess other more desirable characteristics than others. His results may be of use to forest industry, to the Christmas tree industry and to those looking for ways to grow wood as fuel.

The research at Michigan State has concentrated on using what Dr. Hanover calls "accelerated-growth" technology. By growing young trees—from seedlings or from cuttings—under specially controlled light, temperature, water and nutrient regimes, the researchers have been able to test their theories about certain trees more quickly than could be done the traditional way.

In the Midwest, on the commercial side, only Vans Pines of West Olive, Michigan has been growing container seedlings in large quantities for several years. Vans has produced more than 500,000 pine seedlings in containers annually since the mid-1970's. A real selling job still needs to be done to convince tree planters and landowners in the Midwest that a container-grown tree has advantages over the bare-root seedling. The costs of the container trees are substantially higher than they are for the traditional trees.

Dr. Hanover believes that container rearing techniques will eventually revolutionize the nursery industry. He thinks the costs of starting trees in greenhouses is less per seedling when consideration is given to the facts that less land is tied up, less labor is required, and there is better quality control in the greenhouse. Of course, results are obtained much quicker, too.

Vans Pines and Dr. Hanover make several worthwhile points about accelerated growth seedlings, as follow:

1. The accelerated growth habit is stimulated through the use of controlled, ideal growing conditions, up to 24 hours a day. Seedlings are carefully maintained under strict watering and fertilizing schedules. Temperatures and humidity are carefully regulated, and seedlings are grown in special containers with a measured amount of growing medium for each plant.

2. Seeds used are specially selected from trees with superior genetic traits.

3. The roots of seedlings develop uniformly and are not damaged when shipped. This is one of the main advantages of the container rearing system because the entire root plug is planted intact and the roots never need be touched or exposed to drying out.

Seedlings have heavy caliper and longer, larger needle development. Buds are larger and have a greater quantity of stored nutrients.

The technique is being refined continually and it is inevitable that economies of operation will bring down the presently high cost of container grown seedlings. Experience in the West, the South and Canada has been exceedingly favorable. It can't be too long before the rest of the United States moves heavily in this direction. I've even just begun growing 6,500 seedlings under lights in my basement!

For further details, interested readers will find the best sources are two recent publications that may be obtained free. They are:

HOW TO GROW SEEDLINGS IN CONTAINERS IN GREENHOUSES by R. W. Tinus and S. E. McDonald, 1979, U.S. Forest Service. Rocky Mountain Forest and Range Experiment Station, Fort Collins, Colorado. General Technical Report RM-60. 256p.

GUIDELINES FOR REARING CONTAINERIZED CONIFER SEEDLINGS IN THE PRAIRIE PROVINCES by L.W. Carlson, 1979. Canadian Forestry Service, Northern Forest Research Centre. Edmonton, Alberta. Information Report NOR-X-214. 62 p.

Marion Van Slooten, President of Vans Pines, Inc., examines a crop of pines growing under the accelerated growth program.

CHAPTER 10

FOREST WEED CONTROL PRACTICES FOR SMALL WOODLAND OWNERS

Richard D. Iverson, Forest Weed Control Specialist
Department of Forest Resources
University of Minnesota

Introduction

Minnesota's forests are valuable. They contribute to the states economy through recreation, tourism, and wood-based industries. Approximately one-third of the state's total land area (16.7 million acres) was covered by forests in 1977 and of this amount about 13.7 million acres was classified as commercial forest land (Lewis 1984). Small woodland owners including farmers and private individuals own 38 percent of the commercial forest land in Minnesota (Jakes 1980). Because of this ownership pattern, small woodland owners can have a significant impact on the wood supply and long term productivity of Minnesota's forests.

Productivity can be significantly improved by implementing sound silvicultural practices. Current forest conditions in the Lake States are a result of past forest practices that left non-merchantable trees, brush, and non-stocked areas on many of the lands now owned by small woodland owners. The natural re-establishment of pine was precluded in part by the lack of natural seed sources, poor seedbed conditions, excessive competition from herbs and hardwoods, adverse weather, and frequent wildfires in the late 1800's and early 1900's. It is estimated that an additional 6 million acres once supported red pine stands in the Upper Great Lakes Region (Benzie 1982). Artificial regeneration of conifer species is the best way to insure establishment of plantations and the continued productivity of pine and spruce types. The management of forest weeds is an essential part of the artificial regeneration process.

In the paragraphs which follow, I will describe where forest weed control practices fit in the establishment of conifer plantations, discuss treatment alternatives, and explain the use of herbicides for brush and weed control.

Site Preparation

Site preparation is necessary before planting to insure maximum seedling survival and growth. Site preparation will create additional plantable spots by removing residual debris and vegetation left after logging and if done properly it will control undesirable vegetation that will compete with the seedlings for sunlight, water, and nutrients.

Site preparation can be accomplished by using fire, heavy equipment, herbicides, or a combination of these tools. Fire can be used to reduce slash and debris left on the site after logging, but use of broadcast burning for site preparation in the Lake States is seldom practiced because conditions are rarely suitable to obtain a good burn without risk of causing a wildfire. Fire is often used in conjunction with mechanical methods to eliminate the debris left in windrows and piles after windrowing or brush raking.

Several types of heavy equipment are used to prepare sites for planting. Crawler tractors with blades are often used to remove undesirable brush and hardwoods. Blades are designed to meet varying site conditions and site preparation objectives. The blade rake, for example, promotes more thorough mixing of the soil organic layer with the mineral soil than a straight bulldozer blade. However, the Rome K/G Clearing Blade is advantageous on lowland sites during winter months. Other equipment is available which can be pulled behind a crawler tractor or skidder to plow, disc, or scarify the site. A complete description of all this heavy equipment is contained in a silvicultural equipment reference catalogue for Northern Ontario (Smith 1979). Site preparation with heavy equipment can be done by contractors who usually have the type of equipment that is needed to match conditions on the site.

Herbicides can also be used to prepare sites for planting. Herbicides will not remove physical impediments to planting, but they do an excellent job of controlling potentially competing vegetation. Controlling weeds with herbicides is also easier before planting than afterward because applications can be made without major concern about damage to crop trees.

Site Preparation Herbicides

There are many products registered for forest site preparation in Minnesota. The right product to use will depend on characteristics of the site, species of weeds, and the methods of application. Sometimes products can be combined to improve the effectiveness of the treatment on certain hard-to-kill species such as red maple.

Following is a description of the major herbicides used for site preparation. Table 1 provides information about the critical elements of herbicide prescriptions: 1. Major weeds controlled; 2. Time of application; 3. Rate of application; and, 4. Method of application. This information highlights details contained in the labels and should not be used in lieu of the label. The label should be carefully read before using any pesticide product.

2,4-D

There are a variety of brand names for 2,4-D including Esteron® 99C, Weedone® LV-4,

Formula® 40 and Weed Rhap® . Most formulations of 2,4-D are applied to actively-growing foliage in a mixture with water. Formula 40 is used in undiluted form for tree injection.

The herbicide 2,4-D belongs to a class of herbicides known as growth regulators. Growth regulators affect plant growth and appear to act at the same site as the natural plant auxin IAA (indole acetic acid). Growth regulators are transported within plants by means of the water and food transporting systems. Because of this process, known as translocation, the growth regulators can be effective when only part of the plant is treated and the herbicide can be applied by low pressure and low volume sprays as well as wiping applications. Grasses are generally resistant to growth regulators.

2,4-D + 2,4-DP

The combination of 2,4-D and 2,4-DP is sold under the brand name Weedone® 170. Dichlorprop is a common chemical name for 2,4-DP and it is also a growth regulator. The combination of 2,4-D and 2,4-DP is more effective on hard-to-kill species such as oak than is either herbicide used alone.

Glyphosate

Glyphosate, sold under the brand names Rodeo® and Roundup® , affects plants by inhibiting the synthesis of chemicals within the plant essential for development (the amino acids phenylalanine, tyrosine, and tryptophane). Glyphosate is a broad-spectrum postemergence herbicide that is readily translocated throughout grass and broadleaf plants. Low spray volumes are more effective than higher volumes with glyphosate and injury symptoms are slow to appear. A week or more is required to see control on annual plants and much longer periods for perennials, especially woody plants.

Hexazinone

Hexazinone is available in a liquid formulation under the brand name Velpar® L, a wettable powder formulation under the brand name Velpar, and a granular formulation under the brand names Pronone® 10G and Pronone® 5G. Hexazinone belongs to a class of herbicides that inhibit photosynthesis and it can enter the plant through the foliage and root system depending on the formulation used.

The proper rate of application for hexazinone depends on the soil texture and amount of organic matter where it is applied. Soils with a high amount of clay and organic matter will require higher application rates than light-textured soils with low organic matter. The labels list application rates for various soil types.

Triclopyr

Triclopyr is sold under the brand name Garlon® . It is a growth regulator and therefore has the same mode of action as the phenoxy herbicides, 2,4-D, and 2,4-DP. It is effective on a wide range of broadleaf weeds and hardwoods, but will not affect most sedges or grasses. Garlon 4 is formulated as an emulsifiable concentrate with 4 pounds of triclopyr acid equivalent per gallon and Garlon 3A is formulated as triethylamine salt with 3 pounds of triclopyr acid equivalent per gallon.

Simazine

Simazine is a photosynthetic inhibitor like hexazinone and is sold under several brand names including Princep® . It is available in liquid, wettable powder, and granular formulations. Simazine is most effective if it is applied to bare mineral soil before weeds emerge. It will control a variety of grasses and broadleaf herbaceous weeds.

Sulfometuron Methyl

Oust® is the brand name for sulfometuron methyl. It belongs to a fairly new chemical class of herbicides called the sulfonyl ureas which affect target plants by inhibiting synthesis of chemicals essential for proper plant development (branch chain amino acids, leucine, isoleucine, and valine). Oust is formulated as a water dispersible granule and it will control a variety of annual and perennial grasses and broadleaf weeds. Oust can be applied pre-emergent or post-emergent to weeds.

Amitrole + Simazine

The combination of amitrole + simazine is sold under the brand name Amizine® . Amitrole affects plants by inhibiting formation of pigments in the leaves. New growth on affected plants become almost white. Amitrole is translocated throughout the plant by both the water and food transport systems. All plants are affected by amitrole. The combination of amitrole and simazine, therefore, provides both pre-emergent and post-emergent control of weeds.

Picloram

Picloram is available alone in a pelletized formulation called Tordon® 10K or in a liquid formulation called Tordon® K. Picloram is a growth regulator that is highly mobile in the soil and persists for longer periods than other growth regulators. Consequently, some products with picloram in them have been classified as restricted use pesticides by the Environmental Protection Agency. The restricted use classification for picloram was not declared on the basis of its toxicology because it has low mammalian toxicity. Picloram is often tank mixed with other herbicides to improve effectiveness for site preparation.

Picloram + 2,4-D

The brand name for picloram + 2,4-D is Tordon® 101. Tordon 101 is applied as a broadcast foliar spray and will kill a variety of shrubs, hardwood trees, conifers, and various broadleaf weeds. It is not effective on grasses. Tordon 101 and Garlon 4 can be tank mixed to enhance control in areas where there is a rich mixture of hardwoods.

Dicamba

Dicamba is a growth regulator sold under the trade name Banvel® . Banvel will control many broadleaf weeds, brush species, hardwoods, and conifers. Banvel can be mixed with one or more additional herbicides to control grasses, additional broadleaf weeds, brush species, and trees.

Release

After trees are planted, weed control is often necessary to control weeds that were not controlled by site preparation, or weeds that develop in the favorable environments created by site preparation. Woody competitors such as aspen or hazel will resprout from stumps and roots even after mechanical removal and herbaceous weeds may develop on sites where woody plant herbicides are used.

There is currently no objective quantitative method for deciding whether or not a plantation needs release. Professional foresters use their judgement based on the adequacy of plantation stocking and some criteria of competition severity. Their goal is to treat only those plantations in which the cost of treatment is less than or equal to the value of additional wood volume made possible by the treatment. A quantitative method that would insure reaching this goal involves many variables including the inherent productivity of the site, the cost of the treatment, the response of the conifers to control of various species of herbaceous and woody weeds, and the future value of the wood which is grown. Such a system would be complicated and is needed mainly for decisions in marginal situations.

Studies have shown that conifers released from competing vegetation will produce more wood volume than their counterparts left to grow with competing vegetation (Stewart et.al. 1984). Conifers that are overtopped by hardwood trees and woody shrubs or completely surrounded by dense woody vegetation will generally respond favorably to release treatments. The degree of response depends on the conifer species and its age when treated. Dense grass and herbaceous weeds will also compete with tree seedlings, especially during dry growing seasons. Control of herbaceous weeds may even aid survival of newly planted seedlings.

Release treatments can be accomplished by using herbicides, cutting or removing the competitors, using mulches to control herbaceous weeds, or by controlled grazing and browsing. Woody vegetation can be removed or cut with chain saws and other equipment, but these methods are very time consuming, dangerous to implement, and the effects are short-lived. Woody vegetation that has been cut will often produce more shoots than were present originally and reach heights that existed during treatment in just a few years (Roberts 1980). Mulches are effective, but they are expensive and must be maintained. Grazing can be accomplished with sheep or cattle. However, fences are necessary for cattle and herders may be necessary for sheep. Herbicides are the most common and effective method for accomplishing release objectives.

Herbicides for Release

Most of the herbicides used for site preparation can also be used for release. However, rates and time of application are often different when the herbicides are applied over the tops of conifers. Also, some herbicides can be used in some species of conifers, but not others. Hexazinone, for example, can be used on red pine, but it will damage jack pine. The label or supplements to the label will specify the species of conifers that the herbicide can be used on and it will provide other information about rate and time of application. Following is a list of herbicides that can be used to release conifers from competing vegetation (Information for making prescriptions is contained in Table 1):

2,4-D
2,4-D + 2,4-DP
Glyphosate
Hexazinone
Triclopyr
Simazine
Sulfometuron Methyl

Herbicide Application Equipment

Herbicides can be broadcast, banded, or spotted to the site, injected directly into undesirable hardwood trees, or applied as a spray to the lower stem. The appropriate application method will depend on the age and type of vegetation needing control, the type of herbicide used, and the age and species of the crop tree.

Spot application of herbicides are made with hand sprayers to control individual weeds in a forest plantation, or to control weeds within a specified radius of each crop tree. Spot applications use less herbicide than broadcast, or band applications, but take more labor to apply.

Band applications are made over rows of seedlings with hand sprayers, or with mechanized equipment at variable widths. Less herbicide is needed (probably at less cost) in a band rather than a broadcast application.

Broadcast applications over large acreage needs to be done by motorized ground equipment or aircraft. Broadcast applications are usually necessary for control of multi-layered woody competition. Aerial application of foliar-active herbicides is necessary for control of weed trees beyond the reach of ground equipment.

Basal spray applications, injection, and stump treatment are used to control individual stems or clumps of woody species. Basal sprays are applied to the bottom 12 to 15 inches of the hardwood stem. Injections are made by applying the herbicide in cut surfaces around the trunk, or by using special tools. Stump sprouting can be controlled by treating the cut surface with a selected herbicide. This can be done at any time of year while a foliar application is limited to the growing season.

Herbicide granules can be applied from the ground or the air. Aerial applications are usually made by helicopters equipped with buckets or aerial seeding equipment. Ground applications can be made from skidders equipped with rotary-type application buckets, or with machinery manufactured by Omni Spray, Inc. The Omni Spray equipment uses air-driven streams to distribute the granules through tubes. Granules can also be applied by hand using rotary-type applicators commonly used for grass seeding or fertilization of lawns.

Calibration of Equipment

The calibration of application equipment is an essential part of sound forest weed control practices. The application of dosages lower than those recommended on the label may not adequately control weeds and may create the need for another application. The application of dosages, higher than those recommended may injure crop trees and at best unnecessarily increase herbicide cost. The cost of poor chemical application is estimated to be $1 billion nation wide. In almost all cases application errors result from poor calibration, or defects in equipment design or operation.

Determining Herbicide Amount

Recommendations are sometimes expressed in terms of the amount of active ingredient needed per acre rather than the amount of product needed per acre. Consequently, it is necessary to convert the recommendation in terms of active ingredient to units of product before the herbicide can be mixed. The herbicide label will specify how much active ingredient is contained in the product. The amount of active ingredient in liquid formulations is expressed in pounds per gallon and the amount of active ingredient in wettable powders and granules is expressed as a percent of product weight. The amount of herbicide needed for application to one acre can be easily calculated for each herbicide formulation.

For liquids:

$$\text{gals/acre} = \frac{\text{lb/acre of active ingredient recommended}}{\text{lb of active ingredient/gallon}}$$

Example: If the recommendation is 2.5 pounds active ingredient per acre and the herbicide contains 4 pounds active ingredient, then the gallons/acre needed of herbicide $= \frac{2.5}{4} = .625$ gal. = 5 pints.

For wettable powders or granular herbicides:

$$\text{Lbs/acre} = \frac{\text{lb/acre of active ingred. recom. X 100}}{\text{percent active ingredient in formulation}}$$

Example: If the recommendation is 5 pounds active ingredient/acre and the herbicide contains 80% active ingredient, then the pounds/acre of herbicide to use $= \frac{5 \text{ X } 100}{80} = 6.25$ pounds.

After determining the amount of herbicide product needed per acre, it is possible to determine the amount of herbicide needed in the spray tank by multiplying the recommended herbicide rate (in terms of product units not active ingredient units) times the acreage covered with a full spray tank:

Amount of herbicide needed = recommended herbicide rate X acreage covered with a full spray tank.

The number of acres covered by a full spray tank will depend on the following four variables: 1. nozzle capacity, 2. pump pressure, 3. sprayer speed, and; 4. effective swath width. The proper combination of variables will deliver the specified amount of solution per acre in a uniform pattern. Spray equipment dealers will be able to design systems that meet your particular needs. Nevertheless, once you have a system, it should be periodically checked to insure you are still getting the prescribed delivery rate.

Calibration

Properly calibrated equipment will deliver a known amount of spray solution to a known amount of spray area. The performance of your sprayer can be checked in two ways. In one method the sprayer is filled with a known volume of water, an area of known size is sprayed, and then the amount of water needed to refill the sprayer to the original capacity is measured. Another way is to fill the sprayer with a known volume of water, spray it on an area in the same manner that you would apply herbicide, then measure how much area the spray covered. Both methods allow you to determine the spray delivery rate in gallons per acre by converting the units used for measurement to gallons and acres.

Example: If it takes 3 pints to refill a 4 gallon backpack sprayer that was used to spray 1000 square feet, the gallons per acre delivered by the sprayer can be calculated as shown below:

1. Gals applied $= \frac{\text{3 pints applied}}{\text{8 pints/gallon}} = .375$ gals applied

2. Acres applied $= \frac{\text{1,000 sq ft}}{\text{43,560 sq ft/acre}} = .023$ acres app

3. Gals/acre $= \frac{\text{.375 gals}}{\text{.023 acres}} = 16.30$ gallons per acre

The delivery rate of granular applicators can be checked by the same methods used for sprayers with the exception that product and delivery rate is measured in terms of weight and not liquid volume. Manufacturers directions must be followed to determine the effective swath width for rotary spreaders.

Sources of Additional Information

Much more information is available on the use of herbicides. Chemical company representatives will have detailed information concerning herbicide use and characteristics of the chemical. An excellent reference which summarizes chemical characteristics, major uses, and toxicology of all the herbicides is **Herbicide Handbook of the Weed Science Society of America.** It is available from: Weed Science Society of America, 309 W. Clark Street, Champaign, Illinois 61820.

WEED AND BRUSH CONTROL: Herbicide Use

Alan C. Jones

Forest Pest Specialist - Minnesota Department of Natural Resources

A herbicide is one of several tools which can be used by a forest landowner to help manage competition. You, as landowner and manager, must ultimately make a decision which tool should be used. Consider the following when deciding which is the best tool:

1. Competition is Present on the Site: This may seem obvious, but make sure you do have a competition problem. Just because there are noncrop species of trees and shrubs on your site does not mean a problem exists. Your plantation does not have to look like a bare ground agricultural field. Try to compare height growth between crop trees growing out in the open and crop trees growing among the noncrop trees or brush. If there is a difference of 50% or more, treatment will be needed. Also, put in a series of circular milacre plots throughout the area you want to treat. With plantations, use the crop trees as plot centers. If 50% or more of the plots have thick grass sod, or all of the plots average at least two overtopping woody stems and at least 50% of the plots have at least one overtopping woody stem, competition is probably severe enough to make treatment necessary.

2. Competition Has Developed: Most herbicides, except for the pre-emergent grass herbicides, work best when there is leaf surface to absorb the herbicide. Even in the case of soilactive herbicides such as picloram (Tordon) and hexazinone (Velpar), leaf uptake is also important. For best results, be sure that the target vegetation is above ground, has fully expanded leaves, is actively growing, and is not under stress.

3. Species Conversion is the Objective: When planting conifers on sites formerly occupied by hardwoods, herbicide use will nearly always be required. The tremendous sprouting capability of most hardwood root systems will quickly reestablish the hardwoods and choke out the conifers. Herbicides used especially during site preparation will help to kill out the root systems and give the newly planted conifers a chance to establish and grow.

4. Large Areas Are To Be Treated: Often, herbicides are the most cost effective treatment, especially when areas are large enough to be aerially treated. Consider using herbicides by aerial application on sites 20 acres or larger in size. On large areas with spotty competition problems, consider treating only those areas with competition rather than aerially spraying the entire site.

5. Area To Be Treated is Remote or Inaccessible: Since herbicides kill both tops and roots, one application is usually sufficient. Remote or inaccessible areas, then, would have to be treated only once and could be done with a helicopter reducing both expense and effort.

Back Pack Mist Blower
Sometimes used for herbicide applications.

6. Funding is Limited: Herbicide costs typically range between $30 and $60/acre; costs for competition reduction with heavy equipment range between $60 and $120/acre; burning costs range between $40 and $140/acre; and costs for utilizing hand labor range between $150 and $200/acre. Heavy equipment, burning and hand labor rarely disturb the root systems, and resprouting, retreating and spending of more funds are the results of not using herbicides.

7. Area is Close to Homes, Water or Neighboring Property: There are no herbicides registered for forestry applications that can be applied into water. Most neighbors do not want herbicides to drift or run onto their homes and properties. If your treatment area is close to any of these features, either use noherbicide-treatment buffer strips or choose an alternate means of treatment.

8. Neighbor Attitudes and/or Local Political Climate is Antiherbicide: Herbicide use is a very controversial issue. If your use of herbicides may lead to legal litigation or hostile neighbor action, consider alternate means of treatment.

9. A Herbicide Can Be Found Which is Legal to Use and Will Accomplish Your Objectives: Be sure you read the label of the specific herbicide you wish to use. Not all herbicides can be used on all sites. For example, there are many formulations of 2,4D but there are only a few that can be used on forestry sites in the Lake States. The label will tell you what kind of site on which to legally use the herbicide, the legal rates to use, the target vegetation, the crop trees on which the herbicides can be safely applied, the time of the year to apply the herbicide, and precautions to be aware of. Read and be familiar with the label before making the final decision to use a herbicide.

To help guide you in your planning and decision-making, the following is a guide for herbicide selection for forestry uses in Minnesota as of January, 1984. Because pesticide registrations and regulations vary from state to state and they are constantly changing, consult state forestry personnel, consulting foresters and/or extension agents before making your final decision.

HERBICIDES IN MINNESOTA FORESTRY

Agricultural Extension Service
University of Minnesota - U.S.D.A.

A variety of herbicides are currently used in forest management activities. Some of the more common substances are 2,4-D, Dicamba, Dichlobenil, Diuron, Glyphosate, MCPA Picloram, Picloram and 2,4-D, Simazine, Trifluraline, Amitrole-T, and others. Regarding the use of 2,4,5-T and Silvex; note the following statement:

"On February 28, 1979, the Environmental Protection Agency announced emergency suspension of uses of 2,4,5-T products on forests, rights-of-way and pastures; suspension of Silvex products registered uses for forests, rights-of-way, pastures and home, aquatic, and recreation areas. Concellation products were initiated at the same time. Decisions on these uses of products will not be known until final actions are taken. During the interim period, these uses are not legal."

The amount of herbicide used in forest management activities is minute when compared to agricultural useage of herbicides. In recent years, less than 10,000 acres of forest land in Minnesota was sprayed annually with herbicides. To put the figure into perspective, this area amounts to less than one-half of a standard township (23,040 acres). Herbicide usage should be considered as a part of a total weed and brush control program.

Weed and Brush Control Programs

Effective weed and brush control usually results from a combination of cultural, mechanical, and chemical practices. The ideal combination for each field will depend on a number of considerations including:

1. the crop grown
2. the kinds of weeds
3. the seriousness of the weed infestation
4. the soil type
5. the cropping system
6. the availability of time and labor

Cultural practices that are optimum for crop growth should be followed. These practices include good seedbed preparation, adequate fertility, optimum stands and row width, and proper seeding or planting dates. Tillage operations should be timed to destroy weeds. Tilling the soil immediately before planting will kill weeds that have germinated, thus giving the crop a competitive advantage and often improving weed control from chemicals that do not control weeds that have germinated.

Early cultivations, when weeds are small, are most effective. Use a rotary hoe, harrow, or cultivator as soon as weeds begin emerging and are in the "white stage", even if herbicides have been applied. Set cultivators for shallow operation to avoid crop root pruning and to reduce the number of weed seeds brought to the surface. Throw enough soil into the row to cover small weeds, but avoid excessive ridging that may encourage erosion or interfere with harvesting. Shallow cultivation should be repeated as necessary to control newly germinated weeds.

Example of Discing for Site Preparation.
August is best month for discing.

Mowing is an effective weed control practice in perennial forage crops, non-cropland areas, and Christmas tree plantations. To be effective, mowing must be done before seeds are formed which means you must mow by the time weeds are in the bud stage or just beginning to bloom. Earlier mowing will reduce weed competition and improve crop yield more than later cutting.

Herbicide Characteristics

The proper selection and use of an herbicide in forestry is influenced by: 1) characteristics of the herbicide, and 2) the treated vegetation, both pests and crop.

Important characteristics of the herbicide include: A) foliage-absorbed or root-absorbed, B) contact or translocated, C) selective or non-selective, and D) persistent or non-persistent.

A) Foliage-Absorbed or Root-Absorbed

Foliage-absorbed herbicides enter the plant through the leaves (post-emergence). Root-absorbed herbicides usually enter the plant through the roots and are generally most effective when applied before the weeds emerge (pre-emergence). Root-absorbed herbicides are often applied in combination with an herbicide having post-emergence activity. Some herbicides can be absorbed by both the foliage and the roots. Site of absorption can be influenced by formulation (granule or pellets versus liquids). Examples: (foliage-absorbed)--2,4-D, dalapon, dicamba, glyphosate, hexazinone, picloram; (root-absorbed)-atrazine, oryzalin, picloram, simazine.

B) Contact or Translocated

Contact herbicides kill only the green portion

of the plants with which they come in contact. Good coverage is necessary because they do not translocate through the plants. Most contact herbicides are non-selective. They are effective in controlling annual weeds, and provide temporary suppression of perennial vegetation. Translocated herbicides move throughout the plant, whether foliar- or root-absorbed. Foliage-absorbed translocated herbicides are especially useful for controlling perennial weeds because they move into the roots. Examples: (contact)--cacodylic acid, diquat, paraquat; (translocated)--amitrole, atrazine, dalapon, 2,4-D, dicamba, dichlorprop, glyphosate, hexazionone, picloram, simazine.

C) Selective or Non-Selective

Selective herbicides control only certain types of plants. When applied to mixed vegetation, some plant types/species will be noticeably unaffected. Non-selective herbicides will generally control most weed species. Selectivity may reflect physiological differences, (broadleaf weed control by 2,4-D, dicamba, and picloram, without injuring grasses), timing (glyphosate resistance by dormant conifers), application rate (control of annual weeds by simazine in black walnut plantations) or method of application (shields or directed sprays or wiping applications to prevent herbicide contact with desired species). Examples: (selective)--2,4-D, dicamba, dichlorprop, glyphosate, hexazinone, picloram; (non-selective)--amitrole, cacodylic acid, diquat, glyphosate, paraquat.

D) Persistent or Non-Persistent

Persistent herbicides remain active in the environment for an extended period of time. Nonpersistent herbicides are relatively short-lived in the environment. The degree of persistence is greatly influenced by temperature, moisture, soil type and rate of application. Persistence is important for residual weed control in new plantations or Christmas trees. Persistence within the plant is important for control of woody species or other perennial weeds. Nonpersistence may be desired when applying herbicides for weed control before planting (dc pending on herbicide used, rate of application and species to be planted). Examples: (persistent)--atrazine, hexazinone, picloram, simazine; (non-persistent)--amitrole, 2,4-D, dalapon, glyphosate, paraquat.

Plant Factors Affecting Chemical Weed Control

A) Plant Types

Most plants are either grasses or broadleaves. Broadleaf plants include herbaceous forbs and many woody plants. The sedges, such as the nutsedges, are neither grasses nor broadleaves. However, they have similar characteristics to grasses and are often listed under grasses on herbicide labels.

Grasses--Grass seedlings have only one leaf as they emerge from the seed. Their leaves are generally narrow and upright with parallel veins. Most grasses have fibrous root systems. Consequently, grasses are very competitive for soil moisture. The growing point on seedling grasses is sheathed and located below the soil surface. The growing point gradually moves above the soil as the plant grows and matures. Examples: foxtail, quackgrass.

Broadleaves (Forbs)--Herbaceous (plants that do not develop persistent woody tissue above ground) broadleaf seedlings have two leaves as they emerge from the seed. Their leaves are generally broad with netlike veins. Broadleaves usually have a taproot and a relatively coarse root system. All actively growing broadleaf plants have exposed growing points at the end of each stem and in each leaf axil. Examples: dandelion, mullein, pigweed, plantain, ragweed.

Woody Plants--Woody plants are those that form wood. They include brush and shrubs, and trees. Brush and shrubs are woody plants that have several stems and are less than 10 feet tall. When trees are present, brush or shrubs may be called understory. Trees are woody plants which usually have a single stem (trunk) and are over 10 feet tall. Examples: hickories, oaks, pines.

B) Plants' Life Cycles

A plant's life cycle is either annual, biennial, or perennial.

Annuals--Annuals complete all four stages of development in less than 12 months. There are two distinct kinds of annuals–winter annuals and summer annuals.

Winter Annuals germinate in the fall, overwinter, mature, set seed, and die in the spring. For best results, control winter annuals at or soon after germination in the fall. Examples: cheat, downy brome, henbit, little barley, mayweed, wild mustard.

Summer Annuals germinate in the spring, grow, set seed, and die in the fall. For best results, control summer annuals at or soon after germination in the seedling stage of growth. Examples: common ragweed, crabgrass, foxtail, giant ragweed, marijuana, partridge pea, pigweed.

Some weeds are specifically winter or summer annuals. Other species can be germinated and grow in either the fall or spring.

Biennials--Biennials complete their life cycle in two years. Biennial plants complete the seedling and vegetative stages of growth in the first year and the seed production and maturity stages in the second year. Biennial weeds are most easily controlled in their first year of growth. Some herbicide labels identify biennials as annuals. Examples: bull thistle, common evening primrose, common burdock, common mullein, teasel, wild carrot, wild parsnip.

Perennials--Perennials may complete all four stages in the first year and then repeat the vegetative, seed production, and maturity stages for several following years, or the seed production and maturity stages may be delayed for several years. Some perennial plants die back in the maturity stage each winter; others, such as trees, may lose their leaves but do not die back to the ground. They all reproduce by seed but many are able to spread and reproduce vegetatively. Perennials are difficult to control due to the persistent root system.

Simple Perennials spread by seed, crown buds, and cut root segments. Most have a large, fleshy taproot. Examples: chicory, curly dock, dandelion, gray goldenrod, plantain, spiderwort, white heath aster, white oak.

Creeping Perennials spread vegetatively with stolons or rhizomes as well as by seed. Examples: big bluestem, common milkweed, hemp dogbane, horsenettle, horsetail, poison ivy, Virginia creeper and yellow toadflax.

C) **Developmental Stages**

All plants have four stages of development.

Seedling--small, vulnerable plantlets; seed leaves still present; seed leaves lost, true leaves present.

Vegetative--rapid growth of stems, roots, and foliage. Uptake of water and nutrients is rapid.

Seed Production--little or no growth; production of fruit. Uptake and movement of water and nutrients slow and directed mainly to reproductive parts--flowers, fruits and seeds.

Maturity--little or no growth; movement of water and nutrients in plant is slow.

Best control of annual and perennial weeds is best obtained when the herbicide is applied at the seedling stage of growth. As an annual plant matures control becomes more difficult. Another effective period for perennial weeds is the vegetative or regrowth stage. This causes the greatest drain on the underground food reserves. Treatment during early seed production stage is also effective for perennial weed control. When perennials reach full flower control becomes more difficult.

With these factors in mind consider two areas of forest management in which herbicides play an important role; plantation establishment and timber stand improvement.

Selecting Chemicals

Selection of an appropriate chemical or combination of chemicals should be based on consideration of the following factors:

1) label approval for use
2) use of the crop
3) crop and variety tolerance
4) potential for soil residues that may affect following crops
5) kinds of weeds or brush
6) soil texture
7) pH of soil
8) amount of organic matter in the soil
9) formulation of the chemical
10) application equipment available
11) potential for drift problems

Proper application of chemicals is essential for obtaining satisfactory results. Follow carefully the suggested rates on labels for specific soil and weed situations. Apply herbicides at the times specified. Delayed applications usually result in poorer control and may injure other crops.

Weather conditions will affect herbicide performance. Weed control from soil-applied herbicides may be poor if there is insufficient rain soon after treatment, which will make timely cultivation necessary to control emerging weeds. If rainfall is very heavy, some herbicides may move downward in the soil, resulting in poor weed control and/or crop injury. Temperature and moisture conditions affect the weed control and crop injury resulting from herbicides applied postemergence. Observe label precautions regarding weather conditions and crop size when applying herbicides on growing crops.

Granular Versus Spray Forms of Herbicides

Several herbicides are available in formulations to be applied as dry granules or as sprays. With a few exceptions, approximately the same weed control can be expected from either form. The cost of granules is usually higher than the cost of an equal amount of the spray form. Distribution of chemicals with granule applicators is sometimes not as uniform as with sprayers, especially on rough ground. In some instances poor distribution has resulted in variable weed control. Dry granular formulations of triazine herbicides require more rain and may result in more soil residues than the spray formulations. Chemicals that cause irritation are less irritating in the granular form than in the spray form.

Herbicide Mixtures

Herbicide mixtures are used to overcome limitations of single chemicals. Certain mixtures may:

1) control more kinds of weeds
2) give more consistent performance with different soils and weather conditions
3) lessen soil residue problems
4) increase persistence enough to give full-season weed control
5) reduce crop injury

Only those mixtures that have been field tested under local conditions and registered by EPA and State Agencies should be used. Use of some mixtures may result in poor weed control or crop injury. Growers or applicators may be responsible for chemical residues in crops, crop injury or lack of weed control resulting from use of unlabeled mixtures.

HERBICIDES AND USE

For Forest Management in Minnesota

Alvin A. Alm and Richard D. Iverson*

Background information

Chemicals used for killing weeds or interrupting plant growth are commonly referred to as herbicides: a useful tool for preparing planting sites and releasing crop trees from vegetative competition. Herbicides are effective and safe when used according to instructions on the label.

Prior to herbicide selection, the problem weed species should be determined. The herbicide label provides information on weeds that are controlled and specifies appropriate application methods, herbicide rates, and when to apply. The herbicide label should always be read carefully before applying any herbicide. It is illegal to use a herbicide at a higher rate or for a purpose other than labeled.

It is also important to become familiar with the appropriate application technique. Herbicides can be broadcast, banded, or spotted to the site, injected directly into undesirable hardwood trees, or applied as a spray to the lower stem. The appropriate application method will depend on the age and type of vegetation needing control, the type of herbicide used, and the age and species of the crop tree.

Spot applications of herbicides often are made with hand sprayers to control individual weeds in a forest plantation or to control weeds within a specified radius of each crop tree. Spot applications use less herbicide than broadcast or band applications but take more labor to apply.

Band applications are made over rows of seedlings with hand sprayers or with mechanized equipment at variable widths. Less herbicide is needed (probably at less cost) in a band rather than a broadcast application.

Broadcast applications over large acreage need to be done by motorized ground equipment or aircraft. Broadcast applications are usually necessary for control of multi-layered woody competition. Aerial application of foliar-active herbicides is necessary for control of weed trees beyond the reach of ground equipment.

*Alvin A. Alm is a silviculturist and Richard D. Iverson, a forest weed control specialist, both with the College of Forestry, Department of Forest Resources, University of Minnesota.

Basal spray applications, injection, and stump treatment are used to control individual stems or clumps of woody species. Basal sprays are applied to the bottom 12 to 15 inches of the hardwood stem. Injections are made by applying the herbicide in cut surfaces around the trunk or by using special tools. Stump sprouting can be controlled by treating the cut surface with a selected herbicide. This can be done at any time of year while a foliar application is limited to the growing season.

TERMS AND DEFINITIONS

The following definitions are for terms likely to be encountered on herbicide labels or in discussions about them:

acid equivalent (a.e.) = the amount of active ingredient expressed in terms of the parent acid.

active ingredient (a.i.) = the chemical compound identified on the label in a herbicide formulation that is responsible for the killing of or disruption of plant tissue.

adjuvant = any additive that activates a herbicide or that facilitates or modifies characteristics of herbicide formulations or spray solutions. Examples of herbicide adjuvants include surface-active agents (surfactants), anti-foam materials, and anti-drift materials.

backflash = herbicide kill that results from root contact of trees within the treated area. Desirable trees can be killed accidentally so care must be taken when desirable trees are near those being treated.

contact herbicides = kill only the portion of the plant they contact, while *translocated herbicides* are taken up by the plant and then moved into the stem and roots. Most herbicides are the translocated type.

formulation = how the active herbicide ingredients are packaged. Herbicides are available in both liquid and dry formulations.

frill application = cutting continuous slits around a tree and filling them with chemical.

LD_{50} value = the single dose of a chemical required to kill 50 percent of the test animals when given orally. This is expressed in milligrams (mg) of chemical per kilogram (kg) of body weight of the test animal. *Remember that the* *higher the LD_{50} value the less toxic the chemical;* for example, Velpar® L has a relatively low toxicity level with an LD_{50} of 7,500.

non-selective herbicide = chemicals that will kill or suppress plant life regardless of species.

restricted-use pesticide = a pesticide which can be sold to and applied only by certified applicators or persons under their direct supervision. Every pesticide label must state use: either general use or restricted use.

selective herbicide = a chemical that will kill or suppress some plants but not others; for example, 2, 4-D will kill most woody shrubs but not grasses. Herbicides used for release are selective, meaning they will not kill most conifers if used according to label instructions.

soil-applied herbicide = herbicides that are taken up by target weeds primarily through the root system. Soil texture, percent organic matter, and soil pH are factors which determine application rate and effectiveness of soil-applied herbicides.

volatility = a compound is volatile when it evaporates or vaporizes when exposed to air; it is important to read the label instructions since some herbicide formulations should not be sprayed in hot weather.

Calibration of equipment

Herbicide equipment must be calibrated to insure application of the prescribed rate of chemical. Lower dosages than label recommendations may not adequately control weeds and higher dosages than label recommendations could damage crop trees and unnecessarily increase application costs. The amount of herbicide product needed and a method for calibrating equipment are explained in the following sections.

- Mixing liquid herbicides:
  To determine the number of gallons of herbicide to mix with water for spraying an acre:

$$\text{gallon of herbicide} = \frac{\text{lb/acre of a.i. recommended}}{\text{lb of a.i. gal}}$$

Example: If the recommendation is 2.5 pounds a.i. per acre and the herbicide contains 4 pounds a.i., then the gallons/acre needed of herbicide =

$$\frac{2.5}{4} = .625 \text{ gallon} = 5 \text{ pints}$$

- Mixing wettable powders:
  To determine the amount of wettable powder to mix with water or the amount of granules to apply to an acre:

$$\text{lb/acre} = \frac{\text{lb/acre of a.i.} \times 100}{\text{\% a.i. in formulation}}$$

Example: If the recommendation is 5 pounds a.i./acre and the herbicide contains 80% a.i., then the pounds /acre of herbicide to use = $\frac{5}{80} \times 100 = 6.25$ pounds.

Calibrating a sprayer

The first step in sprayer calibration is to determine the amount of spray delivered by the sprayer to a known area. One way to do this is to fill the sprayer with a known volume of water, spray an area of known size, then measure the amount of water needed to refill the sprayer. Another way is to fill the sprayer with a volume of water, spray it on an area in the *same manner* that you would apply herbicide, then measure how much area the spray covered.

- After this determination, use the following formula:
  amount of herbicide needed = recommended herbicide rate (pounds/acre) x acreage covered with known volume of water.

Example: if the spray covers .0125/acre with 1 gallon of liquid and the recommended herbicide application rate is 3 pounds/acre (48 ounces/acre) then:

amount of herbicide needed = (48 ounces/acre) x (.0125) = .6 ounces. In this example, .6 ounces of herbicide should be mixed with each gallon of water.

Example: if the sprayer covers 1.1 acre with 10 gallons of water and the recommended application rate is 2 pounds /acre then:

amount of herbicide needed = (32 ounces/acre) x (1.1 acre) = 35.2 ounces = 2.2 pounds/10 gallons of water or .22 pound (7 ounces)/gallon of water.

If narrow strips or swaths (bands) are being sprayed, the following formula can be used to calculate how many feet must be covered with a given swath width to cover an acre:

$$\text{feet} = \frac{\text{43,560 square feet/acre}}{\text{width of swath in feet}}$$

Example: a 6-foot wide band would require 7,260 feet of coverage for an entire acre.

When calibrating equipment, size and type of nozzle, and speed of walking or driving will determine rate of herbicide delivery. Sprayers must be recalibrated if nozzles are changed or if speed is varied.

Useful units of measure

1 acre = 43,560 square feet
1 mile = 5,280 feet
1 pound = 16 ounces
1 gallon = 4 quarts = 8 pints = 128 liquid ounces

Herbicides labeled both for site preparation and release

(can be sprayed safely over conifers other than larches if directions are followed — table 1).

2,4-D: sold under a variety of brand names such as ESTERON® 99C, WEEDONE® LV-4 and WEED RHAP®; used for control of woody species such as aspen, hazel, willow, and alder; recommended rate of 1 to 3 quarts/acre; applied to foliage in mid-summer after conifers have set overwintering buds; not effective for grass control. The 2,4-D product FORMULA 40® can also be used in undiluted form for tree injection; effective on hard to kill species such as oak, elm, hickory, ash, and maple; can be used any season, but is least effective during winter months and spring sap flow.

2, 4-D+2, 4-DP: brand name is WEEDONE® 170; similar to 2, 4-D but is more effective on hard-to-kill shrubs such as oak.

Glyphosate: sold as ROUNDUP®; is less selective than the compounds just discussed because it will kill woody species, grasses and an assortment of broad-leaved weeds; recommended rate is 1½ to 2 quarts/acre; should be sprayed late in growing season after conifer growth has ended to avoid damage to conifers; can be used effectively for site preparation if sprayed when target species are in full leaf; there is no residual effect so areas can be planted about seven days after application.

Hexazinone: brand names are VELPAR® L, PRONONE® 10G, PRONONE® 5G, and GRANVAL®; these formulations kill a wide range of plants including woody shrubs, grasses and broad-leaved weeds; the liquid (VELPAR® L) should be broadcast sprayed or spot sprayed early in the growing season at the rate of 1 to 2 gallons/acre; PRONONE® 10G, PRONONE® 5G and GRANVAL® are granular products used at rates varying from 5 to 40 pounds/acre depending on soil texture and age of crop trees for release; they should be applied before or during the period of active growth of brush species; these formulations have a residual effect lasting more than one season; label should be checked for which species these herbicides can be used on without damage since species such as jack pine and white pine can be severely damaged; rainfall is necessary to effectively move these products into the soil to be absorbed by the root system.

Triclopyr: sold under the brand name GARLON® 4; a foliar herbicide effective on a wide range of hardwoods including ash and oak and broadleaf weeds but not most grasses or sedge; the recommended application rate for release is 1½ to 3 quarts/acre; depending on application rate used, planting should be delayed one or two months after application; for release, application should be in late summer or early fall during full leaf but prior to leaf coloration; conifers should have set overwintering buds; rates should be decreased for jack pine; this product can be used undiluted or mixed with oil, or oil and water for basal stem application.

Simazine: brand name PRINCEP®; this is a pre-emergent herbicide that will kill grasses and other weeds if applied early in the season before growth begins; used at rates ranging from 2½ to 5 pounds/acre; formulations are wettable powder, liquid or granular; can be sprayed over conifer seedlings at least three years old.

Sulfometuron methyl: or OUST® as of October 1984 has only an experimental use permit for forestry purposes. A label for forestry use is expected in 1985. It is a dispersible granule used on non-cropland areas for control of annual and perennial grasses and broadleaf weeds; it can be applied pre-emergent or post-emergent at rates ranging from 2 to 8 ounces/acre depending on target species and desired residual effect.

Herbicides used for site preparation *(should not be sprayed over crop trees — table 2).*

Amitrole + Simazine: brand name is AMIZINE®, this is a combination pre-emergent and post-emergent herbicide that kills a variety of grasses and non-woody plants; can be applied after growth begins; recommended rate is 7 pounds /acre; directed spray should be kept away from crop trees.

Picloram + 2, 4-D: sold as TORDON® 101; used as a broadcast, foliar spray; will kill a variety of shrubs and hardwood and conifer trees as well as a number of broadleaf weeds, but not effective on grasses; applied at rates varying from 1 to 4 gallons/acre; has a relatively long residual effect in the soil so planting of conifers should be delayed until 8 to 9 months after application; TORDON® RTU and TORDON® 101R can be effectively used on individual large hardwood stems by injection or used for treatment of cut surfaces.

Picloram: sold as TORDON® K, this product can be tank mixed with a number of other herbicides to increase effectiveness for site preparation.

Labels and additional information on all these herbicides can be obtained from the following addresses:

ESTERON® 99C, FORMULA 40®, WEED RHAP®
Vertac Chemical Corporation
5100 Poplar, Suite 2414
Memphis, TN 38137

WEEDONE® LV-4, WEEDONE® 170
Union Carbide Corporation
T.W. Alexander Drive
P.O. Box 12014
Research Triangle Park, NC 27709

ROUNDUP®
Monsanto Agricultural Products Company
800 N. Lindbergh Boulevard
St. Louis, MO 63167

VELPAR® L, OUST®
E.L. duPont de Nemours and Company
Biochemicals Department
Wilmington, DE 19898

PRONONE® 10G, PRONONE® 5G
PRO-SERVE, Inc.
400 E. Brooks Road
Memphis, TN 38116

TORDON® RTU, TORDON® 101, TORDON® K, GARLON® 4
The Dow Chemical Company
P.O. Box 1706
Midland, MI 48640

PRINCEP®
CIBA-GEIGY Corporation
Agricultural Division
P.O. Box 11422
Greensboro, NC 27409

GRANVAL®
HABCO, Inc.
1418-5th St. So.
Hopkins, MN 55343

These lists are not all inclusive. The herbicides included are those most frequently used for weed control in Minnesota forest situations. *None of the herbicides listed should be used without thoroughly reading the label.* Herbicide use often brings on controversy regarding safety. Most of the problems relate to misuse or misinformation or both.

If questions in the use of herbicides arise, a knowledgeable person should be contacted, such as an ag-chemical or industrial chemical representative, through the local dealer selling the chemical.

An excellent reference for information on herbicides including use and toxicology is *Herbicide Handbook of the Weed Science Society of America,* 5th edition 1983. It is available from: Weed Science Society of America, 309 W. Clark Street, Champaign, IL 61820.

Table 1. Herbicides most commonly used for site preparation and release

Chemical[1]	Brand names[1]	Crop trees	Important target weeds[2]	Rate per acre	Application time	Application methods
2,4-D	Esteron® 99C Weedone® LV-4 Weed Rhap®	jack pine, red pine, white spruce, black spruce	alder, aspen, birch, hazel, willow	1 to 3 qt-release 3 to 4 qt-site preparation	mid-July (after conifers set buds)	broadcast, tree injection
2, 4-D + 2,4-DP	Weedone®170	jack pine, red pine, white pine, scots pine, black spruce, white spruce	similar to 2,4-D but is more effective on oaks	2 to 4 qt-release 4 qt-site preparation	mid-July (after conifers set buds)	broadcast, tree injection
Glyphosate	Roundup®	pine species except E. white, spruce species, fir species	aspen, cherry hazel, certain forbs and perennial grasses such as quack grass	1½ to 2 qt-release 3 to 4 qt-site preparation	fall—for shrubs, after conifers have set bud but before first killing frost; spring — before buds swell on conifers unless they are protected	broadcast, tree injection, spot to avoid conifers if applied in spring for herbaceous weed and grass control
Hexazinone	Velpar® L Pronone® 10G Pronone® 5G Granval®	balsam fir, red pine, black spruce, white spruce	aspen, birch, raspberries, oaks, pin cherry certain forbs and perennial grasses such as quackgrass	1 to 2 gal Velpar® L for red pine release (2 to 4 lb a.i.) ½ to 1 gal Velpar® L for release of other conifer crop trees, (1 to 2 lbs. a.i.); 1 to 3 gal Valpar® for site preparation (2 to 6 lb. a.i.)	early in the growing season near conifer budbreak until early summer	broadcast, spot, injection
Triclopyr	Garlon® 4	spruce species, balsam fir, red pine, jack pine	aspen, birch, maple, cherry, willow, oak, hazel, rubus sp. and certain broadleaf weeds	1½ to 3 qt-release 4 to 8 qt-site preparation	late summer or early fall after conifers have formed over-wintering buds for release, when plants are actively growing for site preparation	broadcast, basal bark treatment
Simazine	Princep® (several different formulations including Caliber® 90 Princep® 4G Princep® 4L Princep® 80W)	black walnut, balsam fir, red pine, white pine, scotch pine, white spruce	controls a broad range of annual weeds and grasses when used at selective rates	2 to 4 lbs a.i. depending on soil texture for control of annual weeds in new plantations (amount of 80W needed would be between 2½ and 5 lbs product)	weed growth should be removed before application, apply before weeds emerge	broadcast, spot
Sulfometuron methyl	Oust®	label for Lake States Conifers expected in 1985	annual and perennial grasses and broadleaf weeds	2 to 8 oz dependent on new label	early pre-emergence or post-emergence to weeds in the spring and either after or before planting	broadcast, spot

1. Read labels for detailed information on each product.
2. List is not all inclusive.

CHAPTER 11 — MEASUREMENTS

Minnesota 4-H Bulletin No. 87 and (or)
Forestry For Minnesota Schools—North Print Co.

An organized system of land measurement is a basic requirement for any phase of forest management whether it is evaluation for timber, water, recreation, wildlife, or forage. Forest management requires a system of land measurement and an organized way of measuring and comparing individual trees and forest stands. First, let us examine the way land is measured.

MEASURING LAND

The United States basic unit of land measurement is the foot (12 inches). We speak of a man being 6 feet tall; it is 340 feet from home plate to the right field fence; and a mile is 5,280 feet long. In Europe, the metric system is used with 1 meter equaling 100 centimeters or slightly longer than 3 feet. The United States may adopt the metric system soon because of its convenience and world-wide use.

In measuring the earth's surface, we use the square foot, which is 1 foot of ground, 12 inches on a side. This is a very small area to man, but a large area to an ant, earthworm, or a beetle. We might think there are not very many things in 1 square foot of earth surface, but imagine the number of blades of grass in 1 square foot of lawn.

A common unit of measurement in any management plan is the acre: 43,560 square feet. We refer to the number of bushels of corn per acre, the number of board feet per acre, or the average number of acres for each man, woman, or child living in the United States or one of its cities. One square acre of land is slightly over 208 feet on each side. A football field is 300 feet long and 150 feet wide, or 45,000 square feet, slightly over an acre in size. Not to a human being, but to smaller animals, an acre is a huge domain. It is not unusual to have a population of 400,000 to 500,000 earthworms in 1 acre.

Consider now, measuring the earth's surface, the terms used to describe where we are located, and the tools and the methods used in measuring and describing land areas.

Land is measured by surveying. Surveying means measuring and locating points, lines, angles, and elevations on and within the surface of the earth or on bodies of water.

Early in United States history, land was measured in "metes and bounds." This was convenient where boundary lines of private property consisted primarily of natural features such as tidewater shorelines, roads, fences, trees, and stones. Since these natural features could be destroyed or changed, many errors and arguments occurred in the description of boundary or property lines.

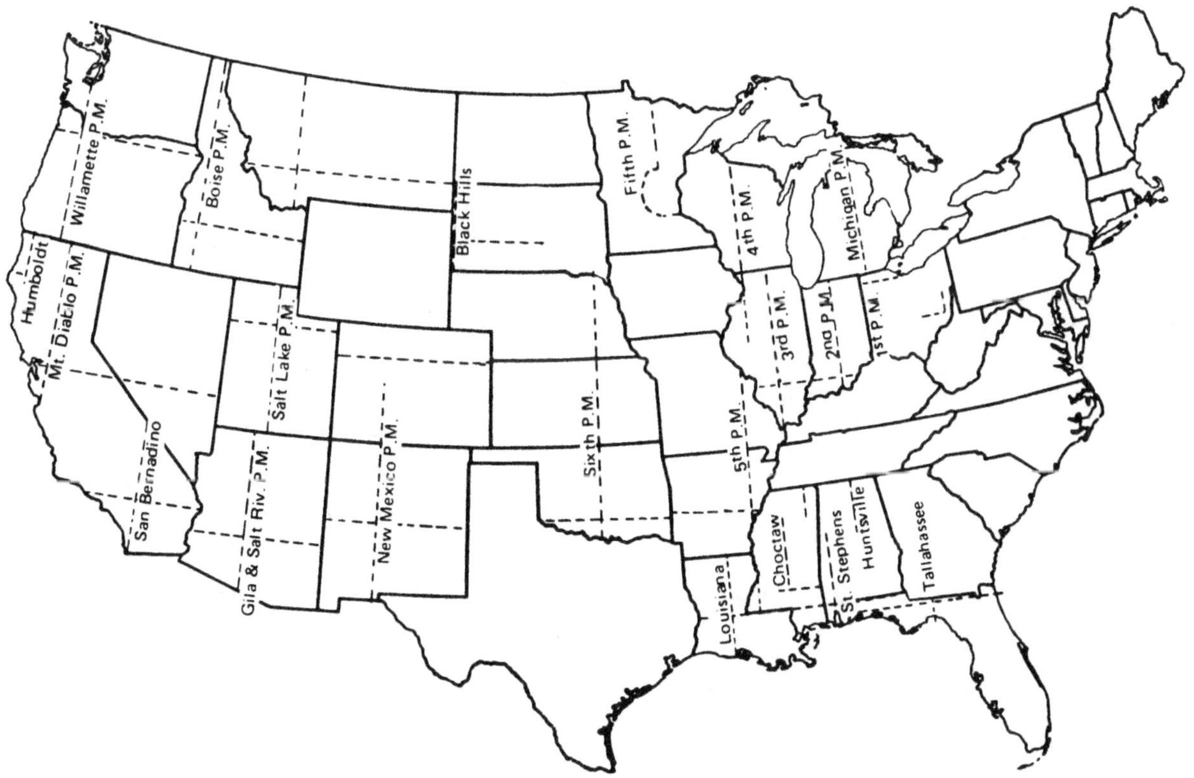

An organized system of describing the land using square-mile sections was started by the Continental Congress in 1785. This system became known as the "Cadastral Survey" and is the official system of land measurement. This rectangular land survey method is found in 31 states, primarily west of the original thirteen colonies.

West of Pennsylvania, initial surveying points were chosen at irregular locations extending from the east to the west. From these initial points a base line, running east and west, and a principal meridian, running north and south, were established (figure 1). Lines were then run at 6-mile intervals parallel to the meridian and the base line.

East-west lines were called township lines and numbered consecutively north and south of the base line; for example, Township 1 North (T.1 N.) was the first row above the base line and Township 1 South (T.1 S.) was the first row below the base line.

North-south lines were called range lines and numbered consecutively east and west of the meridian lines; for example, Range 1 East (R.1 E.) is the first row east of the meridian line and Range 1 West (R.1 W.) is the first row to the west of the meridian line (figure 2).

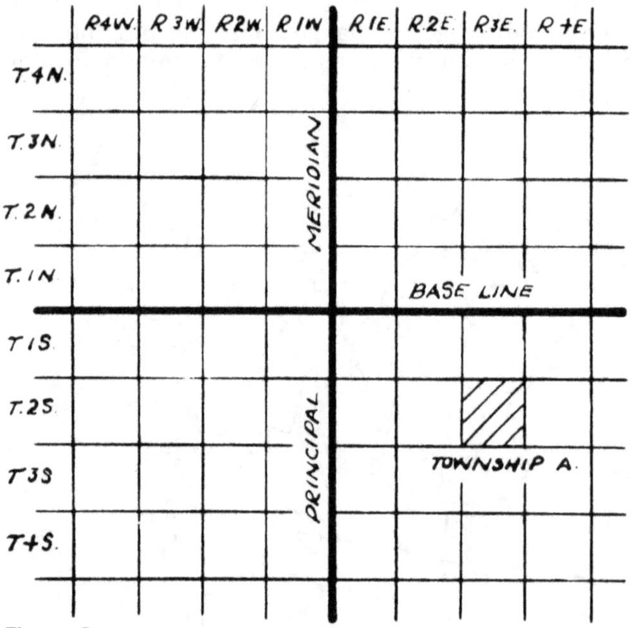

Figure 2.

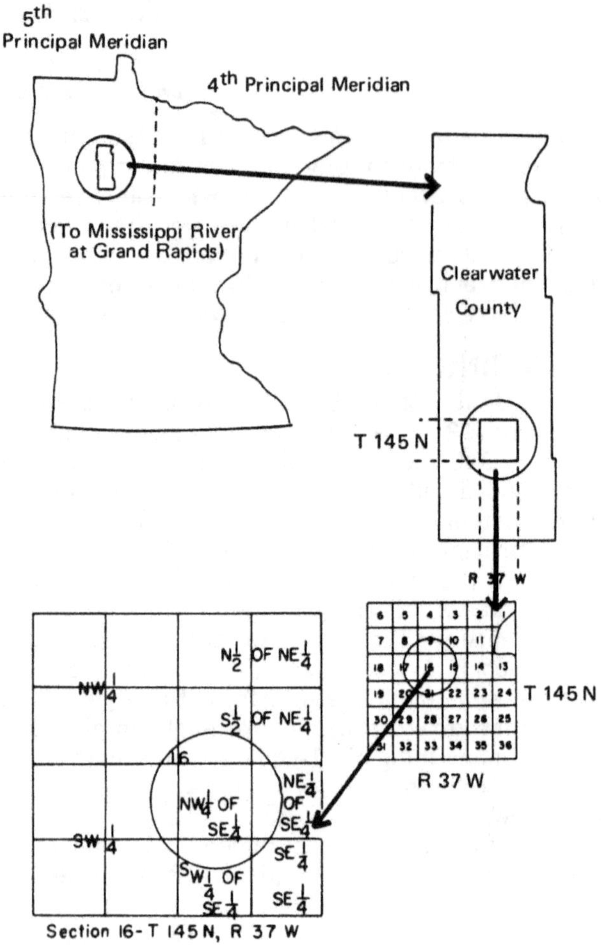

Figure 3.

In Minnesota, township and range lines are measured from the 4th or 5th Principal Meridians. In these illustrations the township and range lines are indicated as they would be numbered from the initial point with Township "A" being T.2 S., R.3 E. This land measurement system conveniently divides the land into smaller subunits or townships.

Each township (6 miles on a side) was divided into 36 sections, each 1 mile square. Each section contains 640 acres. Figure 3 illustrates the division of the township into 36 sections and shows the system of numbering these sections.

Each section was then divided into four parts of 160 acres each. These quarter-sections are further subdivided into quarter-quarter-sections, commonly referred to as "forties," each 1,320 feet on a side.

For legal purposes each tract of land must have a description so that it will not be confused with another tract of land. This legal description has been designated by the original land survey and is on record in that county's courthouse. For example, the description of the tract of land in figure 3 is legally designated as the NW¼, SE¼ of Sec. 16, T.145 N., R.37 W. of the 5th Principal Meridian, Clearwater County, Minnesota.

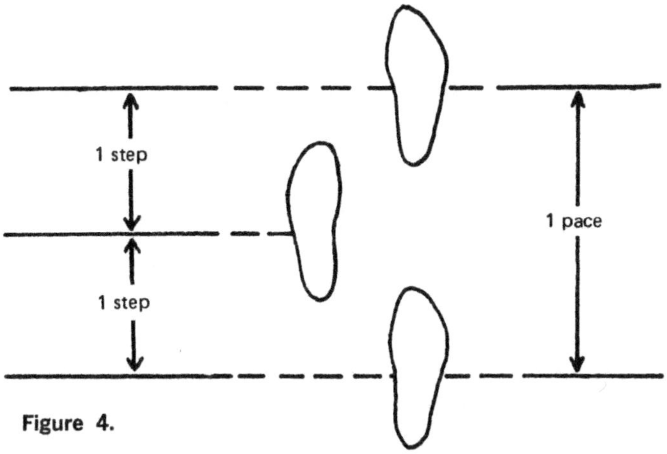

Figure 4.

There are two important and widely used measurement concepts when considering forested tracts of land: distance and direction. Forested areas are often laid out with a tape, but since a tape can be very cumbersome in a wooded area, many forest cruisers measure distance by pacing. Accurate pacing makes it easy to lay out forested areas in which you plan to apply various management practices.

A pace is two steps. Stepping off with your right foot, count a pace each time your right foot strikes the ground (figure 4).

To measure the length of your pace, a 500 foot course can be measured out along a forested road with a tape, marking the beginning and the end. Then pace the distance using your normal stride, keeping count of the number of paces, dividing this number into the length of the measured course to determine the number of feet in each pace. You can now measure distances in the field or woods. A measured course could also be laid out through the woods to get a comparison of pacing where it may be rougher going because of hills, trees, or brush. You can determine the number of acres in an area of a rectangular forest tract of land by multiplying the length of the area by its width and dividing by 43,560.

Foresters commonly use the unit of measurement called a chain, which is 66 feet long, because it is easier to simply count the number of chains than to total the paces. There are four rods (each 16½ feet long) in each chain. A mile is 80 chains long (80 x 66 = 5,280 feet). Ten square chains equal 1 acre. It does not matter whether you use chains or feet to measure your forested acreage.

Direction is the other important land measurement concept. Some people have an inherent sense of direction, but most of us must rely on other means to establish where we are going. Many objects used as references are present in nature; for example, the sun, moon, and the stars.

Use Your Wristwatch

At noon the sun is always to the south in the northern hemisphere. By using a watch a person can lay out a north-south line any time during the day, providing the sun is shining. Place a small stick or pin standing upright on the center of your watch dial so that it will cast a shadow on the dial (figure 5). Rotate the watch so that the shadow will fall halfway between the hour hand and twelve. When you look from six to twelve on your watch, you will be looking north. It is important to remember that this method is more accurate in the late morning or early afternoon and that this method must always be used with standard time and not daylight saving time.

Figure 5.

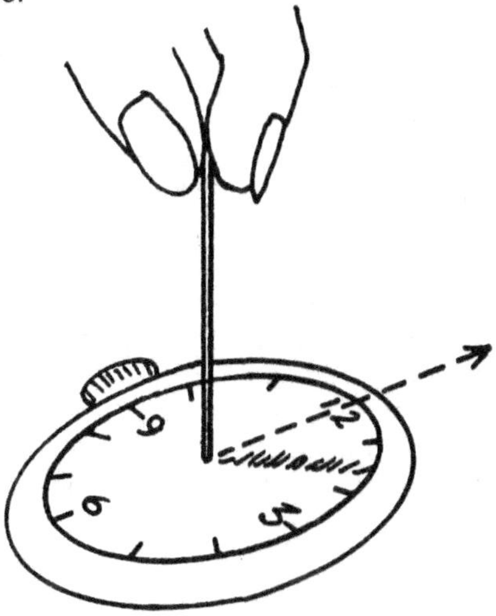

Stars

Another method of establishing a north-south line is by using the stars at night. The most accurate way is to use the North Star, Polaris. First, locate the Big Dipper, Ursa Major. Using the pointer star of the Dipper, the star on the top outer edge of the Dipper, the North Star will lie in a northwest direction (figure 6). Next place a long stick in the ground and with a shorter

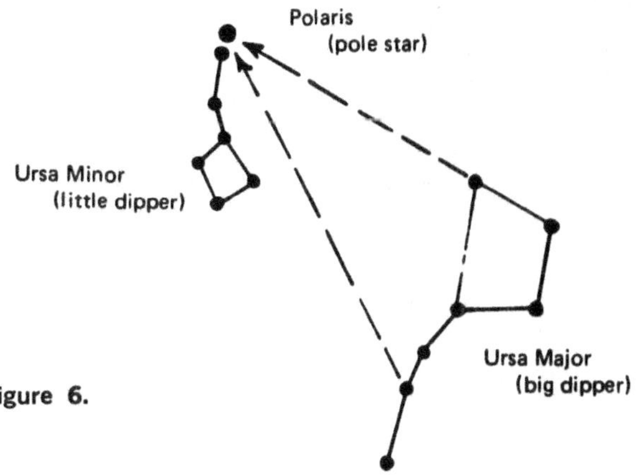

Figure 6.

stick, sight across the top of the two sticks at the North Star as if you were aiming a gun. When the two sticks and the North Star are in line, push the short stick into the ground. By connecting the short and the long stick with an imaginary line, a piece of string, or another stick, you have established a north-south line which serves as a reference point for other directions.

Compass

Clouds, tree cover, or other physical barriers often make it inconvenient or impossible to use these natural methods, so foresters use a compass.

A compass is an instrument used to determine directions on the earth's surface by means of a magnetic needle which swings on a pivot and points to the magnetic North Pole. There are many types of compasses in a varied price range but a simple directional compass such as the Silva is adequate for most directional referencing (figure 7). Cost of the compass makes little difference. The important thing is to know how to use it.

One of the first things you should know about a compass is which end of the compass needle points north. Usually this is indicated by an arrow or by the letter "N" on the needle. The compass needle points to the magnetic north pole, but your directions are based on true north. The difference between magnetic north and true north is called magnetic declination (figure 8). The declination varies with the locality. In Minnesota

Figure 7.

Figure 8.

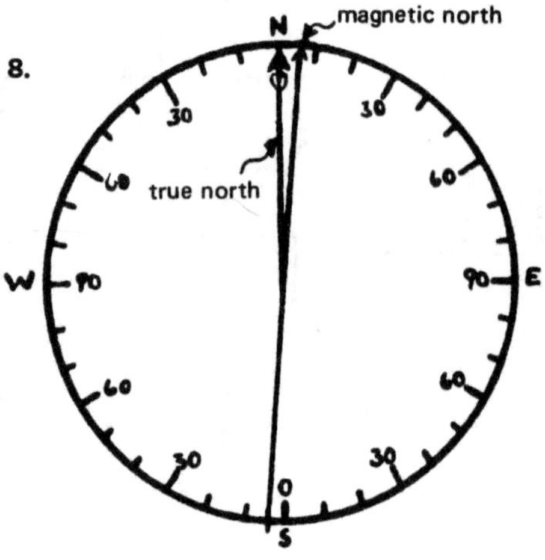

Figure 9.

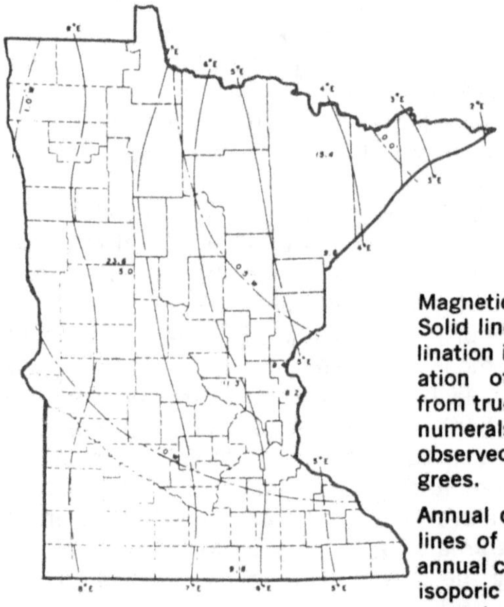

Magnetic declination—Solid lines of equal declination indicate the variation of the compass from true north. Isolated numerals are abnormal observed values in degrees.

Annual change—Broken lines of equal westward annual change are called isoporic lines and the annual rate of motion is given in minutes. (1960 data from the Isogonic Chart of the U.S. published by the U.S.C. & G.S.)

this declination varies from about 2 degrees east of true north to more than 8 degrees east of true north (figure 9). For example, the line for 4 degrees east magnetic declination is very near Duluth, Minnesota. This means that when you are in that vicinity and have a reading of 0 degrees on the compass, the true north line of sight would be 4 degrees west.

Figure 10.

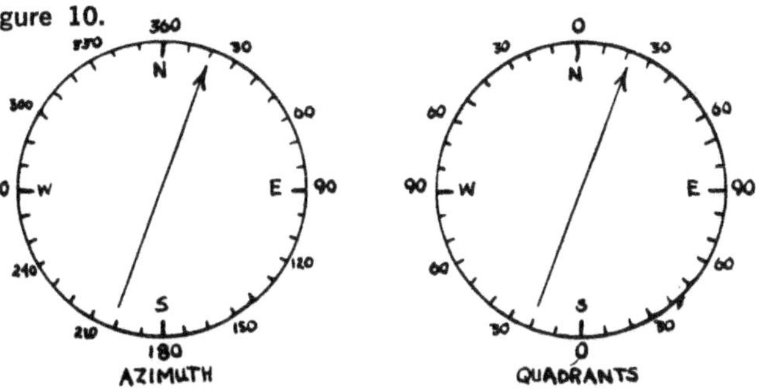

Some compasses are graduated from 0° to 360° clockwise, called azimuths (figure 10) so that north is either 0° (or 360°) and south is 180°. The majority of compasses are graduated by dividing the compass into four quarters, called quadrants (figure 10) and representing northeast, southeast, southwest, and northwest. There are 90° in each quadrant or quarter. In reading a bearing (your location) start with either north or south, depending on which half of the compass you are concerned with. You will end with either east or west. For example, the direction northwest would be read on the compass as north 45° west, or the direction southeast as south 45° east.

Always stand directly behind the compass so you can sight in the direction you wish to go (figure 11). In other words, move your body, not the compass.

Figure 11.

To Use a Hand Compass:

1. Hold firmly in both hands with elbows against body (far enough from magnetic objects such as jackknife, keys, etc.).
2. Sight toward the object or the direction wanted. (Always read the north end of needle. The south end of the needle usually has a wrapping of fine wire.)
3. Make sure that the compass is level.
4. Long 'shots' (objects farther away from the compassman) are better than short 'shots' (objects close to the compassman).
5. Be sure to sight on an object which can be seen and readily identified at all times.
6. If in doubt as to the location of your target, take another sight.
7. Rely upon the compass.

Upon occasion, in areas of iron ore deposits, "local attraction" is encountered. When using a compass in these areas, the needle will not "settle down" and will give an incorrect reading. To check against local attraction, take both foresight and backsight readings (backsight readings are exactly opposite or 180 degrees plus the foresight reading). When it is proven that there is local attraction, and it is necessary to continue your line, do it by means of projecting a straight line, or use the sun as a guide.

Pacing

Pacing is the least accurate means of measurement, although it is indispensable in timber cruising and reconnaissance surveys. A few simple rules are:

1. Be systematic. Record all paces, and check constantly against known distances. After experience, a person should have a pacing accuracy of 1/100.
2. Count *double steps only.* As used here, *a pace means two steps.*
3. Adopt a pace which can be held all day. Pace repeatedly over a measured course on such ground until a variation of less than three paces in 100 is attained.
4. Attempt to adopt a pace which will be an average of all conditions (open field, gently rolling terrain, heavy brush, etc.). Remember, when pacing up steep hills, you take short steps, therefore you need to have more paces per chain. Going down hill, the reverse is true. Consistent accuracy in pacing is obtained only by practice.

Measuring Individual Trees

Determining the Height of a Tree

Learning the total height of a standing tree is usually of little value when one is interested primarily in learning how much of the total length is usable. Some trees have a short stem and a tall crown, while others may have a long stem or trunk and a relatively short crown. It is possible to make close estimates of the usable, or merchantable length, or height of a standing tree by any one of three units of measure. It should be kept in mind that by merchantable or usable length, we refer to that part of the stem or trunk which has value for processing into some useful product such as lumber, railroad ties, pulpwood, box boards, etc.

Total Merchantable Length in Feet

The most important factor in arriving at the merchantable length in feet is the smallest diameter of log or bolt which will be accepted, and then estimating on the trunk of the tree where this diameter occurs. For logs, this smallest diameter is eight inches—pulpwood it is four inches. One of the simplest of all methods to determine the total length, after the point of minimum diameter is estimated, is to use a Merritt Hypsometer.

This is a wooden or metal stick, usually about 36 inches long that enables the user to measure the merchantable length of a standing tree. Graduations on the stick are based on the length-of-arm class of the user, the distance from your eye to your index finger pointed up. Learning this length, you start at the left end of the stick, considered "zero", and measure off intervals corresponding to your arm reach.

For example, a 25" arm reach will have a 6.2" interval between each 16 feet of height. Each successive interval should be marked and numbered as to total height. Remember, each interval equals 16 feet of tree height.

Merritt Hypsometer Length-of-Arm Class

Length-of-Arm Class (Inches)	23	25
Graduating Interval (Inches)	5.4	6.2

The Hypsometer is used for height determinations by standing 66 feet from the base of the tree. By grasping the stick vertically near the zero end, you extend your arm fully and make certain the stick is perpendicular to the ground. Sight along the bottom of the stick to the base of the tree. Then, **without moving your head,** glance up toward the merchantable top of the tree (8" minimum diameter or at the base of a crotch) and note the point on the stick crossed by your line of sight. This reading gives you the height of the tree in number of 16-foot legs. You should make your estimate to the nearest one-half log or 8 feet of height. The Hypsometer is on the back side of the Biltmore Stick.

USE OF THE HYPSOMETER

STEPS IN USE OF HYPSOMETER
(An Alternate Method)

1. Place 8' mark on tree.
2. Back off and hold stick perpendicular.
3. Holding perpendicular, move stick forward or backwards till bottom of stick at base of tree and the first 3" interval mark lines up with 8' mark on tree.
4. Without moving head, sight across stick to top of usable log.
5. Count the number of marks and multiply by 8.

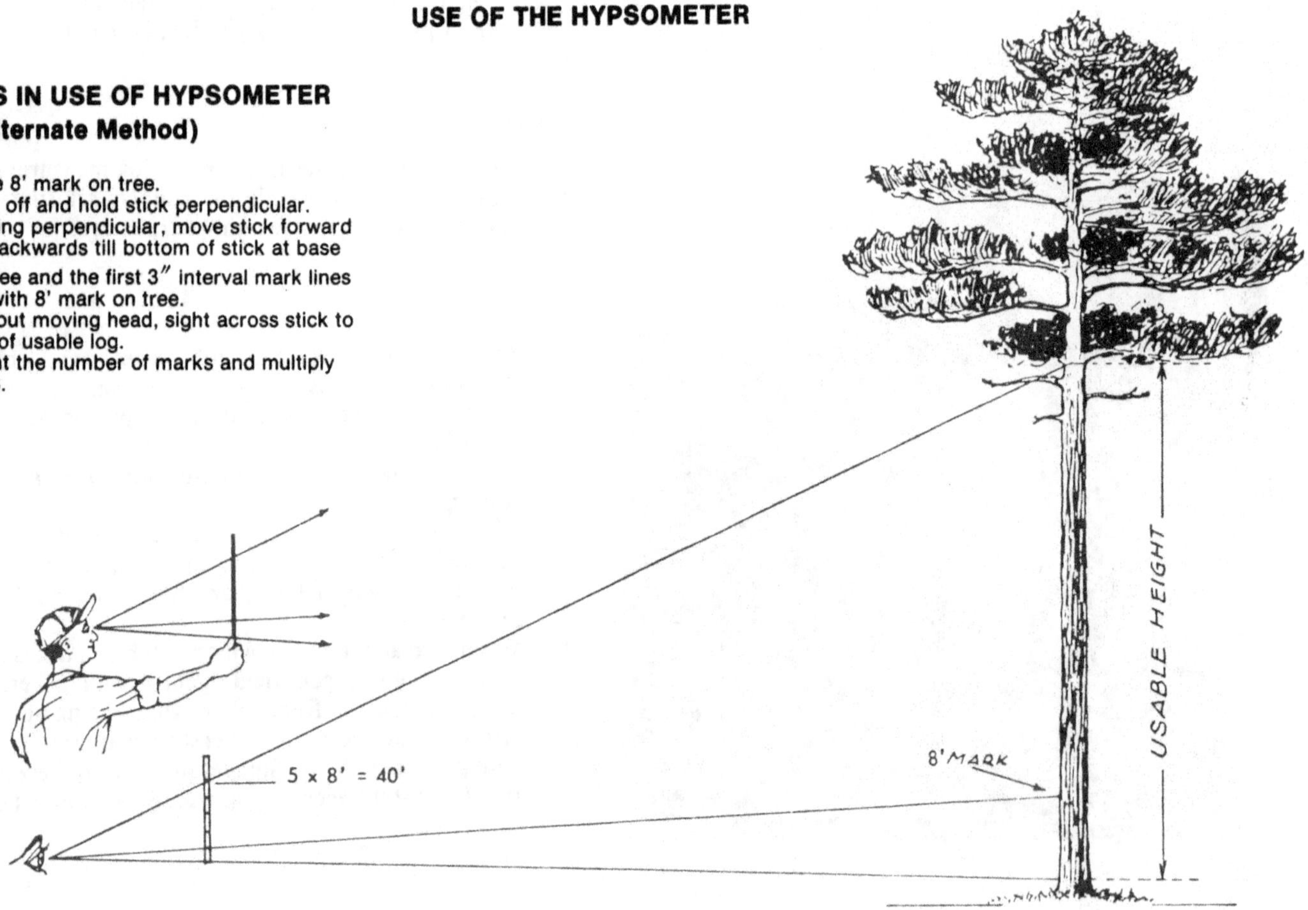

Determining the Diameter of a Tree

The diameter of a tree is measured in inches at a point 4-1/2 feet above the ground. This is known as diameter breast height, abbreviated D.B.H. It is a measurement of the outside bark diameter of a tree.

As in measuring tree height, the other side of the cruiser stick used as a Merritt Hypsometer can be used to measure diameters. You can make your own diameter or Biltmore stick by following these instructions. You must find your length-of-arm class as you did in making the Hypsometer—by measuring the distance in inches from your eye to your index finger with your arm fully extended. With this measurement determined, measure from the left end of the stick the distances given in Table 1. Label each mark with the diameter that is indicated. For example, if your arm length is 25 inches, the table indicates that you would measure 5.37 inches from the left end of the stick and mark it 6, then measure 7 inches and mark it 8 and so on. Divide the space between the successive numbers with a long line. This will give you the odd-numbered diameters.

To use your Biltmore stick for measuring tree diameter, grasp the stick firmly with your arm extended at a right angle to your body. The stick should be at a right angle to your arm and just touching the tree at D. B. H. Glance toward the left side of the tree and move the stick until the left end lines up with the outer edge of the tree. Then, *without moving your head,* glance toward the right edge of the tree and observe the point on the stick crossed by the line of sight. This reading gives the diameter of the tree. Most trees are not truly round so take two readings and average them. The second reading should be made at right angles to the first.

Tree diameters are usually grouped in 2-inch diameter classes. This means that the 14-inch class would include all trees measuring between 13 to 15 inches. The 16-inch class would include all trees from 15 to 17 inches in diameter, and so on.

Two measurements of a standing tree are needed in order to estimate the volume of a tree.

1. The diameter of a tree at 4 1/2 feet above ground (D.B.H.).
2. Usable height of a tree, usually in units of 16-foot logs.

Table 1

Diameter of tree - inches	Interval to be marked on stick	
	23" arm reach	25" arm reach
3	2.82	2.83
5	4.53	4.56
6	5.34	5.39
7	6.13	6.19
8	6.81	6.96
9	7.63	7.72
10	8.35	8.45
11	9.05	9.17
12	9.73	9.86
13	10.39	10.54
14	11.03	11.21
15	11.67	11.86
16	12.29	12.50
17	12.89	13.12
18	13.49	13.73
19	14.06	14.32
20	14.63	14.91
21	15.18	15.48
22	15.72	16.05
23	16.26	16.60
24	16.79	17.14
25	17.31	17.68
26	17.81	18.20
27	18.31	18.72
28	18.80	19.23
29	19.29	19.73
30	19.76	20.22

Measuring the diameter in inches outside bark of a tree at a height of 4½ feet above ground. (Known as diameter breast high, abbreviated D.B.H.)

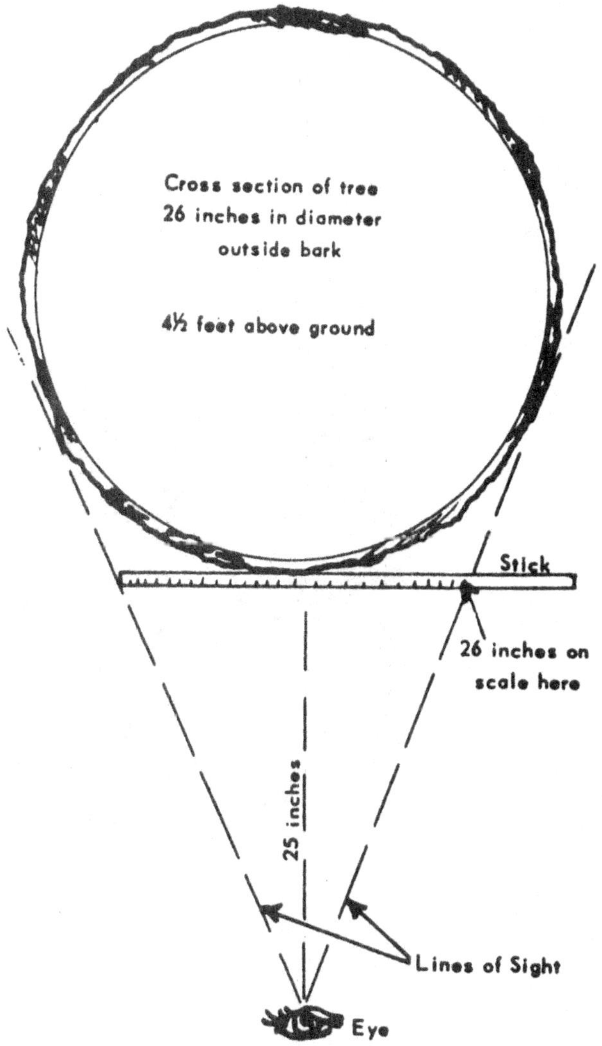

Estimating Volumes

Volumes are usually stated in board feet, cords or cubic feet. Cubic foot volume tables are used primarily in research, therefore they will not be discussed further.

To estimate the board foot volume in a standing tree, first measure the D.B.H. with the Biltmore stick. Next, measure off 66 feet from the tree base and estimate the merchantable height with the Merritt Hypsometer. Usable heights are estimated to an 8" minimum diameter or to a fork in the merchantable trunk. From a volume table, you can determine the tree volume. Example: a tree measuring 20" D.B.H. and 3 logs in usable height has a board foot volume of 364. (Table 4)

Cordwood size trees are measured for diameter and estimated for height similar to the way in which sawtimber trees are measured. Instead of measuring height in log lengths, it is estimated in *8-foot bolt* lengths. This is actually the 1/2 log interval marked on the stick. The same principles apply as in sawtimber trees except that bolt lengths are not divided into half lengths but are kept in even bolts, 1, 2, and so on. Thus, a tree with 24 feet of usable length is classed as a 3-bolt tree. Usable height is estimated to a 4" minimum top diameter. After diameter and height are determined, the volume can be obtained by using a cordwood volume table. Example: a tree measuring 8" D.B.H. and 4 bolts in usable height would have a volume of 0.09 cord. In other words, it would take about 11 trees of this size to make one cord of pulpwood.

Through practice, you can estimate tree volumes quite accurately with the cruiser stick.

Pulpwood

A pulpwood cord is composed of bolts, or logs, 100 inches long, placed in a pile four feet wide and four feet high. This represents a gross cubic volume of 133 cu. ft.

Board Foot

A board foot is represented as 144 cubic inches of lumber. As an example it could be a board 1" thick 12" long and 12" wide.

How to Measure 66-Foot Distance.

It is not necessary to use a tape measure to find the correct distance from the tree each time you estimate a tree height. Measurements can be made quickly and quite accurately by pacing. A pace is equal to *two ordinary steps.*

Lay off a 66-foot distance with a tape measure and mark the ends. Check the number of paces you need to take to cover this distance. If you start your stride with your *right* foot, the second step will be when you *left* foot

A PULPWOOD CORD

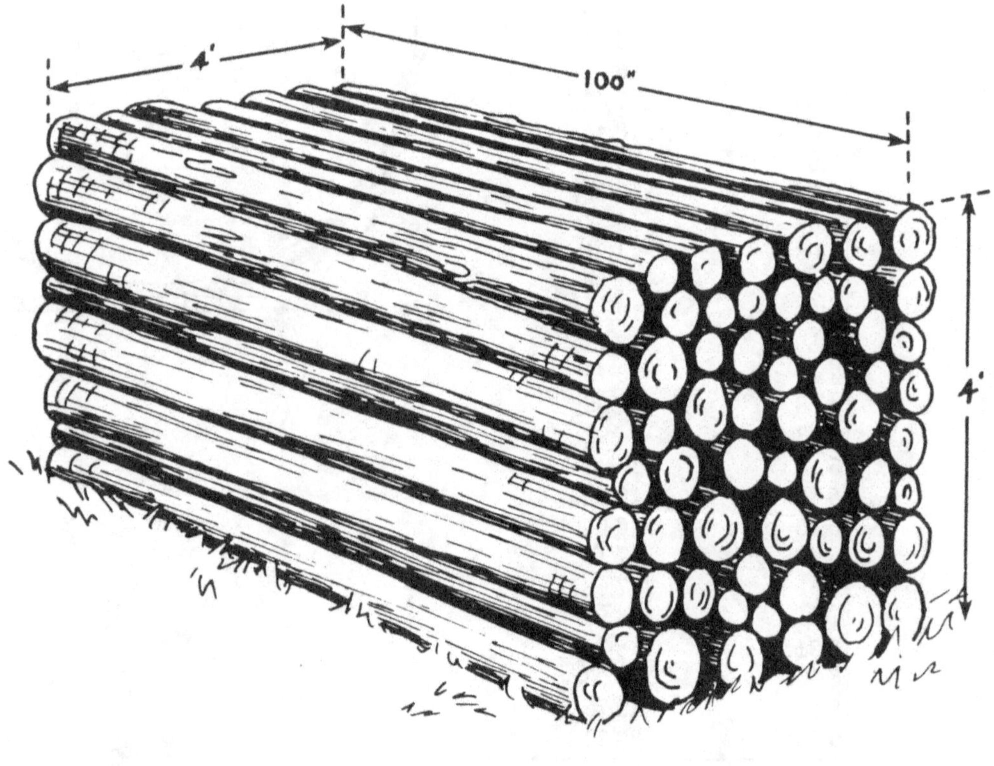

hits the ground. This is one pace. Count only when your left foot touches the ground. This will give you the total paces you need to walk the 66 feet.

The average person takes about 12 paces for the 66-foot distance, but this will vary according to your *natural* stride, the topography and the amount of vegetation that you pace through.

Tallying Trees on Timber Cruiser Form

You can keep track of each tree you measure for D.B.H. and height by a simple *dot-dash* system. This will enable you to keep a neat and readable record of your field work even though writing space is limited and your pencil gets a bit dull.

1 2 3 4 5 6 7 8 9 10

The Tree Volume Table.

Since the forms of trees are extremely variable, as influenced by differences in tree species, the rate at which they grew, and the degree to which they were crowded while growing, there are literally hundreds of tree volume tables. The volume table printed on this stick represents an average table for trees whose usable stems contain 78 per cent of the volume of a cylinder having a length equal to the usable length of the tree, and a diameter equal to the diameter of the tree.

Using the Tree Volume Table (Table 4.)

Let us assume you have measured the diameter of a tree at breast height (DBH) and found it to be 20 inches, and that in estimating the height you determined it had a merchantable length of 48 feet, or three 16-foot logs if utilized to a top diameter of eight inches. You now want to get the volume in board feet. On the scale below "Diameter of Trees—in Inches" you will note volumes by

COMPOSITE VOLUME TABLE FOR STANDING TREES

(from U.S. Forest Service Administrative Handbook)

Tree Diameter in Inches	Volume - Board Feet - Scribner Rule - by No. of Usable 16-foot Logs				
	1	1-1/2	2	2-1/2	3
10	28	36	44	48	52
11	38	49	60	67	74
12	47	61	75	85	95
13	58	76	94	107	120
14	69	92	114	130	146
15	82	109	136	157	178
16	95	127	159	185	211
17	109	146	184	215	246
18	123	166	209	244	280
19	140	190	240	281	322
20	157	214	270	317	364
21	176	240	304	358	411
22	194	266	338	398	458
23	214	294	374	441	508
24	234	322	409	484	558
25	258	355	452	534	617
26	281	388	494	585	676
27	304	420	536	636	736
28	327	452	578	686	795
29	354	491	628	746	864
30	382	530	678	806	933

Steps

1. Obtain diameter and usable height.
2. Consult table for volume of tree.
3. Add the volumes for all tallied trees for the gross volume of area cruised.

diameters of trees containing one, two, or three 16-foot logs. Since our tree had an estimated 48 feet of merchantable length, we would refer to the volume under 20 inches for a three-log tree, which gives a gross volume of 364 board feet. For trees with one and one-half, two and one-half, or three and one-half log length, volume is determined by interpolation. For example, a tree 20 inches DBH, containing one and one-half logs merchantable length (24 feet) will have a gross volume of...

$$\frac{157 + 270}{2}$$

...or 214 board feet. Note that 157 feet is the volume of a one-log tree; 270 feet is the volume of a two-log tree, 20 inches in diameter. It should also be noted that the volume table is constructed for trees shown at two-inch intervals in diameter. This is satisfactory when large numbers of trees are measured, and the diameters are rounded off to the nearest even number. In like manner, the volume of a tree with an odd numbered diameter can be had by averaging the volumes of trees in the even numbered diameters preceeding and following. Hence, a 19-inch three-log tree will have a volume equal to 280 + 364 ÷ 2, or 322 board feet. (Table 4 includes volumes for odd numbered inches of diameter and half logs. Some volume tables do not.)

Composition of Log Rules and Volume Tables

It is well to emphasize the fundamental differences between a log rule (regardless of name) and a volume table:

1. A log rule gives board foot volumes of *logs* only; a volume table gives board foot, and may give cubic foot or cordwood content of *standing* trees.
2. The diameter column in a log rule refers to diameter at the *small end* of a log; the diameter column in a volume table refers to the diameter at breast height of a standing tree.
3. The several columns indicating lengths in a log rule refer to length of logs; in a volume table these columns refer to usable lengths of trees, generally in terms of 16-foot logs.

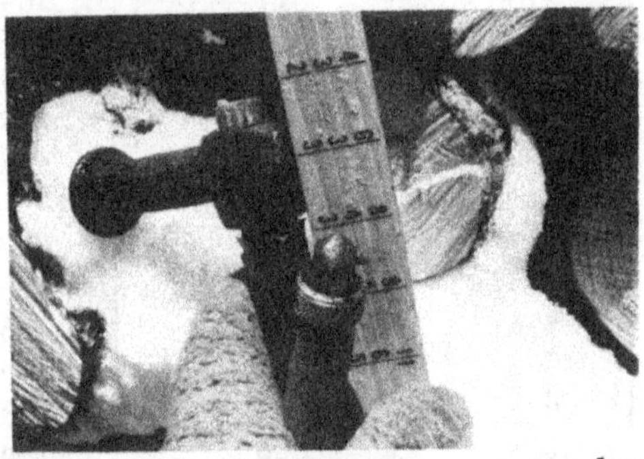

Common scaling tools include a stamp hammer, logging chalk and a scale stick.

Scaling Cut Products

By definition, scaling is the term used in estimating the lumber content of a log or a group of logs considered together, based on a specific log rule. It is a system of measurement of forest products.

Logs

Logs or timber intended for sawing into boards, planks or other construction materials are known as sawlogs. In Wisconsin, the more important species are elm, maple, oak, basswood, walnut, ash, white birch, white and red pine, and jack pine. Trees to be cut for sawlogs should be straight, sound, and at least 16 feet to the nearest branch. After a tree is felled, it is usually cut into 8-, 10-, 12-, 14-, or 16-foot lengths plus an additional 4 to 6 inches for trimming and squaring the ends of the resulting board. Knowledge of the length desired by the ultimate user is essential before felled trees are cut. The board-foot content or scale of a log is determined by measuring the average diameter inside the bark at the small end of the log in feet, and referring to a table indicating the board feet. A board foot is equal to a piece of wood 12 inches wide x 1 inch thick x 1 foot in length.

In Wisconsin, the Scribner Decimal C scale is used in measuring logs cut from the national state and county forests. Walnut logs are usually scaled on the "Doyle Rule" which has inaccuracies when applied to small logs.

The International Rule is the most accurate rule devised but, unfortunately, it is not widely used. The figures given will come quite close to the actual board feet obtained by most sawmills in operation in the state.

When many logs are measured, scale sticks giving direct readings in board feet using the desired log rule, such as Scribner, International or Doyle, are often used. If logs are defective, the board-foot contents are reduced accordingly.

Posts, Poles and Pilings

Material used for these purposes are usually sold by the piece. The standard unit of measure is the linear foot. Sizes for the various products vary, depending use. Posts, used for fencing, are usually 5 to 7 feet in length and 3 to 8 inches at the small end. Piling-size range is 15 to 60 or more feet in length and between an 8-inch tip and 20-inch butt.

Poles used in utility work have a length requirement of 30 to 45 feet and a diameter inside bark at the small end ranging 5 to 7 inches.

In most cases, the product is scaled or measured as to length in feet and diameter inside bark at the small end.

Variable Radius Plots
(Point Sampling)

In 1948, Walter Bitterlich of Austria described a new and very rapid method of determining basal area per acre by point-sampling. In 1952, L. R. Grosenbaugh and others translated and expanded on this method to make it applicable to forestry in this country. Once basal area is known, it can be correlated to tree and stand volumes.

Basal area, as applied to tree measurement, is the area in square feet of the cross section of a tree at breast height. The total area in square feet of the cross sections of all trees on an acre is referred to as "basal area per acre." Basal area per acre can be easily determined with the aid of several different instrument using Bitterlich's principle. Such instruments are the wedge prism, stick-type angle-gauge, Panama angle-gauge, Spiegel Relaskop and various improved gauges.

In using these instruments, a cruiser counts the number of trees around a sampling point whose diameters at breast height appear larger than the crossarm of the angle-gauge (Figure 1.) or that are not cut in two when viewed through a wedge prism (Figure 2.). When tree count is multiplied by a predetermined factor (basal area factor), the basal area per acre is obtained around the sampling point. A number of points must be taken and averaged to obtain basal area and volume estimates.

It is recommended that instruments used in timber stands of Wisconsin have a basal area factor of 10. With a factor of 10, the ratio of width of crossarm to length of stick (angle-gauge) or the displacement by the wedge prism is 1:33. That is, a tree one foot in diameter would be counted as an "in" tree up to 33 feet from the point center. In all references to the basal area factor in this exercise, the factor will be assumed to be 10.

The probability of a tree being counted as "in" is proportionate to its size (Figure 2.). A tree twice the diameter of another and at the same distance from the point center has four times the probability of being counted as would the smaller. Larger trees, which are more variable in volume and quality, are sampled in greater proportion than smaller trees which are less variable and of less importance.

Figure 1
POINT SAMPLING WITH ANGLE GUAGE

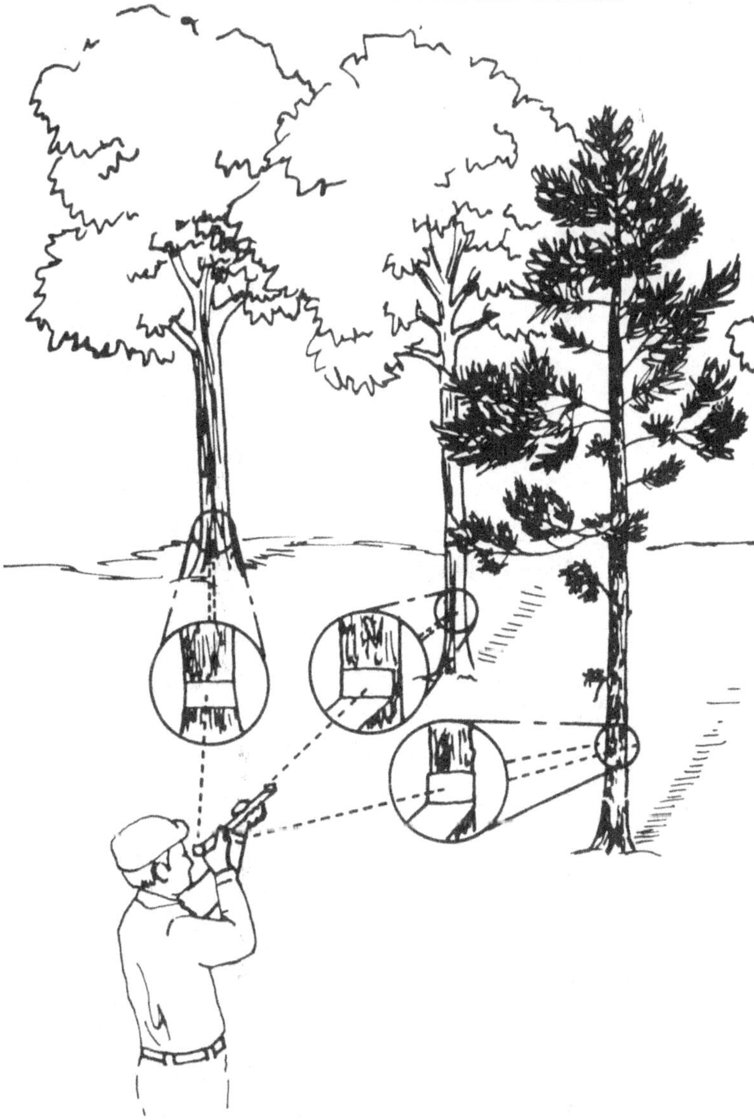

Figure 2
POINT SAMPLING WITH WEDGE PRISM

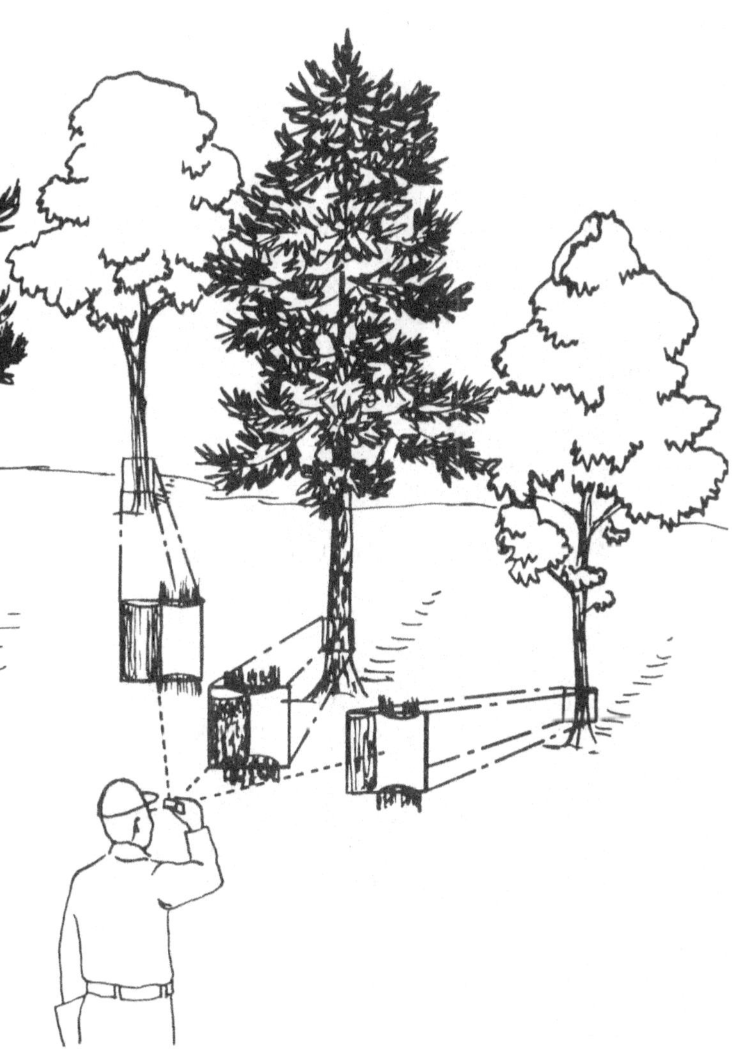

Figure 3
RELATIONSHIP BETWEEN TREE DIAMETER AND PLOT DISTANCE. BAF 10

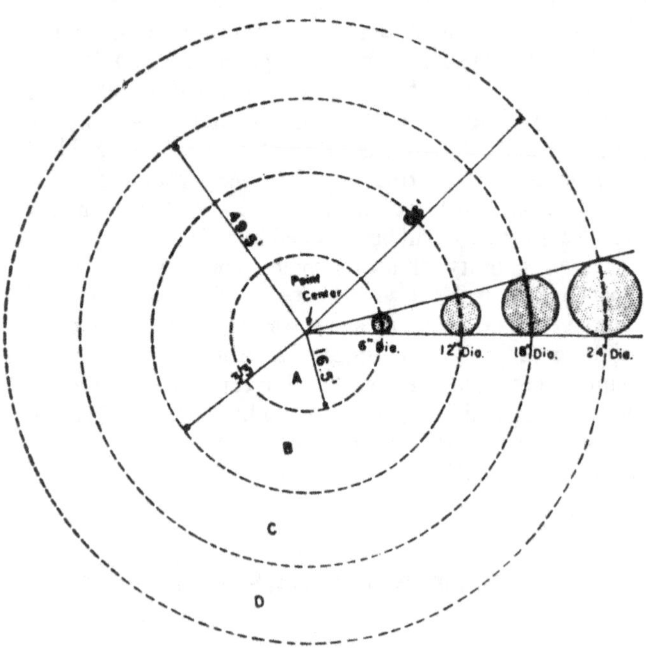

Relationship between tree diameter and "plot" distance. BAF 10.

Each counted tree represents 10 square feet of basal area per acre. It also represents a certain part of an acre and a specific number of trees per acre. A small tree represents a small area but a large number of trees. A large tree represents a large area but a small number of trees.

The use of the point-sampling system requires considerable care in the determination of "in" and "out" trees. The following precautions must be taken:

1. Be certain all trees are counted. Watch for the hidden ones.
2. Count only the merchantable trees. Do not count dead or cull trees.
3. Do not stray off point center.
4. Hold angle-gauge close to eye.
5. Hold wedge prism over point center in a vertical position and at right angles to the line of sight.
6. On steep slopes a correction factor must be applied.

This correction for tree count, basal area and volume per acre is true only when it is assumed that in this sampling procedure a proportionate number of trees will be added when the slope distance is corrected to the true horizontal distance.

Figure 4
CONSTRUCTION OF ANGLE GAUGE

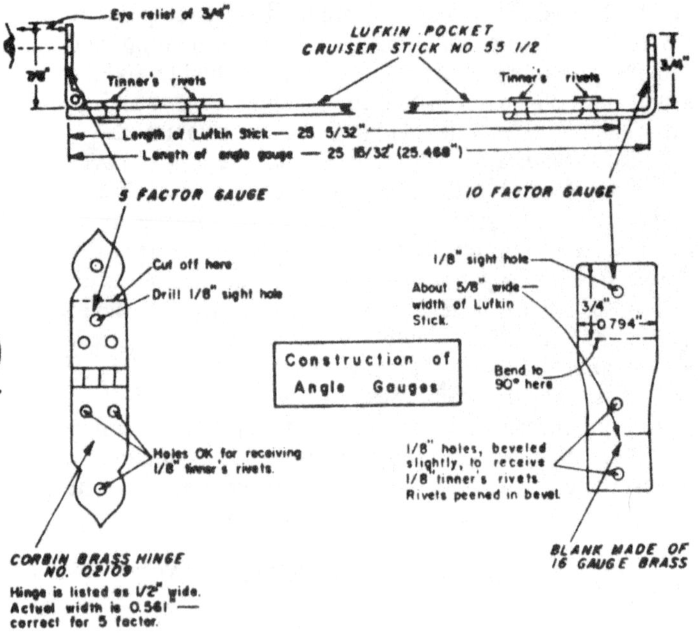

Computations for Gauge Widths

Total length of Bitterlick Stick	25.468"
Eye relief displacement	+0.750
Effective length of Bitterlick Stick	26.218

	Effective length	x	Width of gauge per 1" of stick	=	Gauge width
10 FACTOR WIDTH	26.218	x	0.0303	=	0.7944"
5 FACTOR WIDTH	26.218	x	0.0214	=	0.5611"

Tomahawk 1962

Construction of angle gauges.

Table 1
RELATIONSHIP BETWEEN CROSSARM AND LENGTH OF ANGLE-GAUGE

Width of Crossarm (Inches)	Length of Angle-Gauge—Inches		
	BAF 5	BAF 10	BAF 20
.25	11.67	8.25	5.83
.30	14.00	9.90	7.00
.35	16.33	11.55	8.17
.40	18.67	13.20	9.33
.45	21.00	14.85	10.50
.50	23.34	16.50	11.67
.55	25.67	18.15	12.83
.60	28.00	19.80	14.00
.65	30.34	21.45	15.16
.70	32.67	23.10	16.33
.75	35.00	24.75	17.50
.80	37.34	26.40	18.66
.85	39.67	28.05	19.83
.90	42.00	29.70	21.00
.95	44.34	31.35	22.16
1.00	46.67	33.00	23.33

BAF 5 has a ratio of 1/46.67; therefore length of stick is 4.667 inches per .1 inch of crossarm.

BAF 10 and 20 have ratios of 1/33 and 1/23.33; therefore length of stick is 3.300 inches and 2.333 inches per .1 inch of crossarm, respectively.

Fixed Radius Plots

Allen Wickman

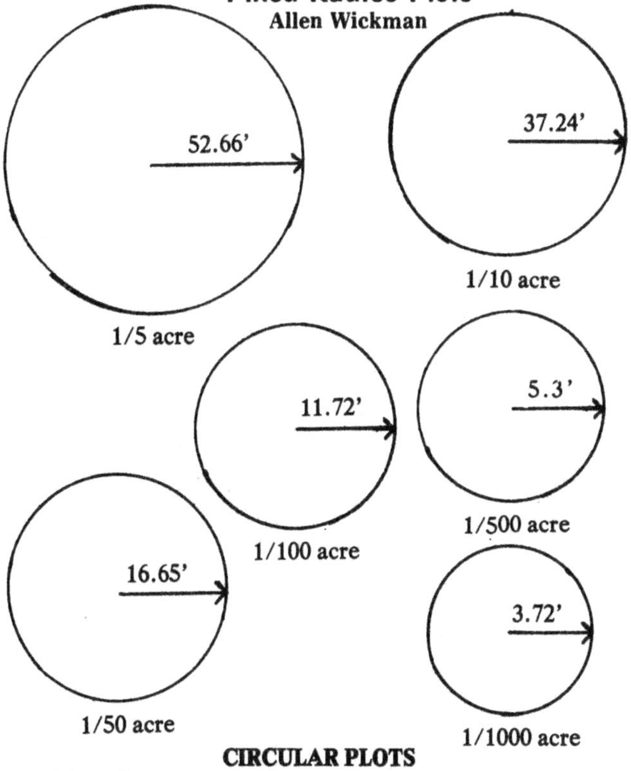

CIRCULAR PLOTS

Many foresters use circular plots to help determine the numbers and volumes of trees per acre. For example, if we were to make a circle with a radius of 37.2 feet (74.4' diameter) we would have a circle equal in size to 1/10 acre. If we were to count all the trees inside that circle and multiply by 10, we would have an **estimate** of the number of trees we have per acre. Each measurement as such is called a **plot.** The more plots taken the more accurate the estimate (normally).

When foresters count trees in a plot they often **"tally up"** (add up) the volumes and sizes for each species. This estimate is used for making timber sales.

Circular plots are frequently used to determine stocking levels in plantations too. Stocking levels can tell us if a plantation is too thin or too thick. In Hubbard County we often give landowners the following set of instructions to see if plantation thinning is necessary. If thinning is necessary, we advise which trees to cut. The recommended trees per acre depends on their average diameter.

GUIDE TO THINNING AN EVEN AGE NORWAY PINE PLANTATION

STEP 1

Determine the desirable number of trees per acre. Use an applicable chart, table or consult a forester. Divide by 100 to find recommended number of trees on a 1/100 plot.

STEP 2

Make a measuring stick that is 11.8' long. An 11.8' radius of a circle will give readings on 1/100 of an acre.

STEP 3

Find a dense clump in your plantation. Stand in the middle of the clump and use yourself as the center of an imaginary circle with an 11.8' radius. Count all the trees in your circle. If unsure whether a tree is "in or out", use your measuring stick.

STEP 4

Compare your count total with the recommended amount for a 1/100 acre plot determined in Step 1. If you have 12 trees and the recommended is 8—then plan to cut 4. Several plots such as this should be taken in a plantation to give a good overall picture of its needs.

STEP 5

Which trees should be cut?

- Cut the smallest trees.
- Cut trees with deformed trunks.
- Cut trees with double or deformed tops.
- Cut trees growing very close together (especially 2' apart or less—cut only one).

Which trees should be saved?

- Save the trees of largest diameter.
- Save the tallest trees.
- Save the straightest trees.
- Save trees not crowded by others.

These trees have the potential for highest value.

SPACING CHART FOR EVEN AGED NORWAY PINE

AVERAGE D.B.H.	RECOMMENDED TREES PER ACRE
0-4½"	700 to 900
5"	650
6"	450
7"	350
8"	250
9"	200
10"	170

BASAL AREA
WHY IS IT IMPORTANT?

What Is It?

Suppose we were to measure off one acre and then cut down a tree on it at 4½ feet above the ground. Suppose we measured the surface area of that stump and found it to be 144 square inches (1 square foot). 1 square foot would be the basal area of that stump.

Now suppose we cut down all the trees an that acre at 4½ feet above the ground and found the total surface area of all stumps - we would then have a figure giving us the basal area for that acre.

By now you must be saying to yourself that there must be a better way to measure basal area. There is! Special tools can be made or bought that will give us a good approximation of basal area. These tools can be found in any forestry supply catalog.

What Is It For?

Basal area measurements are quite common to foresters because they can tell how "thick" a stand of trees is. Stands that are too thick should be thinned to improve growing conditions for the remaining trees. For example: A forester may feel that a particular 50 year old norway pine stand should have a basal area of 90 square feet for optimum growing conditions. His measurements may indicate a basal area of 130. He would then make plans to thin down this stand by selective cutting. 40 square feet of basal area would be removed from each acre.

Many foresters rely quite heavily on basal area measurements to determine what should be done with a stand of trees. The "ideal" basal area may vary depending on the age or species of a group of trees.

A full page chart in the appendix can help tell us how many trees per acre we should have if we know the average diameter at breast height and the "ideal" basal area. It's an excellent chart for those who know how to use it.

What Is a Site Index?

How Fast Do Trees Grow

By Allen Wickman

Site index is very important when planting trees—yet thousands of people who plant trees every year have never heard of it!

What is it? Simply stated, site index is an estimation of a tree's height when it is 50 years old. If a tree has a site index of 60, we would expect that tree to be 60 feet tall when it is 50 years old. A tree with a site index of 40 would be 40 feet tall after 50 years.

Site index depends on the characteristics of the soil and is a measurement of a soils capacity of growing a particular species of tree. (Of course, climate also has something to do with that!)

Each soil type can grow some species of trees better than others. We try to match the tree to the soil for maximum productivity in our forests. Site indexes usually range from 30 to 80 for different species of trees on the same soil type. We try to plant trees with a site index of 55 or more. The higher the site index the better.

Trees with a very low site index grow very slowly and we say that these trees are "off site". This means that the trees would be much better off growing somewhere else (on a better site). When we find trees growing off site we often consider removing these trees and planting a tree better suited to the conditions.

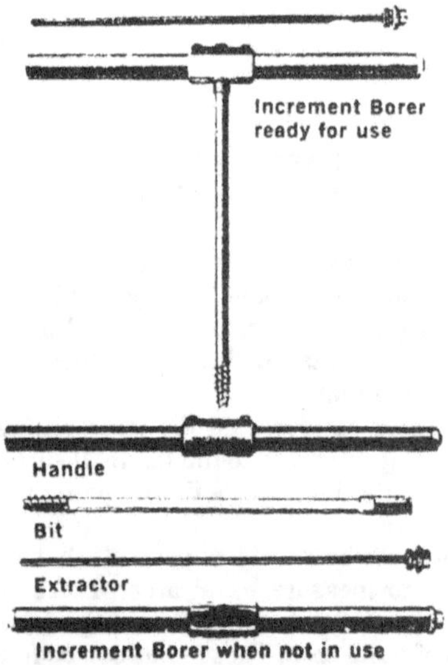

The increment borer is an expensive little tool ($60-$70) used by foresters to determine a tree's age.

Increment borer being screwed into tree.

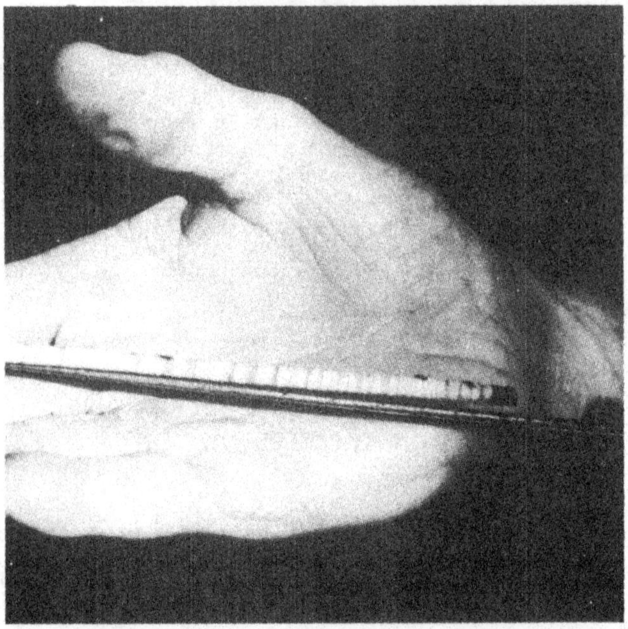

Core sample — Each dark ring normally represents one year's growth and is formed by "summer wood" which is more dense than faster growing "spring wood".

Why Is Site Index Important?

Trees growing on soil where they have a high site index will grow faster, yield more timber (or other products) and are more resistant to insects, disease, drought and other problems.

A yield table for Norway pine will tell us that a 40 year old stand of Norway pine on site index 75 soil will yield 67% more wood than a similar stand on site index 45 soil. A site index of 75 not only yields more volume, but the wood will be of much higher quality and value.

How Is Site Index Determined?

It is necessary to determine a tree's height and age to find out its site index. A tree's height also depends on the amount of sunlight it has received during its life. A tree growing in heavy shade will not give us a good measurement of the true site index. Lack of sufficient sunlight will retard its growth.

Consequently, we try to select a tree that appears to be getting plenty of sunlight. We select one of the taller trees in a grove or a tree growing out in the open.

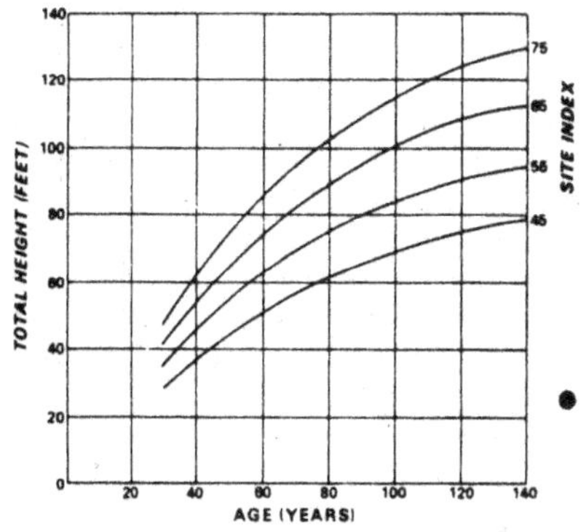

This is a red pine site index curve. Determine a red pine's height and age then refer to this chart to find its site index. For example: a 100 year old tree that is 80 feet tall would fall closest to the 55 site index curve. Its height at 50 years would be about 55 feet.

Core sample being extracted from increment borer.

The second step is to bore into the tree with a special tool we call an increment borer (see photo). The core sample extracted will have a number of dark colored rings. Each ring normally represents one year of a tree's life. To determine the age, we simply count the rings. The first ring counted should be at the center of the tree. Each ring from the center to the outer edge is counted.

Once we have a tree's age it is necessary to compute its height. Several forestry tools are available for this purpose. Once we know a tree's height and age we can refer to a table

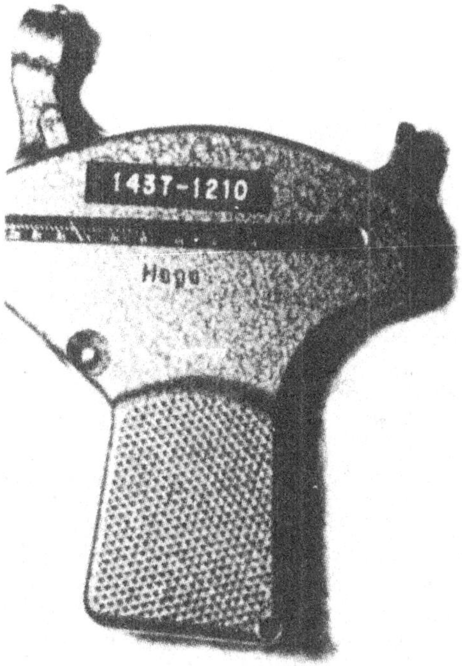

The haga altimeter is used by foresters to determine the height of a tree.

(see chart) which will indicate its height at age 50—this is the site index for that particular tree.

Since most people can't afford a $60 increment borer to find out a tree's age, there are a few easier routes to follow. Watch for trees being cut in nearby localities with similar soils. After a tree is cut, measure its length and count the rings in the stump. Use these measurements to determine the site index. Better yet, contact a forester or county agent to find out which trees grow best in your particular locality.

What is site index? To a landowner it can be a matter of profit or loss. To a tree it can be a matter of life or death. Matching the proper tree to your soil can be a wise decision.

A GUIDE TO SELLING AND SCALING SAWLOGS

Introduction

On occasion forestland owners find it appropriate to sell timber. Even nonindustrial private woodland owners including farmers and nonfarm rural landowners may decide to harvest their woods. Reasons for logging a woodland include benefits from additional income and, if properly planned and conducted, silvicultural improvement of forest stands.

When harvesting one's own woods, in contrast to selling stand timber (stumpage) to a logging or sawmill business, marketing the products becomes the landowner's responsibility. Selling harvested hardwood sawlogs will be either at roadside adjacent to a woodland or delivered to a sawmill. **It is advisable before harvesting begins to have secured appropriate markets for the different products, including pulpwood, firewood, and logs.**

Marketing hardwood sawlogs profitably requires a knowledge of: (1) buyers, (2) log specifications established by different sawmills, and (3) values of different tree species for such products as veneer, baseball bats, pulpwood, utility poles, fence posts, furniture grade lumber, and construction or pallet lumber. In addition, an understanding of log grading, log scaling, and log rules is important.

Log value factors

The value of logs sold to a sawmill is determined by the following factors:

1. end-use products which are obtained from logs,
2. local supply and demand situation relative to logs,
3. quality of a log which determines value of final products, and
4. volume of a log which determines yield of wood products.

Tree species

Values of different species of logs of the same quality and volume may vary due to relative differences in prices for end-use products. For instance, beech is of lower value as a species than white ash or black cherry. The more valuable species of logs, in recent years, have been red oak, white oak, black cherry, sugar maple, white ash, and yellow birch. However, it should be noted that the higher prices paid for these species do not apply to low quality logs. And, although there may be exceptions relatively low priced species of logs include red spruce, American beech, aspen (popple) and hemlock.

Log length

Log lengths affect lumber value, logging and sawmilling costs, and therefore log values. Short logs of 8 feet normally are not sought except for specialty products; but 10-foot and 12-foot logs may be preferred over 14-foot and 16-foot logs depending on the sawmill and its markets. Log lengths typically are specified in even-foot units from 8 to 16 feet, with about 4 inches of trim allowance required in addition to the specified log length. Thus, an 8-foot log should be cut 8 feet 4 inches in length and a 16-foot log will measure 16 feet 4 inches to meet requirements of sawmills. Many sawmills purchase logs in even lengths of 8 ft., 10 ft., 12 ft., 14 ft., and 16 ft.; and they will determine volume of a log to the nearest even foot which is less than actual log lengths. For example, logs measuring 13 feet 4 inches and 13 feet 11 inches would be scaled for volume as 12 foot logs. Sometimes, however, sawmills will purchase logs in odd-foot lengths, but the seller should determine the log-length scaling procedures before cutting (bucking) logs to length for the marketplace.

It is easy to lose value by miscutting logs. Logs which are too short are reduced in scale. Logs which are too long waste timber because the purchase price for logs is based on standard lengths specified by the purchaser. Logs are normally cut from tree butt to top, and if logs are bucked overlength, the last log may be shorter than necessary or possibly not even long enough to meet the minimum length of 8 feet 4 inches required for a sawlog. Additionally, the scaling diameter for an excessively long log as well as for the logs which occur above it in a tree is reduced due to the natural taper of a tree from stump to top.

Log diameter

Log diameters also affect log values. Normally, small diameter logs are more costly to harvest, truck, and saw into lumber than large diameter logs. For logs between 9 and 18 inches in diameter, measured inside the bark (d.i.b.) on the small end, it is usual for logging and sawmilling costs to increase as log diameters decrease. In addition, the more valuable lumber grades are usually yielded by logs of at least 13 inches in scaling diameter (measured d.i.b. on the small end of the log).

Log grades

Hardwood log grades are used to predict value of lumber recovery from sawmilling and to separate logs of one species into different purchase price categories based on factors such as the following:

- log length
- log diameter
- knots or brand stubs
- bumps which indicate internal wood irregularities or overgrown knots
- seams with ingrown bark
- frost cracks
- splits
- ring shake or separation of growth rings on end of log
- bird peck
- worm holes
- rot or declay
- gum spots (in black cherry)
- sweep which is a continuous deviation of a log's length from a straight line
- crook which is an abrubt deviation of a segment of a log's length from a straight line.

The highest grade of hardwood log is normally a veneer log for slicing. And the highest priced veneer logs are commonly exported to Europe. High grade veneer logs for rotary turning are valuable, but container or construction grade veneer logs are of less value. The highest grade of sawlog is commonly called "prime" although sometimes individual sawmills specify the highest grades as "veneer," "select," or "No. 1." After these grades the next lower grades in succession from best to poorest are frequently identified as No. 2, No. 3, and cull.

The mill-delivered prices for logs graded as No. 3 may be so low that they are not profitable to produce. It is common for cull logs to be priced at less than the cost of logging and trucking because sawmill businesses do not want to be bothered with processing them due to high production costs and unprofitable value of lumber recovered.

Log prices

Prices for logs are quoted per thousand board feet (MBF) of logs delivered to a mill or sold at roadside next to a woodlot. Sometimes a "woods run" price is given for all logs removed from a woodland with a requirement that no logs be sold to other buyers. However, in most cases prices are quoted accordingly to log grades which vary with each individual sawmill. It is common for the price range per thousand board feet of sawlogs to vary as much as $100 to $200 for valuable species such as black cherry, white ash, sugar maple, red oak, and white oak based on different log grades.

Since different log rules give significantly different board-foot volumes for the same log size, one must ensure that price comparisons are based on the same log rule. Also, one must ensure that scalding techniques and log grades are similar when comparing prices quoted by different sawmills.

Log rules

In New York State a number of different log rules may be used to measure the volume of board feet in logs. Since different log rules give significantly different volumes for logs of the same diameter and length, one must know what log rule is being used when prices are quoted and logs are measured in board-foot units. For example a 12-foot long, 14-inch (d.i.b.) log contains 100 board feet on the International ¼-inch Log Rule, 90 board feet by the Scribner Decimal C rule and only 75 board feet on the Doyle Log Rule.

According to the USDA Forest Service there are over 95 recognized log rules bearing about 185 names which have been developed in the United States and Canada. However, in New York only three log rules are commonly used. These are the International ¼-inch, Doyle, and Scribner Decimal C. In addition, the Vermont, Roy, and Ontario log rules may be used on occasion.

The International ¼-inch Log Rule is based on a mathematical formula (V equals $0.199D^2 - 0.642D$; where V equals volume in board feet for 4-foot section and D equals d.i.b. on small end of a 4-foot length). The formula is changed for long lengths of 8, 10, 12, 14, and 16 feet in order to account for log taper. This log rule is quite accurate but may result in less lumber being sawn than predicted by log scale so an underrun of 5 percent in actual lumber recovery compared to log scale may occur.

The Doyle Log Rule is based on a mathematical formula (V equals $(d-4)^2$ L/16; where V equals volume in board feet, d equals scaling diameter in inches inside the bark on small end of log, and L equals length of log in feet) which is especially inaccurate on small logs because of its excessive allowance of 4 inches for slabs and edgings. The Doyle log rule under scales log volumes for small diameter logs; and for these logs it provides an overrun of lumber sawn in comparison to volume of logs scaled. This tends to compensate the sawmill business for the inefficiencies of handling small logs; and in some cases motivates loggers and woodland owners to leave relatively small sawtimber trees to grow. From a forest management standpoint, the Doyle log rule sometimes is considered a disincentive to harvesting small diameter, immature trees.

For example, the volume of an 8-inch (d.i.b.), 10-foot log on the Doyle log rule is 10 board feet, only 50 percent of the 20-board-foot volume yielded by the International ¼-inch log rule for the same size sawlog. In general, the under scaling of logs by the Doyle log rule increases as log diameters decrease from 23 inches (d.i.b.) to 8 inches. Also, for the same diameter logs between 8 inches and 23 inches, 16-foot logs are under scaled about 5 to 15 percent more than 8-foot logs when comparisons are made with the International ¼-inch log rule. Therefore it is advantageous to sell 8-foot long logs rather than 16-foot logs when the Doyle Rule is used, if the buyer permits it.

The Scribner Log Rule is based on log diagrams indicating sawing patterns for logs of exact inch diameters (measured on the small end of logs) with no allowance made for log taper. The Scribner Decimal C Log Rule, which is commonly employed, is a modification of the Scribner log rule where board-foot volumes are rounded off to the nearest 10 feet and the cipher is dropped. Therefore 2 on the Scribner Decimal C log rule represents 20 board feet and 16 represents 160 board feet. The Scribner log rule commonly gives log volumes which are less than the International ¼-inch log rule and more than the Doyle log rule for logs between 8 and 21 inches in diameter (d.i.b.).

Determining Log Volume

To determine the gross board-foot volume of a log, measure the average diameter inside the bark to the nearest inch on the small end of a log and measure the log length in feet. Remember, log lengths typically are measured in even-foot lengths of 8, 10, 12, 14, and 16 feet with the aforementioned trim allowance of about 4 additional inches required for each log length.

Gross log volume in board feet is determined by using volumes estimated by any of the recognized log rules. The volume tables for different log rules are provided in tables appended to this publication for gross volume determination. A log scaling stick may be used if you prefer. Scaling sticks which are used to measure logs read directly in board-foot volumes for logs of different diameters and lengths.

To determine the actual or net board-foot content of a log, deductions are made from the gross volume for scaling defects which reduce the amount of lumber that can be sawn from a log. The scaling defects which must be subtracted from gross log scale account for losses in lumber yield due to holes, decay, splits, spiral seams, crook, and sweep in logs.

Common scaling practices involve rules of thumb to obtain a log's net volume by reducing gross log scale for scaling defects. Reducing the scaling diameter of a log or reducing the length of a log are typical methods of accounting for scaling defects. However, the recommended procedures for determining scaling deductions to account for holes, splits, sweep, crook, and decay are depicted in Figure 1.

Also, it should be noted that actual log scaling practices of sawmills vary, especially with the Doyle log rule. Because of competition in the log-buying market and due to under scalding of actual log volumes by the Doyle rule, some sawmills scale logs outside the bark so both bark thicknesses are included in the log diameter used to determine log volume in board feet. Other scaling practices include measuring the large end of small diameter logs and including one bark thickness but not the other when measuring the small end of a log. Of course there is only one prescribed method of measuring a log's diameter, and this is to determine the average diameter inside the bark on the small end of the log. **But due to the variety of log scaling practices, a seller of logs should know and understand the scaling methods of log buyers because these practices affect log volume which with log grade (and associated price) determines log value.**

Determining Log Grade

Since every sawmill in New York State uses its own log grading specifications based on a combination of quality of logs available, particular needs of the sawmill, and competition or custom in the area, **it is important to evaluate the potential for markets before producing and marketing logs.** In general, the higher log grades are free from or have fewer scaling and grading defects because sweep, crook, knots, surface bumps, and decay in the quality zone (all of which are grading defects) reduce the grade of lumber sawn and therefore lumber value. Holes and decay in the low quality

Figure 1. Methods of determining scaling deductions.*

(Examples based on a 16-foot log with 20-inch scaling diameter)

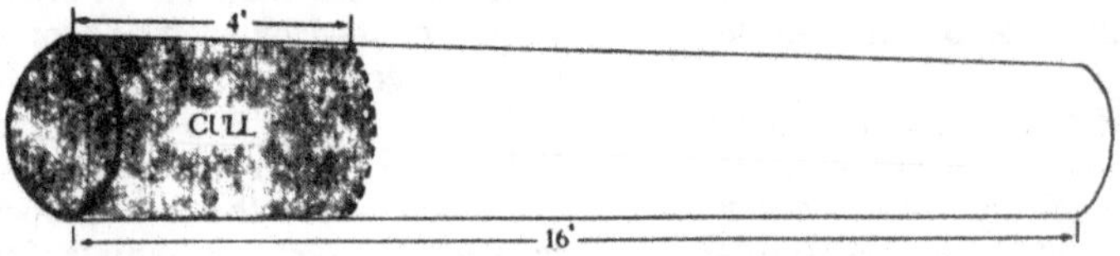

Defect section (rule 1):

$$\text{Percent deduction} = \frac{4}{16} = 25\%$$

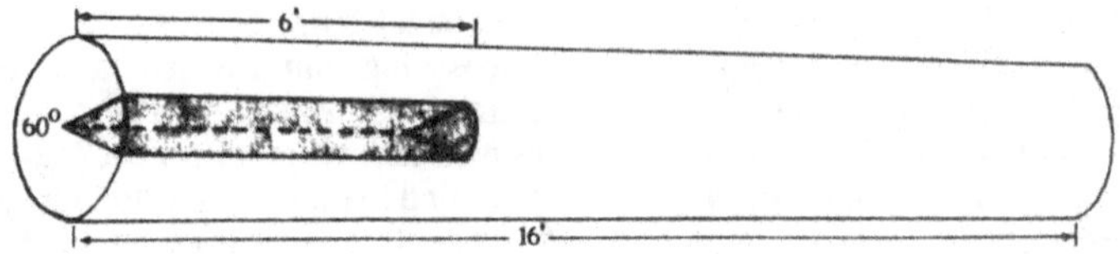

Defect section (rule 2):

$$\text{Percent deduction} = \left(\frac{6}{16}\right)\left(\frac{60}{360}\right) = 6\text{-}1/4\%$$

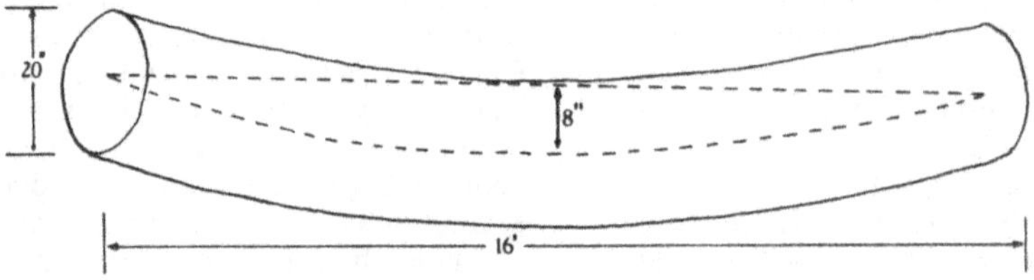

Sweep (rule 3):

$$\text{Percent deduction} = \frac{8-2}{20} = 30\%$$

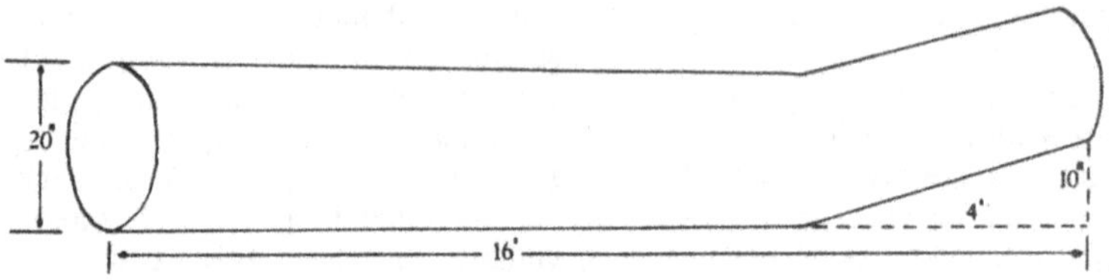

Crook (rule 4):

$$\text{Percent deduction} = \left(\frac{10}{20}\right)\left(\frac{4}{16}\right) = 12\text{-}1/2\%$$

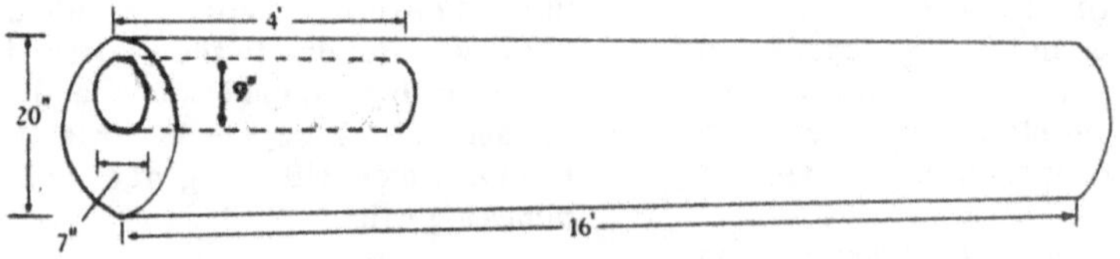

Interior defect (rule 5):

$$\text{Percent deduction} = \frac{(8)\ (10)}{(20-1)^2} \times \frac{4}{16} = 5\text{-}5/9\%$$

In practice each elipse axis can be divided by (20 — 1) and rounded to nearest tenth if desired.

Thus $\frac{8}{19} = .4,\ \frac{10}{19} = .5$, and $(.4)\ (.5)\left(\frac{4}{16}\right) = 5\%$

*Source: "A guide to Hardwood Log Grading," U.S.D.A. Forest Service, General Technical Report NE-1, 1973, by Everette D. Rast, David L. Sonderman, and Glenn L. Gammon.

heart center (Figure 2) of a log are scaling defects which reduce log volume but not log grade according to the U.S.D.A. Forest Service's hardwood log grading procedures.

A hardwood sawlog's "quality zone", as defined by the U.S.D.A. Forest Service, is measured inside the bark from the circumference of a log on the small end toward the log's center with a radius equalling 30 percent of the log's scaling diameter (Figure 2). This quality zone contains higher grades of lumber than the heart center which is defined on the small end of a log as the area measured from the center of the log to a radius of 20 percent of the log's scaling diameter.

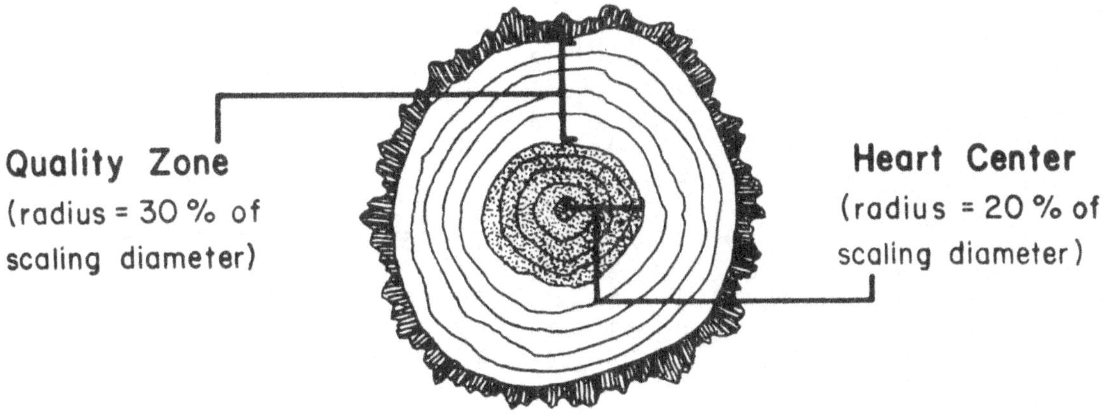

Figure 2. Heart center and quality zone of sawlogs.

Although sawmills may not use the term "quality zone" in defining their log grades, the higher grades of hardwood lumber are produced from this region of a log. Therefore, decay or a hole in the heart center does not significantly reduce the production of high grade lumber from a log; but, of course, it does reduce volume of lumber sawn.

Industry commonly defines high grade hardwood logs as those free of surface defects such as knots, bumps, splits, and spiral seams on the outside of a log. Also, some sawmills require high grade logs to be sound (free of decay and holes) even in the heart center. Of course veneer logs for rotary turning must be sound in the heart center so veneer lathe chucks can be used.

A helpful reference for those interested in log grading to predict lumber grade yields is "A Guide to Hardwood Log Grading," U.S.D.A. Forest Service, General Technical Report NE-1, dated 1973.

Forest Management

Log grades and values are influenced by forest management practices. Species composition of a woodland, volume of stocking rate of tree growth, and quality of trees are determifned by the practice of silviculture and careful logging. As a foundation for forest management, silviculture is concerned with the establishment, cultivation, and reproduction of forest stands.

Logging can be employed solely for immediate economic benefit, possibly destroying or reducing future timber values. Perhaps a more sound approach is to integrate timber harvesting with silviculture to provide both immediate income and improved forest growth, thereby maximizing timber production.

Managing forest stands includes harvesting timber and establishing tree reproduction as well as protecting soil from erosion in skid trails and roads, and favoring some high vigor, dominant, codominant, and what appear to be genetically superior trees for future growth.

Only a small percentage of all hardwood logs produced in New York State are of high grade. They are obtained from the very small percentage of all trees which are of highest quality. And most high grade logs are the first (butt) log of a tree. However, proper forest management and good timber harvesting practices will increase the quantity and quality of valuable trees. Improvement cuttings for firewood and commercial timber harvesting for pulpwood and sawlogs can be silvicultural practices providing increased future returns from forests while meeting present needs for income.

An Extension Publication of the Department of Natural Resources, New York State College of Agriculture and Life Sciences, A Statutory College of the State University at Cornell University, Ithaca, New York.

BOARD FOOT CONTENT OF LOGS
Length of Log In Feet

Diameter inside bark small end (inches)	Scribner decimal C. rule (in tens) [1]/						Doyle rule[2]/					International rule[3]/				
	6	8	10	12	14	16	8	10	12	14	16	8	10	12	14	16
	Bd. ft.	Bd. ft.	Bd. ft.	Bd. ft.	Bd. ft.	Bd. ft.	Bd. ft.	Bd. ft.	Bd. ft.	Bd. ft.	Bd. ft.	Bd. ft.	Bd. ft.	Bd. ft.	Bd. ft.	Bd. ft.
6	0.5	0.5	1	1	1	2	2	3	3	4	4	10	10	15	15	20
7	.5	1	1	2	2	3	5	6	7	8	9	10	15	20	25	30
8	1	1	2	2	2	3	8	10	12	14	16	15	20	25	35	40
9	1	2	3	3	3	4	13	16	19	22	25	20	30	35	45	50
10	2	3	3	3	4	6	18	23	27	32	36	30	35	45	55	65
11	2	3	4	5	5	7	25	31	37	43	49	35	45	55	70	80
12	3	4	5	6	7	8	32	40	48	56	64	45	55	70	85	95
13	4	5	6	7	8	10	41	51	61	71	81	55	70	85	100	115
14	4	6	7	9	10	11	50	63	75	88	100	65	80	100	115	135
15	5	7	9	11	12	14	61	76	91	106	121	75	95	115	135	160
16	6	8	10	12	14	16	72	90	108	126	144	85	110	130	155	180
17	7	9	12	14	16	18	85	106	127	148	169	95	125	150	180	205
18	8	11	13	16	19	21	98	123	147	172	196	110	140	170	200	230
19	9	12	15	18	21	24	113	141	169	197	225	125	155	190	225	260
20	11	14	17	21	24	28	128	160	192	224	256	135	175	210	250	290
21	12	15	19	23	27	30	145	181	217	253	289	155	195	235	280	320
22	13	17	21	25	29	33	162	203	243	284	324	170	215	260	305	355
23	14	19	23	28	33	38	181	226	271	316	361	185	235	285	335	390
24	15	21	25	30	35	40	200	250	300	350	400	205	255	310	370	425
25	17	23	29	34	40	46	221	276	331	386	441	220	280	340	400	460
26	19	25	31	37	44	50	242	303	363	424	484	240	305	370	435	500
27	21	27	34	41	48	55	265	331	397	463	529	260	330	400	470	540
28	22	29	36	44	51	58	288	360	432	504	576	280	355	430	510	585
29	23	31	38	46	53	61	313	391	469	547	625	305	385	465	545	630
30	25	33	41	49	57	66	338	423	507	592	676	325	410	495	585	675

In the eastern United States the board Foot contents of logs are measured by one of three different log rules or scales.

[1]THE SCRIBNER RULE is based on diagrams which intend to show the actual board feet that can be sawed from logs of varying lengths and diameters.

The original rule was published over a century ago, but subsequently the last digit was dropped and the figures rounded off to the nearest 10 board feet. The rule is in general use in Wisconsin and Minnesota and in national forests. It is now known as "Decimal C" as given above.

[2]THE DOYLE RULE is based on the formula $\frac{(D\text{-}4)^2}{4}$ L= Board feet, in which D is the diameter of the small end of the log, inside the bark in inches, and L is the length in feet. In spite of its inaccuracies when applied to small logs, this rule is used almost exclusively in scaling hardwoods of the Mississippi and Ohio Valleys, and in thc southern pine region. It underscales average-sized logs by about 30 board feet.

[3]THE INTERNATIONAL RULE is the most accurate rule yet devised but unfortunately does not have the widespread use of the others. The figures given will come quite close to actual board feet that can be derived from sound logs by saws which cut 1/4 inches of kerf (saw-dust). It has been adopted by statute in Michigan. The formula, .8D(D-2) will give values closely approximating the board feet in 16-foot logs.

*Taken from Forestry Handbook for Upper Mississippi Region. U.S. Soil Conservation Service

Basic Volume Tables

The basic tables used to calculate volumes are the "composite volume tables." Use of a single table for many species is based on the fact that where the same standards of utilization are employed, differences in volume for most species are of no practical consequence. Errors in determining usable height and DBH outweigh the slight volume variations between species. International ¼-inch, Scribner Decimal C, Doyle and Cordwood Composite Volume tables have been developed through research to meet most of the needs of forestry. The International ¼-Inch Rule is illustrated in Table 1.

Composite Table: Gross Volume[1] in Board Feet (International ¼-Inch Rule) by Number of 16-Foot Logs

Diameter breast high (inches)	Volume when number of 16-foot logs is								
	½	1	1½	2	2½	3	3½	4	5
	Bd. Feet	Bd. Feet	Bd. Feet	Bd. Feet	Bd. Feet	Bd. Feet	Bd. Feet	Bd. Feet	Bd. Feet
8	15	24	35	46					
9	18	32	44	54	63				
10	21	39	54	68	76	81			
11	25	48	68	82	91	98			
12	30	57	80	100	114	124	130		
13	36	68	96	118	134	149	161	171	
14	42	79	110	140	163	184	194	205	
15	50	92	128	160	188	214	232	250	
16	59	105	147	180	213	247	274	295	326
17	66	118	166	208	245	281	314	340	378
18	74	135	188	235	278	320	360	400	440
19	83	152	212	265	314	360	405	450	500
20	92	170	236	295	350	400	450	500	570
21	102	189	262	328	390	450	505	550	635
22	112	209	290	362	430	495	555	610	715
23	122	228	316	396	470	540	610	680	800
24	133	252	346	430	510	595	670	740	870
25	145	275	376	470	555	645	730	810	950
26	158	300	410	510	605	700	790	880	1,020
27	172	325	440	550	650	760	850	950	1,100
28	187	348	480	595	700	810	920	1,020	1,190
29	203	378	515	640	760	870	990	1,100	1,280
30	220	410	550	685	810	930	1,060	1,180	1,360
31	237	440	595	740	870	1,000	1,140	1,260	1,450
32	254	470	635	790	930	1,070	1,210	1,350	1,550
33	270	500	680	840	990	1,140	1,290	1,440	1,650
34	291	530	725	900	1,060	1,210	1,380	1,530	1,760
35	311	565	770	950	1,120	1,290	1,460	1,630	1,880
36	333	600	820	1,010	1,190	1,370	1,550	1,725	2,000
37	353	635	860	1,070	1,260	1,450	1,640	1,830	2,120
38	374	670	910	1,120	1,330	1,530	1,730	1,930	2,240
39	394	705	960	1,180	1,400	1,620	1,830	2,040	2,360
40	415	745	1,010	1,250	1,480	1,700	1,930	2,160	2,480

Table 1

[1] The boxed-in figures in the upper portion of the table show volume to a top diameter of 6.0 or more, but less than 8.0 inches, and hence are applicable only to softwoods.

CORDWOOD VOLUME FOR TREES

(from Technical Note 241 - Lake States Forest Experiment Station)

Diameter Breast High (DBH) Inches	8' (1 stick)*	16' (2 sticks)	24' (3 sticks)	32' (4 sticks)	40' (5 sticks)	48' (6 sticks)
	Tree Volume in Cords					
6	.02	.03	.04	.06		
8	.03	.05	.07	.09	.12	.14
10	.05	.07	.10	.13	.17	.20
12	.07	.10	.14	.18	.22	.27
14	.10	.13	.18	.23	.29	.35
16	.12	.17	.22	.29	.36	.44

Standard Cord is 4' x 4' x 8'. To find approximate volume of peeled wood, subtract 12 per cent.

Steps

1. Obtain diameter and usable height.
2. Consult table for volume of tree.
3. Add the volume for all tallied trees for the gross volume of area cruised.

....

*Note: The terms stick and 8 foot bolt are synonymous.

CONVERSION FACTORS FOR CUBIC MEASUREMENT OF WOOD PRODUCTS

Unit of Measure	Board Feet 1/	Cubic Feet 2/	Cords	Cunits 2/	Cubic Meters 2/
One Board Foot 1/	1	.1886792	.0023883	.0018868	.0053428
One Cubic Foot	5.3	1	.0126582	.01	.0283168
One Cord (4x4x8)	418.7	79	1	.79	2.2370309
One Cunit	530	100	1.2658228	1	2.8316847
One Cubic Meter	187.1677335	35.3146667	.4470211	.3531467	1

1/ For 14" DBH trees. (Scribner DEC C Log Rule, board foot per cubic foot ratio increases as tree size increases and ranges from 4.4 board feet per cubic foot for 10" DBH trees to 6.2 board feet per cubic foot for 24" DBH trees.)

2/ Cubic feet, Cunits, and Cubic Meters are units of measure for solid wood excluding bark and air space. A cunit is a 100 cubic feet of wood.

Metric Conversion Table

TO METRIC MEASURES

Approximate Conversions

	SYMBOL	WHEN YOU KNOW	MULTIPLY BY	TO FIND	SYMBOL
length	in	inches	2.5	centimeters	cm
	ft	feet	30	centimeters	cm
	yd	yards	0.9	meters	m
	mi	miles	1.6	kilometers	km
area	in^2	square inches	6.5	square centimeters	cm^2
	ft^2	square feet	0.09	square meters	m^2
	yd^2	square yards	0.8	square meters	m^2
	mi^2	square miles	2.6	square kilometers	km^2
		acres	0.4	hectares	ha
mass (weight)	oz	ounces	28	grams	g
	lb	pounds	0.45	kilograms	kg
		short tons (2000 lb.)	0.9	tonnes	t
volume	tsp	teaspoons	5	milliliters	ml
	Tbsp	tablespoons	15	milliliters	ml
	fl oz	fluid ounces	30	milliliters	ml
	c	cups	0.24	liters	l
	pt	pints	0.47	liters	l
	qt	quarts	0.95	liters	l
	gal	gallons	3.8	liters	l
	ft^3	cubic feet	0.03	cubic meters	m^3
	yd^3	cubic yards	0.76	cubic meters	m^3
temperature (exact)	°F	Fahrenheit temperature	5/9 (after subtracting 32)	Celsius temperature	°C

FROM METRIC MEASURES

Approximate Conversions

	SYMBOL	WHEN YOU KNOW	MULTIPLY BY	TO FIND	SYMBOL
length	mm	millimeters	0.04	inches	in
	cm	centimeters	0.4	inches	in
	m	meters	3.3	feet	ft
	m	meters	1.1	yards	yd
	km	kilometers	0.6	miles	mi
area	cm^2	square centimeters	0.16	square inches	in^2
	m^2	square meters	1.2	square yards	yd^2
	km^2	square kilometers	0.4	square miles	mi^2
	ha	hectares (10.000m^2)	2.5	acres	
mass (weight)	g	grams	0.035	ounces	oz
	kg	kilograms	2.2	pounds	lb
	t	tonnes (1000 kg)	1.1	short tons	
volume	ml	milliliters	0.03	fluid ounces	fl oz
	l	liters	2.1	pints	pt
	l	liters	1.06	quarts	qt
	l	liters	0.26	gallons	gal
	m^3	cubic meters	35	cubic feet	ft^3
	m^3	cubic meters	1.3	cubic yards	yd^3
temperature (exact)	°C	Celsius temperature	9/5 (then add 32)	Fahrenheit temperature	°F

MAXIMUM MEAN ANNUAL GROWTH BY SITE AND SPECIES

Minn. D.N.R. Forestry Div.

(To convert cords to ft., multiply the cord value by 82)

Site	OAKS		ASPEN		BIRCH		HARD MAPLE		RED PINE		WHITE PINE		JACK PINE	
Index	Bd. ft.	Cords	Bd. ft.	Cords	Bd. ft.	Cords	Bd. ft.	Cords	Bd. ft.	Cords	Bd. ft.	Cords	Bd. ft.	Cords
40	50	0.20	—	0.10	—	.10	100	0.25	140	0.40	100	0.25	160	0.40
50	100	0.40	120	0.30	120	.30	160	0.40	240	0.60	200	0.50	200	0.50
60	200	0.50	200	0.50	200	.50	215	0.50	360	0.90	400	1.00	240	0.60
70	290	0.70	280	0.70	280	.70	260	0.65	520	1.30	590	1.45	300	0.75
80	370	0.90	340	0.85	320	.80	310	0.75	720	1.80	700	1.75	360	0.90

Site	White Spruce		Black Spruce		Balsam Fir		White Cedar		Tamarack	
Index	Bd. ft.	Cords	Bd. ft.	Cords	Bd. ft.	Cords	Bd. ft.	Cords	Bd. ft.	Cords
40	—	0.25	—	0.45	—	0.35	—	0.45	—	0.30
50	—	0.50	—	0.75	—	0.50	—	0.65	—	0.45
60	—	0.90	—	1.00	—	0.70	—	0.75	—	0.65
70	—	1.10	—	1.25	—	0.85	—	0.95	—	0.70
80	—	1.40	—	—	—	1.05	—	—	—	—

STOCKING GUIDES — EVEN AGED

Avg. Dia.	Oak	N.Hwds.	Red Pine	Pine-Spruce-Fir
6"	D+7	D+6	D+4	D+3
8"	D+8	D+7	D+5	D+4
10"	D+9	D+8	D+5	D+4
12"	D+10	D+9	D+5	D+4
14"	D+11	D+9	D+5	D+4

CHART SHOWING CORDS PER TREE

DIAMETER	Ht. to 4" Top	16'	24'	32'	40'	48'
	6"	.03	.04			
	8"	.05	.07	.09	.11	
	10"	.08	.11	.13	.16	.19
	12"	.12	.16	.20	.23	.26
	14"	.17	.23	.27	.31	.35
	16"	.22	.30	.37	.42	.47

SHORTCUT FORMULAS

for

Cruising with 10-factor prism or gauge

PULPWOOD

(CORDS)

1. $$\frac{\textbf{Number of 8-foot sticks in countable trees + number of countable trees}}{\text{2 x number of point samples}} = \text{Cords per acre}$$

2. If trees average 2 or 3 sticks:

$$\frac{\textbf{Number of 8-foot sticks in countable trees x .7}}{\text{Number of point samples}} = \text{Cords per acre}$$

If trees average 4 sticks or more:

$$\frac{\textbf{Number of 8-foot sticks in countable trees x .6}}{\text{Number of point samples}} = \text{Cords per acre}$$

BASAL AREA TABLE

Diameter (Inches)	Area (Sq. Ft.)	Diameter (Inches)	Area (Sq. Ft.)
1	.006	21	2.405
2	.022	22	2.640
3	.049	23	2.885
4	.087	24	3.142
5	.136	25	3.41
6	.196	26	3.69
7	.267	27	3.98
8	.349	28	4.28
9	.442	29	4.59
10	.545	30	4.91
11	.660	31	5.24
12	.785	32	5.59
13	.922	33	5.94
14	1.069	34	6.30
15	1.227	35	6.68
16	1.396	36	7.07
17	1.576	37	7.47
18	1.767	38	7.88
19	1.969	39	8.30
20	2.181	40	8.73

POUNDS PER CORD

Species	Air-Dry	Green
Hickory	4,600	5,700
White Oak	4,300	5,600
Sugar Maple	3,900	5,000
Red Oak	3,900	5,800
Beech	3,900	5,000
Yellow Birch	4,000	5,100
Paper Birch	3,800	—
White Ash	3,800	4,300
Red Maple	3,200	4,700
Gray Birch	3,500	—
Elm	3,100	4,400
Pitch Pine	3,200	4,900
Norway Pine	2,800	3,500
Red Spruce	2,600	3,000
Aspen	2,400	4,200
White Pine	2,700	3,500
Hemlock	2,600	4,400
Balsam	2,200	3,700

UNITS OF MEASUREMENT IN LAND SURVEYING

1 square mile	=	640 acres
1 township	=	23,040 acres (more or less)
1 township	=	36 square miles
1 mile	=	80 chains
1 chain	=	66 feet
1 chain	=	100 links
1 link	=	7.92 inches
1 rod	=	16.5 feet
4 rods	=	1 chain
1 pace	=	2 steps
1 acre	=	10 square chains
1 acre	=	43,560 sqare feet
1 tally	=	5 chains =330 ft.

PULPWOOD MEASUREMENT

(Standing Trees)

CORDS

DBH	8-Foot Sticks 1	2	3	4	5	6	7
6"	.018	.030	.043				
8"	.032	.050	.070	.092			
10"	.049	.074	.101	.132	.167	.200	.239
12"	.070	.100	.138	.180	.225	.271	.324
14"	.095	.134	.179	.233	.291	.351	.419
16"	.120	.168	.222	.290	.361	.437	.521
18"	.146	.208	.270	.350	.439	.531	.634
20"	.186	.246	.320	.416	.522	.632	.755
22"	.220	.294	.374	.494	.612	.739	.833

TONS*

DBH	8-Foot Sticks 1	2	3	4	5	6	7
6"	.045	.075	.108				
8"	.080	.125	.175	.230			
10"	.123	.185	.253	.330	.418	.500	.598
12"	.175	.250	.345	.450	.563	.678	.810
14"	.238	.335	.448	.583	.728	.878	1.048
16"	.300	.420	.555	.725	.903	1.093	1.303
18"	.365	.520	.675	.875	1.098	1.328	1.585
20"	.465	.615	.800	1.040	1.305	1.580	1.888
22"	.550	.735	.935	1.235	1.530	1.848	2.208

*Conversion Factor: 1 Standard Cord = 2.5 Tons.

To find weight for a certain species: (1) determine cords; (2) find species in table on page 19; (3) multiply cords by ratio for that species.

HANDY FORMULAS

Area of Circle = 0.7854 x $(\text{Dia.})^2$

Circumference of Circle = 3.14 x Dia.

Area of Sphere = 3.14 x $(\text{Dia.})^2$

Volume of Sphere = 0.524 x $(\text{Dia.})^3$

Temperatures:

°F to °C: °C = 5/9 (°F-32)

°C to °F: °F = 9/5 °C + 32

SHORTCUT CALCULATIONS FOR GROWTH

1. Rule-of-Thumb for Central Hardwoods:

   Anual growth rate (board feet) = No. of trees per acre above 9" x 5 — good site / 4 — average site / 3 — poor site

2. Quick Growth Computation for an Individual Tree:

   Annual growth = **Future volume after increment of 2" — present volume** / Growth rate (RPI)

3. Shortcut Growth Formulas:

   a. $\frac{4}{\text{DBH x RPI}}$ x 100 = % Growth

   b. $\frac{\text{Vol. Per Acre x 4}}{\text{DBH x RPI}}$ = Growth/Acre/Year (Volume)

   c. $\frac{\text{Basal Area per Acre x 4}}{\text{DBH x RPI}}$ = Growth/Acre/Year (Basal Area)

4. Ingrowth Computation When Using Point Sample Cruising

   $\frac{\text{Basal Area in 10" Trees x 91.5}}{\text{Rings Per Inch}}$ = Board Feet/Acre

CONES PER BUSHEL

SPECIES	Number of Closed Cones Per Bushel	Collection Minimum Cones Per Tree
Red Pine	1,300-1,800	300
White Pine	500-700	200
Jack Pine	2,500-3,200	300
Shortleaf Pine	1,400-2,500	300
Loblolly Pine	400-1,100	150
White Spruce	6,500-8,000	1,500
Black Spruce	7,000-16,000	1,500

NUMBER OF TREES PER ACRE BY VARIOUS METHODS OF SPACING

Spacing (ft.)	Trees number	Spacing (ft.)	Trees number	Spacing (ft.)	Trees number
2x2	10,890	7x9	691	12x15	242
3x3	4,840	7x10	622	12x18	202
4x4	2,722	7x12	519	12x20	182
4x5	2,178	7x15	415	12x25	145
4x6	1,815	8x8	681	13x13	258
4x7	1,556	8x9	605	13x15	223
4x8	1,361	8x10	544	13x20	168
4x9	1,210	8x12	454	13x25	134
4x10	1,089	8x15	363	14x14	222
5x5	1,742	8x25	218	14x15	207
5x6	1,452	9x9	538	14x20	156
5x7	1,245	9x10	484	14x25	124
5x8	1,089	9x12	403	15x15	194
5x9	968	9x15	323	15x20	145
5x10	871	10x10	436	15x25	116
6x6	1,210	10x12	363	16x16	170
6x7	1,037	10x15	290	16x20	136
6x8	908	10x18	242	16x25	109
6x9	807	11x11	360	18x18	134
6x10	726	11x12	330	18x20	121
6x12	605	11x15	264	18x25	97
6x15	484	11x20	198	20x20	109
7x7	889	11x25	158	20x25	87
7x8	778	12x12	302	25x25	70

WEIGHT TABLE FOR VARIOUS WOODS

	Weight Per Standard Cord (Pounds)	Tons Per Standard Cord Ratio	Weight Per 1000 B.F. of Green Lumber (Pounds)
Ash, white	4300	2.2	4000
Aspen	3900	2.0	3600
Basswood	3800	1.9	3500
Beech	4900	2.5	4500
Birch, yellow	5100	2.6	4800
Birch, white	4500	2.3	4200
Cedar, red	3300	1.7	3100
Cedar, N. white	2500	1.3	2300
Cherry, black	4000	2.0	3800
Cottonwood	4400	2.2	4100
Elm	5000	2.5	4600
Gum, red & black	4500	2.3	4200
Hackberry	4500	2.3	4200
Hemlock	4500	2.3	4200
Hickory	5700	2.9	5300
Locust, black	5200	2.6	4800
Maple, hard	5300	2.7	4600
Maple, soft	4300	2.2	3900
Oak, red	5700	2.9	5200
Oak, white	5600	2.8	5200
Pine, red	3800	1.9	3500
Pine, shortleaf	4700	2.4	4300
Pine, white	3200	1.6	3000
Poplar, yellow	3400	1.7	3200
Spruce	3000	1.5	2800
Sycamore	4700	2.4	4300
Tamarack	4200	2.1	3900
Walnut, black	5200	2.6	4800

CENTRAL HARDWOOD SPACING TABLE

Diameter (Inches)	Maxium Stoocking "A"-Level Spacing (feet)	Minimum Stocking "B"-Level Spacing (feet)
2	4.6	6.5
3	6.1	8.2
4	7.6	9.9
5	9.0	11.6
6	10.3	13.4
7	11.6	15.0
8	13.0	17.0
9	14.3	18.7
10	15.6	20.4
11	17.0	22.1
12	18.1	23.8
13	19.4	25.6
14	20.8	27.2
15	21.9	29.0

RULE-OF-THUMB:

For "A"-level stocking spacing, multiply diameter by 1.5

For "B"-level stocking spacing, multiply diameter by 2.0

FIELD REFERENCE HANDBOOK
For Service Foresters
Northeastern Area
State & Private Forest Service
U.S. Department of Agriculture

U. S. FOREST SERVICE

RESEARCH NOTE LS-67

• LAKE STATES FOREST EXPERIMENT STATION • U. S. DEPARTMENT OF AGRICULTURE

Alinement Chart for Numbers of Trees - Diameters - Basal Areas

The chart given here eliminates or simplifies many calculations involving relationships among numbers of trees, diameters, and basal areas.[1] If any two of these variables are known or can be estimated, the other can be found. This chart has many applications in forestry, both in the field and in the office. Some examples are given below.

Estimating Average Stand Diameters. — If the number of trees and basal area per unit of area are known (the unit of area can be an acre, a 10th-acre plot, or any other sized area), the diameter of the tree of average (mean) basal area can be simply estimated.

Place one end of a straightedge on the NUMBER OF TREES scale, for example 220 trees per acre, and the other end on the BASAL AREA scale, say on 90 square feet. The average diameter is read off the DIAMETER scale at the point where the straightedge crosses — 8.7 inches in this example.

Estimating Basal Areas. — If the number of trees of a given average diameter is known, their basal area is readily found. Aline the straightedge on the NUMBER OF TREES and DIAMETER scales and read off the BASAL AREA scale. The basal area for 400 trees averaging 6 inches in diameter is 78.5 square feet. Note that this same answer tells us that the basal area in 40 trees is 7.85 square feet; in 4 trees, 0.785 square feet. For the basal area of one 3.9-inch tree, find the basal area for 1,000 trees, 83 square feet, and divide by 1,000. Thus, one 3.9-inch tree has a basal area of 0.083 square feet. Any convenient number of trees can be used in this manner to find the basal area of a single tree or a group of trees of any diameter.

Estimating Diameter and Basal Area Growth. — If future values of any two variables can be estimated, then the third can be found. One use of this is to estimate stand diameter growth (the increase in diameter of the tree of average basal area). For example, after thinning a red pine stand to 90 square feet of basal area we have 220 trees averaging 8.7 inches in diameter. Over the next 10 years these 220 trees are expected to grow in basal area to 135 square feet, with no mortality. Thus, in 10 years the average diameter would be 10.6 inches (220 trees, 135 square feet), a growth of 1.9 inches.

Conversely, past ring counts may have indicated a future 10-year average diameter growth of 1.9 inches in this stand, from 8.7 to 10.6 inches. Knowing the present number of trees (220) would enable us to estimate the future basal area as 135 square feet if there is no mortality, a growth of 45 square feet (135-90).

This technique only approximates actual stand diameter and basal area growth. It applies best to stands with a narrow and even distribution of diameter classes, such as a well-spaced plantation.

Estimating Numbers of Trees To Be Cut In Thinning. — Another use of this chart is to estimate how many trees must be cut to thin to a given basal area (assuming a cut from above and below that does not change the diameter of the tree of average basal area by cutting). To illustrate, a stand averaging 10.6 inches in diameter with 220 trees and 135 square feet of basal area per acre is to be thinned back to 90 square feet. A stand with 90 square feet in 10.6-inch trees would have 147 trees. Thus, about 73 trees per acre would have to be cut to thin the stand back to 90 square feet.

Other applications of this chart may come to mind, but these examples illustrate its use in several problems commonly encountered in timber management.

---

[1] *The formula used in constructing this chart was:*

$$BA = \frac{\pi}{144} \left(\frac{D}{2}\right)^2 N$$

$$BA = 0.005454\ D^2N$$

where BA *is total basal area in square feet,* D *is tree diameter breast high in inches, and* N *is number of trees on an area of known size.*

ALLEN L. LUNDGREN
Economist

OCTOBER 1965

MAINTAINED AT ST. PAUL 1, MINNESOTA, IN COOPERATION WITH THE UNIVERSITY OF MINNESOTA

NUMBER OF TREES - DIAMETER - BASAL AREA CHART

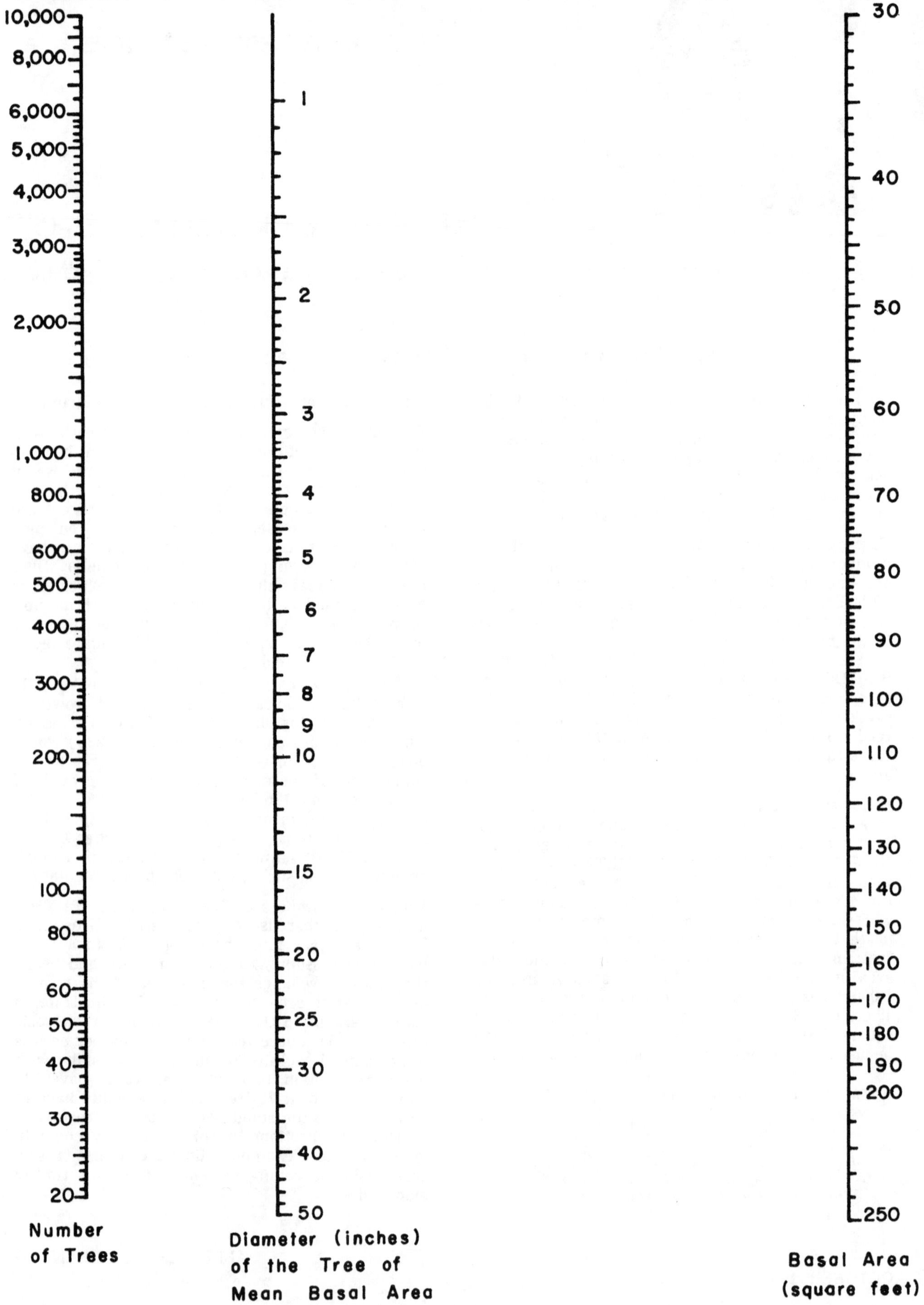

CHAPTER 12
PEST MANAGEMENT FOR WOODLAND OWNERS

Michael R. Carroll
Forest Pest Specialist
Minnesota Department of Natural Resources
Brainerd, Minnesota

and

Marguerita A. Palmer
Research Plant Pathologist
North Central Forest Experiment Station
St. Paul, Minnesota

INTRODUCTION

Any effort to grow trees for wood products, Christmas trees or green space value will be influenced by insects, diseases and adverse enviromental conditions. A woodland owner should become familiar with the pests that commonly affect Minnesota trees (see Table 1) and develop management strategies to avoid pest buildup and widespread economic losses.

Why worry about pests in Minnesota forests?

There are many insect & disease pests that cause serious losses in forest tree plantations. Pest agents can cause needle loss, shoot & branch mortality, deformed stems and whole tree death. You can get an idea of what pest agents are costing you by comparing the cost of establishing the crops trees (i.e. site preparation, planting, and maintenance) with the ideal monetary return from harvest. Any reduction in growth, product quality or overall yield caused by a pest agent will reduce the dollar return. A management program emphasizing early detection & control of pest problems will help you avoid these losses from pest outbreaks.

Here are some guidelines you can use to develop your own pest management program:

I. PLAN AHEAD

The easiest and most cost effective way to control pest outbreaks is to **prevent** pest outbreaks. By careful planning, you can avoid creating situations that invite past outbreaks. When planning your tree planting consider the following:

1. Select species or varieties resistant to pests abundant in your area.
2. Select tree species compatible with the site you have selected.
3. Prepare the site properly to avoid stressing trees through competition with weeds and brush.
4. Avoid introducing pests into your plantation by inspecting seedlings thoroughly before planting for signs and symptoms of insects & diseases.
5. Use good planting techniques.
6. Recognize that future control operations or stand improvements may require specialized equipment and proper application techniques. Budget accordingly and maintain access lanes to allow equipment operation with minimal damage to trees.

II CONDUCT FREQUENT SURVEYS

It is important to check your insect and disease problems throughout your woodland ownership. Establish survey locations to check all the different species and age classes of trees you are growing. Be familiar with the life cycles of important pests in your area to time your periodic inspections to detect the damaging stage of the pest before widespread damage has occurred. Keep records of your surveys and make annual comparisons of population or damage levels.

III. LEARN TO RECOGNIZE MAJOR PESTS IN YOUR AREA

Before starting any control measures, be sure to postitively identify the problem-causing agent. Controls for different pests vary considerably, and you can waste valuable time and money by using incorrect control methods. Many times, with the help of reference materials, (e.g. Pest Alerts, How-to's) you can correctly identify pests yourself. However, if you are not sure of your diagnosis, have a pest specialist for identification. When collecting plant material for mailing, try to keep the material as fresh as possible. Include both affected and non-affected material. A handout is available on the display table giving exact instructions for collecting and shipping samples.

IV. DECIDE IF CONTROL IS NECESSARY

In order to properly evaluate pest outbreaks it is important to ask yourself the following questions:

1. How widespread is the pest on my and adjacent ownerships?
2. Has the pest caused serious problems in my area in the past?
3. What is the potential for continued buildup or collapse?
4. What will the damage do to my product goals?
5. Will the value of trees saved be greater than control costs?

Depending on your answers to these questions, you may or may not decide to start control measures. Use the next section to help you select the best controls for your specific pest problem.

V. CHOOSE CONTROL METHODS

A good pest management program uses several types of controls to reduce the impact of insects and diseases. Many different types of controls are available:

1. **Genetic controls** use insect & disease resistant species or varieties to reduce pest impact.
2. **Legal controls** support quarantines used to prevent introduction of a pest into a area where it has not been found.
3. **Biological controls** use organisms, such as viruses and bacteria to reduce pest populations.
4. **Cultural controls** are usually directed at preventing pest outbreaks. They can be easily integrated into the normal plantation management routine. Examples of cultural controls are: eliminating alternate hosts of rust fungi, site preparation to reduce weed competition, and slash treatments to eliminate insect buildup.
5. **CHEMICAL CONTROLS** should only be used as one part of an integrated pest management program. They can be very effective in reducing pest population for a short period of time if used properly.

VI. FOLLOW UP ON CONTROL EFFORTS

It is essential to keep good records to evaluate the success of your control efforts. Was the pest controlled? Was it cost effective? How can the control operation be improved? Anwering these questions will help plan future pest management strategies.

SUMMARY

Insect and disease control is an important, ongoing part of woodland management. By using the guidelines provided above, you can develop a pest management program to reduce pest impact and meet the objectives of your particular woodland.

TABLE 1 REOCCURING PEST PROBLEMS OF COMMON TREES IN MINNESOTA

SPECIES	PEST	DAMAGE
Aspen	Forest Tent Caterpillar	Defoliation and reduced growth
	Popple Borer & Gall makers	Structural damage and open wounds
	Hypoxylon canker	Sapling Mortality and stocking reduction
	White Rot	Fiber deterioration in older trees
Oaks	Two lined chestnut borer	Top kill and tree decline
	Armillaria root rot	Mortality - replanting consideration
	Leaf spot	Mortality - legal shipping restrictions
Walnuts		Defoliation and reduced growth
		Deformed to dead leaders
		Defoliation and reduced growth
Birch	Bronze birch borer	Top kill of stressed trees
Spruces	Yellowheaded Sawfly	Defoliation and mortality
	Rhizophaera needle cast	Needle loss and lower branch mortality
White Pine	Tip weevil	Dead leaders and branchy form
	Blister Rust	Branch and stem mortality
	Animal Damage	Seedling mortality from browzing or root feeding, sap sucker
	Air pollution	Needle loss, stunting, mortality

SPECIES	PEST	DAMAGE
Red Pine	Shoot blights (Diplodia or Siroccocus)	Dead current years growth, twig and branch mortality
	Sclerroderris canker	Branch flagging and mortality
	Needle Rust	Needle loss
	Bark Beetles	Tree mortality
	Saratoga spittlebug	Branch mortality and stem deformation
Jack Pine	Tussock moth and budworn	Defoliation and top kill
	Stem rust	Branch or whole tree mortality
Scots Pine	Winter burn	Needle discoloration and loss
	Gall rust	Branch or whole tree mortality
	Needlecasts	Needle discoloration and loss
	Weevil complex	Tip mortality and root collar girdling
Austrian Pine	Diplodia Tip Blight	Shoot and tip mortality
	Dothistroma Needle Blight	Needle discoloration and loss
	Winter Burn	Needle discoloration and loss
Ponderosa Pine	Winter Burn	Needle discoloration and loss
	Dothistroma Needle Blight	Needle discoloration and loss

Defoliators — The entire leaf or needle may be consumed and the tree completely defoliated, or only a part of the leaf destroyed depending upon the insect species concerned and their abundance. The seriousness of defoliation depends upon the tree species attacked, the completeness of defoliation, and the frequency at which it occurs. A single complete stripping of pines, spruces, firs, and cedars (conifers) usually results in their death. Partial defoliation is reflected in growth reduction. Broadleaf trees, such as maple, oak and aspen, are more capable of withstanding foliage loss and mortality does not occur until stripping has been repeated annually for several seasons. Partial defoliation of both conifers and hardwoods also results in the increased susceptibility of the trees to bark and root inhabiting insects. This is the most serious effect of non-fatal defoliation. The forest tent caterpillar, jack pine and spruce budworms, and the sawfly larvae are typical defoliators.

Leaf Miners - The leaf miners feed between the upper and lower surfaces of leaves or inside the needles of conifers. When the attack on the latter is severe growth reduction and mortality may result; on hardwoods, resistance to other insects and disease is reduced. An example of this type of injury is produced by the activities of the birch leaf miner.

Sucking Insects — Sucking insects reduce tree vigor by the withdrawal of sap to the extent that the tree or a portion of it dies. These insects may introduce toxic substances which cause growth malformations, or their feeding punctures may serve as points of entry for disease organisms. Many insects of this group also injure trees by means of punctures made during egg laying activities. Among the sucking insects causing economic loss in forest stands are the plant lice (aphids), scales and spittlebugs.

Twig and Stem Feeders - Although these insects seldom kill trees, they cause stunting and distortion of both coniferous and hardwood trees by boring into the buds, twigs and small branches. In addition, height growth is frequently reduced to such extent that tree harvest is prolonged several years beyond the planned rotation. The white pine weevil, the European pine shoot moth, and the oak twig pruner are examples of insects in this group.

Bark Beetles — The bark beetles must be considered among the most dangerous of the forest insects. By tunneling into the bark and cambium they are capable of killing healthy trees as well as trees weakened by some other cause. In Wisconsin forests, their greatest threat is to trees which would otherwise recover from the effects of defoliation, sap withdrawal or drought. In addition to consuming wood, the bark beetles introduce wood staining and wood rotting fungi which further reduce the value of the wood for paper and lumber. The most common Wisconsin species include the pine engraver and the red turpentine beetle.

Wood Borers — Like the bark beetles, these insects are more important as enemies of weakened trees, or logs stored in the forest. The borers can be further subdivided as follows: (a) the flatheaded or metalic wood borers, (b) the roundheaded or longhorned wood borers, and (c) the ambrosia beetles. The young grubs of the borers feed in the cambium region for a time and later penetrate the solid wood. Wood thus infested is rendered unfit for lumber. Substantial volume losses also occur in wood used for pulp and paper. Ambrosia beetles bore in the sapwood of weakened and recently cut trees, and in unseasoned wood stacked in lumber yards. The wood surrounding the tunnels is stained by the ambrosia fungi, which the beetles cultivate in the burrows. Infected wood is weak, and often useless as lumber and tight cooperage. The bronze birch borer typifies the flatheaded wood borers, while the sugar maple borer and locust borer are examples of the longhorned beetles.

Gall Formers — Seldom do the gall insects cause the death of trees, but indirectly reduce the growth or stunt otherwise healthy trees. Galls may be formed on leaves, small as well as large twigs, and even on the main stem and roots. This ability to irritate certain parts of trees and cause gall formation is not limited to any specific group of insects or mites, and only rarely are control measures necessary.

Cone, Seed and Nut Insects — The flowers, fruits, nuts, and seeds of practically every forest tree species may be attacked and destroyed by insects. Since reforestation and the future supplies of timber depends on an ample supply of fertile seeds, this group of insects is of prime importance; however, comparatively little is known about them. The acorn weevil, cone moths and beetles, and the seed chalcids are responsible for the loss of many cones or seeds.

Root and Root Collar Feeders — Trees, particularly nursery stock, may be killed, stunted, or lowered in vigor by such insects as white grubs, wireworms, root collar weevils, and cutworms. Many mealy bugs and aphids also feed on root systems.

Carriers of Disease — Insects play a very important role as carriers of disease pathogens. They may accidently introduce them into wounds caused by some other agent, or carry the organism into their own burrows and feeding areas. In a few cases, the insect may be dependent on the disease-causing fungus for food while the fungus is dependent on the insect for its dissemination. Long distance transmission of oak wilt occurs when an insect feeds on a fungal mat and is then attacted to a fresh wound in a healthy tree. Dutch elm disease also is spread by insects, particularly the smaller European elm bark beetle.

MAJOR INSECT PROBLEMS IN MINNESOTA

Minnesota DNR

Borers or Tunneling Insects
1. Bronze Birch Borer
2. Poplar Borer
3. Bark Beetles (Elm & Pine)
4. Pine Root Collar Weevil
5. Zimmerman Pine Moth
6. European Pine Shoot Moth
7. White Pine Weevil

Defoliating Insects
8. Cankerworm
9. Forest Tent Caterpillar
10. Large Aspen Tortrix
11. Walking Stick
12. White Grub
13. Orange-humped Mapleworm
14. Red-humped Oakworm
15. Variable Oakleaf Caterpillar

Insects Attacking Needleleaved Trees (conifers)
16. Jack Pine Budworm
17. Pine Tossock Moth
18. Introduced Pine Sawfly
19. Jack Pine Sawfly
20. Larch Sawfly
21. Red-headed Pine Sawfly
22. Yellow-headed Spruce Sawfly
23. Spruce Budworm

Sap-sucking Insects
24. White Pine Aphid
25. Pine Bark Aphid
26. Pine Needle Scale
27. Pine Tortoise Scale
28. Pine Spittlebug
29. Saratoga Spittlebug
30. Spruce Spider Mite
31. European Red Mite
32. Two-spotted Spider Mite

MAJOR TREE DISEASE PROBLEMS IN MINNESOTA

33. Hypoxylon Canker
34. White Trunk Rot
35. Northern Hardwood Decay
36. Sirococcus Shoot Blight
37. Diplodia Tip Blight
38. Sclerodorris Canker
39. White Pine Blister Rust
40. Armillaria Root Rot
41. Sweetfern Rust
42. Oak Wilt
43. Dutch Elm Disease
44. Lophdermium Needle Cast
45. Brown Spot
46. Pine Needle Rust

BORERS OR TUNNELING INSECTS

1. The Bronze Birch Borer

The bronze birch borer attacks birches, aspen, poplar, and beech. The damage is caused by the feeding of the larvae just beneath the bark resulting in girdling and death of a limb. The damage is recognized by the death of branches usually at the top of the tree.

Bronze Birch Borer eggs are laid on the bark in June and the larvae chew into the inner bark and overwinter under it. The following April or early May the larva, about ¾" long, changes to pupal stage. The adults emerge shortly after, through characteristic D-shaped exit holes. The adult is a slender, olive-bronze beetle about 3/8" long and does little damage.

2. The Poplar Borer

The poplar borer attacks poplar, aspen, cottonwood, and willow. The attack is evidenced by irregular scars and holes on the bark of large branches and the trunk from which sap and wood borings are extruded. The borer weakens or kills trees and provides an entry for many wood-destroying rots.

Poplar Borer adults emerge in July and August and cut small holes in the bark in which to lay their eggs. The larvae hatch and bore into the inner bark and sapwood where they spend two winters and a summer, mining and making galleries. Late in the second summer, the larva, which is 1¼" long and cream-colored, changes to pupa. After about 30 days, the adult emerges--a long-horned beetle (antennae are almost as long as the body), 1¼" long, gray in color with irregular, yellow, elongated spots sprinkled with small black dots.

3. The Bark Beetles

The Elm Bark Beetles. All elm species are attacked by bark beetles. Two kinds of elm bark beetles are of economic importance--the native elm bark beetle, and the smaller European elm bark beetle. These beetles are small, dark brown, make shot holes in the bark, and tunnel beneath the bark into the sapwood. The two kinds of beetles can be distinguished by the type of egg galleries and fan-shaped larval feeding tunnels. The egg gallery of the native elm bark beetle is formed across the grain of the wood (typical horizontal gallery), whereas the smaller European elm bark beetle forms its egg gallery along the grain of the wood (typical vertical gallery). Adults feed in twig crotches, cause twig casting, and transmit spores of Dutch elm disease fungus.

Bark Beetles typically pass the winter as larvae in the bark. The native elm bark beetle may also overwinter as an adult. Populations that overwinter as adults one year overwinter as larvae the next, and vice versa. The adults, after feeding in the crotches of living elm trees bore through dead, dying, recently cut, or otherwise devitalized elm material, and form brood galleries. The female places her eggs along the sides of the gallery. After hatching, the larvae form frass-packed tunnels that fan out characteristically from the gallery and penetrate the inner bark. Pupation occurs in cells formed at the end of larval tunnels. Emergence of second generation adults begins in August and continues until late fall.

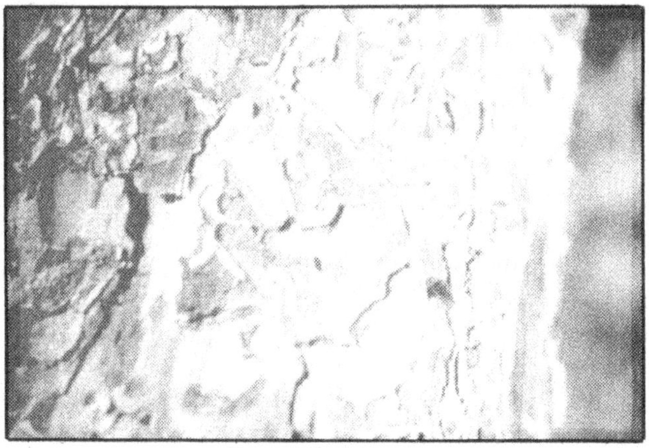

Typical Bark Beetle
Galleries (tunnel) found in cambium layer beneath bark.

Pine Bark Beetles

Bark beetles are commonly found in the pine forests of Northern Minnesota. The beetle is rarely seen but the small round BB sized exit holes they make are easy to see in the bark. Sometimes saw-dust can be seen in or near these holes.

The important thing to know about bark beetles is that they have the ability to kill pine trees - especially large ones. Norway pine and jack pine seem to be the ones most often affected.

A healthy pine tree can normally tolerate these beetles that burrow around in the cambium layer just under the tree bark. When the pine is in a weakened condition and there are above normal populations of beetles it is not uncommon to see the beetles getting the upper hand resulting in the death of the trees.

Why would a pine tree be in a weakened condition? The most likely cause is drought. Old age can also weaken large pine as well as a host of other problems such as insects, disease, wind damage, root rot and logging damages.

Here is an example:

Diplodia Tip Blight is a fungus disease of Norway pine needles that is quite common in Minnesota. It can be identified by new spring growth that suddenly curls and dies. These dead areas can be seen at the "tips" of the branches. Tip Blight can be fatal to small Norway pine - larger ones will rarely succumb. However, the combination of a pine weakened by Tip Blight with a follow up attack by bark beetles may be enough to kill it.

Dead and dry pine are not attractive to bark beetles except for recently cut or killed pine. Beetle populations build up the fastest in dying weakened pine and in pine cut between March through June.

There are two generations of bark beetles per year. The first generation starts in early spring and is considered the "building generation". They usually attack downed trees from April to mid-June. The second generation is considered the "killing generation" and will attack standing trees from mid-June to September. The adult beetles over winter in duff near infected trees.

Control of bark beetles is based on their preference for weakened, dying and recently killed trees. The most common recommendation is to cut and burn all stems 2" and larger. This should be done as soon as it is certain that a tree is on the way out. When dying trees are left, large populations of bark beetles can build up. These beetles will soon move to nearby trees - (even healthy ones) and - may kill normally healthy trees that cannot withstand the sudden large influx of beetles.

Any pine killed in the spring or early summer should immediately be cut down and destroyed - it should not be saved for firewood since it is prime beetle territory. The one exception to this would be if the bark were stripped from the trunk and all larger branches. Stripping the bark will promote rapid drying of the cambium layer. Once the cambium layer is dry beetles loose their interest.

4. The Pine Root Collar Weevil

The pine root collar weevil attacks jack, Scotch, red, Austrian, and occasionally eastern white pines. The damage is caused mostly by the larva feeding in the root collar region that results in the separation of the bark and wood, reduction in stem diameter with structural weakening, and callus formation just above the root collar. Trees between 1 and 4" in diameter are most susceptible. The weevil larval attack can be recognized by black masses of pitch, combined with soil containing galleries.

The pine root collar weevil overwinters in the adult, larval, and sometimes pupal stages. Eggs are laid in mid-June in bark wounds at the root collar or in adjacent soil up to 3" from the stem. The larvae overwinter until the following June, then feed on the inner bark and surface of the wood. Pupal chambers are constructed in resin-soaked soil or in the bark. Adults emerge in July or August, feed on the bottommost branches during the day and on the bark of the branches in the crown at night.

5. The Zimmerman Pine Moth

The Zimmerman pine moth is a native of North America. This insect infests the tips of pine branches and the main stem, thus retarding and deforming the tree. Young trees in plantations are most vulnerable. In Minnesota, the Zimmerman pine moth attacks jack pine, red pine, Scotch pine, white pine, and Austrian pine. Jack pine is particularly susceptible. When this insect attacks the main trunk of pines, it generally is found in the pitch near wounds. The larvae tunnel into the wounded area, causing the trunk to enlarge and a copious amount of pitch to flow and harden on the trunk. The larvae that attacks shoots and twigs tunnel into them causing the tips to turn brown and break off.

Zimmerman Pine Moth. Adults emerge between the end of June and mid-September. They are gray with a wing span of 1 to 1½ inches; forewings are gray with a reddish tinge, and with zig-zag markings across wings. The hindwings are yellowish-white. Eggs are laid on the bark or on branch tips in late summer; they average 20 and hatch in 7 to 10 days. Some eggs overwinter. A mature larva, ¾- to 1-inch long, is a dirty white to reddish-yellow or green caterpillar, with chestnut brown head. On the body are a series of black dots, from each of which arises a single bristle. The larva matures from June to August, changes to pupa near the surface of the tunnel, then to an adult.

6. European Pine Shoot Moth

These moths mainly attack Red pine but will occasionally be found on Scotch pine. They feed in May and August. Their identifying features are pitch at the base of buds and production of "post-horns" which are shoots which are crooked.

7. White Pine Weevil

Prefers white pine but will attack other pines and spruces. Feed from early May to mid-July, and August to mid-September. Identifying features are injury caused forking of the top or multiple leaders; early in the growing season the terminal will bend and display a "shepherd's crook".

DEFOLIATING INSECTS

A INSECTS ATTACKING BROAD-LEAVED TREES

8. Cankerworm

Cankerworms are known also as loopers, inchworms, and measuring worms. There are two species, the fall cankerworm and the spring cankerworm, so called because of their season of adult moth emergence. Both species prefer foliage of apple, basswood, boxelder, and elm. They also attack ash, beech, cherry, hickory, maple, oak, fruit and ornamental trees and shrubs. Cankerworms destroy all but the midribs and larger veins of leaves. The crowns of trees attacked often appear ragged.

Cankerworm eggs are laid on the bark of twigs and branches; eggs in a cluster range from 50-100. Fall cankerworms overwinter in the egg stage, unlike spring cankerworms which overwinter as pupa. The larvae of both species feed about the same time. The fall cankerworms are light to dark green with dark stripes down their backs. The spring cankerworms are light green to brown or black with pale, whitish lines down their backs. The full grown larvae, ¾-1" long, feed for 3 to 4 weeks. They then enter the soil and go through a resting or pupal stage, are about 1/3" long and brown. Adult moths of fall cankerworms emerge after the first frost in the fall; moths of the spring species emerge early in the spring, shortly after the first thaw. Female moths of both species are wingless, so they crawl up tree trunks to lay eggs.

9. The Forest Tent Caterpillar

The larvae of the forest tent caterpillar unlike the eastern tent caterpillar do not construct a tent, but feed openly, often in large groups or clusters. Because of the nuisance created by the migrating larvae, the insect is often called the armyworm of the forest, and can have a serious impact on camping in state and national parks. Trembling aspen is the principal host tree, but larvae also feed on alder, ash, basswood, beech, birch, maple, oak, willow, and most any broadleaved forest trees. **Forest Tent Caterpillar**. The moths appear in late June or early July, and eggs are laid in bands that encircle small twigs of host trees. Fully-formed larvae develop within the eggs in about 3 weeks and overwinter in this form. The hatching of eggs usually coincides with the swelling and breaking of buds in early spring. With favorable weather and development of foliage, the larvae feed voraciously and mature by mid-June. The larvae attain a length of 1½ to 2 inches, are bluish with a row of key-hole-shaped white marks down the middle of the back. Cocoons are spun among the leaves of trees, shrubs, or other vegetation, and the moths emerge about 8 to 12 days later.

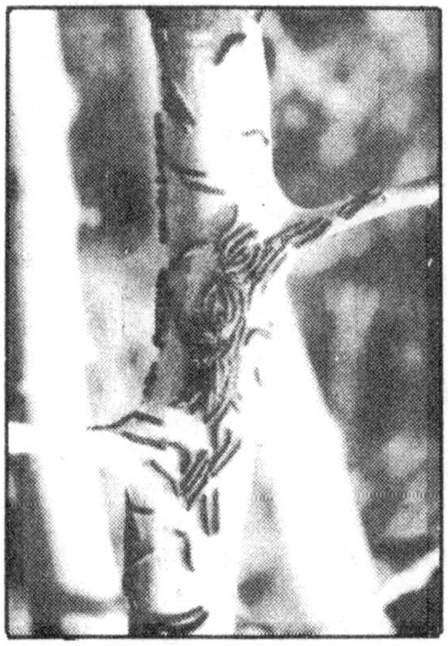

Forest Tent Caterpillars on Aspen

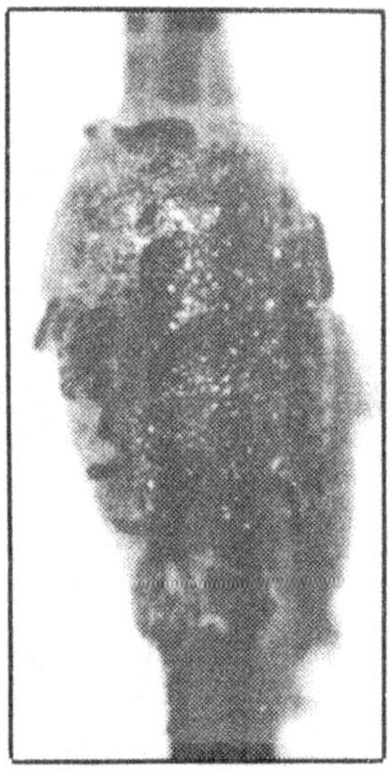

Forest Tent Caterpillar Egg Mass

10. The Large Aspen Tortrix

The tortrix becomes a problem only where quaking aspen is a major component of the forest stand. During epidemic conditions it will also feed on alders, balsam poplar, black cottonwood, chokecherry, white birch, and willow. The larvae also construct heavy webbing and damage understory plants.

Eggs of the large Aspen Tortrix are laid in masses; the individual egg is scale-like, oval, and pale green, 60 to 400 eggs per egg mass, with egg rows overlapping like fish scales. The young larva which is ½- to 1-inch long, is dark green to almost black. The gregarious first instar larvae web 2 leaves together in July and August and skeletonize the touching surfaces. They overwinter as second larval instar under loose bark or bark crevices. Activity is resumed in the spring and, as solitary feeders, they bore into the buds and feed on the tissues of expanding leaves. Pupation takes place in the rolled leaves. The moths emerge in late June.

11. The Walkingstick

Basswood, cherry, and oaks of the red oak group are the preferred hosts. Other trees, such as apple, quaking aspen, paper birch, hickory, and locust, are fed upon if they are in stands with the preferred hosts. Defoliation occurs every other year unless both the even-year and odd-year broods are present in a particular locality.

Two years are required to complete development. Development occurs through incomplete metamorphosis, there being no change in body form. The nymphs hatch from eggs in May or early June and feed on preferred hosts. Adults, 2½- to 3½-inches long, are formed by July or early August. Both sexes are slender and wingless, with long thin legs and antennae. Mating and egg laying continues until October. The eggs are dropped to the ground from wherever the female happens to be on the tree. The eggs overwinter in the leaf litter, and most remain unhatched throughout the following summer and winter. Nymphs emerge the next spring. In Minnesota walkingsticks are found during odd-numbered years.

12. White Grub

White grubs are the larvae of several species of beetles. As larvae, they seriously damage root systems of trees, particularly after transplant or in the nursery. As adults, they feed on foliage of broadleaved trees, especially the oaks.

Adult white grub beetles vary in color from light brown to black, ½- to 1¼-inches long. The typical life cycle takes 2 to 4 years. The eggs are laid in the soil. They hatch into the characteristic C-shaped tiny grubs in 3 to 4 weeks, and feed on decaying or living vegetable matter in the soil. Later, they molt and eventually burrow deeper into the ground to overwinter. In the spring, they return close to the surface and feed voraciously on the roots of plants. About mid-October they burrow back into the soil to spend the second winter. In the following spring they move to the surface once more and feed for a month or two on the roots. About the middle of June they move downward in the soil and change to the pupal stage. After a month as pupae, they change to adults but remain in the pupal chamber throughout the fall and winter, and emerge as adults the following May or June. For this reason, the adults are often called "May" or "June" beetles.

THE FALL DEFOLIATOR COMPLEX:
13. Orange-humped Mapleworm
14. Red-humped Oakworm
15. Variable Oakleaf Caterpillar

Oak, chiefly bur oak and white oak; basswood; beech; birch; elm; sugar maple; and other hardwoods. The early larval stages are gregarious and skeletonize the leaves. As the caterpillars develop, they scatter and consume all but the midribs of leaves. Generally, infestations are localized and may persist for several years, affecting tree vigor and growth.

Normally there is one generation per year. Eggs are laid in small clusters or singly on the underside of the leaves. The larvae feed from June through October and reach a length of 1¾ inches. The full-grown caterpillars are yellowish; the distinguishing characteristic is one reddish-brown band bordered by a yellow stripe along sides for the variable oakleaf caterpillar, three black dorsal lines for the mapleworm, and five black dorsal lines for the oakworm. About mid-September to early October the larvae cease feeding, move to the ground, and spin their cocoons. The larva pupates and overwinters in the cocoon. Moths begin to emerge near the end of May or early June; they are grayish in color with a wing span of 2 inches.

INSECTS ATTACKING NEEDLELEAVED TREES (Conifers)

16. The Jack Pine Budworm

This defoliator occurs throughout the range of jack pine in North America. Jack pine is the principal host tree, but red pine, Scotch pine and occasionally eastern white pine are severely defoliated when they occur with jack pine in natural stands or in plantations. The larvae feed chiefly on new growth, some mine staminate (male) cone clusters or new needle fascicles. Feeding sites are formed within a loose webbing which ties clipped needles and adjacent shoots together. The accumulation of this material imparts a distinctive reddish appearance to attacked trees.

Jackpine Budworm - Adult Moth

Jack Pine Budworm. The moths appear from July to early August and deposit 10 to 60 eggs in clusters on the underside of pine needles. Hatching occurs in 7 to 10 days, from late July to mid-August. First instar larvae crawl to hibernation sites under bark scales or between the needles, spin overwintering protective coverings called hibernacula, and molt to the second instar without feeding. Emergence in spring coincides closely with the shedding of pollen from staminate jack pine flowers, usually in the latter part of May. Larvae feed for about 6 weeks. A mature larva is about 1-inch long, with a shiny light brown to black head and a dark brown "collar" separated from the head by a narrow white band. The body is reddish brown with yellowish sides. There are 2 rows of white dots on the back, arranged in groups of 4. Pupation begins from late June to late July, usually among needles or between webbed shoots. One generation of budworms occurs per season.

17. The Pine Tussock Moth

The pine tussock moth has a history of infrequent outbreaks in east central Minnesota and northwestern Wisconsin. In Minnesota, jack pine, red pine, eastern white pine, white spruce, and occasionally tamarack are the host trees. Jack pine is the preferred host. In early spring, the larvae prefer to feed on the developing staminate cones, but later their activities are confined to the needles, which they consume down to the needle sheath, leaving a small stub.

Pine Tussock Moth - Larva

The pine tussock moth has one generation per year. Moths are gray-brown with irregular cross stripes on the forewing, and have a wing spread of 1½ inches. Females lay eggs usually on the needles, twigs, branches, or trunk of the tree, in late July to early August. An average egg mass contains 25 eggs in 2 or 3 rows. The tiny larvae hatch in 8 to 12 days, feed until late August, and hibernate under the bark for the winter. Larval activity is resumed in June. The mature larva is a gray-brown hairy caterpillar, 1- to 1½-inches long, with a "toothbrush" of four tufts of grayish or brownish hairs on the back, a pair of pencils of black hair at the front end, and 3 pencils of black hair on the rear end. The characteristic caterpillar, upon completion of feeding, forms a gray-brown cocoon composed of silk and hairs from its body. This cocoon is usually attached to the underside of a twig. Pupation occurs about the first week in July, with the moths emerging 10 days later.

SAWFLIES:

Among the insects known as sawflies are several species which are considered as the most serious threat to the successful growing of conifers (needleleaved trees) in Minnesota. These defoliators have destroyed vast areas of vigorously growing trees and have wiped out both mature stands and newly established plantations. The name sawfly is derived from the saw-like structure used for egg laying, which is located at the tip of the abdomen of the female. The hairless caterpillars have a body with 3 pairs of legs toward the head end and 7 pairs of short, fleshy legs on the abdomen. A few of the sawflies of economic importance in Minnesota are identified below:

18. The Introduced Pine Sawfly

In North America the introduced pine sawfly was first discovered in 1914 in New Haven, Connecticut. Its present distribution range includes Minnesota. White pine is the preferred host; other common host trees are jack pine, red pine, and Scotch pine. Older larvae are known to feed on Austrian pine. Old needles are eaten before the current season's growth is attacked, but the new needles are also consumed after they have hardened.

Introduced Pine Sawfly

Normally the introduced pine sawfly has one-and one-half to two generations per year. Eggs are deposited from early to mid-May, in slits cut along the edges of old needles, about 10 eggs per needle. Larvae emerge in 10 to 14 days, at first feed gregariously but later become solitary feeders. The mature larva is about 1-inch long, has a dull gray body with black legs, and a shiny black head; a dark brown and black double stripe runs along the back, with wide, irregular, yellow stripe broken into oval, yellow patches along the sides. Cocoon spinning begins in early July, frequently on the host among the needles, or at the bases of small branches. Pupation occurs shortly after. Adults emerge in early August, with egg laying commencing shortly thereafter. Some sawflies do not emerge but remain in the cocoon as prepupae in a resting stage, called diapause, for 1 to 3 seasons. The second generation larvae emerge in 7 to 8 days. In September, the mature larvae drop to the ground and spin their cocoons among the dead leaves and litter on the topsoil. Most of them overwinter as prepupae in the cocoons and pupate the following season. Adults emerge from April until June.

19. The Jack Pine Sawfly

The jack pine sawfly is a serious defoliator of jack pine in Minnesota. Injurious infestations periodically develop in natural stands and plantations. Occasional damage is done to red pine, and Scotch pine found in infested jack pine stands. The larvae characteristically devour only the old needles.

This sawfly has one generation per season in Minnesota. Eggs are laid in the jack pine needles. The sawfly spends the winter in the egg stage. Young caterpillars hatch in early June. Larvae are light green to whitish in color, with the full-grown larva about 7/8-inch long. The head of the larva is shiny black, the body has two pale greenish-gray lines extending along the back, with a single row of 11 black spots on either side. The mature larvae drop to the ground during early to mid-July, spin cocoons and pupate. Adults emerge from the duff under infested trees during September.

20. The Larch Sawfly

The larch sawfly develops and completes its life cycle on larch species (tamarack, western larch, European larch, Japanese larch, and Siberian larch). The larch sawfly has been a pest in the United States since 1880, with Minnesota reporting its presence in 1908. Evidence of infestation is revealed by the twigs developing a characteristic curl, which remains for several years. If the attack is heavy, trees are completely stripped. Successive defoliations drastically affect the growth and size of trees.

Adult sawflies emerge from the middle of May to August. Eggs are laid in the new shoots, and the newly hatched larvae move to the needles of the older shoots and feed on them for about 3 weeks. The mature larvae fall to the ground, spin cocoons, and spend 1 or 2 winters as prepupae. A full-grown larva is about 5/8-inch long, its head becomes jet black, and the body is whitish beneath the gray-green along the back. Pupation occurs from late spring to mid-summer.

21. The Red-headed Pine Sawfly

The red-headed pine sawfly is an important defoliator of ornamental, natural growing, and plantation pines. In Minnesota, jack pine, red pine, and Scotch pine are heavily attacked. White pine, Norway spruce, and tamarack are occasionally attacked. Old needles are eaten first, followed by complete defoliation. Evidence of feeding is seen in the reddish-brown, straw-like remains of needles.

The Red-headed Pine Sawfly has a single generation per year in Minnesota. In June, each female deposits about 120 eggs on adjacent needles, in the current or previous growth. The eggs hatch in 3 to 5 weeks. The larva, about 1-inch long, has a bright reddish orange head, hairless elongated bright yellow body, covered with 6 rows of dark black spots. A conspicuous black spot is present at the tail end. The larval stage lasts about 30 days. Pupation occurs in cocoons in the litter or soil. The sawfly overwinters as prepupae, some of which may remain in diapause for several years.

22. The Yellow-headed Spruce Sawfly

The yellow-headed spruce sawfly is widely known as a destructive pest of various species of spruces. It attacks white, black, Norway and blue spruces. The larvae prefer new foliage, but later consume both new and old foliage. Trees attacked by the sawfly have a ragged appearance early in the season; later, complete defoliation may occur. This species, although commonly found on spruce in natural stands, prefers trees growing in windbreaks, plantations, and ornamentals. Young plantation trees usually become susceptible from the third to fifth year after planting.

The female yellow-headed spruce sawflies emerge from the soil about the time the scales fall from spruce buds. Eggs are laid on new growth in May or early June. The larvae hatch in mid-June. The larva, about ¾-inch long, has a chestnut brown or reddish-yellow head, a body yellowish green above and lighter beneath, with a double row of broad, brown to olive green stripes along the back, and a broad stripe on the sides. Cocoons are spun in the soil in late July or early August in which the larvae spend the winter. A single generation occurs per season.

23. The Spruce Budworm

The spruce budworm is the most widely distributed destructive forest insect in North America. It has several varieties or biological races which attack a wide range of host plants, killing or stunting them through defoliation. The spruce budworm defoliates balsam fir, white spruce, and black spruce, which are preferred hosts in Minnesota. Spruce/fir forests, attacked severely by the budworm have, at first, a scorched appearance. Later, the crowns of trees become grayish and bare as the dead, webbed foliage disappears. Defoliation is usually heaviest in the upper part of the tree crown. The budworm is a wasteful feeder and often clips needles off at the base and webs them together without consuming them.

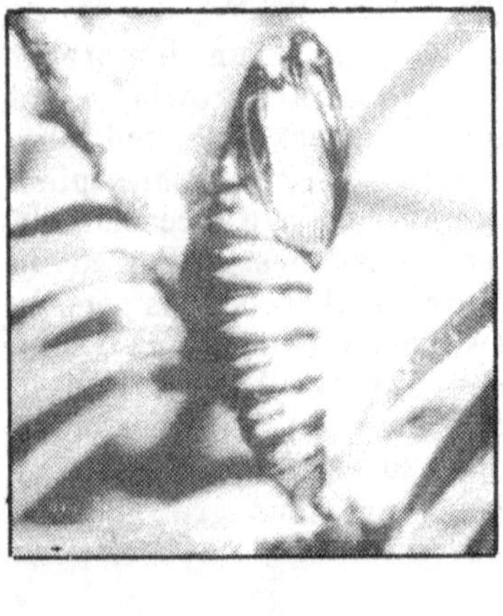

Spruce Budworm - Pupae Stage

The life cycle of the spruce budworm is quite similar to the jack pine budworm. The moths appear in July or early August and deposit eggs in masses on the underside of needles. The eggs hatch in 10 days, with the tiny larvae overwintering in webbed mats, the hibernacula. They emerge in May to early June, about 2 weeks before the buds begin to open on balsam fir. Feeding and molting continues throughout June. Mature larvae are about 1" long and are very similar to the jack pine budworm. The larval head is shiny black; the body is dark brown to dark reddish-brown with 2 rows of conspicuous white dots arranged in pairs, 1 pair on each side of a body segment. Pupation occurs during mid to late July.

Major factors and processes affecting the spruce budworm. Studies conducted during an outbreak of the spruce budworm showed that several factors are involved in making mature and overmature balsam fir stands more favorable than immature stands for population increases and the development of outbreaks. Some of the factors include:

(a) Stands providing warm, dry conditions in comparison with the cooler, more humid climates of dense, young stands promote a rapid development of all life stages, including the overwintering larvae before the onset of cold, lethal temperatures in the fall.

(b) Dry and sunny weather favors the normal flight activity of moths and their dispersal through wind transport. Wind dispersal, however, is not considered an important factor influencing population trends in any given area.

(c) The production of staminate flowers is a characteristic of mature and overmature balsam fir trees, a good source of food for the budworm. The periodicity of flowering is set basically by the climate of the region. In Minnesota, balsam fir and spruces normally produce staminate flowers in alternate years. Short term changes in climate can cause breaks in the normal flowering cycle, resulting in continuous flowering or nonflowering years.

SAP-SUCKING INSECTS

Diagnosis of tree damage by sucking insects generally depends on pest identification. Frequently a waxy or wool-like substance covers the pest and is useful in identification. In scale insects, the characteristic color and shape of "shells" will aid in identification. The reaction of the tree to damage by sucking insects is usually characterized by wilted leaves, dying branches, reduced growth, etc. Sometimes the causal agents, like aphids, spend only a part of their lives on the primary host and the rest on an alternate host-usually specific shrubs or herbs.

24. The White Pine Aphid

The white pine aphid causes injury by sucking the sap from white pine twigs. If severe, young trees and individual branches on larger trees are killed. The honeydew produced by the aphid may allow sooty molds to develop. Symptoms of attack are either yellowing and fall of needles, or blackening of needles and twigs.

Winged adults appear in the fall, are up to ¼" long, black or dark brown with conspicuous white lines down the center of the body, and with spots on the abdomen. The wingless forms are similarly colored. Eggs are laid in rows on white pine needles; at first they are pale, but soon turn jet black, oval in shape and look like a row of tiny footballs. The insect overwinters in the egg stage. Hatching occurs in the spring. The species then reproduces by the wingless forms producing living young. After a variable number of generations, winged females are produced which migrate and produce young.

25. The Pine Bark Aphid

The insect usually attacks white pine, and sometimes red pine. The evidence of the attack is seen in patches of white, cotton-like masses, usually observed along the trunk, especially in well-shaded places. Purplish to yellow, soft-bodied insects are found under the waxy material.

In general, resembles the white pine aphid.

26. The Pine Needle Scale

The pine needle scale attacks pines and spruces primarily, but also will attack firs and is especially injurious to ornamentals like mugho pine, and windbreaks. The elongated white scale with the yellowish cast skin at the narrow end stands out in sharp contrast to green needles. The feeding causes discoloration of needles and sometimes may result in needle drop and weakening of the tree.

Pine needle scale eggs are laid under the scale in the fall, and the insect overwinters in the egg stage. In the spring, hatching begins at about the time the new needles appear, and continues for a period of 2-3 weeks. The reddish crawlers move about for a few days, settle down and commence feeding. They turn yellow, become flattened, and molt after about 2 weeks. The females molt again after about 3 weeks but still lack the white scale. Males, however, after the first molt, develop a scale that has parallel ridges running the length of the scale. At about the time the females molt or shortly thereafter, the adult males, which are winged, emerge and fertilize the females, and the females begin to develop their scales. The scale is formed of wax produced by glands at the rear end of the body. Usually there is one generation per year.

27. The Pine Tortoise Scale

Jack pine, Scotch pine, red pine, and Austrian pine are the main host trees. The presence of the pine tortoise scale is indicated by branches or trees becoming black due to sooty mold or by yellowing of the needles (terminals). The scale insect withdraws sap from the twigs and produces copious amounts of honeydew. When the branches are examined, chestnut brown, hemispherical bodies up to ¼" long may be seen in clusters.

Up to about 1,500 eggs are produced by the adult female in June. Crawlers hatch from the eggs in a few hours and are dispersed by the wind. They settle down on new growth, begin to feed, and develop into scales. Males, which look like tiny gnats, fertilize the females about mid-August and die afterwards. Females continue to feed until cold weather sets in and overwinter.

SPITTLEBUGS:

Two native species, the pine spittlebug and the Saratoga spittlebug, are common in Minnesota. Their feeding habits are somewhat different. Nymphs of pine spittlebug develop on pines, while those of the Saratoga spittlebug develop on sweet-fern.

28. The Pine Spittlebug

Small nursery stock to mature forest trees are attacked by the pine spittlebug. Scotch pine, jack pine, and white pine are very susceptible to injury. Also, Norway, white, and black spruces, and hemlock are commonly attacked. Early evidence of this pest is the presence of froth or spittle among the needles or on the small twigs and branches. Trees injured by this insect have discolored brown foliage and stunted appearance.

There is one generation per season. The spittlebug passes winter as an egg, laid in late July and August, in the tissues of the pine twig. In early May, the eggs hatch and the tiny nymphs begin to feed on pine twigs and branches. These nymphs are alligator-like in shape, with a slender, shiny black body; beneath it they are pale orange, with purple sides. When nymphs are fully grown by the end of July, they leave the spittle mass, migrate to the needles and molt to winged adults. Adults are about 3/8" long, boat-shaped, with a pointed rear end. They are light brown, dotted with pale and dark spots forming transverse bands. More damage is done by adult feeding than by nymphs. The adults feed during July and August.

29. The Saratoga Spittlebug

The Saratoga spittlebug attacks young red pine and jack pine plantations in Minnesota. Red pine is the preferred host. Occasionally, white pine, white spruce, and Norway spruce are attacked. The first evidence of injury, usually caused by adults, occurs in the late fall and early winter as a slight yellowing of the needles at the branch tips. By spring, pronounced discoloration is present. Severely damaged trees present a scorched appearance. The spittlebug injury can be checked by removing bark from twigs, and finding brown to tannish flecks in the wood--signs of spittlebug feeding. The combined effect of feeding and the dehydration in the plant results in dead tissue at feeding sites and subsequent girdling of twigs. This creates an entry point for fungus diseases.

Life history: Very similar to the pine spittlebug with the exception noted earlier under "SPITTLEBUGS".

SPIDER MITES:

30. Spruce Spider Mite

31. European Red Mite

32. Two-spotted Spider Mite

The tiny mites are not insects but are close relatives of spiders. They feed on the needles of arborvitae, juniper, larch, pine, hemlock, and various species of spruce. Spruce trees are favored. By sucking the sap from foliage, they may severely devitalize or even kill trees. Their attack may be recognized by speckled yellowish appearance of the foliage. Closer examination reveals numerous, small, light-colored to brown spots, representing mite feeding sites. The mites also produce silken webs on the needles. The best way to test for mites is to hold a piece of stiff white paper under a twig and jar the foliage against it a number of times. The mites will appear as tiny, moving, reddish or dark specks.

The mites spend the winter as eggs, except the two-spotted spider mite, which overwinters as an adult. They are tucked under bud scales and at the bases of twigs. The brown-colored eggs hatch in the spring. The young look like the adults, begin feeding, and molt 3 times before becoming adults with 4 pairs of legs. The young mites are pale green, while the adults are dark green to nearly black. Probably 10 generations occur in a year. All stages can be found on most host plants during the summer months.

WOODLAND DISEASES

33. Hypoxylon Canker

Hypoxylon canker is the most serious killing disease of aspen. The disease starts when fungal spores infect wounds in the bark, such as, those produced by insects on small branches.

Any size tree can become infected. Cankering is commonly associated with dead branches. At first, yellowish-orange areas appear on the bark. After a couple of years, the outer bark layers become papery and peel off to reveal a dark, carbonaceous appearing material where new spores are produced. At the edge of the canker, white fan-like growths of the fungus can be found.

Hypoxylon canker kills trees thereby reducing stand density. Decay fungi and insects can also enter through the cankers, which weakens the stem and subjects them to wind breakage.

Hypoxylon canker cannot be prevented, however, proper management can minimize its impact. The amount of Hypoxylon canker infection is related to species, clonal differences and stand characteristics. Of the native poplars, quaking aspen is most susceptible, bigtooth is moderately susceptible and balsam poplar is rarely infected. Within each species, clonal susceptibility varies. Infection and canker formation are favored by open stands, poor stocking and along stand edges. Infection can be prevented by maintaining dense stands with closed canopies.

Aspen stands should be checked periodically for Hypoxylon canker. If 15-25% of the trees are cankered, harvest the stand by age 40 in a manner which will encourage dense aspen reproduction. If more than 25% of the trees are cankered, harvest as soon as possible and convert to another species. If the stand is lightly infected, it may be safe to manage it on a rotation longer than 40 years.

34. White Trunk Rot

White trunk rot causes the most volume loss in aspen. It becomes more common as the stand ages.

Conks, the fruiting body of **Phellinus tremulae**, are the most obvious indicators of infection and decay. The conks are hoof-shaped with a dark, rough upper surface and a tan to white lower surface. Conks are often found on the trunk at the bases of the dead branches. White trunk rot is a white, spongy decay of the aspen wood. Black zone lines surround the decayed wood.

A single conk on a 16 foot aspen log means the log is probably not useable for lumber. Trees may be infected and have decay in them for a number of years before conks are produced. To estimate the amount of rot in your aspen stand, determine the percentage of trees with conks on them and double that percentage. Almost twice as many aspen trees have decay as have conks.

Protect your trees from fires and other stem wounds. Wounds often become infected with decay organisms, such as **Phellinus tremulae.** Maintain dense stands with closed canopies. Harvest your stand before decay becomes severe, generally by age 40.

35. Northern Hardwoods

This type includes sugar maple, basswood, yellow birch, paper birch, oaks, elms, ashes, etc. The net annual growth of hardwood sawtimber is greater than removals, yet the demand for high quality hardwoods exceeds the supply. The difference is due to losses to decay. In general, decay fungi cause about 80% of all volume losses due to disease in hardwoods.

Decay Fungi are the major cause of decay in trees. Decays can be grouped by where they occur on trees, such as, root rot, butt rot and trunk rot. Spores are spread by wind and rain. Most fungal spores must enter the tree through wounds or branch stubs to become established and to cause damage.

External indicators of decay are: rotten branches, cracks, seams, fire scars, logging scars, butt bulges, burls, conks and cankers. Although cankers are caused by fungi that primarily kill the tree cambium, they are wounds through which decay fungi can enter.

Trees have no wound healing process. Trees compartmentalize the injured wood and may generate new tissues over the injured areas, but they can never replace or repair the injured tissues. It is not possible to eliminate decay from stands, however, decay can be reduced by preventing wounds, thinning sprout stands properly and using timber stand improvement techniques.

Prevent wounds by:

1. Protecting stands from fire.
2. Reducing logging wounds by better design of skid trails, reducing felling damage and harvesting damaged residual trees.
3. Maintain stocking levels to encourage self-pruning of small branches and rapid wound closure.
4. Proper pruning. Pruning can be the best thing you can do for your trees, or the worst, depending on how you do it. It's best to prune most trees when they're young and during the dormant season. In live branch pruning, only branches that will leave wounds two inches or less in width should be pruned. Find the bark ridge and cut downward and slightly outward. Do not injure or remove the branch collar. Do not paint the wound except for cosmetic purposes, and if you do, use a very thin coat of some commercial material (not house paint).

Many high value hardwood species are vigorous sprouters. Stands developing after harvest are likely to contain many stems of sprout origin. Sprouts are generally more susceptible to decay than are seedlings. Decay can be reduced in sprout origin stands by:

1. Favoring seedlings, seedling sprouts or sprouts from small, decay-free stumps.
2. Removing excess stems in sprout clumps when stands are young, generally less than 20 years old.
3. Favoring sprouts of low origin on the stump over those of high origin.

In established stands, proper thinnings and improvement cuts can reduce the number of decayed trees. Sound management practices include:

1. Prevent wounds, as discussed above.
2. Remove cull, conky, scarred, or root-diseased trees. Such trees are of low value and are taking up valuable growing space and resources. They also serve as sources of insects and diseases for surrounding healthy trees.
3. Avoid thinning in the spring when bark is easily knocked off the tree.

36. Sirococcus Shoot Blight

37. Diplodia Tip Blight

38. Sclcroderris Canker

A number of different fungi cause blights and cankers on red pine; three of them will be discussed briefly. They are **Sirococcus shoot blight** caused by Sirococcus strobilinius, **Diplodia tip blight** caused by Sphaeropsis ellisii and **Scleroderris canker** caused by Gremmeniella lagerbergii.

In the forest situation, it is often difficult to differentiate the three diseases. In general, the symptoms are needle discoloration and death or needle droop or premature needle drop and shoot mortality. Branch and stem cankers caused by Diplodia are resinous and dark colored below the bark. Scleroderris cankers causes a greenish discoloration beneath the bark of dead branches.

Any one of these diseases, under the proper conditions, can destroy a red pine plantation. Damage is cumulative and heavily infected trees can die, especially seedlings and saplings. Damage can be more severe if the tree is under stress from drought, J-rooting or being planted offsite.

Proper management techniques are:

1. Remove all overstory red and jack pine from a site scheduled for red pine planting as they are potential disease carriers.
2. New plantings should not be established adjacent to older pine plantations with existing disease outbreaks without treatment of those problems.
3. Do not plant red pine in frost pockets or off sites. Avoid J-rooting of seedlings.
4. Do not replant old Scleroderris infection sites to red pine.
5. To reduce the amount of infection on a site, infected branches can be pruned and burned along with infected or dead trees. Do not prune when the trees are candling or while wet.
6. Do not plant infected nursery stock.

39. White Pine Blister Rust

White pine blister rust, caused by Cronartium ribicola, was introduced into the USA about 1900. Rust diseases are unique in that two hosts are required for completion of their life cycle. In this case, gooseberry and currant are the alternate hosts. Only spores produced on gooseberry or currant can infect pine needles. The fungus kills branches and trees by girdling them.

The Lake States have been broken up into four hazard zones. Zone 1, southern MN and WI, has the lowest hazard of rust infection and Zone 4, northern MN and WI, has the highest hazard.

By the end of the first year, needle infections have spread to the branch or main stem where the bark becomes brown, bordered by yellow. By end of the second year, the branch becomes swollen and spindle shaped. In subsequent years, cankers enlarge and take on a sunken appearance. Abundant resin flow from the canker is an obvious symptom.

This is the most serious disease on white pine and has severely limited its planting. White pine can be grown in the southern parts of MN and WI with no significant mortality from blister rust. However, open field plantings in the northern part of the States generally will not produce productive plantations and should be avoided.

In Zones 3 and 4 (Northern MN & WI) establishment should only be attempted in an understory situation where it is possible to control the overstory through gradual removals. Prune the lower branches off the white pine to 50% of live crown at age 5-7 and continuing every two years until there are no live branches within at least 9' of the ground level. Consider planting resistant stock when it becomes available. In central and southern MN & WI (Zones 1 & 2), open field planting is acceptable. Prune as described above. Managed areas should not be located in areas where cold air collects at night or in or on the edges of small forest openings.

40. Armillaria Root Rot

Infects most all of the trees of importance in northern Minnesota, but of main concern are red and jack pines planted in areas that were covered with hardwoods.

Roots and the collar area of the tree become decayed; if the bark is peeled away, white fungal mats can be observed. Young pine infected in the spring will suddenly turn red and the newly elongating shoots will all wilt. In September edible honey-colored mushrooms can be found at the base of the trees.

41. Sweetfern Rust

Attacks jack pine and sometimes red pine.

An elongated swelling on the stem first appears on seedlings; as the trees grow, the stems show long, deep vertical fissures which develop into "catfaces".

42. Oak Wilt

Affects primarily oaks belonging to the red oak group including northern red oak and pin oak; rarely the white oak group (including bur oak) can become infected.

Leaves turn brown and droop, then fall; often one branch or a section of the tree will show wilting symptoms first; death usually occurs in one year. Usually one to a few trees in a group will become infected; widespread leaf loss in oak is not oak wilt. (This disease has not been found north of Brainerd, and it is thought that unfavorable environmental conditions will keep the disease out of northern Minnesota).

43. Dutch Elm Disease

Attacks all native species of elm, but American or white elm is the most susceptible. Chinese or Siberian elms are not susceptible.

Symptoms are the same as oak wilt disease but disease occurrence can be widespread. Look for the brown discoloration in the sapwood of twigs that show signs of recent or fresh wilting. This disease is present and spreading in northern Minnesota.

44. Lophdermium Needle-cast

Affects red and Scotch pines.

In the spring the needles show brown spots with yellow margins; entire needles turn brown and drop by mid-summer; black, football-shaped, protruding fruiting bodies appear on the dead needles.

45. Brown Spot

Found on Scotch and red pines, but occasionally white and jack pines will become infected.

Symptoms appear in late August and show up as yellow spots on the needles. The entire needle turns brown from the tip back and drops in late fall. Most infections are confined to the lower half of the tree.

46. Pine Needle Rust

Infects red, jack and Scotch pines.

Most common on young seedlings and symptoms show up best in the spring. Look for a browning of the foliage on the lower half of the tree and the production of orange blisters on the needles.

Diseases—

Prophylactic spray: applied at 7-14 day intervals starting at bud break and ending when leaves are fully extended.

Foliage Diseases	
Anthracnose Leaf Spots Scab	Bordeaux mixture, Zineb, Captan, Benlate, Cyprex, Maneb
Needle Casts	
Lophidermium, Brown Spot, Diplodia, Rhizosphaera	Bravo, Bordeaux mixture
Powdery Mildew	Karathane, Sulfure
Fire Blight	Streptomycin & Resistant Varieties
Rusts	Resistant Varieties & eradicate host
Oak Wilt & DED	Sanitation, Vapam for root grafts
Other Wilts	
Verticillium, Dothiorella, Phloem Necrosis	Main tree vigor (water & fertilize) Prune off affected parts

Note: Use of these chemicals is subject to federal and state regulation. Check all current laws pertaining to these chemicals before use. Consult labels on container to determine if it can be legally used.

Information for the insect and disease section is from publications by the Minnesota Agricultural Extension Service and Minnesota DNR.

Guide to Some Common Insect Pests of Minnesota Conifers

Insect pests	Primary hosts	Diagnosis keys	Chemical recommendations	Control details
Bark Beetles				
Pine engraver	Pines and spruces	Whitish or reddish dust on bark. Shothole exit holes through bark. Needles turn yellow then reddish. Star-shaped galleries under bark.	Carbaryl	Spray bark at 1 gallon solution per 50 sq ft of bark. Prompt removal of cut logs from April 15 to October 15. Sanitation.
Red turpentine beetle	Pole-sized and larger conifers; main damage is to pines	Pitch tubes of mixed resin and boring dust concentrated in basal 6 feet of tree. Wide galleries beneath bark. Needles turn yellow to sorrel to red.	Nothing approved	Reduce tree stress.
Borers				
Pine and fir sawyers	Recently dead pines and firs; logs and pulp sticks	Rough strands of chewed wood and bark extruded from dead tree accumulated at tree base, also on sides and beneath cut logs and pulp stick. Large galleries stuffed with excelsior-like wood below bark. Adults--longhorned beetles, emergence holes round. Larvae naked whitish.	Nothing approved	Prompt removal of cut logs and pulp sticks from April 15 to October 15.
Hemlock borer	Overmature and windthrown hemlock	Wet patches on bark of trees stressed by overmaturity or defoliators. Larvae elongate, whitish.	None	Sanitation; harvest of overmature trees; control of defoliators.
Defoliators				
Spruce budworm	Spruces and balsam fir	Current season and older needles partly or entirely eaten; needles webbed together in May-June. Larvae hairless, 1/2-1 inch in length with 2 pairs of whitish spots per segment.	Carbaryl	Reduce balsam fir component of stand below 50%. Reduce harvest age below 50 years.
Jack pine budworm	Jack and red pines	Current season and older needles partly or entirely eaten; needles webbed together in May-June. Larvae similar to spruce budworm.	Carbaryl	Increase stand stocking to 90-100 sq ft basal area. Remove wolf trees and large crowned trees. Reduce cutting cycle to 45 years.
Pine tussock moth	Pines and Norway spruce	Current needles with surface feeding in late summer; current and older needles completely eaten in May-June. Larvae black to brown with grey tufts of hair. Widspread tree killing.	Carbaryl Methoxychlor	Spray in mid-May to mid-June. No silvicultural techniques known.

Insect pests	Primary hosts	Diagnosis keys	Chemical recommendations	Control details
Larch sawfly	Tamarack and larches	Eggs laid in terminal shoots bearing single needles in early June cause shoot to have hooked appearance. Naked, pale, grey larvae with black heads consume only tufts of needles on old wood buds. Trees completely defoliated in June-July may refoliate in August.	Methoxychlor	Mid-June spraying; avoid applications to water because of toxicity to fish.
Red-headed pine sawfly	2-3 needle pines	Naked yellowish larvae with orange heads and black spots on back and sides. New and old needles often removed during June-July-August results in tree death.	Carbaryl Acephate	Spray foliage in mid June to mid-August.
European pine sawfly	2-3 needle pines and white pine	Naked grey-green larvae with black heads and dark lines on back and sides. Eat only older needles in mid-May to mid-June. Defoliation often extensive; tree death infrequent.	Carbaryl Acephate	Spray in mid-May to mid-June. Toxic to bees. Avoid direct application to water.
Red pine sawfly	2-3 needle pines	Same as above.	Same as above	Same as above
Introduced pine sawfly	2-3 needle pines and white pine	Naked dark green larvae with fine yellow mottling over body. Eat new and old needles from mid-June to mid-September. Trees often killed.	Carbaryl Acephate	Spray June through September. Avoid water and bee areas.
Root Weevils				
Pales weevil	Pine seedlings, especially Scots, red and white	Bark stripped from tender areas of seedlings and trees girdled just beneath the ground. Most serious on pines planted 1-2 years after clear-cutting pine forests. Also serious killing of pines in Christmas tree plantations where partial cutting practiced.	Nothing approved	Avoid replanting cutover areas for 2-3 years; allow land to lie fallow.
Root collar weevil	Pines, especially Scots, and red in 3-6 inch diameter class (5-15 year)	Trees off-color, yellowish or dying. Main stem girdled beneath the soil surface with the area of attack surrounded by pitch-impregnated blackened soil. Trees may be toppled over or broken off at girdle.	Benzene hexachloride (see note below)	Treat soil surrounding infested girdled area of tree. Remove lower 2-3 whorls of branches. Pull away needle accumulation from root collar area.

Insect pests	Primary hosts	Diagnosis keys	Chemical recommendations	Control details
Root tip weevil	Red and jack pines	Gradual deterioration of 20-30-year-old closely-spaced plantations on poor sites. Small (1/2 inch and smaller) roots destroyed. Reproduction in openings may be girdled in addition to having small roots destroyed.	None recommended	Avoid nitrogen depleted soils for establishing pine plantations.
Gall Makers				
Red pine needle midge	Red pine	New needles turn yellow then reddish in mid-October to mid-November. Tiny orange maggot between needles within sheath. Trees stunted in height and branch growth.	None	No silvicultural controls.
Balsam gall midge	Balsam fir	New needles with gall swellings two times width of needle. Galled needles drop in late November; trees frequently unfit for Christmas use.	Diazinon	Spray new foliage in late May or as soon as galls are seen.
Spruce gall aphids	Spruces, especially Norway, Colorado blue, and white	Galls involve swollen needle bases forming pineapple-like structure; eastern spruce gall aphid at twig base; Cooley spruce gall aphid at twig end. Galls open to permit insect escape.	Carbaryl Azinphosmethyl	When insects first seen.
Spittlebugs-Scales				
Saratoga spittlebug	2-3 needle pines	Reddish dead twigs/branches; exposure of wood by stripping bark reveals brownish, pitchimpregnated wood. Adults feed on needle-bearing twigs--no spittle mass formed. Immature insects form spittle mass on sweetfern and hawkweed stems beneath leaf litter. Tree mortality often widespread.	Carbaryl Methoxychlor Chlorpyrifos Malathion	Apply mid-July when adults feeding on twigs. Avoid alternate host areas in planting pines.
Pine spittlebug	Pines, especially white, Scots	Spittle mass on small branches and bark. Adult tree-feeding not in spittle mass. Injury results in stunting and tree death.	Same as above	Apply mid-July to mid-September.
Pine tortoise scale	Scots and jack pines	Female scales 1/16 inch in diameter, maroon color and crowded together on twigs between needles. Whitish empty skins of males may be present. Twigs covered with sooty mold and needles may be yellowed.	Carbaryl Diazinon Acephate Malathion	Spray in early May-June when crawlers emerge.

Insect pests	Primary hosts	Diagnosis keys	Chemical recommendations	Control details
Needles scales	Pines and spruces	Whitish or blackish elongated scales on needles. Sooty mold covering twigs. Needles yellowish.	Diazinon Carbaryl Malathion	Spray when crawlers seen.
Shoot and Twig Borers				
European pine shoot moth	2-3 needle pines	Current shoots mined in late May and June by several larvae.	Dimethoate Azinophosmethyl Carbaryl Malathion	Spray late April/early May; repeat late June and early July. Avoid planting sites adjacent to Lake Michigan.
Zimmerman pine moth (typical)	Pines	Infest main stem where wounds present from mechanical injury such as branch breakage from accumulating snow or ice, wind, and improper thinning. Extruded resin and frass at point of attack.	Endosulfan Naled Trichlorfon	Apply when insects or damage appears.
Zimmerman pine moth (shoot feeder)	Red pine	Infested shoots hollowed and killed turn brown in mid-July. Heaviest infestations on leader and upper branches.	Nothing approved	No effective controls.
White pine weevil	All pines, especially white and jack; Norway spruce	Current year and previous season's leader wilt and turn brown in mid-July. Larvae feeding beneath bark circling the previous season's leader.	Nothing approved	No effective controls.

Note: BHC could still be used for the control of pine root collar weevil, pine tip weevil and balsam gall midge if it had a Minnesota registration. Contact the Minnesota Department of Agriculture or the Department of Minnesota Natural Resources for the most recent information.

Guide to Some Common Insect Pests of Minnesota Hardwoods

Insect pests	Primary hosts	Diagnosis keys	Chemical recommendations	Control details
Bark Beeltes				
European elm bark beetle	American elm	Bark appears as if hit by bird shot; galleries running parallel to vertical wood grain.	Methoxychlor	No forest treatment recommended. A number of University of Wisconsin-Extension bulletins on Dutch elm disease and its control are available from your county Extension office.

Insect pests	Primary hosts	Diagnosis keys	Chemical recommendations	Control details
Native elm bark beetle	Elm species; ash, basswood, wild cherry	Same general appearance as above; galleries running horizontal and at right angle to wood grain.	Methoxychlor	See above.
Shot hole beetles	Most trees	Numerous bird-shot-like holes in bark. Trees dying from top down, and cambial area with numerous galleries.	None recommended	Sanitation.
Borers				
Bronze birch borer	Most birches, especillay serious on paper birch	Stressed trees mainly infested. Tree trunk and larger branches with irregular swellings; "D shaped" holes in bark often associated with rust-colored oozing sap. Zig-zag galleries on surface of wood.	Lindane may be used on noncommercial ornamental trees only	Spray on bark in late May-June and July to protect individual lawn and yard trees. Watering and fertilizing recommended.
Two lined chestnut borer	Oaks	Stressed trees mainly infested. Branches dying from upper crown downward, often leads to entire tree death in a few years. Adults emerge through "D shaped" holes in bark. Zig-zag galleries on surface of wood.	Chlorpyrifos Dimethoate	Spray tree bark in early to mid-summer. Spray logs in early to late summer. Watering and fertilizing recommended.
Locust borer	Black locust	Tree deterioration and branch dying. Bark dead with large tunnels into wood. Galleries on surface of wood, then to heartwood. Adults emerge through round holes. Early spring wet patches on bark surface. Trees on poor sites generally injured.	Carbaryl	Avoid stress of poor sites, grazing cattle and fire. Use resistant varieties. Spray in early May to June.
Defoliators				
Cankerworms	Most tree species	Leaves with holes and edge-feeding to complete defoliation.	Carbaryl Methoxychlor Bacillus thuringiensis	Chemical controls generally not necessary. If needed, apply as foliage is expanding.
Eastern tent caterpillar	Most species of wild Prunus	Small web nests in the crotches of trees. Leaves near nests consumed. Trees may be completely stripped and nest of 6-10 inches in length common. Larvae inside nest during daylight.	Carbaryl Methoxychlor Acephate	Not a forest problem. Most commonly seen on roadside cherry.

Insect pests	Primary hosts	Diagnosis keys	Chemical recommendations	Control details
Forest tent caterpillar	Aspen, birch, ash, maple, oak, willow	No tent constructed. On aspen, feeding begins when leaves unfolding. Larvae amass on trunk during non-feeding period. Widespread stripping of leaves common; forces larvae to wander across roads and open fields. Tree killing rare.	Carbaryl Methoxychlor Acephate	Chemical controls seldom required in forest. If necessary, applications made in early May when feeding begins.
Gypsy moth	Most hardwoods, oaks, birch, poplar, several conifers	Leaves eaten. Complete widespread stripping common. Migrating larvae feed on conifers. Widespread killing of hardwoods not common. Conifer death occurs if trees completely stripped.	Carbaryl Bacillus thuringiensis Acephate Dimilin	Application is timed to coincide with beginning of larval feeding in mid-May.
Walking sticks	Black oaks, elm, basswood, black locust, cherry and many others	Feeding begins on woody shrubs in early May-June then moves upward into nearly trees. Stripping may be widespread by mid-July to August. Because of the 2-year life cycle, with eggs laying dormant on the forest floor for one year, defoliation occurs alternately every other year. Tree killing uncommon.	Carbaryl Methoxychlor	Treat in late July.
Scales/Aphids				
Scales	All tree species	Discolored leaves with round or elongated waxcovered insects. Sooty mold may occur on nearby branches and foliage.	Malathion Diazinon Carbaryl	When crawlers active in spring.
Aphids	All tree species	Generally on tender leaves or newly expanded shoots. Colonies of insects feeding.	Malathion Acephate	When insects become numerous or on majority of new twigs.

University of Minnesota
FACT SHEET
Agricultural Extension Service ● U. S. Department of Agriculture

PLANT PATHOLOGY NO. 12–Revised 1973
WARD C. STIENSTRA

Noninfectious Diseases of Trees

Many tree diseases are caused by fungi, bacteria, viruses, or nematodes (living organisms). Some are not. Noninfectious, or abiotic diseases, are a major cause of tree problems in forests and home landscapes. Abiotic diseases, together with animal damage, have reduced yields in forest stands as much as 20 percent. Recognition of noninfectious tree diseases and diagnosis is difficult because the causal agent often has disappeared. The following categories describe many of the common noninfectious diseases of trees:

HIGH TEMPERATURE DISEASES

Heat Defoliation

This may occur as a result of high temperatures, coupled with severe dry winds.

Winter Sunscald

This is caused by above-freezing temperatures on sunny days combined with freezing temperatures at night. It usually occurs on the south-southwest side of smooth-barked trees (apple, aspen, maple, mountain ash, and white pine) during the late winter or early spring when the daytime sun warms the tree trunk and temperatures are freezing at night. Smooth-barked trees can be protected by wrapping the trunk with any material that shades the stem or reflects sunlight and thus prevants excessive warming of the cambial tissues.

Heat Canker

This is common on young trees when the soil surface temperature rises enough to cause localized killing of cells, which results in a swelling of the stem above the injured tissues. Soil temperatures at the ground line can be as high as 160° to 175° F. and temperatures of 140°F. are not unusual on sunny, summer days. Shading young seedlings during the first and second year prevents heat cankers.

Birch Dieback

Birch is a forest tree. When planted in a lawn, it does not have the leaf and surrounding trees to protect its shallow root system. Without this protection, soil temperature can rise, on hot days during the summer, enough to kill the roots. Root loss leads to dieback of upper branches. The weakened tree is susceptible to attack by the bronze birch borer, which hastens the tree's death. Thick soil mulches help keep soil temperatures lower and thereby prevent dieback. Soaking the ground around the tree during hot weather may give added protection. The river birch is generally longer-lived in Minnesota than imported varieties.

LOW TEMPERATURE DISEASES

Low Temperature

Low temperatures, common in Minnesota winters, may kill tree roots growing close to the soil surface if there is no snow cover. Introduced tree species, newly planted trees, and trees growing in shallow soils are more susceptible to this injury.

Frost Damage

Early fall and late spring frosts can injure or kill woody tissues, buds, and leaves. Damage due to late spring frost is obvious. The day after the frost, leaves and flowers wilt and die. Fall frost damage is usually not noticed until the following spring and, therefore, diagnosis is difficult, in fact, seldom seen until the plant thaws. Fortunately, most trees can survive frost injury although growth rates may be reduced.

Frost Crack

This occurs in both hardwood and softwood trees during a pronounced drop in winter temperature. The outer layers of wood, which are colder, contract more rapidly than the inner layers, which are insulated. The result is the opening of long vertical tissues in the wood and bark. Frost crack may occur in succeeding years in the same place, causing excessive formation of callus tissue or swelling on one side of the tree.

Frost Shake

This is a crack or separation of wood at the growth ring and occurs as a result of the sudden warming of the outer layers of wood while the inner tissues are still very cold. The rapid expansion of the outer layer forces it away from the cold inner tissues.

Winter Browning

This tree injury is common on arborvitae and other species of ornamental conifers grown in northern latitudes. The rapid drop in foliage temperature at sunset (17 degrees per minute) causes sudden freezing of water within the foliage and results in injured tissue, which becomes evident in May or June when it turns brown. This problem occurs predominantly on the southwest and south side of shrubs. Planting susceptible species on the north or east sides of buildings or other locations protected from continuous exposure to sun and wind will reduce this type of injury.

WATER-RELATED DISEASES

Drought

In years of low rainfall, many tree species can be substantially weakened or killed by drought. Deciduous trees appear to die from the top down, have small, off-colored foliage, narrow growth rings, and may be invaded by many secondary fungi and insects, which seem to be causing the tree decline. Pine trees appear to die from the bottom up when subjected to artificial drought. Individual trees vary in their resistance to drought.

Issued in furtherance of cooperative extension work in agriculture and home economics, acts of May 8 and June 30, 1914, in cooperation with the U.S. Department of Agriculture. Roland H. Abraham, Director of Agricultural Extension Service, University of Minnesota, St. Paul, Minnesota 55101. We offer our programs and facilities to all people without regard to race, creed, color, sex, or national origin.

Winter Drying

This differs from winter browning, but the symptoms are similar. Winter drying, primarily a disease of evergreens, occurs when the roots are in frozen ground and warm winds cause excessive transpiration (foliage water loss). Winter drying can be avoided to some extent by protecting evergreens with a mulch to prevent deep freezing. The very susceptible species, northern white cedar, should not be planted on exposed southwest facing slopes. Wrapping susceptible evergreens in burlap is helpful.

Leaf Scorch

A sudden and rapid loss of water from maple leaves during hot, dry weather causes leaf scorch. Affected leaves have yellow or brown margins and may remain on the tree or the affected tree may just lose its leaves prematurely. Watering during hot, dry weather may prevent or alleviate this problem on small trees.

Excessive Water

Tree roots require oxygen for respiration as well as water and will seek a level in the soil that best supplies both these requirements. Spring floods and backfilling around trees will smother the root system of the tree and tree dieback results. Often secondary fungi and insects invade these trees. Symptoms normally do not appear until later in the same year or the year following root damage.

MECHANICAL INJURIES

Bruising

Ice, snow, hail, lightning, wind, and machinery can extensively damage trees. These injuries may or may not kill a tree, but wounds can be entrance points for disease-causing parasites.

Transplant Injuries

Transplant injury is a major cause of tree loss. Many roots may be lost during digging. So, to compensate, a proportional amount of the top of broad-leaved trees is removed. Nurserymen train the tree roots to form more compact masses so that a greater percentage of the roots can be taken up when trees, especially evergreens, are transplanted. Early spring is the preferred time for transplanting trees.

CHEMICAL DAMAGE

Nutrient Lack

Soils deficient in nutrients do not support proper tree growth. Know the fertility of your soil by having it tested and adjust nutrient levels accordingly. Trees grown on nutrient-deficient soils may be more subject to attack by parasites.

Herbicide Damage

Tree damage from herbicide sprays or granules applied to turf underneath trees is common in metropolitan areas. Affected leaves become distorted, twisted, curled or cupped, and margins turn brown. Trees may be killed by this treatment, but more often are stunted for 1 year.

Salt Toxicity

Salt applied to streets in winter is absorbed by roots of boulevard trees during the growing season. This results in varying degrees of marginal browning of leaves when moisture becomes limited. Dieback may occur in trees damaged in successive years; some trees have been killed.

Other Chemical Injuries

Fumes from some industrial processes and incinerators can cause chlorosis, spotting of foliage, and in severe cases, tree defoliation. Dog urine on conifers, especially arborvitae, can result in black needles. Foliar sprays, normally considered non toxic to trees, may cause leaf injury if applied during hot weather. Also, a spray concentration recommended for one species may be toxic to another.

Air Pollution

Injury to vegetation induced by exposure to air pollutants is a popular issue. Air pollution is certainly not new. When plants are injured by an air pollutant, symptoms characteristic of the specific pollutant usually develop. Recognition of pollution-induced effects and identification of the specific causal agent or agents is a complex task, requiring an integrated knowledge of several diverse fields. Since the pollutant itself normally undergoes a chemical change soon after it contacts plant tissue, symptoms are often the only remaining evidence of the pollutant.

The major air pollutants are peroxyacyl nitrates (PAN), ozone, fluoride, sulfur dioxide, and ethylene. Smog symptoms, now recognized as PAN injury, typically develop only on the expanding leaves. The very young and most mature ones are usually resistant. Banding, along with undersurface glazing, bronzing, silvering, and tissue collapse are associated with injury produced by PAN. Ozone symptoms vary and are grouped under four different general types. The most common ozone symptom on many deciduous trees and shrubs and some herbacious species is localized thickening and pigmentation of the cell walls resulting in sharply defined small dot-like lesions. The lesion may be dark brown, black, purple, or red. Small unpigmented necrotic (dead) spots or more general upper surface bleaching is the second common type of injury on most herbacious and many woody species. The individual lesions tend to be irregular in shape and can develop on either leaf surface.

The third symptom involves killing of all of the tissues through a leaf. The upper and lower surfaces are often drawn together, forming a thin, papery lesion. The fourth symptom is a fine chlorotic stipuling of either surface. Sometimes light green chlorotic areas develop with many irregular islands of normal green tissue. The characteristic symptom of fluoride injury on many broad-leaved plant species is necrosis (dying), which occurs predominantly at the leaf tip and margins where the fluoride has accumulated. A dull, gray-green, water-soaked discoloration of tissues along the leaf tip and margins is the first sign of this injury.

Fluorides may also cause chlorosis or yellowing, extending in from the leaf margin between the larger veins. A general chlorotic appearance of the leaf is symptomatic of chronic sulfur dioxide injury. These cells are usually not killed but are bleached of their chlorophyll. In addition to the chlorosis or yellowing of the leaf, a silvering or bronzing of the undersurface may accompany sulfur dioxide damage. Acute sulfur dioxide injury appears as marginal areas of dead tissue which at first have a dull grayish-green, water-soaked appearance. When dry, these areas take on a bleached tan color in most plant species. Ethylene acts as a growth hormone and causes a general reduction in growth. Plant leaves or modifed leaves droop, show chlorosis (yellowing) or necrosis (cells dying). Abscission and failure of flowers to open properly has been reported for many floral crops.

There is a range of plant markings produced by all air pollutants. Some symptoms are typical of a given pollutant and may be used to identify the specific pollutant, while other agents (frost, disease, insect pests, chemical sprays, soil herbicides, nutrient deficiency, etc.) may also produce markings on plants that very closely resemble air pollutant symptoms.

University of Minnesota
FACT SHEET
Agricultural Extension Service ● U. S. Department of Agriculture

FORESTRY NO. 8

Protecting Trees from Animal Damage

WILLIAM R. MILES

Animal damage to trees is apt to be most severe from late fall to early spring. During this period, rabbits, mice, deer, and squirrels turn to trees for food because of the lack of other plant material. Some animals--porcupines, pocket gophers, and dogs--damage trees throughout the year.

The traditional method of preventing tree damage has been to remove the offending animal from the area. Hunting, trapping, and poisoning are effective removal methods but all involve time or skill and present a hazard that may deter the average tree owner from using them.

Other methods that put the tree out of reach or make it unpalatable can be employed. Outlined below are the recommendations for protecting trees from specific animals.

RABBITS

Rabbits girdle trees and sometimes even cut down small seedlings. You can fence trees against rabbits, but a more effective barrier is created by placing cylinders of hardware cloth or mesh screen around the base of each tree. Be sure to wrap the tree high enough so rabbits can't get at it by standing on the snow.

If you have a number of trees to protect, using screens may be too expensive and time consuming. Repellents may be the best solution for you. Remember that a repellent is not a poison; it simply renders the tree undesirable through taste or smell.

You can either spray or paint repellents on trees. Though good repellents can be made at home, the preparation is rather involved. Since many good commercial repellents are available, we recommend that you consider using them.

Effective commercial repellents generally contain one or more of the chemical ingredients abbreviated TNB-A, TMTD, or ZAC. Some recommended repellents for rabbits are Z.I.P., TAT-Go, Selco Rabbit and Deer Repellent, Peter Rabbit Repellent, Chaperone, Arasan 42-S, Magic Circle, and No. 96-A. There may be other effective commercial repellents besides these.

All of the above preparations are effective against rabbits. You may purchase them from hardware stores, garden supply stores, feed and seed stores, or nurseries. Always follow the manufacturer's directions when you use commercial products.

Two mix-your-own repellents are:

(1) Add 1 gallon of asphalt water emulsion to 2 gallons of water. Then stir in 3 pounds of Thiram (75 percent) and mix thoroughly.

(2) Add 1 gallon of asphalt water emulsion to $2\frac{1}{2}$ gallons of water. Then add 1 pound of household detergent and mix until smooth. Add 3 pints of Black Leaf 40 and mix thoroughly.

Asphalt water emulsions usually can be obtained at lumberyards and from other building supply dealers. Thiram (75 percent) is sold as a fungicide under the names Arasan SF-X and Tersan 75 and can be obtained from seed and feed outlets.

MICE

Mice generally damage a tree by gnawing and subsequently girdling its stem. You can effectively protect the bases of trees by placing cylinders of $\frac{1}{4}$-inch mesh hardware cloth around them. Be sure to extend the cylinder deep enough into the soil to prevent mice from getting underneath the screen.

You can insure good tree protection, especially for young fruit and yard plantings, by placing 5-quart oil cans with both ends removed around tree bases.

One of the best deterents against mice is a clean planting, especially in the row. Repellents also are effective; those recommended for rabbits usually will work against mice. Using mouse poisons containing strychnine or zinc phosphide on oats and wheat usually is effective. Caution: Be sure to follow manufacturer's directions when using any commercial poison.

DEER

Deer may feed on and damage the tops and side branches of small trees. Bucks often damage stems by rubbing them with their antlers. The repellents listed for rabbits sometimes are effective against deer. Fencing for deer is very expensive and you should consider it only if you have very high priced plantings. For deer exclosures to be effective, fences must be high and constructed with posts and heavy mesh wire. If damage from

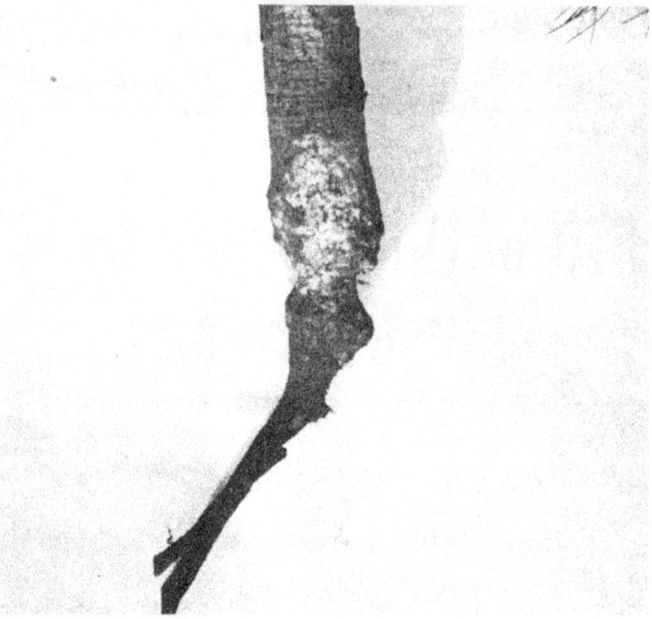

Pocket gopher damage on red pine roots.

deer is severe, notify your local game warden.

SQUIRRELS

Squirrel damage usually is confined to the gnawing and stripping of bark in tree tops. Squirrel damage may be severe in stands of sugar maple. Protect isolated trees by placing a 2-foot band of smooth metal around the tree trunk 6 to 8 feet above the ground. Remove any branches that are less than 6 feet from the ground.

Z.I.P. rabbit repellent is effective against squirrels. Nicotine sulfate spray or a mixture of 1 teaspoon Lysol, 3 ounces Epsom salts, and 1 gallon of water also is effective. Spray every 2 weeks during the growing season and after any hard rain during the season.

PORCUPINES

Porcupines usually cause damage by girdling the tops of large trees. The most effective control is to kill the offending animal. Repellents are not generally recommended, though Tanglefoot may be effective if used on individual trees. To protect individual trees, place a broad band (3 to 4 feet wide) of smooth metal around the tree 3 to 6 feet above the ground. Remove all branches below this band.

According to the National Wildlife Federation, an application of the following formula will keep porcupines from gnawing on paddles, handles, and cabin steps. Stir 1 pound of copper naphthenate into 2½ quarts of mineral spirits and spread on wood as you would paint. This formula gives the wood a greenish tint but discourages animals.

POCKET GOPHERS

The pocket gopher is active year round, but he is the most damaging in the spring and fall. Gophers feed on tree roots below the ground line; the trees eventually die and can be easily pulled up. In plantations, your best and easiest control method is to use a tractor-pulled "burrow-building" machine. This machine deposits poisoned bait in artificially created tunnels that intercept pocket gopher runways. Trapping and poisoning are the most successful methods of getting rid of individual gophers.

For further information on pocket gopher control, contact you county extension agent or Soil Conservation Service office.

DOGS

Damage to small conifers is apparent when patches of foliage near the ground appear brown and eventually die. The male dog is responsible, and this condition is most noticeable in the spring on outer exposed branches. Protect trees with fences or use a dog repellent.

BIRDS

Sapsuckers (commonly but erroneously called woodpeckers) cause the most apparent and severe bird damage to trees. Damage appears as rows of small holes on the tree trunk in neat horizontal and vertical lines. Use a mixture of two parts malathion to one part sugar syrup to discourage sapsuckers. Or you can smear the bark with Tanglefoot in the area where birds are active or wrap this area with hardware cloth or burlap.

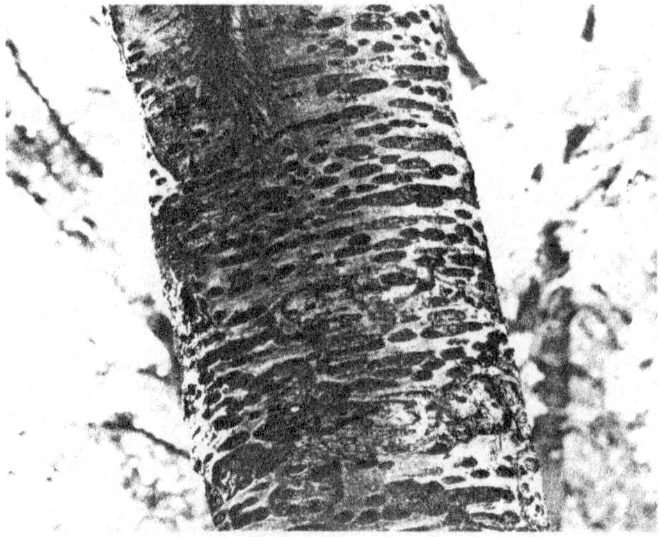

Yellow-bellied sapsucker damage on mountain-ash.

The use of trade names in this publication is solely for the purpose of providing information. Reference to commercial or trademarked materials does not imply endorsement by the Minnesota Agricultural Extension Service; failure to mention a product does not imply criticism.

Issued in furtherance of cooperative extension work in agriculture and home economics, acts of May 8 and June 30, 1914, in cooperation with the U.S. Department of Agriculture. Roland H. Abraham, Director of Agricultural Extension Service, University of Minnesota, St. Paul, Minnesota 55101.

CHAPTER 13
ACCESS ROAD DESIGN, CONSTRUCTION, MAINTENANCE AND RENOVATION

Adrian Hagen
Forester
WI Dept. of Natural Resources

Larry Himanga
Private Forest Mgmt. Spec.
MN Dept. of Natural Resources

One of the most worthwhile investments you can make in your woodland is a system of woods roads or trails. For purposes of this discussion, a woods road is the type of access which a truck or other hauler can use. A trail, on the other hand, is access a farm tractor, pickup or log skidder would use. Their primary forest management purpose is to provide access to aid you in the management and enjoyment of your forest.

Access is essential for the harvesting of wood products. Whether the products are logs, pulpwood or firewood, being able to reach them with equipment adds to their value. Use of roads and trails can also reduce damage that can occur to uncut trees during harvesting.

Roads and/or trails are valuable also for other management activities such as tree planting, thinnings, pruning, fire control and wildlife management. When kept in herbaceous cover, they are a permanent opening in the forest and are a source of food plants otherwise unavailable to wildlife.

This is a firebreak that doubles as an access road and winter ski trail.

Recreational activities such as viewing wildflowers and picking mushrooms in spring, observing fall coloration, hunting, horseback riding, snowmobiling, skiing and hiking are all more enjoyable with a good trail system.

In planning for access, when you have none, there are some important questions to consider. The first question is whether you need permanent access at all. Seemingly contradictory, there are woodlands that merely need some designated travel areas during management but no permanent road. These woodlands are usually level terrain, excessively wet land or excessively steep/minimally productive areas. Level land is easy to use equipment on, wet areas may only be entered while the ground is frozen and steep areas of low productivity will not repay construction costs. Access for enjoyment of woodland amenities is a different story.

Aerial photos or topographic maps are very helpful in layout. Study your woodland, learning where the best growing sites, the most valuable timber and the most difficult road-building areas are. Location decisions should be made with as much information in hand as possible. Avoid flat ridgetops and bottomlands as they are difficult to drain. Build roads on a side-slope and work from the top of the slope to the bottom. Watch for and avoid obstacles such as rock, springs, gullies and streams.

All potential future uses and needs should be considered. Once a "temporary" access has been established, they usually become a permanent fixture of the landscape. If they're not suitable to your needs, resources have been underutilized. Access may be a simple two-track path, a complex system of loops and connector roads or something in between. A small property may need little road and more trails while a large one may need a more complex system. Many roads are built by loggers to suit their needs. Since you, as the owner, may have slightly different needs, these should be worked out with the logger so both parties' needs are satisfied. Road construction costs are normally figured into the sale price of the timber. Therefore, you may have to reinvest some of the sale proceeds to get additional roads constructed. It is an investment in future management, however, that will repay you many times.

Typical logging access trail.

In steep terrain, an important question is whether or not there's enough land to establish the access. Because of the length of slope required to achieve a proper grade, it may become necessary to combine forces with a neighboring owner. Season of construction should also be considered to achieve desirable construction standards. Good planning will provide for all uses, control costs, minimize potential problems and provide a useful, practical access system.

I'd like to discuss briefly the effects of access on forest soil in terms of precipitation, water runoff and erosion. A natural forest soil is deep and porus, that is, filled with air spaces and protected by a litter and humus layer. This litter layer protects the soil surface from the force of the raindrops, allowing water to move into the soil (infiltration). Any action that destroys or removes the litter layer reduces this beneficial action. The act of cutting trees has little effect on runoff as the litter layer remains intact; rapid revegetation replaces the water transpiring activity of the lost tree crowns. Infiltration is also reduced when the soil is compacted. Compaction reduces the pore space in the soil thereby reducing the amount of water that can move into and through it.

Constructing roads or trails both removes the litter layer and compacts the soil, although the latter effect is most pronounced during and immediately after use. Water falling on bare soil dislodges soil particles. On flat land, "puddling" of water occurs in the low spots where soil was compacted by travel. On sloping land, however, water flows and carries soil particles with it, which helps to loosen and carry more soil particles–erosion! The faster the water and soil mix flows, the more erosive it becomes. It's obvious that a bare soil road on a steep grade can be a problem; more on this later.

Erosion is a natural and normal phenomenon, on natural forestland about .2-.3 tons per acre per year (Soil Conservation Service agricultural land erosion tolerance losses vary from 2-5 tons per acre per year in Southwestern Wisconsin). Though tree cutting has little effect on runoff and thus doesn't contribute to erosion, the act of removing cut products can have significant effects. Skidding logs on the ground and use of roads, trails and log landings will increase erosional possibilities. The reasons include exposing the mineral soil, the drying out of the upper soil layers and the loss of soil structure due to compaction. The effect is a tremendous reduction in infiltration (movement of water into the soil), from 8-12 inches per hour to as little as .5 inches. The water then runs off the bare roads and trails carrying the loosened top soil with it (losses up to 30 tons/acre of road have been recorded). Runoff will naturally collect and flow in any wheel tracks increasing water velocity with a correspondingly greater erosive action. These effects, most pronounced during active logging, can be greatly reduced through proper road layout, design and maintenance.

As you have seen, road layout and design is quite important. If a road is designed properly, it can reduce both installation and maintenance costs. In laying out the road, you should choose a location that will require the least amount of earth disturbance, while providing access to the specific areas you designated in planning the road. After you have an idea of the location, walk the entire length of the road and make any changes you feel are necessary. You should take care to avoid areas of poor drainage, spring seeps, or valley bottoms. There are a few other basic points to keep in mind when building a road:

1) Width of road 12 feet, road surface 10 feet
2) Keep the road under 10% grade, if possible
3) Don't ever exceed 20% grade, and hold areas of the road over 10% grade to very short distances
4) Keep road as short as possible
5) Avoid sharp corners unless necessary (switchbacks)
6) Cross ditches at right angles
7) Never allow water to run down the road

For ease of discussion, we will expand on each of the above points separately; however, it should be remembered that they all work together to create a useful, permanent road.

The first and most basic point of road construction is the road width. For most purposes, a 12-foot road with a 10-foot surface will do. The road width is important for maneuverability with equipment. Clearing brush and trees out another foot or two past the road surface can save paint on your truck as well as make the road more useable.

Second, keep the road under 15% grade, if possible. This is fairly simple to do. By angling up the hill, you can decrease the grade, however, at the same time you are increasing the road length. A few factors that could affect the road's grade are the percent of rock, drainage and the investment dollars available. These factors usually work together to determine the grade. The more rock in the soil the greater the construction costs per foot. The amount of money available can then become the limiting factor as to how much rock should be moved to obtain the grade desired. In the interest of keeping costs down, the operator should have some flexibility during the actual construction of the road. If the operator hits rock at one point and he has the flexibility to move a few feet up or down slope, he can save several hundred dollars. At the same time, he may have to increase the slope of the road for a short distance.

Woods roads, unlike high use roads, should not contain long stretches of a continual grade. These stretches allow the buildup of water and increase the potential for erosion. You should use occasional dips to drain off water, with the average grade of 10%.

The third item to keep in mind during construction is keeping the grade below 20%. This is very important since grades above this will have a very limited use period. Immediately after a rain or snow, the grade may be too steep to drive without damaging the road. For the same reason, grades over 10% should be only for short distances - such as crossing a ditch or going up or down slopes to avoid rock.

The fourth point is probably the most fundamental. Keeping the road as short as possible will keep your costs down as well as disturb the least amount of soil. At the same time, you should still keep the grade as gentle as possible.

If you have to build a road on an unusually steep and short hillside, you may need to install a switchback. A switchback is simply a sharp curve in the road to change its direction, thus, allowing you to maintain grade in the given area. Switchbacks should be located in areas with little side slope where the loop can be constructed with the least excavation. The grade of the straight stretches should be reduced into and out of the switchback. This will help maintain the grade throughout the curve. As always, no more excavation should be done than necessary to get the fill for along the lower side of the switchback.

The sixth item of fair importance deals with stream and ditch crossings. When you must cross a drainage system, there are several factors to consider. They are:

1) How large an area does it drain?
2) What is the potential water movement?
3) Cross at a right angle.

If the drainage area is small, the problem may be solved with a rock riprap crossing or small culvert. If the drainage area is larger, the culvert size would increase proportionately. Culvert installation should be done with care to maintain proper alignment with the channel. Good culvert alignment improves its function and helps maintain water quality. When culverts are placed at an angle to the channel, they can cause bank erosion and collect debris. The slope of the culvert should be the same as that of the stream or ditch. The culvert length then would have to be increased accordingly to account for the extra distance.

In areas where there is a very large drainage area with no continual water flow, a ditch crossing may be in order. A ditch crossing consists of removing soil or boulders to expose the parent material of the ditch in question. This will give you a rough but useable road surface that won't wash away in the first large rain.

In some instances, it may be necessary to cross permanent water. In these instances, several other items must be considered:

1) Permits required
2) Minimal disturbance of stream
3) Avoidance of washouts or undercuts

Although the crossing could be made with a culvert or some type of bridge, these items are often quite expensive. The best and usually the least expensive method of crossing a stream or creek is to drive through it at a right angle. Even then, the banks have to be sloped to permit easy entering and exiting of the stream. Decreasing the slope of the stream banks allows for better traction and reduces the chance of tearing up or rutting the banks. If the stream bottom is too rough to drive through, rock may be hauled in and placed in the stream and along the banks. This usually will suffice for traction, yet not greatly affect stream flow. The crossing should be designed such that the stream's natural dimensions are not altered. Do not widen the stream bed itself or greatly decrease the stream depth. Under no circumstances should fish barriers (screens on culverts, etc.) be constructed. After construction, all disturbed areas around the stream should be landscaped and seeded to grasses. A couple basic rules of thumb for timing construction are 1) don't do any fall or winter work on trout streams and 2) don't do any spring work on warmwater streams with gamefish populations (March to June depending upon species of gamefish).

The final and most important point to remember is not to allow water to run down the road. Anytime there is a potential of water moving down the road itself, precautions should be taken to eliminate potential erosion. During construction, the road should be given a 3% outslope to help move water off the road. That is, the inside of the road should be a few inches higher than the outside. Seeding the road impedes a lot of water movement. The type and quantity of seed will depend on your geographical area and soil type. Basically, you prepare the road for seeding much the same as you would for a new lawn. The majority of work should have already been done during construction.

A good seed mix will contain a variety of grass types, usually including Kentucky bluegrass, creeping red fescue, smooth bromegrass and perennial ryegrass. A few other species of grasses will usually be mixed in in small percentages. In our area, the seeding rate is usually around 35 pounds per acre or ¾ pound per 1,000 square feet. After seeding, a mulch should be applied to give the grasses the best start possible.

Although seeding does slow water movement, seeding alone often isn't enough. There are two basic methods that, when combined with seeding, can bring water movement down to an acceptable level. The two methods consist of installations of either waterbars or open top "box" culverts. These structures bring water that is traveling down the road across or off the road surface.

A waterbar is a gradual rise in the road surface with a small dip on the uphill side. These structures are installed during the actual construction of the road surface and should be angled 30° downslope to facilitate water moving across the road as well as smooth out the ride. Spreading the hump and dip out over 6-8 feet will decrease the abruptness of the hump as well as minimize any tendency to undercut.

An open top culvert or box culvert is a three-sided structure creating a permanent trough in the roadway. Water traveling down the road enters the culvert and is diverted off the road. As with waterbars, these structures should be installed at a 30° angle downslope. Although these culverts are easier to drive over, they have special maintenance needs. Leaves, twigs, duff and silt can collect in them if not kept clean. If they are not installed properly and not flush with the road surface, the water may try to undercut the box and create a larger problem than if they were not there.

Either waterbars or box culverts should be installed whenever the slope is greater than 5% depending upon the road surface material. Roughly one structure should be installed every 140 feet of straight stretch of road. As the grade increases, the distance between the stretches should be decreased.

Table 1. Recommended distance between waterbars or open top culverts on woods roads

Grade of Road	Distance Between Waterbars
Percent	Feet
2	250
5	135
10	80
15	60
20	45
25	40
30	35
40	30

Another couple of items to remember when designing and constructing woods roads are sloping the banks and the frequency of turnouts or turnarounds.

Whenever the slope is such that the road requires a lot of cut and fill, the inside bank will slough onto the road if left unattended. These inside slopes should be banked to decrease their slope and generate a bed for seeding to prevent more erosion than necessary.

Whenever the start and end points of a road are a considerable distance apart, turnouts or a turnaround may be beneficial. These are areas where two vehicles can meet or pass each other on the road or one vehicle can turn around.

Having turnouts installed during the construction of the road adds very little to the road cost yet makes the road easier to use and eliminates driving to the top or backing down, in some cases.

A final word of caution on design and construction. In areas where the topography is fairly flat and little or no earth movement is required, don't fight mother nature; that is, do not blade your road with a cat or scraper. Blading the road in these areas will actually lower the level of the road, creating a low spot for water to collect. These areas would be better off having a load or two of fill hauled in to smooth the road surface.

Maintenance is the most critical factor in the long-term usefulness of your access system. After periods of heavy use, make sure the road surface is smooth, free of debris and out sloped slightly (about 3%). Check all culverts and drainage structures to be sure they are operational. This should also be done after heavy rains and annual leaf fall. These actions on your part will assure that water is kept drained off the road with minimal erosion. It is desirable to cut overhanging trees and those close to the road so the sun and wind can help dry the surface. An additional protection was mentioned earlier, that is, the seeding down of roads and trails with grasses. In our temperate climate, no bare soil area will remain that way for long. A grass mixture will help keep out tree seedlings and prevent erosion. Add some white dutch clover (alsike for wet areas) and you'll have a perennial cover of outstanding wildlife benefit! Control the traffic, especially during wet periods, and your access should be there whenever you need it.

If your property has some existing access, there are some steps you can take to improve it. Inventory, and map if possible, what you have and decide which roads are most useable for current and potential uses of the land. Close off those sections that aren't needed (closed roads could be seeded down or re-stocked with trees or even wildlife shrubs, if needed). New roads needed to develop a desirable system can be added at anytime.

Any erosion on the existing roads and trails to be used should be repaired, installing drainage, sloping and seeding as needed. Chuck holes should be dug out and regraded or can be filled with larger road rock, cinders or a sawdust/bark mixture. Avoid selection of roads within 100 feet of streams due to the potential for siltation into the water. Woods roads can approach public roads at any angle (desirable aesthetically) but should tie-in at a right angle. A little additional expense at the outset will cut down on maintenance later.

Carefully planned and maintained access makes your woodland more useful and, therefore, more profitable, whether you count it in dollars, enjoyment or both!

Culvert should cross road at about a 30-degree angle downgrade.

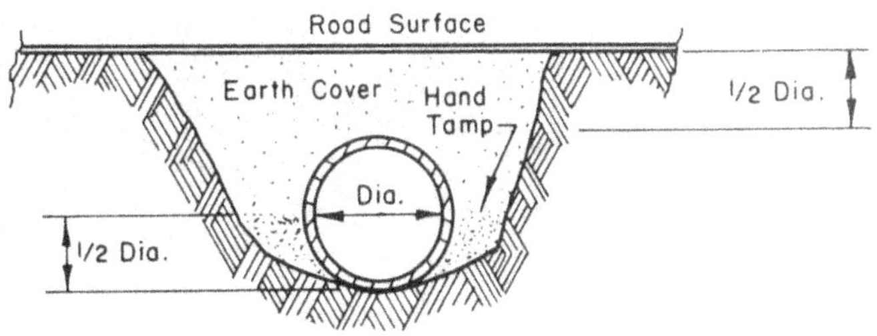

CULVERT INSTALLATION

**From: Woodlands of the Northeast
U.S.D.A. Forest Service**

FOREST ROADS VERSUS WATER

Pipe Culverts For Roads

Pipe culverts—Pipe culverts are usually installed on permanent roads at the time of construction. They are commonly used where vehicle traffic will be relatively heavy following logging activities. Pipe structures are the most expensive type of cross drain but are effective in controlling water.

Culvert should cross road at about a 30-degree angle downgrade.

CULVERT INSTALLATION

Pipe Culverts

A culvert inlet should be placed on the drainage level and as near as possible to the natural channel. In some instances where the culvert level has to be lower than the drainage gradient, a drop box can be constructed. This box is a place for sediment to settle out and needs close maintenance. Pipe culverts are used primarily to channel water across roadways from uphill drainages. They can also be used to divert water collected in the road side ditch across the road. Spacing for this use would be the same as for water breaks.

During construction, seat the culvert on firm ground and compact the earth at least halfway up the side of the pipe to prevent water from leaking around it. Adequate cover is needed, the rule being a minimum of 1-foot or half the culvert diameter, whichever is greater. If adequate cover cannot be achieved, then an arch pipe (squashed pipe) or two smaller culverts should be installed. The cover must also be compacted to prevent settling in the road. If erosion of the inlet end is a problem, a headwall must be provided. Sandbags, with some cement mixed with the sand, durable logs, concrete, or hand-placed riprap are suitable.

It is suggested that 12 inch pipe be the smallest used.

Open Topped Culverts

This type of culvert is used on low cost logging roads, and is usually constructed from lumber, logs, or a combination of the two. When properly installed and maintained, these culverts will adequately drain small sources of water, such as seeps and springs. Log skidding tends to damage and plug them up, making them ineffective for water control on skidroads. They must be cleaned frequently, even on haul roads, to be effective. Open-top culverts should be installed at a 30-degree angle down grade.

The following tabulation presents the recommended spacing for open top wood culverts.

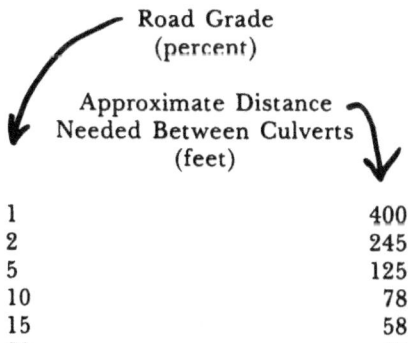

Road Grade (percent)	Approximate Distance Needed Between Culverts (feet)
1	400
2	245
5	125
10	78
15	58
20	47

As with other water diversion methods, the discharge area should be protected to prevent erosion and enable sediment to settle out.

OPEN TOP CULVERTS FOR ROADS

POLE CULVERT

BOX CULVERT

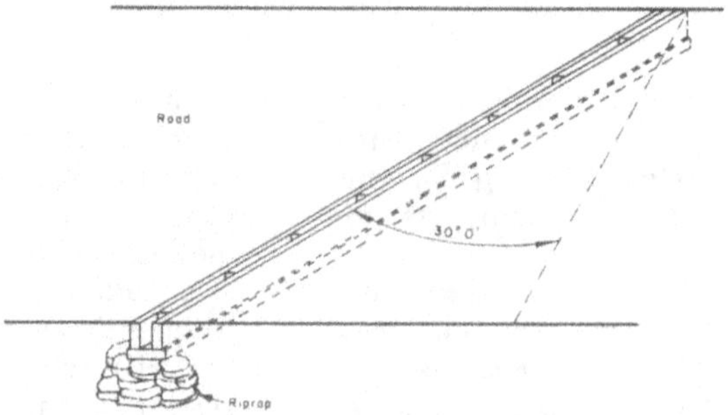

OPEN TOP CULVERT LAYOUT

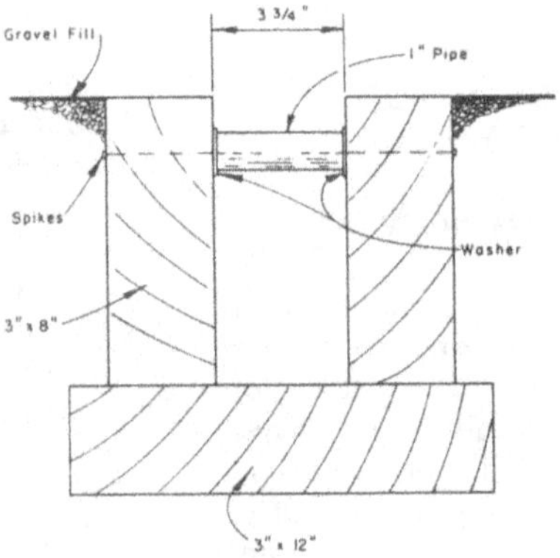

BOX CULVERT DESIGN

Outsloping Of Roads

"Bad" Road sloping

Outsloping is a good way to remove small quantities of water from logging road surfaces. A road that is properly outsloped, slopes slightly from the cutbank to the outside edge of the roadbed. The surface of the road must be smoothed out before outsloping can be effective. Generally this (smoothing and outsloping) is done after logging is finished, and the road surface can then be waterbarred and/or planted to grass.

"Good" Road sloping

A SIMPLE LOGGING ROAD BRIDGE DESIGN

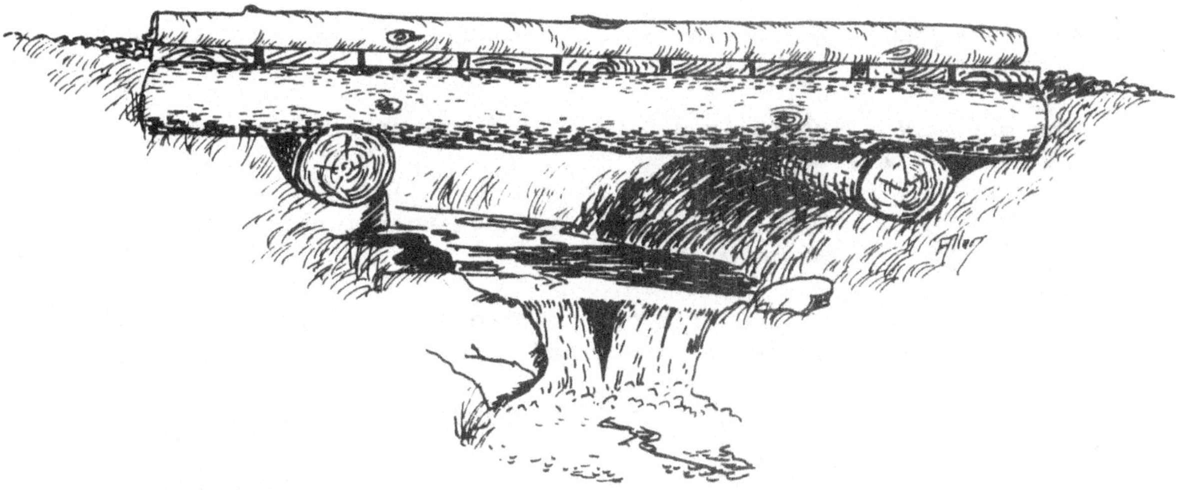

1. Bridge streams crossings if there is a potential sediment or polution source.
2. Use alternate road location to avoid the necessity of a bridge, if possible.
3. Bridge crossing should be located where the stream channel is straight with an unobstructed flow of water.
4. Abutments should be in a direction parallel to the stream flow and imbedded in good foundation material.
5. Locate at right angles to stream flow where approaches are reasonably level for a minimum of 50 feet from both sides.

SUGGESTED METHOD FOR LAYING OUT A CURVE

1. Set a stake at beginning of curvature (end of straightaway)——point A.

2. Decide spacing of stakes in curve (distances from 25 to 100 feet are suitable for logging roads, with the closer spacing applicable to sharp curves).

3. Measure selected distance from A to C, in line with B.

4. At right angles to CA, set stake at D and mark distance CD on measuring stick.

5. Lay off same distance from D to E, and distance AC from A to E. Set temporary stake at E.

6. Double selected distance to F in line with AE and set stake at F.

7. Pull up temporary stake at E and set it at G by using measuring stick and tape in same way as at E.

8. Set stake H in same manner as that by which F was located.

This procedure is repeated around the curve. As long as all measurements are the same each time, the curve will be smooth. If it does not end at the right place for the next straightaway, either increase or decrease length marked on measuring stick and reset the stakes.

By using 25-foot distance A-C, length on measuring stick should not exceed 4 feet if road is to be used by ordinary trucks, or 2 feet if built for truck and trailer traffic. Shortening the distance on stick will increase radius of curve and make driving easier.

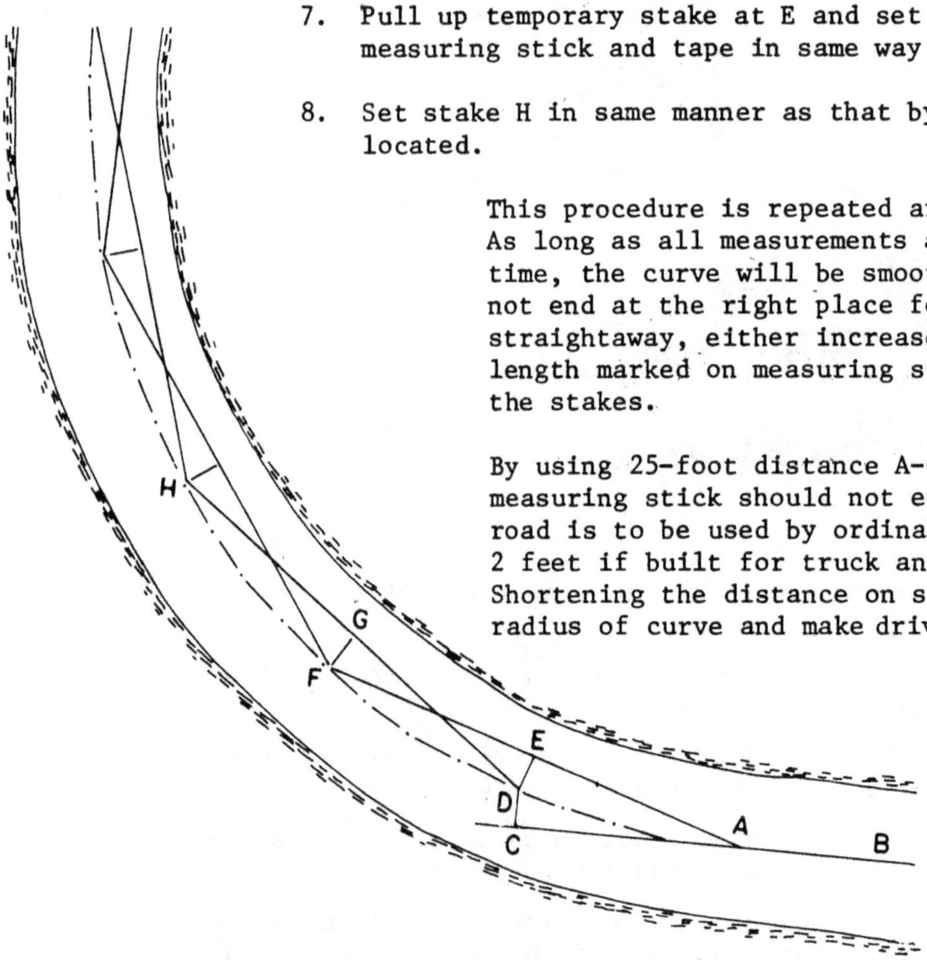

Source: Haussman, R.F., "Permanent Logging Roads for Better Woodlot Management", 1960. USDA, Forest Service, S&PF, Upper Darby, Pa.

CHAPTER 14 — FIRE

Minnesota 4-H Bulletin No. 87 and (or)
Forestry For Minnesota Schools—North Print Co.

Minnesotans depends on 17 million acres of commercial forest land for much of their livelihood and well-being. Fire, insects, disease and animals are the forests' greatest enemies. Since some of these destructive agents are beyond our control, complete protection is not possible. However, we can minimize the loss of growing trees and merchantable timber by applying certain control measures known as forest protection. First, examine how we protect the forest from the most dramatic influence, forest fires.

Forest Fires

Fire is the most spectacular and can be the most damaging enemy of the forest. Anyone who has witnessed a forest fire spreading with explosive violence, roaring through the trees like a thousand locomotives, generating waves of heat and gas that fan the flames to even greater fury, knows that it can be terrifying. It destroys nearly everything in its path. (figure 30).

That was what happened when the Peshtigo Fire in Wisconsin in 1871 wiped out whole settlements, burning 1,280,000 acres of pine, and killing 1,500 people; and when the Hinckley Fire in Minnesota in 1894 burned 160,000 acres killing 418 people. It happened again in 1918 when the Cloquet-Moose Lake Fire burned more than 250,000 acres of timber worth more than $30 million, killed 538 people, and wiped out the town of Cloquet, most of Moose Lake, and more than a dozen smaller villages. In mid-May, 1971, in northern Minnesota, the Little Sioux Fire burned almost 15,000 acres of forest land in 3 days.

Fire tower used for smoke detection.

Surface Fires

Most of the forest fires, especially in the East and South are "surface fires", burning mostly in the duff or leaf litter on the forest floor. These fires, consuming the dry leaves, grass twigs and underbrush on the forest floor, may permanently injure, but not kill, many of the larger trees, but they will kill seedlings and small trees. Fought promptly with adequate manpower and equipment, such fires are fairly easy to control. But nearly every small forest fire has the potential of a big one. Surface fires, with the right combination of dry weather and high winds, may develop into crown fires.

Foresters try to stop forest fires before they become building fires.

Crown Fires

It is usually the "crown fire" or combined surface and crown fires that cause the greatest timber and property damage and loss of human life. Such a fire is usually the result of a surface fire which, driven by a strong wind, leaps into the tree tops and rushes through the timber, often jumping barriers such as open fields or large rivers. Crown fires occur mostly in coniferous forests because the green leaves of hardwoods are not easily ignited. These fires may, however, travel through forests of mixed hardwoods and conifers. Usually they create showers of flying embers or brands which set fires far in advance. Crown fires may kill all the trees over wide areas; they may destroy homes, personal property, and even entire villages and towns.

Ground Fires

Sometimes fires burn below the surface in the thick duff of decayed leaves or needles, or in peat soils that have become dry. Giving off very little smoke between surface outbreaks, such ground fires may smolder for days or weeks before being discovered, and it is difficult to know when they may safely be declared out. Ground fires are common in our northern forest region. These fires usually kill most of the trees in their path, for although they burn slowly, they generate intense heat beneath the surface.

FIRE SUPPRESSION

Factors affecting the behavior of a forest fire.

1. Relative humidity.
2. Winds
3. Forest fire fuels.
4. Topography
5. Temperature
6. Fuel moisture.

Influence of relative humidity—Extremely low humidities will result in high inflamability of forest fuels. The degree of inflamability becomes greater as the low humidity period continues. Fire can start and spread at relatively high humidities. Normally, however, fires spread only slowly when the relative humidity is above 50 per cent. Between 40 and 50 per cent, fires tend to pick up and, with unfavorable conditions, (pertaining to fuels and wind,) may spread rapidly. During the fire seasons, when the relative humidity drops below 40 per cent, the fire danger becomes acute.

Winds—They are of outstanding importance in gauging the rate and direction in which a fire will spread and in determining the inflammability of forest fuels.

Fire fuels—Flash or critical fueisl are those which, under normal conditions, are susceptible to easy ignition and rapid combusion. They include: dead vegetative materials, needles, light litter, small branches, dead leaves, grasses and other light fuels and ferns.

Heavy or slow burning fuels include the remaining dead materials, which by reason of their structure or arrangement, are covered, and will not burn rapidly. In many instances, they do not burn at all until the overlying materials are consumed. They include: humus, heavy limbs, logs and decaying wood.

Green fuel embraces all the growing forest vegetation such as foliage, brush, etc. These materials, while living, have a high moisture content and do not constitute a severe hazard.

Topography—The steepness of the slope is always a big factor in governing the spread of a fire. The steeper the slope, the more rapidly the fire advances. A fire can go up a hill faster than it can go down.

Temperature—Air temperature has a very important effect on the behavior of a forest fire and also upon the moisture content of fire fuels. Fires burn more severely at high temperatures because warm fuels ignite and burn more quickly.

Moisture—The inflammability of all forest materials is governed primarily by their moisture content. This is one of the most important factors in determining the speed with which a fire will burn. It is governed by the amount and circulation of air through the fuels.

What to do before a fire gets started.

Three things a landowner can do to protect his land from fire:

1. Establish firebreaks and access ways.
2. Reduce hazards by disposing of dangerous fuels.
3. Organize fire control crews and locate fire control tools in a handy place.

Persons who start uncontrolled fires are responsible for the damage they cause. The state or the town may recover the costs of suppression.

Firebreaks and Access Roads

Firebreaks are vital to the protection of conniferous plantations. A satisfactory barrier should be made at time of planting. This can be done by plowing a strip about a harrow-section wide. It can be kept open and free of vegetation by subsequent harrowings. One harrowing in the early spring and a second in July or August are usually sufficient to keep the weeds in check and mineral soil exposed. This operation must be repeated yearly to prevent loss of the trees by fire. The plantation should also be established with an "access" route around the entire planting. This route would serve as an aid in future management practices, and would enhance wildlife values. It would also help in the prevention and suppression of destructive forces that are a threat to a plantation. This "access" route should be at least 25 feet in width. Within the plantation, an "access" route of at least 18 feet should be provided at 300-foot intervals.

Fire Prevention

It has been said again and again that there is no honor attached to the detection and suppression of a forest fire if you could have prevented it. You have probably also heard many times, "A fire prevented is a fire that does not have to be suppressed."

Both statements plainly point out that the first step in a forest fire control program is a strong preventive effort. This requires strong public support for the prevention of wildfires.

Much time, energy and money is spent yearly in Minnesota for forest fire prevention. Commonly known practices, such as plowing furrows around forest plantations, are an important part of prevention, but it is only a small part of the total program. Other areas of prevention of importance include:

Debris Burning

Always be sure you are complying with the forest fire laws before burning anything.

The safest time to burn is late in the day after the wind has gone down.

Always have fire fighting tools handy before you start a fire. If possible, plow a fire break around the material to be burned.

Be sure your fire is out.

YOU are responsible if your fire gets out of control. You become liable for all expenses incurred in suppressing the fire, and for all damage caused by it.

Liability for damages includes injury to young tree growth.

Incinerators

Have your incinerator inspected and approved before you use it. It may be necessary to obtain a burning permit. Check with your local forest ranger or other officials responsible for forest fire protection.

Smokers

Be sure your match is out. See that pine ashes and cigar or cigarette stubs are dead before discarding them. Use the ashtray in your car.

Campers

Build campfires only in safe, approved areas, away from flammable forest fuels. Never leave a campfire until it is completely extinguished. Be doubly sure that the last spark is out.

PROTECTING BUILDINGS FROM FOREST FIRES

The following practices are recommended by the Department of Natural Resources to reduce the chance of fires starting near cabins and summer homes in wild land areas. These practices may also protect buildings from wild fires that may occur in the area.

1. Trees should be thinned out in the immediate vicinity of buildings.
2. Existing trees should have branches limbed to prevent their loss in case of fire.
3. Prune tree branches that are very close to buildings.
4. Leaves and flammable materials should be removed from under and around buildings.
5. Grass and small vegetation should be cut or mowed 200 feet around buildings.
6. A screen should be provided for your chimney. Recommendations for screens can be obtained from your local CNR field office.
7. Keep a water supply and shovel handy in case of fire.
8. Purchase or make an incinerator for burning trash. Blueprints for an approved incinerator are available from DNR field offices.
9. Have a ladder available in case of a roof fire.
10. Check if trees are too close to power lines.
11. Pine plantations are safer with a disced firebreak around them.
12. Clear all flammable materials within five feet of outdoor fireplaces. Equip the smoke outlet with a screen and a method of controlling indraft.
13. When constructing access roads to buildings, considerations should be made to allow large fire trucks to drive in easily and turn around.

Loggers

When refueling power saws, do so in a cleared area, free of any flammable material; then move a safe distance away before starting the motor. Flammable material should be moved away from saw cuts. Sawdust should be checked for smoldering embers before moving to the next cut. A fire extinguisher and shovel should be part of your regular equipment when using a power saw in the woods.

HOW FOREST FIRE DANGER IS MEASURED

Most people can tell reasonably well how dry vegetation is by observing some natural signs present. That was once the way the fire fighters determined forest fire danger—but no more.

Now the forester uses such factors as temperature, relative humidity, fuel moisture, precipitation, wind velocity, and condition of ground vegetation—dry, green, or in between these two. Analysis of the interaction of these variables indicates to the fire manager what he has to prepare for in the way of men and equipment to cope with the current fire situation.

Thus he knows what to do about burning permits, detection requirements, and other items of importance relating to his job of preventing and suppressing forest fires.

Information on the current forest fire danger in forest and wild land areas is often given on radio and TV programs. The fire danger rating chart will help you to understand what the situation is where you live.

FIRE DANGER RATING CHART

Explosive

Explosive conditions. Fires start easily, burn fiercely and crown readily. Often very difficult or impossible to control during the day. Burning is not recommended.

High

Dangerous conditions. Fires start readily from a match or sparks, spread quickly, and spot readily. Difficult to control. Burning is not recommended.

Medium

Fires start from a match or burning embers, spread quickly in dry grass or leaves. Burn with extreme caution.

Low

Fires will start from an open flame, spread slowly, and in absence of wind, tend to go out. The safest time to burn.

HOW TO AND WHOM TO REPORT A FOREST FIRE

Fire should be reported to the nearest forester, fire warden or local resident. If you are a stranger in a community, go to the nearest phone, or, if no phone is nearby, notify the nearest farmer, storekeeper or tavernkeeper. The telephone operator will also be able to connect you with the proper authorities.

When reporting a fire, be sure to include this information:

1. Location of fire.
2. Size.
3. What is burning and what is in immediate danger of burning.
4. How hard the fire is burning.
5. How many men, if any, are fighting the fire.

Helicopter with bucket helping douse a fire.

Common fire fighting hand tools and their use

The common tools used in fire fighting are pictured and described as follows:

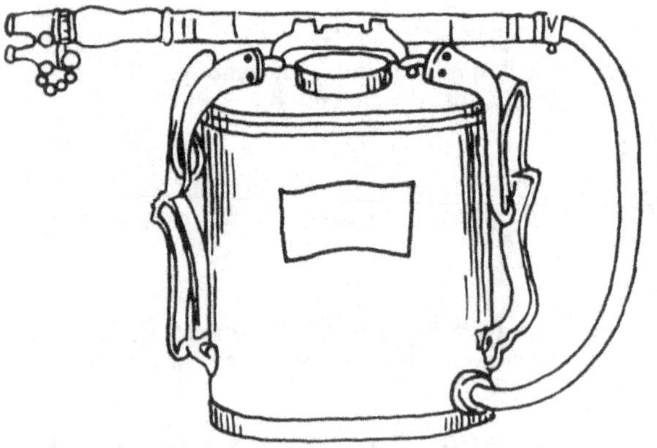

Back-Pack Pump — Carries about five gallons of water used to reduce the fire so that a line can be constructed. It also puts out a fire by placing water in the proper areas.

Shovel — Either short or long handled, are used for fire line construction by shoveling duff and humus to one side until mineral soil is reached in line construction. It is used to follow up plowing or other digging, in order to obtain a clean fire line.

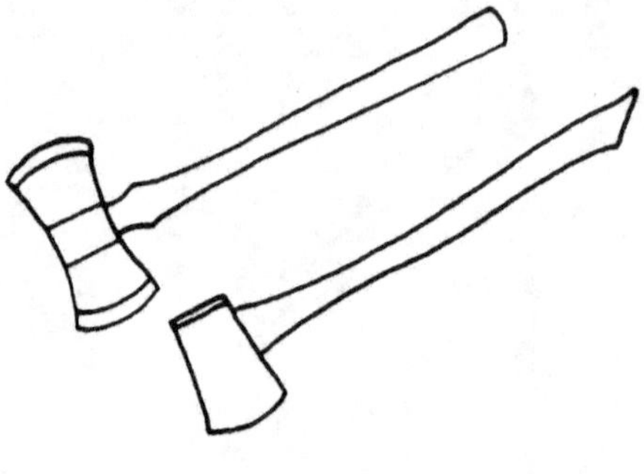

Axe — Used to cut material for fire line clearing, snag felling, etc.

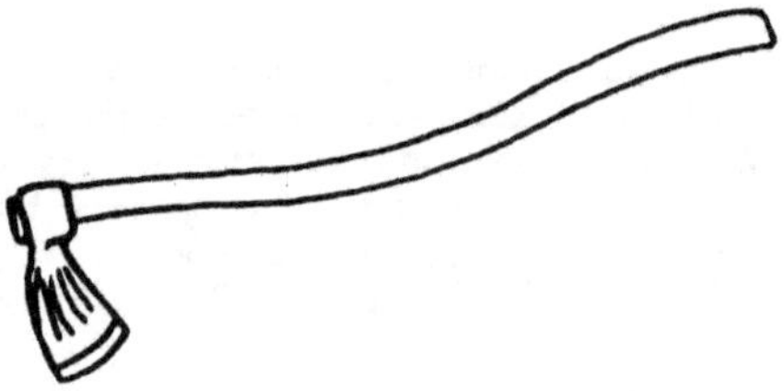

Grub Hoe — Used for digging and grubbing in soil, humus, roots, and decayed logs in constructing fire lines.

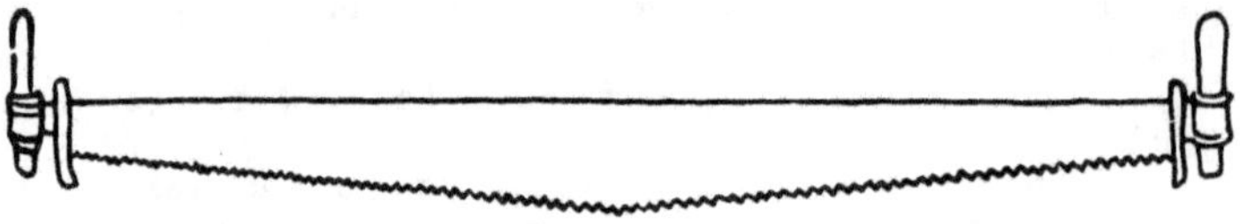

Crosscut Saw or Chain Saw — Used in bucking logs that lie across the fire line. It is also used in felling snags.

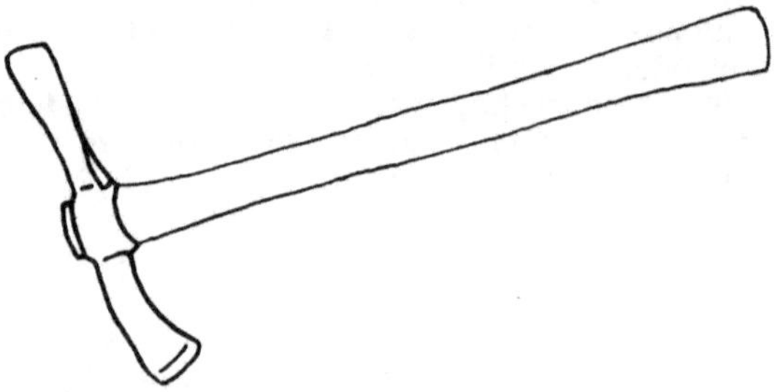

Cutter Mattock — Used much as a combination axe and grub hoe. The sharp edge is used as an axe, while the other edge is used for grubbing in fire line clearing and construction.

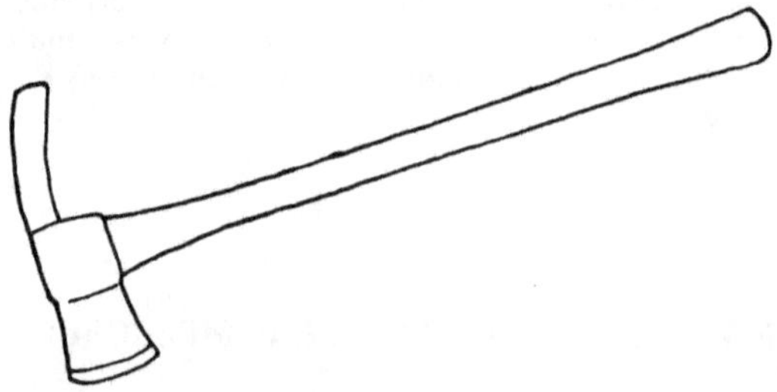

Pulaski Tool — Similar to a cutter mattock except that it is somewhat lighter and sharper and used for lighter work in fire line clearing and construction.

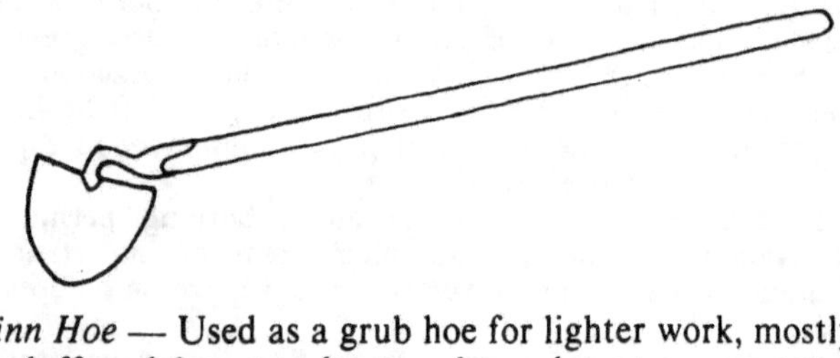

Finn Hoe — Used as a grub hoe for lighter work, mostly in duff and humus where rocks and roots are not too numerous.

Flaming Fury

By Allen Wickman

The Huntersville-Badoura Fire of 1976

Typical Jackpine devastation left in the wake of the Huntersville-Badoura fire.—Photo by Ron Norenberg

On Tuesday, September 7, at 11:28 a.m., the Nimrod fire tower reported a smoke about three miles southeast of Huntersville. The forester in charge immediately started the seven-mile trip to the fire. The station caterpillar tractor and fire plow followed. A 20-man crew stationed nearby was also notified and quickly left for the scene.

Within minutes, the forester knew this could be a bad fire. By radio, he requested helicopter and bomber water drops to slow its spread. An hour later, a bomber made the first drop of fire retardant.

I became aware of the fire at noon. I spotted the column of smoke as I drove to another fire about 20 miles to the northeast. This fire was quickly controlled by a crew from the Badoura Nursery. By 1:30 p.m., I had returned to the Park Rapids area headquarters. For the next two hours, I helped to

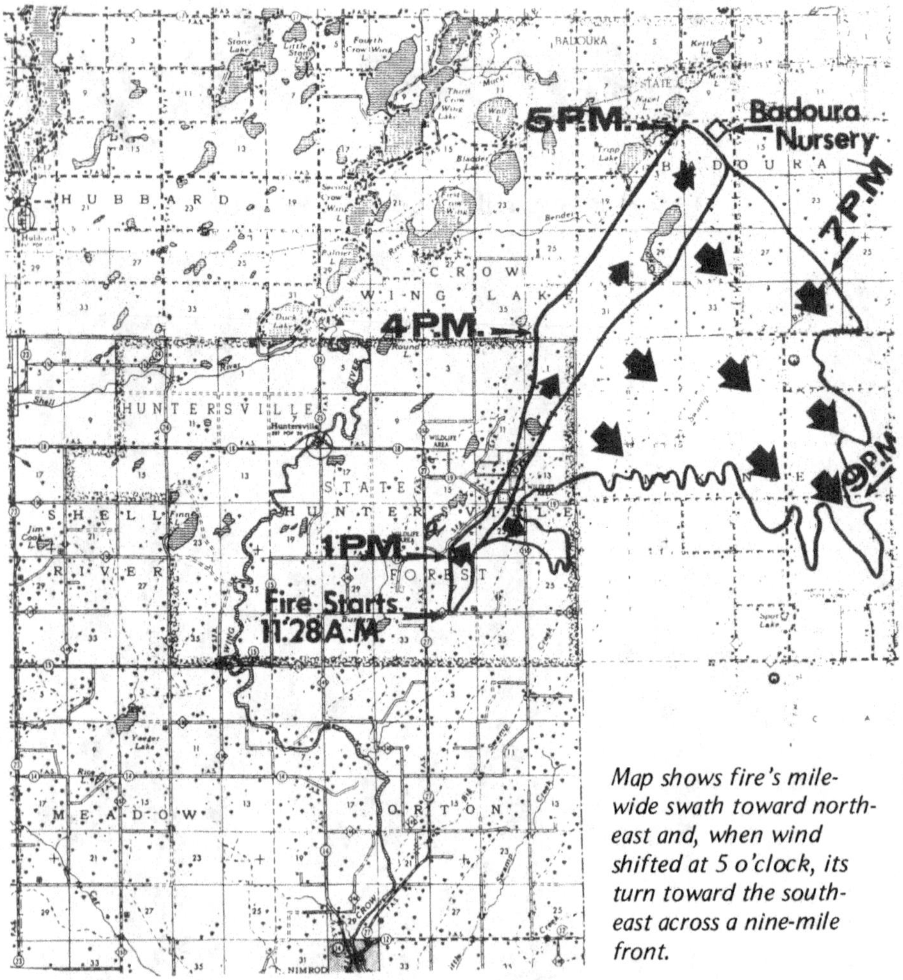

Map shows fire's mile-wide swath toward northeast and, when wind shifted at 5 o'clock, its turn toward the southeast across a nine-mile front.

dispatch men and equipment to the area and answered phone calls from people worried by the spreading Huntersville fire.

Meanwhile, the fire itself had crept into willow brush and was soon marching boldly through a marsh. Next, it hopped into an aspen thicket and began to accelerate.

A farmer was the first man to attack the fire. Using his tractor, he plowed a furrow through a field along the east side of the fire. He then cut in front close enough to scorch his face and singe his hair. But the furrows did not deter the flames. The fire raced over the plowed-up ground as if it wasn't there.

A 20-man crew arrived and began working on the rear and flanks. At its head, no one had a chance of stopping the fire.

Crowns Aflame. By one o'clock, the fire was one mile long and a quarter-mile wide. Leaving the aspen thicket, it began burning faster. It raced through tinder-dry slash, then leaped into the crowns of jack pine clumps.

The Menahga and Sebeka fire trucks arrived, but were helpless. Between 1 and 2 p.m., large cats quickly deployed to construct firebreaks to deprive the fire of fuel. Removing fuel from the path of a fire is usually an effective way to stop it.

By four o'clock the fire had advanced five miles. The breaks constructed by the cats proved to be no more effective than the first break plowed by the farmer. The cat operators tried plowing the breaks wider, but the fire was stronger too. When flames approached a break, they simply flung burning debris across it and leaped.

Sometime after 4 p.m., I arrived on the scene. I would be there for the next four days, but of course I didn't realize it then.

The towering cloud of smoke was visible for miles. It billowed to 4,500 feet. Strong winds kept it from going higher. The top was white; the underside a dirty gray-black. At 10 miles, I saw two huge orange balls glowing at its base.

The smoke reminded me of a wasp nest with planes circling it instead of wasps. Helicopters, World War II bombers (B-26s) and cargo planes (C-119s), and light planes circled and darted in and out of the dense smoke. All seemed as helpless as the men and machines on the ground. The cargo planes with their 2,000 gallons of fire retardant seemed to have about as much chance of putting out the flames as an eye dropper squirting a bonfire.

The planes, men, and equipment concentrated on the flanks of the main blaze and on spot fires. On the flanks, they were in less danger and could slow the fire's spread.

One Hour, Four Miles. Late in the afternoon, the fire jumped a road and moved into dense stands of pine. With this added fuel, it picked up speed. It had taken four hours to go the first five miles. It raced the next four miles in one hour.

As the fire front advanced, the area filled with milling people. Many were just sightseers. They blocked and slowed people working on the fire or trying to flee it. Trucks and cars loaded with personal possessions moved down the road between fire trucks, forestry vehicles, sheriff cars, and trucks hauling heavy equipment.

More than 20 people were caught in the fire. Miraculously, all survived! One cat operator banked up a wall of sand between himself and the advancing fire. A sheet of searing flame passed right over his head. His cat was in a power-line clearing; otherwise he would have fried.

Five other cat operators quickly made a clearing and sweated it out until the fire passed. Another operator

Traffic congestion at Highways 87 and 64 during fire. Sightseers often blocked vehicles such as this National Guard truck carrying men and equipment to fight fire.

Minnesota Army National Guard

bent his throttle as he "opened her up" and narrowly outran the fire.

By 4:30, the fire was one mile wide and 8½ miles long. During the afternoon, the wind slowly shifted from the southwest to the west and finally to the northwest. At five o'clock, the wind's direction had changed the path and intensity of the fire. The fire's head now became its flank and the east flank became the head. Instead of a one-mile front, the fires started galloping toward the southeast across a nine-mile front.

The wind speed and direction changed all afternoon. Its last shift was a fortunate one. When the flames were less than one-half mile from the Badoura Tree Nursery, the wind turned the fire away from millions of tree seedlings. Behind the nursery were hundreds of acres of young jack pine and a large experimental forest planted in the 1930s. All this would have been reduced to charcoal if the wind had held from the southwest for one hour more.

Desperate Escapes. Two foresters saw the flames cross the road ahead but stubbornly decided to proceed anyway. The smoke got so thick they couldn't see the front of their pickup truck. The driver opened his door and looked at the ground. With this "navigational aid," the truck crept through the smoke into cleaner and cooler air.

At the Paul Day farm, 15 people watched the fire at a safe distance to the west. But at five o'clock, when the wind shifted, it came directly toward them. A helicopter flew to the rescue. It hovered just a few feet above them, but the smoke was so thick the pilot couldn't see them—or even the ground.

The people retreated to the center of an oat field and the fire went around them. The field had been cut over and the stubble did not ignite. Their lives were spared.

At times the fire leaped ahead a quarter mile in five minutes or less. One helicopter pilot saw a convection column full of flaming embers swirl ahead of the main wall of flame and shower the ground. The dry, hot ground exploded as if gasoline had been poured on it.

After a helicopter water drop, men on the fire line found small fish flopping on the ground. Evidently the Sikorsky, with its 900-gallon bucket, had scooped the fish from a nearby lake.

One forester let a large, frightened dog into his truck. Smoke came through the window on the dog's side. The animal leaped to the floor onto the accelerator and wouldn't budge. The truck leaped and lurched ahead. The forester had a quick, hectic ride

Above: A World War II bomber drops fire-retardant. First drop was made one hour after fire started. Left: National Guardsmen spray water from portable pump tanks in mop up operation. Below: "Troops" march to fire along forest road behind caterpiller tractor.

Photos Courtesy Minnesota Army National Guard

down the smoking forest road before he got the trembling dog off the accelerator.

The Giant Lulled. By nine that night, the wind slowed, the temperature dropped, and light rain started to fall. This combination lulled the giant fire.

The area looked like a scene from *Gulliver's Travels.* While the giant slumbered, the elves worked hard to tie it down. By morning, it couldn't move. The warmth of the sun tried to rouse it. But try as it might, it couldn't cross the 40-odd mile fire break encircling it. Where it did, its life was very brief. A wave of men, equipment, and aircraft pounced on it and tied it down again.

The fire was contained, but it was far from dead. Inside the line it continued to smolder in the peat, creep through brush, and leap into unburned tree tops.

The smoldering peat posed special problems for the firefighters. It burned underground where it was impossible to detect until it suddenly burst into smoke and flame. Even a slight wind could topple a grove of trees after the fire had eaten the peat which anchored the roots. This peat is certain to burn and smoke until winter.

In spite of the destruction caused by the fire, not one person was killed or seriously burned. One man suffered a chainsaw injury; another had 12 bee stings. Three people were found wandering in a state of shock inside the blaze area. A helicopter rescued them.

While taking off from Fishhook Lake, the Hubbard County sheriff and his pilot crashed. Their plane was mounted with a siren-speaker to warn residents to evacuate. Hot, uneven, gusty winds caused the accident.

Wildlife Loss. The loss of wildlife was significant. Small birds and mammals died by the hundreds. But rabbits seemed to recognize the danger and started to flee long before humans could even smell smoke. During a fire, most animals lose all fear of humans. At one point some distance from the blaze, a rabbit hopped close to me and stood still for more than a minute.

In the Huntersville - Badoura area, the mix of large marshes, scattered pine, and light sandy soils—soil that dries out fast after a rain—have always presented a critical fire situation. I doubt this situation will change. Homes, tree plantations, and natural stands of timber will always be in danger of fire.

In four days, the fire destroyed seven homes. Only the cinderblock foundation and chimney of this home remained.

Allen Wickman

The September fire left a wake of destruction. More than 23,000 acres were inside the fire line and 95 percent of the area burned. About 5,000 acres of pine were destroyed—although some of the larger timber will be salvaged.

Six occupied and several unoccupied summer cabins, hunting shacks, one store, and a few abandoned buildings were destroyed. Many buildings somehow escaped destruction although the fire lapped at their doorsteps and scorched their sides.

Close to $1 million was spent to control this blaze. The 42 cats cost about $150,000, the Sikorskys about $100,000, and the bombers about $25,000.

Newspapers called the Huntersville-Badoura fire the worst since the dust bowl days of the '30s—a distinction we could just as well do without. □

Allen Wickman *is a Private Forest Management (PFM) Forester for the Division of Forestry, MDNR, at Park Rapids.*

Men & Equipment Used on the Fire

Men

[Estimated Partial Listing]
150 Volunteer Firefighters
225 National Guard Troops
89 men—Hot shot crews from Idaho (18), Montana (31), Florida (40)
25 DNR (Department of Natural Resources Personnel)
14 students from University of Minnesota (Cloquet Forestry Session)
40 Prisoners from the Sandstone Penitentiary and the Willow River Detention Center.

Equipment

Quantity	Type	Capacity	Purpose
1	C-119	2,000 gallons	Water & retardent drops
2	B-26	1,200 gallons	Water & retardent drops

Helicopters

Quantity	Type	Capacity	Purpose
2	Sikorsky	900 gallons	Water drops
2	Bell	50 gallons	Water drops
1	Hiller	None	Observation
1	Bell	None	Observation
2	Hueys (Nat. Guard)	12 men	Carry smoke jumpers
1	Jet Ranger (Nat. Guard)	5 men	Carry smoke jumpers to hot spots

Cats

42 cats of various sizes were used on the fire to construct fire breaks of various sizes from 12' to 100' wide.

CHAPTER 15 — WATER

EU-TRO-PHI-CA-TION

A NEW WORD FOR OLD AND AGING LAKES . . . AND MAN'S ACTIVITIES OFTEN BRING ON PREMATURE SENILITY

By John B. Moyle

Every now and then a technical word escapes from the scientific community and becomes part of the everyday language. "Eutrophication" is such a word and like "ecology" and "isotope", it now confronts readers of magazines and the daily paper. It is an important word, for it affects all of us. It refers to conditions that cause lakes to change in character as they grow older.

The causes of such changing conditions in lakes may be natural or may be the result of human activities. The most obvious symptoms of eutrophication are increasing growth of plant life in waters and changes in the chemistry of deeper waters so that they become less suitable for fish.

Natural eutrophication is going on in every lake, for every lake is slowly changing as it grows older. The present concern about eutrophication largely arises because some of our activities can speed up the aging process and shorten a lake's life.

Like a person, each lake has a life history. It is born, most of ours being formed by continental glaciers of about 10,000 years ago; it ages, becoming gradually more productive of plant and animal life of all kinds; and finally a lake may disappear by becoming filled with the remains of the myriads of plants and animals that have lived in it, combined with materials washed in from adjacent lands.

It is the natural fate of all lakes, if given time enough—and this may be thousands and tends of thousands of years—to gradually fill, become a marsh, and eventually dry land. Many Minnesota lakes have already disappeared in this way. If you were to take a core of soil extending from the surface of a bog to the original underlying gravel, clay or sand, you could find in this core of soil the remains of the many kinds of aquatic animals and plants that lived in this former lake during its long life before it became a bog.

The word "eutrophication" and the history of the ideas it represents is an interesting story and one that is becoming increasingly important to Minnesotans who jointly own more than 15,000 lakes. The story begins more than 50 years ago when certain Scandinavian, German and Swiss scientists became curious about what went on in the depths of the lakes of northern Europe. It had long been obvious that lakes differ greatly both in appearance and in the kinds of plants and animals that live in them. For example, certain lakes had trout in their deeper waters while others did not. And no amount of fish stocking could establish trout in the non-trout lakes.

As biologists, chemists, physicists and earth scientists accumulated information on lakes they were able to group most deep northern lakes into two general types. Since there were no names for these, terms were coined for them from Greek word roots. These lake types were called **"oligotrophic"** and **"eutrophic"**; meaning, literally, "poorly nourished" and "well nourished" lakes.

The poorly nourished, or oligotrophic lakes, had clear waters with small crops of water plants and animals in the upper waters. Conditions in the deeper waters were suitable for trout and other cold-water fishes, for throughout the year there was ample oxygen dissolved in the deep water so these fish could live there.

In contrast, in upper waters of well-nourished eutrophic lakes there was an abundance of water plants and animals. In summer, the upper waters often became green with microscopic free-floating algae. Fish were abundant as were water weeds along shores. **But in these eutrophic lakes the depths were dead in summer because the water here contains little or no dissolved oxygen. Why were the deeper waters dead?**

Well, throughout the summer myriads of tiny plants and animals were born, flourished and died in the upper waters. When they died their remains sank and settled in the lower waters. Here they decayed, and the decay used up oxygen leaving none for the fish. With no oxygen, no fish could live in the deeper water.

These early lake scientists, or "limnologists", also found that the water in deep lakes is not uniform in temperature from top to bottom. Rather, in summer the water of a lake separates into three layers of different temperatures and there is little exchange of water between them. Only in spring and fall does all the water of the lake mix. Any skin diver can tell you that there is a warm upper layer extending downward to usually 15 to 25 feet. Below this there is a narrow zone or layer in which the temperature declines rapidly. The technical name for this transition has also become almost a common word. This word is **"thermocline".**

Below the **thermocline**, in the depths of the lake, the water is cold—as much as 25 degrees colder than the surface water. Currents may circulate the water in each of the three layers of this aquatic sandwich but there is little exchange of water between them. This is why, even though there may be ample oxygen supplied by wave action and manufactured by plants in the upper layer, the **lower layer in a well-fed eutrophic lake is usually deficient in this life-giving gas.**

Throughout the summer, no oxygen is added to the deeper waters and much is used up. In contrast, in the poorly-fed oligotrohpic lakes lthere is but light fallout of organic material from the upper waters that will use up oxygen and the oxygen which has been stored in the deeper water in spring never becomes too low for fish.

We have been surveying lakes in Minnesota for more than 30 years. One of the jobs a lake survey crew does is to make a determination of the basic characteristics of each lake. If you were to travel with a lake survey crew, you would find that most of the deep lakes of the poorly-nourished oligotrophic type, and therefore trout waters, are found in the rocky northeastern border country. Here the land is forested and the soils are thin or even lacking.

Northeastern lakes, in consequence, are poorly fed by their watersheds and the lake water is low in dissolved minerals, many of which are plant nutrients. **The water is clear but may be stained brown from drainage from forest soils.** There are often trout in the depths as well as other cold-water fishes, but these fish are comparatively few. **Oxygen is abundant but food is comparatively scarce.**

Elsewhere in Minnesota, where the lakes lie in fertile glacial soils, most of the deep lakes are of the well-fed eutrophic type. They have hard waters because they receive an ample supply of minerals, such as lime, from waters draining glacial soils. Many become quite green in the Dog Days of summer and have weedy margins. Usually there are no trout or other fishes in the depths. **But the warmer upper waters have fish in abundance**, especially along weedy shores. Such lakes contain bass, panfish, northern pike and walleyes.

These southern lakes have aged more rapidly than those in the northeast and are now at a productive middle-age stage of life. The main deficiencies that they have are that the deeper waters are not available to fish during the summer months. They get green in summer and grow crops of submerged water plants along shores. Like the expanding waistline in human middle age, these are natural and common conditions. Many lakes are like this.

Thus far, we have been considering deep lakes that stratify into temperature layers in summer. But there are many lakes, especially the larger shallower lakes, in which all the water in them circulates throughout the summer because of wind and wave action. Here the water becomes fairly warm from top to bottom and the water of the entire lake is well oxygenated. **All of the bottom is available for the feeding of fish even though the water is fairly fertile.**

Our large walleye lakes, such as Mille Lacs and Winnebigoshish, are like this. They are eutrophic in that they are well fed but constant circulation prevents lack of oxygen in the lower waters. They may become somewhat green in summer but because of heavy wave action are not as weedy along shores as neighboring smaller lakes. Then there are also the shallow, non-stratified lakes of the prairie farmland. If very shallow, they may be completely filled with the water weeds that are the food of wild ducks and geese. **If deeper than 12 feet** they support large crops of fish, especially crappies, northern pike, bullheads, carp and other rough fishes.

Such lakes often become quite green in summer. Some have "nuisance" growths of algae—growths or "blooms" so heavy that the shoreline property owners sometimes control them by applying copper sulfate. Such lakes may also "winterkill" in years of heavy snow. At such times, the oxygen stored in the water may be used up by the abundant life in the fertile water and the fish suffocate.

Lakes, therfore, differ greatly. They are oligotrophic, eutrophic and some are in between. Some are better fed and more productive than others. Some are layered or stratified in summer, others are not. All have aquatic life growing in them but some have more than others.

Our lakes, in the thousands of years they have been in existence, have continually received dissolved minerals and plant nutrients from their watersheds. Yet, most are not excessively fertile. **Why is this?** And what has happened to the dissolved minerals and other material they have received? Part of this material leaves the lake in water flowing from it, but most does not. **Much of the material a lake receives from its watershed becomes part of the bottom soil.** Here, in the lake bottom, the minerals, including plant nutrients, are continually being stored and "locked up" indefinitely.

This is part of the slow filling process—part of aging. If you continue to add sugar to a cup of coffee, some settles out in the bottom of the cup. It is somewhat the same with lakes. Lakes are, among other things, a complex biological and chemical system whereby the materials they receive from the land are removed from the water and stored in the bottom soils. It is an efficient system but one on which depends the growing crops of water weeds, algae, and aquatic animals.

These "crops" both concentrate minerals in their remains and by their growth change conditions in the water so some plant nutrients (such as phosphates and lime) are precipitated chemically. We cannot object to the results of this process for it really works, but we may object to the water plants and algae that cause it to work.

But what about the effects of human activities on the eutrophication of waters? Aquatic plant crops tend to increase if a lake is articially fertilized by the activities of man—for the plant crops are then being better fed. Such added fertility can be nutrient salts, such as phosphates and nitrates, from tilled lands or from manure produced by livestock. It can come from sewage or purified effluents from sewage treatment plants, **for the plant nutrients in sewage are, for the most part, not removed by usual sewage treatment.**

Runoff from fertilized lawsn and gardens can also fertilize lakes as can drainage of seepage from cesspools. Wash water containing detergents adds phosphates.

All of these, when added to a lake, provide more food than it would normally have. This causes more waterweeds to grow and the lake to be greener in summer. If the lake was originally of the poorly fed oligotrophic type, oxygen may disappear from the deeper waters and then no longer be suitable for fish. This is what wastes and nutrients from other sources have done to Lake Erie and several large lakes in Europe.

In these lakes, fertilization or artificial enrichment with plant nutrients has been the means whereby eutrophication and the aging of these lakes has been speeded up. **The effects of added fertility is apt to be most noticeable in shallow lakes,** because as has already been pointed out, the water in these circulates throughout the summer and in them there is little temporary summer storage of plant nutrients such as occurs in the stagnant deeper colder waters of lakes that stratify in summer.

To get back to our cup-of-coffee analogy, there will be more sugar in the coffee and less on the bottom of the cup if it is continually stirred. Stirring of the waters of shallow lakes is mostly by wind and waves. But the bottom may also be stirred in other ways and stored nutrients in the soil again made available. Rough fish, such as carp and bullheads, can be a factor. Large outboard motors may also stir up bottom soil in shallow water.

Continued addition of nutrients to shallow lakes throughout the summer is especially apt to produce excessive growths of aquatic plants and algae. The plants are continually fed and nutrients are being added faster than they can be removed and deposited in the bottom soil. **But it should also be remembered that fertilized lakes produce large crops of sportfish, for the plants ultimately feed the fish.**

What can we do about eutrophication of our lakes? First, we should recognize that, to a great extent, this is a natural process that has been going on for a long time. We should not expect our lakes to be crystal clear and free from water plants. The tiny algae that cause lakes to be green in summer are among the oldest plants in the world and there have been waters suitable for these to grow for billions of years. The waterweeds have also provided food and shelter for waterfowl and fish for a long time.

Both the algae and waterweeds which result from eutrophication can be controlled, for a price, if present in excessive amounts. But such control is usually temporary. **And remember that all animal life, including production of game fishes, depends on these plants. We should not expect tap water quality in a lake. Remember the tap water has gone through treatment and filtration.**

We should also distinguish between natural conditions in a lake and those caused by man. The long-term natural changes can be expected to continue. **But we can slow the eutrophication process of waters by holding man-made additions of fertility to a minimum.** Proper soil conservation on watersheds is essential. In more northern areas, good forest management is of great value.

The escape of plant nutrients from watersheds into lakes might be considered a matter of "leakage" from the soils of the watershed. Adequate plant cover, forests or crops, on the tributary watershed will do much to reduce such leakage and reduce the amount of nutrients entering the lake.

Most important of all is keeping from the lake plant nutrients in sewage, sewage effluents, cess pools and wastes of domestic animals. These may be a source of continuous feeding of algae and water weeds. In some places, like Lake Tahoe in California and Lake Washington in Seattle, sewage effluents have been diverted from the watersheds of the lakes. Treatment of sewage effluents to remove plant nutrients is now the subject of much investigation, and is being tried in pilot plant operations.

The Moods of Minnetonka . . . Onset of winter [left] and, above, the tranquil calm of a mid-summer day.

Answers are coming. Control of wastes where domestic stock and poultry are concentrated is also being worked on.

The rate of eutrophication and the speed of aging of our lakes depends on what we are able to do about minimizing the amount of plant nutrients entering them. In proper amounts, these nutrients are not harmful for they are the same substances we purchase in manufactured fertilizers. But a lake should not receive excessive amounts of them because of human activities. Eutrophication is a big and very complicated problem. Eutrophication will be with us for a long time. It is a word to remember.

MINNESOTA FISHING WATER

How much fishing water have we? This is not an easy question to answer, for even the total water area of the state is only approximately known. The usual figure for the Minnesota water area, 2,370,000 acres, is that of the inland meandered lakes.

Then there are the smaller lakes and ponds whose shorelines were not mapped in the original government surveys, and the rivers and streams. In addition, the Minnesota portion of the border waters should be included; such waters as Big Stone, Traverse, Pepin, Lake of the Woods, and the long string of lakes along the Canadian border. Then there is the Minnesota part of Lake Superior.

All in all, it appears that the area of the inland and border waters, not including Lake Superior, is about 3,149,000 acres or **4,900 square miles**. To this add 1,167,000 acres of Lake Superior for a total water area of 4,316,000 acres or 6,740 square miles.

Of the inland and border water, exclusive of Lake Superior, it is estimated on the basis of 15 years of lake surveys, that about 65 per cent or 2,029,000 acres in fish and fishing waters. This is 3,170 square miles, or an area about the size of Koochiching County, Minnesota, or an area as large as the states of Rhode Island and Delaware combined.

WATER TALK

Hydrologists are people who study water as it occurs on the earth's surface and underground. They use words to describe the phenomena they observe that are assigned very specific meanings. These words, used outside the science of hydrology, may have more general meanings. Learn to recognize these terms and to distinguish when they are used in a technical sense.

Surface water is on the surface of the earth. **Ground water** is in the ground beneath the water table. The **water table** is the level below which the ground is saturated with water; this water-saturated zone is called the **zone of saturation**. Above the water table is a zone where some spaces or fractures in the rocks contain air; this zone is the **zone of aeration.**

An **aquifer** is a layer of rock or a body of sand that can store and transmit water in sufficient quantities to supply a well. Pumping from a well removes water from the rocks next to it and causes water to flow toward the well. The result is a depression in the water table called a **cone of depression** because of its shape. **Artesian wells** are wells in which the water level rises above the upper surface of the aquifer penetrated. The water is under pressure. Artesian wells may flow at the surface, though not often.

Surface water is generally thought of in terms of lakes, streams and "wetlands"—such as marshes or swamps. The land area from which water flows toward a lake or stream is called its **watershed**. At some distance from a lake or stream a point is reached where rain falling on the land will divide, part of it flowing in another direction toward another lake or stream. These **divides** separate one watershed from another.

Streams and lakes supplied with too much water may **flood** or overflow their banks. The banks of a stream are usually well defined while a lake may have very gentle slopes. A stream occupies a **channel** between its banks which contains the water most of the time. Bordering a stream on either side is a flat lowland subject to flooding called the **flood plain**. **Stage** is a term used to describe the height of water at which the stream begins to spill over its banks. **Flood stage** is the height of water at which damage begins to occur.

The terms selected are often heard in the news. They are used by hydrologists because they have accepted, precise meanings understood by all hydrologists. When used as technical terms they save time and save the use of many adjectives, and contribute to unity, coherence, and emphasis in scientific thought and writing.

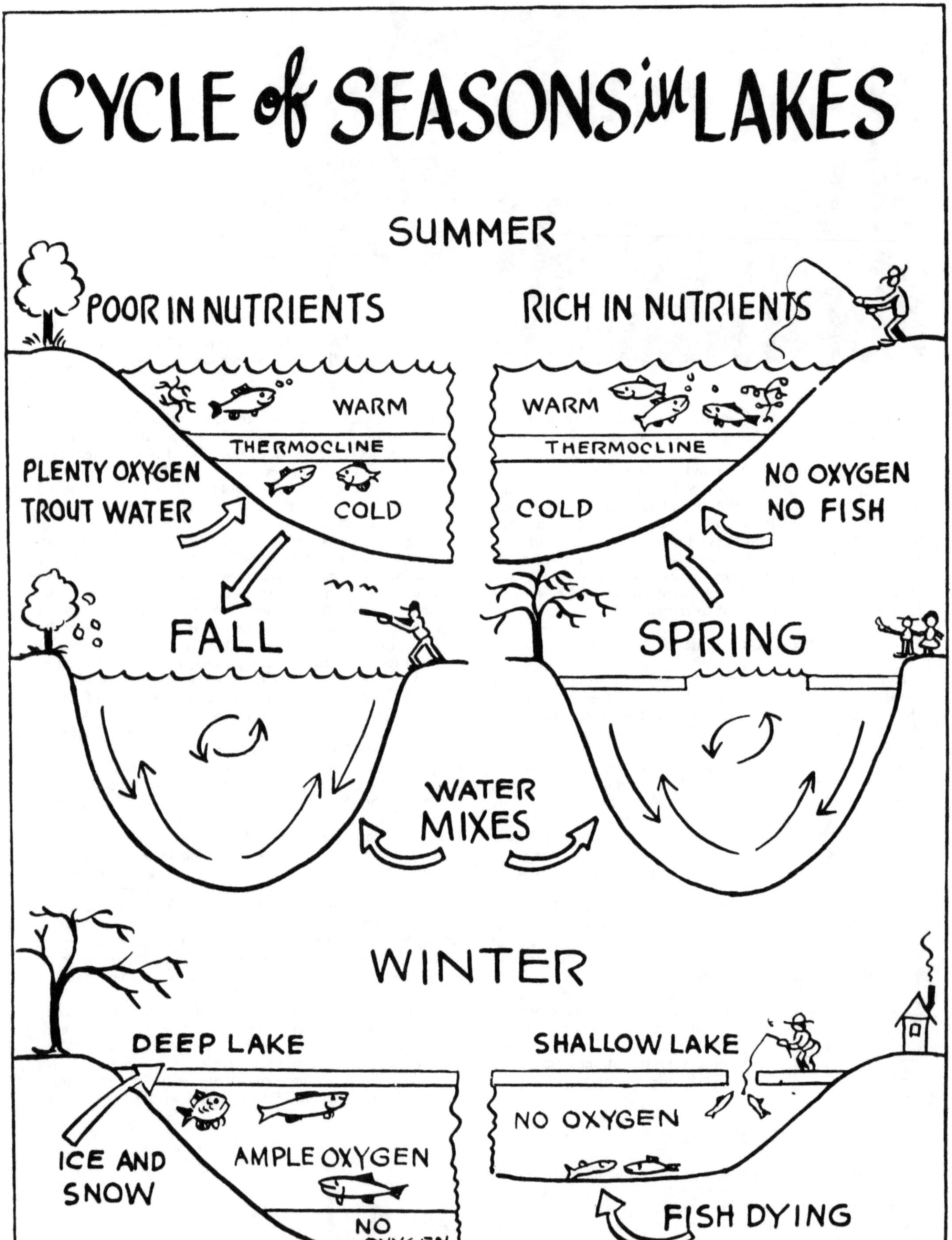
CYCLE of SEASONS in LAKES
SUMMER
POOR IN NUTRIENTS
RICH IN NUTRIENTS
WARM
WARM
THERMOCLINE
THERMOCLINE
PLENTY OXYGEN
TROUT WATER
COLD
COLD
NO OXYGEN
NO FISH
FALL
SPRING
WATER
MIXES
WINTER
DEEP LAKE
SHALLOW LAKE
NO OXYGEN
AMPLE OXYGEN
ICE AND
SNOW
NO
OXYGEN
FISH DYING
The Minnesota Volunteer

FNR-69
April, 1977

forestry & natural resources

WILDLIFE AND FISHERIES

Fish Kills in Indiana—Their Causes and Prevention

by Herbert C. Krauch, Extension Specialist in Wildlife

For anyone who has a stocked pond—and there are thousands in Indiana—fish kills can be quite a problem. The condition called "fish kill" occurs when a number of fish in a given body of water die from a specific cause. Among these causes are chemical pollution, water quality (pH, hardness, oxygen supply, etc.), old age, physical disturbance, oxygen depletion, parasites and/or disease.

Often, fish kills can be anticipated and measures taken to prevent them.

The most common cause of fish kills in Indiana ponds is oxygen depletion. Oxygen depletion results when conditions create a greater demand for oxygen than the aquatic environment can produce. Oxygen depletion can occur in both winter and summer.

Winter kill

The conditions which produce oxygen depletion in winter are quite different from those which result in summer kills. The results of a winter kill are seldom noticed until spring when the ice melts. Then the dead fish, often the larger ones, are seen washing up along the edge. The large fish die first, because they require more oxygen, and will suffocate first.

During winter, most ponds are covered by ice. The oxygen supply under the ice depends on the passage of light through the ice and the resulting production of oxygen by the tiny algae in the water. If the ice is snow-covered, sunlight cannot penetrate and the plants cannot produce oxygen. Just an inch of snow can shut out as much as 90 per cent of the light; and five inches will shut out about 99 per cent.

The existing oxygen is then gradually used up by the respiration of the fish and by natural decay processes. If the snow remains on the ice long enough, the fish will suffocate.

Generally, not all the fish are killed. Some are more resistant to low oxygen levels than others. In addition, poor conditions may not exist uniformly throughout the pond. So, the result may be that the larger fish, such as bass, die, leaving the smaller, immature fish in the pond. This creates conditions further unbalancing the pond population very quickly.

Preventing winter kill

Pond design is important to preventing winter kill. In Indiana, ponds should be at least eight feet deep over at least 25 per cent of the pond area. Greater depths are even better.

If, however, an eight foot depth isn't possible, a six foot depth over at least half the pond is an alternative. Deepening the pond and/or removing build-ups of dead and decaying organic matter from the pond will help preserve the oxygen.

Removing snow cover from at least 50 per cent of the pond surface may help. But, unless the ice is thick enough to hold a person or small tractor, this can be very dangerous. Be sure the ice is sufficiently thick! Chopping holes in the ice won't help.

Summer kill

The most common cause of summer fish die-offs is the same as winter kill—oxygen depletion. But the causes of depletion may be more complex than with winter kills. Summer depletion can result from oxygen removal by excessive quantities of plants, animals or decaying organic matter.

This can be caused by overstocking, overfeeding, over fertilization, pollution from barns, feedlots, improper septic drainage, or chemical treatment of aquatic weeds during the critical months—June through September.

Another common cause of summer fish kills occurs when a dense growth of submerged aquatic plants or algae in a pond dies suddenly from natural causes or from herbicides.

The decay process from the dead plants may use up the oxygen in the water. This type of summer fish kill almost always happens about sunrise, when the dissolved oxygen is at its low point for the day.

PURDUE UNIVERSITY • COOPERATIVE EXTENSION SERVICE • WEST LAFAYETTE, INDIANA 47907

Another condition leading to summer fish kill, and related to the plant die-off, is high water temperatures. During July, and especially in August, water temperatures in ponds may reach 85-95°F. Water can hold much less oxygen when its temperature is above 80°F.

When overcast skies persist for several days, while temperatures are high and winds are calm, a fish kill may occur. Plants cannot produce sufficient oxygen at reduced light levels and calm weather reduces the exposure of water to oxygen in the air. Therefore, the dissolved oxygen may disappear entirely.

Signs of oxygen depletion

1. Large numbers of fish are seen at the water surface gulping air at night or early in the morning. When disturbed they dive, but quickly return to the surface.
2. If oxygen depletion is not severe, fish are at the surface in the early morning but go to deeper water as oxygen builds up during the day. This may continue for several days. If the owner is observant it will give him time to take corrective action.
3. Although feeding is not recommended in Indiana, if fish are being fed they will suddenly stop eating.

Prevention

Conditions leading to fish kills from aquatic plant die-offs and temperatures can be alleviated by controlling rooted aquatic vegetation and algae. When plants are dense, chemically treat only a portion of the pond at one time and allow that part of the vegetation to decay before further treatment.

In new pond construction, a bottom-water overflow outlet will reduce the chances of a die-off. Briefly, this type of outlet releases water from the pond bottom. Contact the local Soil Conservation Service office for details on this construction.

Emergency treatment is suggested if signs of oxygen depletion are observed. Steps should be taken immediately to prevent losses.

Physical treatments

1. Flush the pond with fresh aerated water from a well or adjacent pond.
2. Spray water from a 2 to 3 foot depth into the air with a pump.
3. In small ponds, add oxygen by stirring the surface water vigorously with an anchored outboard motor.

Chemical treatment

There is a chemical treatment to alleviate oxygen depletion temporarily, but this should not be attempted without obtaining advice from a fisheries biologist.

Emergency treatments will help alleviate oxygen depletion. However, the key to preventing oxygen depletion is first, proper pond construction, and secondly, proper pond management. After the emergency has passed, find and eliminate the cause of oxygen shortage.

Other factors causing fish kills

Fish kills can also be caused by organic pollution, chemical run-off and disease. Organic pollution from barnyards, feedlots, and faulty septic system drainage consumes oxygen as it decays and quickly depletes oxygen content in the pond. Fish kills from these sources of pollution often occur after a rain has washed quantities of these materials into the pond.

Steps should be taken to prevent all organic waste from entering ponds. Preventive measures include properly installed tiling, grading, and/or diversion ditches.

Pesticides used on farm crops or in home yards and gardens may wash into a pond during heavy rains and cause a fish kill. Fish may die from the direct effect of the chemical.

Caution should always be exercised in the selection of pesticides and in the time of application.

In the spring or summer a few fish may be found dead along the shoreline. Such mortality is often the result of natural causes. The natural resistance of fish to disease is lower in the early spring than at any time of the year. Larger fish seem to be more susceptible, but, it may also be a simple case of old age. Heavy parasite infestation may also be a cause of death.

There are other less common causes of fish kills, but those dealt with here will account for the great majority.

Restoring the balance after a fish kill

Generally, to correct the population imbalance resulting from a fish die-off, regardless of cause, it is best to renovate the pond. That is, kill the remaining fish, either with a chemical or by draining, and restock.

If there is a question concerning the severity of the kill, contact the district fish management biologist, Indiana Department of Natural Resources, for advice on checking the pond's fish population.

Cooperative Extension Work in Agriculture and Home Economics, State of Indiana, Purdue University and U.S. Department of Agriculture Cooperating. H. G. Diesslin, Director, West Lafayette, Ind. Issued in furtherance of the Acts of May 8 and June 30, 1914. It is the policy of the Cooperative Extension Service of Purdue University that all persons shall have equal opportunity and access to its programs and facilities without regard to race, religion, color, sex or national origin.

Chapter 16

URBAN FORESTRY
Make your town a Tree City USA

TREE CITY USA

A COMMUNITY IMPROVEMENT PROJECT SPONSORED BY THE NATIONAL ARBOR DAY FOUNDATION IN COOPERATION WITH THE U.S. FOREST SERVICE, THE NATIONAL ASSOCIATION OF STATE FORESTERS, THE U.S. CONFERENCE OF MAYORS, AND THE NATIONAL LEAGUE OF CITIES.

Winning communities receive a Tree City USA flag, a walnut-mounted plaque, and signs to place at city entrances.

The National Arbor Day Foundation, in cooperation with the U.S. Forest Service, the National Association of State Foresters, the U.S. Conference of Mayors, and the National League of Cities, is prepared to recognize towns and cities all over America who meet the standards of the TREE CITY USA program.

When you think about it, every community in America is a mini forest. All you have to do is fly from one place to another and you can see that. In many instances, however, on closer observation, you discover that the "forest" isn't being managed as well as it could be. Trees are planted haphazardly. Dead trees aren't being removed. New trees are not being planted or cared for. Generally, about half of all the trees are on public property . . . along streets, in parks and around public buildings. It is important, therefore, that an ongoing community forestry program is initiated.

TREE CITY USA has been designed to recognize those communities that are effectively managing their tree resources. Just as important, it is geared to encourage the implementation of a local tree management program based on the TREE CITY USA Standards through the professional leadership of participating state foresters of the National Association of State Foresters.

Public Law 92-288 of 1972 gives the state foresters authority and responsibility for providing technical services for the "protection, improvement and establishment of trees and shrubs in urban areas, communities and open spaces."

Many states have developed excellent urban and community forestry assistance programs. Contacts may be made with local or state forestry department offices.

A quick check with the city forester of the Mayor's office will tell you whether your community is meeting the standards and therefore is eligible for TREE CITY USA recognition. It could be that you need only have a formal Arbor Day observance or the current local forestry program has to be a little better defined to become eligible. On the other hand, your community may have to make a concerted effort to get an effective management program underway and TREE CITY standards and guidelines will help provide direction.

A word about the standards. They are designed for a TREE CITY USA award to be made to the community that has a workable program in urban/community forestry. The award is not simply for pretty trees, but also for the program that makes them pretty. The standards are also designed to be as objective as possible. Standards 1 and 2 provide for an urban/community forestry program structure in a town or city. Standard 3 requires the program to have demonstrated success based on the judgement of the state forester's office. Standard 4, the actual Arbor Day observance helps create a new awareness and appreciation of trees among all the residents of the community.

State forestry personnel will evaluate applications from individual cities and forward such evaluations to the National TREE CITY USA committee no later than December 31 of each calendar year. Communities that will receive the award will be notified before February 15 by either the Arbor Day Foundation or the state forester.

It is suggested that the actual TREE CITY USA Award be presented during your annual Arbor Day celebration.

Because TREE CITY USA is an ongoing community improvement program, your community can and should re-apply for TREE CITY USA designation each year.

Once a community has received recognition and has met the standards, it can contact the state forester at the end of each calendar year and apply for continued recognition. Communities are encouraged to continue to develop their urban forestry programs beyond the standards set forth.

The Four Standards . . .

A Legally Constituted Municipal Tree Body
The first step in a municipal forestry program is the formation of a local organization. This can be department, board, commission, or other authority. In towns of perhaps 10,000 population and over, city forestry departments with salaried employees are often feasible. These departments may or may not be supported by advisory boards or administrative commissions. In smaller towns, city tree boards are recommended. They should have legal status and be charged with the responsibility for the development and administration of a comprehensive city forestry program. In many small towns, tree boards not only plan to program, but physically carry it out. They thus function as both a board and department. It is recognized that successful tree projects are often completed by beautification committees, civic and service clubs, etc. However, these efforts generally lack the continuity from year to year to provide for a town's total tree program needs.

Adoption of City Tree Ordinance
A city tree ordinance designates the department or board responsible for managing the city's trees. Such a legally constituted body will assure continuity in implementing a local forestry program. The ordinance should specify the number and qualifications of members, their terms of office, and duties and responsibilities. The ordinance should include sections on the operation of the Board, list tree species to be planted, note spacing and planting location requirements, consider planting as it effects utilities and determine public tree care policies.

A Comprehensive Community Forestry Program
Prior to the adoption of a program, an inventory is often taken of all trees growing on public property and their condition noted (healthy, needs pruning, should be removed, etc.). After the inventory data is compiled, a written report should be prepared for presentation and approval by the City Council. The report should be an objective analysis of the town's present tree situation with recommendations for future needs.

There are three general activities in a municipal forestry program: planting, maintenance and tree removal. Priorities must be determined between and within each activity. For example, a community hit hard with Dutch elm disease must at the outset give highest priority to dead and diseased tree removal.

Ideally, a City Tree Board should serve in a planning and advisory capacity with a qualified city employee to physically implement work plans. In small towns, the Tree Board might have to actually administer and implement the program. Board members would purchase and distribute trees, arrange publicity, mark trees for removal, stake planting sites, etc.

The question of financing a Municipal Forestry Program must be met head on and dealt with realistically. The Arbor Day Foundation has set a minimum $1.00 per capita for a community to be eligible for TREE CITY USA recognition, believing that this is a minimum amount to begin to underwrite an effective local program. With the exception of federal funds for some parks, the cost of tree projects must be borne by the residents of a community - either through taxes or by group or individual efforts. Thus, in developing programs, the Tree Board is urged to look at three areas: (1) what can the individual property owner do; (2) what can be accomplished by community action projects by civic clubs, youth organizations, etc.; and (3) what funds are required to implement the program from the city budget.

Arbor Day Observance
To create an appreciation of trees throughout the community, it is appropriate that an Arbor Day proclamation is made annually and a commemorative tree planting take place. This can be done on the date designated for the observance of Arbor Day in you state or on an alternate date convenient to your local tree planting season.

The sample ordinance was designed for use in midwestern communities of average population. The ordinance that your community ultimately develops should be designed to fit its specific needs.

SAMPLE

CITY

TREE ORDINANCE

Be it ordained by the City Commission of the City of____________________________,
State______________________________.

Section 1. Definitions

Street trees: "Street trees" are herein defined as trees, shrubs, bushes, and all other woody vegetation on land lying between property lines on either side of all streets, avenues, or ways within the City.

Park Trees: "Park trees" are herein defined as trees, shrubs, bushes and all other woody vegetation in public parks having individual names, and all areas owned by the City, or to which the public has free access as a park.

Section 2. Creation and Establishment of a City Tree Board.

There is hereby created and established a City Tree Board for the City of____, *(state)* which shall consist of five members, citizens and residents of this city, who shall be appointed by the mayor with the approval of the Commission.

Section 3. Term of Office

The term of the five persons to be appointed by the mayor shall be three years except that the term of two of the members appointed to the first board shall be for only one year and the term of two members of the first board shall be for two years. In the event that a vacancy shall occur during the term of any member, his successor shall be appointed for the unexpired portion of the term.

Section 4. Compensation

Members of the Board shall serve without compensation.

Section 5. Duties and Responsibilities

It shall be the responsibility of the Board to study, investigate, council and develop and/or update annually, and administer a written plan for the care, preservation, pruning, planting, replanting, removal or disposition of trees and shrubs in parks, along streets and in other public areas. Such plan will be presented annually to the City Commission and upon their acceptance and approval shall constitute the official comprehensive city tree plan for the City of _____ ________________, State_____________.

The Board, when requested by the City Commission, shall consider, investigate, make finding, report and recommend upon any special matter of question coming within the scope of its work.

Section 6. Operation

The Board shall choose its own officers, make its own rules and regulations and keep a journal of its proceedings. A majority of the members shall be a quorum for the transaction of business.

Section 7. Street Tree Species to be Planted

The following list constitutes the official Street Tree species for_____________________, State__________________. No species other than those included in this list may be planted as Street Trees without written permission of the City Tree Board.

* **Small Trees**	**Medium Trees**	**Large Trees**
Apricot	Ash, Green	Coffeetree, Kentucky
Crabapple, Flowering (sp)	Hackberry	Maple, Silver
Golden Rain Tree	Honeylocust (thornless)	Maple, Sugar
Hawthorne (sp.)	Linden or Basswood (sp.)	Oak, Bur
Pear, Bradford	Mulberry, Red (fruitless, male)	Sycamore
Redbud	Oak, English	Sycamore, London plantree
Soapberry	Oak, Red	Cottonwood (Cottonless, male)
Lilac, Jap. Tree	Pagodatree, Japanese	
Peach, Flowering	Pecan	
Plum, Purpleleaf	Birch, River	
Serviceberry	Osageorange (Male, thornless)	
	Persimmon	
	Poplar, White	
	Sassafras	

Section 8. Spacing

The spacing of Street Trees will be in accordance with the three species size classes listed in Section 7 of this ordinance, and no trees may be planted closer together than the following: Small Trees, 30 feet; Medium Trees, 40 feet; and Large Trees, 50 feet; except in special plantings designed or approved by a landscape architect.

Section 9. Distance from Curb and Sidewalk

The distance trees may be planted from curbs or curblines and sidewalks will be in accordance with the three species size classes listed in Section 7 of this ordinance, and no trees may be planted closer to any curb or sidewalk than the following: Small Trees, 2 feet; Medium Trees, 3 feet; and Large Trees, 4 feet.

Section 10. Distance from Street Corners and Fireplugs

No Street Tree shall be planted closer than 35 feet of any street corner, measured from the point of nearest intersecting curbs or curblines. No Street Tree shall be planted closer than 10 feet of any fireplug.

Section 11. Utilities

No Street Trees other than those species listed as Small Trees in Section 7 of this ordinance may be planted under or within 10 lateral feet of any overhead utility wire, or over or within 5 lateral feet of any underground water line, sewer line, transmission line or other utility.

Section 12. Public Tree Care

The City shall have the right to plant, prune, maintain and remove trees, plants and shrubs within the lines of all streets, alleys, avenues, lanes, squares and public grounds, as may be necessary to insure public safety or to preserve or enhance the symmetry and beauty of such public grounds.

The City Tree Board may remove or cause or order to be removed, any tree or part thereof which is in an unsafe condition or which by reason of its nature is injurious to sewers, electric power lines, gas lines, water lines, or other public improvements, or is affected with any injurious fungus, insect or other pest. This Section does not prohibit the planting of Street Trees by adjacent property owners providing that the selection and location of said trees is in accordance with Sections 7 through 11 of this ordinance.

Section 13. Tree Topping

It shall be unlawful as a normal practice for any person, firm, or city department to top any Street Tree, Park Tree, or other tree on public property. Topping is defined as the severe cutting back of limbs to stubs larger than three inches in diameter within the tree's crown to such a degree so as to remove the normal canopy and disfigure the tree. Trees severely damaged by storms or other causes, or certain trees under utility wires or other obstructions where other pruning practices are impractical may be exempted from this ordinance at the determination of the City Tree Board.

Section 14. Pruning, Corner Clearance

Every owner of any tree overhanging any street or right-of-way within the City shall prune the branches so that such branches shall not obstruct the light from any street lamp or obstruct the view of any street intersection and so that there shall be a clear space of eight feet (8') above the surface of the street or sidewalk. Said owners shall remove all dead, diseased or dangerous trees, or broken or decayed limbs which constitute a menace to the safety of the public. The City shall have the right to prune any tree or shrub on private property when it interferes with the proper spread of light along the street from a street light or interferes with visibility of any traffic control device or sign.

Section 15. Dead or Diseased Tree Removal on Private Property

The City shall have the right to cause the removal of any dead or diseased trees on private property within the city, when such trees constitute a hazard to life and property, or habor insects or disease which constitute a potential threat to other trees within the city. The City Tree Board will notify in writing the owners of such trees. Removal shall be done by said owners at their own expense within sixty days after the date of service of notice. In the event of failure of owners to comply with such provisions, the City shall have the authority to remove such trees and charge the cost of removal on the owners property tax notice.

Section 16. Removal of Stumps

All stumps of street and park trees shall be removed below the surface of the ground so that the top of the stump shall not project above the surface of the ground.

Section 17. Interference with City Tree Board

It shall be unlawful for any person to prevent, delay or interfere with the City Tree Board, or any of its agents, while engaging in and about the planting, cultivating, mulching, pruning, spraying, or removing of any Street Trees, Park Trees, or trees on private grounds, as authorized in this ordinance.

Section 18. Arborists License and Bond

It shall be unlawful for any person or firm to engage in the business or occupation of pruning, treating, or removing street or park trees within the City without first applying for and procuring a license. The license fee shall be $25 annually in advance; provided, however, that no license shall be required of any public service company or City employee doing such work in the pursuit of their public service endeavors. Before any license shall be issued, each applicant shall first file evidence of possession of liability insurance in the minimum amounts of $50,000 for bodily injury and $100,000 property damage indemnifying the City or any person injured or damaged resulting from the pursuit of such endeavors as herein described.

Section 19. Review by City Commission

The City Commission shall have the right to review the conduct, acts and decisions of the City Tree Board. Any person may appeal from any ruling or order of the City Tree Board to the City Commission who may hear the matter and make final decision.

Section 20. Penalty

Any person violating any provision of this ordinance shall be, upon conviction or a plea of guilty, subject to a fine not to exceed $____________.

*Please note: The above species are offered as size-class examples only and may not be suitable for planting in your area. Please check with local sources to develop a species list for your area.

URBAN TREE MANAGEMENT

Alan C. Jones
Minnesota Department of Natural Resources
Bemidji, MN

Mankind benefits from and is dependent on the beauty and utility of trees. Despite this close relationship with trees, we see dead, diseased and damaged trees all around us. The worst enemies of trees are not insects, diseases, weather, or air pollution; trees' worst enemies are people--you and I! Eighty percent of urban and backyard tree problems are caused by humans.

A recent survey conducted by a national tree care company identified 4 major areas of tree abuse:

The first abuse is in the **SELECTION** of plant material. Poor quality planting material is selected, or the planting material is poorly suited to the site or geographic region. This abuse often results from poor planning. Many trees are planted by people who do not do their homework. Trees often are selected on the basis of cost or aesthetics and not on quality, size, and suitability. Losing a tree later on because of poor quality planting stock becomes much more expensive than paying initially for good quality stock.

The second tree abuse is poor **PLANTING** techniques. This abuse causes the greatest number of problems and leads to the greatest monetary losses. Poor planting techniques include soil preparation, care and handling of the planting stock, the actual planting, and care immediately after planting.

The third area of tree abuse is a general lack of good **MAINTENANCE** practices. Many people think that a tree will take care of itself. Plant it and step back! We often take trees out of their native habitat and place them into something totally unfamiliar such as a city street or a backyard. Even though it may be the right tree for the right spot, the spot may be foreign to the tree. Like any other living organism, a tree needs to be taken care of and nurtured. All too often maintenance is done by well meaning individuals equipped with Dark Ages' knowledge. This kind of maintenance often may cause greater harm than no maintenance at all.

The last area of tree abuse is lack of adequate **PROTECTION.** Trees need protection from wounding, construction activities, insects and diseases, animals, weather, and chemicals. Trees need to be protected from us because many things we do place trees under stress and make them more susceptible to natural pests. Trees need our protection because they cannot get up and move away from potential injury-causing situations.

The paper that follows is a discussion of these 4 abuses and what can be done to minimize these abuses. If we are responsible for putting our trees at risk, then we should be able to learn what is proper so that our trees give us many years of beauty and utility.

SELECTION

Before selecting the tree to plant, you need to do some planning and research. First **determine** your species options, **evaluate** those options to select the best species to fit your needs and conditions, and then **obtain** the best planting stock.

Determining Species Options

Determine your options by finding out what species will grow in your particular geographic area, and what species are readily available. Gather this information by consulting with your local County Extension Agent and/or your local State, County or Federal forestry personnel. If you don't have access to these individuals, try local nursery or farm and garden store personnel. Some of these places may even have a landscape architect on their staff who could give you invaluable advice. There is also a wealth of published information from government pamphlets to landscaping books and gardening magazines which can help you gather the information you need to develop a species options list. Don't overlook the possibilities of consulting with a homeowner who has a tree or landscaping idea you particularly like. Most homeowners are proud of what they have done and are willing to give free advise to anyone who asks.

Evaluate Your Options

Once you have a list of possible options that will grow in your area, make the selection based on the following criteria:

1. What is the purpose of planting your tree? Are you planting a tree for shade, for wind protection, to attract birds, as a sound and/or visual barrier, or simply as an ornamental to make your property look good and you feel good about where you live? Determine the purpose for the tree and then limit your species options to those having the physical attributes which will meet your objectives. For example, you would not choose a Norway pine as a shade tree, but you may choose a Norway pine as a visual barrier.

2. Where will you plant the tree? Are your species options compatible with the soil characteristics of the planting location? If there are species which are not suited to the soil conditions, eliminate them from consideration. At this point you may need to do some more homework. Get a soil test or ask someone who knows soil properties to look at the planting location. Compare the soil conditions with the silvics or ecology of your species options to make sure they are compatible.

3. Will the tree fit the space? More important is the space big enough to accommodate the tree when it attains a full grown size? When we plant a tree, we simply cannot believe that the tree will ever outgrow the space even though we water and fertilize it with the hope that it will grow and get big. The tree is so small when we plant it that we actually "cheat" and crowd the tree because it looks so small and dwarfed. A common abuse is severely pruning a tree to refit it into the space it once fit when newly planted. These pruning techniques are better known as "mutilation".

This space problem can also apply to the roots. Usually roots will spread out even further than the top. Plan on root growth! Lack of rooting space will cause the tree to become unbalanced with more top than root system. An unbalanced tree will decline and die. Avoid planting close to underground items such as drain fields, gas tanks, or pipelines. Both you and the tree may suffer for your miscalculation when planting.

To avoid future problems, be sure you know the potential height and crown spread of the tree when fully grown. Use half the crown diameter as a guide in determining the minimum planting distance from all above ground objects. For example if a tree has a 50 foot diameter crown spread, plant the tree at least 25 feet from any above ground obstacles. If the objects are overhead lines, double your planting distance to reduce the chances of having your tree periodically "pruned" by the power company. Use the potential tree height as a guide for providing rooting space. A 100-foot tall tree should have at least a 100-foot diameter space around the trunk clear of underground obstacles which will interfere with root growth. Species on your options list which will not fit your space should be eliminated as viable choices for planting.

Become familiar with the characteristic shape of the type of tree you are considering. Is it a widespread oval crown or is it more pyramidal or columnar? The shape of the crown will help you fit the tree into the space you wish to fill, and the shape should be compatible with the shape of your buildings and other landscaping features.

4. What are the tree's potential problems? Are there recurring insect and disease problems? For example, if you live in a jack pine area, you may not want to plant pine because the jack pine budworm will cycle through the area about every 10 years creating headaches for you. White pine is a sensitive species to salt. Don't choose white pine where salt spray from the highways can easily reach your tree. Oak wilt can be a serious problem in the southern half of the state, and if you live there plant oaks only if you are willing to learn how to protect your trees from this disease. Silver maple is notorious for pushing up sidewalks and getting into sewer pipes. It also suffers from alot of broken branches in wind storms. If these situations are important for you, choose some other species. Sugar maple and paper birch grow where soil temperatures are cool and are not adapted to sites with warm soil temperatures. With these species avoid lawns and areas with lots of blacktop and concrete. And remember, fast growing trees generally are short-lived trees. If you decide on hybrid poplars or other fast growing trees, anticipate outlasting your tree. Plant slow growing replacement species the same time you plant the fast growing tree. Also, have a plan in mind for the removal of the fast growing tree when it becomes old and decadent.

5. What are the tree's undesirable characteristics? Some tree species have male and female trees. Generally, male trees are more desirable because you don't have to contend with the seeds and fruit. If your choice has both male and female flowers, can you live with the seed production? For example, black walnut is not real desirable in yards because the nuts are a nuisance when mowing the lawn. The nuisance of boxelder bugs is one of the reasons boxelders are not popular trees; however, boxelder bugs are seed feeders and are attracted only to the female boxelders. Varieties have been developed for some species which do not produce flowers, and messy seed production is eliminated. An example is the Marshall seedless ash.

Obtaining planting stock

You now have determined the best species to plant, and now you should obtain good quality planting stock. Cheaper is not necessarily better. If you obtain cheaper, lower quality planting stock which is going to give you problems after planting, chances are good you will spend more maintenance money than the difference between high quality and cheaper quality planting stock. With cheap stock you may find yourself having to severely prune the tree, the tree may be deformed from repeated dieback, severely stunted because of poor growth, or the tree will die several years after planting. Start with good quality planting stock which was locally grown, if possible. Some local nurserymen also guarantee their stock for 1 or 2 years after planting. The guarantee may be worth the price, since handling and planting can be extremely stressful activities. However, most trees will "survive" or appear to be green and alive during the first growing season, and it will not be until the second or third growing seasons when planting and handling problems appear. Guarantees which last through at least the second growing season are more desirable than guarantees which last only through the first growing season.

When choosing planting stock, you have a choice between buying the stock or transplanting native trees. When buying planting stock you will be faced with three options: bare root, balled and burlap, or container grown.

Bare root. This stock will be the cheapest but has the greatest limitations. To prepare bare root stock for sale, it has to be dug up, the soil removed from the roots, and then the roots packaged to protect them from drying out. Handling during lifting and planting provide plenty of opportunities for root damage by breakage and dessication. Once the roots dry, they die. Tiny absorbing roots can dessicate and die by being exposed to the air for less than a minute.

To help minimize the risk when planting bare rooted stock, plant as quickly as possible after purchasing. Storing the tree will increase the chances of root damage. If possible, avoid temporary storage by healing the tree into the ground. Healing in and then removing the tree to plant can cause root abrasions and fine root breakage. If you have to heal in the tree because you cannot plant right away, dig a hole or trench large enough so the roots don't have to be jammed into the trench. Gently cover the roots with broken up soil, not clods. When digging up the tree, do not pull it out of the trench or remove the soil from the roots. Dig up both soil and roots. For shorter term storage, the roots can be immersed in water but not for any periods longer than 8-12 hours. Tree storage for periods longer than 8-12 hours can be done by keeping the roots moist in wet peat moss. Plant bare root stock when it is dormant. The tree has broken dormancy if shoots are elongating, new leaves appear, or the roots have white tips. Chances for a successful planting will be greatly reduced under these conditions.

Bare root stock is best for spring planting of large numbers of small seedlings, usually 2-4 years of age. Trees with a stem diameter of over 3 inches should not be transplanted bare rooted. If you wish to plant larger stock, just a few trees, or plant anytime during the growing season, then balled and burlap, container grown, or transplanted trees with the soil attached to the roots are probably better choices.

Balled and burlap. This is a tree dug out of the nursery bed with the soil attached to the roots, and then the roots and soil are wrapped in burlap to form a root ball. The chances for root drying are minimized because the roots are not exposed to the air and the entire ball can be watered. When handling balled and burlap stock, the tree should not be lifted or carried by the tops or branches. This may allow the soil to crack, loosen, or fall away from the stem and roots; if this happens, severe damage to the roots can occur. When planting, the burlap should not be removed, but the intact root ball is placed in the planting hole. The burlap will keep the soil and roots together and make transplanting much easier. Root development will not be restricted by the burlap which will readily deteriorate once in the ground. Be sure to loosen the burlap around the stem so the rope or wire holding the burlap to the tree doesn't girdle the stem. The burlap should not stick up above the surface of the ground because it can act as a wick and draw moisture away from the roots.

Balled and burlap stock is usually cheaper than container grown stock, but more expensive than bare root stock. The additional cost over bare root stock may be worth the price since balled and burlap stock can gain 2 or more growing seasons of top growth over the same size bare root stock. If the decision is made to purchase balled and burlap stock, the factor of cost should be secondary to the appearance of the tree and the relationship between the size of the tree and the size of the balled roots and soil. A general rule of thumb is for trees with a stem caliper (at the surface of the ball) of 1 to 1-1/2 inches, the minimum ball size should be 16 to 18 inches in diameter; 1-1/2 to 2 inch stem, 18-23 inch ball; and 2 to 2-1/2 inch stem, a 22-24 inch ball. As the stem diameter increases, the ball diameter should also increase.

The other advantages of balled and burlap stock are that root disturbance is minimized and planting does not have to be confined to the dormant season. Even though root disturbance is minimized, roots can be damaged if care is not taken during lifting. Often the tap root or long lateral roots have to be severed to get the tree out of the nursery bed. If the nurseryman is not careful during the lifting process and roots are crushed, ripped or mangled, new roots will not generate from the ends of these damaged roots, and root rot organisms may become established. Balled and burlap stock can be planted anytime during the growing season, but if planting is delayed, the ball should be buried in sawdust or wood chips and regularly watered to prevent the roots from drying out. To provide maximum protection for the root system, it is always best to plant the tree as soon as purchased.

Container grown. This stock provides the best chance for a successful planting although the cost usually is the greatest. Container grown stock can be a tree seeded directly into a container, but, more likely, it is a tree that has been transplanted into the container. Make sure the tree has not been transplanted into the container during the same growing season you buy it. Root damage can occur while transplanting into the container. The container grown tree should have at least one full growing season in the container to give it time to generate roots. If the root system has not been damaged or it has been pruned before transplanting into the container, you will be purchasing a tree with a better root system than it had before transplanting.

When planting, simply lift the tree out of the container, and place roots and soil mass into the planting hole without breaking up or disturbing the root ball. You should never leave the tree in the container when planting even if the container is decomposable and the instructions tell you to plant the container with the tree.

A real advantage of container grown stock is that it can be planted anytime during the growing season when you have the time and when the growing conditions are ideal, i.e. cool and moist. Container grown stock can be stored easier and longer than balled and burlap stock. Be cautious about large trees in small containers and small trees in large containers. Your objective should be to try to get a tree which completely occupies the container but has not occupied it to the point of being pot bound. Choose trees in which you can see roots at the drain holes of the containers. This means the roots are fully occupying the pot, and the roots are being air pruned at the drain holes. But you also need to look at the top of the tree because roots at the

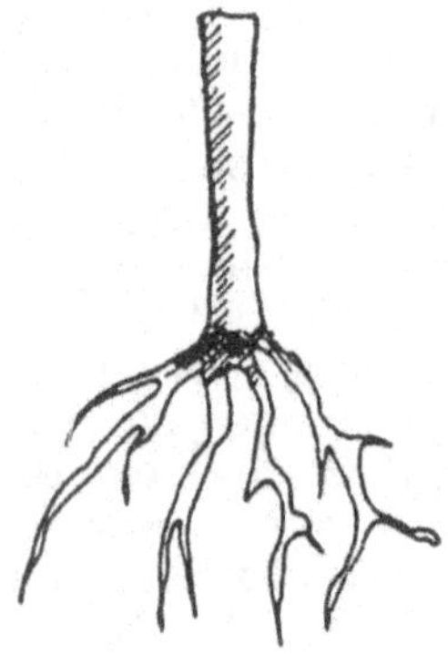

a. Bare root

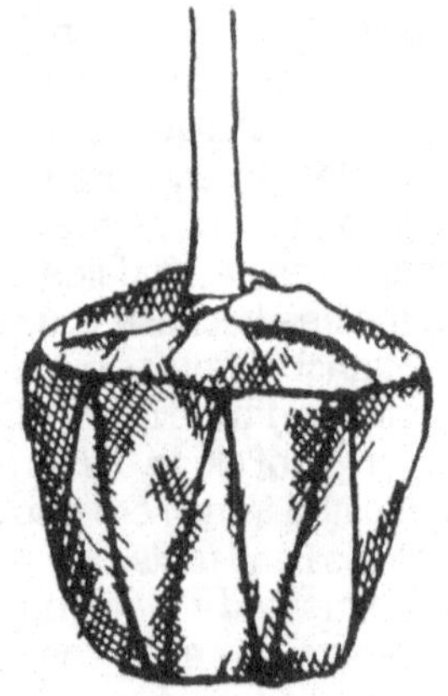

b. Balled and burlapped

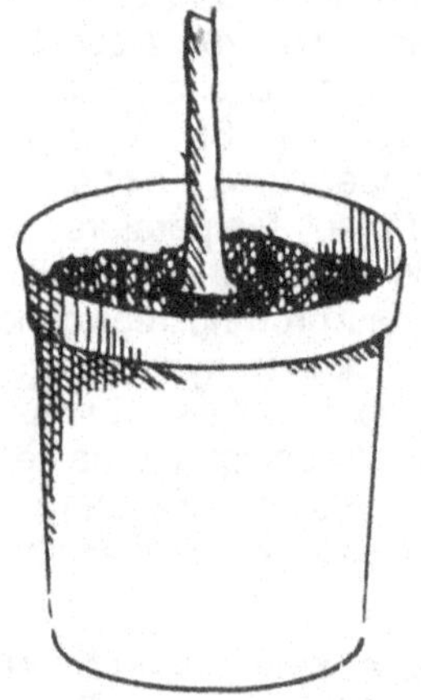

c. Potted and container grown

d. Pre-packaged

drain holes may indicate a tree has been in the container too long and is pot bound. The tree should display good growth, and the foliage color should be a rich green. You can also get an idea of how much of the pot is occupied with roots by wiggling the tree while holding the pot. If you can't see roots, are unsure of how much of the pot is occupied with roots, or the top does not look good, choose another tree. It is important that you screen the trees and choose the best possible specimen.

Transplanting. Rather than purchasing a tree, a wild tree or a plantation grown tree can be transplanted by hand or with a mechanical tree spade. Both methods are fine; obviously, the tree spade is more expensive but can handle much larger trees. Tree spades can handle trees with stems up to 8 inches in diameter. The maximum size tree that can be safely transplanted without a tree spade is a tree with about a 2-1/2 inch diameter stem. The danger of transplanting especially with the tree spade is trying to take too large of a tree. Large trees have such large root systems that too much of the root system has to be severed to move the tree. The tree may survive for a number of years, but the tree often declines in vigor. Top dieback occurs, and/or small and sparse foliage is produced with little growth. These are symptoms of a damaged root system. Choosing a smaller tree that you can handle will pay off in the long run. The tree will survive and get a faster start without an extraordinary effort on your part.

Transplanting is similar to handling balled and burlap stock except transplanting should only be done during the dormant season. Get as much of the root and soil mass as possible, and don't remove any of the soil from the roots. Keep the root and soil ball intact as you would with balled and burlap stock. If long lateral roots or the tap root must be cut, use a sharp shovel or ax to cleanly sever the roots. Avoid stripping bark off of the roots, crushing the ends of roots, and fraying the ends by trying to sever them with a dull shovel. A cleanly severed root will generate new roots and lead to better transplanting success. The planting hole should be prepared before digging up the tree so there is as little time as possible between digging and planting.

The root systems of "wild" trees will often be different than nursery grown trees. The nursery grown tree will have compact, fibrous roots from being pruned and confined. The wild tree will have long lateral roots ranging well beyond the tree's dripline. The more open grown the tree, the more wide-spread the root system. It will be difficult, then, to get adequate roots when transplanting; therefore, prepare for transplanting a year ahead of time by root pruning some of the laterals while the tree is still intact in its original location. During the spring planting season, make cuts straight down into the soil just outside the dripline or outer crown spread of the tree. Do this on all four sides but not completely around the tree. These cuts will loosen the tree by severing some roots while the unsevered roots will anchor the tree. The severed roots will then generate new roots to help offset any damage to roots during transplanting. During the next growing season when transplanting, only the unsevered laterals will have to be cut; the severed laterals will have generated a mass of fine roots which should remain intact on the root system.

Making the final selection. It is important that you carefully inspect the top of the tree before you purchase any tree. Look not only for good twig growth and good foliage color, but make sure that the foliage is proper size for the species you are choosing. Trees with small leaves indicate a root problem and should be avoided. Look for wounds on the branches and main stem. Avoid purchasing trees with wounds; the wounds may attract insects, may be colonized by fungi, and/or may lead to structural problems in the future. Look for pruning wounds, and make sure they have been done correctly. If stubs are left or branches have been cut off flush with the main stem, avoid purchasing trees with these kinds of pruning damage. Look for branch pruning cuts made between nodes or buds. If pruning has occurred between nodes, avoid the tree. Look at the architecture of the crown. If a lot of branches arise at the same point on the main stem, this will spell trouble later, and these kinds of situation should be avoided. If the bark ridges in the branch angles are turning inward, avoid these situations. Finally, avoid trees with cracks in the main stem; and sunken areas, odd looking bark patterns, and/or wet areas on the main stem or branches.

PLANTING

Now you're ready to plant! You know why you want to plant, you have matched your tree's needs to the soil conditions, you know where the tree will be located, you have an idea in your mind what your tree will look like when it gets larger, and you have obtained good quality planting stock. Let's plant.

First, prepare the planting site. To do this you should not only dig a planting hole, but also loosen the soil around the hole at least to the dripline of the tree and preferably beyond the dripline. The soil can be loosened with a shovel or fork, and loosening should be to a depth of at least 8 inches. When digging the planting hole, make the hole about twice the size of the root ball or container. If you are planting into sand or subsoil, you may want to enlarge the planting hole even more.

Use the topsoil that was excavated out of the planting hole as backfill material around the root ball. It is best to use this topsoil unless you are planting into excavated subsoil. The

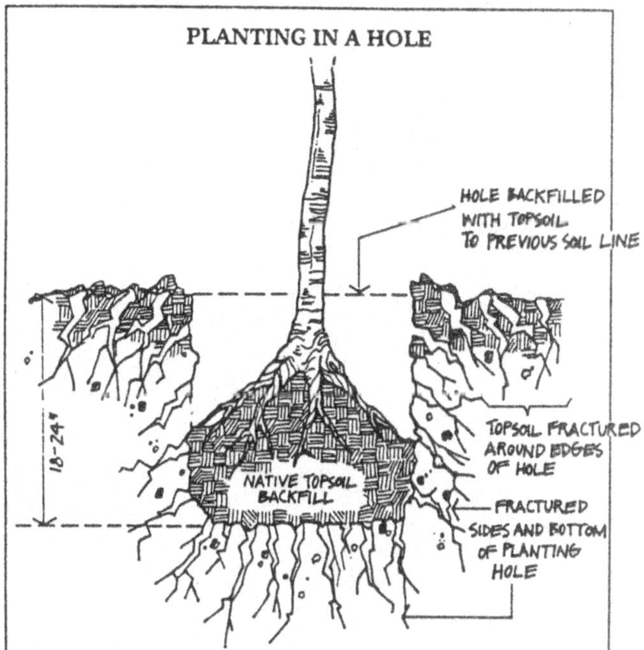

Ground Rules For Planting Trees, By Robert Kourick, Illustrations By Maia Farrell, "Harrowsmith" September-October 1988

tendency is to enrichen the backfill material with "black dirt" hauled in from somewhere else or with peat and other organic amendments. Using too rich of backfill material will create such an ideal environment for the roots that the roots will not expand outward into the surrounding soil. An important key for the tree to survive and grow is a wide ranging root system. Adding rich backfill material into the planting hole will create a "vase" which will confine the root system to the planting hole, and the tree will be limited in its growth and survival. Mixing sand into the backfill material will not "enrichen" the backfill but will change the soil texture and may help drainage in heavier clay and clay loam soils. Some sand can be added when planting on the heavier soils, but your main objective should be to keep the backfill material as similar as possible to the soil where the tree is being planted.

Fill the hole with enough topsoil so the surface of the tree will be at the same depth or slightly higher than it was at the nursery or in the container. The soil should be packed firmly in the hole so the tree doesn't settle. Water the hole before planting to moisten the backfill material. Gently place the roots into the hole. If the hole is too shallow or too deep, take out the tree and adjust the hole rather than trying to force the tree to fit. Hold the tree in place to keep it at the proper depth and to keep it straight. If you are planting bare root stock, gently spread out the root system so no roots are wrapped around the stem but are distributed to grow outward away from the stem. Never bunch up the roots or wrap them around the stem to get the tree into the hole. When the tree is in the correct position and the roots are spread out, partially fill the hole with soil and then water. Repeat this procedure until the hole is filled. Be sure there are no roots sticking out of the ground although the top surface of large buttress roots can show if they were growing over the top of the soil when you bought or dug up the tree.

Leave a small depression around the tree for a catch basin for future waterings. Do not make the catch basin so deep that water will stand for long periods of time in the basin. Eliminate the basin after 2 or 3 growing seasons to avoid pooling of water over the main woody roots. As the tree grows, the location for watering will shift outward, beyond the dripline, and the roots will not be efficiently watered by filling up the catch basin. The catch basin, by pooling water over the woody roots, can lead to root rot and root suffocation.

Pay attention to the planting depth so the depth is not drastically altered from what it was in the nursery. For container grown and balled trees, this depth will be fairly obvious since the trees will already be "planted" in their own soil. Bare root stock may be more difficult to identify the proper planting depth. Look for color differences on the stem which will tell where the soil line was originally located. If planting grafted stock, the grafted union may be swollen or there may be a stub of the old top remaining. This grafted area should be right at the groundline but above the groundline. Sugar maple trees often suffer from collar rot and basal cankers which are associated with being planted too deeply. Maple growing in a natural forested area have many buttress roots growing over the surface of the ground. It is important to plant sugar maple fairly shallow with roots distributed near the surface of the ground.

Mulching. To increase survival chances during the critical first growing season and first winter, apply a mulch from the trunk to at least 1 foot beyond the tree's crown spread or root ball. The mulch should be 4-6 inches thick and can consist of wood chips, bark, grass clippings, leaves, corn stalks, etc. A mulch will help conserve soil moisture by killing out the sod under the tree and by preventing evaporation from the soil surface. Sod can be a major competitor for soil moisture since grass roots are closer to the soil surface and will absorb moisture before any can reach the tree roots. A mulch can help moderate temperatures, protecting the roots from high soil temperatures in the summer and cold temperatures during the winter. Mulch-

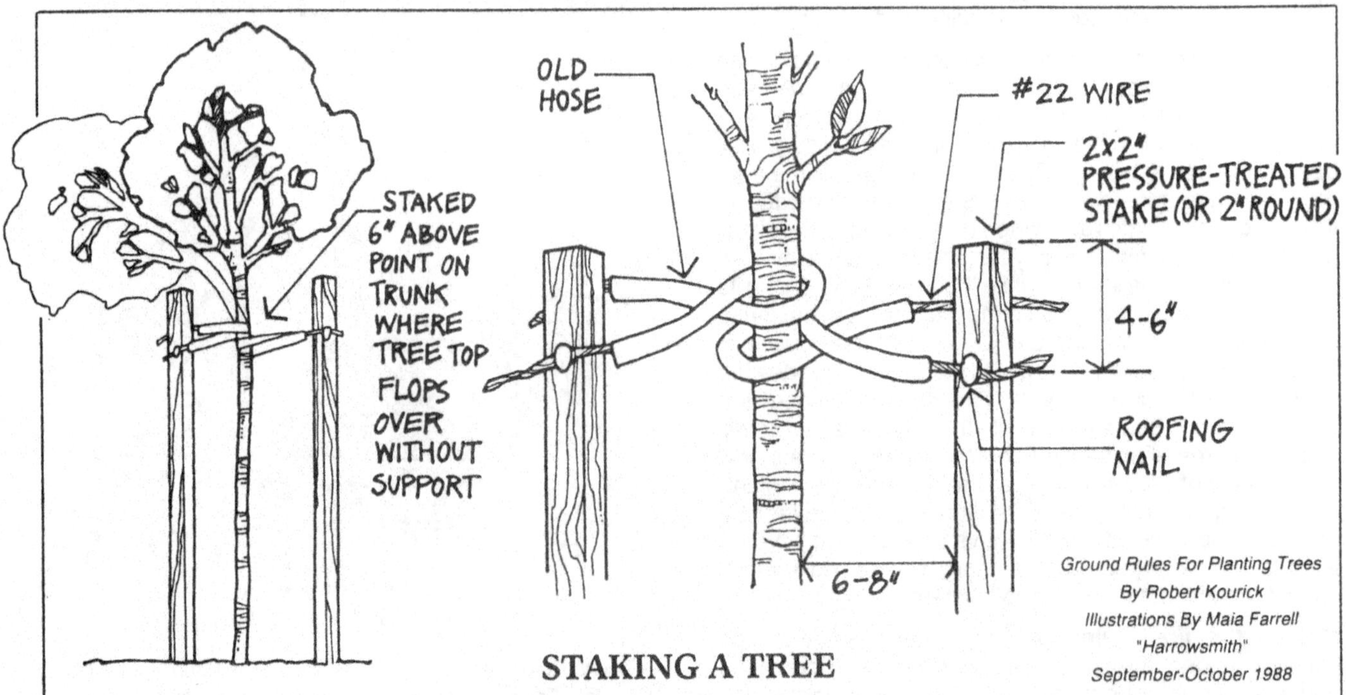

STAKING A TREE

ing can also help reduce mechanical damage to the main stem by keeping lawnmowers away from the trunk. Mulching is particularly recommended when planting in sandy soils, planting species such as paper birch and sugar maple which can't tolerate warm soil temperatures, or planting in the late fall to protect roots from early low temperatures.

Staking. Some bracing of newly planted trees should be done until new root growth can adequately anchor the tree; however, staking is usually overdone. It is important for a tree to develop wind firmness, and it is difficult to do that when a tree is firmly staked for a number of years. Staking should not be used to prevent tree movement or to prevent a thin stem from bending in the wind, but staking should be done only to prevent the tree from tipping out of the ground because roots have not expanded out into the soil. Trees which especially need staking include conifers at least 4 feet in height because their dense foliage will easily catch the wind making them vulnerable to windthrow, broadleaf deciduous trees with large tops in relation to their root masses especially in exposed and windy locations, trees transplanted with a tree spade because these trees are usually very large compared to their root mass, and sugar maples which are shallow planted to prevent collar rot.

Staking can be done with fence posts or with short wooden stakes. Three stakes placed evenly around the tree are ideal for staking. Try to attach the guy lines as low on the trunk as possible so that the top can move about in the wind. The guy lines should be a wide, flat strapping material or nylon stocking so that the stem is not girdled or bruised. Nylon stocking is particularly effective since it will not damage the trunk, but it will stretch and allow tree movement helping the tree develop wind firmness. Ropes and especially wires cushioned with rubber pads or hose sections at their point of attachment to the trunk can injure or girdle the stem. Avoid the use of both rope and wire for guy lines. The tension of the guy lines should be adjusted to allow tree movement. Periodically check the guy lines to adjust the tension and to make sure they are not cutting into the bark. Remove the guy lines by the end of the first or second growing season. By then the tree will have developed enough of a root system to adequately anchor the tree.

Pruning. Removal of 1/3rd to 1/2 of the top of deciduous trees to bring the top into balance with the roots has historically been a recommended pruning practice. **This is no longer recommended.** It is difficult to know how much of the root system has been damaged or lost during the planting process and therefore it is difficult to know how much of the top to prune. You may prune too much, or you may prune the more vigorous stems without knowing. The food that is needed for growth of roots is manufactured by the leaves. Trees need as much leaf area as possible to produce enough food for the roots. Cutting back the top can set back root establishment by reducing the available food supply which will ultimately affect the top growth. If the tree is handled carefully, planted correctly, mulched and watered after planting, a tree even with a serious top/root imbalance should be able to survive. Pruning is also a wound to that tree, and wounds can lead to more serious damage by providing an entry point for harmful fungi, bacteria and insects. The rule now is: DON'T PRUNE BACK A NEWLY PLANTED TREE! If the root system is damaged the tree will shed some of its top to balance its mass with its energy levels. This shedding will be in the form of top dieback, and it is the portions that have died which should be pruned out during the following dormant period.

MAINTENANCE

The tree is now in the ground; you're allowed to step back to admire your handiwork and enjoy its beauty. But don't step back too far and for too long. Remember, your tree is a living organism just like your pets or cattle or bees. They all need your care and attention.

Watering

The most immediate maintenance practice for newly planted trees is watering, but watering is a practice that can be used for all ages and sizes of trees. Always keep trees well watered during the first 2 growing seasons for newly planted trees. Trees with root damage should be watered during the growing season when injury occurred and during the next two growing seasons. Finally, water all trees during periods of prolonged dry weather and for one growing season after the dry period has ended. For newly planted trees, apply water to the area of the root ball every 7 to 10 days unless rainfall has been heavy enough to saturate the soil. Resume watering 7 to 10 days after the end of a saturating rainfall.

Weekly rainfall to benefit any aged tree should be at least 1 to 1-1/2 inches and should be received at a rate of at least 1/2 inch per rainfall. If this amount of rainfall is not received, a regular watering schedule of every 14 to 21 days should be undertaken. Water by allowing a sprinkler to run over the root zone until at least 1-1/2 inches of water is caught in a container placed in the area being watered. Also, every 6 weeks, continuously apply water to the rooting zone for a period of 12 hours, starting at 8:00 PM and ending at 8:00 AM.

Where is the rooting zone? On newly planted trees, the rooting zone is fairly obvious and corresponds to the root ball. On established trees, the rooting zone can extend outward from the trunk to a distance way beyond the dripline of the tree. Open grown trees will have a wider ranging root system than trees growing in groups since root competition from other trees will limit the spread of the roots. Roots tend to grow horizontally away from the trunk rather than downward into the soil. The major woody roots that form the framework of the root system will be found in the top 3 feet of the soil. The fine roots which grow from the woody framework roots tend to grow upward and are found in the surface 6 to 8 inches of the soil. Trees growing in sod will have the majority of their fine roots right under the sod layer. These fine roots will be the important absorbing roots during the growing season, and they will be located from the dripline outward. Watering, then, should take place from about the dripline outward for a distance at least equalling the length of the longest branches.

Water only during the growing season. Stop watering about September 15 unless there is a severe and prolonged drought. Late watering can promote a late season flush of growth or delay the hardening off process. Succulent tissues produced by late season watering will be damaged by sudden drops in temperatures.

Fertilization

Fertilization is another maintenance practice which should be used on a routine basis. Since nitrogen is lost through leaching, volatilization, and bacterial action, it may be necessary to apply nitrogen every year. Phosphorous and potassium applications generally are needed every 3 to 5 years. However, sandy soils and subsoils from the backfill of construction are soil conditions which may necessitate a yearly fertilization of all three minerals.

An important principle of proper tree care is to fertilize only when there is a need to fertilize or when the tree will benefit. Trees have been damaged by too much fertilizers, and many dollars have been thrown away by applying fertilizers which do not benefit the tree. It must be realized that fertilizers are not plant foods. Plant foods are sugars manufactured in the leaves which are then transported to the roots. Fertilizers do not "feed" trees, and the roots are not "feeder" roots. We tend to overfertilize if we think we are feeding the tree by fertilizing. Fertilization is the act of adding elements to the tree which are involved in the physiological processes of the tree but do not directly provide energy for the tree to survive and grow.

A starting point for fertilization is to take a soil sample and get it analyzed. State Universities usually do soil testing with instructions and shipping cartons handled by the local County Agent. Soil and Water Conservation District personnel in each county can also help determine fertilizer needs by helping with a soil test.

A soil test will tell you the quantity in the soil of elements necessary for plant survival and growth. For example, a general rule of thumb is that phosphorous levels above 30 pounds per acre and potassium levels above 160 pounds per acre are usually adequate for the tree's needs. This is a generality and may not be specific enough for your tree which illustrates a potential problem with soil tests. Soil tests may not always tell you everything you need to know. The reasons for this are that fertilizer requirements are not always known for trees, and soil labs may only have experience interpreting soil test results for agricultural crops. Most fertilizer information is based on agricultural crops, and fertilizer recommendations that are best for corn may not be best for your tree.

Another starting point whether or not you have obtained a "believable" soil test is to observe the tree. If the tree has good leaf color and displays good shoot growth, at least 1-14 inches per year, fertilization probably is not necessary. If you do fertilize, look for a positive response. If the leaf color doesn't change by becoming "richer" green in color and if there is no significant change in shoot growth over the next two growing seasons, provided the growing seasons are normal, then the fertilizer application was probably not needed.

The three major elements most needed are nitrogen (N), phosphorous (P), and potassium (K). The amount of each of these elements will be stated in a percentage and will always be stated in the order listed above. For example, a 20-10-5 fertilizer contains 20% nitrogen, 10% phosphorous, and 5% potassium. If using a fertilizer which contains all three elements, commonly referred to as a "complete" fertilizer, the ratio of N-P-K for trees should be 3-1-2 or 3-1-1, and generally the amount to use should be based on the nitrogen content of the fertilizer.

To determine how much fertilizer to use after a soil test is performed, base the fertilizer application on the projected size of the root system. Root systems will usually extend well beyond the crowns, but a good average is to think of root systems being twice as large as the crowns. Measure the crown area by placing stakes in the ground along the dripline of the tree, and average the dripline so that you end up with a circle, square or rectangle formed by the stakes. Calculate the square foot area within the stakes, and double the area to arrive at an area for the root system. The rule of thumb for fertilizing trees is 1-6 pounds of actual nitrogen, 1 to 2-1/2 pounds of actual phosphorous, and 2-4 pounds of actual potassium per 1,000 square feet of rooting area. But remember, nitrogen may be needed yearly, but it is rare when phosphorous and potassium are needed yearly. Normally they may be needed every 3 to 5 years. It is best to use the low figure of the pound ranges for each element and then observe the tree's response. In other words, experiment; fertilize lightly but more often and constantly observe the response of your tree.

Other elements are also necessary for tree survival and growth, but these "microelements" are usually needed in lesser amounts than N-P-K. Often the availability of these other elements is related to the pH of the soil. Trees usually do best growing in soils with a pH range of 5.0 to 6.2. A pH above 6.2 often makes microelements such as iron unavailable. It should be realized that for good lawn growth, a higher pH is desired. Iron chlorosis from high pH have occurred when lawns were limed to raise the pH to grow better grass. Often, grass and trees are not real compatible.

A fertilization routine should be started about the second or third growing season after planting. It is important not to start fertilizing too soon after planting. NEVER place fertilizer into the planting hole or fertilize the same year of planting even with container grown trees. Nitrogen fertilizer will stimulate leaf and shoot growth; a transplanted tree probably has too much top for the root system to supply with water. By fertilizing before the top and roots are in balance, additional stress can be placed on the tree by creating a greater top imbalance. The same principle applies to established trees after root injury has occurred. DO NOT apply any nitrogen fertilizer that growing season. You absolutely do not want to stimulate any top growth. If you feel compelled to fertilize, fertilize with phosphorous and potassium which will help stimulate root generation without affecting top growth.

Apply fertilizers when new growth is desired and at a time when the fertilizer will be absorbed. Applications can be made in both the fall and spring, but a good average time is in the fall. Fall applications are best because the fertilizer is then in place the next spring when the tree resumes activity, there is little danger of burning the grass, and tree roots are undergoing a flurry of growth in fall with some uptake occurring. The danger of a fall application is applying the fertilizer too early in the fall. Warm temperatures may cause the nitrogen to be lost, but more important an uptake of nitrogen in the late summer or early fall may stimulate the tree to produce new and succulent tissues which can be damaged by cold temperatures. Apply fertilizers in the fall from leaf drop to ground freeze up, and a good guide for application is after a killing frost. **Never** apply fertilizers to frozen ground; fertilizers will run off and not benefit the tree but will harm the environment.

Generally, fertilization is done with dry or liquid fertilizers applied on the ground under the tree. Fertilizers should be applied to the area where the major portion of the absorbing roots are located: 3 to 4 feet beyond the dripline of the tree to about one half the distance between the dripline and the trunk. On larger trees it is not necessary to get any closer to the trunk than 3 feet.

Applications can be made by broadcasting on top of the ground and then thoroughly watering so the fertilizer is carried beyond the grass roots to where the tree roots are located. If the fertilizer is not thoroughly watered into the soil, only the grass will benefit. If heavy nitrogen applications are made in spring when the grass is actively growing, some burning of the grass may occur.

To avoid the problems inherent with applying fertilizers to the surface of the ground, fertilizers can be applied below the soil surface. "Fertigation" is an application technique which injects the fertilizer into the soil in a liquid form using a soil probe. Use of this technique may also help loosen the soil. The alternative is to punch holes in the ground and pour either granular or liquid fertilizer into the holes, or place slow release tablets into the holes. The holes should be evenly distributed under the tree, 6 to 8 inches deep, 1-2 inches in diameter, and spaced 2 to 4 feet apart. A crowbar is an effective instrument for making holes, but any instrument used can cause soil compaction around the holes. Be sure the holes are made when the soil is not very wet; this will help to minimize soil compaction.

There are other ways fertilization can be done. Commercially prepared solid forms of fertilizer called tree spikes can be pushed down into the ground around the tree. The use of these would be similar to punching holes in the ground. Since these are specialty items and commercially prepared, they are more expensive. When using them be sure the amount of fertilizer being applied is the recommended amount.

Another method, although not very effective and not used very often is spraying a fertilizer solution onto the leaves. Leaves are not as effective as the roots are in taking in fertilizer. Sometimes fertilizers are mixed with systemic insecticides and both are sprayed together.

Finally, microinjections can be made in the trunk or root flares particularly to apply micronutrients such as iron and magnesium. The advantages of this technique are less fertilizer used, less danger of the fertilizer causing damage to other organisms, and less danger of the fertilizer leaching into ground water since it is injected into the tree itself. The drawbacks include the cost, the need to use specialized equipment and trained personnel, questions concerning the technique's effectiveness, and each injection site is a wound which can be invaded by fungi and bacteria leading to cankers, decay, and wetwood.

Pruning

Another routine maintenance operation is pruning. There are two underlying principles of pruning: (A) Consider the consequences of every pruning cut by visualizing the future shape of the tree **before** the cut is made, and (B) every pruning cut is a wound. These two principles tell us that there should be a definite reason for making every pruning cut. Even though pruning is usually a benefit to the tree, pruning can lead to structural weakness, decay, cankers, wetwood and even tree death. Pruning cuts made on live limbs during the growing season can attract insects

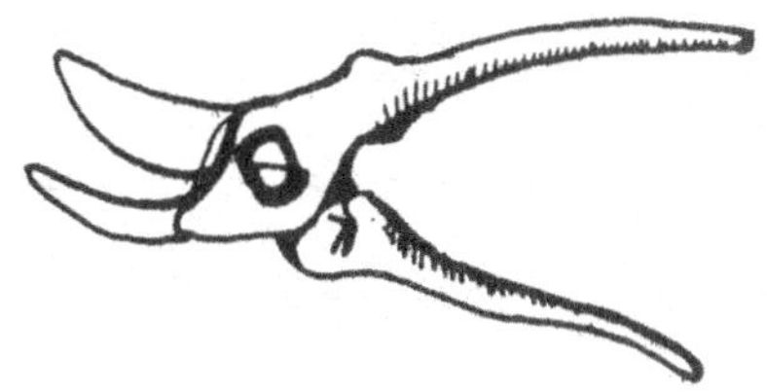

Pruning shears

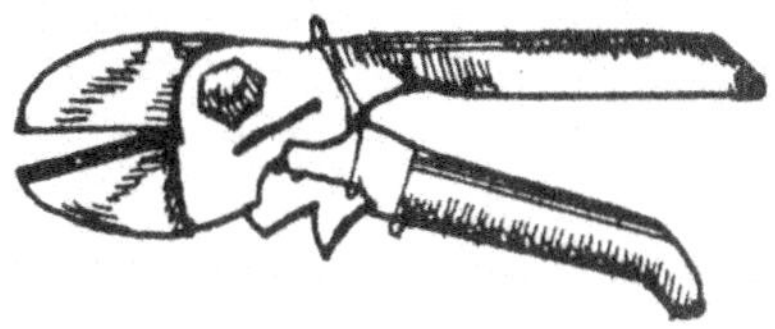

Anvill-type pruning shears

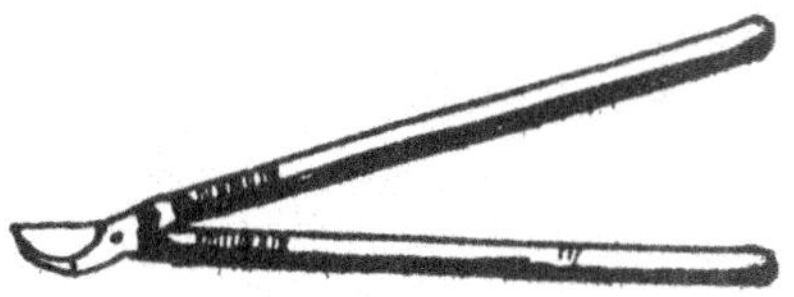

Lopping shears

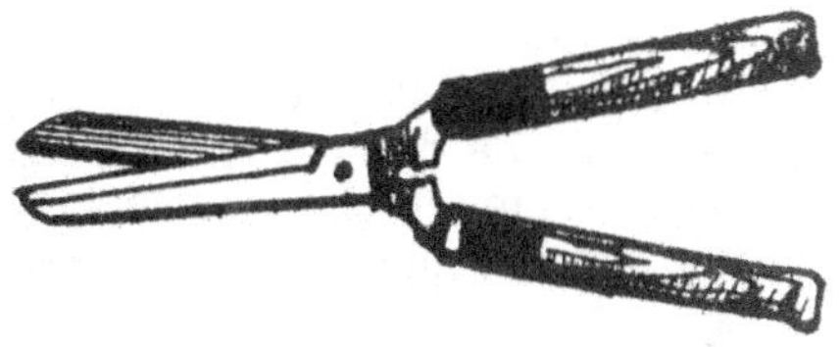

Hedge shears

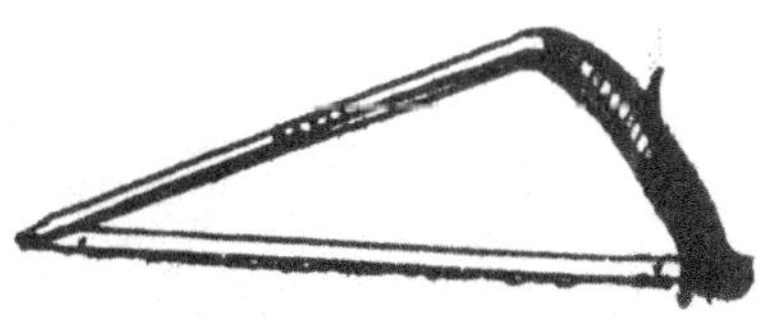

Bow saw

Pruning saw

which spread disease. An example is oak wilt. In oak wilt areas, **never** prune an oak tree during the growing season because the fresh wound will attract sap feeding beetles; these beetles may be carrying the oak wilt fungus. Always have a purpose for pruning, visualize what the tree will look like before you make the cut, carry out that pruning using proper pruning techniques, and prune at the proper time.

Why prune? There are a number of reasons to prune. If you don't have one of the following reasons, don't prune.

1. To alter the appearance of a tree: Remember, don't rely on drastic pruning to try to fit a tree into a space. But there are situations when pruning for appearance may be necessary: a lopsided tree, branches which interfere with human activities, major storm damage, or an inherited problem.

2. To provide safe conditions around the tree: Trees with broken or dead branches and weak crotches pose hazards to people and property. Dead branches can fall anytime, and weak crotches will usually fail when put under stress from wind or snow and ice. Weak crotches are narrow, V-shaped crotches of less than 30°. They are weak crotches not because of the angle but because the narrowness of the crotch forces one side of the crotch to overgrow the bark of the other side of the crotch. The edges of the bark ridge in the crotch will turn downward or one side will be under the other side of the crotch. This ridge is called the "branch bark ridge", and its appearance is a better indicator of crotch strength than crotch angle. Any bark inclusions form weak unions which fail under loads. Strong crotches are broad U-shaped crotches with no bark inclusions, and the branch bark ridge in the crotch is turned upward.

3. To keep the tree healthy: Some disease incidence can be reduced by pruning off the parts of the tree which are most susceptible to the disease, or by pruning off the disease portions before the disease can spread to other parts of the tree and other trees. Examples include removing the basal branches of white pine because infections occur more commonly on lower branches, removing a cankered branch of white pine before the fungus has moved into the main stem, and pruning off the black fruiting bodies of black knot of cherry during the winter before they can fruit in the spring and spread the disease. Other times to prune for health reasons include removing suckers and water sprouts from the root collar and main stems, removing branches which are rubbing together, and making clean cuts on broken branch stubs to promote more rapid healing.

4. To rejuvenate the tree: This type of pruning consists of removing the dead wood, removing branches in the interior parts of the crown, or by cutting off the tree at the groundline. New growth should be directed to the outside of the crown where there is better utilization of sunlight. Cutting off the entire tree at the groundline can rejuvenate a poorly developed tree if the tree is known to be a sprouter. The new sprout is easier to work with, a better formed tree is produced, and the original tree height is attained quite rapidly when rejuvenating young trees.

5. To increase production: This type of pruning primarily applies to fruit production. If you want to produce fruit, you must prune so the plant will be stimulated to blossom and set fruit.

6. To develop special products: This kind of pruning would be used in the Christmas tree industry where shearing produces a full, uniformly tapered crown. Forest management practices to produce a clear bole for high quality lumber or a straight bole for higher quality wood products would also fit into this category.

When to prune. Once you have determined why you want to prune, then you must decide when to prune. Generally, late in the dormant season just before bud swell is the best time to prune live branches. Dead branches can be pruned anytime of the year. During the dormant season, the branching structure can easily be seen, access to branches to prune is easier, and the branches weigh less. Tree bleeding is not a problem during the dormant season although late dormant season pruning will produce heavy sap flows on species such as maple and birch. Sap flow or bleeding from pruning cuts will gum up pruning equipment, hands, and clothing; however, unless it attracts disease carrying insects, this sap flow will not harm the tree. Pruning late in the dormant season produces wounds which will have less exposure to fungal spores. The wounds will be overgrown more quickly by callus and sealed off. In the late winter and early spring there are fewer fungal spores in the air than during the fall. Since the tree will produce little callus in the fall, the wound is exposed longer to a greater spore load than if pruned in late winter. **Do not** prune in the spring when the buds are flushing because energy to produce tissue to overgrow the wound is going into growth. The moist, sap-filled bark is loose in the spring and can easily be stripped off the trunk and branches causing great injury.

The exception to dormant season pruning is shearing of pine Christmas trees and repairing storm damaged trees. Pine shearing has to be done when the new growth is elongating and before it hardens off and sets buds. Shearing usually takes place in June. Storm damage often occurs during the growing season when the crowns are heavy with leaves. Storm damage demands immediate attention by pruning out broken branches, removing branch stubs left behind from broken branches, and making clean straight cuts to otherwise jagged breaks.

How to prune. Make clean cuts without torn or rough edges. To do this, keep pruning tools sharp. When making cuts, take care that no branch stubs are left. Branch stubs will die back; healing will be retarded and the pruning cut will be vulnerable longer as an infection site for fungal diseases. Leaving a small branch stub, however, is not as bad as cutting flush against the trunk. In between the main stem and the branch is a swollen collar called the "branch collar", and it separates the branch from the tree. This should not be damaged by cutting. Remove the branch to the branch collar, but never beyond. Cutting into the branch collar will cause a larger and more serious wound. Be careful that the cut is completely through the branch so the bark is not stripped down the stem.

Begin a pruning program when the tree is small so wounds are small. On larger trees try to keep the wounds as small as possible by confining your pruning to small branches. If large limbs must be pruned, plan on making 2 cuts before the final cut to prevent bark ripping. Make the first cut a partial cut through the branch on the underside about 12 inches from the main trunk. The second cut then should be about 1 inch beyond the first cut, and it should be started on top of the branch until the branch gives way. Now that most of the weight of the large branch has been removed, make the final cut next to the branch collar.

When partially pruning back branches for shape or flower production, always select a healthy bud to prune to. NEVER PRUNE BETWEEN BUDS! Be sure the bud is on the outside and on the side to which you want the tree to grow toward. Cut off the stem right above the bud with a downward angled cut. If the tree has opposite branching, i.e., two buds at the same point on the stem, remove one of the buds. If both buds are left on, forking will occur at this point since both buds will produce shoots which will compete for dominance. Most flowers and fruit are produced on previous year's shoots. If pruning for flowers or fruit, don't remove alot of the past year's shoots.

Wound dressings. The question always arises concerning treating the pruning wound with a wound dressing. This is no longer advised unless you live in an oak wilt area and a fresh wound or pruning cut has been made during the growing season. It is important to cover the wound as quickly as possible so it is no longer attractive to sap feeding insects. In all other situations, wound dressings are not needed because only the tree can protect itself by overgrowing the wound. Wound dressings or paints do not help the tree but can retard the growth process by killing the new tissues formed after wounding. Spores from fungi which cause decay and cankers are so common in the air, that it is difficult to seal the wound quickly enough to prevent spore contact with the wound. The wound dressing does not seal long enough to prevent fungal spores from landing on the wound. Keep the wounds small, smooth, and clean, and try to prune in late winter. These practices will help minimize diseases and later structural weaknesses associated with pruning wounds much more than applying wound dressings.

Wound dressings do not stop rot.

Do not apply house paints or wood preservatives nor apply heavy coats of any material. Research shows that wound dressings do not stop decay or stall rot. Trees have been responding effectively to their wounds for over 200 million years. Do not interfere with this natural process. Keep your tree healthy and it will take care of its wounds. In a short time the wound surface will blend perfectly with the tree bark.

Other maintenance activities

There are times and situations in the life of your tree when you may have to act as a tree surgeon. Large wounds may occur from an automobile or large piece of machinery hitting the tree, rot and cavities may occur which weaken the tree, despite your best pruning efforts weak and split crotches or major cracks in the trunk may develop, or most likely you have inherited someone else's tree abuses. You need to take drastic actions in these cases.

Large trunk wounds, no matter how bad they look, should not be covered over with wound dressings. If the wound has jagged edges and bark flaps around the edges, these can be removed. Your objective should be the same as it was for pruning: you want a small and smooth-edged wound. Use a sharp knife to cut off the bark edges and jagged pieces of wood, and try to make the margins of the wound smooth. Don't cut into undamaged wood; this action will only enlarge the wound. Simply remove the jagged bark and wood; then, step back and allow the tree to do the rest.

Weak or split crotches and major cracks in the trunk can be strengthened by running a bolt through the tree **below** the weak or damaged area, and combining this with cabling above the weak area. A hole is drilled through the trunk below the crotch, and a bolt is put completely through. Use large round or oval washers on both ends of the bolt to anchor it to the tree. These washers should be seated up against the outside of the tree, and not driven into the tree. The bolt can be left in place and the tree allowed to grow around the washers and bolt. Cabling should be done with wide, flat material to prevent as long as possible over-

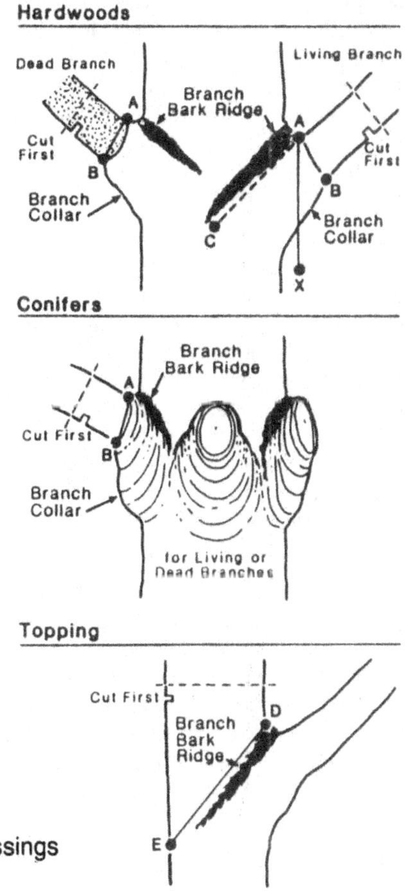

Wound dressings

growth around the cabling material. Cabling and bracing is a drastic action to be taken, and more harm than good will be done if not done properly. Consult a professional, but make sure the professional knows to bolt below the crotch.

PROTECTION

If you cannot or care not to do the routine maintenance practices just described, at least give your tree a measure of protection. Protection is needed from insects and diseases, animals, weather, destructive cultural practices, and chemicals. Even though the tree may look invincible and has stood its ground longer than you can remember, it still needs your protection if you want it to stand for many more years. Without your protection, your tree is living on borrowed time.

Wounding

Almost all protection centers around protection from wounding. A wound is any injury which damages **living** bark, cambium, or wood. When a tree is wounded, it does not react to that wound as you and I would react. The difference in the reaction is important to know. We heal a wound by replacing the injured cells and tissues. A tree cannot replace damaged cells or tissues but "heals" a wound by separating itself from the wound and by overgrowing the wound. It doesn't regenerate damaged tissue; it grows new tissue in a different spatial arrangement than the original tissue.

A tree is a highly compartmentalized organism, and when a tree is wounded, the wound is compartmentalized and shut off from the new parts of the tree formed after the wound was made. All parts of the tree present at the time of wounding stand a chance to be invaded by destructive microorganisms such as decay fungi. New wood formed after wounding is separated from the old wood by a barrier often referred to as the "barrier zone". The barrier is both a chemical barrier and an anatomical, physical barrier of specialized cells produced by the cambium in response to the wound. The barrier zone confines the damage to the wood present at the time of wounding which is why it is good to prune when the tree is small.

The second way the tree reacts to a wound is to grow over it with callus tissue or wound wood. Callus is soft, undifferentiated tissue; tissues formed later which are hard and impervious and produces the characteristic roll around a wound is called wound wood. Callus and wound wood are not replacement tissues; they are new tissues which overgrow the wound to seal it off. As the wound wood continues to develop and the roll gets larger, the wound will be completely overgrown and sealed off. This helps prevent fungi and bacteria invasion. A tree is not completely immune to microorganism invasion until the wound is completely covered over with new, living tissues.

Both barrier zone and wound wood formation are influenced by the vigor and genetics of the tree. A young, vigorously growing tree will react and seal a wound much quicker than an older tree. A wound made immediately before the spring growth period closes much more rapidly because the tree has a great store of food reserves and is about to undergo a period of vigorous growth. Wound closure is the least rapid during bud and leaf expansion since the tree's energy is being used for leaf flush.

Genetic control is the biggest factor influencing barrier zone formation and wound closure. Work is currently being done to identify individuals, clones, and varieties which show a strong genetic ability to form barriers in response to wounding. Young trees are wounded, and after a period of time they are split open at the point of the wound to be inspected for discoloration. Little discoloration associated with the wound is an indication that the tree has a stong capability of compartmentalizing the damage. That capability is carried in the tree's genes. The trees showing the strongest tendency to compartmentalize are the trees to propagate and manage.

A tree's ability to wall off a wound is both a blessing and a curse. The barrier zone between the wound and new, healthy wood is a plane of stress and tension. It is along this barrier zone that additional damage may occur. Tree cracks start along this zone because of the stress and tension. Cold weather may put the additional stress on the tree which will force the crack outward to the tree's surface, but it is the barrier zone which starts the crack. In the same way ring shake occurs. Ring shake is separation of the growth rings and appears as concentric cracks around the stem. Both stem cracks and ring shake occur because the tree has been wounded sometime in the past and a barrier zone has formed. Since the barrier zone is formed along the entire length of the tree when wounded, cracks and shake can occur long distances from the wounds. For example, hemlock will experience ring shake in the trunk from sapsucker feeding in the top of the tree. Reduce the amount of wounding to the tree, and cracks and ring shake occurrence will be reduced.

The formation of the barrier zone also shuts off vital food or energy storage tissues. The barrier zone separates the tissues present at the time of wounding from any new tissues formed after wounding. When a tree is wounded, much of the trees food storage capacity is shut off from the new functioning part of the tree. As the tree is repeatedly wounded and energy storage and reserves are shut off, the tree's vigor and ability to resist additional stresses are reduced. The formation of the barrier zone also requires energy which otherwise could be used for growth and seed production. Each new wound creates a new barrier zone, and multiple wounds create multiple barrier zones. Each zone shuts off reserves and uses up energy while it is being formed. Continual wounding leads to a steady decline in the tree. Wounds may seem innocuous particularly when they are small, but over time they take their toll and can lead to tree death.

Lawnmower injury. Next to pruning, the lawnmower is probably responsible for more tree wounds than anything else. Even the most careful person rarely avoids hitting the tree when mowing. The wound is low on the trunk and often leads to tree decay because the greater humidity near the groundline creates ideal conditions for fungal spore survival and infection. It is not unusual for lawnmower wounds to occur several times during the mowing season and year after year. The chances of decay getting started are increased because of the frequency of wounding. Trees which break off at or near the groundline or trees with dead bark lesions close to the ground have been wounded by lawnmowers, and the wounds have been successfully colonized by decay or canker-causing fungi. Prevent lawnmower injury by not allowing grass to grow right up to the trunk. Establish a flower bed around the trunk, maintain a mulch under the tree, or put an area of rocks or bark around the trunk so that mowing close to the trunk is not necessary. Grass is also a serious competitor with the tree for moisture and minerals. Reducing the grass not only protects the trunk from physical injury but also reduces the potential moisture stress that grass can cause.

Construction Damage

An area of tree damage which often can be overlooked is damage done during building construction, underground line installation, and road or driveway construction and maintenance. We are aware of the obvious wounds such as a gouge or breakage from heavy equipment but overlook root damage from exposure, compaction, severing, or changing grade.

The objective should be to **prevent** construction injury, not try to cure a tree damaged from construction activities. Often it is impossible to save a construction damaged tree, or at the least construction damage can seriously change a tree's appearance making it no longer desirable to have around. The keys to preventing construction injury are COMMUNICATION and CONTROL. You must communicate with the contractor to make him understand the value of your trees, to educate him as to the vulnerability of your trees, to inform him of the restrictions you will be imposing on his activities to protect your trees, and finally to specifically spell out the penalties that will be imposed if the contractor damages any trees. After communicating all of these things, you then need to be present at all times, or have a representative present at all times to control the contractor's activities. The contractor may understand your point of view, but unless you are out there enforcing it, the contractor may well ignore your point of view.

It takes planning on your part long before the contractor even bids on the project. You must realize not all trees can be saved. It is more costly in the long run to try to build in among trees, damage those trees, and then have to pay very expensive fees to remove those trees which die 2-8 years after the construction is completed. If parts of the root system are damaged, there is a good chance the tree will die. Excavation obviously will sever roots and cause massive root injury; trees close to the excavation hole should be removed as part of the site preparation activities. A "no-impact" zone should be established around each tree, and this no-impact zone should be the area around the tree where no construction activity or vehicle travel occur. A rule of thumb for determining the size of the no-impact zone is 1 foot of horizontal distance from the trunk for every inch of trunk diameter. For example, a 20 inch tree should have a "no-impact" zone of 20 feet around that tree. During excavation, if roots larger than 2 inches in diameter are severed, then the tree should be removed. There is a strong possibility that the tree will die after losing large roots.

Make a site map with the trees plotted on the map, and draw the no-impact zone around each tree. This will allow you to visualize where construction can safely occur, where an access road can be routed, and which trees need to be removed. Lay out the access road by flagging a route, and try to use the same access road for ingress and egress. Make it clear to the contractor that only the access road can be used; he is not to drive **or** park any vehicles or equipment except on the access road. Any tree that is being saved and is close to the construction activity should have a fence erected around the tree with the fence located at the outer edge of the no-impact zone. Use snow fence for this protective fence. Snow fencing is readily available, it will reasonably limit the contractor from getting into the no-impact zone, and it will be readily apparent if the snow fence has been altered.

Contractors will have a need to store equipment and building materials on the site, and they will have to park their private vehicles somewhere on the site. Unfortunately, contractors will tend to store their tools, etc. under trees. This will compact the soil and should not be allowed. Plan for a storage/parking area before the contractor arrives, and keep it away from any trees you want to save by using the rule of thumb for the no-impact zone. The same is true for temporary soil storage from the excavation. Even a 30-day temporary storage of soil close to a tree can cause root damage. Plan on soil storage away from the trees you want to save.

All of these restrictions will cost you money, but it will be an investment in the value of your building site. It will save trees, and trees will have a very definite impact on the value of your site. But if the trees are really important to you and you are willing to invest even more, there are other practices which can be used to provide more protection or allow you to save more trees. Some of these practices include building on pilings rather than excavating, bridging access routes across wet areas so that drainage is not altered, restricting when the contractor can be active to avoid the most damaging times of the year such as the wet times and the spring of the year when the trees are most vulnerable, using cement conveyors or pumps rather than having heavy cement trucks drive onto your land, and then finally altering your house plan to fit the site rather than altering the site to fit your house plan. Take advantage of what you have on the site and build without changing the existing grade and drainage patterns. This may make the house plan more expensive, but the expense will be an investment in your trees.

Construction activity injures trees in a number of ways. The most obvious injury is mechanical damage to the tops of the trees. Gouges and scrapes on the main trunk can be common if activity is not restricted, and breakage of low hanging limbs can also be a problem. Before activity occurs, prune low hanging branches to raise the level of the crown so that no equipment will hit the crown. Pruning can reduce the amount of injury even though pruning is a wound. With pruning you control the wounds, you make the wounds outside the branch bark collar, or you make the pruning cuts at nodes. These proper pruning techniques will allow the tree's defenses to compensate for the wounds. Without pruning, equipment will break off limbs, leave stubs and jagged edges, or rip bark into the trunk. These kinds of injuries will result in future cracking and mechanical failures in the tree. Even if the tree survives the construction activities, it may be structurally unsound, and become a serious hazard to you and your home. Wounds made on the trunk should not be painted or covered, and they should not be trimmed up so that additional wounding occurs. You can carefully remove the jagged wood and bark edges, but then allow the tree to produce callus and wound wood to seal the wound. Again, DO NOT PAINT THE WOUND OR USE ANY WOUND DRESSINGS!

Less obvious damage from construction injury is root damage. Roots can be damaged by severing the roots, exposing the roots, and smothering the roots by raising the grade or changing the drainage patterns so that the soil becomes wetter. Severing and exposing the roots usually are quite obvious. Any roots that have to be severed that are over 2 inches in diameter spells disaster for the tree. When excavating, you or an arborist should be present to prune any roots which are exposed. If the roots can be cleanly cut before the equipment severs them, you will be helping your tree immensely. A good clean cut will produce callus tissue on the end of the cut, and this callus tissue will produce a mass of roots while sealing off the cut. If the roots are not pruned they will be crushed, torn, and frayed by the equipment. Roots damaged in this way will die back, not produce the proliferation of new roots, and root rot organisms will become established more easily.

A trenching machine probably is the most destructive piece of construction equipment. It generally looks innocuous because it produces only a narrow trench, but the trench destroys alot of roots. The severed roots are not always obvious in the trench. Again direct the trenches away from the no-impact zone, and try to combine all of the utilities in the same trench. All too often, several trenches along different routes are made by different utility companies. Consider tunneling rather than trenching particularly when you cannot stay outside of the no-impact zone. Root systems are rather shallow, especially if you are building in an undisturbed wooded area. Tunneling at a depth of 2 or 3 feet will miss most of the major woody roots and all of the short-lived, fine absorbing roots.

The construction damage which is most difficult to understand and recognize is damage to the roots from compaction and raising of the grade. The roots need oxygen to live, and oxygen is obtained by the roots from the pores and voids in the soils. Oxygen diffuses into these soil spaces from the surface atmosphere. Compaction reduces the size and number of air spaces in the soil, and raising the grade reduces the oxygen infiltration rate into the soil. Both compaction and grade change can affect the water infiltration rates and water holding capacity of the soil. All of these conditions can stress the roots and cause root death.

Be aware of the vulnerable rooting area, the no-impact zone, and keep heavy equipment away from this zone, particularly when the soils are wet. Wet soils compact much easier than dry soils. It is important to know your soils. A fine-textured soil will compact more easily than a coarse-textured soil. Laying down a 4-6 inch layer of wood chips **before** construction can help reduce compaction, but it isn't a guarantee that compaction will not occur. Again, the best way to avoid compaction is to avoid the rooting zone.

Perhaps the least recognized construction damage to roots is changing the grade. For sensitive species such as maple, oak, and beech, adding 2 to 4 inches of fill around the tree will cause root death. If a grade has to be changed, build a retaining wall as far away from the trunk as possible. The retaining wall should keep back the new fill so the old grade is maintained. From this retaining wall extend tile outward in a spoke fashion and install vertical tile every 8 to 12 feet along the spokes. The vertical tile then should extend above the new grade. These vertical and horizontal tiles will allow oxygen to diffuse downward and out through the rooting area. Before adding soil to establish the new grade, put down a layer of straw or coarse gravel to help the oxygen and water diffuse outward from the tile.

Any root damage will be evidenced by either a sudden or gradual crown dieback. A gradual thinning and dieback of the crown is the more common situation. In extreme cases of root death, the crown may be limited to a few small leaves on the trunk. It is important to realize, then, that construction injury often causes a slow tree death. Contractors will comment that all of the trees were green and "healthy" when they left the site. That statement is correct. Sometimes the only evidence of construction injury during the same growing season as the construction may be premature fall coloration in the crown. Crown dieback may not occur for 2 or 3 years after construction, and tree death may take 8 years. But the fact is that it was the contractor and the contruction activities that caused the tree death or injury.

Construction injury is difficult to reverse. When construction activities are completed, begin a program of watering even without any evidences of crown damage. Chances are great that some portions of the root systems have been damaged, and your immediate objectives should be to reduce any additional stress on the trees by watering. DO NOT fertilize with any nitrogen fertilizer that will stimulate top growth. The top will be out of balance with the root system, and the root system needs to be rebuilt. A fertilizer without nitrogen but one that contains phosphorous and potassium can be used to help stimulate root growth, but it is best to delay any fertilization until the following dormant season. Then apply a light application of a 10-10-10 fertilizer. A light application would be 1/2 to 1 pound of actual nitrogen per 1,000 square feet of rooting area.

DO NOT reduce the crown if you suspect you have root damage. How much should you reduce the crown? What does a 20% root loss mean in terms of the amount of crown to remove? Which portion of the crown corresponds to the part of the root system that was damaged? These are difficult questions to answer, and therefore crown reduction should not be done. Allow the tree itself to adjust to a damaged root system by allowing it to naturally shed the parts it can't maintain. You then can prune out the dead wood.

If compaction has occurred, try to aerate the rooting zone, but you must realize it will be difficult. Aeration can be done with commercial equipment that employs a hydraulic probe inserted into the ground and a jet of air is sent through the probe to physically loosen the soil. This equipment is limited in finer textured soils. Another aeration technique is to auger a series of 2-inch diameter holes throughout the rooting area of the trees. The holes should be 8 to 12 inches deep or deep enough to go completely through the new fill and into the rooting area. The holes should be spaced 8 to 12 inches apart and extend away from the trunk at least throughout the no-impact zone, i.e., 1 foot horizontal distance for each inch of stem diameter. The holes then should be filled with coarse gravel, pea rock, organic matter, or anything which doesn't compact but which allows water and oxygen to freely infiltrate into the holes. This is an expensive treatment, but it may save the tree. This again, emphasizes the fact that remedial treatments, if they work at all, will be more expensive than preventative treatments. And, with remedial treatments, there is no guarantee that they will work and no guarantee that the tree hasn't been damaged so badly that any remedial treatment will be fruitless.

Insects and Diseases

When we think of protecting trees, we often think of insects and diseases first. It is true that insects and diseases do kill trees, but insects and diseases are often secondary, taking advantage of the tree's weakened condition. Insects and diseases take advantage of poor cultural practices, wounds, adverse weather, and our general lack of tree care. Pests are always present waiting for opportunities to parasitize trees, and if they can't do it on their own, we often help by innoculating our own trees. Examples include spreading fire blight bacteria on our pruning tools, pruning oak during the growing season in an oak wilt area, and shoestring root rot successfully colonizing and then killing trees when the trees are put under stress either from weather extremes or from our mismanagement practices.

To protect a tree from insects and diseases, you must be able to identify the pest, know its life cycle, and understand when it is most susceptible to control efforts. Many insects undergo a metamorphosis in which their body structure and function may change 4 times during their life cycle. Generally, only the larval or eating stage is vulnerable to controls. You need to know when and where the laval stage will be in order to control the insect. Many diseases are inside the tree's tissues and therefore are protected by the tree itself. You must know when the disease agent is present on the outside of the tissues, or you must know what part of the tree to destroy in order to control the disease. Finally, you must learn to evaluate the seriousness of the insect and disease pest. The presence of a pest does not mean that it will cause damage. A control action should not be initiated because of pests occurrence; it should be initiated when serious and irreversible damage is likely to occur.

The most common types of pests you will likely encounter will be leaf-eating caterpillars (defoliators), insects and fungi which cause swellings (galls) on leaves and branches, sucking insects (scales and aphids) on branches and leaves, and fungi which cause defoliation. All of these pests are nuisances, but they are not always serious. A tree under stress will be more easily damaged by these kinds of pests. Newly planted trees should be considered under stress, and some control action should be taken when any pests are encountered. When the trees have been in the ground for at least two full growing seasons and/or the length of the new shoots are at least 6 inches long, the tree probably is not under stress and control actions may not be needed.

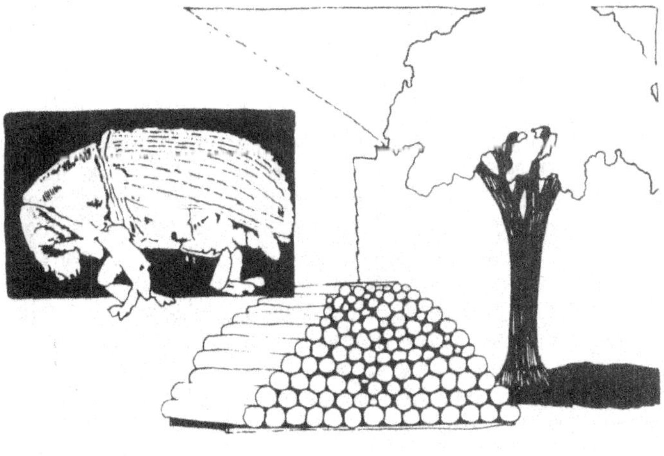

Gall-making insects rarely cause damage enough to warrant any control. If galls are on branches, try to prune off the galls and destroy them as soon as you see them. Galls noticed during winter will rarely contain any insects, and pruning at this time will not control the insects. Swellings caused by fungi, may produce spores in the spring and reinfect the tree. Pruning these galls in the winter is advisable. Sucking and defoliating insects pose a greater risk, but damage from sucking insects such as aphids and scales usually occur from heavy infestations in successive years.

The chances of your tree sustaining damage is greatest from both the fungal and insect defoliators. Serious, long-term damage from defoliation is dependent on the condition of the tree, on the time of year of defoliation, and on the type of tree. Since broadleaf trees naturally lose their leaves every fall (deciduous trees), defoliation usually is not as serious as defoliation of evergreens. One complete defoliation of an evergreen may cause tree death, but it may take 2 to 5 defoliations of a deciduous tree to cause tree death. Plan on taking some control action the first growing season insect defoliators are found on evergreens, but do not take any action on deciduous tree defoliators until the second growing season and then only if defoliation occurred prior to August and resulted in a loss of at least 50% of the leaf area the first year.

The other factor influencing defoliator damage to deciduous trees is the timing of the defoliation. Generally, defoliation early in the growing season soon after the tree has flushed and expanded its leaves is more serious than later in the growing season when the tree is preparing to shed its leaves for winter. Food reserves in the roots are low after leaf flush, and defoliation at this time may cause branch dieback because food reserves in the roots are not adequate to sustain all of the branches while energy is being used for refoliation. If defoliation occurs before August, consider taking a control action. Defoliation occurring during August and September probably will not require any control action because the roots have adequate food reserves.

Fungal defoliators are more difficult to control than insect defoliators. By the time you see the symptoms of a fungal infection, i.e., leaf spots and blotches, curled leaves, or falling leaves, the fungus is inside the leaf tissue and protected by the leaf tissue. The best control strategy for fungal defoliators on evergreens is to identify the fungus, learn its life cycles, and then apply a protective fungicide to the tree the following growing season when spores are being produced. For deciduous trees, rake up and burn the leaves in the fall since most fungal defoliators survive in the leaves and reinfect the tree the next spring from the fallen leaves. These fungi also are very dependent on moisture, and if the spring and early summer are characterized by rainy days and /or many days of high humidity and overcast skies, protective fungicides also should be used as the leaves are expanding. To control foliar fungi on both evergreens and deciduous trees, 2 to 6 fungicide applications spaced 7-21 days apart will be needed during the time when the fungus is producing the spores which infect the leaves.

For small numbers of trees in your backyard, try to hand pick the insects and prune off the galls, cankers, and fungal fruiting bodies. Practice good sanitation in the fall and winter. Use chemical pesticides only as a last resort, but don't totally disregard their use because they do have a place in any control program. The danger with chemical pesticides is the tendency to over use them. We choose chemicals without really knowing what kind of pest we have and whether or not the chemical will do the job. Be sure you know what you are controlling, and be sure that it is in a life stage that is vulnerable to the pesticide being used. Carefully read the label to find the pest and situation that confronts you. If you are using the chemical for some other pest or in some other way than what is stated on the label, you are illegally using the chemical. Do not use a shotgun approach by applying an all purpose mixture of insecticides, miticides, and fungicides. You are wasting your money and effort with this kind of approach, and you may be unnecessarily contributing to pollution of the environment by applying pesticides when not needed.

Whether or not you take a direct control action against the insect or disease pest, you should always try to maintain the tree in a good healthy vigorous growing condition. Make sure the tree is not under any additional stress from lack of moisture and/or nutrients. An application of fertilizer to help stimulate foliage is a way to try to get the tree back into a healthy condition. Remember, the leaves are the food and energy manufacturing machinery of your tree, and the only way the tree can regenerate new roots, branches, and leaves is for the tree to first manufacture food. But fertilizers are a two-edged sword: Overfertilization can cause the production of succulent shoots which will be more attractive to sucking insects such as aphids, and overfertilization can increase the soil microorganisms and lead to more root rot.

Animals

Deer. In the rural backyards, deer are probably the number one trouble-some animal pest. Deer will browse off the buds, shoots, and foliage, and will scrape off the bark with their antlers. The most critical type of damage is the removal of the terminal bud. This retards height growth and changes tree shape. The continual removal of the leader's bud keeps the buds, new shoots and foliage low to the ground where the deer can continually feed each year. The best way to minimize deer damage is to do everything legal in your power to separate the deer from the terminal bud. Once the tree starts to outgrow the deer's reach the deer generally leave it alone. There may be some lower branch pruning during severe winters, but you and your tree can tolerate the damage.

Most bud damage occurs during the dormant season, especially very early and very late in the dormant season. Bud protection should start soon after the first frost and last until the buds begin to elongate. Protection can be accomplished with the use of repellants, cages, or bud wrappings. Repellants offer varying success but have to be continually reapplied to be effective. Their effectiveness is related to weather conditions; the more moisture, the shorter period of effectiveness. Commercial repellants are expensive, and you generally need spray equipment to use them.

There are a number of home remedies which usually are cheaper than commercially made repellants. One homemade repellant is human hair. Fresh cut human hair is collected from beauty shops and barbers; small amounts are placed in bags made from nylon stockings and hung on the trees. When the oil and the odor are leached out of the hair, the repellant effect is lost. The main advantages of hair are the low cost and continual supply. Other home remedies include hanging bars of fragrant soap on the trees, wrapping the buds in Christmas tree tinsel, and spraying the buds with a mixture of eggs and red peppers which are blended into a liquid concoction to be applied with a sprayer.

Cages or netting are longer lasting than repellants but can be more expensive. They also are slower to apply. Some netting or cage material is biodegradable. Our experience with cage material, however, casts doubt on the material's biodegradable nature. The material has not degraded, and when the tree grows up through the cage, the leaders become deformed and curled. If you use cages or leader covers, plan on removing them each spring before growth begins.

Bud wrappings are inexpensive and effective alternatives to repellants and cages. These wrappings are simply a physical impediment to deer browsing. They can consist of paper or cloth stapled or tied around the terminal bud. Use small squares of kraft tree wrap stapled around the buds. Tree wrap is treated to withstand moisture and will last throughout the dormant season. Bud wrappings **must** be removed in the spring right before bud break so you don't damage and deform the new terminal.

Lower stem-chewing animals. Rabbits, mice and rats will chew on the stem and girdle the tree. Young, thin-bark trees are particularly attractive and vulnerable. Once the bark thickens and roughens, these animals rarely bother the tree. The damage is almost always done during the dormant season, but rats have been observed climbing trees and eating the tree's fruit during the growing season. Most damage is close to the groundline, but in the winter it will occur at the snowline which may be several feet above the ground.

Clearing away grass and other vegetation around the base of the tree should be your first action. Tall and thick vegetation around the tree will provide protective habitat for the rabbits and mice. Mice do not like to venture out in the open where they are vulnerable to predators. If the ground can be exposed or the vegetation cut short from the trunk out beyond the dripline, you stand a better chance of reducing gnawing damage. If you mulch your trees mice may live under a heavy mulch and feed on your trees from this protective cover. Pull the mulch away from the main stem so that you have a bare space from the trunk to half way to the dripline. Always use a trunk protector when using mulch.

Whether or not you mulch, consider using some form of trunk protector to prevent the animals from getting to the trunk. Build a loosely fitting fence around the trunk. It can be made of any material that rabbits and mice cannot chew through. Mice, however, have been known to chew through soft bricks, aluminum, and electrical wiring. Tin cans or cardboard oil cans with the ends removed may provide enough protection, but hardware cloth with a mesh no larger than 1/4 inch is probably the best exclusion device. Plastic sleeves can also be purchased to go around the trunk. An additional advantage of these devices is that they help protect the trunk from sunscald damage. Their disadvantage is that they are more expensive than homemade fences.

The trunk protector must be high enough to stick up above the anticipated snowline and should go down into the ground so mice can't burrow underneath. If the fence around the trunk has a loose enough fit, it can be left on for several years. Be sure to check it periodically so that it doesn't girdle the tree. Erect the trunk protector at the end of the growing season when the ground is soft enough to partially bury it and before the mice and rabbits have to use the tree as a food source.

There are rabbit repellants on the market which also work against mice, but their disadvantage is their short life. A last resort approach is to trap or shoot the rabbits and poison the mice and rats. In order to have an effective poisoning program, you must know whether you are dealing with mice or rats. Mice have a smaller range than rats, they tend to be curious about strange objects, and they are constant nibblers. Rats on the otherhand cover more territory, are more suspicious, and will gorge themselves in one eating. Because mice graze or nibble constantly, the poison used for mice control has to be stronger than that used to control rats.

Most of the poisons used are anticoagulants which cause internal hemorrhaging. With anticoagulants, the rodents must feed several times to get a lethal enough dose to kill them. A new poison now on the market has an active ingredient of Vitamin D3. This causes a calcification of the circulatory system which leads to death by heart attack. The advantage of this new poison is that rodents discontinue feeding after initially taking in the poison.

Poison placement is as important as the choice of poisons. Oats and wheat are usually used to carry the poison, and it's these grains which attract the mice to the poison. When using treated grain, the real danger is birds and other non-target animals eating the treated grain. Place the grain in the mice runways, where you see the mice, and under the canopy of the tree. Be sure the grain is covered over with vegetation or something artificial such as pieces of plywood or cardboard so only mice can get to the grain. Periodically check the grain to monitor its use.

Upper stem-chewing animals. Squirrels and porcupines will strip and chew bark off of the upper branches and the upper trunk area. Squirrels favor maple trees particularly when the sap is running. They will strip large areas of bark off small branches to get at the sap. Trees in stands where the squirrels can go from tree to tree through the crowns are impossible to protect. Isolated trees which have to be climbed from the ground can be protected by placing a band of smooth metal, 2-3 feet wide around the trunk about 6-8 feet above the ground. Remove all branches that are closer to the ground than 6 feet.

Porcupines feed all year long and usually feed on the main trunk, girdling the top. On pine trees they will feed on the trunk in the interbranch area, 2 to 4 whorls from the top. They feed only enough to girdle the top and then move to another tree. The best control is killing the animal before it girdles the top or before it can move to another tree. Tanglefoot, which is a sticky material used to prevent insects from crawling up the trunk, sometimes will discourage porcupines from climbing the tree to feed. Try using a smooth metal band similar to that recommended for rabbit control.

Sapsuckers. Sapsuckers, often mistaken for woodpeckers, can cause serious tree wounds and death. The sapsuckers will drill holes into the main stem, mix ants in the sap which fills the holes, and feed the ant/sap mixture to their young. The damage appears as multiple parallel rows of holes on the main trunk. The tree can be girdled at the point of hole drilling, but more often the wounds lead to ring shake and streaks of discoloration. Because the holes are wounds, a barrier zone is formed and the wood will often shake along these barrier zones. Sapsuckers are protected by federal law; so, their elimination is not advisable. The use of Tanglefoot, hardware cloth or burlap placed over the holes can discourage the sapsuckers. However, the area of drilling is usually high enough on the tree that most people cannot reach it to apply the protective covering. Sapsuckers seem to return year after year to the same trees. Try to keep the tree alive as long as possible and "donate" it to the sapsuckers. As long as they have some favorite trees, they usually will not go after other trees. Once their favorite tree dies or you remove it because it looks bad, the sapsuckers will seek out other trees.

Burrowing animals. Pocket gophers, moles, and ground squirrels can all cause serious tree damage and death. Pocket gophers are probably the most common and destructive of the burrowing animals. Gophers feed on roots and are particularly destructive during the spring and fall. Young trees are vulnerable because the root systems are not very extensive and are composed primarily of fine, small roots which are attractive to the gophers. Older woody roots are rarely attacked. When gophers eat the roots, the tree turns red in the case of a conifer or drops its leaves in the case of a deciduous broadleaf. The tree may also tip over because there are no roots to anchor the tree. To identify gopher damage, pull or dig up the tree and examine the root system. Missing roots usually mean gopher damage. Sometimes the entire root system will be gone, and only a stub will be left.

Pocket gophers have the capability of constructing an extensive underground tunnel or runway system 3-12 inches below the surface. As much as 800 feet of tunnels per acre can be constructed. Gophers push the excavated dirt to the surface, and circular or heartshaped piles of fine soil and narrow, sinuous mounds are noticeable on the surface.

The only control is trapping or poisoning. The spring and fall seasons are the most effective times of the year to trap and poison. Fresh mounds are easily found, and gophers are very active. Both poisons and traps must be placed in the runways; so, fresh evidences of activity are important to be successful in controlling gophers. Proper placement of trap and poison without significantly disrupting the runways is the key to control. The University Extension Service has an excellent publication describing control methods, and this publication can be obtained through your local County Agent.

Weather

Severe and extreme weather conditions can cause serious problems for your tree. Too little precipitation and too much precipitation can cause problems; low temperatures, bright sunlight and drying winds during the winter can also cause problems. Watering and mulching can help reduce drought stress, but conditions too wet can rarely be changed. Adverse winter conditions can't be changed, but you can do some things to help the tree resist these conditions.

Winter sunscald. This condition can be a serious problem on young deciduous trees and deciduous trees with thin, smooth bark. When the bark thickens and roughens, sunscald is rarely a problem. This condition is sometimes called "southwest injury" because it is frequently found on the southwest side of the tree. The afternoon sun warms up the bark, and cambial temperatures climb above freezing. Nighttime temperatures then drop below freezing, and the cambial area freezes killing the bark. This dead area appears discolored and can be covered with shrinking and cracking bark. Open grown trees are more commonly affected by sunscald because they are not protected by shading from surrounding trees. Sunscald commonly occurs in late winter when the days are getting longer and warmer, and the sun is at a higher angle.

You can prevent this kind of injury by wrapping the trunk. Use the plastic sleeve mentioned to protect the trunk from chewing animals, or use a paper tree wrap. This is a treated kraft paper which withstands moisture and will last until the following growing season. Wrap the tree in late fall and keep the wrap on until the start of the growing season. Do not leave the wrap on during the growing season. Wrapping the tree will insulate the trunk and prevent temperature fluctuations responsible for sunscald injury.

Winter browning. This is a common injury to the foliage of arborvitae and other ornamental conifers not totally suited to northern Minnesota. The sun is warm enough in late winter to thaw out the foliage, and when the temperature drops in the evening with the setting sun, moisture in the leaves freezes. The foliage turns red or brown. Often the browning foliage is not evident until the following spring. Like sunscald, symptoms are more pronounced on the south and southwest sides of the tree.

Control for this condition is not to plant these susceptible species on the west and south sides of buildings where temperatures will fluctuate much more than on the north and east side of buildings. Construct a shade barrier on the south and west sides of exposed trees. Anything you can do to protect the tree from the winter sun will help minimize damage. Withholding fall watering so the tree has adequate time to dry out and harden its tissues will prevent winter browning.

Winter burn. This condition is similar to winter browning but happens to all conifers, even the well-adapted native conifers. Rather than a freezing of moisture in the leaves after being warmed up, it is a drying out of the leaf tissues. Winter winds suck moisture out of the leaves, and because the ground is frozen, no water is available to replace the lost moisture. The needles turn red. Injury is usually on the side which faces the predominant wind. It is also on the most exposed parts of the tree, the outer and upper needles. If there is snow cover, the needles below the snow are protected from drying out and remain green. The pattern on the tree during years of bad winter burning is red on top of the tree down to the snowline, and then green below that.

Conditions leading to winter burn are not yet fully understood. It is advisable to prepare for winter burning possibilities every year by protecting your tree from direct exposure to the drying wind. Make sure the tree has had adequate moisture going into the dormant season, but do not water too late in the fall. Establish a thick mulch well beyond the dripline to prevent the frost from going very deeply into the ground. If the ground remains unfrozen, the roots should be able to replace some of the lost moisture. When establishing trees, be careful where you put them. Trees susceptible to winter burning should not be put on the sides of your building where they will get the full force of the wind. Some varieties of conifers are not very winter hardy. The short needled varieties of Scots pine, for example, tend to winter burn every year. Plant only those trees that you know are not particularly susceptible to winter burn.

Chemicals

Lawn trees are exposed to harmful chemicals more than any other kind of tree. This is due to the large amount of chemicals homeowners use on their lawns. The chemicals most harmful to trees are herbicides. Everytime you weed your lawn of dandelions or other undesirable broadleaf plants, you are using herbicides which can damage your tree. Herbicides most commonly used for broadleaf weed control in lawns are 2,4-D and dicamba. These are chemical names and are not the brand names. Be sure to read the label to determine which chemicals you are using.

Deciduous broadleaf trees and conifers actively growing can be damaged by both of these herbicides. Be careful when using these herbicides; stay at least 10 feet beyond the dripline. The herbicides can be absorbed through the roots, and some herbicides, particularly dicamba, are volatile and will move through the air on hot days. The greatest danger of herbicide use is during spring and early summer when the trees are actively growing. Avoid herbicide use around the trees during this time if the temperature is above 80°F or the winds are over 10 MPH.

Lawn care companies will produce nice lawns, but they can damage your tree. These companies rely on heavy doses of fertilizers, lime, and pesticides to produce green grass. Heavy applications of any of these chemicals can affect your tree. Heavy applications of lime can raise the soil pH which will tie up the iron and make it unavailable to the tree.

Tree foliage then becomes light green (chlorotic) between the veins. Use all things in moderation, and be careful to stay away from your tree. Remember, the root system extends beyond the crown of the tree.

HELP FOR TREES

Backyard tree care requires **your** active involvement! Tree abuse can be reduced, but it does take an effort. Remember, your tree is a living organism which does need your constant attention to give you life-long enjoyment. To maximize your enjoyment, be sure you select the proper tree for the planning site; use good quality planting stock; know how to take care of the tree until it is planted; understand proper planting techniques; adequately and properly mulch, water, fertilize, and prune your tree; and finally, be vigilant by protecting your tree from wounding, weather, chemicals, and insects and diseases. These actions will not only produce a tree you are sure to enjoy but will allow you to gain a new understanding and appreciation of your tree. You will develop an intense pride in your tree through this closer "relationship", and you will reap the benefits of many years of enjoying your tree's utility and beauty.

LITERATURE CONSULTED

Appleton, B.L. 1988. Fertilizing Landscape Trees and Shrubs: The Revised Version. Arbor Age, Vol. 8, NO. 10: 12-14.

Anonymous. 1986. Killing Rats and Mice That Pose Hazards to the Life of Trees. Arbor Age, Vol. 6, No. 11: 30-32.

Anonymous. 1986. Trees' Worst Enemies. Arbor Age, Vol. 6, No. 11.

Anonymous. 1988. Wildlife and Insect Damage and Hunter Abuses. The American Tree Farmer. March-April.

Ball, J. 1988. Backyard Tree Care. From a paper prepared for the Woodland Owners' Conference. February. St. Paul, MN.

Dornholt, P., and M. Schmidlin. 1986. How to Prune Trees. Minnesota Forests. MN For. Assoc. Newsletter, Fall 1986: 10-11.

Elfring, C. Winter Burn on Trees. Univ. of WI Ext. Serv. IN The Forest Management Digest, 5th Ed. ed. H. Allen Wickman. Forestry Farmers of MN, Inc., Park Rapids, MN: 216-217.

French, D.W., and W.C. Stienstra. 1975. Oak Wilt Disease. Univ. of MN Agric. Ext. Serv. St. Paul, MN. Ext. Folder 310-1975. 5p.

Goehring, III, J.J., and J.L. Goehring. 1988. Fertilization and Tree Health. Arbor Age, Vol. 8, No. 3: 24-25.

Hamilton, W.D. 1986. Tree Wounds: Their Causes and Treatment. Arbor Age, Vol. 6, No. 10: 36-39.

Hasselkus, E.R., and E.E. Schulte. 1986. Tree and Shrub Fertilization. Urban Phytonarian Series. WI Ext. Serv., Madison, WI. A2308. 2p.

Miles, W.R. Protecting Trees From Animal Damage. Univ. of MN Fact Sheet. Agr. Ext. Serv. St. Paul, MN. Forestry No. 8. IN The Forest Management Digest, 5th ED. ed. H. Allen Wickman. Forestry Farmers of Mn, Inc. Park Rapids, MN: 216-217.

Morrison, J. 1986. Trees Never Heal--They Just Seal. Arbor Age Vol. 6, No. 11: 24-28.

Ostry, M.E., and T.H. Nicholls. 1976. How to Identify and Control Sapsucker Injury on Trees. USDA For. Serv. NC For. Exp. Sta. St. Paul, MN. 5p.

Parten, H.L., and R.L. Isaac. Pocket Gopher Control. Univ. of MN Agr. Ext. Serv. IN The Forest Management Digest, 5th Ed. ed. H. Allen Wickman.

Rideout, R. 1978. Fertilizing Shade Trees. Univ. of MN Tree Line. Agr. Ext. Serv. St. Paul, MN. No. 15. IN The Forest Management Digest, 5th Ed. ed. H. Allen Wickman. Forestry Farmers of MN, Inc. Park Rapids, MN: 151-152.

Shigo, A.L. 1986. A New Tree Biology: Facts, Photos, Philosophies on Trees and Their Problems and Proper Care. Shigo and Trees, Associates. Durham, NH. 595 p.

Shigo, A.L. 1986. A New Tree Biology Dictionary: Terms, Topics and Treatments for Trees and Their Problems and Proper Care. Shigo and Trees, Associates. Durham, NH. 132 p.

Shigo, A.L. 1985. Wounded Forests, Starving Trees. Journal of Forestry Vol. 83 No. 11: 668-673.

Shigo, A.L. 1982. Tree Decay in Our Urban Forests: What Can Be Done About It? Plant Disease Vol. 66, No. 9: 763-768.

Stienstra, W.C. 1973. Noninfectious Diseases of Trees. Univ. of MN Fact Sheet. Agr. Ext. Serv. No. 12. IN The Forest Management Digest, 5th Ed. ed. H. Allen Wickman. Forestry Farmers of MN, Inc. Park Rapids, MN: 213-214.

Suihra, P. 1987. When to Fertilize Deciduous Woody Ornamentals. Arbor Age, Vol. 7, No. 6: 26.

Tattar, T.A. People-Pressure Diseases of Trees (PPD). Cooperative Extension Service, Univ. of Massachusetts, Shade Tree Laboratories. C-134. 7 p.

Watt, R.F. 1978. Shade-Tree Selection. IN Urban Foresters Notebook. ed. Silas Little. USDA For. Serv. General Technical Report NE-49. NE For. Exp. Sta. Broomall, PA: 79-86.

Worf, G.L., J.E. Kuntz, and M.J. Drilias. Maple (Acer) Decline: Collar Rot and Basal Canker Complex. Urban Phytonarian Series. WI Ext. Serv. Madison, WI A2532. 3p.

AGRICULTURAL EXTENSION SERVICE
UNIVERSITY OF MINNESOTA

HORTICULTURE
FACT SHEET No. 22—Revised 1977
M.C. EISEL

Street Trees for Minnesota

Trees can add infinite beauty to a community and create a feeling of peace and tranquility. The flowers, the cool green foliage, the form, the fruits, the bark, the branching habit, and the autumn color add beauty to our cities and towns. They help achieve continuity in the community by tying together varied styles of architecture.

Since Dutch elm disease is taking its toll in Minnesota, people in towns and communities are becoming interested in other trees they can plant. In selecting trees for a community today, several different kinds should be considered. A good plan is to use different kinds of trees on different streets rather than a random planting of many different kinds of trees on the same street. Variety is needed to prevent a situation where a single tree species may be injured or killed by a disease or pest.

There are a number of important considerations when selecting trees. Due to the severity of our climate, hardiness is of primary importance. The map at right shows hardiness zones in Minnesota. Microclimates within the hardiness zones can be more favorable or more severe than the general climate of the area. These must be considered when selecting trees. Winter temperatures often are not as severe in urban areas as in the surrounding country. Plants grown in a lawn area are likely to "harden off" in the fall and winter over better than those grown under clean cultivation.

Rainfall, wind, and soil pH must be considered. Along the Red River Valley in northwest Minnesota trees must be tolerant of a high pH.

The tree's mature size should be considered when street trees are chosen. Selecting trees that will not outgrow their allotted space will save maintenance time and eliminate the necessity of removing trees before they reach maturity. Many trees are being brutally pruned because attention was not given to their mature size when they were planted. It would be better if such trees were removed and replaced by smaller ones.

If trees are planted beneath utility wires, their mature height should be below the wires, or the tree should be planted on the homeowner's property on the house side of the sidewalk. Where salt poses a threat, setting trees back from the street will minimize the problem. Trees should not be planted near corners where they are likely to obstruct vision, traffic control lights and signs, or street lights.

Some trees litter sidewalks and streets with large and messy fruits, twigs, and leaves that can plug sewer entrances. The best trees have deep root systems, since the roots of some trees can clog and damage sewer lines. Check the following list of undesirable trees. Be certain of where underground utility wires and gas lines are located before you start digging. Selecting trees that have few insect and disease problems will minimize the amount of spraying required. All trees require some maintenance, but those on the recommended list require a minimum. Trees should be able to grow in existing soils, or the soils should be modified before trees are planted. Trees must be able to tolerate city air and the reflected heat from streets and concrete buildings.

All trees require care; even the best may not survive without it. Trees should be carefully planted, immediately watered, and carefully pruned. Water should be provided during dry spells at least through the first two growing seasons. Staking usually is necessary for trees larger than 6 feet or trees located in a windswept area. Weed competition must be controlled either by mulching or hand hoeing. A guard, mulch, or cultivated area around the base of the tree will reduce the chances of damaging it with lawn mowers and other equipment. Trees planted in parks or along roadways where there is tall grass should be protected from rodents with a cylinder of quarter-inch hardware cloth. To prevent sunscald, thin, smooth-barked trees must be wrapped in the fall for several years. The proper training and pruning of young trees will eliminate many potential problems and reduce maintenance costs. Street trees should be pruned to remove lower branches that interfere with pedestrian and vehicular traffic.

Most of the trees on the list on the back of this fact sheet can be seen at the Minnesota Landscape Arboretum at Chaska.

TREES NOT SUITED FOR STREET TREES

(Some of these trees may be useful for other purposes.)

Boxelder *(Acer negundo)*—Brittle wood; has boxelder bugs on female trees.
Silver Maple *(Acer saccharinum)*—Weak wood; shallow roots; gets too large.
Birch *(Betula spp.)*—Due to unfavorable conditions, susceptible to die-back and birch borers.
Catalpa *(Catalpa speciosa)*—Messy leaves, flowers, and fruits.
Black Walnut *(Juglans nigra)*—Often defoliates early in the season; messy; slow to leaf out in spring.
Mulberry *(Morus alba tatarica)*—Fruits messy.
Poplars *(Populus spp.)*—Short-lived; brittle wood; many get too large; some have suckering roots.
Black Locust *(Robinia pseudocasia)*—Subject to borer damage; has suckering roots.
Willows *(Salix spp.)*—Weak wood; has numerous twigs.
Elms *(Ulmus spp.)*—Susceptible to Dutch elm disease; some have brittle wood.

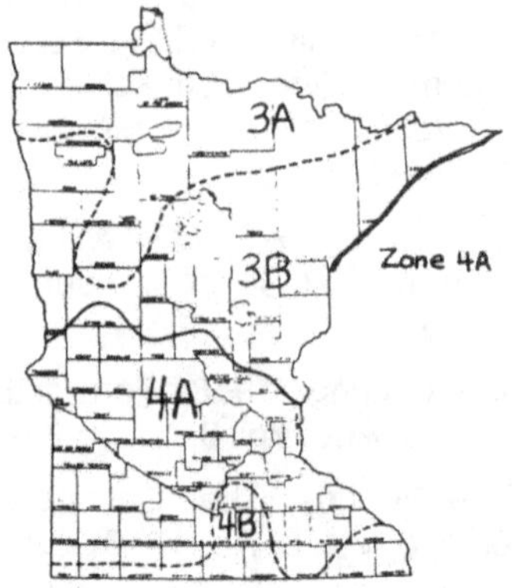

The zones approximate the hardiness zones taken from the plant hardiness zone maps prepared by the U.S. National Arboretum, Agricultural Research Service, U.S. Department of Agriculture, in cooperation with the American Horticultural Society.

Issued in furtherance of cooperative extension work in agriculture and home economics, acts of May 8 and June 30, 1914, in cooperation with the U.S. Department of Agriculture. Roland H. Abraham, Director of Agricultural Extension Service, University of Minnesota, St. Paul, Minnesota 55108. We offer our programs and facilities to all persons without regard to race, creed, color, sex, or national origin.

RECOMMENDED STREET TREES †

Tree	Height	Spread	Growth rate	Hardiness zones 3a	3b	4a	4b	Remarks
Norway Maple *(Acer platanoides)*	M	W	M			X	*	Protect all maples from sunscald.
Cleveland Norway Maple *(Acer platanoides 'Cleveland')*	M	I	M			X	*	One of the best maples.
Emerald Queen Norway Maple *(Acer platanoides 'Emerald Queen')*	M	W	M			X	*	
Schwedler Norway Maple *(Acer platanoides 'Schwedleri')*	M	W	M			X	*	
Summer Shade Norway Maple *(Acer platanoides 'Summer Shade')*	M	W	M			X	*	
Red Maple *(Acer rubrum)*	M	W	M	*	*	*	*	Best on acid soils.
Sugar Maple *(Acer saccharum)*	L	W	S-M	*	*	*	*	Heavy, moist soils required.
Ohio Buckeye *(Aesculus glabra)*	M	W	S-M	*	*	*	*	
Hackberry *(Celtis occidentalis)*	L	W	M	T	*	*	*	
Horse Chestnut *(Aesculus hippocastanum)*	L	W	M				*	
Russian Olive *(Elaeagnus angustifolia)*	S-M	M	M	T	*	*	*	
White Ash *(Fraxinus americana)*	L	M	M-F		*	*	*	Native source recommended.
Green Ash *(Fraxinus pensylvanica)*	L	W	F	*	*	*	*	
Marshall Seedless Green Ash *(Fraxinus pensylvanica 'Marshall Seedless')*	L	W	F	*	*	*	*	Broader than Summit.
Summit Green Ash *(Fraxinus pensylvanica 'Summit')*	L	W	F	*	*	*	*	
Blue Ash *(Fraxinus quadrangulata)*	M	W	S			X	*	
Ginkgo *(Ginkgo biloba)*	M	W	S			X	*	
Imperial Honeylocust *(Gledtsia thriacanthus 'Imperial')*	L	W	F			X	*	
Skyline *(Gledtsia thriacanthus 'Skyline')*	L	W	F			X	*	
Kentucky Coffeetree *(Gymnocladus dioicus)*	M-L	W	S			*	*	Slow to leaf out in spring.
Flame Flowering Crabapple *(Malus 'Flame')*	S	I	M	*	*	*	*	
Radiant Flowering Crabapple *(Malus 'Radiant')*	S	I	M	*	*	*	*	
Red Splendor Flowering Crabapple *(Malus 'Red Splendor')*	S	I	M	*	*	*	*	
Vanguard Flowering Crabapple *(Malus 'Vanguard')*	S	N	M	*	*	*	*	
Ironwood *(Ostrya virginiana)*	M	I	M	*	*	*	*	
Amur Corktree *(Phellodendron amurense)*	M	W	M			T	*	
Swamp White Oak *(Quercus bicolor)*	L	W	S-M			T	*	
Pin Oak *(Quercus palustris)*	M	I	M		T	*	*	Requires acid soil.
Sovereign Pin Oak *(Quercus palustris 'Sovereign')*	M	I	M		T	*	*	Requires acid soil.
Showy Mountain *(Sorbus decora)*	S	I	M	*	*	*	*	Eastern counties only. Protect from sunscald.
Japanese Tree Lilac *(Syringa amurensis japonica)*	S	W	S	*	*	*	*	
American Linden *(Tilia americana)*	L	W	M	*	*	*	*	
Littleleaf Linden *(Tilia cordata)*	M	W	S		T	T	*	
Greenspire Linden *(Tilia cordata 'Greenspire')*	M	I	S		T	T	*	
Redmond Linden *(Tilia x euchlora 'Redmond')*	L	I	S		T	T	*	

† Height
S = small (25 ft. and smaller)
M = medium (25-50 ft.)
L = large (50 ft. and over)

Spread
N = narrow (15 ft. and less)
I = intermediate (15-25 ft.)
W = wide (25 ft. and wider)

Growth rate
F = fast
M = medium
S = slow

* = recommended
T = trial planting (protected sites)
X = recommended in southern part of zone, trial in northern part of zone on protected sites.

WINTER BURN ON TREES

CHRIS ELFRING, UW-Extension Writer

Man's reaction to cold is immediate — we bundle up, move indoors, or suffer the consequences with alarming rapidity.

Trees may look contentedly inactive all winter, but they can suffer. Damage may not show until spring, but nonetheless winter is a test of strength for trees just as it is for other living creatures.

Low temperature though, isn't the sole culprit. Cold, bright sun, wind and frost combine to take their toll in the forest. Come spring, the appearance of brown, dead branches, and whole trees as well, will attest to the severity of another Wisconsin winter.

"The problem is called 'winter burn'," says Gordon Cunningham, University of Wisconsin-Extension forester. "During the deep freeze of mid-winter, moisture is removed or transpired from tree bark and foliage faster than the roots can replace it."

Many factors contribute to this dehydration. All plants lose water through their leaves or needles. Transpiration is a physiological cooling mechanism similar to perspiration. The winter burn problem begins when, despite the cold air, sunlight warms the foliage and branches enough to increase transpiration.

More warming can occur when snow reflects even more sunlight onto the plant. This reflected sunlight and wind, which worsens the dehydration by carrying moisture away more rapidly, can further increase the amount of water lost.

Winter burn actually is caused by more than the rapid loss of moisture through transpiration. At the same time the tree is losing essential moisture to the air, it is also unable to replace it because the roots are threaded through frozen ground. Ice simply is not a good source of water for plants — it cannot be readily absorbed. And in cold weather, what liquid water there is moves more slowly into the tree and through its cells. Like oil in a car, the colder the weather, the slower the movement.

"Sunlight, wind and radiation off the snow combine to cause excess transpiration, and these set the stage for winter burn," Cunningham says. But the extent of the burn also depends on how deeply the ground is frozen and how deep the tree's root system extends.

Frost depth varies with the severity of the winter and snow cover. The long-term average is 7.5 inches.

The extreme cold often associated with bright, sunny weather aggravates the winter burn problem.

Winter burn is not very visible in winter. It is in early spring, when damaged foliage starts turning brown, twigs fail to bud, or dead bark becomes noticeable.

When spring arrives, the dying foliage and branches will be most obvious on the tree's south and west faces — the sides where winter sunlight is most intense. Severe winter burn can kill an entire tree, but more often damages only portions.

"Sometimes you can judge how deep the snow cover was in a spot by examining low evergreens for the extent of winter burn," says Cunningham. "Snow is a good insulator, so the parts of the evergreen that were buried will be alive and green. But there can be a sharp line above which needles are brown—these were above the snow and subjected to wind, sun and harsh weather."

Winter burn is harder to spot in bare, deciduous trees than in conifers. Thin-barked trees like maple will show "frost cracks." These are small cracks in the bark that form when a cold spell freezes sun-warmed water in the underbark. But the extent of the damage to a deciduous tree also will not be known until spring, when some branches blossom and others do not.

For forest trees, winter burn is just one of nature's tests, a factor in the process of natural selection by which the more resistant survive to reproduce.

You can help lawn shrubbery and small trees by protecting them. Burlap shields will block both wind and sun — around the south and west sides of susceptible plants.

"You'll find more damage to shady-site trees like yews. These will winter burn more severely than sun-tolerant species like junipers", Cunningham says. This is because yew foliage has a thinner cuticle, or outer skin, which loses moisture more easily than waxy juniper needles.

One of the best ways to avoid losing trees to winter burn is to keep site-preference in mind when planting. In Wisconsin's severe winter climate, yews do better when they are protected from the sun, perhaps in the shade of buildings, while junipers are able to survive sunny southwestern exposures.

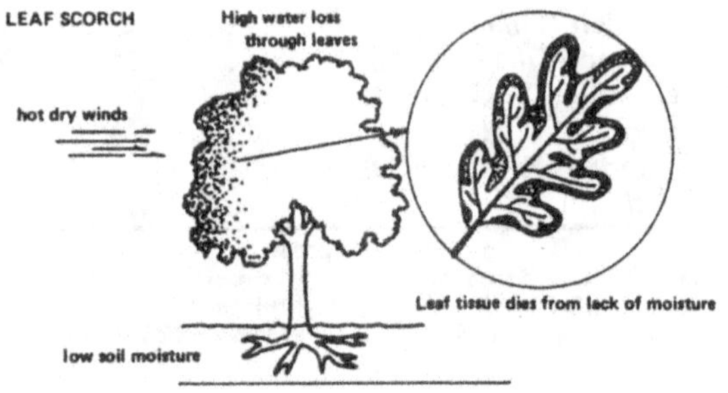

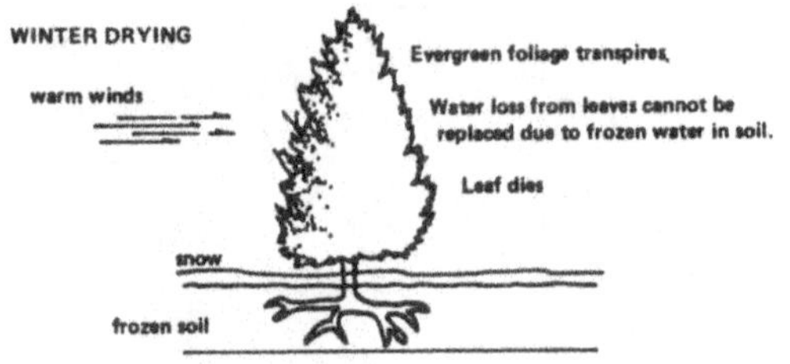

MINNESOTA TREE LINE Agricultural Extension Service
University of Minnesota

Protecting Trees and Shrubs Against Winter Damage

No. 17–1978
By Richard Rideout

Minnesota's harsh climate is responsible for a lot of damage to plants in the landscape. Winter sun, wind, and cold temperatures can burn evergreens, damage bark and kill branches, flowerbuds and roots, while snow and ice can break branches and topple entire trees. Salt used for deicing pavements is harmful to landscape plantings. Winter food shortages force rodents and deer to feed on bark, twigs, and foliage injuring and sometimes killing trees and shrubs. All is not bleak, however, for there are some things we can do to protect our landscape plants and help minimize any injury.

COLD DAMAGE

Cold temperatures can cause damage in several ways. Plants that are not hardy in Minnesota will be killed by early frost or low temperatures, but plants that normally grow here may be injured as well.

Sun Scald

Sun scald is characterized by elongated, sunken, dried or cracked areas of dead bark, usually on the south or southwest side of a tree. On a cold winter day the sun can heat up bark to the point where cellular activity begins. When the sun goes behind a cloud or building, the bark temperature drops rapidly, killing the active tissue.

Young trees, newly planted trees, and thin barked trees (cherry, crabapple, honeylocust, linden, maple, mountain ash, plum) are most susceptible to sun scald. Trees that have been heavily pruned or transplanted from a shady to a sunny area are also sensitive because of the loss of shading from other branches. Older trees are less subject to sun scald because the thicker bark can insulate the dormant tissue from the sun's heat.

Sun scald can be prevented by wrapping the trunk with a commercial tree wrap paper, plastic tree guards, or any other light colored material. The wrap will reflect the sun and keep the bark at a more constant temperature. Put the wrap on in the fall and remove it in the spring after the last frost. Newly planted trees should be wrapped for at least 2 winters and thin barked species up to 5 winters or more.

To repair sun scald damage, cut the dead bark back to live tissue with a sharp knife, forming the wound in the general shape of an elipse with rounded ends to facilitate healing (figure 1). Wrap the trunk in subsequent winters to prevent further damage.

Figure 1. Repairing sun scald damage.

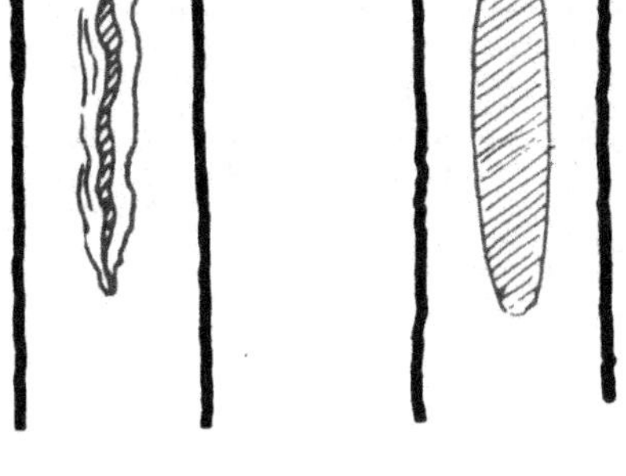

Winter Browning of Evergreens

Browning of evergreen foliage occurs when winter sun and wind cause excessive transpiration (foliage water loss) while the roots in frozen soil are unable to replace lost water. Browning, like sun scald, can also be caused by rapid drops in foliage temperature resulting when the sun drops out of sight on a cold winter day. Damage normally occurs on the south, southwest, and windward side of the plant, but in severe cases the whole plant may be affected. Yew, arborvitae and hemlock are the most susceptible, but winter browning can affect all evergreens. New transplants are particularly sensitive.

There are several ways to minimize winter browning damage. The first is proper placement of evergreens in the landscape. Yew, hemlock, and arborvitae should not be planted on south or southwest sides of buildings or in highly exposed places. A second way to reduce damage is to prop pine boughs or Christmas tree greens against or over evergreens to protect them from wind and sun and to catch more snow for natural cover.

Winter browning can be prevented by constructing a barrier of burlap or similar material on the south, west and windward sides of the evergreen (figure 2). If a plant has been showing injury on all sides, surround it with a barrier but leave the top open to allow for some air and light penetration. Keeping evergreens well watered throughout the growing season and into the fall is another way to reduce winter burn. Watering only in the late fall does not help reduce injury.

Anti-desiccant and anti-transpirant sprays are often recommended to prevent winter burn. Most studies, however, have not shown them to be effective and in some cases there has been more damage with them than without them.

If an evergreen has suffered winter burn, wait until late spring before doing any repair. The brown foliage is dead and will not green up, but the buds, which are more protected than the foliage, will often grow out and cover up the brown. If the buds have not survived, prune off the dead branches, and fertilize in early spring. You should water the tree well throughout the season and provide some protection the following winter.

Figure 2. Protecting evergreens from winter burn.

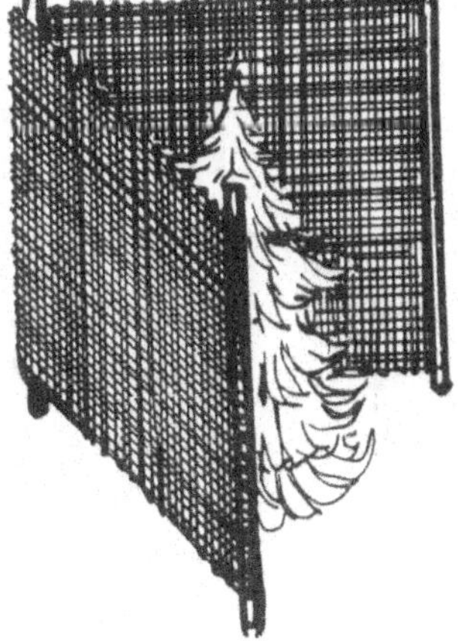

Die Back

Deciduous trees and shrubs can suffer from winter shoot die back and bud death. Flower buds are the most susceptible to injury. A good example of this is forsythia, which grows well, but often flowers very poorly.

There is little you can do to protect trees and shrubs from winter die back. Plants that are not completely hardy should be planted in a sheltered location. Plants in a vigorous growing condition late in the fall are more likely to suffer winter die back, so avoid late summer pruning and fertilizing. Apply fertilizer in the spring or in the fall after the leaves have dropped.

Root Injury

Roots do not become dormant in the winter as do the stem, branches and buds and so they are less hardy than the tops. The roots of most trees and shrubs that grow in this state will be killed at temperatures below 0° to +10°F. These plants survive in Minnesota because the soil temperature normally is much higher than the air temperature.

Many things influence soil temperature. Moist soil holds more heat than dry, so in a sandy soil or during a dry year, frost penetration will be deeper and soil temperatures colder. Snow cover acts as an insulator as does mulch, keeping the soil temperature higher. With newly planted trees, any cracks in the planting hole will allow the cold air to penetrate into the root zone.

To reduce root kill, mulch new trees and shrubs with 4 to 6 inches of a material, such as wood chips or straw if snow cover is unreliable. If the fall has been dry, water heavily before the ground freezes to reduce frost penetration. Check new plantings for cracks in the planting holes and fill them with soil. Check stakes and guy wires to be sure they are tight to prevent further cracking.

Heaving

Repeated freezing and thawing in the spring causes soil to expand and contract which can result in heaving shrubs and new plantings out of the ground. A 2- to 6-inch layer of mulch will prevent heaving by maintaining a more constant soil temperature.

ICE AND SNOW DAMAGE

Heavy snow and ice storms cause damage by bending and breaking branches. Multiple leader upright evergreens, such as arborvitae and juniper, and multiple leader or clump trees, such as birch, are most subject to damage. Relatively small trees can be wrapped together with heavy twine or the leaders tied with strips of carpet, strong cloth, or nylon stockings two-thirds of the way above the crotch (figure 3). These wrappings must be removed in the spring to prevent girdling, and allow free movement of the stem. For trees with large wide-spreading branches or large multi-stemmed trees, the branches should be cabled together by a professional arborist.

Figure 3. Protecting trees from snow or ice damage.

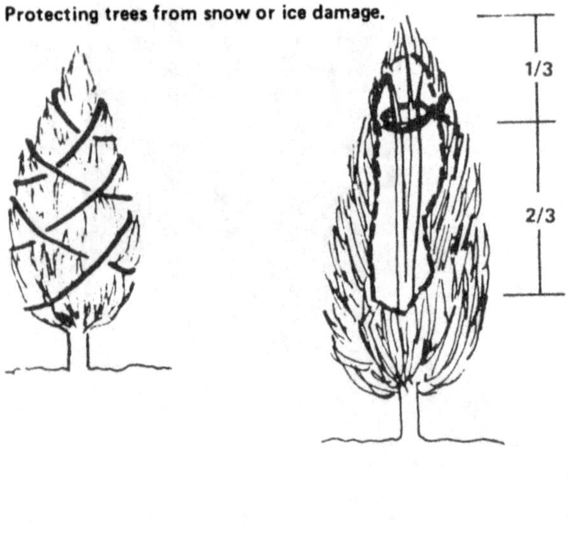

SALT DAMAGE

Salt used for deicing roads in the winter can cause or aggrevate winter burn and die back. In addition, salt runoff can injure roots and be absorbed by the plant, ultimately damaging the foliage.

To prevent salt damage, do not plant trees and shrubs where salt is present. Avoid areas where salty runoff collects and where salt spray kicked up by cars may land on plants. Burlap barriers (figure 2) will protect plants from salt spray.

ANIMAL DAMAGE

Mice, rabbits and deer are the main culprits causing damage in the winter. These animals feed on the tender twigs and foliage of landscape plants. They can girdle trees or eat shrubs to the ground line.

Rodents

Trees can be protected from rodent damage by placing a cylinder of ¼-inch mesh hardware cloth around the trunk. The cylinder should extend 2 to 3 inches below the ground line for mice and 18 to 24 inches above the anticipated snow line for rabbit protection (figure 4). The hardware cloth can be left on year around, but it should be larger than the trunk to allow for growth. For small trees, plastic tree guards are also effective. You can protect shrub beds from rabbits by fencing the beds with chicken wire.

Figure 4. Protecting trees from rodents.

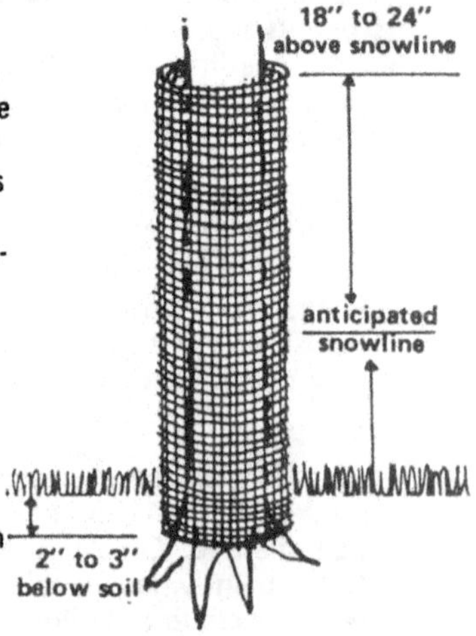

If you have many trees or shrubs to protect, using screens may be too expensive and time consuming. In such a situation, repellents may be the best solution. Remember that a repellent is not a poison; it simply renders the plant undesirable through taste or smell.

The most effective repellents for rodents are those containing thiram, a common fungicide. You can either spray or paint these on trees and shrubs. Repeat applications may be necessary periodically and after heavy precipitation because of their limited lasting ability.

If these methods do not control the rodents, commercial baits containing poisoned grain are effective. However, baits are hazardous to humans, pets, and beneficial wildlife. Sickness or death can result in animals that eat the bait directly or consume bait-killed rodents. Use baits only as a last resort. Trapping and shooting, where legal, will also control rodents.

Deer

Deer feed on and damage tops and side branches of small trees and shrubs. Repellents containing thiram give effective control. For deer exclosures to be effective, fences must be high and constructed with posts and heavy wire mesh. However, this is usually not feasible. If deer are starving there is little you can do to prevent feeding. Providing a more palatable forage may help, but it may also attract more deer.

Issued in furtherance of cooperative extension work in agriculture and home economics, acts of May 8 and June 30, 1914, in cooperation with the U.S. Department of Agriculture. Roland H. Abraham, Director of Agricultural Extension Service, University of Minnesota, St. Paul, Minnesota 55108. The University of Minnesota, including the Agricultural Extension Service, is committed to the policy that all persons shall have equal access to its programs, facilities, and employment without regard to race, creed, color, sex, national origin, or handicap.

CHAPTER 17 —ASSISTANCE FOR THE FOREST LANDOWNER WHAT'S C.F.M.?

The Management Plan

By Terry Helbig

CFM, M & U, I&D, TF, ACP, FIP, TSI, and 3e are abbreviations for programs that Department of Natural Resources foresters use to help private landowners manage their woodlots.

These programs are important because of two-thirds of commercial forest land in the U.S. and seven million of the 18 million acres of commercial forest land in Minnesota are private woodlots/parcels of less than 1,000 acres.

Last year, Minnesota's 150,000 small woodlot owners produced almost all the state's Christmas trees, maple syrup, and quality hardwood. A good share of the pulp and saw timber also came from their lands.

Because of the importance of the woodlot owner, federal and state governments and private industry have joined together in programs to assist landowners to increase tree production on these lands. Collectively, these programs are known as Cooperative Forest Managment.

CFM programs fall into two major groups: management assistance and financial assistance.

Management Assistance

This group of programs covers a broad range of services ranging from individual tree inspection to examinations of intire forests. Perhaps the best way to illustrate the varieties of management advice available to private landowners through these programs is to use the case of a hypothetical person, Joe Landowner.

In December 1970, Joe bought 40 acres of land 90 miles from his home for recreation, hunting, investment, and as a possible site for a future retirement home. The land consisted of five acres of tilled field, seven acres of abandoned field, 13 acres of mature woods and 15 acres of immature woods.

Joe had many ideas for developing his land, but was unsure how to proceed. He was sure of one thing, though. He wanted to plant the abandoned fields with trees. However, before ordering trees, he had the foresight to contact his local DNR forester for assistance.

The forester welcomed the opportunity to work with another absentee landowner under the CFM program. Although Joe didn't know it, he was about to receive the first of many kinds of advice from his forester.

In his first talk with Joe, the forester was able to accomplish many things. In addition to getting a general idea about Joe's objectives for managing his land, he recommended trees best suited for the soil, advised on ordering trees at cost from DNR nurseries, pointed out proper planting techniques, and suggested methods of weed and brush control. He also encouraged Joe to plant trees over several years and to do a thorough job rather than try to plant all his land the first year.

Spring 1971 came and so did Joe's trees. Following the forester's advice, he planted the trees carefully and knew the satisfaction of doing the job right. The planting behind him, Joe then became interested in managing the forest already growing on his land. To develop a plan, he again made an appointment with the forester.

On the appointed day, the two men met and started through the woods. While gathering data on various timber types, the forester and Joe discussed many items related to woodland management. When the examination was over, the forester left with a better idea of what Joe hoped to accomplish by managing his woodland. Joe in turn gained an appreciation of the ecology of his land.

Upon returning to his office, the forester developed a management plan for the Joe Landowner woods. The plan included a description of timber types and management recommendations, wildlife management information, and suggestions for various miscellaneous activities, such as Christmas tree culture and maple syrup production. All the forester's recommendations were tailored to Joe's management goals.

Joe read his plan and began formulating a time schedule to complete the work. Some of the plan required no effort on Joe's part, such as leaving a portion of crops as wildlife food. Other parts of the plan, such as sanitation cuttng (removal of high-risk trees), required considerable work.

One recommendation the forester made was to harvest the

mature timber. Joe decided to have this done during the winter of 1971-72. Needing advice on prices to expect for timber and who possible loggers might be, Joe again contacted the forester for Marketing and Utilization advice (M&U).

The majority of the mature timber was clearcut, that is, all trees were harvested. However, three acres made up a mixed hardwood forest and this timber, on the forester's recommendation, was partial cut. The trees left standing would provide food for wildlife. With this stand, the forester, for a charge of 25 cents per thousand board feet and 15 cents per cord, marked the timber needing removal.

Over the winter, all timber harvesting was completed and Joe received a nice return for the logs cut.

Joe ordered more trees, and the spring of 1972 found him busy planting a section of an abandoned field. One day in late spring he walked through the cutover area. Nature had already begun renewing the area. Stump sprouting was starting in the clearcut areas where aspen was dominant before harvest.

On another walk later that summer, Joe noticed that leaves were turning brown on several trees. Fearing the start of an epidemic, Joe called the forester. As it turned out, the problem was an insect, the leaf miner, eating chlorophyll out of leaves. Since that insect seldome reached epidemic numbers, the forester recommended no action.

Many woodlot owners find, however that Insects & Disease (I&D) problems are much more severe than the one Joe experienced. As such, an important part of the CFM program involves identification of I&D problems and, where possible, recommending cures and preventive measures.

In the winter of 1972-73, because of Joe's interest in proper management of his woodland, the forester recommended that he become a **tree farmer**. The forester's recommendation was approved by the State Tree Farm Committee and Joe's woodland was certified as a Minnesota Tree Farm.

The Tree Farm (TF) program is sponsored by the nation's forest industries through the American Forest Institute (AFI). The purpose of the program is to recognize and assist landowners who are managing their woodlands for timber production and other forest uses.

To be eligible for the program, a landowner must meet standards for protecting his trees from fire and disease, for managing "crops" of timber, and for harvesting under approved methods.

While contributing to the nation's future timber needs, tree farmers contribute to their own financial security.

Financial Assistance

For his 1973 spring planting, Joe signed up for cost-sharing under the ACP (Agricultural Conservation Program).

The ACP and the more recent FIP (Forestry Incentive Program) are sponsored by the federal government. These programs were enacted by Congress because tree planting is a long-term investment and, thus, unattractive to many landowners. Cost-sharing is designed to lessen the amount of investment for a landowner.

There are two cost-sharing programs available to woodlot owners/tree planting and TSI. To be eligible for tree planting funds, Joe had to plant between 800 and 1,200 trees per acre. He also had to plant the trees in a solid block, i.e., windbreak and landscape plantings were not eligible. Weed and brush control around seedlings were essential. In addition, tree species planted had to be approved by the local forester. This assured that the right trees were planted on the right soil so optimum growth could be attained.

The second cost-sharing program, TSI (Timber Stand Improvement), provided funds for Joe to improve his 15-acre stand of immature trees. The forester pointed out proper ways to prune these trees (not more than 150 trees per acre) and thin the stand (removing weaker ones).

Site preparation for natural reseeding is another feature of the TSI program. Site preparation consists of removing or breaking up brush or dense litter (the leaf layer) to allow the area to regenerate naturally.

Spring 1974 saw Joe continuing his planting program. It also brought a property tax bill that he thought was too high for land dedicated to growing trees. He again contacted the DNR forester and was informed of special property tax breaks for woodlot owners in many counties.

Each County Board can adopt one of two special laws to reduce property taxes for forest landowners. The first is known as the 3e law which provides that forested land used exclusively for growing trees can be assessed at the homestead rate (20 percent of value).

The second law is the Tree Growth Tax Law. This places the tax for a piece of woodland at 30 percent of the average growth rate per acre multiplied by the average stumpage price per cord for that timber type paid over the last two year period.

Because valuations of forested land vary from county to county, it's impractical here to show tax savings on a "typical" piece of woodland. It is sufficient to say, however, that property tax savings are often significant. Joe found it worthwhile to apply to the County Board to place his land under one of the special taxes noted above.

In spring 1975, Joe Landowner tapped maple trees on his woodland and made his own syrup. In early summer, he began a Christmas tree shearing program on a stand of pines he had planted previously. When cut, the Christmas trees will serve as a thinning.

What will the future hold for Joe Landowner's woods? It's impossible to predict. Natural disasters, fire and a host of other problems may force Joe to change his managment plans completely. Whatever happens, though, Joe can call the DNR forester anytime for professional advice on managing his woodlot-and so can you.

WHAT IS A TREE FARM?

Left: Dick White 1988 Region 1 Tree Farmer of the Year,
Right: Allen Wickman Minnesota DNR

A Tree Farm is a tract of privately owned forest managed to produce continuous crops of trees with added benefits of improved wildlife habitat, watershed protection, outdoor recreation and aesthetic value.

BENEFITS

- All new Tree Farm members receive a certificate suitable for framing.
- Tree Farmers receive an attractive green and white Tree Farm sign that must be properly displayed in a scenic and conspicuous place on the Landowner's property.
- A colorful and informative magazine entitled "The American Tree Farmer" is mailed to members four times a year.
- Tree Farmers normally continue to receive Forestry assistance in order to follow their forest management plan.
- There is no charge for Tree Farm membership.

QUALIFICATIONS

How does a person become a Tree Farmer? The main requirements are listed below:

- A Landowner must own 10 or more wooded acres that are not being grazed.
- The trees on at least 10 acres must have been growing for a minimum of three years.
- The land must have been in the same family for at least one year.
- The property must have a forest management plan made or approved by a Forester.
- The Landowner must be following the recommendations listed in the plan. All forest land must be under management.
- The Landowner must request Tree Farm status. Normally this request is made to the Forester who made up the forest management plan.
- The Forester, with assistance from the Landowner, must make up a Tree Farm inspection report on the property.
- The Forester must recommend Tree Farm status. This recommendation must be approved by the Regional Tree Farm Committee.
- Regional approvals must be signed by the State Tree Farm Chairman.
- The Regional Tree Farm Committee is composed of public, private and industry foresters and normally meets two to four times a year. This committee considers new applications and reinspections (done on each Tree Farm every 5 years). Once a year the committee selects a Regional Tree Farmer of the year. Regional winners are entered in the state contest.

GETTING STARTED

Solid forest management begins with determining your objectives. Determining what you have on your lands and developing a forest management plan. A professional forester can answer your questions. A free inspection of your woodlands can be arranged by your state forest service, state forestry association or state Tree Farm Committee. Also, your county agent can recommend a Forester to inspect your property.

PIONEER TREE FARM PROGRAM

If you're just discovering the benefits of forest management and haven't met the qualifications for full Tree Farm certification, you may be eligible as a Pioneer Tree Farmer. By developing written management recommendations and showing a desire to follow them, you'll be on your way toward full certification. It costs nothing to join and doesn't obligate you in any way, other than your own personal dedication to good forest management.

COST SHARING FOR WOODLAND MANAGEMENT

What is cost sharing?
Cost sharing is a general name given to federal programs in which the cost of a practice is shared between the federal government and the landowner.

Why does the federal government cost-share?
The federal government believes forest resources are vital to the nation. Over half the timberlands in the U.S. are in small, private ownerships. Future wood supplies for our nation will have to come from these lands if the demand for wood is to be met. Therefore, to encourage private woodland management, the government will share some of the costs to the landowner. This provides the landowner with an incentive to manage his woodlands properly.

What forestry practices can be cost-shared?
1). **Timber stand improvement**
   a. **Thinning.** To improve individual tree quality and growth and have a good variety of trees. This is accomplished by removing trees of poor form, undesirable species, and even good trees if there are too many. The trees to be removed may be marked by your local forester.
   b. **Pruning.** To improve the quality of the bottom log of selected trees. At present, only black walnut, red pine, and white pine can be cost-shared, although other species would benefit from pruning. Pruning must be done according to the specifications on file at the A.S.C.S. Office.
   c. **Grape vine removal.** Grape vines are a serious threat to a woodlot. Grape vines compete with trees for light. They can kill or deform trees. And they increase the damages which result from ice storms.

2). **Planting.** Cost-sharing is available for planting labor, the cost of the trees, and weed and grass control for only one year on the same acreage.
This practice is for establishing trees for long range forest products only. Plantings for Christmas trees or ornamental purposes will not be cost-shared. Also the trees species must be approved by the local Forester and planted according to his recommendations.

3). Cost-sharing can be obtained for windbreaks, and the technical advice is given by the Soil Conservation Service. To qualify for any cost-sharing, woodlots or planting sites must be protected from grazing.

4. **Fencing** is required if there are grazing animals because grazing is damaging to woodlots. Only interior fencing can be cost-shared. Fencing must be permanent and the minimum is 3 strands with a maximum distance between posts of 16½ feet. Treated wooden posts (except for cedar) or steel can be used.

Where Can I Sign Up For Cost-Sharing?
Cost-sharing is administered by the Agricultural Stabilization and Conservation Service. Their offices are located in the county seat and listed under U.S. Dept. of Agriculture.

How Much Can Cost-Sharing Amount To?
Cost-sharing varies from 50% to 90%. See your local A.S.C.S. office for details.

Who Determines What The Cost-Share will Be?
Some of the practices, like planting, may have fixed cost-sharing rates. Thinning, grape vine removal, and pruning costs are usually determined by the local DNR forester.

If I Want To Plant Trees, Where Can I Get Them?
Trees can be obtained from State, Soil Conservation District, and Private Nurseries. See your local forester or the local Soil Conservation District for more details and assistance.

If I Sign Up For Thinning My Woodlot, Who Does The Cutting?
You can cut the trees yourself or hire it done. Your local forester can supply you with a list of cutters.

What Time Limit Is There, If Any, To Complete A Practice?
You have one year to complete your practice although extensions may be granted.

Must I Thin My Whole Woodlot In A Year?
You must accomplish whatever you signed up for.

Example No. 1: Let's say you sign up to thin 20 acres, but the local forester determines only 10 acres need thinning. You can either thin the whole 10 acres or inform the A.S.C.S. that you want to do a lesser amount this year. But you will not get paid for more than 10 acres or the number of acres you do.

Example No. 2: If 20 acres of your woodlot needs thinning, you can sign up to do only 5 or 10 acres this year. The remainder may be done in succeeding years.

Cost-sharing must be requested by filing an application with the County A.S.C.S. Office. Written approval from the County A.S.C.S. Committee must be received before work is started on the practice.

For further details on cost sharing, consult your A.S.C.S. Office or local DNR forester.

Region III Forest Management, Michigan Department of Natural Resources

Acreage Conservation Reserve Program (ACR)

Reduces the acreage of specified crops at a specified percentage to increase crop values and reduce crop surplus. The acreage taken from production must be eligible cropland devoted to conservation uses to protect the soil from wind and water erosion. More information can be obtained by contacting the local **USDA — Agricultural Stabilization and Conservation Service (ASCS) office.**

Agricultural Conservation Program (ACP)

Provides cost sharing with farmers and ranchers to carry out conservation and environmental protection practices on agricultural land that result in long-term public benefits. More information can be obtained by contacting the **local USDA — Agricultural Stabilization and Conservation Service (ASCS) office.**

Conservation Reserve Program (CRP)

Pays farmers, based on bids, to discontinue growing crops on highly eroding cropland and plant it to grass and trees. Ten-year contracts are developed on the enrolled acres with annual payments to the landowner. More information can be obtained by contacting the local **USDA — Agricultural Stabilization and Conservation Service (ASCS) office.**

DNR Gift Program

Provides landowners with tax savings for donating full or a partial interest (easement) in property to the DNR or to a tax-exempt public foundation. Additionally, the Reinvest in Minnesota Resources Act of 1986 (RIM) provides a special gift feature whereby contributions to the fund are "matched" by state funds. For more information contact the **Minnesota DNR Section of Wildlife, 500 Lafayette Road, St. Paul, MN 55155-4007.**

Emergency Conservation Program (ECP)

Provides emergency funds for sharing the cost of rehabilitating farmland damaged by wind erosion, floods, or other natural disasters and for carrying out emergency water conservation measures during periods of severe drought. More information can be obtained by contacting the **local USDA — Agricultural Stabilization and Conservation Service (ASCS) office.**

Federal Waterbank Program

Authorizes 10-year lease contracts with landowners to protect qualifying wetlands. Agreements require landowners not to drain, burn, fill, or otherwise destroy the wetland. Thirty-eight Minnesota counties are eligible for this program. Payment rates typically are $10 per acre per year for wetlands and may range from $20 to $55 per acre for adjacent upland. Upland payment rates are based on cropland capability classes and a percent of the documented corn yield. Upland acres are planted to permanent grass-legume cover. Contact the **local UDSA — Agricultural Stabilization and Conservation Service (ASCS) office.**

Forestry Incentive Program (FIP)

Encourages landowners to plant trees on suitable open lands or cutover areas, and to perform timber stand improvement work for production of timber and other related forest resources. The FIP shares the cost of these practices with private landowners. For more information contact the **local USDA — Agricultural Stabilization and Conservation Service (ASCS) office or your local DNR Forestry office.**

Minnesota Forestry Incentives Program (MFIP)

Provides cost sharing for forestry related practices not covered by other state and federal programs such as pest control, fire break establishment, forest road construction, etc. More information can be obtained by contacting the **local DNR Forestry office or local Soil and Water Conservation District office.**

Minnesota Waterfowl Association's Habitat Development Program

Provides cost sharing to landowners to restore or create shallow wetlands and other wildlife habitat development projects. For more information contact the **Minnesota Waterfowl Association, 3701 Normandale Road, Minneapolis, MN 55424.**

Native Prairie Tax Credit and Exemption Program

Exempts approved native prairie from property taxes and provides property tax credit. Landowners can obtain application forms from their county assessors. The credit is equal to 1.5% of the average estimated market value of an acre of tillable land in the township in which the approved tract is located, multiplied by the number of acres in the tract. Contact your county assessor's office for application forms. For information on native prairie, write the **Minnesota DNR, Natural Heritage Program, 500 Lafayette Road, St. Paul, MN 55155-4007.**

Nongame Wildlife Program (NWP)

Provides written materials and advice to private landowners on projects to enhance their property for nongame wildlife. A book "Landscaping for Wildlife" will be available in summer 1987. Staff will prepare habitat management plans for bald eagle nest sites on private lands. Write the **Minnesota DNR, Nongame Wildlife Program, 500 Lafayette Road, St. Paul, MN 55155-4007.**

Pheasants Forever Cost-Share Program

Provides cost-sharing to establish 10-16 row shelterbelts, plant grasslands, or maintain corn or sorghum food plots over winter. Pheasants Forever cost-share funds are often blended with public funds to reduce the landowner's cost. For more information contact **Pheasants Forever, P.O. Box 75473, St. Paul, MN 55175.**

Private Forest Management (PFM)

Provides technical assistance to landowner participants in state and federal cost-share programs and state tax laws. This assistance includes inventory, multiple use management planning, timber-stand-improvement, timber harvesting and restoration. In addition, tree seedlings may be purchased from the DNR Nurseries. Write **Minnesota DNR, PFM Program, 500 Lafayette Road, St. Paul, MN 55155-4044 or your local Area or District DNR Forestry office.**

Private Lands Wildlife Habitat Improvement Program (WHIP)

Provides technical and cost-share assistance to landowners to help plant 10-16 row woody cover shelterbelts, restore or create small wetlands, establish over-winter food plots and maintain grassland nesting cover. Because funding for the program is mostly from pheasant stamps receipts, pheasant habitat development is emphasized. Waterfowl habitat and deer food projects are also cost-shared. Write **Minnesota DNR, Private Lands Program, 500 Lafayette Road, St. Paul, MN 55155-4007.**

Roadside/Wildlife Program (RWP)

Offers technical assistance to road officials, conservation groups, and landowners to improve the quality and/or management of roadsides or other grassland sites for the benefit of nesting farmland wildlife. This includes recommendations for seedings, renovation techniques, equipment, and vegetation maintenance. Write **Minnesota DNR, Roadsides for Wildlife Program, P.O. Box 756, Highway 15 So., New Uln, MN 56073.**

Save the Wetlands Program

Acquiries wetlands and adjacent uplands. Since the program's inception, 1000 Wildlife Management Areas (WMAs) have been acquired encompassing 530,000 acres. WMAs provide critical habitat for hundreds of plants and animal species. If your property is adjacent to an existing project, purchase may be a high priority. For more information contact the **Minnesota DNR, Section of Wildlife, Box 7, 500 Lafayette Road, St. Paul, MN 55155-4007.**

State Cost-Share Program

Assists landowners in the installation of permanent non-production oriented soil and water conservation practices. Cooperators are eligible to receive up to 75 percent cost sharing of specified erosion control and water quality practices. Technical assistance to landowners is also available through the program. Contact your **local Soil and Water Conservation District (SWCD) office or the Minnesota Soil and Water Conservation Board, 90 W. Plato Blvd., St. Paul, MN 55107.**

State Waterbank Program

Compensates farmers for not converting qualifying wetlands to cropland. Payments are based on appraised land values, providing incentives to keep qualifying wetlands in their natural state. The landowner must have been denied permission to drain the wetland and must show that drainage of the area would not violate any property agreements, that outlet rights can be obtained, that the proposed drainage would be profitable and that the area, if drained, would make high quality cropland. For more information contact the **Minnesota DNR Division of Waters, 500 Lafayette Road, St. Paul, MN 55155-4032.**

Streambank, Lakeshore, and Roadside Erosion Control Program (SLR)

Provides financial assistance to local units of government for the control of erosion along streambanks, lakeshores, and roadsides. Cost sharing is available for up to 50 percent of the local share of installing erosion control practices. Contact your **local Soil and Water Conservation District (SWCD) office or the Minnesota Soil and Water Conservation Board, 90 W. Plato Blvd., St. Paul, MN 55107.**

Wetland Acquisition Program (WAP)

The federal WAP uses two methods of acquisition, fee title and easement. Specified counties in northwestern, west central and southern Minnesota are eligible for this program. Eligible wetlands are primarily prairie wetlands with associated uplands managed to provide waterfowl habitat. More information can be obtained by contacting your local **U.S. Department of Interior Fish and Wildlife Service office.**

Wetlands Tax Credit and Exemption Program (WTCEP)

Exempts eligible wetlands from taxation and provides credit on wetland acres enrolled. Qualifying wetlands include land that is mostly under water, produces little, if any income, and is important to wildlife, water conservation and drainage must be legal, feasible and economically practical. All protected wetlands qualify. Counties qualifying for exemption and credit are those where wetland drainage for agricultural production is a common practice. Contact your **county assessor's office in participating counties or Department of Revenue, 658 Cedar Street, St. Paul, MN 55145.**

TREE PLANTING
BID INVITATION AND CONTRACT

You are invited to bid on a tree planting job to be done on the property of ______________________

______________________ (landowner's name). This bid will be considered as the planting contract if signed by the landowner and vendor.

PLANTING DESCRIPTION

Acres to Plant ______________ Total Number of Trees to Plant ______________

☐ See Attached Map for Location ☐ See Attached Instructions

Type of Planting: ☐ Hand ☐ Machine ☐ Other

Access to Plant Site is By: ______________________

Soils and Topography: ______________________

Type of Site Preparation: ______________________

Planting Specifications and Special Instructions:

Job to be Supervised By: ______________ Date All Trees to be Planted By: ______________

Party in Charge of Picking Up Trees: ______________________

Trees Coming From ______________ Nursery. Tree Order Number ______________

Tree planting must be approved by forester before payment will be made ☐ yes ☐ no

PLANTING SPECIFICATIONS

(A) Tree roots must be kept moist until trees are planted.
(B) Tree bundles must be protected from the drying effects of the sun and wind.
(C) Tree bundles must be well ventilated and not left in closed car trunks, vans, etc. which may heat up and kill the seedlings.
(D) Trees must be planted so that there are no "J" roots. Roots should ALL be below ground and not twisted or balled up - a good root spread is essential to the survival of the tree.
(E) Tree roots may not be pruned.
(F) Trees must be planted firmly in the ground.
(G) No trees should be planted in unsatisfactory spots such as water filled depressions, rock piles, rotted logs, heavy duff, brush piles, etc.
(H) For each work day required to complete the contract after the termination date, any extension, thereof, a sum of $ ________ will be deducted from the bid price.

THE CONTRACTOR MUST:

(A) Report any leftover trees to the landowner or his designated representative within 24 hours after the job is finished. The location of any leftover trees must be clearly stated.
(B) Submit a bill to the landowner or written notification that the job has been finished.
(C) Clean up all debris resulting from the planting job such as tree bundle wrappers and sticks, lunch wrappers, pop cans, etc.
(D) Repair, replace or pay for damage done to any property in the performance of the contract, beyond ordinary wear or tear.
(E) Provide all necessary equipment, labor and supplies for planting trees in a satisfactory manner.
(F) Notify landowner immediately if tree roots appear to be excessively dry as trees are removed from the bundle, or if trees appear to be defective.
(G) Inform landowner when starting project and make daily reports to owner while the job is in progress.

THE LANDOWNER MUST:

(A) Pay the tree planters within 30 days after receipt of written notification that the job has been completed.

(B) Provide all trees necessary to finish job.

(C) Provide rights of ingress and egress to the planting site for all men, materials and equipment necessary to perform the planting job.

PENALTIES

A 5% non-compliance to the above specifications is allowed. Non-compliance will be penalized at a rate equal to double the value of the trees improperly planted.

Non-compliance for the care and handling of the seedlings as stated shall be penalized at the rate of $50.00 per violation.

Where differences occur between the first and second parties on the amounts of penalties to be assessed, they shall be resolved by a mutually acceptable third party, who shall act as an arbiter and whose decision shall be final. Each party shall pay one-half his fee.

LIABILITY CLAUSE

No liability shall be imposed upon or incurred by the landowner or any of his officers, agents or employees, officially or personally, on account of the granting hereof or on account of any injury or damage to persons or property, whether avoidable or not, resulting from any of his agents, employees or contractors relating to any matter hereunder. Vendor shall indemnify and hold harmless the landowners from all claims arising out of the performance of this agreement. This agreement shall not be construed as estopping or limiting any action of the landowner against the vendor, their agents, employees or contractors for violation of or failure to comply with the provisions of this agreement or applicable provisions of law.

PLANTING BID

I bid $________ per 100 trees to be planted as specified in this bid invitation. This bid includes my transportation of trees as specified in the bid.

Name ____

Address ____

City ______________ State __________ Zip __________ Phone ______________

NOTE: Bidder must sign at bottom.

Submit your bid to:

Name ____

Address ____

City ______________ State __________ Zip __________ Phone ______________

This bid will be opened at ____ (place)

on: __________ (hour) ______________ (day) ______________ (month)

THE OWNER HAS THE RIGHT TO REJECT ANY AND ALL BIDS

NOTICE OF ACCEPTANCE*

LANDOWNER, sign and return one copy to Bidder of your choice as soon as possible.

YOUR BID HAS BEEN ACCEPTED

Landowner (signature) ______________________ (date) ______________

Bidder (signature) ______________________ (date) ______________

*** Contract must be signed by Bidder and Landowner - one copy to each**

TIMBER BID INVITATION

NOTE

"F.F.M., Inc." cannot guarantee the legality of any bid or contract form in this book. They are only to show what types of information might be included and are samples of forms that have been used. If you want a legal form see a lawyer or timber company.

Dear ....:

You are invited to place a bid for the purchase of timber stumpage on my property as described below:

Landowners name .... Phone ....

Address .... Zip ....

LOCATION ....

....

(A map showing location, sale area and access is attached.)

SPECIES, ESTIMATED VOLUME AND PRODUCTS ....

....

....

....

....

....

SALE SIZE .... Acres

ACCESS TO CUTTING AREA IS BY: ....

....

....

SLASH REGULATIONS ....

....

....

HARVEST TIME ALLOWED ....

LANDINGS ARE TO BE LOCATED: ....

SCALING: ( ) None ( ) Consumer Scale ( ) On Permit Area ( ) Other ....

....

Other Specifications: ....

....

....

....

....

....

....

PAYMENT: ( ) Lump Sum ( ) Based on volume estimate with final settlement to depend on total wood scale.

( ) other ....

............% of payment is due within .... days after approval of bid. The balance if any is due as follows: ....

....

TO LOOK AT WOOD: ....

....

BID WILL BE OPENED AT: (time, date, & place) ....

....

TIMBER BID

Name .... Signature ....

Company Represented (if any) ....

Address .... Zip ....

Phone .... Date ....

I bid $.................... for all of the timber as described (lump sum & sold as appraised sales).

I bid as follows: (for products that must be measured)

SPECIES	UNIT OF MEASUREMENT	PRICE PER UNIT
....	....	....
....	....	....
....	....	....
....	....	....
....	....	....
....	....	....
....	....	....
....	....	....

Return this bid to:

Name ....

Address ....

....

Mark your envelope; **TIMBER BID**
LAND OWNER HAS THE RIGHT TO REJECT ANY OR ALL BIDS.
All bidders will be notified within seven days after bids are let.

Timber Sale Contract

Agreement entered into this ____________ day of ____________________19______ by and between ____________ ______________________________ of ____, hereinafter called the Purchaser, and ______________________________________ of ______________________________________, hereinafter called the Seller.

SECTION I - Location:
Whereas the Seller has the right to sell, and warrants to the Purchaser that the timber and property are free and clear of all encumbrances, and agrees to sell to the Purchaser, forest products herein described on certain lands located in the ______________________________________, Section(s) ____________________ Twp. ____________, Range ____________, __________________________ County, Minnesota.

SECTION II - Terms of Contract:
Time is of the essence and unless an extension of time is granted in writing, all forest products must be paid for, cut and removed on or before ______________________________19______. All rights to the remaining forest products, cut or uncut, revert to the Seller. In the event that logging operations must be suspended due to adverse weather conditions, adverse ground conditions, fire, flood, strike, or other circumstances beyond the control of the Purchaser; this contract is extended an equal number of days.

SECTION III - Payment Schedule and Bond:

(1) Method and time of payment shall be as follows:

(2) Title to all forest products or standing timber included in this contract shall be and remain the property of the Seller until paid for.

(3) To insure performance of this contract by the Purchaser, a performance bond, surety deposit or irrevocable letter of credit from an approved financial institution for the amount of $____________________ is required to be presented to the Seller or his Agent at ______________________________________.

(4) Upon the completion of this sale, all bonds, surety deposits, or letter(s) of credit will be returned, refunded or canceled forthwith.

SECTION IV - Harvest Specifications:

SECTION V - Products and Prices:

Species	Product	Quantity	Price/unit	Total Price

SECTION VI - Other Provisions:

(1) The Seller's Agent shall be notified five working days in advance of each phase of the logging operation.

(2) Spring poles and "hung" trees shall be brought to ground level.

(3) The Purchaser shall have the right of ingress and egress to and from the lands over any other property which the Seller owns, including the right to use roads located thereon for the purpose of completing this sale.

(4) Stump heights will be as close to the ground as practical; however, they are not to exceed their diameter in height unless multiple stemmed, scarred, cat faced, etc.

(5) Location of landings, haul roads, and decking areas to be mutually agreed to by the Purchaser and the Seller's Agent.

(6) The Seller or his agent shall establish sale boundary lines and timber markings and does assume responsibility for their accuracy. The Purchaser is not responsible or liable for errors resulting from the Seller or his agent's incorrect marking of sale boundary lines or individual trees.

(7) The Seller is responsible for providing all necessary ingress and egress for the Purchaser to complete the terms and obligations of this contract.

(8) All debris generated by the Purchaser such as machine parts, oil cans, lunch papers, etc. will be removed from the sale area by the Purchaser.

(9) Trails and haul roads must be opened prior to skidding operations. All roads, landings, and decking areas shall be cleared of debris and bladed smooth by the Purchaser at the conclusion of the operation. Tops and limbs falling outside of the sale boundaries, water courses, or woodlands will be pulled within.

(10) Damaged and/or incidental residual trees removed to facilitate the logging operation will be charged to the Purchaser at the Seller's Agent's scale according to the price of the species in this contract.

(11) The Purchaser shall do all in his power to prevent and supress any and all forest fires. In the event that he or his employees cause fire damage, the value of the timber lost shall be charged to the Purchaser.

(12) The Purchaser shall execute reasonable care to protect the residual stand from damage. Excessive damage will be assessed by the Seller's Agent and be charged to the Purchaser.

(13) This contract is binding on all heirs, executors, administrators, successors, and assigns of all parties.

(14) This contract shall be construed according to the laws of the State of Minnesota.

(15) It is agreed between the parties that the Purchaser will not sub-contract out this agreement without the express prior written consent of the Seller.

SECTION VII - Arbitration:
In the event that a conflict arises regarding this contract, and cannot be resolved through discussion by the parties as a first step; then either party may request in writing to the other party that the area of disagreement be presented to an arbitration panel chosen as follows: one member to be selected by each party, and a mutually agreed to third party. All costs of the arbitration shall be charged against the party determined by the arbitration panel to be at fault. The rules of the American Arbitration Association shall govern the arbitration of said dispute. The determination of the arbitration panel shall be final and binding on the Purchaser and the Seller.

SECTION VIII - Liability:
The Seller or his Agent shall not be liable for injury or damage suffered by the Purchaser or his employees or contractors resulting from the use of the premises herein described or arising out of the performance of any act or thing required by the terms of this contract. The Purchaser does hereby release the Seller and his Agent from any claims or demands that he may have of whatsoever kind and nature in any manner arising out of any matter or thing related to this contract or the premises described; and the Purchaser covenants and agrees with the Seller that he will indemnify and save harmless the Seller and his Agent from any claims and demands that any person may have or make in any manner arising out of any matter or thing related to this contract or the premises described.

Workman's Compensation Insurance will be provided by the Purchaser during the term of the contract. Proof of insurance will be provided to the Seller's Agent prior to the commencement of operations.

General Liability Insurance in the amount of $______________ will be provided by the Purchaser during the term of this contract. Said insurance policy shall be provided to the Seller's Agent prior to commencement of operations.

SECTION IX - Special Provisions:

In witness hereof the parties hereto set their hands and seals this _______ day of _______________19____.

SELLER: ______________________________ DATE: ______________
SELLER: ______________________________ DATE: ______________

PURCHASER: ______________________________ DATE: ______________
SELLER'S AGENT: ______________________________ DATE: ______________

SELLER'S ADDRESS & PHONE: ______________________________
____

PURCHASER'S ADDRESS & PHONE: ______________________________
____

This contract was prepared by the Michigan Association of Consulting Foresters and The Michigan Association of Timbermen.

TIMBER STAND IMPROVEMENT BID

Landowner Name:________________________________ Phone:____________________

Address: ____

________________________________ Zip:____________________

LEGAL DESCRIPTION:____

____

CURRENT STAND CONDITIONS AND DESCRIPTION OF WORK TO BE DONE:

____

____

____

____

____

This job must be started by:________________________

This job must be completed by:________________________

SPECIAL REQUIREMENTS: ____

____

A map showing the location of the project area is ☐ is not ☐ attached.

Work must be checked and approved by:________________________________before payment will be made.

BID OPENING: TIME____________ DATE____________ PLACE____________

SUBMIT YOUR BIDS TO: (Name)____

(Address)____ Zip____________

BID FORM

I bid a total of $____________for the job as described above.

Vendor Name:____________________ Address:____________________________

____________________________ Zip:______________ Phone:______________

Please mark your envelope "T.S.I. Bid" in red.

Owner reserves the right to reject any or all bids.

Owner will notify all bidders of results.

TIMBER STAND IMPROVEMENT CONTRACT

This agreement is between____

hereinafter called the vendor, and____
hereinafter called the landowner.

A map of the area is attached showing (a) legal description, (b) approximate acres, and (c) boundaries.

The exact acreage will be determined after the job has been completed by:

____

Payment will be made on a (per acre) (per hour) basis as follows:____________________

____

____

Type of timber stand improvement to be done, equipment to be used and special instructions to follow:____

____

____

____

The Vendor will submit his bill in duplicate.

The Vendor will save and hold the landowner harmless from any and all liability arising from the Vendor's action for use and occupancy of Seller's property.

Signature of Vendor____

Date________________________________ Phone_________________________

Address ____

Witness ____

Signature of Landowner____

Date________________________________ Phone_________________________

Address ____

Witness ____

SITE PREPARATION BID INVITATION

You are hereby invited to bid on the site preparation work to be done on my land as described below.

Legal Description ______________________________

Acres ______________________________ (Estimated)

Equipment to be used ______________________________

____

Terrain Level ________% Rolling ________ % Hilly ________ %

Vegetative Cover ______________________________

____

Rockiness ______________________________

Soil Type ______________________________

Access to use ______________________________

____

Special Requests ______________________________

____

____

____

Type of site preparation desired ______________________________

____

____

Special limitations (work must be done by...) ______________________________

____

____

Please submit your bid to:

Name:

Address:

City ______________________________ State ________ Zip ________ Phone ________

This bid will be opened at (place) ______________________________

on (hour) ________________ (day) ________________ (month) ________________

Owner has the right to refuse any or all bids.

Note: Please submit your bid on a per acre basis.

SITE PREPARATION CONTRACT

This agreement is between ____________________

hereinafter called the Vendor, and ____________________
hereinafter called the Landowner.

A map of the area is attached showing: (a) legal description, (b) approximate acres, and (c) boundaries.
The exact acreage will be determined by ____________________
____________________ after the job has been completed.

Payment will be made on a (per acre) (per hour) basis and will be made as follows: ____________________
____
____

Type of site preparation to be done and equipment to be used: ____________________
____
____
____

The Vendor will save and hold the landowner harmless from any and all liability arising from the Vendor's action for use and occupancy of seller's property.

Special instructions from landowner (trees to reserve, direction of windows, (location of piles, etc.): ____________________
____
____
____

The vendor will submit his bill in duplicate.

In Witness whereof, the Parties hereto have hereunto set their hands and seals on this ________ day of ________________, 19____.

____________________	____________________
Witness	Vendor
____________________	____________________
Witness	Landowner

Vendor's Address ____________________

Phone ____________________

Landowner's Address ____________________

Phone ____________________

TREE PLANTING WORKSHEET

Use this worksheet when planning your tree planting projects - call a forester for advise and assistance.

Name (First, M.I., Last)				Home Telephone Number	Work Telephone Number
Address					
Legal Description					
Section	Township	Range	County	Total Plant Acres	Plant Year Season

Cost share program (check one) ☐ ACP ☐ FIP ☐ CRP ☐ RIM ☐ Other______ ☐ None

Foresters Name - Address - Phone

Ground Cover ________________

Soil Type ________________

Rocks?________ Standing water In spring? ________

Topography

Site Preparation Options

☐ Establish a cover crop of 3/4 bushel oats and 3 pounds of timothy per acre.

☐ Seed 3 pounds of timothy into small grain stubble.

☐ Eliminate noxious weeds before planting or seeding.

☐ Herbicide control of grass or weeds as follows:

____

____

____

☐ Existing cover crop acceptable.

☐ Mechanical site preparation as follows:

____

____

____

Tree Planting Options

☐ Hand planting to be done by ☐ vendor ☐ landowner.

☐ Machine planting to be done by ☐ vendor ☐ landowner.

☐ Note: use planting machine 45 horse power or larger.

☐ Landowner to bill A.S.C.S. at the rate of _____ ¢ per tree.

☐ Landowner will provide tractor and driver and will hire vendor with planter.

☐ Landowner to charge _____ ¢ a tree, vendor to charge _____ ¢ a tree.

☐ Landowner to hire a vendor to do all work at the rate of _____ ¢ a tree.

Tree Ordering Options

☐ Trees to be ordered on assignment.

☐ Check to be sent with tree order.

Tree ordering notes on orders from the state nursery

☐ Be sure to sign your name and fill in your address <u>twice</u>. The second signature is so that you do not have to pay sales tax on anything but shrubs.

☐ Landowner must add 6% sales tax on all shrubs. Make out check to the Minnesota DNR <u>or</u> take your completed tree order to your county A.S.C.S. and ask to order your trees on assignment.

☐ What is assignment? The A.S.C.S. will order your trees and will pay for the tree with part of your cost share funding. They cannot pay for your trees until the project is completed. The balance of the cost share funding is then sent to the landowner. Landowner must go in to A.S.C.S. office to sign up for assignment. A tree order must be filled out.

Firebreak Options

- ☐ A "green" firebreak will be constructed (moved and raked in Fall).
- ☐ A "black" firebreak will be constructed (disced in Fall).
- ☐ The firebreak will be ☐ around the plantation border ☐ between any road and the plantation.

Tree Pick Up Options

- ☐ Trees to be delivered to landowner via UPS.
- ☐ Landowner to pick up trees at the tree nursery.
- ☐ Vendor to pick up the trees at no extra charge.
- ☐ Vendor to pick up trees at the rate of $ ______per ______

  Trees to be delivered to ☐ planting site ☐ central drop off point ☐ other
- ☐ Landowner add cost of tree pick up when billing the A.S.C.S.

Contract Options

- ☐ A contract will be made up between vendor and landowner.
- ☐ No contract will be used - oral agreement only.

☐ If doing your own planting, be sure the Forester is aware of when you are planting.

☐ This plantation may require chemical release to ensure plantation survival. This release will only be done to comply with DNR Forestry recommendations.

☐ If hiring a vendor, don't pay the vendor until a Forester has checked and approved the planting.

- ☐ Plantings must be protected from destructive fire and grazing.
- ☐ Chemicals used in performing this practice must have Federal and State registration and must be applied strictly in accordance with authorized registration uses, directions on the label, and other Federal and State Policies and requirements.
- ☐ Consideration must be given to preserving and improving the environment.
- ☐ The practice shall be maintained for a minimum of ________ years after the calendar year of installation or establishment.
- ☐ Shearing for Christmas tree production is not allowed during the practice lifespan.
- ☐ All plantings should be as early as possible in the spring, but no later than May 31, unless approved for a later date by the Division of Forestry representatives. Seeding must be done during the time of the year as designated by the forester.
- ☐ Cost-sharing is limited to _______ trees per acre.

Spacing Options

600 trees per acre	6' x 12' (605)	7' x 10' (622)	8' x 9' (605)
700 trees per acre	6' x 10' (726)	7' x 9' (691)	8' x 8' (681)
800 trees per acre	6' x 9' (807)	7' x 8' (778)	
900 trees per acre	6' x 8' (908)	5' x 10' (871)	
1,000 trees per acre	6' x 7' (1,037)	5' x 8' (1,089)	
1,200 trees per acre	6' x 6' (1,210)	5' x 7' (1, 245)	

Remarks:

Replant

☐ This is a replant recommended by the Forester.

Species	Trees/Ac.	Spacing	Quantity	Cost (Include Tax)	State or Private Stock
Norway Pine					

COST SHARE BILLING*

Name ______________________________
Address ______________________________
City ________________ State _____ Zip ________
Phone ______________________________

Practice ________________
Acres Done ________________
Date Completed ____________

		Total Costs
HIRED LABOR (Be sure to include copy of bills.)		$ ________
LANDOWNER LABOR Hours _____	Rate Per Hour ________	$ ________
TREES PLANTED ____________	Rate Per Tree ________	$ ________
HIRED EQUIPMENT (Be sure to include copy of bills.)		$ ________
LANDOWNER EQUIPMENT Hours _____	Rate Per/Hr ________	$ ________
TREE PURCHASE (Include bill.)		$ ________
TREE HAULING (Include copy of bills.)		$ ________
HERBICIDE EXPENSES (Include copy of bills.)		$ ________
Misc. Expenses (Explain Below)		$ ________
	TOTAL - (All of the Above)	$ ________

Misc. Expense

____
____
____
____
____

*Return this form to your County A.S.C.S. with the **RE-245** (Project Approval and Payment Application form), in order to be eligible for payment on your project.

I have checked this project and find that it meets specifications. ______________________
Forester Signature

Landowner Sign Here ______________________________ Date ____________

WOODLANDS FOR WILDLIFE

DECIDING WHAT YOU WANT

by
Glenn R. Dudderar
Extension Specialist
Department of Fisheries and Wildlife
Michigan State University

About a year ago at a conference for woodland landowners, a woodland owner asked me if he should leave den trees for wildlife. "What kind of wildlife are you interested in?" I asked.

"Deer and grouse" he replied.

"For what reason?" I asked, and he replied, "Mainly for hunting by me and my friends."

I then recommended that he not leave any den trees in his woods for wildlife, since den trees have little benefit for deer and they can be detrimental to the number of ruffed grouse because den trees are occupied by squirrel and raccoons, which are predators on grouse eggs. The landowner was somewhat surprised that I had recommended the removal of den trees because he had been told that the leaving of den trees was good wildlife management in forests. Obviously, den trees are beneficial if a landowner is interested in a variety of wildlife, particularly those that utilize the cavities in den trees (such as raccoons, squirrels, porcupines, woodpeckers, woodducks, and other species of cavity nesting birds). Had the landowner expressed an interest in those species, then I would have recommended that he leave den trees on his property.

If you landowners are to get the maximum of benefits that you desire from your woodlands, then you must decide what those benefits are. In addition, you must be fairly precise in deciding what those benefits should be. Then professional foresters or wildlife biologists can help you create a forest management plan that produces those benefits.

For example, if a landowner asked for a woodland management plan that would produce the greatest variety of birds possible then I would recommend a tree cutting and planting program that would create as dense an understory (shrubs and small trees growing beneath large trees) as possible. If, however, a landowner said that he wanted his woods to produce a variety of birds so that he could observe them, my management recommendations would be entirely different. In some places my plan would create a very dense understory, but in other places there would be very little understory. I would also recommend access trails, probably a two-track, on which both or at least one side was kept free of shrubs and small trees. These recommendations would produce the same variety of birds as the first recommendation, but would definitely not produce as many birds. More important, however, these recommendations would make the observation of birds much more likely and far easier, which is of course what the landowner requested.

Most landowners, however, desire multiple benefits from their woodlands. In the surveys that I have taken of woodland landowners' desires, most landowners say they want the benefits of natural beauty, outdoor recreation and wildlife. However, it may not be possible to achieve all of those benefits equally from every acre of land at the same time. For example, access routes, which are unnatural, may be necessary in order for the owner to appreciate and protect the natural beauty of wildlife that is there. For another example, aspen which are over-mature, dying and breaking up may be natural but not very beautiful in the landowner's eyes, and certainly of limited benefit to wildlife. Thus you, the landowner, must be willing to rank desired benefits in order of importance so that if necessary, a particular benefit will be achieved in lesser amounts than desired in order to achieve some other benefits.

In addition, the benefits that a landowner chooses must be economically and biologically compatible with the other benefits chosen and with the condition of the woodlands that he or she may own. For example, the management methods for aspen that would produce the greatest economic benefits are very good for wildlife as well. However, maximum wildlife benefits usually require that the aspen be cut in smaller size areas than would be done for maximum economic benefit. This difference can often be mitigated by making the area cut long and narrow with irregularly shaped edges. Further, more wildlife will be produced if mature aspen is cut in areas somewhat larger than desired than if no aspen is cut at all.

Many woodland owners express a desire to keep the woods exactly as they are for wildlife. Biologically this is not possible because trees grow old and die. Second, wildlife may or may not benefit from this process, depending on the species of wildlife and the trees involved. Allowing aspen to become overmature and decedent benefits only woodpeckers, and only then only for a short period of time. Allowing trees like oak, beech, maple and others to become over mature and decedent may be beneficial for squirrels, turkeys, bears and woodpeckers, but may be very detrimental to grouse, deer, snowshoe hare, and many species of songbirds.

Sometimes landowners attempt to achieve greater wildlife benefits without any timber cuttings, utilizing methods such as food plantings, artificial nest box construction and artificial feeding. Though these methods can be very successful, they are very expensive and sometimes pose hazards to the wildlife. The irony of the situation is that such methods become more expensive the more successful they are, when in most cases a good timber management plan could produce equal or greater benefits and a profit to the landowner.

Finally, when managing for wildlife, you must consider what is happening on your property. If there is extensive timber cutting on surrounding properties, you may wish to delay cutting on your property as long as biologically and economically possible, because your mature timber will provide the variety that many wildlife species need. Conversely, if there is little or no timber cutting on neighboring properties and you desire greater wildlife benefits, you may wish to do more timber cutting than desirable to provide the same variety of wildlife.

Do not hesitate to ask for professional advice on deciding which benefits you would like to derive from your woods. A professional forester or wildlife biologist may be able to suggest benefits or a combination of benefits that you were unaware of. Keep in mind that as in medicine, law and finance a second opinion is highly advisable. Beware of popular trends, slogans and catchwords. They can be absolutely wrong or very misleading. The following list points out the shortcomings of some of these items:

- The natural way is the best way – It depends. Remember, nature destroys over-mature trees by fire, severe weather, insects, and disease, often en mass. Such a process may destroy the things you value most. A good woodland management plan produces the same benefits of these natural processes, while eliminating the short term intense destructiveness.
- Selective cutting of single trees disturbs the forest least – Not always, especially over time. Selective cutting of single trees will assuredly destroy a stand of aspen. In a mixed stand of maple, oak, beech and basswood, single tree selection is very likely to gradually eliminate the oak, which would be detrimental to wildlife.
- Eliminate undesirable species – It depends on your objective. Ironwood and hickory are not undesirable species if your objective is wildlife.
- Clean up your woods – Great, if your intention is to maximize the revenue from your trees; but such a process will assuredly reduce the number of cavity-nesting species of wildlife.
- Plant food plots for wildlife – This is expensive and often less beneficial than a good woodland management plan.

What Can I Do For Farm Wildlife?

Many modern farming practices that have been a boon to agricultural production—have been a bust for wildlife.

Removal of abandoned farm groves; drainage of marshes; elimination of fence lines and odd brushy areas; fall plowing; channelization of natural streams; mowing, burning, plowing and driving in roadsides; and planting of fewer small grains (oats and wheat) - all of these practices have benefited farmers and virtually every American because of increased production of food and materials.

But what about wildlife? As farming operations have steadily improved crop production, wildlife habitat in the intensively farmed areas of Minnesota has steadily diminished.

Hold Out For Habitat

Fortunately, wildlife habitat can be re-established or maintained. You can help bring back wildlife by following any of the suggested habitat improvement recommendations provided in this pamphlet. And remember, a project doesn't need to encompass thousands of acreas and cost large sums of money. Even one acre of good habitat will help to increase wildlife numbers.

It should be emphasized that pheasant populations have not responded to stocking, refuges and/or closed seasons. Studies to date have failed to demonstrate that any one of the three practices will increase pheasant numbers. Here's why.

Adding pen-raised birds to wild populations increases competition for available nesting cover, and winter food and cover. Of course, the much hardier wild birds can survive this strain, but stocked birds do not.

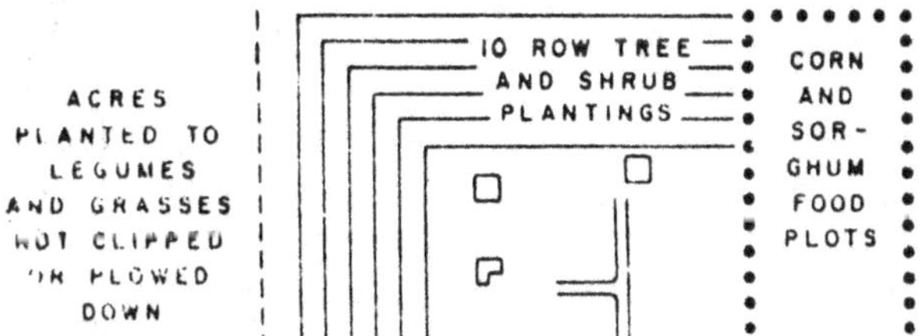

Pheasants and other birds have a high reproductive rate, but their life spans are quite short—less than one year in most cases. Thus, hunting has little effect on the abundance of game birds, because the birds will produce a large number of new chicks the following spring. The elimination of hunting only serves to needlessly deprive hunters of recreational opportunities.

WILDLIFE

The key to good populations of farmland wildlife is to provide high-quality nesting cover, winter cover, and winter food next to winter cover.

Nesting Cover

Safe nesting cover is rapidly disappearing and poses a serious threat to the future of ground-nesting farmland wildlife such as rabbits, meadowlarks, hungarian partridge and ring-necked pheasants.

Legumes or native and tame grasses provide cover for nesting, brooding, loafing and escaping enemies. Early nesting pheasants usually establish nests in dry grasses and sedges, which carry over from the previous fall.

Later in spring, as soon as new plant growth is six to eight inches high, most hens nest in alfalfa fields, grain fields, pastures, roadsides and other grassy-type cover. Accordingly, nesting cover may be planted on retired acres, roadsides, odd pieces of cropland and other areas.

The most important source of pheasant nesting cover in many areas of Minnesota is the unmowed roadside. Studies have shown that unmowed roadsides contain more than twice as many nests per acre than other sources of cover.

It is important, however, that nesting sites remain undisturbed for 35 to 40 days. Nesting cover sites should not be moved until after July 31st.

Winter Cover

Most resident farmland wildlife can resist disease and survive for a period of time without food. However, wildlife with little or no winter cover does not last long in the face of a severe winter storm.

Cattail marshes provide good cover during most winters. Other good cover sources include small woodlots, farm shelterbelts, brushy bottom lands and other areas of dense cover.

Many large cattail marshes have been drained or filled. So, pheasants and other farmland wildlife have been forced to depend more and more on farmstead windbreaks and other woody cover for winter protection. Well-designed plantings consisting of ten or more rows of trees and shrubs, of which at least four rows should be conifers, are excellent sources of cover. These plantings are large enough to provide shelter from the wind, and they remain relatively free of snow. During severe storms, narrower plantings can provide temporary false security and become a death trap to wildlife.

Shelterbelts are very important to many species of farmland wildlife. Rabbits and squirrels make shelterbelts their year-around home. Mourning doves, robins and numerous songbirds are tree-nesting birds which use shelterbelts for nesting sites.

Food Plots

Winter food, or the lack of it, can be a factor limiting pheasant populations during severe winters especially in areas where fall plowing is extensive.

Row crops such as corn, soybeans or sorghum provide an excellent source of winter food when left standing until April 1st of the year following the initial seeding.

Food plots should meet the following requirements:

1. The plots must be adjacent to good winter cover.
2. One or more kinds of wildlife should be benefited.
3. The plot must be in an area where the food supply would otherwise be insufficient.

Pheasant Egg Salvage

Numerous pheasant nests are destroyed during mowing of hay crops each year. Individuals are encouraged to salvage and hatch pheasant eggs that would otherwise be wasted as a result of the mowing operation. Persons desiring to participate in such a project should secure a dependable incubator prior to the hatching season. Let farmers in your area know that you will be hatching eggs and they can mark and report destroyed nests. Other nests must be left alone.

For More Information . . .

For more information on possible cost-sharing for winter cover, nesting cover and food plots contact the Department of Natural Resources, Section of Wildlife, 390 Centennial Building, St. Paul, Minnesota 55155. The Minnesota Department of Natural Resources, Bureau of Information and Education, 350 Centennial Building, St. Paul, Minnesota 55155 also has brochures and fact sheets on game birds and other wildlife.

What Can I do? For Forest Wildlife

Major forest-game species in Minnesota include the White-tailed Deer and the Ruffed Grouse. These species are most commonly found in the mixed conifer-hardwood forests of our state.

Much of our northern forest has been reaching maturity since the early 1950's. The tree canopy of this maturing forest is shading out low-growing plants and shrubs.

In addition to a maturing forest, Minnesota is faced with a natural succession toward a spruce-fir forest in some counties. These areas are poor producers of forest wildlife.

Forest wildlife management is the art of managing a forest to produce annual crops of wildlife. The habitat phases of wildlife management are almost entirely a matter of managing forest vegetation. This includes cutting the right amount of trees in the right places, and at the right time.

To do this effectively, a management plan must be developed so that desirable species of food and cover plants are available to provide for the needs of wildlife throughout the year.

This plan should be developed with assistance from State Foresters and Wildlife Managers of the Department of Natural Resources.

Determine Objectives

The first step in management is to inventory the resource. Second, determine your objectives. This will involve setting up some land use priorities. The third step will be habitat development to improve the carrying capacity of the land. Remember, ideal habitat conditions are difficult to attain. If your objective is primarily *timber production* or *cattle grazing*, little will be accomplished for deer, grouse or other forest wildlife.

How To Improve Deer Habitat

The White-Tailed Deer is most abundant in brushy habitat that is largely composed of young deciduous trees and shrubs. Numerous small openings containing grasses and herbs should be well interspersed throughout the area. Such conditions represent early stages of forest development, and support a large number of wildlife species.

Deer tend to avoid large single-species tracts of vegetation and prefer a blending of various types of cover and openings. They are primarily browsers feeding on the leaves and succulent stems of trees and shrubs. This diet is supplemented with grasses and herbs. Aquatic plants, fungi, nuts, and a variety of other plants are also eaten. Food studies indicate that deer require about five pounds of browse per hundred pounds of body weight per day.

Beneficial Practices

The following forest manipulation practices are beneficial to deer.

*Avoid development of large areas of a single tree species. Clearcutting should be in small tracts of five to fifty acres. Small blocks of different timber types provide more and better food and cover for deer.

*Preserve and maintain openings. Openings or clear-cut tracts of five acres or less should not be planted to woody cover. Clover may be introduced to help maintain openings along trails, firebreaks and clearings.

*When re-planting clear-cut tracts of five to ten acres, leave openings one chain wide (66 feet) between the planted area and the existing forest. When re-planting openings over 10 acres, numerous small clearings should be scattered throughout the plantation.

Winter logging is desirable because it provides supplemental deer browse during the most critical period of the year. Plan winter cutting so it is done when and where needed. It is important to note that cutting browse solely for deer food should be employed only when wildlife officials have declared an emergency situation.

In Minnesota, there are sizeable stands of closed-canopy, pole-sized timber with little wildlife value. These stands can be greatly improved by cutting, crushing, dozing, shearing or burning. These practices should be employed only upon recommendation of a forester or wildlife manager.

How To Improve Grouse Habitat

Grouse are primarily browsers subsisting on buds, twigs, leaves, and fruits of various forest plants. They are non-migratory and spend most of their lifetime within a relatively small area. The abundance of grouse depends on the quantity and the quality of available food.

Working with the vegetation already present is by far the surest, quickest and least expensive way of producing more grouse.

Uneven-aged timber stands resulting from selective cutting on a sustained yield basis are much more productive than even-aged stands resulting from clear-cutting large areas. Openings created by clear-cutting can be valuable to grouse but in any cutting operation these should be kept as small as possible. Large areas of the same tree species should be avoided. Grouse prefer areas with a variety of plants.

Suggested Practices

Suggested practices beneficial to grouse may include one or more of the following:

*Existing openings of five acres or less should be maintained as such.
*A long narrow clear-cut provides a maximum amount of edge. The total width of the opening should be 1½ to 2 times the average height of adjoining uncut trees.
*Predator losses can be minimized by removing slash, brush piles and other litter accumulation.
*Planting conifers, especially pine, in grouse habitat is not recommended.
*Shearing, crushing or controlled burning will stimulate the growth of preferred grouse foods. Prescribed burning should be done only under the direction of foresters or wildlife managers.
*During winter, grouse feed primarily in aspen over 30 years of age. When large stands of trees are clear-cut, clones or groups of mature male aspen should be left standing.
*Hardwood stands lacking a good distribution of aspen or sites predominated by pine, spruce and fir are regarded as marginal ruffed grouse habitats. Treatment should include methods to encourage aspen.
*Although ruffed grouse eat the fruits of many wild plants, they do not appear to be essential food items. The planting of berry-producing shrub seems to have little value.

How To Improve Habitat For Other Forest Wildlife

The practices suggested to benefit deer and grouse will create a diversity of forest cover that has the potential for supporting the greatest variety of wildlife species.

MINNESOTA DEPARTMENT OF NATURAL RESOURCES
DIVISION OF FISH & WILDLIFE

What Can I Do? For Wetland Wildlife

The great flights of waterfowl, stretching endlessly from horizon to horizon, are gone. But the stirring sound of migrating geese, the raucous quacking of a mallard as it leaps off the water can remain with us for many generations to come. This is where you come in!

Western and much of southern Minnesota is part of the prairie-pothole region extending through the Dakotas and north into south-central Canada. Here, many of the ducks in the Mississippi Flyway are raised.

Development of agriculture and other activities of man have adversely affected these prairie wetlands. Millions of acres have been drained, eliminating much needed habitat for waterfowl, muskrats, mink, herons and a host of other wetland wildlife species. Modern farming practices and conversion to row crops have reduced the amount of safe nesting cover for ducks, geese, and other ground nesting wildlife.

In 1951, the "Save Minnesota's Wetlands" program was started in an attempt to rescue prairie marshes from drainage. Since then, 200,000 acres of prime marshland have been purchased by the Department of Natural Resources that otherwise would have been drained. These wetlands, called Wildlife Management Areas, provide valuable breeding habitat for waterfowl and other wildlife.

Though many wetlands have been saved, still more are needed to bring about any increase in waterfowl numbers. With your participation and the involvement of thousands of other conscientious Minnesotans, our waterfowl will have a fighting chance.

Many Minnesotans, including hunters, bird-watchers, amateur photogaphers and the thousands who just enjoy the sights and sounds of waterfowl on the wing will benefit from your project, no matter how insignificant it may seem to you. This we guarantee!

Waterfowl Impoundments

Shallow impounded water is generally more attractive to waterfowl than other man-made wetlands because the result more closely resembles a natural prairie marsh. Marshes can be developed, improved or restored by the construction of a dike or dam—preferably with a water control structure—at less cost per acre than construction which entails removal of earth.

Impoundment sites should have gradual slopes of less than one per cent. Sites with slopes greater than that should be considered only on individual merit because of the extra height and cost of a structure required to flood even relatively small areas.

The marsh basin must have sufficient water to assure a year-round supply. A good watershed-marsh ratio is approximately 12 to 1. For every acre of marsh covered by water there should be 12 acres of watershed providing run-off into the marsh.

Dabbling ducks, such as the mallard, prefer shallow water. For these ducks, a marsh with 75 per cent of the basin less than two feet deep is ideal.

Nesting Boxes

Wood Ducks and Goldeneyes usually nest in natural tree cavities. Where natural sites are not available these ducks will accept and use artificial nest boxes.

Plans for nesting box construction are available from the Minnesota Department of Natural Resources, Bureau of Information and Education, 350 Centennial Bldg., St. Paul, Minnesota 55155.

Loafing Sites

Waterfowl are attracted to marshes that have loafing sites where they may get out of the water to rest and preen their feathers in safety. Such sites include open shoreline, floating logs, rafts and small islands.

Assistance

Technical and possible financial assistance is available from both the U.S. Department of Agriculture and the Minnesota Department of Natural Resources for:

1. Impoundments
2. Dugouts
3. Fencing

Please contact your local Area Wildlife Manager (DNR), District Conservationist (SCS) or County Executive Director (ASCS) for more information.

Waterfowl Dugouts

Dugout construction is a popular method of improving waterfowl habitat. With this type of development, it is possible to open up shallow wetlands which have grown over with vegetation.

The greatest production from dugouts is in areas having high waterfowl densities where breeding and courting sites are at a premium. These small water areas are most valuable during the breeding season. Generally, ducks nesting near a dugout will move their young to a larger, permanent body of water shortly after the young are hatched.

Open water may be created in dense marsh vegetation using a bulldozer or dragline.

Dugouts with the greatest attraction for waterfowl have the following features:

Maximum depth less than four feet.
Average depth two to three feet.
Minimum size - 1,000 square feet.
Dugouts that are grouped together.
Mineral soil which is exposed at the marsh bottom.

Nesting Cover

Waterfowl must have adequate undisturbed cover in Spring and early Summer to nest and bring off broods of young. The most valuable nesting cover is a mixture of legumes and grasses. To be of greatest value for ducks, nesting areas should be less than one-half mile from permanent marshes. Sites of this kind should be protected from grazing and mowing.

Nesting cover should be established for a period of two years or longer since perennial legume-grass plantings produce early nesting cover. Such cover crops are soil conserving measures and should be encouraged under land retirement programs.

WOODLANDS AND WILDLIFE
How Do I Get Started?

By Mike Bolin and T.W. Steele

To start off blindly without a course of action is a mistake in anything you attempt. You must first establish your wildlife goals and - in all fairness to yourself - they must be realistic. Remember that while it is doubtful you will attract rare or unusual wildlife to your woodlot, you are assured of attracting species that live nearby. The key to the types of wildlife that will be attracted to your woodlot is the habitat conditions that surround it.

What are your objectives? Do you want to manage your woodland for one specific species, or do you want a forest that is capable of supporting a variety of animals? The professional who assists you in developing a management plan for your property will need to know your objective.

Property Inventory

It is important to know what your property contains and its relative condition. This can be accomplished with a simple inventory of the property. You want to include a list of all wildlife that currently visit your property, noting the time of year and the duration of the visit. You will also need to construct a map showing your woodland and the resources on adjoining acreage.

The map should be divided into simple blocks or compartments defined by areas of similar vegetation. Examples would be a stand of mature oak-hickory timber, a recently clearcut area with shrubs and brush, a pine plantation, a young stand of pole-size timber, or a wet, marshy area. Within each compartment, it is important to note what is found growing there and the area's general condition. A recent aerial photograph of your property will help you develop your map. They are available for a slight charge from your county U.S. Soil Conservation Service Office.

Developing a Management Plan

With this information, you are now ready to contact a forester or wildlife biologist for assistance in developing a management plan - the first step in meeting your woodland and wildlife objectives. The challenge of wildlife management lies in manipulating the woodland and its surroundings to create an inviting mix of cover, food and water. However, be prepared to make some trade-offs!

General Thoughts

Remembering that vegetative diversity is the key to variety and abundance of wildlife in your woodland, a professional forester or wildlife biologist most likely will suggest that it contain a mix of habitats (described below). His recommendations will be based on your wildlife objectives, the life

requirements of the species, and the present conditions that exist in your woodland.

Herbaceous Openings

Herbaceous openings are areas where the ground is covered with a mixture of grasses and other non-woody plants; there are few or no trees present in such an area. These openings are particularly important to forest wildlife for nesting and brood rearing. They provide food in the form of herbs, grasses and insects eaten by a variety of birds and mammals.

The openings should be at least one-quarter acre in size, with no more than one to three acres needed for each 50 acres of woodland. Herbaceous cover can be established either by seeding or by allowing native plants to invade the site. The opening will need site preparation. All woody vegetation must be removed except understory species near the perimeter of the opening. The site should be disturbed, exposing bare mineral soil. If nutrient levels are deficient, soil fertility levels should be built up.

Annual plants should be favored over perennials in herbaceous openings. Since annuals rely heavily on seed to reproduce each year, they will be far more attractive to upland game birds and seed-eating songbirds. To maintain an annual plant cover crop, the opening must be tilled or disked at least every three years. Otherwise, perennial plants will overtake the site and its attractiveness to wildlife will be reduced. If the opening is intended to provide forage for deer, the site should be mowed periodically during the growing season to promote succulent growth of grasses and herbaceous plants.

Brushy Areas

Approximately 10 percent of your woodland should be in permanent brush cover. This can be in the form of (1) young trees of the seedling or sapling stage, or (2) shrubs and vines. Brushy areas are valuable to wildlife for nesting cover, escape cover, and food.

Most brushy areas require nearly full sunlight in order to maintain themselves. Openings ranging from one half to two acres in size should be created. This means that a few trees may have to be removed in various locations, or a brushy border adjoining the woodland's boundary might be maintained. However, within 10 years, the dominant tree species will begin to outgrow the lower undergrowth and as increased shade falls on those lower plants, they begin to pass out of the picture.

Therefore, to maintain brush patches in your woodland, these areas will have to be cut with a brush-hog every five to six years. This is best done during the winter months, which protects the nesting season. Native shrubs such as dogwood, elderberry, greenbriar, hawthorn, witchhazel, honeysuckle, and viburnum are considered excellent sources of food for wildlife and planting is encouraged where none exist.

Sapling-Pole Stands

A landowner with a woodland comprised mainly of sapling and pole-size trees 4-12" in diameter will have a frustrating time managing for wildlife because, at this age, the woodland has little to offer in the way of habitat or food for most wildlife. The foliage generally is too high to be of any value for browse, and the dense canopy shades out most understory plants.

The best recommendation for this type of woodland is to get it to the next stage as quickly as possible. This calls for repeated Timber Stand Improvement practices such as thinning and improvement cuttings. Periodic thinnings which favor the selected crop trees will provide an important source of browse to deer if it is done in the winter and the tops are left on the ground. It is also advisable to create a few herbaceous and brushy openings to make this type of woodland more attractive to wildlife.

Mature and Overmature Timber

Mature timber is a primary source for all types of seeds and mast (nuts and acorns) and therefore plays a very important role in managing timber for wildlife. Overmature trees are also important, for they provide much-needed cavity nesting sites in their rotten centers. These trees are called den trees. Snags are dead trees that have damaged crowns and often have hollow centers.

Since mature timber has its greatest value as lumber or veneer, a professional forester should be consulted prior to harvesting any trees from the woodland. Since mast production is a direct function of the tree's crown size and the amount of light it receives, he may recommend leaving some large trees within the woodland to insure a continuing supply of mast for wildlife during the winter. Generally, den trees and snags have far more value for wildlife than they do for the limited products that can be cut from them. A good guideline is to leave approximately five den trees or snags for each acre of your woodland. Den trees can be girdled or poisoned with an herbicide so they will not retard the development of reproduction in the understory.

Conifer Plantings

Conifers or evergreens such as spruce, pine, arborvitae, red cedar, or hemlock provide excellent and often-needed cover for wildlife during the winter months. They are primary nesting sites for some bird species and provide shelter and safety from attacking prey. Most hardwood forests would be better off from a wildlife standpoint if a few patches of conifers were interplanted. This can be done easily by clearing small areas and planting conifers at a 10x10 foot spacing.

Establishing a C-, J-, or S-shaped planting will be more beneficial to wildlife than a rectangular block planting since these shapes create more edge effect. Conifer plantings should be located next to herbaceous or brush openings, if possible. No more than five acres of conifers for each 100 acres of hardwood timber is necessary for improved habitat.

Water Reservoirs

A source of good quality water is important to wildlife and they will migrate toward it. In general, shallow streams, potholes, and very small ponds with water depth of less than 12 inches are much more valuable to wildlife than one large pond located in or near your woodland. If you are considering the construction of a watering site in your woodland, it would be best to discuss your intentions with a wildlife biologist before you proceed.

Keeping A Journal

Once you have developed and implemented a management plan, keep a journal of your activities and observations. Not only will the journal serve as a record of your successes, it will also be a living history of your woodland that you can look back on and enjoy.

WILDLIFE ASSISTANCE PROGRAM FOR PRIVATE LANDOWNERS

Timothy Bremicker, Private Lands Development Coordinator
Section of Wildlife
Minnesota Department of Natural Resources

Goals for Wildlife Assistance Program

The Section of Wildlife's goal is to improve wildlife populations on private lands. This effort requires a multi-faceted approach because private land use decisions affecting wildlife are highly integrated with economic and social pressures. Relying upon one program initiative would undoubtably prove unsuccessful and certainly counterproductive to the ultimate goal; more habitat thus more wildlife.

In Minnesota, wildlife assistance programs are offered by several private organizations and public agencies that provide tax incentives, cost share assistance, technical advice, tax and estate planning, policy and regulatory control, tax disincentives, commodity reduction or land retirement policies and often habitat land rental practices.

TAX INCENTIVES AND DISINCENTIVES

Wildlife is definitely affected by taxing authorities. Investment credits and depreciation schedules for ditch and tile lines are good examples of adverse affects. Thankfully, however, the tax system also offers advantages that can provide very positive benefits for wildlife.

Income Deductions

Soil and water conservation expenditures, are deductible expenses. Allowable deductions on water courses, wind breaks and other conservation investments such as ACP conservation practices on lands used for farming are permitted to reach at least 25% of the adjusted growth income. Unused deduction can be carried over the following year.

Property Tax Credits

Minnesota's Wetland Tax Credit and Exemption program, grants land owners with qualifying wetlands a sizable credit deducted from their property taxes for the preservation of a wetland. Acreage qualifying for the credit is also exempt from the property taxes. The exemption and credit cost are reimbursed from the state general revenue fund to the county level of government. Qualifying wetlands are those defined as "lands mostly under water, which produce little or any income and have no use except for wildlife or water conservation purposes". Wetlands should also be preserved in a "natural condition" and "draining should be legal, feasible and economically practicable", and after drainage, "suitable for the production of livestock dairy animals, poultry, food, vegetables, forage and grains". Generally these wetlands are those types that have open water mixed with emergent vegetation. The enabling statute specifically excludes woody swamps containing trees or shrubs, wet meadows meandered waters, streams, rivers flood plains or river bottoms. Qualifying wetlands are often generally located south and west of the Mississippi River where drainage pressures have been chronic. **All protected wetlands** (MS Chapter 105.341) can be enrolled in a wetland tax credit and exemption unless these basins are enrolled in another protection program. Last year over 70,000 acres were enrolled. The DNR estimates that another 200,000 acres are also eligible. The Native Prairie Tax Credit and Exemptions is very similar and applies to native unpastured prairie.

Gifts

Another unique way that wildlife can be benefited through the tax system statutes is gifting of full or partial interest in real property. The U.S. Internal Revenue codes permit the gifting of fee or partial interest (conservation restriction) in a real property if the gift is for a benefit or legitimate conservation purpose. A legitimate purpose is probably best first defined as a public agency or qualifying public trust's willingness to accept the gift. The gift must offer conservations values that the agency or trust is already trying to preserve.

Full fee interest conservation gifts are readily accepted by organizations such as Minnesota Wildlife Heritage Foundation, and the Department of Natural Resources. The Nature Conservancy, however, prefers to accept specialized land values such as native prairie. Land values gifted can be deducted between 20% and 50% from an individuals adjusted gross income. Gifts to publicly supported charities or agencies that are real estate assets of long term capital gain value can be deducted up to 30% of your adjusted gross. A carry over deduction can, however, be applied for the next six years, permitting the donor to take a full deduction for the next six years, permitting the donor to take a full deduction for the gift. Gifts to private foundations are limited to 20% of your adjusted gross income. Perpetual easements for conservation purposes,

such as restrictions against draining or filling of wetlands, are permitted by Internal Revenue Codes if the easement meets specific conservation goals that are being protected on similar properties sponsored by specific public programs. Scenic easements along established Scenic Rivers or a perpetual wetland protection easements are good examples of a qualifying conservation restriction. If public agencies are acquiring easements to protect unique resources for the broader public good, open space easements would be acceptable.

Congress is now considering a bill authored by Wyoming Senator Malcom Wallop and co-sponsored by Senator David Durenburger that would substantially improve charitable donations, particularly those pertaining to partial interest. Appreciated capital gain gifts for easements or full fee would receive allowable deductions of 50% of the adjusted gross income versus the current 30% limit. Gifted easements also would be permitted as an acceptable means to pay state taxes. A very beneficial option for a cash poor rich landowner.

Cosh Share and Technical Assistance Programs

Wildlife assistance programs in Minnesota also provide landowners cost share and technical assistance to improve or provide wildlife habitat on private land. Minnesota's Department of Natural Resource's Wildlife Habitat Improvement Program (cost share) is designed to meet what Wildlife Managers like to refer to as the critical four; more secure nesting and roosting cover, more wetlands, more food resources and more woody cover. These are land cover types that agricultural pressures have diminished in quantity and quality. Consequently any attempt to improve wildlife populations in agricultural regions must include or address crucial habitat needs. The Wildlife Habitat Improvement Program was funded in 1984 by $120,000 from the Game and Fish fund, $374,000 in Pheasant Stamp revenues, and $150,000 from the Deer Habitat Improvement account.

The section of Wildlife cooperates closely with private groups such as Pheasants Forever and the Minnesota's Waterfowl Association. Blending cost sharing with these groups offers another $200,000 to habitat development. The Section of Wildlife attempts to concentrate work and money in counties where we feel the greatest opportunity exist for habitat and populations improvements. For example, counties within the pheasant range may receive an "A" priority because of historically and currently high pheasant populations, due to more conductive private land use and public ownership. "A" priority counties receive $13,000 annually whereas counties with very limited opportunities receive $1,000. If A priority counties have an active conservation organization, the Area Wildlife Manager might also expect to receive another $20 - 30,000 from the group annually for habitat practices.

WHIP attempts to cost share habitat improvement practices at a 75% of cost. Food plots, however, that benefit deer and pheasant can be purchased at 100% of the cooperators cash cost if necessary. Because standing corn is often crucial wintering deer herds. Area Managers are authorized to pay 125% of the establishment cost, and further justified a land rental rate may be included. Other food plot options proving extremely beneficial for pheasants and popular with landowners are the sorghum food plots established on diverted lands.

Wood cover or shelter belts cost share practices are encouraged wherever winter cover is lacking or needs improvement. Shelter belts are cost shared at 75% requiring a minimum of ten rows, of which four rows must be conifer. To accelerate growth and survival of the conifer stock, we conifers must be potted or container grown. A typical woody cover practice involves ten to sixteen rows and is cost shared up to $1,000 total. Shelterbelt improvements are also authorized if the practice represents an enlargement of the planting. Farmsteads, feed lots and field block plantings are all permissible.

Cost share practices for wetlands improvements are limited to the restoration of drained basins or sites where water level manipulation would prove very beneficial. We do not, except in rare instances or very specific locales, recommend dugouts. Blastouts are even less preferred. Neither receive technical or cost share assistance. The Section will do its absolute best to provide the landowner with 100% cost reimbursement for approved project. The WHIP program typically offers 75%, while the Minnesota Waterfowl Association will contribute another 25% of the cost. Cost sharing is blended with ACP or SWCD practices that accomplishes the same type of restoration if possible. Restored basins qualify for the Wetlands Tax and Credit and Exemption Program if they are Type 3,4,5 wetlands after restoration and located in an area where drainage is a persistent problem.

Nesting cover practices, the most critical, and unfortunately, the most expensive habitat component, are concentrated around tracts of public ownership throughout the better pheasant and waterfowl producing counties. Where pasture lands can be improved, WHIP cost shares the establishment of nesting cover at 75%. Landowners are sometimes offered a land rental payment equal to the mean county cash rent for the low quality pasture land. If a rental payment is offered, cattle grazing is not permitted. Nesting cover agreements are at a five year period. Approximately 40% of the Pheasant Stamp funds go to this practice. WHIP nesting cover agreements are piggybacked onto federal ACP or land retirement programs if at all possible.

PONDS FOR WILDLIFE

Wetlands and wildlife. An abundance of one means an abundance of the other.

With each wetland drained in Minnesota these days, the remaining, undreained wetlands take on ever-increasing value. Unfortunately for wildlife, the loss of another wetland is often passed off as an inevitable event that must take place if Minnesota's economy is to prosper. In 1964, a 19-county survey showed 64,275 wetlands basins. In 1974, a second survey showed 38,271 basins remaining, a 40 percent loss in only ten years.

Luckly, more and more private landowners recognize the vital role of wetlands in maintaining an abundance of wildlife, game and non-game alike. And they are doing something about it: constructing small artificial ponds on unused portions of their property.

Interest in constructing private ponds is increasing, according to U.S. Soil Conservation Service (SCS) officials. In 1974, more than 2,000 such ponds were designed for Minnesota landowners by SCS engineers. Of this total, more than half were intended for wildlife.

Artificial ponds could create a significant increase in Minnesota's waterfowl population by providing a network of breeding and courting areas where they are currently nonexistent or in short supply. Natural potholes of less than one acre serve this important function, but are the first casualties in any drainage project. Untold numbers of natural potholes are now raising corn instead of ducks.

How is a pond constructed? It usually involves more than providing a basin that will retain surface and/or ground water — water that would otherwise be discharged from the land. However, constructing a pond that will give maximum benefit to wildlife is a different story. The applicaiton of both biological and engineering expertise is essential, both of which are readily available at no cost to the landowner.

The object is to provide a water supply that will support both emergent and submerged aquatic vegetation. In other words, to duplicate wetland habitat — "the natural pothole" — the most diverse and biologically productive environment in Minnesota. A pond equally blessed with cattail and patches of open water will attract a menagerie of wildlife, including waterfowl, songbirds, upland game, furbearers, and even deer.

A single pond can provide other benefits: fish, wildlife, recreation, water supply, and erosion control, to name a few. but a small pond (two acres or less) cannot be all things to all people. Many pond owners get discouraged when they realize their fish pond won't attract a pair of ducks. A parallel to this is the lakeshore owner who wants his lake to provdie him with muskies, walleyes, trout, waterfowl, waterskiing, canoeing, picturesque scenery, warm water, plus a beautiful weed-free beach. It just doesn't work that way.

Private ponds constructed at the beginning of the SCS program provided soil-erosion control and water to reverse effects of the 1930's drought.

By accident, many of these ponds attracted wildlife. The Dakotas have experienced this spin-off. A 1958 Bureau of Sport Fisheries and Wildlife survey showed that 2.75 breeding pairs of waterfowl were using each of 89,000 stock ponds in South Dakota.

Two basic designs are used in pond construction — the impoundment and the dugout. Both types have certain advantages. The suitability of each depends upon site conditions.

Most common is the dugout, which is usually constructed in a low area where the water table is within one foot of the surface. While shape can vary, most dugouts are either round, rectangular, or L-shaped and have a water-surface area from one-tenth to two-tenths of an acre (from about 4,500 to 9,000 square feet). Generally, several dugouts with a total surface area of two-tenths of an acre will attract more wildlife than a single pond of the same area. Cost depends upon the yardage of soil removed.

"A Natural Pond"

The impoundment pond is preferred by most wildlife specialists, but is limited because of topography; therefore, it involves the geatest amount of engineering. It consists essentially of a low dyke or dam with an outlet — usually a culvert — to handle excess run-off from an average storm.

The dyke is installed in a portion of a small watershed, wherever the ratio of watershed area to pond surface area is at least 10-to-1. This ratio varies according to soil type, slope of land, land use, and expected rainfall. Depending upon the size of the watershed, an emergency spillway may also be constructed to handle excess flows. This is simply a grass waterway of a size, shape, and slope capable of withstanding the torrential runoff from the type of severe storm that may occur once every ten, 25, or 50 years.

The principal spillway of an impoundment permits either partial or complete draining of the pond. Wildlife specialists have found that periodic exposure of the pond bottom is necessary to sustain aquatic vegetation beneficial to wildlife. all the Department of Natural Resources' larger, controlled-wildlife management areas have this feature. Dugout ponds lack this feature. However, this deficiency can be compensated for if 50 percent of the shoreline is sloped gradually at a 5-to-1 ratio (five feet forward to each one-foot drop). Normal fluctuation of the groundwater table will then expose larger portions of the pond bottom.

Remember, the growth of aquatic vegetation is paramount to the success or failure of a wildlife pond. Vegetation also removes nutrients and sediments that would otherwise be discharged into lakes and streams. Research has shown that species such as cattail and bulrush remove large amounts of phosphorous from water. Since phosphorous is one of the major causes of algal blooms, a pond owner can expect good correlation between the amount of emergent vegetation growing within his pond and the clarity of the pond water.

Important to the success of any wildlife pond is a livestock-proof fence around all or a large portion of a pond. The fence should be at least 50 feet from the perimeter of the pond. If

"Potential Pond Site"

allowed to roam at will, livestock will destroy both the aquatic and semi-aquatic vegetation on the pond's periphery. Puddled soils will form and vegetation growth will be stifled. Soil erosion, polluted water, and algal blooms will result.

Fencing will also maintain the lush growth of upland grasses which is essential for nesting birds. Light grazing within the fenced portion is sometimes permissible, usually after mid-July, when most nesting is completed. Sometimes light grazing can be beneficial if controlled burning or mowing is not practiced on a periodic basis. Your DNR area wildlife specialist should be consulted before you undertake such grazing.

Owners of ponds constructed in cooperation with the SCS are required to initiate erosion control practices on those portions of watershed contributing water to the pond. Usual methods of erosion include grass waterways, minimum or reduced tillage, and contour-strip planting. These methods can save as much as five tons of soil per acre per year. Watershed treatment is especially important where impoundments are constructed; it reduces the burden of the natural water filter downstream and extends the pond's life span.

Many other land management practices can be added to the pond and its surroundings once it has been created. In some areas of Minnesota, especially in the south and west, it is beneficial to plant rows of shrubs around most of the pond, preferably next to the surrounding fence. The shrubs not only enhance esthetics of the pond, but also provide shelter to upland wildlife in winter.

To attract waterfowl, loafing sites can be provided within the pond. These can be made by anchoring log rafts or large, single logs in open water, or simply by placing several old hay bales in shallow areas. When ducks aren't using them, turtles or a family of muskrats will take in the commanding view provided by these structures.

A small island should be considered when planning a wildlife pond. Islands are especially well-suited for dugouts because they provide a predator-resistant nesting area, while breaking up the man-made appearance of the final product. Grasses and legumes should be seeded shortly after construction to stabilize the soil and insure growth of desirable nesting cover.

How much does all this cost? Because of state and federal cost-sharing programs, costs to the landowner can be remarkably low. In fact, there have been some instances where landowners have qualified for both state and federal funds, thus receiving 100 percent reimbursement. Unfortunately, during tight-money periods, government spending is cut back; conservation programs are the first to feel the pinch.

A current estimate to construct a one-tenth to two-tenths acre dugout pond, without cost-sharing, is from $400 to $800. Cost of an impoundment varies considerably. However, depending upon site conditions, cost per acre of surface water created can be lower than that of the dugout.

Most cost-sharing for pond construction is processed through the U.S. Department of Agriculture's Agricultural Conservation Program (ACP). Landowners must apply for financing at their Agricultural Stabilization and Conservation Service (ASCS) office in the county where the project is located. Amount of funding varies from county to county depending upon ACP funds available for the year in question. Some landowners have been reimbursed for 80 percent of construction costs, but 50 percent is more common.

State funding for private wildlife ponds is another possibility. It is sometimes available through the DNR's Private Land Development Program. For assistance, contact an area wildlife manager of the DNR Section of Wildlife, Centennial Office Building, St. Paul, MN 55155. Reimbursement from the DNR varies according to a pond's wildlife potential.

Landowners should check with the DNR to see if a special permit is required for their project. Generally, a permit is required for pond construction which would affect the course, current, or cross-section of any natural watercourse, wetland, or lake which is classified as "public water". For information, consult your DNR regional hydrologist at either Bemidji, Grand Rapids, Brainerd, New Ulm, Rochester, or St. Paul.

With continued interest by private landowners and financial support provided by state and federal programs, the future for wildlife-pond development in Minnesota is bright. Increased numbers of private ponds are needed to help make Minnesota the waterfowl producing state it once was.

Text for this leaflet was prepared by Bruce Sandstrom, DNR Division of Waters. DNR Report Number 51.

Nitro blast in progress, note soil lifted into air.

Dug water hole 30', 120' long and 3' deep. Used extensively by nesting ducks.

Wildlife hole blown in peat using amonium nitrate. All photos this page provided by U.S.D.A. - Soil Conservation Service.

Charles E. Schafer
Michigan Wildlife Sketches

NORTHERN MINNESOTA IS MOOSE COUNTRY

A full-grown moose weighs as much as four or five full-grown deer. Hunting it requires careful preparation.

Moose is an Algonquin Indian word meaning "twig-eater" or "he who eats off," and indeed the moose is a browser by habitat. In the summertime, the moose lives where lakes, rivers, and bogs are abundant. Like deer, moose prefer shrub and second-growth forest, but are more suited to larger areas of cutover forest than are deer.

DESCRIPTION

The moose is the largest member of the deer family. While a large Alaskan bull moose may weigh up to 1,800 pounds, those found in Minnesota and adjoining Canadian provinces seldom exceed 1,200 pounds.

Its coat is usually dark brown to almost black. The mature moose has a shoulder hump and a flap of skin hanging below the throat (called a bell or dewlap) that may be up to two feet in length.

Bulls carry large antlers which become palmated (flat and extended) after the third year. Antlers begin to grow in April and, after the velvet covering is rubbed off in late August or September, the polished antlers are carried until late December or early January. Antlers achieve maximum size around the fifth year and generally decrease after the tenth year.

Long legs and splayed hooves support the moose's weight in both marshy areas where it feeds during summer and in deep winter snow.

Moose have weak eyesight, but an acute sense of smell and hearing. They are adept swimmers and are known to submerge entirely to feed on aquatic plants.

RANGE AND POPULATION TRENDS.

Before people settled in northern Minnesota, moose were the most common big game animal, except in the extreme northeast where woodland caribou were more common. Logging, fires, settlements, and the resulting second-growth forests created ideal habitat for white-tailed deer. This spelled immediate trouble for the moose because deer carry a parasite that is harmless to them, but fatal to moose. Consequently, moose populations declined sharply.

In Minnesota, moose thrive in two different habitats: the northwest Red Lake region in the flat, low-lying willow, aspen, and jack-pine area. These two major ranges contain about 8,600 square miles. Between the two is a secondary range — 15,000 square miles — where hunters occasionally see moose.

In Minnesota, moose populations have changed from year to year. The brainworm, a parasite, caused numerous deaths among moose in the mid-1920s and again in the mid 1930s. Since the 1940s a gradual increase in moose numbers and the area they occupied occurred. The moose population in Minnesota reached its peak by the mid-1960s. Since then, it has remained fairly stable at slightly reduced numbers. In 1977, the state-wide moose population was estimated at 9,000 to 12,000 animals.

In the northeast, the population increase was due to pulpwood harvests that allowed second-growth forests to develop as moose habitat. A decreasing deer population at the same time lessened the incidence of brainworm infection. In the northwest, a decrease in agriculture, and an increase in brush land provided good moose habitat.

BEHAVIOR

Moose are not gregarious, though bulls occasionally band together or cows and mature offspring form small groups of two to five. Except for mating season, moose tend to be solitary.

Although they look formidable, moose are generally not aggressive, except when a cow feels her calf is in danger, or when a bull is in rut (the breeding season which occurs from mid-September to mid-October).

Eight months after mating, cows give birth to one or two calves in May or June. A calf weighs 25 to 35 pounds at birth. Twin births are common in good range. Within a week of birth, the calf is strong enough to accompany the cow both on land and in water. Calves remain with the cow until the following year's calving, and may congregate with her after this.

Moose are most active at daybreak and sunset, although they may feed continuously through the day. During the summer months, they spend a considerable portion of their time in or near water, but in the winter they head for browse and cover on higher ground. As the snow piles up, the moose will remain in one region for several weeks, bedding and feeding in the same general areas.

MANAGEMENT

At the present time, moose management in Minnesota consists of aerial surveys over major moose range every winter to count animals and to estimate the survival of calves. A summer survey assesses the proportion of newborn calves in the population.

Habitat for moose in northeastern Minnesota is largely a product of forest management, particularly timber harvesting. In portions of the northwest, moose habitat, along with that of other wildlife species, is disappearing as timber harvesting declines.

Two parasites common to the white-tailed deer may infest moose when the two animals share the same range. The first and most deadly, a parasitic roundworm, called the brainworm, is normally harmless to deer. Moose are infected when they accidently eat snails that feed on deer feces. The second, liver flukes, though not necessarily fatal, are definitely debilitating. Liver flukes may also be transmitted by snails.

The only predator in Minnesota capable of killing a moose is the timber wolf. The primary food of wolves in Minnesota is the white-tailed deer. However, wolves do prey on moose — particularly calves — at various times of the year.

HUNTING

Moose hunting in Minnesota was a way of life with Native Americans and early settlers. The first hunting regulations came into effect in 1887 when the season was restricted to the month of November. Because of continued declines in moose populations, the DNR further restricted hunting to bulls only in 1903. Seasons were closed completely after 1922. The moose season remained closed from 1922 to 1971. In that year, the legislature authorized a controlled hunt.

Current moose seasons limit the number of hunters, the area in which they hunt, and season length. Because there are more moose hunters than permits, hunters must apply in parties of four before the season opens. The DNR holds a drawing to determine the successful applicants. During 1976, the DNR granted 930 hunting permits. Successful applicants then had to wait two subsequent seasons to apply again.

The season is about three weeks long, depending on the area. Success is about 90% due to the length of time. Hours are from sunrise to sunset.

There are several things the prospective moose hunter should consider:

•Be in good physical condition. A full-grown moose weighs as much as four or five very large deer. The largest recorded dressed moose weighed 910 pounds, which came to 227 pounds of meat for each of four hunters to carry.

•Use a rifle of sufficient caliber and be sure to sight it in beforehand.

•Observe the ethics of a good hunter.

Dressing out, quartering, and transporting the heavy meat and head is easier with the right equipment. Pulley hoists, back-packing equipment, hatchet, knife, sharpening stones, wrapping cloths, and bags should be part of hunters' equipment on a moose hunt.

Copy for this article was prepared by the DNR Bureau of Information and Education. DNR Report Number 6.

White Tail Deer
Spring
Summer
Autumn
Winter
Charles E. Schafer
Michigan Wildlife Sketches
Schafer

WHITE-TAILED DEER

Managing Minnesota's deer herd means regulating hunting seasons and maintaining suitable habitat. Knowing the white-tail's habits increases a hunter's chances of success.

The white-tailed deer, named for its distinctive tail or "flag," is Minnesota's most common big game animal. Today it is found in every county, though its major range is the forested regions of the state.

LIFE HISTORY

The deer's life history divides into four periods. These periods coincide with the four seasons of the year.

Spring. As snow melts from fields and meadows, deer move out of forest stands into clearings to feed on succulent green growth. The hardships of winter are over; food once again is plentiful.

Antler growth on buck deer begins in spring. Does, after a pregnancy close to seven months, begin fawning in late May. Most fawns are born from early to mid-June.

The fawning site may be in a field, marsh, or woods. Often, twin fawns are born, especially among healthy does over two years old. Healthy fawns weigh about eight pounds at birth. Adult buck deer take no part in rearing fawns.

Within an hour after birth, fawns are up and about. When not nursing, they remain nearly motionless, camouflaged by white spots on their sides and backs which break up the solid coloration. The doe usually does not move too far from her fawns.

A few weeks after birth, fawns begin to supplement their milk diet with leafy greens — although they continue to nurse for about four months.

In the fall, the knobs or "buttons" of the male fawn appear under the dense hair on the top of the skull. Maximum antler development usually comes when a buck is in his prime. Proper diet is essential for good antler growth. Underdeveloped antlers are usually associated with poor range conditions or old age.

Summer. These are the months of plenty. Twigs and leaves are lush and aquatic plants thrive in shallow ponds and streams. Deer regain weight lost during winter. Reddish-tan coats become sleek and shining; underparts turn white.

Fall. The velvet has cropped from bucks' antlers, leaving them hard. A heavy coat of gray has replaced summer red on both bucks and does. Now weaned, fawns have lost their white spots.

Bucks begin "rut" activity in late October by sparring with small trees and pawing "scrapes" on the ground. These marks may serve as signposts of a dominant buck's territory. The scrapes may also assist bucks and does in locating each other during the breeding period. In Minnesota this period generally reaches its peak during the first half of November.

Winter. This is often the season of hardship for deer. Snow, plunging temperatures, and cold winds cause deer to seek shelter in preference to areas which provide good food but no shelter.

With rutting season over, the antlers have served their purpose and are now useless weight. Bucks shed their antlers anytime from December to March.

As snow depths increase beyond 18 inches, deer movement becomes restricted. Moving through deep snow often drains more of a deer's energy than available food can replace. Nor is winter food nutritionally adequate in most areas to maintain proper body temperatures and weight. Even in mild winters, deer lose weight.

Malnutrition caused by severe winter or late green-up in spring may cause starvation deaths. Even without deaths there are other consequences: In spring, many does may not bear fawns, or their fawns may be born so weak that they cannot survive, or many does may give birth to only one fawn rather than twins. These reproductive losses are the primary cause for lower deer numbers in northern Minnesota.

DEER MANAGEMENT

The Department of Natural Resources maintains the state's deer population by setting hunting regulations that increase or decrease the deer harvest, and by improving habitat to increase the capacity of land to support deer.

The size of the deer herd must be a compromise between the minimum number of deer desired for public enjoyment, and the maximum number the land can support without damage to crops and tree reproduction.

In farm country, crops provide much deer food. But there is a limit to how much "freeloading" farmers will allow. A large number of deer can damage crops severely, especially orchards or standing corn. In farm areas, therefore, deer managers maintain the deer herd with farmers as well as sportsmen and wildlife enthusiasts in mind.

HUNTING REGULATION

From 1957 through 1965, the annual deer harvest increased from 67,000 to 127,000 animals. Despite more hunters — from 180,000 in 1957 to 290,000 in 1965 — hunting success generally held above 40 percent because of a seven-year succession of winters that favored deer reproduction.

Although the number of hunters continued to increase — 307,000 licenses were sold in 1967, the peak year prior to 1975 — the effects of severe winters in 1964-65 and 1965-66 began to show declining harvests and in lower hunter success. The harsh winter of 1968-69 was reflected in the harvest of only 68,000 deer in the 1969 season.

The 1970 season was reduced to two days to prevent taking too many deer. Despite the brief season, 157,000 hunters harvested 50,000 white-tails. Another severe winter followed. Anticipating poor reproduction, the DNR closed the 1971 season to allow the herd to rebuild.

In 1972, following legislation to establish a deer season framework between November 1 and December 15, a new approach to deer hunting began.

Hunters were given the opportunity to choose their own November hunting period — two and three days early in the month or five or six days in the later part. This approach reduced the number of hunters in the woods at one time. Hunting quality improved and the deer harvest was reduced from what it might have been under the old system. A few years' experience showed that even more conservative measures would have to be employed to rebuild the deer population and hunting was then restricted to bucks-only in much of the state.

Since 1975 the DNR has made increasing use of the antlerless quota system. Under this system a quota of does and fawns may be taken by special permits in addition to the harvest of buck deer in any management unit.

The deer population of any management unit can be more efficiently increased, decreased, or held stable by the antlerless quota system. This system can assure that neither too few nor too many does are taken from an area. Without this ability to control the doe harvest, that is, if deer are harvested indiscriminately over broad areas, an overharvest occurs in some areas and an under-harvest in others. Controlling the deer population through doe quotas will eliminate the need to change season dates and zone lines for the purpose of managing the deer population.

HABITAT IMPROVEMENT

Both habitat improvement and proper hunting regulations are necessary to maintain a satisfactory deer population.

Large deer herds can be supported only where a large portion of the forest is in a brushy, reproduction stage. Many aspen and other hardwood stands have grown to the "pole" stage in which the "closed canopy" shades out desirable food plant and shrubs.

In addition to maturing forests, northeastern Minnesota is faced with natural plant succession toward a spruce-fir forest. Spruce-fir, poor habitat for deer, is the climax forest in that region. (A climax forest is a self-perpetuating group of trees that have adapted to the climate and soil of a particular region.) This natural succession toward evergreens is already far advanced in Cook, Lake, St. Louis, Koochiching, Itasca, and Lake of the Woods counties.

To restore these areas to good deer habitat, many sites where balsam fir is invading aspen stands must be clear-cut — all trees cut to ground level — to set growth back to the aspen-sprout stage which produces more deer food.

This can be accomplished by commercial timber cutting or by special deer habitat projects. Timber sales are more desirable, but their volume and distribution are not sufficient to meet deer management objectives. The increased use of wood products, however, will require more timber and thus bring about the growth of younger forests. At present, it appears that bulldozing, prescribed burning, and the development of forest openings to provide early supplies of nutritious green food will be a vital addition to commercial timber sales for maintaining adequate deer habitat. Minnesota wildlife managers and foresters are now working together toward this goal.

HUNTING TIPS

Every hunter should familiarize himself with the hunting regulations printed annually by the Department of Natural Resources.

Scouting and learning the country is very important. Both hunters and non-hunters will find that some fundamental knowledge of deer habits is very rewarding.

Deer move early in the morning. Movements dwindle by mid-morning, then increase again shortly before dark. Trail sitting during these hours, reserving the middle of the day for resting, tracking, stalking, or driving, is good strategy. Anticipate movements brought about by the presence of other hunters. Locate and study trails, feeding areas, and places where tracks are found. The number of tracks, freshness, and the direction of prevailing travel are also important.

White-tails eat leaves and ends of twigs, plus grass and forbs. Important feeding areas are located where food is abundant in brushy or open stands where sunlight gets to small trees and shrubs. Large stretches of grass with no shrubs are seldom used during daylight hours. Deer use thick stands of timber for bedding down and hiding, but not extensively for feeding. Wait on a runway near heavy cover where the white-tail may move, or be driven from feeding places.

Snow greatly increases visibility and is important to hunting success. A hunter sitting motionless beside a well-used runway sees even the smallest movement. Over a distance and without snow, that movement would be impossible to see.

FOREST OPENINGS: KEY TO DEER SURVIVAL

By Jay Jenecek

Forest clearings 'set a table' for deer. In spring, they provide protein-rich greens and, come fall, high-energy foods till the snow flies.

Minnesota's population of white-tailed deer could certainly use a few more breaks. Not more luck, mind you, but more breaks in the vast forest canopy that blankets our north-central and northeastern counties.

For years, forest openings were appreciated only by the deer themselves. In fact, many Minnesota landowners planted old fields with pines, thinking they were improving the forest and its wildlife. But within the last decade, wildlife researchers have learned that without forest openings, an important part of the white-tail's diet will be lacking.

Clearings are beneficial because they provide a steady food supply in spring and fall, two critical periods for deer. In spring, young grass shoots and other herbaceous plants appear in forest openings before they are available to deer elsewhere. This new growth provides high-protein, low-fiber food at a time when deer badly need it. Does in their last months of pregnancy urgently require this nutritional boost to produce healthy fawns.

In the fall, after frosts have killed herbs on the forest floor and once-succulent twigs have hardened, the grasses and hardier herbs of openings remain green and nutritious. Then, as in spring, deer feed in clearings to extract the most in nutrition available from their habitat.

Settlement Days. During Minnesota's early settlement, countless openings punctuated northern forests. Settlers cleared the land for crops and pastures. Meanwhile, loggers removed the large timber, leaving farmers to finish the job of clearing brush and pulling stumps.

Today, however, most of these forest clearings are gone. The settlers have long-since departed and the open country they maintained has gradually reverted to forest. Even where agriculture persists, farming methods are more intense and on a larger scale. The fields and openings on farmlands are generally more concentrated rather than scattered throughout the forest.

The disappearance of life-giving openings has been unfortunate for deer. Now that most clearings have grown over with trees and brush, the grasses and herbs have been shaded out and early spring greens are not so readily available.

A survey on state land in northern Minnesota revealed that openings make up less than one percent of the area. Deer habitat investigations indicate that **at least** five percent of a deer's home range should be in openings.

Forest openings in northern Minnesota are of a good size to benefit deer — averaging about three acres — but there are too few of them. Moreover, nearly all of these openings are surrounded by steadily-encroaching forest.

To wildlife managers concerned with maintaining our deer population, the challenge is clear: forest openings must be preserved and increased.

Meeting this challenge is difficult—and costly. Cutting and burning are effective techniques for keeping openings free of undesirable trees. However, these actions usually produce vigorous new sprouting of trees, making it necessary to cut and burn several times (at great expense) before a more permanent clearing is created.

The cheapest and most effective means of maintaining openings is with chemicals. One chemical which produced good results is picloram. It is applied as dry pellets around species such as aspen, hazel and willow. By treating these plants individually, other desirable species—including oaks, blueberries, highbush cranberry, cherries, etc.—can be retained in the opening. The dry chemical has other advantages. It is easily carried in the field and does not have to be mixed with water. Unlike spray-on liquid chemicals, it will not air-drift and affect non-target plants.

New Openings. Creating wildlife openings is every bit as expensive as maintaining them. Removal of timber, stumps and debris—in addition to seeding the bare earth—add to the cost.

One way to defray costs is to use openings created by other activities in the forest. Commercial timber sales, for example, create landing sites where logs are piled and loaded onto trucks, chipper sites where large mechanical tree harvesters operate, and turnouts in access roads. All hold promise for being maintained as openings. Most are free of timber and require only the removal of debris, followed by seeding, to "set the table" for deer. Both wildlife managers and foresters of the Department of Natural Resources develop forest openings this way whenever practical.

Openings scattered throughout a forest provide diverse habitat which supports a variety of species other than game animals. A clearing is a bird water's paradise and a nice place for hikers to stop for a rest and enjoy a cool breeze. It could hold the only good blueberry patch in the area and juneberries and chokecherries might thrive.

A forest opening is many things to many people, but to all it is a source of satisfaction. To date, hunters' dollars from license fees are being spent to preserve and perpetuate forest clearings. But the need is greater than these funds can provide. Once Minnesotans are aware of the value of openings to wildlife, other sources of funding will no doubt be identified. Hopefully, this awareness will not be too long in coming.

SUMMER FOODS OF DEER

By Keith R. McCaffery

In northern Wisconsin, deer prefer non-woody foods when they can get them, from spring through fall. Aspen leaves are favorites—and aspen areas can support sizeable deer populations.

When we think of deer food, most of us are inclined to visualize the woody stems and twigs of trees and shrubs. Deer are considered to be browsers and the common definition is that browse means twigs. However, the technical definition of browse also includes *leaves* of woody plants. We tend to overlook the importance of leaves and herbaceous plants in the day-to-day diet of deer.

There is speculation regarding what deer do or don't eat in summer. Recent research on deer food habits indicates that, although deer are classified as browsers, they apparently would prefer to be grazers!

We began a study of the summer food habits of northern Wisconsin deer in 1969. Stomach (rumen) contents were collected from 76 road-killed deer—a job that was more than occasionally messy.

Collections were made between April 15 and November 15. The collected material was preserved in one-quart jars and sent to Professor Forest Stearns at the University of Wisconsin—Milwaukee. There, two of his students, John Tranetzki and James Pietchura, Jr., conducted the sample analyses.

Each quart sample was washed over screens to separate "mush" from the plant fragments. After washing, individual pieces had to be identified. Identifying these small fragments required a great amount of skill and knowledge of plant structure. These botanists had it. Results of all the samples were subsequently tabulated to determine the annual and seasonal importance of specific foods. The findings were a little surprising.

Woody twigs and stems, which we traditionally consider to be browse, comprised only 3 per cent of all food eaten. Only about one-third of the food eaten would have fallen into the broader technical definition of browse.

The bulk of the food eaten was made up of leaves and herbs. This herbaceous material comprised 87 per cent of the volume eaten by the 76 deer. The remaining 10 per cent of the food volume was composed of fruits and flowers which included mainly acorns, mushrooms and blackberries.

The small percentage of woody browse found indicates that deer definitely prefer leafy foods. It suggests further that deer eat quantities of woody twigs only when forced by snow accumulation to eat this food resource.

Aspen (popple) leaves were by far the most important single source of food for the deer sampled. Although more than 70 different plant types (genera and families) were identified in the rumen samples, aspen alone contributed 16 per cent to the aggregate volume. Aspen leaves were most heavily used in early summer (June-July) when they contributed almost 30 per cent of the food eaten.

Grasses were the second most important food in terms of the volume eaten. They were particularly important in spring and fall when they amounted to 17 to 19 per cent of the food eaten in those seasons.

Other important food plants are included in the accompanying table. The plants are listed in descending order of importance and encompass the period from April 15 to November 15.

The small percentage of woody browse found indicates that deer definitely prefer leafy foods. It suggests further that deer eat quantities of woody twigs only when forced by snow accumulation to eat this food resource.

Aspen (popple) leaves were by far the most important single source of food for the deer sampled. Although more than 70 different plant types (genera and families) were identified in the rumen samples, aspen alone contributed 16 per cent to the aggregate volume. Aspen leaves were most heavily used in early summer (June-July) when they contributed almost 30 per cent of the food eaten.

Grasses were the second most important food in terms of the volume eaten. They were particularly important in spring and fall when they amounted to 17 to 19 per cent of the food eaten in those seasons.

Other important food plants are included in the accompanying table. The plants are listed in descending order of importance and encompass the period from April 15 to November 15.

TEN OF THE MOST IMPORTANT FOODS FOR DEER IN NORTHERN WISCONSIN

Food Plants	*Seasons or Subseason of Heaviest Use*
Aspen Leaves	June to Sept.
Grasses	Spring and Fall
False Strawberry	Spring
Asters	Late-Summer & Fall
Bush-honeysuckle	Summer
Strawberry	Spring and Summer
Cherry	Late-Summer & Fall
Oak Acorns	Fall
Wintergreen	Spring and Fall
Clover	Summer

This list is based on the volume of food contributed by each plant. The real importance of the individual foods to deer may not be accurately represented. Acorns, for example, occur well down the list. This is mainly because they are only locally available for a short time beginning in fall. However, whenever and wherever available, acorns are highly preferred. This rich carbohydrate helps to lay on the fat that, during severe winters, may mean the difference between death and survival for some deer.

Overall, forest openings and aspen forests stand out as the most important sources of summer nutrition for deer living in the heavily forested areas of northern Wisconsin. Openings contain the grasses, false strawberry and other plants that provide transition food stores between the austere winter diet and leaf-out and between the first killing frost and snow accumulation. Aspen leaves provide a protein-rich food from fawning time until the rut. Aspen leaves appear to be the primary and preferred source of summer nutrition.

We have conclusive evidence that aspen areas sustain much higher and deer populations than most other timber types. This is supported by numbers of deer shot in aspen areas and surveys of deer use based on counts of pellet groups, tracks and trails.

While a food habits study seems quite far removed from the hunting season, the maintenance of habitat types that produce year-round deer food will ultimately mean better production, better winter survival and more venison for you.

Copy for this article was prepared by the DNR Division of Fish and Wildlife. DNR Report No. 76.

BLACK
BEAR
Schafer
Charles E. Schafer
Michigan Wildlife Sketches

MEET THE BLACK BEAR

Campers and hunters in Minnesota's woods benefit from knowing the black bear's habits.

Black bears inhabited most of Minnesota at the time it was first settled. Presently, bears are found in the northeastern one-third of the state, or generally north and east of an imaginary line from Roseau to Pine City. The resident black bear range covers about 30,000 square miles.

Research in northeastern Minnesota has shown that adult female bears establish home territories about four square miles in size. Adult male bears normally range over an area of about 10 square miles. Female bears defend their territories from other adult females. Adult males do not defend territories.

Bears prefer to remain solitary except during the breeding season. However, several bears may congregate at locations of abundant good supply such as berry patches, garbage dumps, and along streams.

The black bear *(Ursus americanus)* is an omnivorous animal. It eats anything that resembles food in looks, smell, or taste. Studies made on the food habits of bears indicate they eat primarily vegetable materials consisting of grasses, fruits, berries, buds, leaves, nuts, and aspen catkins.

Bears also eat animal matter consisting of carrion, insects and their larvae, and small animals. Occasionally, an individual bear may develop a taste for domestic livestock such as sheep. A bear food study conducted in Maine, however, found that only 8.1 percent of bears' year around diet was animal matter including livestock and game animals.

DESCRIPTION

The average female black bear weighs 150 pounds; males average about 250 pounds. Bears stand between two and three feet high at the shoulders and are light brown to black in color.

The black bear does not truly hibernate. Instead it goes through a period of dormancy during which normal functions such as eating and excreting are discontinued.

The bear may select a den site under an upturned tree, a log, a brush pile or in a natural cavity under a rock ledge. It generally does not breed until about 3½ years of age and normally every other year thereafter. In January or February, one to four young are born after a gestation period of seven to 7½ months.

When born, cubs weigh three-quarters of a pound and are hairless. They grow rapidly and follow the sow when she leaves the den.

Cubs usually stay with the sow until their second summer. But they are able to fend for themselves come their first autumn if something happens to the sow. In Minnesota, bears usually stay in the den until April, although they may leave for an occasional foray during mild winter days.

MANAGEMENT

Minnesota's black bear population is a valuable wildlife resource. Previously unprotected in most of Minnesota, black bears were given overdue recognition as game animals in 1971. Their harvest has since been regulated to provide a continuing source of recreation. The increased status and protection is likely to provide more bears for enjoyment by hikers, campers, and other visitors to Minnesota's forest country.

The secretive nature of black bears and the dense cover where they are found makes them difficult to census. Population trends have been based on estimates from field personnel. These estimates were based on observation of "bear signs", i.e., tracks, stump and log workings, scats, and tree markings.

Occasionally, individual bears kill livestock or damage crops and apiaries. Damage complaints, especially those involving cabins, campgrounds, and garbage collection sites increase greatly in years when berry and nut crops fall. This is not a reliable sign that the bear population is increasing; it means only that bears are hungrier than normal. Such bear problems should be handled on an individual basis rather than by attempting to reduce the entire population. The food supply for bears is dependent upon variations of temperature and precipitation and cannot be ensured by reducing the number of bears.

The problem of bears damaging camps can often be eliminated by following a few simple procedures. Wash all dishes and pans promptly. Put all garbage and food scraps in a trash barrel, or burn or bury garbage away from camp. Empty trash cans frequently and deodorize them. Store food in odor-proof containers, or hang it from a tree limb where a bear can't reach it. And, remember! Bears are powerful, wild animals, not pets!

HUNTING

Hunters generally agree that the black bear is a difficult animal to hunt, especially without the aid of dogs. In his wild habitat, it is easy alarmed. It is capable of traveling great distances across country.

Sighting a bear is more a matter of chance than plan. Deer hunters, because of their large numbers, may get more opportunities to see bear while hunting deer than a bear hunter gets. This is not to imply, however, that the hunter who concentrates on getting a bear should be discouraged. Beginning in 1979, bear may not be taken during the deer season.

Certain factors are in the bear hunter's favor. During the fall hunting season, the black bear is usually actively searching for food. Hunting where natural food is abundant should be the hunter's first tactic.

Signs of bear feeding are easy to find. Look for broken tops of any fruiting plants and examine carrion for signs of bear visits. Once these signs are located, it is best to approach the area early in the morning and always up-wind.

The successful bear hunter should field dress and *skin* his bear promptly to prevent meat spoilage. A 50-pound sack of livestock salt should be taken on a bear hunting trip so the hide can be salted down prior to delivering it to a taxidermist. A bear skin, when properly cared for, makes an excellent trophy. The meat can be very tasty. Bear grease has long been prized by bakers for making pasteries.

A precautionary note. Bear meat should be cooked thoroughly the same way pork is cooked. In rare or raw bear meat, trichinosis may be present.

The hunter who expects to bag a bear in Minnesota must allow more time to locate his quarry than do deer hunters. But the pursuit of the black bear adds an exciting new dimension to hunting opportunities in Minnesota.

Copy for this report was prepared by the DNR Division of Fish and Wildlife, Section of Wildlife. DNR Report Number 89.

Charles E. Schafer
Michigan Wildlife Sketches

WILDCAT!

Though seldom seen, three species of wildcats stalk our fields and forests.

Most every barn has its complement of feline mouse catchers. But in the winter of 1974, a wildlife researcher discovered a different breed of cat in an Itasca County barn. He was ground-tracking a bobcat which had been equipped with a radio-transmitter collar the previous fall. To his surprise, the signal led him out of the woods and toward a farmyard. He alerted the farmer and together they located the wildcat's den beneath the barn foundation. The farmer was astonished. He hadn't seen the bobcat, though the animal had been living there for several weeks.

Many people live in "cat country" and never see one. Secretiveness and elusiveness are typical of bobcats and the other two species of wildcats that occur in Minnesota, the lynx and mountain lion.

Only a handful of the contiguous 48 states can lay claim to having three species of wildcats. Minnesota's nothernmost tier of counties is considered the southern boundary of the lynx and the northernmost range of the bobcat. But the mountain lion, or cougar, only rarely leaves its saucer-size prints on Minnesota soil.

These predators are valuable because they fill a significant niche in nature's chain of life. The efficient hunting of predators helps to shape the character of their prey, just as the predators are shaped by the prey. Therefore, we should appreciate these wild creatures not only for their beauty and unique evolution, but because they are essential to the balance of nature.

BOBCATS AND LYNX

Most of our bobcats live in the vast, cedar-black spruce bogs in the northern third of the state. They prefer heavy brush areas typical of these densely-timbered swamps.

The Canada lynx haunts a somewhat different habitat, the mature forests of northern Minnesota, habitat of its primary prey, the snowshoe hare.

Bobcats and lynx usually afford the observer only a fleeting glimpse. From a distance the two cats appear quite similar, though closer inspection reveals several distinct differences.

An adult bobcat stands about 20 to 30 inches at the shoulder and weighs 15 to 40 pounds (the Minnesota record is 50 pounds). The bobcat's body is longer than that of a domestic cat and its paws are more massive, with distinct five-sided heel pads.

Short black tufts, up to an inch long, decorate the bobcat's pointed ears. Extending from the ears to the chin is a white, black, and gray ruff. Its pelage (coat) is light fawn to rust brown, boldly streaded, and spotted with black. Its common name is derived from its six-inch "bobtail," which is barred in black three or four times across the top and white underneath. The main distinction between a bobcat and a lynx is that the tip of a bobcat's tail is black above and white below. The tip of a lynx's tail is black all the way around.

The Canada lynx looks oriental. Its golden eyes have a slanted, oval appearance unlike the bobcat's wide, round eyes. Longer legs and a lankier body make the lynx appear larger than the bobcat, though it stands only two feet at the shoulder, and averages 15-30 pounds.

The lynx also has an ear-to-chin ruff, but it is not distinctly marked. Its most identifiable feature is its long, feather-like ear tufts. Its fur, more tawny than the bobcat's coat, is long, fine, and colored pale gray or buff with indistinct streaks of brown.

Rather then having naked heel and toe pads, the lynx has thick coarse hairs over its pads. These act as natural snowshoes, enabling the wildcat to move easily through deep snow.

Typical of most cats, lynx and bobcats have five toes in front and four in back, all with retractable claws. Both cats are efficient at killing small prey. They use their front claws to hold their prey, while the hind feet rowel (rake) the victim. The phrase "lick his weight in wildcats" originated as a tribute to the fighting ability of hunting hounds which survived frays with bobcats.

BEHAVIOR

Bobcats and lynx frequent large territories or home ranges, often covering eight to ten square miles. They normally hunt alone, crossing and re-crossing portions of their territories, investigating small animal tracks and scents while keeping ears and eyes open for any sign of life. Although primarily nocturnal, bobcats and lynx will hunt during daytime in remote areas. Long whiskers act as feelers in the brush during nighttime hunts.

Usually wildcats stalk their prey or pounce upon a victim after one or two, eight-foot leaps. Seldom do they pursue an animal to any length. Instead, they give up the chase to seek easier quarry.

Breeding for both species occurs sometime between January and June. During this time lynx and bobcat may range farther afield in search of a mate. Often a cat will "mark" a tree with urine, returning frequently to this "rendezvous point" in hopes that a cat of the opposite sex has found the spot. The urine contains scent from the anal gland, which is common to both cats.

During gestation, which may last from 50 to 70 days, a female wildcat seeks out an appropriate den—an abandoned den of another animal, a windfall with tangled branches or roots, a rock crevice, or even the foundation of a vacated building.

Kittens are born blind but furry measuring about ten inches and weighing about 12 ounces. Litters average from two to four young. Claws are funtional at birth. In about nine to 11 days the kittens' eyes open and they begin to venture out of the den.

Play for young cats is a learning activity. It serves to hone their hunting skills. They jump at anything that moves — butterflies and small animals their mother brings back alive. While young are nursing, the male may bring food for the female. She keeps her mate away from the kittens, however, for he may kill them.

The young are weaned in two to three months, though they remain with the female while learning the finer points of hunting. By late fall, the kittens are nearly full grown. Bobcat young probably disperse — seek out their own territories — in early winter. Adolescent lynx may stay with their mother until the next spring breeding season before they disperse.

FOOD HABITS

Studies have shown that lynx prey almost exclusively on snowshoe hare. Levels of lynx populations are regulated by kitten survival (low during years when snowshoe populations are low) and by an apparent lack of breeding during low periods of the hare cycle.

The bobcat is also affected by the snowshoe hare cycle, but not to such a great extent. An ongoing study by the DNR Wildlife Research Station in Grand Rapids, headed by wildlife biologist Bill Berg, found that the highest percentage of the northern bobcat's diet is hare, closely followed by deer.

Bobcats are opportunists — they eat whatever is available. For example, in 1975-76 bobcats preyed heavily on porcupine.

But during 1976-77, they seldom dined on "Porkies." This could mean a decline in the porcupine population or an increase in other prey species. When food is plentiful, bobcats may gorge themselves and then not feed again for days. Occasionally they return to a kill for a second meal, especially if it is a large animal such as a deer, but rarely more than once. Generally, wildcats do not eat carrion.

In Minnesota, livestock predation by lynx and bobcats is not a problem. Nor do the two cats affect deer or small game numbers to any noticeable extent.

COUGAR

While bobcat and lynx seldom raid domestic stock, the cougar, at least in years past, was quite fond of horseflesh. The cougar's power and agility allow it to bring down a horse, elk, or other animal five or six times heavier than itself.

A cougar is much bigger than a bobcat, but about half the size of an African lion. A large adult measures up to nine feet in length, one-third of which is its long dark-tipped tail. It stands about 26 to 29 inches at the shoulder and weighs from 175 to 200 pounds.

The head of the cougar is shaped much like a female African lion's, but without the face ruff. Its coat is tawny — void of streaks or spots — and the underparts are white. Ears are small and rounded other than pointed like a bobcat's.

Cougar are typically solitary. But sometimes two females, or a female with full-grown cubs, have been seen. During mating, pairs travel and hunt together.

A cougar will consume two to three large animals each week as it roams over its 30 to 50 square-mile home range. It is adaptable to boreal and hardwood forests and also to transition lands where farm land and forest come together.

Canadian wildlife biologist Robert Nero has documented the presence of cougar in nearby Manitoba. However, it is highly unlikely that a breeding pair exists in Minnesota. Timber wolf expert L. David Mech, for example, has spent thousands of hours in the air and on the ground tracking wolves in Superior National Forest. He has never seen a live cougar or even a track.

Still, sightings by Minnesotans persist. The last authenticated sighting — one verified by a DNR wildlife biologist — was in 1975 in extreme nothwesern Minnesota. There, six sightings were recorded, resumably by several people seeing the same cougar. Elsewhere, mountain lion have been sighted north of Duluth, near International Falls, and in and around Itasca State Park.

In Manitoba, Nero has recorded an increase in cougar sightings in recent years. He believes the animals may be increasing in numbers in the province because of their protected status.

Nero is convinced that the continued abundance of important cougar prey species — mule deer and elk — and the influx of white-tail deer into the mule deer-elk range, has allowed the cougar to expand its range.

Nero estimates a total population of 50 mountain lions in Manitoba. The fact that cougar range through southern Manitoba keeps alive the hope that North America's largest cat may once again roam the forests of Minnesota.

FUTURE

Minnesota's population of bobcats is estimated to be from 1,000 to 3,000 animals. The number of lynx is more difficult to pin down because their populations fluctuate. Researchers use scent posts to determine bobcat and other wildlife populations. A scent post is a circle of dirt — about three feet in diameter — with a pellet infused with egg and coyote urine placed in the center. Biologists check footprints in the dirt each day to identify species that have visited the posts. Scent posts are placed every one-third mile over a 15-mile transect.

In 1976, four out of ten scent post routes showed bobcat signs. In 1977, half of 18 had been visited by bobcats. These findings, correlated with data obtained from radio tracking of animals (wildcats rigged with a collar and small radio signal transmitter) and ground tracking during winter, will enable the DNR to follow any fluctuations in bobcat and lynx populations.

In 1977, the legislature granted the bobcat and Canada lynx furbearer and game status. This enables the DNR to regulate the harvest of bobcats according to population variations. Meanwhile, the predictability of lynx population cycles will help wildlfe managers set reasonable trapping laws to insure that enough breeding animals survive.

These controls, combined with more research into wildcat populations dynamics and inaccesibility of their habitat, is assurance that our native wildcats will be with us for may years to come.

This article was prepared by the DNR Bureau of Information & Education. DNR Report Number 73.

THE LYNX LINK

The 1977 deer season was only a few hours old when a Coon Rapids hunter bagged a spike buck in Aitkin County. After dragging the deer back to his stand, he looked up to see a great cat. The short tail, long legs, and huge paws were evident as the Canada lynx walked out on a sandbar and waded across the small stream toward the fallen deer and spellbound hunter. Suddenly the animal sensed danger and vanished into the forest, leaving the amazed hunter shaking with excitement.

Perhaps the most intriguing aspect about the Canada lynx is that whole populations of these northwoods cats can vanish as mysteriously as they appear. Periodic fluctuations of lynx numbers are called cycles. Each cycle is characterized by about three years of super-abundance and seven years of lower numbers. A period of abundance is called an "irruption," a population explosion.

Irruptions of Canada lynx were documented by fur buyers of Canada's Hudson's Bay Company as early as 1829. Wallace B. Grange wrote in The Way to Game Abundance (1949) that many wildlife species of the boreal forest are cyclic, including Canada lynx, showshoe hares, ruffed grouse, voles, and lemmings.

While particular years of abundance in each decade may vary somewhat, Grange noted a phenomenal regularity and pattern in wildlife cycles. He wrote that, across the northern region of North America, cyclic wildlife species had reached peak numbers from 1852 to 1942 in each decade in years ending in "2." In individual locales, the peak varied as much as two years before or after. Conversely, the greatest scarcity of cyclic species in each decade from 1857 to 1947 has occurred in years ending in "7," give or take two years.

Grange predicted a high for 1952. In Minnesota, both snowshoe hares and lynx peaked that year. Ruffed grouse peaked in 1951.

During the next decade, the Canada lynx peaked in 1962, snowshoes reached their greatest numbers in 1960, and the ruffed grouse harvest was higher in 1961 than in any other year from 1960 to 1966.

In the 1970's the pattern identified by Grange continued. Canada lynx were most numerous in 1973, and both ruffed grouse and snowshoe hares were most abundant in 1971.

During 1975 and 1976, few lynx were harvested in Minnesota, and ruffed grouse hunters had limited success. Hare numbers also were low.

Ruffed grouse and snowshoe hare numbers showed an encouraging increase in 1977, and a few lynx have again shown up. This suggests that the current cycle has "bottomed out" and continued increases can be expected for the next few years. The next cyclic high for lynx, hares, and grouse should occur between 1980 and 1984.

The number of Canada lynx in Minnesota is reflected by the number taken by hunters and trappers each year. The average annual harvest in Minnesota is reflected by the number taken by hunters and trappers each year. The average annual harvest in Minnesota for the past 47 years is 177 animals. Harvest totals have varied widely. During cyclic lows, 25 or fewer lynx were estimated trapped in Minnesota in 1944, 1948, 1950, 1958, 1959, 1968, 1970, 1971, 1975, 1976 and 1977.

LYNX IMMIGRATION

During irruptions of lynx, there may be more than 300 trapped in a single year. This occurred in 1940, 1952, 1962, and 1973. Where do all these lynx come from? A few lynx are probably resident animals occupying remote northern forests along the Canadian border. These are rarely seen and rarely trapped because of the remoteness of their habitat.

Most lynx taken during irruptions probably have wandered into Minnesota from adjacent Manitoba and Ontario. Lynx are great travelers. Dr. L. David Meck marked a lynx in northeastern Minnesota in 1974, and it was taken three years later nearly 300 miles away in Ontario.

Most of these wanderers are young lynx which have been displaced from occupied territories in Canada. Younger animals can be identified by weight — from 15 to 20 pounds — and their sharp teeth. In contrast, older lynx will sometimes weigh over 40 pounds and the teeth are well worn.

These young animals may show up nearly anywhere, but most occur in the northeastern third of the state. During the most recent irruption which bagan in 1971 and subsided by 1974, the most unusual occurrences were in Lincoln, Wilkin, Meeker, Wright, and Anoka counties.

The normally secretive behavior of lynx may be absent during irruptions. They seem quite naive, or fearless, of man. Although not dangerous, random wanderings take them into such unlikely places as Duluth's residential areas and farmyards in Todd County. Frequently they do not flee at the sight of man, and they may get into fights with family dogs when they enter backyards. Others will stand along roads and watch passing traffic.

These lynx experience a high mortality rate. They are displaced animals in habitat that is either marginal or unsuitable for their survival. This mortality does not affect the species' survival, however, because the main range in Manitoba and Ontario remains occupied by lynx.

HARE APPARENT

The factor which apparently contributes most to the welfare of lynx populations is the status of the snowshoe hare. In simple terms, when hares are abundant, lynx become abundant. When snowshoe hares are scarce the physical condition of breeding-age lynx is poorer. The litter size will be about three, and few kittens will survive.

During these lean years, the existing lynx population grows older so that most animals become four years or more in age. Therefore, when snowshoe hares irrupt, the lynx population is comprised mainly of adults capable of achieving their full reproductive potential of five kittens per litter. In addition, the survival rate of the kittens is high because of the abundance of food.

This explains how lynx can make such remarkable population increases from one year to the next. For example, in 1971 Minnesota's estimated lynx harvest was 14. In 1972 it was 215, and in 1973 it was 691.

The dissappearance of lynx on Minnesota's landscape can be as sudden and remarkable as an irruption. In 1974, an estimated 88 lynx were harvested, but none was recorded the following two years. This is, of course, considered a short-term, common cyclic phenomenon, not to be confused with long-term declines.

After reaching a mid-decade low, snowshoe hares begin increasing in the years ending in "8," "9" and "0," and peak during the years ending in "1" and "2." Lynx populations typically show a peak in the years ending in "2" and "3." There is a one-year lag between the occurrence of high hare numbers and an improvment in the physical condition and reproductive success of lynx. Statistically, there is a 75 percent correlation between the snowshoe hare harvest one year and the lynx harvest the following year.

When snowshoe hares decrease, lynx do not adapt well to other food supplies. So, the number of lynx decrease to a level which is capable of being sustained by the low hare population.

WHY CYCLES

All this information does not answer the obvious questions of why there are cycles. There is no clear-cut answer. Various theories suggest complex interrelationship of predators, prey, disease, weather, and parasitism.

Until an explanation is found, the biological pendulum causing cycles will continue to swing, adding a special mystique to the wildlife of our northwoods.

This article was prepared by Carrol Henderson, coordinator of the DNR's non-game wildlife program. DNR Report Number 74.

Coyote
Timber
Wolf

THE COYOTE

Minnesota is one of only several states where both timber wolves and coyotes can be found. There are basic differences between the two animals, yet many people are unaware of them.

Most wolves are found in Minnesota's northernmost counties, though a few occur across the northern two-thirds of the state. Most coyotes are in the central and northeastern parts of the state.

Like the timber wolf, the coyote is a creature of controversy because it preys upon game an livestock. However, a study of coyote food habits and behavior, conducted by the Department of Natural Resources Wildlife Research Division in Grand Rapids, has revealed many popular misconseptions about the coyote.

DESCRIPTION

The coyote resembles a small, shy, German shepherd with a bottle-shaped tail. It has large golden eyes and round pupils. The coyote's coat is long, coarse, and heavy. Typically, its underparts are light gray to dull-yellow with the outer hairs broadly tipped with black. This black tipping creates wavy bands on the back and sides. The top of the tail is colored like the back; below, it is whitish near the base and black-tipped. The front legs are whitish, while the hind legs are redish on the outside and white on the inside. Throat and belly are usually white or pale gray.

The coyote has small erect ears and carries its tail quite low when running, in contrast to the wolf which carries its tail elevated. The coyote's skull is much smaller and narrower than the wolf's and it has small feet — almost fox-sized.

Adult timber wolves weigh between 60 and 120 pounds. Coyotes average 25 pounds for females and 30 for males. (These figures are based on a six-year study by DNR naturalist Robert Chesness.)

FOOD HABITS

After six years of study, the DNR has learned important facts about the coyote's diet. Research includes: examining 925 coyote stomachs (carcasses were obtained from fur buyers, trappers, and hunters), following coyote winter trails totalling 381 miles, air tracking coyotes with radio collars, and investigating every suspected coyote-killed deer.

Forty to 50 percent of the coyote stomachs contained deer remains, but this percentage is misleading. From tracking coyotes, biologists found that most deer eaten by coyotes are already dead. These animals probably died as a result of hunting, poaching, car-kills, or winter malnutrition.

Of the 381 miles of winter coyote trails, one definite and one probable instance were found where coyotes had killed deer, both fawns. And of 17 possible deer kills by coyotes, 11 were fawns. Most of the deer were killed in March and April when they are normally in their poorest physical condition and snow is crusted enough to support a coyote, but not a deer.

It is noteworthy that many reported coyote-killed deer were in fact killed by roaming dogs.

The study also showed that coyote predation in winter was not a serious factor affecting overall deer mortality. It is likely that, during many winters, those deer killed by coyotes would otherwise die from malnutrition and exposure. The impact of coyotes on each new crop of fawns is next to be studied.

One of the surprising finds in the food study was that 24 percent of the coyote stomachs contained porcupine remains. Ground-tracking revealed that a coyote kills a porcupine by flipping it over and latching onto its throat. The coyote then rolls the quilled skin inside out to peel off the flesh, leaving only the quills behind.

Snowshoe hare and ruffed grouse are common prey for coyotes, depending upon their abundance. Neither species, however, comprises a large percentage of the coyote's diet. Coyotes eat small mammals such as mice, voles, and red squirrels, as well as skunks, raccoons, and muskrats.

Basically, the coyote is an opportunist. It eats whatever is available, even a meal of grass or berries. Occasionally coyotes will kill sheep, especially during the spring lambing season. But biologists note that farmers who practice good husbandry, i.e., penning stock at lambing time and disposing of diseased carcasses, have few predation problems.

LIFE HISTORY

The breeding season begins in January and lasts through February. There is evidence that coyotes may pair for life. Adults do not den the rest of the year but sleep on the ground in a protected place. But after breeding they seek a den site, either digging a den in loose soil or enlarging the den of another animal.

Gestation is from 58 to 63 days. Pups, an average of five to seven to a litter, are born in April. at birth the young are blind and helpless, covered with brownish-gray and woolly fur. Pups venture from the den for the first time at three weeks of age, but do not stay out for long periods until they are six or seven weeks old.

Both parents care for the young. When the pups are newborn, the male assists by bringing food for the female. Later he brings food which the female tears into peices to feed the pups. When the pups are weaned, at eight weeks of age, the parents carry partially digested food in their stomachs, then disgorge it in front of the den. The front of the den is usually clean, except at weaning time when the disgorged food may putrify and attract birds and insects.

When pups are eight to 12 weeks old, the parents teach them to hunt. At that time, the entire family leaves the den area. From fall until mid-winter, the pups are moving away from the home area, or "dispersing." DNR biologists have tracked radioed pups for 45 miles. Once dispersed, the young coyotes are often loners as yearlings, usually not mating until two years of age. Then they settle down and form territories of their own.

RANGE AND HABITAT

Coyotes usually live in transitional lands, a combination of farmlands and forest. But they are so adaptable that they often live close to human populations.

The size of a coyote's home territory varies with the animal's age and sex and with food availability. Adult females consistently have home ranges of about six square miles. Adult males, however, are usually more nomadic than females, ranging up to 36 square miles. Females show strong "home" tendencies with no overlapping territories. Males may wander within two or more female territories and may overlap with ranges of other males.

Home ranges are restricted during deep snow periods, and females restrict their movement during the spring denning period. During April and May when pups are near the den, females and pups do not wander far in search of food. During late summer and fall, pups move about within two to three square miles.This range increases as winter approaches.

HABITS

Unlike timber wolves, which are pack animals, coyotes tend to be either solitary or in small family groups. Aerial observations reveal that about 90 percent of the coyote sightings are of lone animals or an often separated, mated pair. Only on a few occasions did the naturalists observe three or four coyotes "packed" together after mid-winter.

There is some evidence that the coyote may have been a pack animal at one time because of the complexity of its facial expressions, which are similar to those of wolves. But the shift to non-pack orientation has allowed the coyote to survive in much greater numbers than the timber wolf.

Coyotes normally move two and one-half to three miles per day. They are primarily crepuscular — active during the early night and dawn hours.

During fall, coyotes are often heard howling; they are seldom as noisy in spring. They may bark alone or together. Often one starts and others take up the call until it becomes a chorus. Their songs often carry for several miles. Young have higher pitched voices than adults.

Coyotes are good swimmers and can run as fast as 45 mph for a short distance. They are curious and playful. Sometimes they initiate play with other species.

FURBEARER

Coyotes are not protected in Minnesota and may be taken by any legal method, such as hunting or trapping. One to two thousand are trapped each fall and winter, but their numbers have remained quite stable. Prime coyote pelts now bring about $40 each. The fur is commonly used as trim for parkas.

POPULATION TRENDS

The Minnesota coyote has an omnivorous diet and can readily adapt to changing habitat conditions. Although there are no figures on Minnesota coyote populations, they are apparently stable in northern regions and probably increasing in southern agricultural areas.

This article was prepared by the DNR Bureau of Information and Education. DNR Report Number 24.

How delicate is the

BALANCE OF NATURE?

A leading expert explains why wolf-control programs are sometimes necessary

By L. David Mech

Something was amiss. For six months, my research assistants and I had been studying a pack of four wolves in the remote northeast corner of Minnesota. We had attached radio collars to two of the wolves, and as the group made its daily winter forays through the snowbound region we followed the twin signals with a receiver and an airplane. But now, our receiver told a strange tale: a week after the group had strayed out of its territory into that of a neighboring pack, the signal from the collar on the alpha—or dominant—male became stationary while the other continued to move. From our bushplane, we could see that three members of the pack had returned to their native territory. But what had happened to the missing male?

We turned the plane and headed over the snowbound landscape toward the site where the other radio collar was emitting its motionless beeping. Then, after setting our bushplane on a frozen lake, we set out hiking on snowshoes through a forest of birches and balsams. After only a half mile afoot, we found the answer to our question: the body of the alpha male lay atop the wind-blown snow. The wolf had been badly chewed and there was a red-rimmed puncture wound in his chest. The tracks told the story: the trespassers had met the home pack and the two groups had squared off against each other. As soon as the host pack had downed the intruding alpha male, the rest of the invaders fled.

As I autopsied the hapless wolf, I thought about the meaning of the event. Under normal circumstances, wolves rarely enter the domains of other wolves. Apparently, this pack had done so out of stress, after killing off almost all of the prey animals within its home range. By conventional thinking, the balance of nature should have prevented that—allowing an equilibrium to exist between the wolves and their prey so that each could maintain a stable population. But here, as elsewhere, the balance of nature seemed to have gone awry, at least in the short run.

During two decades of wolf research, conducting studies in northern Minnesota and on Isle Royale in Michigan, I have learned that, far from always being "balanced," ratios of wolves and prey animals can fluctuate wildly—and sometimes catastrophically. Wolves may actually starve after killing off almost all the moose and deer in an area. This explains why wolf-control programs may sometimes ensure greater and more stable numbers of both wolves and the animals they hunt.

Most people realize that wolves have to kill deer, moose, caribou, elk and other large animals in order to survive. The predators live in family groups called packs, usually containing 6 to 12 members, and it takes a lot of meat to feed them. The pack is well organized, with each wolf occupying its own place in the social ladder. Each pack possesses a territory large enough to encompass hundreds of prey animals and delineates that territory with urine marks and howling.

It has been my good fortune to have spent most of my career researching such details of the wolf's life and of the creature's interactions with its prey. One of my studies that helped fix the balance-of-nature idea in the public mind was the Isle Royale wolf-moose research that I conducted as a doctoral candidate from 1958 through 1962. Flying over the snow-covered, 210-square-mile national park in Lake Superior each winter in a small ski-plane, I learned that there were 20 to 25 wolves on the island and approximately 600 moose. The wolves were harvesting the old and sick moose and the surplus calves, and both predator and prey numbers seemed stable.

It is this type of stability that is often referred to as the "balance of nature". Prey animals superbly adapted for escaping wolves, and wolves are well suited for catching prey, and the result is a rough balance between the two. Unfortunately, it doesn't always work so smoothly over the short run. Human interference, unusual weather, or other "outside" factors can cause disruptions in the predator-prey relationship. For this reason, scientists are increasingly hesitant to use the word "balance." Many of us now prefer the phrase "dynamic equilibrium," which better describes the phenomenon.

In the late 1960s and early 1970s I witnessed fluctuations caused by adverse conditions: a series of severe winters that struck across North America. My students and I were studying wolves and deer in northeastern Minnesota at the time, using aerial radio-tracking. One of my bush pilots wondered why we were doing the study. Noting that wolf-control programs had generally been curtailed in the area, he stated that, "Everyone knows the wolves will wipe out the deer."

It looked like he was right. Year after year, we watched the wolves decimate an overwintering herd of white-tailed deer throughout a 1,500-square-mile region. Almost all the deer were inaccessible during the hunting season, so wolves caused most of their mortality.

But the wolves had help: the severe winter weather. Winter severity played a dual role in the deer decline. First, the deep snow made adult deer easier to kill. Thus in winter 1968-69, wolves even took more deer than they were able to consume. Each time we found a wolf kill, we would land our plane on a nearby frozen lake and snowshoe to the carcass. We saw several that had been killed and left with little or nothing eaten. That meant fewer deer the following summer to produce new fawns.

Secondly, the fawns that were produced that summer were in trouble from the start. Fawns born after severe winters are lighter in weight and have a lower survival rate than those born following normal or mild winters, regardless of whether predators are present. Where predators are present, they find the weakened fawn crop to be a bonanza. It was not surprising to learn that most of the fawns that were born during 1969 failed to survive.

Any prey population can safely sustain intermittent fawn or calf crop loss. However, our herd in northern Minnesota was hit with a series of seven severe winters, from 1966 through 1972, while wolf numbers were high. The result was that wolves killed an unusual number of adult deer each winter, and a high percentage of fawns each summer. Meanwhile, even the surviving deer in prime shape were aging and becoming easier to kill. After about five years of age, deer become much more vulnerable under any conditions. As the deer herd grew older and fewer fawns survived to replace the adults killed, the population declined.

Initially, during this deer decline, wolf numbers actually increased—a fact I couldn't account for until I realized that they were cashing in on the increased vulnerability of the deer. The wolves' increase hastened the deer decline. Each winter I peered out of my ski-plane window to more and more trackless wilderness.

Wolves, therefore, were suddenly faced with a severe food shortage. They slept more, preyed more on moose and beaver and trespassed into other pack territories where there were more deer. After a few more years these tactics were not enough. Wolf pups starved and litter sizes dropped. Then, one day, I found the wolf that had been driven from its territory by hunger and then been killed by another pack. He was the first of several wolves we found killed by their neighbors.

The same severe winters battered Isle Royale, which lies only 20 miles from Minnesota. From 1959, when I first censused the island's wolves, through 1973, wolf numbers remained between 17 and 28. Meanwhile, the moose population on the island about doubled. To me, this showed that a prey herd can escape the control of its wolf population. Then the severe winters struck, weakening even moose. Wolves began killing more than they needed, demonstrating that the reason they did not kill more earlier was that they found healthy moose too difficult to subdue.

Rolf Peterson, a biologist with Purdue University, continued the Isle Royale study during the severe winters. He found that the calves that did survive became "stunted" yearlings and two-year-olds. These animals fell easy prey to the wolves. Thus the Isle Royale moose population began crashing just as the northeastern Minnesota deer herd had. So, too, the island's wolves increased while their prey dropped. Wolf numbers doubled to 50 by 1980. This had the same accelerating effect on the moose decline as I was witnessing between wolves and deer in Minnesota. Then the population of Isle Royale wolves crashed as expected.

To me, these events were extremely interesting. I began to realize that claiming that "the wolves will wipe out the deer" was like trying to tell time with a broken clock: eventually you will be correct—for awhile. My job was to find out why the claim was correct at this particular time, and I was beginning to understand. The effect of wolves on their prey and vice-versa can differ depending on other important factors. Probably the most usual influencing natural factor is winter weather, but there may be others. Wolves are sometimes over-harvested by humans. In addition, the ability of animal populations to recover can be thwarted by human development; man-made structures can pose barriers that keep newcomers from moving in and repopulating a depleted animal stock.

Thus, the more we carve up wildlife habitat with roads, railroads, powerlines, pipelines, open-pit mines and settlements, the more we encourage fluctuations in wildlife populations—and the more we decrease the abilities of these animal systems to automatically and naturally readjust themselves within their habitats.

However, there is little disputing the results of a recent, well-controlled experiment in central Alaska. Some 38 to 60 percent of the wolves were removed each year from a test area while wolves were not controlled in several adjoining areas. Moose and caribou calves and yearlings increased two- to four-fold where wolves had been taken compared with their numbers before wolf control and were consistently higher than in the areas with no wolf removal. Actual moose and caribou herd sizes followed the same trends.

What would have happened if wolves had not been controlled? Because the herds had been declining before the experiment, I expect that they either would have continued to decrease, would have remained stable but low, or might have increased only slowly. Meanwhile, from what I saw in northern Minnesota, wolf pups would have starved to death, wolf productivity would have declined, and adult wolves would have killed each other. Control programs allowed recovery of both prey and wolves so that more of each could live over a longer period. It is something I am reminded of every time I fly over my Minnesota study area and look at lakeshores that were speckled with deer and wolves in the late 1960s, and that now lie empty.

When prey herds are low for whatever reason, wolf control is often proposed as a ready means of relieving pressure on them. The non-hunting public then usually responds with cries of indignation. It looks like wolves are being used as scapegoats. Because many of these people view wolf-prey systems as constantly in balance, they fail to understand how wolf control can aid prey recovery.

Male Timber Wolf by John Ebeling of Lake Hubert, MN 56459

RED FOX

Charles E. Schafer
Michigan Wildlife Sketches

MINNESOTA'S FOXES

Strong survival talents and DNR harvest regulations assure that Minnesota's two fox species will continue as members of the state's wildlife family.

Two species of foxes are commonly found in Minnesota, the red and gray. Both are members of the dog family (Canidae) and relatives of the coyote, wolf, and domestic dog. Foxes are primarily carnivorous, but also eat some plant material. The males of both species are generally larger than the females. They are efficient predators because of their acute senses of hearing, smell, and sight. Like most canids, they have one litter of young per year.

RED FOX

DESCRIPTION

The red fox is the largest of Minnesota's two fox species. Adults weigh eight to 15 pounds and measure 36 to 44 inches, including a 14-to 16-inch tail.

Ordinarily, red fox have a thick pelage that is sunset red to sandy gold, with a lighter belly and black feet. The backside of the ears are black. The animal has a white throat and chin, and usually a white-tipped tail.

Variations in color include black (all black except for a white-tipped tail), silver (black with white-tipped guard hairs, giving the overall appearance of silver), and cross (red with a dark patch running along the back and across the shoulders). All variations may appear in the same litter.

HABITS

The red fox is the traditional symbol of craftiness and cleverness. To be called sly as a fox is usually a compliment. There are endless tales of the red fox or "Reynard" outfoxing both hunters and dogs with elaborate backtracking and hiding schemes.

Sometimes a fox may deliberately initiate a chase with a pack of dogs. This may appear to be purely for fun, but more likely it is a male fox trying to lead the dogs away from a female with pups.

The red fox does not normally use a den except when raising its young. In the winter, a red fox will curl up on the snow and use its tail to cover its nose and feet.

The red fox is primarily nocturnal, though it is most active at dawn and dusk (crepuscular activity). Unlike other members of the dog family, the red fox's pupils become elliptical when they contract, like those of a cat. This elliptical pupil compensates for the extremes of light and also provides for a greater depth of field.

RANGE AND HABITAT

The red fox is extremely adaptable and thrives under a variety of conditions, including proximity to humans. It is common throughout Minneosta, but most abundant in the mixed forest and farmland of the transitsion zone.

The average hoem range of a red fox is about one or two square miles, though range varies accoring to the availability of food and cover.

FOOD

The red fox, like most predators, is an opportunist who is quick to take advantage of any food it can get. A hungry fox can consume up to one-fifth fo its own body weight in food. This is important because many predators live in a feast or famine situaion and must take advantage of a good food source when available.

Samll mammals such as mice, rabbits, and ground squirrels comprise the bulk of the red fox's diet. Foxes also eat birds and their eggs, snakes, frogs, fish, and carrion. It is also not unusual for them to eat a vegetarian meal, such as corn, berries, grape seeds, and nuts.

If a red fox cannot readily eat everything, the remainder is cached or buried under litter or in ahole to be eaten later. It may urinate n the area, possibly to mark its territory or as a signal to other foxes.

LIFE HISTORY

In Minnesota, foxes mate in midwinter, from late January to early February. After mating the pair seeks out a den. Often it is a renovated den of another animal such as a woodchuck or badger. It may be located anywhere, in the woods or a field, in a dense forest, or under a farm building. Red foxes seem to prefer a den site which is well drained and elevated. The entrance is usually on the southerly side of a slope.

The den itself usually has more than one entrance and is quite large, 30 or more feet long and six to eight feet deep — depending on topography and drainage.

Sometimes more than one fox family will use the same den. A DNR biologist once observed three litters using the same one. A more unusual joint occupancy occurred when a woodchuck, rabbit and fox occupied the same den complex at one time!

Gestation lasts about 51 days and the young are born (whelped) in late March or early April. An average litter contains five pups. Litter size may increase if food becomes more available or fox numbers become less dense.

When born, red fox pups are grayish brown and weigh about 3½ ounces. Eyes open between the eighth and tenth days. When pups are newborn, the male (dog fox) brings food to the den for the female (vixen).

If disturbed or if food is no longer available, the vixen may move the pups to a new den location.

Pups do no venture outside the den until they become four or five weeks old. For several days, they spend most of their time playing near the den. Adults bring live prey home so the pups can learn how to hunt and kill in the safety of the den area.

The young are weaned at eight to ten weeks, about the time the parents take them on their first hunting foray. At 16 to 20 weeks, the milk teeth are replaced by permanent teeth. By this time the den has less importance and the family will begin using certain rendezvous points, eventually abandoning the den altogether. The family remains together throughout the summer.

In late summer or early fall, the young disperse or move to new areas. Yong males often travel 35 miles or more; young females travel 15 miles or more. Several foxes are known to have wandered over a hundred miles. This is nature's way of re-populating an area to assure that animals do not crowd a given area (predators need space to hunt in), and to mix different strains (gene pool) to assure a strong species.

GRAY FOX

DESCRIPTION

The gray fox is sometimes mistaken for a red fox because its coat also has reddish hair. The overall salt-and-pepper pelage of the gray fox might also be confused with the silver phase of the red fox. However, the two foxes are distinctly different species and do not interbreed.

The coat of the gray fox is colored by alternate bands of black and white on the guard hairs. There is reddish brown fur on the sides of the neck, in the throat area, on the flanks, back of the ears, inside the legs, the chest and the undersurface of the tail. The bushy tail is gray with a ridge of coarse, black hair along the top, and with a black tip. The cheeks, throat, insides of the ears, and the belly are whitish. Its fur is coarser than that of the red fox and is not as highly prized.

The gray fox's body is gernerally more compact than the red fox and its legs are shorter. It weighs six to 13 pounds and measures 32 to 40 inches, including the 11 to 16-inch tail.

HABITS

Unlike any other member of the dog family, the gray fox readily climbs trees. It usually hunts and is most active at night. During the day it rests in dense thickets.

Gray foxes usually den in cavities and crevices in rocks, caves, and even in hollow trees or logs.

FOOD

Although the gray fox has a diet similar to that of the red fox, it does eat more plant material. It avoids open areas, preferring thick, brushy cover. The home range of the gray is smaller than the red's. The gray fox occurs in the lower two-thirds of the state.

LIFE HISTORY

Mating occurs between gray foxes at the same time of the year as red foxes (late January to early February). The pups, usually three or four, are born after a 63-day gestation period.

Gray foxes care for their young much like red foxes. More is known about the red, however, possibly because grays are more secretive.

Young gray foxes disperse in late summer and fall, but do not travel as far as young red foxes.

STATUS OF MINNESOTA FOXES

Foxes were unprotected in Minnesota until 1977 when they were placed on the protected list. The DNR now regulates the taking of foxes by establishing annual hunting and trapping seasons.

The red and gray foxes are sought after for their pelts, though fur of the red is worth considerably more. In 1979, a prime red fox pelt brought $80 or more on the market and a prime gray fox pelt up to $50.

Both red and gray foxes are resilient species, able to rebound from low population levels. Their high reproductive potential and strong dispersal tendencies enable them to readily re-populate vacant areas. They will be part of Minnesota's wildlife heritage for the enjoyment and benefit of future generations.

This article was prepared by the DNR Division of Fish and Wildlife. DNR Report No. 79.

BADGER

Charles E. Schafer
Michigan Wildlife Sketches

BADGER & WOLVERINE

THE BADGER is a medium-sized, heavy-bodied mammal. Its neck is short and thick — the same width as its head. It has a wide body, short legs, and a short, bushy tail. Five toes and five prominent claws — those on the front feet are over an inch in length provide the badger's main digging tools, and add to its slightly pigeon-toed gait.

The distinctive black "badge-like" marks on the badger's white cheeks probably gave the animal its name. Its coat is generally gray with a slight yellowish cast. A white stripe runs from the nose to the crown of the head, tapering off at the neck. Underparts are predominently yellowish. Males and females look alike.

The average adult badger is about two feet long and about nine inches high at the shoulder. Males average about 24 pounds; females about 17 pounds.

Behavior

During the day, chances are poor for sighting a badger. It is probably underground — except in regions where people are scarce. Then badgers are active during the day as well.

Digging is the badger's occupation, and its body is built for the task. While a badger is hardly fleet-footed, it can easily out-dig a man with a shovel. When digging, the badger will send dirt flying out of the hole four to five feet in the air.

Badgers are effective at fighting and defending themselves. When a dog disturbs one, chances are the dog will lose the fight.

In Minnesota, the northern sector of the badger's range, it stores body fat during late summer. This serves as insulation and an energy reserve during the winter months. While the badger does not actually hibernate — its body temperature, heartbeat, and respiration remain close to normal — it does spend the winter sleeping underground, maybe coming out on an especially warm day.

The badger digs a burrow, uses it for a time, and then moves on. However it returns occasionally, just to check on new residents which may have taken over the hole, such as the ground squirrel. In this way the badger "lays a trap" for its prey.

Coyotes take advantage of the badgers' hunting habits and occasionally the two animals are seen together. While the badger raids a den, the coyote hangs around and nabs any rodents that might slip by the less agile badger. Hawks will also follow badgers, probably on the lookout for furry escapes.

The badger will eat insects, including grasshoppers, crickets, and beetles. The badger has a special taste for bumble bees. It eats the larvae and the nest as well as the adults. The badger also eats lizards, moles, snakes, turtle eggs, and carrion. It even eats hibernating rattlesnakes, all but the head.

Often the badger drags its prey back to its burrow for a leisurely feast. If a burrow isn't close by, the badger digs one for the occasion.

Badgers can be tolerant animals, sometimes sharing their dens with families of foxes. Former badger dens also become valuable as den sites for foxes, woodchucks, and skunks.

Life History

Mating takes place in August or September. The female bears young in a den two to six feet below ground, then lines it with dry grass. Like many members of the weasel family, the young are not born until the following spring because of "delayed implantation." After fertilization of the eggs, the embryos develop slightly and then become dormant for several months. In late winter embryos become implanted in the uterus and develop in about five or six weeks.

A single litter is born between March and June. An average litter numbers three, with extremes of one to seven. Young are furred but blind at birth. Their eyes open between four to six weeks and they are weaned about the same time. The female continues to bring them food until late summer when she begins to teach them hunting. They stay in the female's den until fall when they disperse.

Range and Habitat

The badger lives throughout most of Minnesota. Its home range is a small area which changes as the food supply dwindles. Badgers prefer open prairie land where there are numerous burrowing rodents, but they live in mixed farmland areas as well.

Management

Badgers are not protected by Minnesota law and may be taken by any method except with the aid of artificial lights or by poison.

THE WOLVERINE is the largest member of the weasel family. Known to the Indians as "Devil Beast," the wolverine is the subject of many stories about its penchant for destroying traplines and, generally, for its cunning nature.

Short-legged and heavy-bodied, ounce for ounce the wolverine is the strongest of all North American animals. Wide-round head, powerful jaws, and small round ears combined with its stocky body gives the wolverine the appearance of a small bear.

Most people would not recognize the wolverine if they were lucky enough to see one. Its coat is thick and long, mostly dark brown except for white patches on the chest, a white band above the eyes, and a distinctive yellowish horseshoe from shoulder to rump. The wolverine has a bushy, medium length tail.

Behavior

In recent years the wolverine has been showing up in Minnesota, wreaking havoc with a few traplines. Apparently some animals learn to follow a trapline, eating animals that have been caught. They also are reputed to set off unsprung traps and generally undo trappers' work. However, wolverines are so rare that the occasional sacrifice may be worthwhile for so unique a creature.

The ferocity of the wolverine is also worthy of campfire stories and tales. In one authenticated incident, two wolverines were seen driving a 500-pound cinnamon bear from an elk carcass. Another case tells of a 20-pound wolverine chasing a pair of cougars away from a deer the big cats had killed!

Wolverines have a peculiar habit of shading their eyes when trying to see something in the distance. The animal holds a paw to shield the sun as it squints through its myopic pupils. The wolverines' scientific name **Gulo luscus**, means half-blind glutton.

And indeed, besides being near-sighted, the wolverine is a voracious eater. It feeds on anything from meat, larvae, and eggs, to berries and carrion. It is solitary, active day and night, and travels great distances despite its slow gait.

Anything is fair game for the wolverine, large or small. After making a kill, it may drag the carcass up to a mile if need be. Sometimes a wolverine will track its prey until the victim is exhausted. Other times it climbs a tree to wait for an unsuspecting animal.

Life History

Spring mating is the one time these anti-social animals get

WOLVERINE
FLOUR
Schafer
Marten
Fisher
Charles E. Schafer
Michigan Wildlife Sketches

together. In May or June two or three woolly young are born. They stay with the female until fall.

Management

Minnesota may have as viable a wolverine population as any of the contiguous 48 states. Whatever the reason for the wolverines tenuous return here, such as increased habitat as Minnesota's northern forest continues to mature, we can consider ourselves lucky to be living with this rare and fascinating mammal.

Wolverines are protected under Minnesota law and may not be taken.

This article was prepared by the DNR Bureau of Information and Education. DNR report No. 23

The Fisher: Strong, Swift, Elusive

Like a streak of dark lightning, the fisher pounds through the pines, scoops a paw under a porcupine, flips it over and disembowels it in a matter of seconds. Not an uncommon occurrence. Unlike most animals, the fisher, because of its speed, does not shun the little prickler. Before a porcupine can raise a quill, the swift fisher has already pounced. Its digestive system isn't bothered by the quills.

Rarely seen, the fisher is prized for its luxurious, silky fur. Trappers often complain that this wily hunter snitches bait from their traps. It may even follow a trap line and eat trapped animals.

DESCRIPTION

Known variously as black cat, black fox, pekan, Pennant's cat (after T. Pennant, an early authority on North American furbearers), and the fisher marten, the fisher is best described by the Chippewa Indian name *tha-cho* — big marten.

Short powerful legs, round ears, and a pointy, whiskered nose identify the fisher as a member of the weasel clan.

Its fur, dark brown with white-tipped guard hairs, shades to jet black over the rump. Females are usually darker than males and their fur is more highly prized in the fur trade.

Like other weasels, the fisher has a musky-smell, though not as strong as some others, such as the mink.

Fishers are one of the fastest tree-climbing mammals in the world. They can out-climb the marten, their half-size cousin — and the marten can outclimb red and grey squirrels! Foresters lucky enough to catch a glimpse of this elusive creature say it can climb a tree faster than other animals can fall out of one.

Not many animals in the woods dare tangle with a fisher. In a fight, the fisher can best a coyote or bobcat. Everything else is fair game. Fishers run down hares, chipmunks, squirrels, mink, weasels, martens, muskrats, birds, and even deer mired in deep snow, Frogs pose no challenge to the fisher's speed.

Fishers also eat insects and berries. In spite of its name, the fisher does not like to get its feet wet, and does not usually eat fish, except those washed up on shore.

Although the fisher may kill more than it can eat at one time, it saves the surplus in its den for a later meal.

LIFE HISTORY

The female is receptive to mating only during one or two days in spring, shortly after her litter is born. The unusually long gestation period, 350 days, is possible because the fertilized egg does not become attached to the uterine wall until the proper number of months remain for the embryo to develop and be born in the spring — when survival chances are best for the young.

Hollow trees, logs, or rock cavities with entrance holes no wider than four or five inches are favorite nesting spots for bearing and rearing young.

Born in April, the young fisher's eyes don't open for seven weeks. The mother takes her infants hunting for the first time when they reach three months. Between mid-summer and early fall, the young mature and leave their mother.

Fishers are territorial — they defend and hunt a given area ten to 20 miles in diameter. Covering their territory may take them a week.

RANGE

Fishers once ranged throughout the northern United States and into Canada. Lumbering and lack of management contributed to a severe decline in population. Today, fishers are making a come-back in several northern states.

In Minnesota, fishers occur mainly in the northern third of the state in the coniferous forests.

DNR PROGRAMS

In Minnesota, fishers are currently increasing because the coniferous forest in the north is recovering from earlier, excessive lumbering and the 1930's drought.

Until 1977, the fisher was protected by Minnesota law. Then the legislature nominated it as a game species, after the DNR had verified that the fisher population was increasing. The DNR designated the first season for December 1 through January 31, with a limit of three animals per trapper.

Before this season, conservation officers trapped fishers, particularly in the Ely and Grand Marais region, because of the damage they did to trap lines.

In 1961, the United States Forest Service obtained 44 fishers from Minnesota for transplanting to the Ottawa National Forest in Michigan and the Nicollet National Forest in Wisconsin. Game biologists hoped that the fishers would control the porcupine population.

This article was prepared by the DNR Bureau of Information and Education. DNR Report No. 9.

Charles E. Schafer
Michigan Wildlife Sketches

SQUIRRELS

All kinds of squirrels-including some that "fly" - live in Minnesota. Here's a rundown on the state's squirrel family.

Pressing tightly against the trunk of the oak tree, the gray squirrel changes into an inanimate lump that goes unrecognized by a passing predator. This disappearing act is a common occurrence among Minnesota's tree squirrels — the fox, gray, red, and flying squirrels.

Description

The fox squirrel's coat varies in color from yellowish-brown to black. It is the biggest tree squirrel, 19 to 30 inches long and up to three pounds. The top of its head is dark and its ears are light.

The gray squirrel is not always gray. It may be black or reddish brown during the summer. (There are numerous black squirrels in the St. Cloud area.)

In Minnesota, the gray squirrel is about 18 inches long — including its graceful seven to ten-inch plumed tail — and weighs one and one-half pounds. The gray's tail is trimmed with white tipped fur, whereas the fox squirrel's tail is trimmed in rusty yellow.

The red squirrel is still smaller. In summer it is rusty red. With colder weather, it turns brown and retains only a streak of red down its back. Red measures up to 14 inches, including its perky six-inch tail, and weighs seven to nine ounces.

The smallest, most elusive, and probably the most beautiful squirrel is the flying squirrel. Most Minnesotans have never seen this little creature because of its nocturnal habits. Its coat is silky, reddish-brown above, and creamy white underneath. It is only ten inches long, including its four-inch tail, and weighs two to three ounces.

Behavior

Gray and fox squirrels are noisy and active throughout the day. They prefer to nest in tree cavities, but if these are unavailable, they build leaf nests. These flimsy constructions are usually built in the crotch of two tree branches. Starting with a platform of leafy twigs, the squirrel adds grass and small twigs.

Fox squirrels are slap-dash about their summer nests, which disintegrate in a short while. They take a little more care with winter nests because, though active throughout the cold months, they like warm dens.

Because of its size, the fox squirrel is not as agile as the gray. Sometimes it falls, but rarely hurts itself.

In agricultural regions, fox squirrels use corn cribs as their main food supply.

The fox squirrel doesn't begin hunting for food until the sun is high, but it feeds later in the evening than the gray.

The gray spends most of its time on the ground digging holes to cache nuts for winter. It usually stores nuts inside a hollow tree and beneath the leaf mold. Contrary to popular notion, squirrels are able to locate nuts they bury. Naturalists have seen them dig through a foot of snow to locate a particular cache.

The gray squirrel uses its bushy tail as an unbrella on rainy days, a favorite time for foraging. When the tail gets saturated, a couple of quick twitches dry it out. They sometimes eat baby birds and often insects, especially grubs.

Gray squirrels are the most highstrung of the three, but in cities, they become quite accustomed to people and traffic — sometimes to their disadvantage — and they are naturally curious and friendly.

The red squirrel is every bit as pugnacious as its two larger cousins. Everything and everyone that passes is exposed to the red's scolding, chattering, and whistling. Its high pitched chirp sounds like a bird. L. L. Rue, in his book "Mammals of North America," tells of a red squirrel raiding his lunch while he sat on a log.

Besides an occasional lunch, red squirrels feed on pine cones, acorns, mushrooms — which they dry in the sun to leech out the poison — hickory nuts, beechnuts, walnuts, and maple seeds. During fall gathering, the red's coat becomes caked with pine gum. It likes to eat in the same spots, and leaves large piles of discarded hulls and cones, sometimes over a foot high.

Airborne

Flying squirrels prefer to nest in tree cavities or leaf nests, but even a chimney will do. Unlike the solitary gray and fox squirrels, flying squirrels often live in groups.

Actually, the flying squirrel doesn't fly, it glides. Between its wrists and ankles are loose folds of skin, like webs. Viewed from underneath, a gliding squirrel appears almost square because of its 50 square inches of air surface.

Before taking off from a tree-top, the flying squirrel takes two bearings by bending left and right to determine a landing site, possibly by triangulation.

Next, it coils and kicks off with its hind feet. It quickly spreads out to tighten the membranes into a flat surface. The squirrel controls its direction by the tension of its membranes and by its tail. To land, it lowers its tail and then jerks it up so that it lands nearly vertical. The membranes then act like brakes, providing for a soft landing.

Flying squirrels eat all kinds of nuts except walnuts, which are too thick to crack.

All four squirrels fall victim to numerous predators. House cats, bobcats, snakes, martens, foxes, owls, fishers, hawks, and dogs all take their toll.

Life History

Male squirrels are polygamous and provide no help in raising young. Breeding takes place in January or February and sometimes a second time during summer. Females bear young in about six or seven weeks. Blind, hairless, and helpless, the little ones require plenty of care. Litters usually consist of one or two young.

At six weeks, gray, red, and fox squirrels open their eyes and begin exploring outside the nest. By this time, they are fully furred. The same process takes about 28 days for the flying squirrel, though it does not reach full growth for 18 months. By two months, young fox and gray can feed on nut meats, but only after two more weeks can they crack shells for themselves.

When the babies are threatened, the mother rescues them. They curl around her neck for a quick ride to the nest.

If they manage to elude all their enemies, gray squirrels can live 15 years, red and fox squirrels 10 years, and flying squirrels about five yers.

During cold months, squirrels may stay denned up for several days. They come out on bright, warm, winter days and perch near their nest, or forage for food on the ground.

Range

Gray squirrels prefer mature woodland, but can adapt to second-growth forest. Fox squirrels usually live near oak trees where their principle food, acorns, is available. They also thrive in smaller woods and farmland. The same applies to the flying squirrel.

All four kinds are generally found, in good numbers, south of a line from Duluth to Roseau.

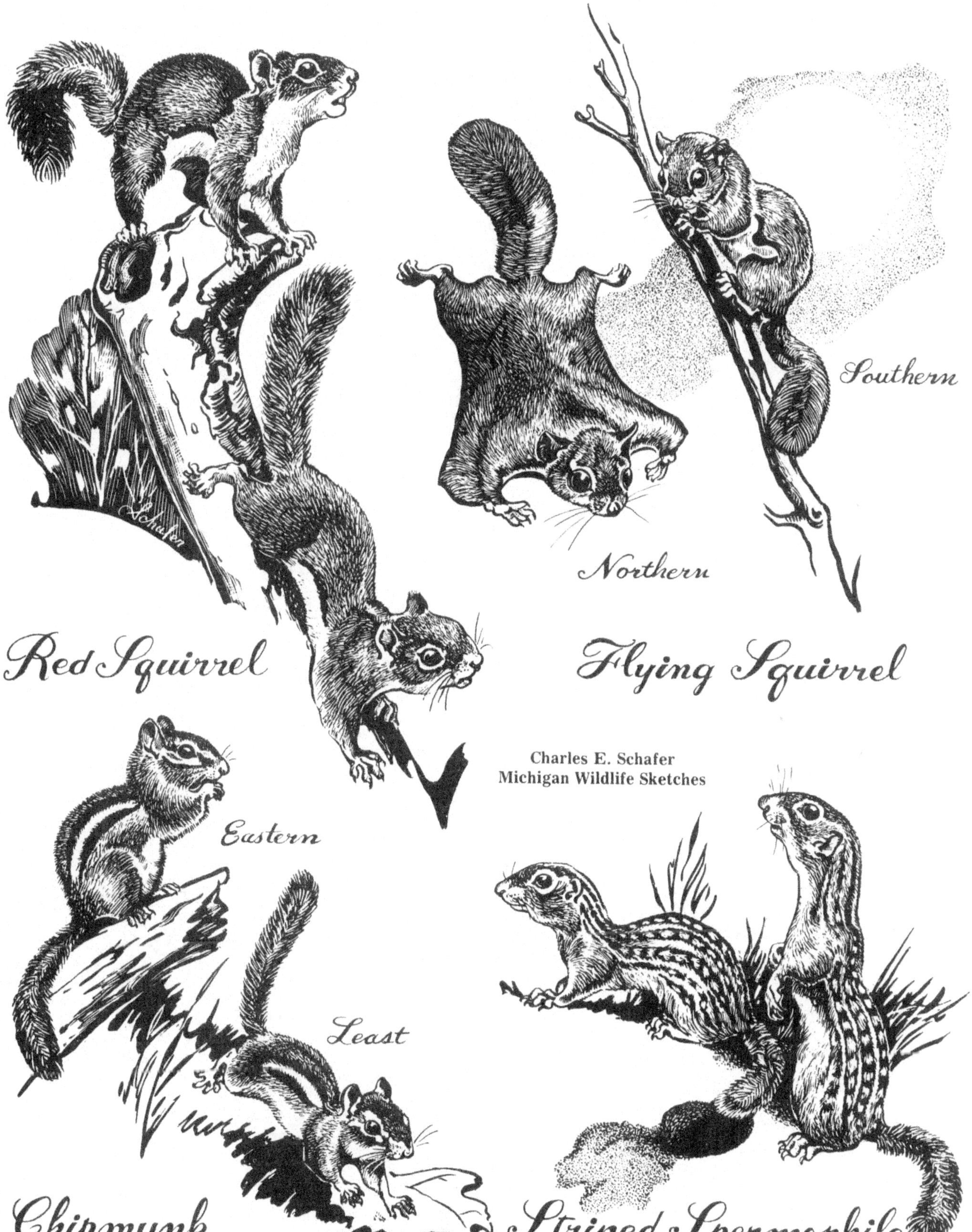

Charles E. Schafer
Michigan Wildlife Sketches

Hunting

Squirrel hunting is a popular sport. Each year many hunters, young and old, test their skills against squirrels. Once hunted, squirrels become wary. They keep the tree between themselves and the hunter. If there are two hunters, the squirrel flattens itself along a branch and becomes virtually invisible.

The flying and the red squirrel are not game species in Minnesota. The gray squirrel accounts for about two-thirds of the total harvest of squirrels. In recent years, the statewide annual harvest has been about 220,000 gray squirrels and 160,000 fox squirrels.

Squirrel season usually opens around October and lasts through December with a limit of seven to ten per day. Department of Natural Resources game biologists use food availability to determine the necessary harvest limits. Every four or five years there is a widespread crop failure — because of late spring freeze. The shortage of acorns that results can limit squirrel populations.

This article was prepared by the DNR Bureau of Information and Education. DNR Report No. 25.

Charles E. Schafer
Michigan Wildlife Sketches

RABBITS & HARES

There's more to know about these unusual creatures than most people realize. The prey of many predators, they have developed uniqure survival tactics.

Many people confuse rabbits and hares and use the names interchangeably. But they are scientifically classified as two different animals. They are members of the same order, Lagomorpha, and both belong to the family Leporidae. But the rabbit is of the genus Sylvilagus and the hare of the genus Lepus.

A true rabbit is born blind, hairless, unable to run, and in a fur-lined nest. In contrast, a hare is born fully furred and can run in a matter of hours.

The hare has long legs. It bounds, not runs. The rabbit has shorter legs. It runs, not leaps.

Finally, a hare has longer ears.

Of the three Minnesota species, the name rabbit may only be applied to the cottontail. Jackrabbits and snowshoes are both hares. They are discussed in the second half of this article.

COTTONTAILS are small animals, measuring only 12 to 16 inches including their fluffy two-inch tail. The female is slightly larger than the male. Both weigh two to three pounds.

The cottontail's fur is grayish-brown with light-tipped guard hairs. The belly, chin, and insides of the legs are white or buff. Some cottontails have a white forehead blaze. Although cottontail rabbits shed fur throughout the year, they do not turn white in the winter as snowshoe hares do.

Watching a startled cottontail go bounding off makes it easy to understand how it got its name. Its white tail looks just like a cottonball.

Habitat

Brush piles, high-tangled grass, thickets, or any obstacle to the cottontail's enemies is a likely home for this little rabbit. The best homesites provide cover close to open grassy areas where the rabbit feeds. The current trend toward clean farming has severely limited rabbit habitat in many areas.

Cottontails scout their small territory - seldom over five acres - until they know every inch of ground. They use the same routes over and over until the area is criss-crossed with trails. This gives them an advantage when pursued by a predator.

Rabbits have powerful legs and claws. They could dig burrows, but don't. Instead, they adopt the deserted home of a woodchuck, badger, or other burrowing animal.

Behavior

The cottontail is most active in night and early morning. During the day, it often loafs in a "form", an arch of loosely-matted grasses or twigs. When no predators are around, it may take a sun-bath or stretch out in the shade. Often it will feed for a few hours in the late afternoon, until about an hour after sunset.

The cottontail does not hibernate, but if the temperature drops much below zero, it may stay in its burrow for long periods.

Kicking with its big hind feet, the cottontail delivers vicious blows, especially against another rabbit. When a mink corners the cottontail, this timid rabbit leaps into the air and runs right over the top of the mink, striking with its feet and claws.

The rabbit's two main defenses are to remain motionless, sometimes for 15 minutes or longer, and its sudden bursts of speed. A cottontail will "freeze" if it sees a predator. But if the predator gets too close for comfort, the cottontail will bounce off on a zig-zag course, dashing through the cover it knows so well.

The cottontail's speed is deceptive. When running through thick cover, it twists and turns. It seems to be running much faster than its top speed of 20 mph.

Normally, when feeding and moving around its territory, the cottontail takes short hops, stopping frequently to sit up on its haunches and survey the surroundings. With its large, protruding eyes, a cottontail has an almost 360-degree range of vision.

Like a cat, the cottontail is picky about its appearance and grooms itself daily.

Cottontails are high-strung animals and may die of shock if they are handled or caged. They are next to impossible to tame.

As vegetarians, cottontails choose from a variety of foods. They eat almost any green plant - whatever is available in their home range. In winter, cottontails have a more limited diet. They eat the bark of trees and fulfill their requirements for water by eating snow. Cottontails also eat their own droppings to obtain recycled protein or intestinal bacteria. They actually select only those pellets which are necessary for their nutrition. These are different from ordinary rabbit pellets. This phenomenon is called coprophagy and is common to hares also.

Life History

In the spring, cottontails become irritable and fights are common. Males compete for access to does. The rest of the year, they tolerate closer contact without aggression.

Rabbits are extremely active during mating from early spring through August. Males chase females, leaping as high as they can. Sometimes one cottontail will jump straight into the air and another will dash underneath.

Gestation is from 26 to 28 days. While pregnant, the female can nurse a previous litter. Larger litters are born sooner than smaller ones.

The female or doe cottontail digs a nest in preparation for her young. This saucer-like depression is about three or four inches deep, six or seven inches long, and five inches wide.

Then the doe plucks fur from her belly and mixes it with dead grass to line the nest and make a "blanket" that she brushes over the babies when she leaves the nest.

A new-born cottontail weighs about one ounce and is the size of a man's thumb. An average litter is five or six babies. They are usually born blind and naked. In smaller litters, the young are further developed at birth and are covered with fine, soft, gray fur.

The doe nurses her young at dawn and dusk. The rest of the time she covers them with the fur and grass blanket, and leaves to forage. Sometimes she may have to move the babies to a new nest because of disturbance. Perhaps a farmer is plowing the field, or children have discovered the nest and handled the young. She carries them one at a time in her mouth to a new nest.

Baby cottontails develop rapidly. They are fully furred by the end of the first week and can wash themselves, though they are still blind. At the end of the second week, their eyes have opened and they are ready to venture outside the nest. As soon as they leave the nest they must be able to stand motionless. Those that can't seldom survive the watchful eyes of predators.

By the end of the third week, the young leave the nest for good. The female breeds shortly after the birth of her litter. She may have as many as five litters in a nesting season.

Predation

Cottontails are an important part of the food chain because practically every predator that walks, crawls, or flies eats

SNOWSHOE
VARYING HARE
Schafer
Charles E. Schafer
Michigan Wildlife Sketches

them. Snakes, especially constrictors and rattlesnakes, snare rabbits. Owls hunt them by night, hawks by day. Minks, foxes, bobcats, fishers, weasels, coyotes, dogs, cats, skunks, raccoons, and wolves — all prey on cottontails.

The rabbit survives in the face of all this predation by producing a great surplus of young. Studies show that, each year, about 80 percent of rabbits will perish in one way or another. The remaining 20 percent set about repopulating the countryside. A single doe and her female offspring can produce 40 rabbits a season.

Hunting

The cottontail is an important game animal in Minnesota. Rabbits provide thousands of hours of hunting recreation for young and old alike. Cottontails are usually taken by "still hunting" — searching through fall or winter cover with a shotgun or rifle.

In Minnesota, the hunting and trapping season is from September 17 to February 28, from sunrise to sunset. There is a limit of ten per day.

Cottontail meat tastes much like ring-necked pheasant. Young adults are excellent table fare.

Disease

Many hunters are concerned about a disease called tularemia. This disease is transmitted from one animal to another by fleas, ticks, or other insects. It is present in a large number of animals besides rabbits. A diseased rabbit is sluggish and sometimes unable to run when disturbed. Its liver is covered with tiny white spots.

The disease is most common in the fall. The first cold weather usually kills insects and diseased animals and eliminates any danger to humans.

Symptoms of the disease in people include chills, headache, fever, and pain throughout the body. It is rarely fatal today and can be cured with mycin-type drugs. To avoid contacting the disease, hunters should wear rubber gloves when cleaning a rabbit.

MINNESOTA has two kinds of hares, the whitetail jackrabbit and the varying hare or snowshoe hare. The varying hare got its first name because it grows two coats of fur, white in winter and brown in summer. Come winter, the snowshoe also grows thick, coarse hairs on the sides of its hind feet that double its foot surface and make it easier for the snowshoe to travel through deep snow.

Description

The snowshoe hare is larger than the cottontail. It measures about 18 inches, and weighs four or five pounds. Females usually weigh more than males.

Temperature does not cause the varying hare's coat to change color. Take one into the house and the hare will go right on turning white as winter approaches — if the house has windows. This is because the hare has an eye mechanism that works like a photo-electric cell. As the days get shorter, the eye picks up less light, causing the hare's pituitary gland to become inactive. The pituitary gland is responsible for pigment production. When it is inactive, hair turns white. This same process is reversed in the spring. As the days get longer, the pituitary gland begins to function again and pigment is produced. The black tips of the hare's ears are the only part that remain dark throughout the year.

The whitetail jackrabbit is the largest hare species. A mature adult weighs from five to eight pounds and measures 22 to 26 inches.

The jackrabbit is usually light brownish-gray on the back and sides, lighter on the belly, and has the same black tipped ears as the snowshoe. Like the varying hare, the jackrabbit's coat changes color in winter, though it becomes more grayish-white than true white.

Both the jackrabbit and the varying hare have large, powerful hind legs. The jackrabbit has longer ears - six to seven inches.

Those magnificent ears, like twin antennaes, are always moving. Along with the hare's keen sense of smell, eyes on the side of its head give it a 360° field of vision to spot predators.

In their natural habitats snowshoes and jackrabbits are well camouflaged. They sit perfectly still and often a predator passes by at close range without noticing them.

Habitat and Range

The snowshoe hares live in the north and central forest and in brushy areas of Minnesota. It spends its life in an area less than 100 acres.

It does not burrow, but seeks out a sheltered spot beneath the branches of a snow-laden spruce, in the bushes of a thicket, or even crouches down in the open.

The jackrabbit lives on the open prairie, in areas of low cover. When it is not feeding, it is sitting motionless in a shallow burrow in a plowed field or short grass meadow or sprinting through tall grasses, wheat, or alfalfa.

Its home range varies according to food availability. In areas of sparse grasses and little agriculture, its home range may be one or two square miles. Where there is mixed agriculture and grassland, sites which provide both food and nesting close together, the home range may be only a couple acres.

Behavior

The jackrabbit has an easy, bounding gait.

At a leisurely pace, it covers about five to ten feet with each jump. Every four or five jumps the jackrabbit leaps a slightly higher leap, above the grass, so it can check for predators. When pressed, jackrabbits can leap 15 to 20 feet and run 45 mph, easily outdistancing a coyote.

Even so, coyotes are its major predator. Coyotes hunt in pairs. One chases while the other waits to intercept the hare.

The snowshoe is not far behind the jackrabbit for speed. It can dash through the well-known paths of its home range at 35 mph.

Like the cottontail, the jackrabbit and the snowshoe employ a "freeze and bolt" method of defense. With ears laid flat against the back and body pulled in tight, they hold absolutely still until the last moment. If the predator gets too close, they bound off — out of sight in a matter of seconds.

Jackrabbits usually feed at night. They eat almost any type of vegetation, including wild grasses, seeds, and shrubs, and also domestic plants. They are especially fond of alfalfa. In times of food shortage, they may turn to tree bark and crops.

Jackrabbits seldom drink water. They get enough moisture from plants and from occasionally licking dew.

The varying hare eats grasses, herbs, and other tender plants growing in forest openings. During the winter, it strips the bark of aspen, young pine, spruce, birch, and willow.

Both jackrabbits and hares practice natural recycling. They normally have two kinds of droppings, dry fibrous ones, and soft mucous-coated ones. The hare deposits the first kind as it moves around, and eats the second kind as soon as they are passed. They are formed several hours after the hare has eaten vegetation. Soft droppings are high in protein and B vitamins which are formed by bacteria in the intestine. These substances are used when the food is passed through the body a second time. The mother feeds her droppings to her young and innoculates them with essential bacteria.

Both jackrabbits and snowshoes are fighters, especially during the mating season. Jackrabbits rear up and then "box" with their front feet. The snowshoe kicks with its powerful hind feet. Males sometimes fight to the death, making deep, ranking cuts with their long, strong claws.

The courtship antics of a hare deserve the expression "mad as a March hare." Males pursue females through the woods or across the fields at top speed. Then they stop, leap straight into the air and go careening off again.

Like rabbits, hares twist and turn when they run, making it difficult for a predator to follow.

Population

Hare populations fluctuate by cycles which follow a ten year schedule. At peak times, researchers have recorded as many as 3,400 hares per square mile. Then suddenly, they mysteriously die-off, almost entirely.

This cyclic decline is to the advantage of both the hare and the environment. With the die-off, the vegetation grows up again and provides the smaller population with adequate food. The die-off also weeds out all but the strongest animals and continually upgrades the species.

There are no accurate estimates of hare populations in Minnesota, but as long as there is available habitat, these prolific creatures will continue to thrive. The number of snowshoes appears to have increased somewhat in recent years.

The whitetail jackrabbit population is much lower today because the grasslands that it needs for nesting have been considerably reduced by farming.

Life History

Both the whitetail jackrabbit and the snowshoe hare mate in the spring, usually in March or April. Gestation is 30 to 38 days for the snowshoe and 41 to 47 days for the jackrabbit.

Both species usually have one to three litters per year with two to four young per litter. Depending on the mother's health, young hares weigh from two to six ounces at birth. They are furry, able to see, and have already cut their incisor teeth. They are tiny replicas of the adult, unlike young rabbits (see first section).

The female scratches out a form, up to eight inches deep, well concealed in thick grass. She may have several nests and keep one or two babies in each. She visits each nest a number of times throughout the day. The whitetail takes care of its young for about a month. Researchers know little about the snowshoe's nesting habits, but believe that it may take care of its young for only three or four days.

Although they have reached adult size by two months of age, most young hares do not breed until the following spring.

Predation

Coyotes are the number one predator of hares, but a multitude of enemies stalk the jackrabbit and varying hare. Bobcats, foxes, lynx, dogs, domestic cats, birds of prey — especially the great horned owl — fishers, and snakes, all try for a meal of hare. The only ones that are consistently successful, however, are birds of prey and the wily coyote.

Hunting

As it does with cottontails, the DNR regulates hunting and trapping of jackrabbits and hares. The season is usually from September 17 through February 28, from sunrise to sunset, with a limit of ten per day. Hunters and trappers harvest surplus animals that normally die-off each year.

The young adults of both species are tender and taste much like other wild game, but the adults tend to be tougher.

The text of this article was prepared by the DNR Bureau of Information and Education. DNR Report No. 57.

Charles E. Schafer
Michigan Wildlife Sketches

THE INDUSTRIOUS BEAVER

The beaver is indeed an active woodcutter, though he is far from being the skillful builder most of us see him as.

Among animals, only the industrious beaver manipulats the environment to suit its needs. And where beaver and people occupy the same environment, the beaver's industry is not always appreciated.

Description

The beaver, largest member of the rodent family in North America, is descended from ancestors that lived more than a million years ago. Those prehistoric giants measured up to eight feet in length and weighed 700 pounds or more.

Our modern version averages 40 to 50 pounds and seldom attains a length of more than five feet. It continues to grow throughout its 12-year life span. Minnesota's record beaver weighed 74 pounds. In South Dakota and Wisconsin, however, trappers have taken animals surpassing 100 pounds.

The beaver is well-adapted to an aquatic life style. Its ears and nose have valves which close when the animal submerges. Furred lips also close behind the incisor teeth so the beaver can carry a branch underwater or gnaw on a woody tidbit without drowning.

The wide paddle-shaped tail acts as a rudder and propeller when swimming, as a balance when walking, and as a prop when the beaver cuts a tree. Slapping the tail against the water startles an intruder and warns other beavers. But tail-slapping is not an automatic alarm response. Beaver have been observed slapping the water when no danger was apparent and, conversely, sinking quietly out of sight when a human intruder was just scant feet away.

While the hind feet are webbed - both for swimming and for supporting its heavy body in soft mud - the front feet are like little hands. This allows the beaver to handle food and construction materials, as well as to groom itself. Two inner claws on the hind foot are used as coarse combs to spread oil from abdominal glands over the fur. This oil keeps it sleek and water repellent.

Internally, the beaver has large lungs and a large liver to store oxygen and oxygenated blood for lengthy underwater excursions, commonly up to 15 minutes.

Up front, the beaver is equipped with strong jaws and chisel-like teeth capable of gnawing through a six-inch tree in 15 minutes. Coated with a hard orange enamel, these incisors grow throughout the animal's life and must be constantly worn down by gnawning. Otherwise the growing teeth would prop the beaver's mouth open.

Beaver make a variety of sounds - churrs, mumbles, whines, snorts, hisses - as well as slapping the water to get attention.

Behavior

Beaver are not the precision lumberjacks many see them as. Trees are felled in any direction - occasionally, a beaver is killed before it can scuttle out of the way. Because shoreline trees generally slant toward water, however, they usually fall in that direction, fortunately for the beaver.

But contrary to what most people believe, the beaver doesn't construct its dams through careful planning. Rather, its dam-building ability is probably a marvelously refined instinct, a conclusion suggested by the beaver's tendency to build too many dams that are unnecessarily large.

The beaver is a night worker. It toils from dusk to early morning, constantly building and repairing its lodge or dam or gathering food for an approaching winter.

Beaver dams vary from several hundred to as long as 2,000 feet in length. They are constructed of branches woven through tree trunks and from rocks carried up fron the stream bed and from shore. Sticks are first lodged into the stream bottom and then covered with mud. Layer upon layer, vegetation s added. Construction in this manner; dams can usually withstand even the strongest spring torrents. Dams often outlast their builders, standing long after the beaver have moved to a new region.

After three to five years in an area, beaver move on because they have exhausted its food supply. Sometimes they build secondary dams to enlarge a pond and thus extend their range.

Like the dam, the lodge is composed of sticks packed with mud. Often it is constructed on top of an island or over a bank burrow. Lodges vary in size, ranging from six to 40 feet in diameter and sometimes extending eight feet above water. Size of the lodge is determined by its age and the number of animals in the colony.

The lodge may have one or several underwater entrances and living chambers. The chamber is usually one and one-half feet wide, two feet high, and three or four feet long.

Most lodges have a small opening at the top for ventilation. Early Indians saw steam rising from these 'wigwams' of wood during winter months, leading them to believe that beaver had once been people who had changed into animals. Legend has it that these "little men of the woods" would someday return to human form as reward for their hard work.

As many as nine beaver share a lodge, living together as a colony. This family will remain together from about September through April. All members, except young kits, participate in gathering and storing food and working on the structure.

Beginning in October, beaver establish a winter pantry by heaping branches (usually aspen) in the water next to the lodge. Eventually this storage pile may protrude several feet above the water.

A typical wintering beaver colony will consist of the two adults, their offspring of the previous spring, and often yearling animals. The number of animals in a colony is limited by the number of offspring produced by the female, ages of offspring, and by the existing food supply.

During winter, beaver remain active under the ice, playing, feeding, or digging new burrows. They may breathe air bubbles under the ice or return to the lodge for air.

Life History

Beaver begin mating when they are one and one-half years old. Breeding season begins in late January and most young are born in May or June.

Kits weigh less than a pound at birth, but can follow their mother underwater before they are a day old. Litter size varies anywhere from three to as many as nine kits. Interestingly, average litter size in Minnesota is about five compared to averages of three or four in many other states.

The chances of youngsters reaching maturity and raising their own young are excellent. Predation on beaver has been greatly over-exaggerated, though large predators such as coyotes, bear, timber wolves, fox, bobcat, and lynx may dispatch the occasional drifter wandering too far from its pond. Disease outbeaks, however, such as an epidemic of tularemia which resulted in widespread die-offs in 1950-52, have led to serious reduction in beaver population.

Management

Unless properly controlled through trapping, the prolific beaver can cause damage in forest and farmland. (Current trapping regulations may be obtained from the Department of Natural Resources, Wildlife Section.) When it depletes its food supply, such as aspen and other typical stream growth, the beaver may forage in orchards or corn fields. Where beaver occur in agricultural areas, they may even construct dams of cornstalks.

Beaver dams ruin trout streams, but only a small percentage of the total trout streams in Minnesota. But their activities often provide benefits. Beaver can transform a small creek into a series of ponds beneficial to countless forms of wildlife. After they cut aspen, new tree growth sprouts to provide more food and cover for wild birds and mammals. Removal of aspen may also result in the growth of trees of greater commercial value.

Beaver ponds control spring runoff, lessening the possibility of downstream flooding. These reserviors are vital for water retention. When abandoned dams eventually break and the water drains, old pond areas become lush green meadows.

The Minnesota DNR's policy advocates management of individual beaver colonies. In areas where beaver create a problem, they may be trapped and moved elsewhere. In other areas, however, the beaver is encouraged so the vital and complex interrelationship between the beaver, the forest, and other wildlife is perpetuated — a harmonious system which has existed over eons of time.

This leaflet was prepared by the DNR Bureau of Information and Education. DNR Report No. 30.

Charles E. Schafer
Michigan Wildlife Sketches

DO SKUNKS LIKE THE WAY THEY SMELL?

The skunk is a comical fellow, but beware. Its odorous spray is accurate to 15 feet. Facts and useful information about the skunk in Minnesota.

THERE IS NO safe way to approach a skunk. One game biologist thought he knew skunks. He reasoned that approaching from the front was the best way to avoid a spraying. His experience had been limited to the larger striped skunk. Unfamiliar to him was the small, agile, spotted skunk he found ensnared in his trap. Just as he approached, the little acrobat did a handstand on its forepaws, flipped its posterior into position, and let loose. Claims the biologist: "There still isn't a safe way to approach a skunk."

Back in the 1930s and '40s, skunks were plentiful, but money wasn't. Many schoolboys made their spending money by snatching skunks from culverts and ditches and selling the skins. In fact, a boy simply wasn't one of the gang until he came to school smelling distinctly skunk-like, and was promptly sent home.

The best remedy for a drenching is a sponge bath of tomato juice. Some victims claim a vinegar wash helps. Skunks carefully avoid getting any spray on themselves. Apparently, they don't like the smell.

In Minnesota, the large, striped skunk if found statewide while the small, spotted (or civet cat) skunk inhabits woods and farmlands in the southern half. Corn cribs, the foundations of barns, and hen coops are their favorite nesting spots.

Farmers call them thieves, but the skunk's reputation for the theft isn't entirely deserved. Although they filch a few eggs, chicks, and grain, they furnish a valuable service in exchange. Hungry skunks eat rats, mice, and insects. In fields, skunks consume small rodents, like field mice, insects — especially grub worms — and carrion. Bird eggs, frogs, and crayfish are also favorite prey.

Skunks are notoriously smelly. Twin nozzles just inside the anal tract can spray an attacker four to six times (accurate up to 15 feet) with a bitter, stinging, yellowish fluid. The sulfur content causes an unpleasant smell.

Skunks do not wantonly waste their ammunition. Most animals know enough to back off slowly — or freeze — until the skunk has time to amble away. But four feet on the ground, tail erect, and a well-aimed posterior mean someone will spend the night in the barn.

Coyotes and other predators will risk an occasional drenching when food is scarce, but the great horned owl is partial to skunk meat. It also has a poor sense of smell. Both owl and skunk hunt at night. The skunk evolved its black and white color to elude the owl's night vision. Great horned owls frequently smell like skunk musk.

Most people are familiar with the striped skunk. Black with a white nape, it has two broad white stripes down its back and a luxurious plumed tail. Adults vary in length from 24 to 30 inches and weigh four to ten pounds. Department of Natural Resources conservation officers once trapped a fourteen-pounder.

Spotted skunks are more stealthy than their larger cousin. Small (one to three pounds) with distinctive white stripes and spots decorating their black coats, they usually live undiscovered on or near farms. This nimble fellow climbs trees and can sneak eggs out from under a setting hen.

Skunks hole-up during winter. They do not hibernate, but are dormant (low metabolism). Mid-winter thaws bring them out for hunting forays. Spotted skunks sometimes stay active throughout the winter.

Young skunks, an average of six per litter, are born in early May. From birth until fall, they follow behind their mother learning hunting skills. Skunks are capable of spraying from birth on.

In Minnesota, no estimates can be made of skunk numbers. Since 1946, low prices and few trappers have drastically reduced the harvest. The 1963 price per pelt was about $1.00. Recent increases to $5.00 or more may revive interest in trapping this fur-bearer, perhaps even revive the old tradition for school children.

Skunk populations fluctuate dramatically. In some years, they become numerous, but then disease, harsh weather, and scarce food decrease their numbers sharply after a year or two.

Apart from research studies in limited areas, the DNR has not attempted management programs on skunks in Minnesota.

Skunks are rabies carriers, a fact which should discourage people from handling them. Because skunks are nocturnal, any skunk observed in the daytime should be suspected of rabies and either avoided or destroyed. For assistance, call the DNR.

Copy for this pamphlet was prepared by the DNR Bureau of Information and Education. DNR Report No. 4.

Animal Tracks

TRACKS are often the only evidence of the presence of wild animals. Such tracks can provide information about their abundance, range, and habits.

The majority of such animals are most active during twilight hours or at night, a principal reason why they are not seen more often.

To "know" an animal is to know the location of his home, his food, his enemies, and his habits.

Tracks are a record of activity and by close study an observer can become intimately acquainted with an animal without ever having seen it.

Tracks of many animals are similar in appearance; for instance, the tracks of dogs, foxes, coyotes and wolves. It is a help in the identification of such tracks to know which of these animals frequent the locality in which the tracks are found.

Tracks are best observed after rains, as in mud along stream banks, or following light falls of wet snow.

Tracks shown on the following pages are more perfect than are usually found under natural conditions.

Schafer

Charles E. Schafer
Michigan Wildlife Sketches

SHORT-TAILED SHREW

WEASEL

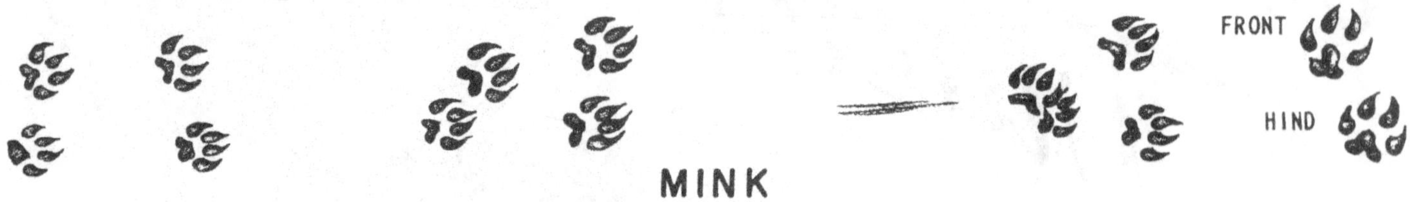

MINK

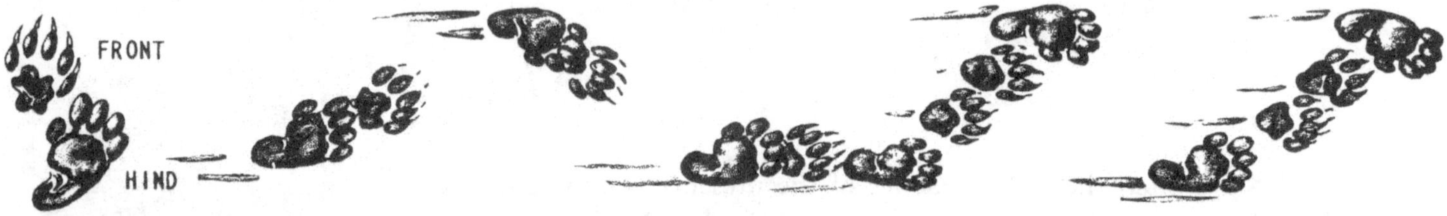

SKUNK

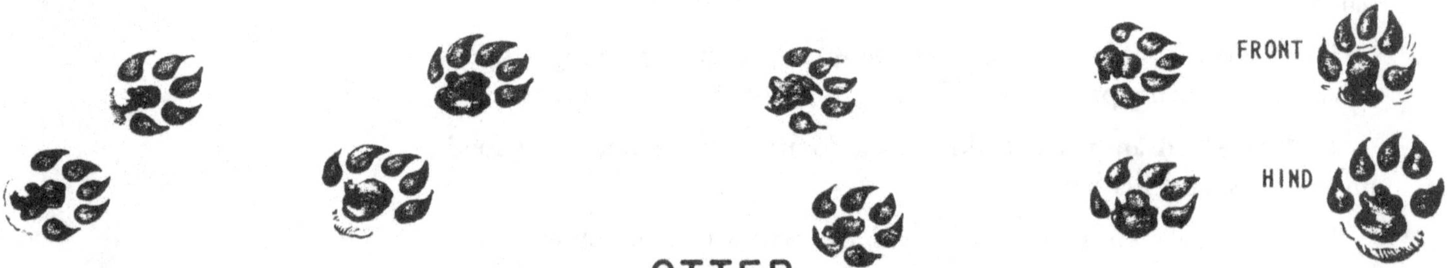

OTTER

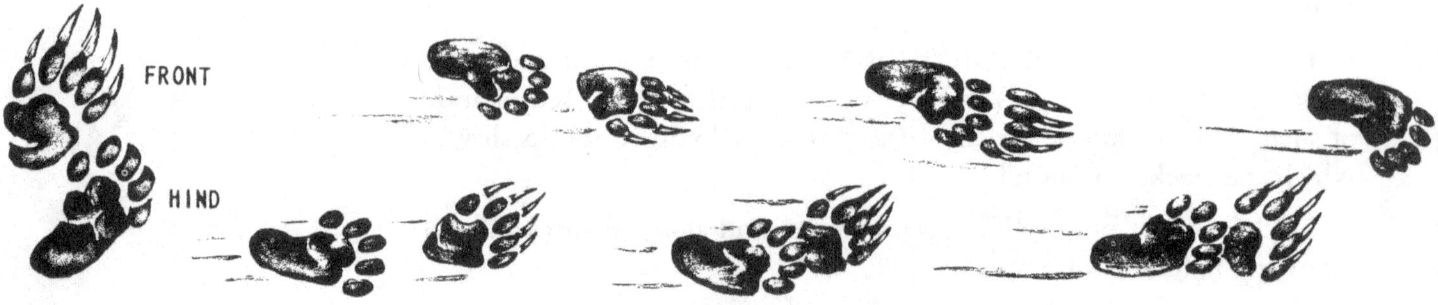

BADGER

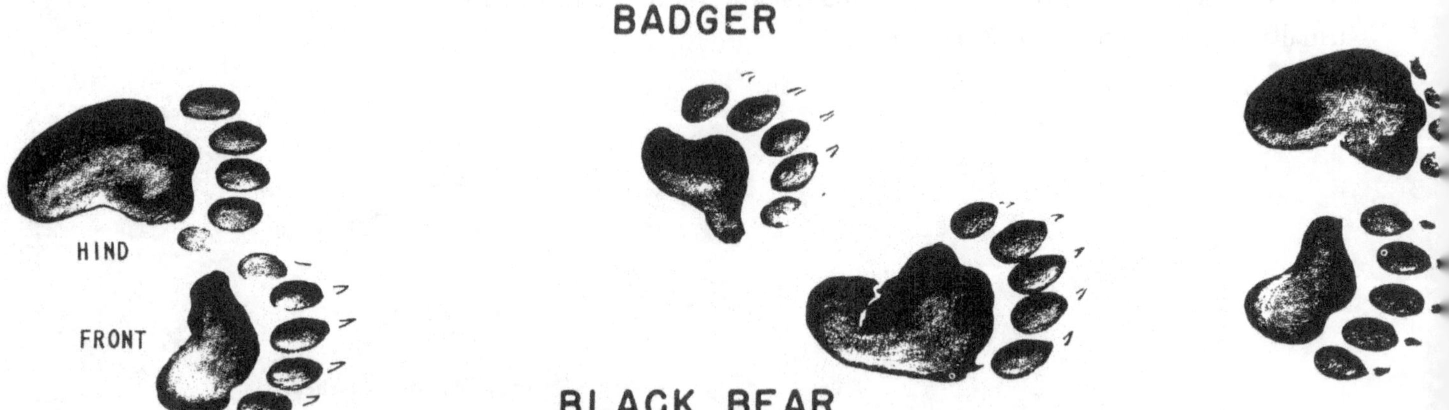

BLACK BEAR

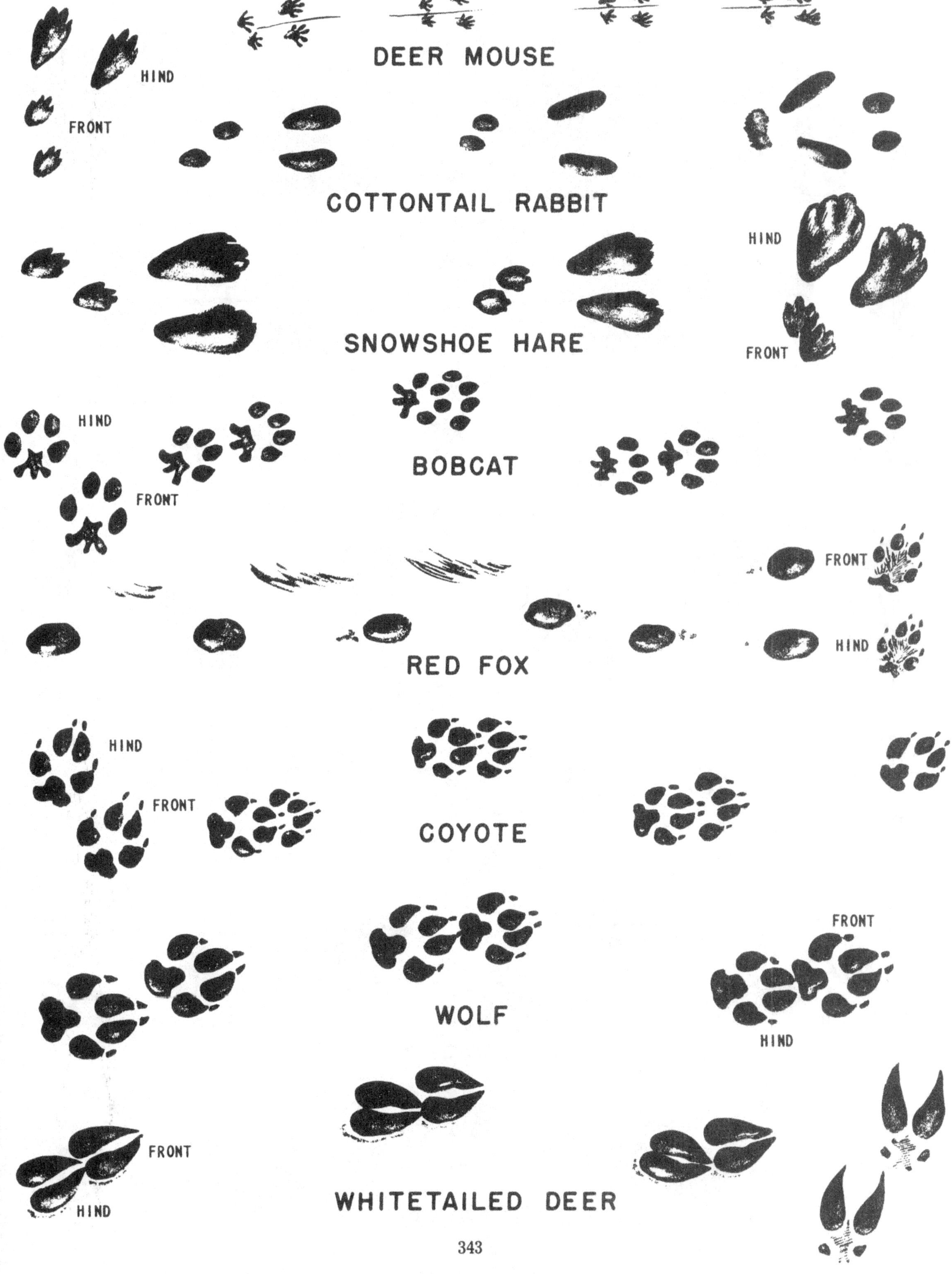
DEER MOUSE
HIND
FRONT
COTTONTAIL RABBIT
HIND
SNOWSHOE HARE
FRONT
HIND
BOBCAT
FRONT
FRONT
RED FOX
HIND
HIND
FRONT
COYOTE
FRONT
WOLF
HIND
FRONT
WHITETAILED DEER
HIND

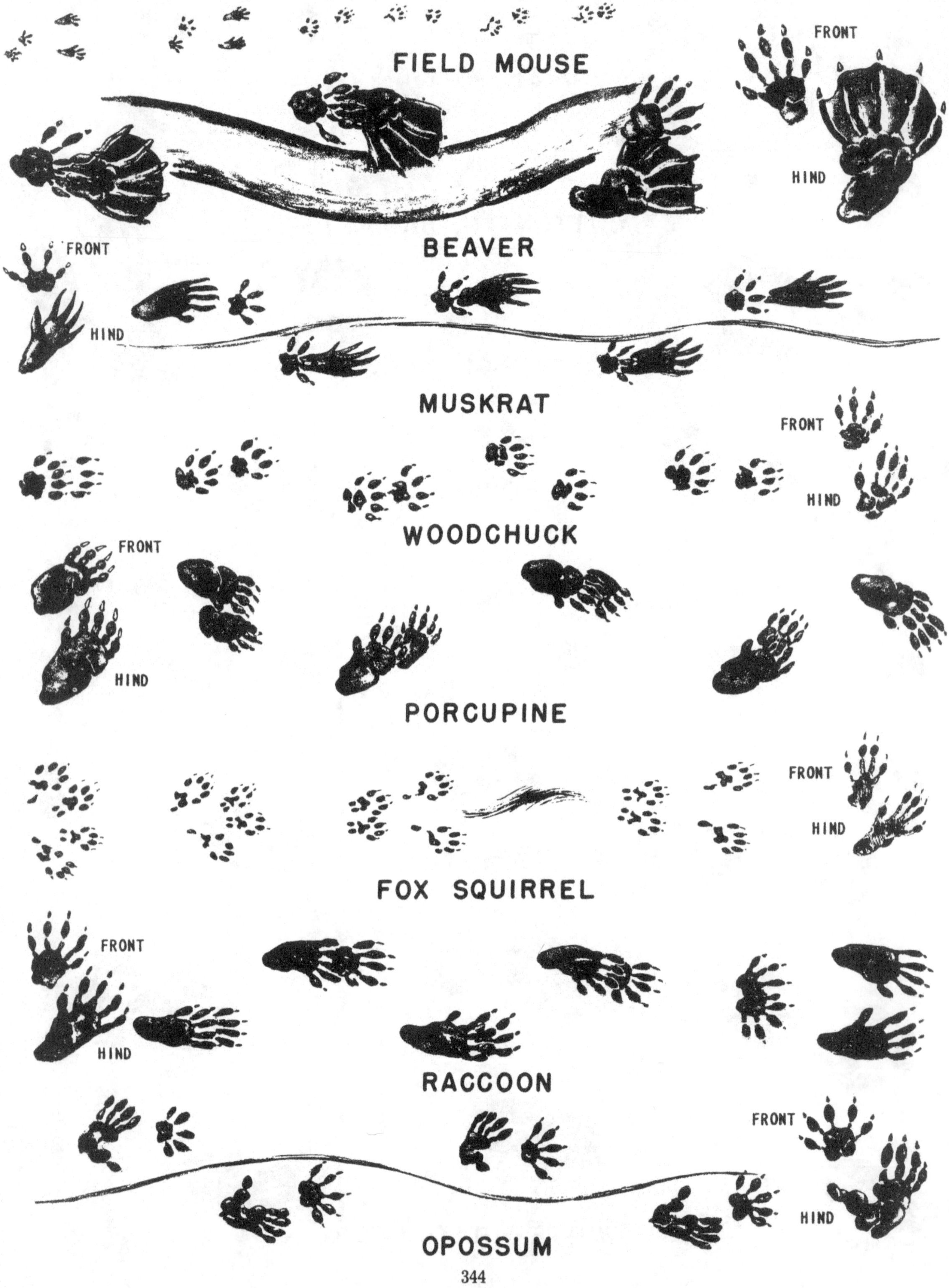
FIELD MOUSE
FRONT
HIND
BEAVER
FRONT
HIND
MUSKRAT
FRONT
HIND
WOODCHUCK
FRONT
HIND
PORCUPINE
FRONT
HIND
FOX SQUIRREL
FRONT
HIND
RACCOON
FRONT
HIND
OPOSSUM

RACCOON

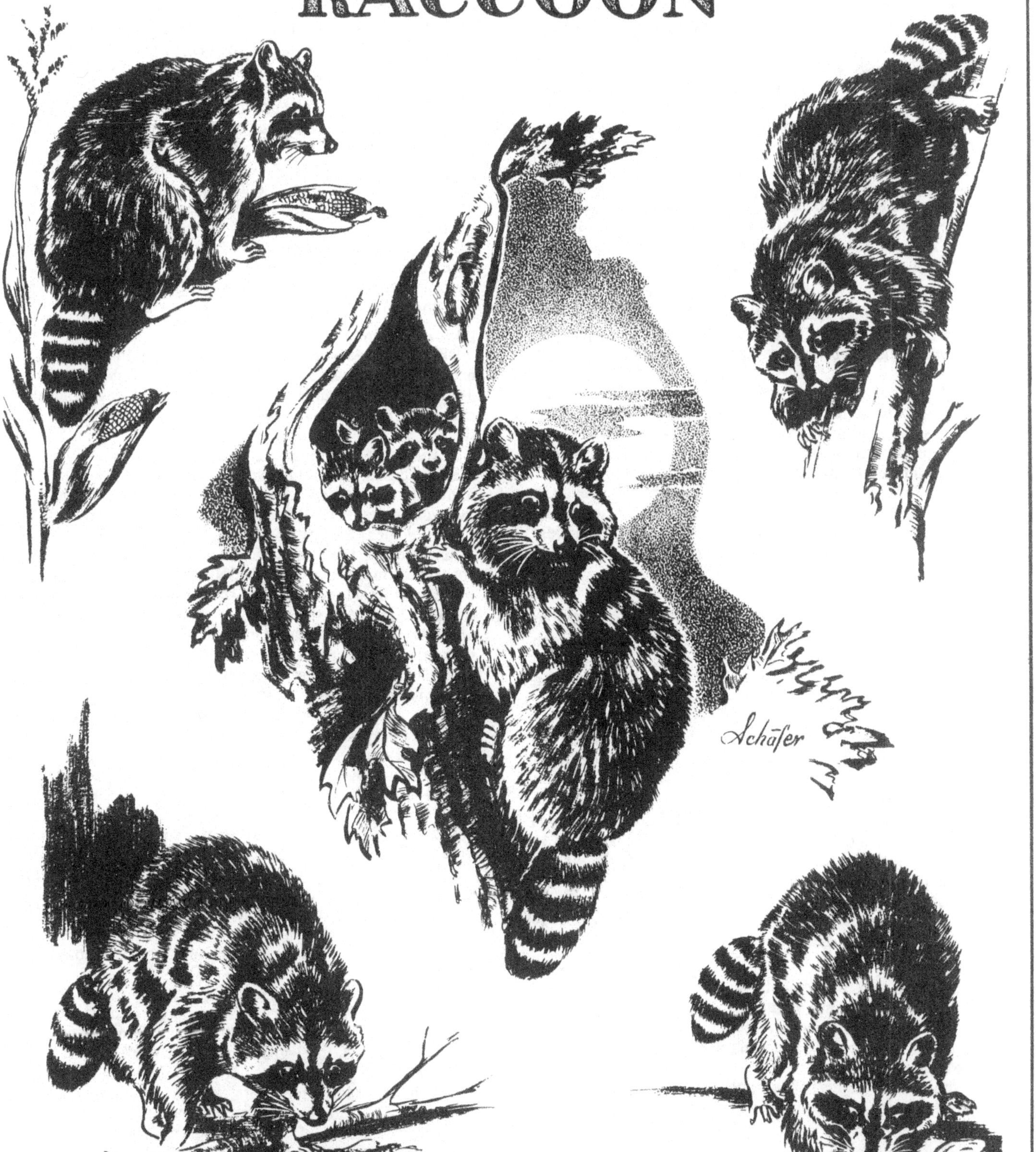

Charles E. Schafer
Michigan Wildlife Sketches

RACCOON

Do rising prices for raccoon pelts mean the animal may once again become scarce? So far, no evidence supports this suggestion.

MISCHIEVIOUS, friendly, intelligent, agile, curious, and ferocious — these are traits which describe the raccoon. To many people, the raccoon is endearing because of its handlike paws and habit of washing food before it eats. When it was legal to possess wild raccoons, children kept them as pets. There was no end to the trouble this masked bandit got into.

Adaptble to many environments, the raccoon exists almost anywhere, from Texas plains to Oregon forests. They occur throughout Minnesota.

During the 1920s when raccoon coats were the fashion, this luxurious-coated fur bearer became scarce. Since then, however, except during a brief revival of coon-skin caps around 1955, the raccoon has flourished. Today, an estimated 60,000 to 150,000 animals live in Minnesota.

Description

The average adult weighs between 15 and 20 pounds. The largest raccoon trapped in Minnesota weighed 36 pounds. Raccoons weighing 60 pounds have been captured in other states.

The Algonquin Indians called raccoons *arakunem*, "he who scratches with his hands." The coon has paws or "hands" nearly as flexible as those of a monkey. When it hunts for small aquatic animals, particularly crayfish, the raccoon sifts through sand and pebbles with its paws. Then it washes off most of the grit. Raccoons continue this practice even in captivity when water is available. This is not a necessary process, for the raccoon has well-developed salivary glands. Perhaps, suggested one wildlife biologist, "they just like to play with their food."

Diet

Raccoons are omnivorous. Crayfish, frogs, insects, fruits, eggs, young birds, small mammals, and sweet corn — which they reach by pulling down stalk after stalk, taking time to devour only half an ear before moving on — all comprise a raccoon's diet. Raccoons also love honey.

It is essential that raccoons consume large amounts of food during the summer and fall. Long Minnesota winters kill them off through starvation, particularly in non-agricultural, northern areas.

Recent studies showed that raccoons are the major predator of wood ducks using wooden nest boxes. Raccoons will prey on wood duck hens in any poorly-made box with an entrance hole four inches or larger. Sometimes they wait in ambush on top of the box and jump the hen when she leaves. Raccoons may even den up for the winter in a wood duck box. Boxes should be placed on poles away from a tree or post so that raccoons cannot get to them.

Family

During the mid to late winter mating season — February and March in southern Minnesota, March and April in the north — the male raccoon covers many miles and mates with many females. Normally raccoons occupy a home range with a one or two mile radius.

Females bear young, about nine weeks after mating, in a hollow tree, log, or other protected den. Litters number from two to six, the average is three or four.

Behavior

Raccoons are born with a climbing reflex. Even before their eyes open — after 19 days — they are climbing spider-like up a tree. It's not uncommon for their mother to rescue them from a tree top. It's easier to go up than down. Adult coons climb trees almost as well as squirrels.

When contented, both adults and young make a churring noise. Adults are even more verbal — churring, hissing, growling, and screeching like an owl.

Bobcats, fox, and great horned owls kill young coons when they get the chance. But a mother coon is a tremendous fighter and predators hesitate to challenge her. The raccoon's only other enemy is the dog. Pound for pound, a raccoon can usually whip a dog. If a dog follows a coon into water, the coon will jump on the dog's head, dig in its claws, and hang on until the dog drowns. Hounds easily catch young coons. But wily old raccoons readily confuse a pursuing dog pack.

In captivity, coons may live to 14 years. In the wild, they are subject to both canine and feline distemper. Starvation and predation also takes their toll, especially during the winter.

Each year, coon hunters take 25,000 to 40,000 and trappers take 15,000 to 25,000 animals. This averages to about one raccoon per square mile throughout their Minnesota range over the last 20 years.

DNR Programs

During the 1930s, raccoons were uncommon in Minnesota except along streams in the southeast. But just a few years, after a severe drought was broken, raccoons flourished everywhere in the state except in three northeastern counties.

In the 1940s, people on the prairies were suddenly aware that raccoons were numerous. The feeling was that the animal had "invaded" the prairies to the west and north of the wooded regions. However, according to Vernon Bailey ("A Biological Survey of North Dakota," North American Fauna No. 49, U.S. Department of Agriculture, 1926), raccoons were numerous in the Red River Valley in 1800.

It seems likely that, through the 19th century, settlers and trappers "living off the land" kept raccoons (and some other animals) at a low ebb. Both the fur and the flesh were valuable. No laws regulated their taking.

In 1909, the Minnesota Legislature considered protecting raccoons between May 1 and November 1. But it was not until 1917 that a closed season was established, and raccoons were protected between March 1 and October 15.

Complete protection was afforded for three years — 1925-1927. From 1928-1952, the open season extended only four to six weeks. This protection, together with diminished pelt values, resulted in an increased abundance of raccoons. No doubt, other conditions such as increased farming (more available food) and weather were also favorable.

Very high fur prices after World War I kept many mammals at low levels. In 1920, raccoon skins sold for as much as $30 each. From that time to 1945, the average price ranged between $1 and $7, but usually it was very low. Prices have been rising in the last few years and a pelt is now worth around $20.

Liberalized hunting and trapping seasons (about 90 to 60 days respectively) under regulations of the Commissioner of Conservation from 1953 to 1464 did not diminish raccoon numbers. In 1965, the legislature removed all protection for raccoon.

Increased values for pelts and a renewed need for management prompted the legislature to give the raccoon full game status as of fall 1975. Current seasons run from mid-October to the end of December.

This leaflet was prepared by the DNR Bureau of Information and Education. DNR Report No. 20.

DNR Reports is a series of free publications on Minnesota's natural resources published and distributed by the Minnesota Department of Natural Resources, Bureau of Information and Education, 350 Centennial Office Building, St. Paul, MN 55155. Please write for a list of titles available in this series. This is title number 20. For this article only.

Charles E. Schafer
Michigan Wildlife Sketches

MUSKRATS THRIVE IN MINNESOTA

This relative of the beaver is our most abundant furbearer.

AMONG WEEDS in ponds and streams, in burrows dug into soft earth banks, hides a shy creature. Hunted by trappers for its pelt and by predators — mink, otter, raccoon, fox, bobcat, great horned owl, snapping turtle, and northern pike — for food, and muskrat tends to stay hidden. Despite the predation, however, the muskrat is thriving in Minnesota. Its high birth rate offsets its high mortality rate and makes it Minnesota's most abundant furbearer.

Description

The muskrat looks like an overgrown field mouse. Its tiny round ears barely protrude beyond its head. It has a stocky body, short legs, and a scaly tail. Its eyes are like glossy black buttons. Its lips close behind the incisor teeth so that the muskrat can eat underwater plants. The small front feet have four clawed toes and thumb; the large hind feet have five clawed toes which are webbed at the base.

Adult muskrats are dark brown. The underpart of the coat is lighter, shading to white under the throat. The coat has a silvery hue from the light underfur showing through the long, glossy overhairs.

Adults lose their fur continuously, but the young ones molt twice during their first year. The muskrat's fur is thickest in late winter or early spring.

The muskrat gets its name from its scent glands that emit a strong, sweet-smelling musk. It uses this musk for trail marking and identification.

Behavior

Muskrats, like beavers, are skillful builders. Although the muskrat prefers to build a den in a soft muddy bank, it will build a lodge in the water if no suitable bank is available.

When the den is on land, the muskrat may build several chambers. In periods of low water, it digs a tunnel out to deeper water.

The lodge, built from aquatic plants, has walls about a foot thick, and members of the muskrat family may construct separate chambers. The soggy walls of vegetation act as insulation and keep the lodge cool in the summer and warm in the winter. The walls also provide food during the winter.

Muskrats are creatures of habit and like to do the same things in the same places. A pile of mussel shells on the bank testifies to numerous muskrat feasts. Nearby there will probably be another pile of droppings. These defecating spots sometimes measure 18 inches in diameter.

Muskrats that live in marshes usually stay within a circular area. Those living in a bank den usually extend their territory out to deep water and along the bank for several hundred feet.

Muskrats are mostly nocturnal, but in the spring they come out in the daytime, too.

On land, the muskrat ambles along, but in the water it is an agile swimmer. Forepaws folded to its chest, the muskrat swims with its webbed hind feet. It uses its tail to paddle and steer. Muskrats sometimes glide along for hours, lazily munching on roots growing along shore.

Muskrats can swim two to three miles per hour. They can swim backward as well as forward. It's not unusual for a muskrat to stay underwater for long periods of time. Biologists timed one that submerged for 17 minutes.

Muskrats are pugnacious and fight among themselves, especially when food supplies are low. This keeps the population down and prevents food shortage. The defeated animals leave; the victorious remain.

Most of the year, muskrats live with their mates or alone, but during the winter, several may get together and share a den. They may warn each other of danger by slapping their tails on the water.

The muskrat's diet depends on the area in which it lives; whatever is available gets eaten. Usually the muskrat eats aquatic plants such as cattails, pond weeds and water lilies. Fresh water clams (mussels), snails, crayfish, fish, frogs, reptiles, birds, and even dead muskrats will serve as muskrat food. But muskrats are primarily vegetarian.

When a muskrat eats cattails, it stands up on its hind feet, balancing with its tial, and cuts off sections of the stem in five to ten inch pieces.

During dry periods, adults can exist without water to drink, providing there are sufficient succulent plants.

Life History

In Minnesota, the breeding season for muskrats is April to September. During this time the female may bear a litter every 30 days under the best conditions, but most young are born in the spring. Two to three litters per season is average. The litter may have from one to 11 young. The average is six to eight.

Males are polygamous and do not help to raise the young.

Measuring four inches and weighing about three-quarters of an ounce, the tiny, blind, and nearly hairless baby muskrats can swim within a week, even though they can't see where they're going. After another week their eyes open. By three weeks they start feeding on vegetation and, soon after, they leave the den.

Young muskrats sometimes wander through the summer. They may cover great distances — for a muskrat. Biologists have seen tagged animals travel up to 20 miles. At the end of the summer, a marsh may be overloaded with muskrats, so they scatter. Then muskrats may turn up almost anywhere. One man told of trapping them in his basement for several days. They had traveled up a drainage tile and out through the drain opening.

Usually, though, the young locate within ten to 60 yards of their mother's den, either in an abandoned den or on heaps of floating vegetation.

Young grow to the same size as adults in six months. Although muskrats born in the early spring can bear offspring by late fall, they usually do not have their first litters until the following spring.

Besides being an important source of food for many predators, the muskrat is subject to numerous diseases. Biolgists have seen muskrats striken with leukemia, gallstones, tularemia, and a fungus disease.

Muskrats also get parasites. Mites, fleas, roundworms, flukes, and tapeworms all plague the muskrat.

Finally, drowning takes a major toll of muskrats. Rising water levels trap them in their homes, particularly the young. Severe cold which freezes water can prevent the muskrat from foraging on trap it under the ice.

Range and Population

Muskrats live throughout Minnesota, wherever there is enough food and water to support them. It is difficult for game biologists to census the muskrat population accurately but the trapping harvest is one indication of population trends.

Until the early 1970s, the muskrat harvest was declining — drought, drained farmlands (which destroyed the muskrat habitat), and low pelt prices limited the number trapped.

Then increases in pelt prices, from less than a dollar to three dollars, renewed interest in trapping. The average annual harvest since then has been 612,000.

In recent years, revenue from muskrat pelts has averaged more than one and one-half million dollars.

DNR Programs

A fisheries biologist said that msukrats are a frequent problem in fish-stocking ponds.

Artificial ponds are divided by dikes wide enough for a truck to drive on, but also narrow enough for a muskrat to burrow through. Pretty soon the ponds drain dry. "We trap a lot of muskrats," said the fisheries man.

"They also chew holes in our nets when we are trying to take a sample of fish populations. They get caught and can't surface, so they gnaw right through."

Currently the DNR commissioner's office sets the muskrat trapping season. It runs between November 1 and April 30, and does not exceed 60 days.

To obtain detailed information on trapping seasons and regulations, contact the Wildlife Section of the DNR.

Copy for this leaflet was prepared by the DNR Bureau of Information and Education. DNR Report No. 8.

Mink
Schafer

MINK

Sought after for its glossy, durable fur, the mink is a three-pound scrapper who holds its own against larger predators. Wild minks wouldn't recognize their 'cousins' raised on mink ranches.

ELUSIVE and catlike, the mink bounds along the edges of streams in a graceful, undulating gait. Canoeists delight in catching a glimpse of one. Trappers find it a difficult quarry.

Ever inquisitive, the mink pokes its nose into everything in its territory. Now and then, it sits up on its haunches and scans the world. The mink is aware of minute changes in its environment.

The female usually remains close to her den, and hunts a territory that she can cover in two or three days. Just as much of a wanderer as an old tomcat, the male may cover an area that takes him ten days to circle.

Description

Wild mink are dark brown and black, usually with a white spot under the chin. Another wild variety, called a cotton mink, has light underfur which is worthless in the fur trade.

Adult males weight about three pounds — females half that — and measure about 20 to 30 inches, including an eight-inch tail.

Mink, like most other members of the weasel family, have long necks, small round ears, pointed snouts, long bodies, and short legs.

Mink have scent glands and can release a sickening odor when alarmed or injured. Skunks hesitate to spray; mink don't. They normally discharge their scent during courtship displays or to mark territory.

Behavior

Mink are spunky fighters and seem to look for trouble. A Department of Natural Resources wildlife specialist was canoeing a Minnesota stream when he was challenged by a mink from a nearby den. The little fighter swam after the canoe in hot pursuit. With just its small brown head sticking above the water, it hissed, dove under the canoe, and surfaced hissing on the other side. Back and forth it went for some time, until the man decided to leave. He was ushered out of the area by the mink, which was probably saisfied that it had chased off the offensive canoe. He figured it must have been a female with young.

Mink den in the same areas as beaver, muskrat, and otter, sometimes taking over muskrat dens after they kill the occupants.

Muskrats are a favorite food, but most muskrats mink kill are those at a biological disadvantage — old, young, injured, or displaced from a home range.

Mink eat fish, frogs, salamanders, snakes, and waterfowl and their eggs. Rabits, mice, rats, chipmunks, and other small mammals are also preyed upon. Mink kill more than they need to live, sometimes destroying all of a farmer's chickens. They cache some surplus food in their dens.

Aided by a webbing of stiff hairs between their hind toes, mink can outswim some fish species. Young ride piggy back in the water until they can swim.

Mink are normally active day and night, but in areas that are settled or heavily trapped, they become exclusively nocturnal.

Oftentimes mink are targets for great horned owls on nightly patrols. Bobcats and fox also prey on mink, but they get plenty of resistance and aren't always successful.

Life History

Both male and female mink are indiscriminate during mating. In February or March, after the male impregnates every available female, he settles down with one to help her raise the young. Gestation is 31 days. The fertilized egg doesn't attach to the uterine wall until the weather is favorable to the litter's survival. Five or six young, blind but furred, are born around May 1.

The entire family stays together throughout the summer, but in the fall the members go their separate ways.

Trapping

The mink wears one of the most sought-after furs in the world. It takes 75 mink pelts to make a full-length coat. Its glossy and durable fur sends trappers sloshing into marshes throughout the northern states.

As with many animals, soil fertility affects mink size. In Minnesota, the largest mink live in prairie counties, second largest live in the northwest, and the smallest come from the southeast.

Mink populations depend on available water. During dry years, the decrease in food and living space reduces production and survival of young. Mink are sometimes their own worst enemies, overcrowding an area and killing each other. Flooding may also decrease populations by drowning young in burrows.

The highest mink harvest in Minnesota took place during 1945 and 1946 — very wet years — when 100,000 mink were harvested.

Today, wetland drainage is making the mink scarce. The huge demand for pelts must be met by commercial mink ranches. They supply five million pelts per year to the fur industry. Mink ranches have used selective breeding to produce many new varieties of minks. Wild mink would hardly recognize their white or gray-blue cousins living on a mink ranch.

Minnesota contributes its share to the wild pelt harvest. During the past decade, the annual harvest has varied from 24,000 to 47,000. Fewer trappers, shorter seasons, and shrinking mink habitat have gradually reduced the harvest since the 1940s.

Until 1944, the season for mink trapping was usually more than 100 days. Mink were completely unprotected between 1924 and 1933. After 1944, the Legislature gave authority for setting seasons to the Commissioner of Conservation; however, the law limits the season to not more than 90 days between November 1 and the following April 30. In the past ten years, seasons have been from 25 to 59 days in length. The season usually opens near November 1, but in some years it has opened as late as November 14.

This leaflet was prepared by the DNR Bureau of Information and Education. DNR Report No. 22.

Charles E. Schafer
Michigan Wildlife Sketches

THE PORCUPINE: NATURE'S LIVING PINCUSHION

A young porcupine is armed with hundreds of sharp quills within minutes of being born.

CALL it quill pig, porky, or porcupine, this is the only mammal in North America whose defense is long, needle-sharp spines. Any animal which has the misfortune to tangle with the porcupine will attest to the effectiveness of its pincushion defense.

Description

The porcupine has a blackish coat with an overlay of long yellow-tipped guard hairs. An adult may have as many as 30,000 hollow quills which extend from above the eyes almost to the tip of the tail. The greatest concentration of quills are on the tail and back.

The porky has a heavy body. It weights from 10 to 30 pounds and measures 25 to 35 inches, including its seven-to-nine-inch tail. It is well adapted to foraging in trees, but is slow and ungainly on the ground. Even if the situation calls for fast action, porky's top speed is an awkward waddle.

Porcupines have hearing that is about equal to that of a human's. That is to say, it is far inferior to that of most animals. And they see only moving objects at close range. They are almost totally unable to recognize stationary objects. One naturalist imitated the female's grunting call. A male porcupine walked up and sniffed around his feet. He seemed puzzled about where the female was.

Defense

Normally the porcupine's quills lie almost flat within the guard hairs. But when the animal senses danger, it quickly raises the quills. This is possible because each quill is under muscular control. The porcupine whips around, displays a prickly rear end and clatters its teeth. if the predator gets too close, porky whips its tail back and forth — this will drive the quills deep if the tail makes contact. Because the quills are loosely attached to the body, this flailing action often sends them flying, leading to the mistaken impression that the porcupine "shoots" its quills.

The quills are engineering marvels. At the tip, each quill is needle sharp. The rest of the shaft is covered with minute barbs shaped like fishhooks. This makes extraction impossible without literally tearing the quill out. Pliers are often the only reliable tool.

A pet owner would be wise to make a quick trip to the vet if his dog runs into a quill pig. Dogs and wild animals sometimes die from encounters with porcupines. Muscular action works the quills deeper and deeper into the body. Near Duluth, for example, a magnificent golden eglc was found dead. The bird had attached a porcupine and received more than 200 quills in its feet and head.

The porcupine is preyed upon by only a few predators. The fisher, a lightning-fast member of the weasel family, is adept at flipping the porcupine over and slashing open the unprotected belly. Supposedly, wolves and coyotes hold onto the porky's nose and shake it until it dies.

Behavior

Porcupines are excellent tree climbers. They are also good swimmers — all those hollow quills give them excellent buoyancy. Left alone, porky ambles along, usually spending part of the day sleeping under a log or in a crevice. In the summer the porcupine eats a wide variety of vegatation — shrubs, clover, alfalfa, dandelions, lily pads, and even corn — so it rarely goes hungry. When the winter winds blow, it heads for the trees. The bark of pine, willow, and aspen provide excellent forage.

David Costello, in his book *"The World of the Porcupine,"* tells how one animal spent the entire winter in one tree, starting at the top and nibbling its way down. In three or so months, the porky strikpped 20 feet of bark off the trunk. Sometimes porky drops twigs from its lofty perches which provide forage for other animals such as deer and elk. It also controls mistletoe, a parasite of some trees.

But this habit of eating bark has given the porcupine a bad name in some circles. While today, most naturalists downplay porcupine damage to timber trees, there have been occasions even as late as 1966 that bounties were offered for the quill pig. The problem is most likely regional and not as serious as once thought. In Wisconsin, fishers have been imported from Minnesota to control the porcupine population.

The porcupine is also known to crave salt. Anything with a residue of salt, either from cooking or from sweat in hands, is a preferred food for the porcupine. Such items include: axe handles, rope, camp tables, lether boots, and toilet seats. One naturalist told of an incident in which he was awakened in the night to find a porcupine resolutely chewing away at the leg of his cot. The cot had been stored in a smokehouse where the man cured meat.

Life History

Porcupines are solitary beasts; rarely do they travel together, except during mating season. Beginning in September the female goes into heat every 30 days until mating occurs. While she awaits courting, the female gnaws on cans and other inedible objects, emitting squeals that echo through the woods as far as a quarter-mile.

Meanwhile, the male is traveling farther than usual, looking for female scent and ready to respond to a female's call. When the two meet, they sit on their haunches, touch forepaws, and clatter their teeth. Somehow they manage to avoid injury while mating in the same fashion as other four-footed mammals.

Gestation lasts for about seven months, and the single baby (occasionally two) is born in April or May. At birth, young porcupines weigh 12 to 20 ounces, and are ten inches long. Their eyes are open and their bodies are covered with a coat of dense black fur. Within a half hour the hundreds of quills that were soft and wet at birth have hardened, and the little porky is armed for combat.

For ten days, the female nurses her prickly baby, but soon after that, the little quill pig starts to eat the fresh greens of spring and is weaned. It can walk shortly after birth, and from the second day of life the young animal is able to climb — but only the lower branches at first. At six months, it can climb as well as the female and has since parted company with her.

Habitat, Range, and Population

Porcupines prefer open forest with plenty of new growth on the forest floor.

In Minnesota, porcupines amble through forests in the upper two-thirds of the state. The survival rate for porcupine young is high, and the porcupine is able to adapt to many human incursions on its habitat. Consequently the porcupine should continue to thrive in Minnesota.

DNR

Porcupines are not protected by Minnesota state law. They may be taken either in daytime or at night in any manner, except with the aid of artificial lights or by poison.

This leaflet was prepared by the DNR Bureau of Information and Education. DNR Report No. 54.

Otter
Schafer
Charles E. Schafer
Michigan Wildlife Sketches

THE RIVER OTTER

The playful otter is well adapted to aquatic life. Webbed feet make it the fastest swimmer among freshwater animals.

Boaters in Minnesota have watched otters bound up a river or stream bank, tuck in its paws, and belly-flop down the bank and into the water. And do it again and again. Otters are one of the few mammals that spend much of their time in play-like activity.

In the winter, otters leave distinctive "dot-dash" trails across the countryside. They travel across land by running a few steps and then sliding on the snow for ten to 20 feet. In this way they can attain speeds of 15 to 18 mph.

Otters often toss rocks and clams into the water and dive for them. In contrast to other members of the weasel family, otters are gregarious and generally live in family groups.

Description

The river otter is larger than a mink. The largest males measure up to 54 inches and weigh about 25 pounds. Females are smaller. An otter has a broad flat head, a wide nose, long whiskers, and small round ears. Its tail is thick at the base and tapers to a point. It has short, powerful legs, and five webbed toes on each foot.

In the water, the otter is a picture of grace. It is well adapted to aquatic life. A streamlined body, a long tapering tail, and webbed feet make it the fastest swimmer among freshwater mammals.

To swim underwater, otters close their ears and nostrils. Its eyes are near the top of its head so that, when it swims, only a small portion of its head shows above water. A thick layer of fat and dense, oily fur insulate the otter, even in icy winter water. Although otters do not have good eyesight or hearing, their sense of smell is acute. Sensitive whiskers also help them find their way in murky water and in dark dens.

The otter's upper coat is deep brown. Underneath, the fur is pale brown or gray. The muzzle is buff or silvery.

Behavior

Otters require a large living space. The otter family may have a total shoreline range of 50 to 100 miles. During one season, the otter will probably stay within an area of three to ten miles. It moves on when the food supply begins to diminish.

The otter has a bounding, rolling gait on land. Even though dogs and predators can easily overtake an otter, it is a scrappy fighter. For its size, it makes a formidable opponent. It is doubtful though that large Minnesota predators seek out otters in preference to other prey.

When frightened, the otter emits a strong musk that probably serves as a warning to other otters. The rest of the time this scent is used to mark territory and identify family members.

Otters have "pulling-out" places along a stream bank where they leave the water for the shore. On land, they roll in grass, leaves, or dust to dry off. Sometimes they wallow in mud.

Otters prefer to stay far from human habitats. Normally, they are active both day and night. If people are around, however, they become strictly nocturnal.

When the otter hunts fish, it glides along the surface, gently paddling with its front feet. Upon seeing a fish, it arches it back and dives, slicing the water with powerful strokes and outmaneuvering its quarry in seconds. Rough fish, not game fish, comprise most of its diet.

In shallow water, the otter catches crayfish, frogs, turtles, muskrats, and small reptiles. It will also eat waterfowl. Earthworms, snails, and insect larvae round out the otter's diet and occasionally carrion.

Otters make a variety of sounds — chrips, grunts, snarls, whistles, and squeals. When a family swims along a stream, members call back and forth frequently with bird-like chirps. When surfacing, the otter blows and sniffs loudly.

Life History

Males sometimes mate with several females, but usually spend most of their time with one. Otters mate in spring, sometime after the birth of young. Because of delayed implantation, the embryos do not attach to the uterus until seven to ten months have elapsed, so that the young will be born in the spring when their chances of survival are best. This means that gestation is from 9½ to 12 months long.

In Minnesota, the female bears her single litter around April, usually about two or three young. The nest may be in tangled shore vegetation, an abandoned muskrat or beaver lodge, bank cavity, or hollow tree.

After the cubs are born, the females curls around them, covering them with her head so they can nurse and protecting them from chilly spring air.

The newborn kits are dark brown, blind, and toothless.

Their eyes open after about 35 days, but they don't leave the nst for ten or 12 weeks. After weaning — at four months or so — the male helps teach the young how to swim and hunt. Both adults carry the kits on their backs in the water and coax them to try swimming. The young stay with their parents through the winter, but leave the following spring before the new litter is born.

Range and Habitat

Otters primarily inhabit the forested regions of northern Minnesota and are expanding their range southeastward along the Mississippi River valley to the Iowa border.

Otters usually live in streams, rivers, and lakes where these waterways are bordered with timber. Otters rarely build their own dens, but adopt those abandoned by other aquatic mammals. During summer, the den has an above-water entrance. In winter, the only opening to the den is under water. The floor of the den may be loose dirt, packed mud, or scatterd grass and leaves.

Trapping

The otter population is probably increasing in Minnesota and their range is expanding southeastward. A yearly trapping season lasts for 15 days. Each trapper is allowed to take three animals. The otter's silky-thick fur is highly prized.

Text for this leaflet was prepared by the DNR Bureau of Information and Education.

Weasels
Long-tailed Weasel
Schafer
Short-tailed Weasel
Least Weasel
Charles E. Schafer
Michigan Wildlife Sketches

THE WEASEL —

NEMESIS OF NATURE'S LITTLE PEOPLE

By G. W. Bradt

Slender, lithe, sharp of tooth and claw, secretively colored, richly endowed with strength, speed, cleverness, and indomitable courage, weasels are the scourge of all small ground-living animals.

There are several species of weasels — three in Michigan — which differ somewhat in size, length of tail, and other minor characteristics, but all are similar in food habits, ferocity and manner of life, so that a single description will suffice for all.

The cunning savagery of the weasel is reflected in his physical appearance, which at times is peculiarly ferocious. The forehead is low, and the nose sharp; the eyes are small, penetrating, cunning, and seem to glitter with a baleful green light. The jaws are worked by muscles which cover the whole side of the skull. This face is set on a long, slender neck in such a way that it can be held at a right angle with neck axis. When we suddenly come upon a weasel, its neck stretched up, flat triangular head swaying from one side to the other, little beady eyes gleaming, we are reminded that ounce for ounce he's one of the greatest little fighters in North America.

Kills More Than He Can Eat

The weasel preys on every living creature which he can catch and kill, and this includes pretty nearly everything from a turkey to a sparrow, and from a rabbit to a shrew. Swift and surefooted, he pursues and runs down his prey; keen of scent, he tracks them; slim and sinuous he follows them into their burrows, and in the end kills them even in their secret nests. Perhaps he does not kill for the sheer love of killing, as some allege, but at least his desire to kill sometimes seems to get the better of him, and he kills far more than he can possibly eat.

We should not too bitterly condemn him for this: he is merely "accumulating a surplus for a possible period of future scarcity," as economists describe similar actions among our own species.

When a weasel finds himself in the presence of a number of creatures which he considers his natural prey, as for example, in a poultry yard, or a mouse nest, he cannot restrain his fierce desires, but kills them all. This wholesale destruction we view with horror in the poultry yard and with delight in the mouse nest. The weasel is perhaps the most efficient destroyer of small rodents known. He kills them by day and by night, in their snug retreats and in the open, in all seasons of the year.

The weasel's killing seems to be regulated only by the scarcity or abundance of prey, and by the limitations of time and space. Yet regulated it must be, for the creatures never multiply until we have a plague of weasels, nor do they ever exterminate their prey over any considerable area. Occasionally a weasel or a weasel family will settle in an old barn teeming with rats and remain until not a rat can be found. But even here rats reappear after weasels have gone, so extermination is not accomplished.

Is Efficient "Control Agent"

The law of diminishing returns operates in nature as well as in the realm of human economics, and when one type of prey becomes scarce, more attention is turned to other types. Hence the weasel is one of the most efficient agents in maintaining the much discussed "balance of nature," since he tends to control certain species without exterminating them. It is generally conceded that weasels do more good than harm from a human viewpoint, since the occasional raid on the poultry yard or the occasional killing of a rabbit, grouse or pheasant is more than offset by the incessant destruction of mice, rats, and other small rodents which are important pests in agricultural areas.

A characteristic of the weasel is his habit of storing food supplies. This is the more surprising since the weasel craves fresh hot blood from living prey, and seldom, if ever, eats from its storage pile of carcasses. Yet a weasel has been known to accumulate as many as a hundred carcasses of mice or rats, dragging some of them a considerable distance to augment the heap.

It is not often realized that weasels can climb trees, but they climb easily and rapidly, and are said to catch chipmunks in that way, although red squirrels easily elude them.

Even larger squirrels may at times be caught in their tree dens, although squirrel catching seems the exception rather than the rule. Unlike his larger cousin the mink, the weasel is not an expert swimmer, and does not pursue aquatic prey.

Observations on the family life of the weasels are scarce, and our knowledge of the subject is, therefore, limited. The gestation period of the long-tailed weasel, our most common species in Michigan, is apparently about 130 days, and five to eight young are born in late April or early May. The male is believed to remain with the family for a time and to assist in gathering food for them. The nest may be in a hollow log, a stone pile, or under an old barn or brushpile, and usually shows remains of mice or other prey along with the grass and hair of the original construction. The young leave the nest at about six weeks of age, and wander with the mother for the remainder of the summer. They seem to become solitary during the fall and early winter, pairing off some time in midwinter.

The long-tailed weasel, also called the New York weasel, is the largest as well as the most abundant species in this state. Its upper parts are dark brown, while its under parts are white, washed with dull yellow. The white does not usually continue on the feet. The tail is tipped with black in both the winter and the summer pelage. This weasel normally turns white in winter in northern regions, and becomes the "ermine," but may not do so in the south. In southern Michigan some turn white, some remain brown, and some become mixed in color during winter.

The short-tailed, or Bonaparte, weasel is somewhat smaller than the long-tailed weasel, but is quite similar in color, except that the feet and toes usually show white. The tail is much shorter than that of the long-tailed weasel, which together with the white on the feet, enables the observer to readily separate the two species.

A third species, the least weasel, is little known and seldom seen by Michigan residents, yet it is probably found in all parts of the state. This weasel, smallest of our carnivores, is only about six to seven inches long, including the short tail. The body is about the thickness of a man's thumb. General coloration and color changes are like that of the other species, but the tail is not tipped with black in either color phase. Food habits are thought to be similar to other weasels, but the small size probably restricts the diet to mice, insects, and possibly small birds. While records of these little weasels are few, it is believed that they are often taken by trappers and dismissed as "young weasels" without further ado, as their tiny pelts would scarcely be worth the skinning.

Fairly Abundant in Michigan

Since weasels are so fierce, active and adaptable, the question arises as to why they do not multiply until they destroy their own livelihood by devouring all available prey. This may occasionally actually occur in a limited area, where mice and other small mammals become reduced to such a degree that weasels must migrate or starve. It is known that bloody fights between individual weasels frequently take place, and the vanquished animal often lies dead on the battlefield. Male weasels are considerably larger and heavier than females, and since these males are quite lacking in chivalry outside of the mating and early family period, a losing fight for the female means no family thereafter. This may constitute a ruthless but effective check on weasel multiplication.

In addition to this, weasels are known to suffer from various diseases which affect mink, so disease may also play a part in controlling their numbers.

Weasels are fairly abundant in Michigan, but few trappers or hunters make any direct attempts to take them, and those taken are shot or trapped incidentally to other activities.

From Michigan Wildlife Sketches, Michigan Department of Natural Resource.

WOODCHUCK

Charles E. Schafer
Michigan Wildlife Sketches

THE WOODCHUCK
Professional Excavator

By G. W. Bradt

The Woodchuck is known by many local names, among the best known being ground hog, whistle pig, and eastern marmot.

As is indicated by its scientific name, *Marmota manox*, the woodchuck is a member of the widely distributed marmot clan, and is actually a big, clumsy, ground dwelling member of the *Sciuridae*, or squirrel family. It seems somewhat odd to associate the small-tailed, heavy-bodied, burrow-digging woodchuck with the large-tailed, slender, tree-climbing fox or red squirrel, but if we remember also the various chipmunks, gophers, and ground squirrels the transition is easier to understand. The woodchuck occasionally reminds us rather sharply of his squirrel ancestry by climbing a tree to escape dogs or to keep a better lookout over his home territory.

Woodchucks Typical Hibernators

Although the habitat of the eastern woochuck, which we have in Michigan, differs widely from that of the western marmots of the great mountains, their habits are remarkably similar. The marmots are all vegetarians, all live in burrows, all stay pretty close to home, all tend to whistle sharply when surprised, and all go in winter hibernation.

Woodchucks may be called typical hibernators, or winter sleepers. When the temperature begins to drop in late fall, about November here in Michigan, the woodchuck, fat as a Thanksgiving turkey, crawls down to his underground hole, pulls the hole in after him by caving in the dirt in the passage to the nest, and goes to sleep for the winter.

During this period of hibernation, nothing is eaten, so that no fuel is received to be converted into bodily heat by the oxidation of the blood through breathing; on the contrary, the lungs almost cease to work. There is an occasional respiration — a sighing inhalation of breath — but most of the time the only oxygen in the lungs is the trifle reaching them by the slow heart beats, and by the diffusion of gases. A mirror held before the mouth and nostrils is not clouded by breath. The animal may be placed under water, or in a jar of carbonic acid gas, for an hour or more, without drowning or suffocating; nevertheless, respiration and other bodily functions do not entirely cease.

Their awakening must be gentle and gradual, as the sudden forcible arousing of hibernators often causes death. Here a curious fact may be mentioned: in spite of their comatose condition, which has been likened to human beings in a state of trance, they are sensitive to the slightest touch; merely blowing on the hair will induce an instantenous response, although not in the least arousing the sleeper. The period of hibernation varies not only with different kinds of animals, but with the same kinds under different or even similar conditions.

That this winter sleep is of a highly protective character cannot be doubted, since the true hibernators are those whose food fails altogether in winter, and is of such a nature that it cannot be stored up in sufficient quantity.

Contrary to tradition, Michigan woodchucks do not emerge from their winter sleep promptly at sunrise of February second, to watch for their shadows on the snow or earth as the case may be. They do seem to become restless and awaken more and more frequently during late winter and early spring, finally coming out to stay when the snow is gone and tender shoots are starting in the fields.

The mating season is believed to be in late March, and the young are born in late April or early May. Very small and undeveloped at birth, probaby blind for about a month, they do not venture from the den until they are six or seven weeks old. They begin to eat solid food in the form of grass and clover as soon as they are able to get out to find it. The mother, or both parents, attends them carefully at first, showing them where and when to eat, as well as some of the dangers of the outside world.

At times the young and adventurous 'chucks cause damage to gardens and truck patches. As the end of August approaches the animals are nearly full grown, and scatter to dig homes of their own.

May Dig Fifty-foot Burrows

Woodchuck burrows may be found in woods or fields, but perhaps the most favored location is at the edge of a wood or swale adjoining a field of clover or alfalfa. One entrance to the burrows is marked by a heap of dirt thrown out during the excavation, but one or more other entrances without external evidences are often present also. These concealed openings are used as lookouts or spy holes, and here the 'chuck watches intruders whose attention may be attracted by the obvious pile of dirt, while the 'chuck himself remains undetected. The length of the burrow varies according to the type of soil and surface topography. The total length may be somewhere between seven and 50 feet, and the depth up to six or seven feet.

Disease does not seem to play a prominent part in the life history of woodchucks, since there is no record of plague or epidemic, and few woodchucks have been found dead from disease or parasitic infection. Man, dogs, and foxes are probably his chief enemies in the order stated, although hawks may occasionally get a young 'chuck during its first days above ground.

Red foxes and badgers are reported to dig out woodchucks under certain circumstances, although this is probably not a common occurrence.

While young woodchucks are good to eat, and frequently are eaten by farmers and sportsmen, the primary economic importance of the woodchuck lies in his digging propensities. Farmers do not like the heaps of dirt piled up by the 'chuck at the entrance to his den in clover and alfalfa fields, nor the concealed almost vertical hole used as a lookout. The cutter bars of mowers are dulled and the blades nicked in the dirt. Occasionally a farm horse broke or twisted his leg by stepping in a 'chuck hole. Nowadays tractors don't have this problem.

Burrows Help Other Wildlife

The other side of the woodchuck hole situation and probably the more important one, is the value of these holes as dens and refuges for other wildlife. Some wildlife investigators maintain that numbers and distribution of the woodchuck are among the more important ecological factors influencing the numbers and distribution of the cottontail rabbit.

Rabbits do not dig their own burrows, or construct their own hiding places. Old farm buildings, brush piles, junk heaps, stone piles and the burrows of other animals are their only safe retreats, and man has a tendency toward eliminating many of these. Buildings are razed, brushpiles and tangled fence rows burned, and stone piles removed.

The recognized importance of the woodchuck as a game animal and a provider of dens for cottontails has brought about a change of public opinion and a change in the legal status of the animals.

No counties and few townships now offer bounties for woodchuck scalps, formerly a common practice.

Woodchucks are given legal protection in Michigan and may be hunted only after the opening of the October hunting season. Since woodchucks are in hibernation during most of the period after late october, this, in effect, closes the season to woodchuck hunting the year around. Farmers are allowed to kill woodchucks doing damage on their property at anytime, and there is no closed season in the Northern Peninsula.

In recent years woodchucks have been holding their own after making a comeback in numbers as compared to their scarcity in the late 30s.

From "Michigan Wildlife Sketches" Michigan Department of Natural Resources.

THE COMMON LOON

David H. "DJ" Johnson
Minnesota DNR Forestry

LOON CALLS

It is the call of the loon on a lonely lake that gives the bird distinction. The loons' calls, once heard, can never be mistaken for anything else. There are four basic calls the loon makes, each described below.

The "wail"

The wail sounds something like "ah-oooo-ooo-ahh". This call carries a general message, and is given when one bird wants to interact in some way with another bird, or when an adult is looking for its mate or a lost chick.

The "tremelo"

This the "laugh" call so commonly heard. This call is given when the bird is disturbed or frightened — such as when a boat comes too close. There are three slightly different variations of this call, but they all sound basically the same. The tremelo is also given when the bird is in flight (in fact, except for the "hoot", the tremelo is the only call given by loons in flight).

The "yodel"

This is by far the most complex call given by loons. It is given only by males, and is usually heard in spring and early summer. An aggressive call, the yodel is given when a bird is trying to establish or maintain its' territory. The call is given for example, at border confrontations between neighboring males, or when a resident male confronts an intruding loon. The yodel begins with a number of tones that gradually rise in pitch; the call then continues with a phrase that is repeated a number of times. The number of repeated phrases can vary from one to more than nine. The next time you hear a yodel, count the number of repeat phrases, this will give you an idea of how excited or angry the bird is, the greater the number, the more excited the bird. Besides communicating aggression, the yodel also contains information on the identity of the bird giving the call, kind of a "name tag". Research has found that a bird's yodel does not change from one year to the next, and by recording yodels each spring, it is possible to determine if the same males are returning.

The "hoot"

The hoot is a simple call which frequently resembles a "hoot" in sound. When the bird is in flight, the hoot may sound more like "gek - gek". The call is given so as to keep in contact with other family members or other birds when in small flocks.

LOONS' LIFE HISTORY

The Common Loon is one of several fish-eating diving birds of the genus Gavia, order of Gaviiformes, found in the Northern Hemisphere. Allied species include the Yellow-billed Loon, the Arctic Loon, and the Red-throated Loon. Only a handful of sighting records exist in Minnesota for the Yellow-billed and Arctic Loon. The Red-throated Loon is an uncommon spring and fall migrant through the eastern part of the state. If you are interested in seeing one of these loons, you'd have the best chance if you look around the Duluth

area, in Lake Superior.

Range of the Common Loon finds this bird breeding from Iceland, central Greenland, the Arctic Islands, and northwest Alaska; to Massachusetts, and in suitable habitat across the northern tier of states. It spends the winters along the coastal areas from Alaska and Newfoundland to Mexico. It is assumed that most of Minnesota's loons spend their winters along the gulf coast.

Primarily a fish-eater, the loon will take some shellfish and only very rarely, vegetation. Unlike the gull, it never has been observed eating dead fish. Loon depredations against the fish population — despite a voracious appetite — are no cause for alarm for the simple reason that a loon is not a prolific bird. A prime nesting area in northern Minnesota showed only 150 loons in a 60-square mile area.

The loon has difficulty in becoming airborne — it can rise only from water — because it has a relatively heavy body (an adult male can weigh up to nine pounds) and small wings. The labored takeoff, half running and half flying over the water, may cover a quarter of a mile. The popular notion that a loon cannot fly without a breeze is a fallacy. Once free of the water, the loon makes several ever-widening circles to gain altitude but seldom is seen flying above 400 feet. Yet they have speed, having been timed at up to 60 miles per hour.

The displays of the loon have long fascinated observers. During courtship, the loon and its mate approach each other slowly and begin a rapid dipping of their bills in and out of the water. Short, quick dives are made, followed by exaggerated preening and stretching. Suddenly both race off across the water, striking the surface with powerful wing beats. They follow a curving course back to the starting place, piercing the northern air with various calls. Because loons are fairly long-lived, they are not prolific, and a young loon does not reach full breeding plumage until its third year.

Nesting takes place soon after courtship and 2, (frequently 1 and rarely 3) greenish or brownish olive eggs are laid in the 2 foot plus diameter nest. The nest consists of a mass of reeds, rushes, grasses, and twigs which may be placed on bare ground, a floating bog, a muskrat house, or attached to the shoreline vegetation. Nesting sites vary, but small wooded islands are favored; usually protected from prevailing winds and wave action. Many nesting sites are used year after year. In an experiment here in Minnesota, 6 of 8 man-made islands of native vegetation were used as nesting sites by loons. Incubation of the eggs is done by both sexes, but mostly by the female; starting with the first egg laid — it is commonly 29 days long.

PROBLEMS LOONS FACE

Many states are experiencing a declining loon population (some as much as 35% or more) with bird numbers dropping as human activity increases near loon nesting and rearing sites. It is sometimes thought that because loons prefer more wilderness-type habitats, they "withdraw" to more remote areas when disturbed. Actually, the remote areas always have had their own populations, and the disturbed birds do not withdraw, but cling to their territories and attempt to nest year after year, even under adverse and impossible conditions. In some areas loons have nested at the same site for eight years, without ever producing chicks.

The disappearance of loons from previously occupied lakes is generally not due to either deliberate human persecution, or to lake desertion by the adults. It is the inability of the birds to rear young over a span of many years that eventually leaves the lake unoccupied. Some birds are lost during migration or during the wintering season on the ocean, a loss which may be increasing because of oil spills and other man-made pollution. But even without winter losses the adults would eventually reach the end of their life spans and fail one spring to return to their nesting territory. In the absence of their own progeny, or populations from nearby lakes, the territory may be permanently unoccupied.

MAJOR CAUSES OF NEST FAILURE

Human disturbance — The most conspicuous conflict between people and loons is the pursuit of birds by speedboat. It is obvious, illegal, and practiced by only a few cruel adults or careless children. More subtle and less observed is the effect of quiet boaters — canoeists and fishermen. Because loon nests are always close to the water's edge, any boat approaching along the shore may flush the bird from its nest. Usually the incubating loon will slip off the nest quietly; to surface at a distance and begin the famous display antics to attract the intruder away from the nest. This may succeed at times, but if the fisherman/canoeist has picked that spot to fish or watch the loons' antics or other natural attractions for any length of time, the eggs in the nest may chill, causing the embroys to die. Repeated disturbances can cause the total abandonment of the nest by the adult birds for that year.

Loon nests are occasionally destroyed by raccoons, gulls, or ravens. Once in a great while young loons are taken by snapping turtles. These predators are not new to the loon but have been around for thousands of years. But some of man's activities, which frighten the loons off their nests, increase the chances of the nest being destroyed by these other wildlife species. We should not chastise these predators (they need to eat too) but direct our efforts at curtailing man's negative impacts on the loons.

Since loon nests are usually built within five feet of the shoreline, they are highly susceptible to flooding or water level subsidence. Wave action caused by speedboats in high water times has caused nest destruction. Although possible, natural wave action probably has minimal effect on loon nests as the birds normally build their nests in areas protected from wind and wave action. Drops in water level after nesting of 14 inches or more has caused nesting failure. It may sound funny, but because of the physical structure of the loon, they are unable to make the climb up the steep incline to the top of the nest. Drops in water level also may leave the nest high and dry or open up passageways for terrestrial predators.

Sometimes when a loon is approached, it will stand up on the water with its wings folded like a penguin. The 'penguin dance' is a sign of extreme distress, and is usually accompanied by frantic calls. It is a desperate effort by the loons to protect their nests or families. If you cause such a response, you should leave the area immediately. Watercraft operators should be aware that loons can die of exhaustion if pursued by motorboat, and chicks can drown in the boats' wake. Monofilament fishing line, improperly discarded into lakes often entangles loons and other birds, drowning them.

WHAT YOU CAN DO TO HELP

The bits of land, islands or marshy shores, where loon nests are still being built today are priceless assets which must be preserved if the loons are to survive. Loons don't need much space. A single strip of quiet shore, 150 feet long will suffice if the neighbors and observers respect the privacy of the area during the critical nesting cycle.

Report harassment or killing of loons to the local conservation officer — loons are completely protected under state law.

SO, IN A NUTSHELL

Loons are very sensitive to human disturbance during the nesting season, so please respect them and their need for privacy. If we want the loon to continue to be a part of our north country — to continue to be a symbol of the genuine wilderness that, deep down, most of us would like to stay a part of — then we must make some simple common-sense efforts to insure the survival of our state bird — the Common Loon.

Kestrel

MINNESOTA'S BIRDS OF PREY

by David H. Johnson

Hawk-Owl

What are the birds of prey?

Birds of prey — raptors — include the owls, eagles, falcons, hawks, vultures, and the osprey. All of these birds are completely protected by law and cannot be taken, transported, or possessed without a special permit.

How many species does Minnesota have?

To date, a total of 35 different species of raptors have been seen in Minnesota. This total is comprised of 12 owls, 2 eagles, 1 vulture, 1 osprey, 5 falcons, and 14 hawks. Twenty-six of these species have nested here at some point in time, with the other 9 seen in winter or migration.

What good are they?

Predation has long been recognized as a necessary and essential element of the natural environment by which life supports life. Given a predator-free environment, prey populations would soon overload the carrying-capacity of their habitats — only to succumb to starvation, disease, or a host of other factors. Predators have evolved with prey, to keep their populations in check. Raptors, being predatory birds (except for the vultures which eat carrion), have an obvious value in that they help to keep in balance populations of their prey — namely small mammals, birds and insects.

Because raptors are on the top of their food chains, they act as indicators of the quality of the environment. The nesting success — on lack of success — by these birds can indicate either a healthy environment, or one which may be having problems. Species like the Short-Eared Owl, Burrowing Owl, Red-Shouldered Hawk, and others have narrow habitat requirements which make them very susceptable to even minor changes in environmental quality. A good example of raptors as indicators came when it was discovered that the insecticide DDT was effectively eliminating the Peregrine Falcon, Bald Eagle, and Osprey. These birds served as warning signals which eventually led to the banning of DDT and other harmful insecticides which are dangerous to humans.

Raptors also have an aesthetic value. For example, watching an osprey hover, plunge into the water, and rise with a fish in its talon is a spectacular sight indeed! Seeing the big, round eyes and silent, steady gaze of an owl is quite a thrill. To hear its call in the evening hours offers a uniqueness that cannot be derived from any other animal. It is impossible to place a dollar value on the howl of a Timber Wolf or the call of the Common Loon. But imagine how dismal this world would be without them! They are indeed of value and well deserve our appreciation and concern.

Raptors are broken down into the following categories:

Vultures:

The only vulture in Minnesota is the Turkey Vulture. A large (wingspan 72"), blackish, broad-winged bird with a naked head. Adults' head is red while the immatures' are black. These birds are scavengers, eating carrion such as car-killed animals and dead fish along lakeshores. While most people think the turkey vulture is an ugly bird, this bird becomes a thing of beauty when it spreads its wings and rises on thermals. Most researchers agree that vultures find their food using their keen eyesight, however there is some evidence that, unlike most raptors, they may also have a highly developed sense of smell.

Turkey Vulture

Turkey Vultures have been found to nest in old dilapitated buildings, on cliff ledges, on the top of large broken-off dead trees, on the ground in hollow logs and under piles of windrowed timber.

Kites

Kites are medium-sized birds with pointed wings. Their food is mainly invertebrates, frogs, and snakes. Three different species of kites have been seen in Minnesota; the White-tailed Kite (seen only once), the Mississippi Kite (seen three times) and the Swallow-tailed Kite. The Swallow-tailed once nested regularly in the State, but disappeared rapidly after the turn of the century. It was last reported during the breeding season of about 1907. Since 1907 the Swallow-tailed has been seen 14 times. It is now receiving consideration for reintroduction efforts using Broad-winged Hawks for cross-fostering (replacing Broad-wing eggs with Swallow-tail eggs under incubating adult Broad-winged Hawks).

Sharp-shinned Hawk

Accipiters

Accipiters are medium to small woodland hawks and are strong sprint-fliers with short rounded wings and long tails. Their main diet is birds and small mammals. Minnesota's species include the Goshawk, Cooper's Hawk, and the Sharp-shinned Hawk. As is true with nearly all raptor species, the female is larger than the male. In the case of accipiters, sexes look alike and species nearly overlap (for instance, a female Sharp-shin is very close to the same size as a male Cooper's), so as to make correct identification a bit tricky.

The crow-sized **Goshawk** (pronounced gos-hawk) is the largest of the accipiters and is readily distinguished by its heavily-built appearance and dark facial markings. Rabbits, hares, squirrels, ruffed grouse, and crows are common prey. Although nowhere abundant, Goshawks nest in the heavier forested areas of northern Minnesota. At approximately 10 year intervals, Goshawks can be seen migrating south in higher numbers. Many feel this migration is closely linked with the cyclic nature of the prey populations, namely the snowshoe hare and ruffed grouse.

The **Cooper's Hawk** is a medium-sized accipiter. It nests in woodlands and is especially susceptible to pesticide contamination because it preys on small birds, such as Robins and Starlings, that may feed on insects carrying insecticides. Birds sick and weak from insecticides are easier prey for the hawk. The concentration of insecticide builds to a dangerous level which may not kill the hawk outright, but results in sterility or thin-shelled eggs that do not last through incubation.

The **Sharp-shinned Hawk** is the smallest accipiter and is about the size of a Kestrel (Sparrow Hawk) or Robin. Sharp-shins feed on small birds, and most commonly nest in association with dense conifers. Although upwards of 10,000 plus Sharp-shins are seen at Hawk Ridge Nature Reserve near Duluth each fall, only 9 nests have been recorded in the State. This may seem hard to believe, that is, until you ever try to find a nest!

Harriers

The **Northern Harrier** (also called the Marsh Hawk) is the only Harrier found in Minnesota. It is slim, medium-sized, mouse-eating hawk with long wings and tail. In flight, a distinctive white rump patch is visible. The numbers of Harriers have dwindled and are still declining due to extensive drainage of marshland, their primary nesting and feeding ground.

Harrier (Marsh Hawk)

Buteos

Largely rodent-eaters, buteos are medium to very large soaring hawks with broad wings, husky body, and a fairly short tail. Minnesota has eight species — Red-tailed Hawk, Red-shouldered Hawk, Broad-winged Hawk, Swainson's Hawk, Rough-legged Hawk, Ferruginous Hawk, Bald Eagle and Golden Eagle.

The **Red-tailed Hawk** is a fairly large buteo, and is one of the more common raptors found in Minnesota. It feeds primarily on small mammals, but will also take reptiles, amphibians, insects, and birds. Adults of the species are identified by their brick-red tail. Young have a brownish tail with black barring across it. Three subspecies have been seen in Minnesota — the "Krider's", the "Harlans", and the "Western" Red-tails. The "Eastern" race is the breeding bird in the State. Coloration in Red-tails is quite variable, which makes identification of the subspecies a big challenging.

The **Red-shouldered Hawk** nests primarily in large, contiguous tracts of mature lowland hardwood forests near watercourses. These habitats are diminishing due to forest fragmentation by timber harvest, urganization, large pasture development, and flooding caused by dam construction. Researchers have found that this hawk has declined and is still in trouble throughout the nation and Ontario. It feeds on earthworms, insects, frogs, snakes, small birds, and small mammals. Its loud call heard mainly in spring and early summer, sounds similar to "kee-yer, kee-yer, kee-yer".

The **Broad-winged Hawk** is smaller than other buteos. It inhabits forests throughout Minnesota, but is most common in the northern forests. When hunting, the Broad-wing sits quietly along forest openings and woodland trails on a perch until it sights prey. It begins to sway back and forth excitedly, then flashes into action. It feeds on large insects, frogs, snakes, small birds, mice, and other small mammals. The adult Broad-wing is distinguished by its short, wide tail with black and white bands, two of each. Thousands of these hawks are seen in Duluth during fall migration. In migration to its wintering grounds in South America, they often ride the thermals in large circling flocks called "kettles".

A dark-breasted buteo of the prairies and plains is the **Swainson's Hawk**. Minnesota is on the eastern edge of this species' North American range. The Swainson's nests in low numbers along the western and southern portions of the State. It feeds largely on gophers, rats, and grasshoppers. This species glides with its wings slightly uptilted, and like the Broad-wing, travels in flocks to its South American wintering range.

During winter, the **Rough-legged Hawk** is a frequent visitor to Minnesota. Its feet are small and well-adapted to catching mice — its main prey. In pursuit of prey, it can be often seen hovering over fields. Of the larger-than-crow size buteos, the Rough-leg, Ferruginous Hawk, Osprey, and more rarely the Red-tail are the only ones that hover. Identifying marks on most of the Rough-legs are the dark "wrist" spots, dark chest band, and white on the base of the tail feathers. As with other buteos there is a good deal of individual color variation, and melanistic (darkness of feathers resulting from pigmentation) Rough-legs are not uncommon.

The **Ferruginous Hawk** occurs in small numbers in Minnesota. It is a bird of open dry country where it perches on badger mounds, hillocks, or posts. It feeds primarily on small rodents and jack rabbits. No Ferruginous nests have been found in Minnesota, although some birds have been sighted during the summer. When seen, it is usually west of the Mississippi in the spring or fall.

The **Bald Eagle** represented power and majesty for thousands of years before John Adams, Thomas Jefferson, and the other founding fathers chose it as our national symbol in 1782. At the same time our republic was formed, as many as 25,000 Bald Eagles soared through the skies over the timber-bordered lakes, streams, and coastlines of what is now the lower 48 states. Today the Bald Eagle is listed as a "threatened and endangered species" south of the Canadian border. The population is estimated at little more than 1,100 breeding pairs; a frightening drop from the thousands which inhabited the American wilderness. As the American wilderness began to vanish — so did the eagles.

Lakes and rivers that were once bordered by eagles' nests are now surrounded by vacation homes. As leisure time increased Americans flocked to the Nation's waterways for recreation.

As we cleared land for farms and roadways, and cut down forests to get lumber for houses, we took the trees in which eagles built their nests. "Progress" created other hazards as well. Eagles eat mostly fish, but also take an occasional waterfowl or small mammal. They're at the end of a complex food chain that is based on water. Many of our waters have been contaminated over the years by agricultural and industrial chemicals such as DDT, dieldrin, PCB's, and mercury. Fish and waterfowl which live in this polluted environment absorb these poisons by eating chemical-laden water plants and smaller animal organisms. Bald Eagles feed on the fish and waterfowl and become victims of what has turned into a pollution chain.

Eagles also fall victim to: (a) the ingestion of lead shot, such as can be found in crippled or sick waterfowl; (b) the deliberate shooting of birds by irresponsible and misinformed persons (some are even killed for their feathers, which are sold on the black market!), and (c) the innocent, but well-meaning, but uninformed bird-watchers and wildlife photographers.

Bald Eagle

There is a slightly optimistic side to the Bald Eagle story. Research studies and management surveys have generated the kind of intimate understanding that's necessary to save this magnificent bird for future generations. Starting in January 1979, the Annual Midwinter Bald Eagle Survey was begun. This survey, (usually held around mid-January), is designed to show the number of eagles that winter in the lower 48 states and eastern Canada. In 1980, 12,323 eagles were reported. In 1981 some 13,709 were spotted during the survey — a 5 percent increase over 1980. In 1979, Minnesota had 111 producing nest sites, in 1981 this author revealed some 132 producing nest sites. Restoration programs in various eastern states are showing signs of success. Overall, the Bald Eagle situation appears to be improving. Although this is encouraging, it should be viewed with caution since little is known about the mortality rates of young eagles after they leave the nest. Lakeshore development and other people-related problems will no doubt present a limiting factor in the long run.

The **Golden Eagle** is a bird of prairies and mountains. They do not nest in Minnesota, but are seen here during migration and in the winter. Aeries — nests — are usually located on cliffs or in large trees. Both adults and immatures have the rich dark brown body plumage. The golden neck feathers can only be seen at close range. The broad white tail band and white wing of the immatures are good field marks. The Golden Eagle feeds mainly on rodents, but will also take larger prey. This eagle is also a victim of habitat destruction, shooting, and pesticide poisoning. It is becoming increasingly rare.

Osprey

In 1981, Minnesota had some 160 producing **Osprey** nest sites. Pesticides such as DDT has reaked havoc on the Osprey population much like the Bald Eagle, Peregrine Falcon, and other bird populations. The Osprey is a fish eater, and shares many of the same lakes and rivers with the Bald Eagle. Ospreys may nest in the same place for years, but their habitat of building nests in the tops of rotting trees does not assure the nests' longevity. It is also of interest to note that data for Minnesota shows that some 25 producing nest sites are on artificial structures such as wood and metal power poles. This is encouraging, in that it means that if left alone during the critical stages of the nesting cycle, this somewhat adaptable bird can co-exist with man without any major complications.

Falcons

Five falcon species have been seen in Minnesota. Three species nest here. Falcons are streamlined hawks with long pointed wings, large heads, and tails that narrow at the tip. They are rapid fliers, with a direct, choppy, powerful flight. They sometimes soar with the tail spread open.

The largest falcon, the **Gyrfalcon**, is an arctic bird, which rarely wanders south of Canada. A few make it as far south as Minnesota, and can be seen anytime from September to March. Gyrfalcons most often are grayish birds, which are immatures. Their main prey is birds and rodents.

The **Prairie Falcon** is more lightly built than the Peregrine and more pale in color. The black axillars (base of underside of wings) help identify this bird. Flight is strong, rapid, and usually low. Prairie Falcons do not nest in Minnesota, but are seen in small numbers with some regularity (especially in the fall), in the western part of the State. The normal range of this bird is in the western U.S. and Canada.

The **Peregrine Falcon** is a crow-sized, grayish-bluebird that is best identified by the falcon shape, facial pattern, and dark cap. The Peregrine feeds almost entirely on birds. Although it no longer nests in Minnesota, a few can been seen passing through the state in migration. A 1942 research publication, estimated some 350 breeding pairs in the eastern U.S.

Peregrine Falcon

Surveys done in 1964 proved the general suspicion of raptor biologists to be true — that there were no longer any breeding Peregrines or occupied nest sites in the eastern U.S. It is now known that DDE, a metabolite of DDT, causes eggshell thinning. DDT is highly residual; once in the food chain, it is transferred upward until it reaches the top level where predators such as the Peregrine receive accumulative dosages. Chlorinated hydrocarbons, used so indiscriminately to control insects, eliminated the Peregrine Falcon in the eastern U.S. Unsuccessful attempts at reintroducing this species into Minnesota were made in 1976 and 1977. Another reintroduction attempt will be made in 1982. The outlook is very optimistic. Someday soon, the Peregrine Falcon may again grace the skies of this state.

The Merlin (also called the Pigeon Hawk) is an uncommon blue jay-sized falcon that is identified by its dark back and wings, barred tail, pointed wings, and its absence of facial markings. It often captures small birds, mice, insects, and shore birds. In Minnesota, the Merlin has been found nesting in the Northeastern region of the state, and has been sighted during the summer (although quite sparingly) in the other regions of the northern forests.

The Kestrel (or Sparrow Hawk) is the smallest falcon, being not much larger than a Robin. It has dark mustaches on each side of its face and a chestnut colored back and tail. The male has bluish wings. Kestrels are fairly common in Minnesota. They prefer somewhat open country where they maneuver back and forth over fields, hover for a time, then drop onto their prey. They are often seen while hunting from perches such as power and telephone lines along roadsides. Mice and insects, particularly grasshoppers, are its mainstays. This attractive little falcon nests in cavities, and will readily accept manmade nest boxes placed in suitable habitat. Because of fuelwood harvest, agricultural expansion, etc., trees with suitable nest cavities are becoming more difficult to find. Nesting sites are of **critical importance** for Kestrels, Screech Owls, Bluebirds, and others — if they are to remain a part of our wildlife scene.

Owls

Twelve species have been found in Minnesota. While Owls are noted for their silent flight and their nightime activities, some are daytime or crepuscular (active early morning and at twilight) hunters. The Snowy, Short-eared, and Hawk-owls are examples of crepuscular types. An owl's eyes are so efficient that it can discern objects in less than 1/10th the light a human eye requires. In some species, almost 1/100th the amount is sufficient. Unlike our eyes, the owl's eyes are fixed in their sockets. In order to change its field of vision the owl must turn or move its head. Often seen is the bobbing motion of the head, which enables the bird to get better depth perception on an object.

The owl has a second prey-locating adaptation — the placement and design of its ears. Its ears are not flesh appendages, like a human's, but are holes covered by a flap. These are locations at the edges of the facial disk. One ear is slightly larger and is located either higher or lower than the other. This allows the bird to pinpoint the source of a sound. The stiff feathers of the facial disk catch sounds and funnel them to the ears.

Owls are predatory birds that substitute patience for flight. Although the owls' (and for that matter all raptors) taloned feet are quite strong, the beaks are a fair bit weaker than imagined. In all owls except the Burrowing and Barn owls, feathers extend all the way down to the feet and to the tops of the toes. These feathers offer warmth and ensure silent flight when hunting prey. Owls (as with other raptors) eat their prey head first, swallowing smaller prey whole. All undigestable material, (such as bone, feathers, or fur) is later regurgitated in the form of a pellet. Owls do not build their own nests, but use abandoned hawk, crow, raven, or squirrel nests. Many owls nest in tree cavities. It is of interest to note that many owls are prey "specialists" — being dependent on one particular type of prey. This prey is usually small mammals such as mice, voles, shrews, and lemmings. Prey specialists include Short-eared, Long-eared, Great-gray, Barn, and to a greater extent the Snowy, Boreal, Saw-whet, Hawk-owl, and Burrowing owls. The prey specialists living and nesting success depend on adquate numbers of mice or mice-like prey.

The Barn Owl is on the northern edge of its range and is quite rare in Minnesota. This medium-sized owl, has small dark eyes, long legs, and a white, heart-shaped face. It nests in tree cavities and in buildings such as barns and silos. Only four nests have been recorded in Minnesota, the last being in 1963. Because this species has approximately 8 percent less body fat than other owls, it literally can freeze to death in colder climates. The Barn Owl is a very efficient mouser, and has been found to catch prey in total darkness, hunting by sound alone. A study of 200 pellets of a pair of Barn Owls that nested in the rafters of the Smithsonian Institute in Washington, D.C., revealed the remains of 225 field mice, 179 house mice, 20 rats, and 20 shrews. Due to the loss of nesting and hunting habitats, this bird is having trouble throughout much of its North American range.

The **Great Horned Owl** is our most common owl. It is considered a prey "generalist" — taking whatever prey is most available. Its prey includes small mammals, birds, large insects, reptiles, amphibians, or even other owls or hawks. Great Horned owls are found throughout Minnesota and for the most part, are non-migratory. The "horns" are actually tufts of feathers and no explanation has yet been found for them. They may possibly help to camouflage the bird. As with nearly all raptors, the female is larger than the male. The female Great Horned has a wingspan up to 62" (males up to 56") and may weight up to 4¼ pounds (males up to 3½ pounds).

The **Barred Owl** is named for the brownish streaks or "bars" on its chest. Its dark brown eyes and main call (which sounds like "who-cooks-for-you. . . who-cooks-for-you-all") are distinctive. Major habitat for this owl is a mature lowland hardwood forest near a stream, river, lake, or large pond. It nests in tree cavities, and a research study by this author has found them to nest in man-made nest boxes as well as older woodduck boxes without tops. The main prey items of the Barred are small mammals, small birds, minnows, crawfish, large insects, and frogs.

Few people get a chance to view the **Long-eared Owl** because it is strongly nocturnal. Its coloration and long, slender body make it difficult to locate when perched. In winter, Long-ears generally move southward, although some remain within the State. Communal roosts containing up to 20 birds are not uncommon in winter. These roosts are usually located in dense hardwood, or more likely, a dense conifer thicket. Of 2,763 prey items identified, 98.7 percent were small mammals (mice, shrews, voles, etc.), and 1.3 percent were birds. These owls are not known to nest in the prairie regions of the State. They can be found in the more heavily wooded portions of the State. Females may be up to 13" tall and may weigh up to ¾ of a pound. Again, the ears or horns are merely long feathers.

The Short-eared Owl is an open-country bird of the marsh and prairie. Like the Long-eared, it regularly migrates south for the winter and forms communal roosts. This owl has a wingspan of up to 3½ feet and is about 12 inches tall. It's "ears" are small, barely visable feather tufts that either stand up or can be flattened to blend with rest of its plumage. Short-ears perch on the ground or on fence posts in the midst of a marsh or prairie. If flushed, they fly a short distance and land again. Their wing beat is similar to a harriers. Short-ears, as already mentioned, are prey "specialists". The populations of their main prey, small mammals (particularly voles) are cyclic in nature. Cyclic populations tend to build in numbers for a few years, and then will "crash" to a low point, only to start building again. In spring, Short-ears arrive on their breeding grounds. If there is a "low" in the prey population, a few of the owls may remain behind and attempt to nest, or they all may go onward — searching for areas with higher numbers of prey. This behavior appears to be somewhat true of all the small mammal specialists. Due to rapidly diminishing marshlands and prairie areas, this owl is listed as "threatened" by the Minnesota Heritage program.

The **Screech Owl**, a small raptor, is about 8 inches tall, with a 22 inch wingspan. Its body is gray, brown or red. It's feather tufts are pominent. It nests in cavities — either in trees or in man-made structures such as woodduck boxes. The Screech Owl is found mostly in the Southern half of the state in small woodlots, orchard, on the outskirts of small towns, and even in some larger city parks. It's most often heard call — one long, descending quavering note — is distinctive. This little owl eats insects, small rodents, and small birds.

Our smallest owl, the **Saw-whet Owl** is more common than most people realize, although its nocturnal habits make it seldom seen. The Saw-whet is a fairly tame owl and can usually be approached rather closely. Its diet is mainly insects and small mammals. It nests in tree cavities but also has been found nesting in woodduck boxes. Normal nesting range is the northern forests, but has also been seen during the summer months in the southeastern part of the state. Common call is a long series of short whistles. The Saw-whet is somewhat migratory, although some are found throughout the year in Minnesota.

Slightly larger than the Saw-whet, the **Boreal Owl** is normally a rare winter visitor, although it has been found nesting twice in the Gunflint Trail area of northeastern Minnesota. It can be distinguished from the Saw-whet by its light bill, black facial border, and chocolate streaking of its underparts. As its name implies, it prefers the coniferous forests of the north. Like the Saw-whet, it nests in cavities, most often those made by woodpeckers.

The **Hawk-owl** has been found nesting five times in Minnesota. It is a tame diurnal (active in the daytime) owl of the damper northern coniferous forests. The long slender tail gives this bird a falcon-like appearance. It often sits on the tip of a tree and jerks its tail like a Kestrel; raising it and slowly lowering it. Hawk-owls are medium-sized owls, with a wingspan of around 33 inches. They nest in tree cavities or on the tops of broken off snags.

The **Burrowing Owl** once was a regular nest on the prairies of western Minnesota. Massive habitat loss due to agricultural expansion has all but eliminated this bird from the state. In a 1965 study, 46 nests in 21 counties were documented from 1881 to 1964. Since 1964 only 4 nests have

Burrowing Owl

been found. These owls are long-legged and active in daytime. They nest in prairie dog or badger holes. Their diet consists of insects, mice, and occasionally small birds. They are migratory, wintering from the southern U.S. all the way down into South America. The Minnesota Heritage program lists this species as "endangered" in the state.

Another prey specialist, **The Great Gray Owl**, hunts for voles and lemmings in the northern Minnesota forests. Large tamarack and ash swamps are its preferred habitats. Although the Great Horned Owl weighs more, the Great Gray is larger (wingspan up to 5 feet). Its bluish-gray body, long tail, and yellow eyes surrounded by prominent gray concentric circles of the facial disk are good identifying field marks. Its voice is a deep, booming series of whoos, each lower in pitch.

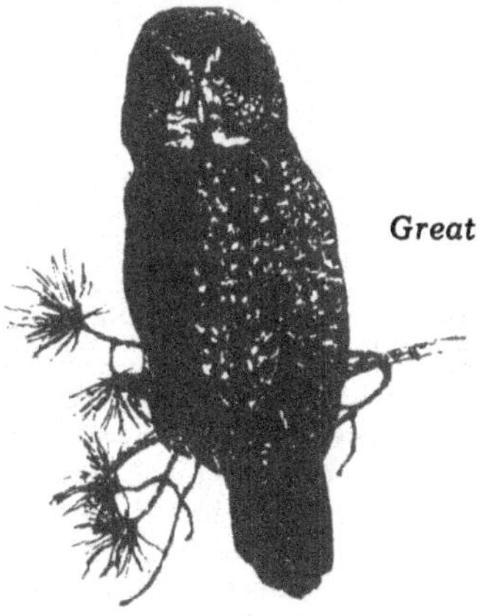

Great Gray Owl

The **Snowy Owl** is the only owl that has not nested in Minnesota. It normally nests on the treeless tundra of the far north. Lemmings (mice-like rodents) are its main diet. Each winter a few of these owls are seen in the state and about every fourth year, when the lemming population crashses, many more owls are forced to move southward. The male Snowy is almost pure white, while the larger female is white heavily marked with black flecks. It perches on or near the ground in open country and often allows birders to approach closely. The large size, whitish plumage, and lack of ear tufts make this beautiful owl easy to identify.

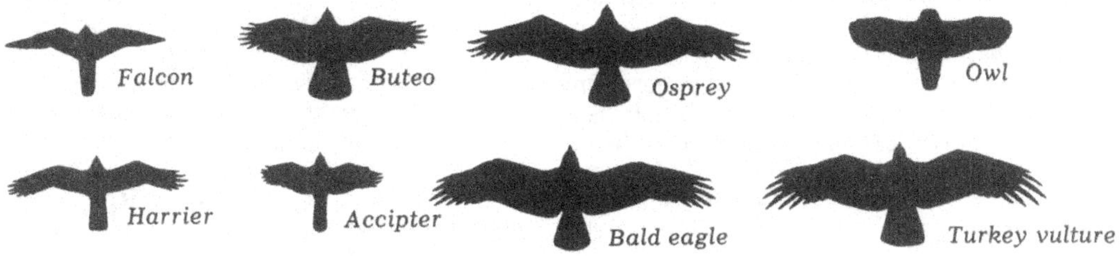

THE AMERICAN BALD EAGLE

by

David H. Johnson

The eagle represented power and majesty for thousands of years before John Adams, Thomas Jefferson, and the other founding fathers chose it as our national symbol in 1782. At the same time our republic was formed, as many as 25,000 soared through the skies over the timber-boarded lakes, streams, coastlines of what is now the lower 48 states. Today the Bald Eagle is listed as a "threatened and endangered species" south of the Canadian border with a population estimated at little more than 1100 breeding pairs — a frightening drop from the thousands which once inhabited American wilderness. As the wilderness began to vanish; so did the eagles.

Before "civilization" moved in, eagles ruled unchalleneged in the wild. As one of the largest and strongest birds of prey, the adult bald eagle has few natural enemies. But the birds do require particular conditions in order to live. In the past, nature has always provided the environment eagles needed to survive, breed, and maintain their numbers. The life cycle of the bald eagle reflects these special needs. This cycle hasn't changed over the years. What has changed is the land — the wild areas and particular habitats the birds need to survive.

Lakes and rivers that were once bordered by eagles' nests became surrounded by vacation homes. As our leisure time increased, we flocked to the Nation's waterways for recreation. And unwittingly, we drove the bald eagle from its home. As we cleared land for farms and roadways, and cut down forests to get lumber for houses, we took the trees in which eagles built their nests.

"Progress" created other hazards. Eagles are raptors — birds of prey that eat mostly fish, but also waterfowl, and small mammals. They're at the end of a complex food chain that's based on water. But many of our waters have been contaminated over the years by agricultural and industrial chemicals such as DDT, Dieldrin, PCB's and Mercury. Fish and waterfowl which live in this polluted environment absorb these poisons by eating chemical-laden water plants and smaller animal organisms. Bald eagles feed on the fish and waterfowl and become victims of what's turned into a pollution chain.

Other very serious situations the eagles are surrounded by are: (a) The ingestion of lead shot, such as can be found in crippled waterfowl — one lead shot pellet if digested can kill an adult eagle, (b) the deliberate shooting of the eagles by irresponsible and misinformed persons, (c) the indiscriminate shooting of flying creatures by "for-the-thrill-of-it" shooters, (d) and the innocent, well-meaning, but uninformed birdwatchers and wildlife photographers.

But there is a slightly optimistic side to the Bald Eagle story. There's an impressive amount of scientific data on bald eagles. Research studies and management surveys have generated the kind of intimate understanding that's necessary to save this magificent bird for future generations.

Today, nesting surveys are conducted periodically by the U.S. Fish and Wildlife Service with assistance of State and other Federal agencies, conservation organizations, and a number of private citizens. Starting in January 1979, the Annual Midwinter Bald Eagle Survey was begun. This survey, usually held around mid-January, is designed to show the number of eagles that winter in the lower 48 states and eastern Canada. In 1980, 12,323 bald eagles were reported. These observations of nesting and wintering eagles provide scientists and researchers with valuable information on population status.

Long before there was a general concern over the eagle's decline, a number of states moved to protect the bird. Then came the Bald Eagle Protection Act of 1940, which made it illegal to disturb, harass, and kill bald eagles. This was the first Federal step taken to preserve the species.

Since 1940, more and more has been learned about the bald eagle. This accumulation of knowledge has paid off. In some areas, bald eagles have recently, increased in numbers. In other regions, the eagle populations have stabilized after long periods of gradual decline. While the overall situation appears optimistic, the picture in some areas is not good — nor is it likely that the eagles can be saved in such areas where man/eagle conflicts are great.

Private organizations such as the National Wildlife Federation and the Nature Conservancy have purchased critical eagle habitat. The government also protects and purchases eagle territory, and many private property owners have successfully maintained eagle habitat and nesting sites.

Some other factors benefiting the eagles are:

—The banning of such pesticides as DDT and Dieldrin and their gradual elimination from the environment.

—Research studies, providing the hard scientific data necessary to influence changes in the laws regulating toxic chemicals.

—The reduction in the amount of trapped and shot birds, through rewards, stiff penalties, and increased public education and understanding.

—Specialized Raptor Rehabilitation clinics, such as the one at the University of Minnesota where injured birds from all over the country are treated and nursed back to health. Often, when the injuries aren't too servere, the rehabilitated birds are returned to the wild.

—Captive breeding programs and transplanting of captive or wild-reared eggs or young eagles to suitable areas elsewhere. Thus far, Minnesota has given 9 young eagles to the state of New York, for repopulation programs there — and its working.

THE EAGLE FACT-FINDER:

—Along with wintering adults, juvenile eagles roost in communes near a good food source.

—As other raptors, the male is smaller than the female. The adult female is about 42 inches long and weighs 10 to 14 pounds. The male is about 35 inches long and weights 8 to 10½ pounds.

—The female's wing-span is about 8 feet; the male's is 6½ feet.

—The name "bald" is a misnomer. It was given to the white-headed eagle during the 17th century. The birds scientific name, Haliaeetus leucocephalus, means "white-headed sea eagle".

—The bald eagle can be found in every state except Hawaii. The largest nesting populations are in the Great Lake States, Florida, and Washington. Alaska is the only state where the bald eagle population is thriving.

—Eagles probably live as long as 30 years in the wild, and up to 50 in captivity.

—Their eyesight is 5 to 6 times sharper than a humans.

—Both members of the eagle pair will alternate in incubating the approximate 35 days it takes for the 1 to 3 eggs to hatch.

—Human disturbance near the nest site may cause the abandonment of the nest, or leave the young vulnerable to severe weather and predators.

—The young eagles develop most of their feathers at 3 to 4 weeks, can walk at 6 to 7 weeks, and begin to fly at 3 months of age.

THE PEREGRINE FALCON
POPULATION STATUS AND LIFE HISTORY

In many regions south of the Arctic, populations of peregrine falcons have declined to a point wherein the species no longer exists as a breeding bird in formerly occupied habitat.

In studies conducted during the late 1930's and reported in the early 1940's, 408 eyries (nesting sites) were recorded in the eastern United States, Canada, Labrador and Greenland. During this period a tentative estimate was made that 350 breeding pairs of peregrines occupied the eastern United States. These nesting birds were historically located in the mountainous regions of the Appalachian Range south to Georgia. Major waterways such as the Upper Mississippi River region also supported local populations of peregrines. The last sites occupied, probably by single birds, were documented in 1963 in Virginia and 1965 in the Upper Mississippi River Region. In these areas the usual nest site consisted of a ledge on a rock, cliff or escarpment that provided a clear and commanding view of the surrounding terrain. The nest was usually a scrape or depression in gravel on the ledge.

Typical of the peregrines' knack for adapting, tree nesting pairs were also reported during the 1800's along the Ohio River and South. On the Sunlife Insurance Building in Montreal, Canada, peregrines nested for many years until the 1940's.

Juvenile

Peregrines are medium-sized hawk-like birds with long, pointed wings and a long tail. They generally reach adulthood at three years of age. Usually, the male arrives first at a cliff site in February or March and goes through a series of aerial acrobatic displays to attract a mate. An average clutch of four eggs is laid in late March and early April, hatching in about 33 days. Peregrines vigorously defend their young and nest site, although, if severely and continuously harrassed, they may attempt to relocate or completely abandon the site. An average of less than two young reach fledging (flying) age per laying pair. Juvenile peregrines are most vulnerable to injury or mortality during their first year of life. During this period, they are developing their flying skills and learning to hunt. As with many other species, first year mortality is much higher than in subsequent years.

Peregrines feed primarily on medium-size bird species such as starlings, blue jays and pigeons. Descending from high above its prey, the peregrine will "stoop" (dive) and strike its quarry in mid-air. Tremendous speeds are achieved during the stoop, and prey is either struck to the ground or killed out-right by the blow of the falcon's talons. Some investigators have estimated the speed of a stooping peregrine to be more than 200 mph.

Adult

According to band-recovery records, the peregrines of southern Canada and the United States to the south are weakly migratory or nonmigratory. The former eastern peregrines were also virtually non-migratory, moving primarily east from the mountains to the Atlantic Coast during the winter. The northern breeding populations from Alaska, Canada, and Greenland are highly migratory, traveling great distances from their breeding sites southward along the east coast and down major drainages to the Gulf of Mexico and into Central and South American countries. Coastal and sand barrier islands which support a food source are essential to fall and winter survival of the peregrine.

REASONS FOR DECLINES

It was not until the late 1940's and 1950's that scientists realized peregrine falcon populations were in serious trouble in North America and Western Europe. In 1965, an international conference was held at the University of Wisconsin to document the extent and examine the causes of the massive failures of nesting peregrines. This conference stimulated much of our present natural interest in and concern for the peregrine.

By the late 1960's the relationship of pesticides to the

decline of the peregrine was becoming clear. Chlorinated hydrocarbon pesticides, particularly DDT and a breakdown product called DDE were causing eggshell thinning. During normal incubation by the adults, eggs were breaking causing mortality of the embryo and completely curtailing the breeding success of peregrine populations. Broad application of these pesticides, used to control insects, eliminated the peregrine falcon in the eastern United States because many of the insects were eaten by birds which in turn were killed and eaten by the falcons.

Use of these pesticides has been reduced in the United States, but those used in other countries are still having an effect on some peregrine populations. Some pesticide monitoring programs indicate a significant reduction of residue levels in the North American environment. Although we are not out of the woods, the indications are good for the future and the time is right for a recovery attempt.

LEGAL PROTECTION

There are a number of federal and state laws protecting raptors. The most important federal laws are the Federal Environmental Pesticide Control Act, the Endangered Species Act, the Migratory Bird Treaty Act, and the Migratory Bird Conservation Act. Raptors, including falcons, were placed on the Migratory Bird Treaty List by amendment in March of 1972. Many states have also enacted their own endangered species laws as well as pesticide regulatory statutes. Presently there stiff fines and jail sentences that can be imposed for taking, harassing, injuring, or killing and endangered species.

RESTORATION

Under the Endangered Species Act of 1973, the Director of the U.S. Fish and Wildlife Service established Recovery Teams to recommend those actions necessary to restore the peregrine to its former range in the United States. To date, there has been a blend of private, public and governmental interest in ensuring that this magnificent global wanderer is guaranteed its right to exist.

Since the 1970's much effort has been devoted to the captive propagation of peregrines as a means to preserve the species and provide young stock to be used for restoration of the falcon to its natural habitats in the United States. By 1976, the means had been achieved to annually produce 50 young peregrines in captivity, and release these young falcons to the wild. Along with providing young in the hope of establishing a wild breeding population, emphasis has also been placed on protecting necessary nesting and migration habitat. In 1980 two pairs of the captive produced falcons produced four young of their own in New Jersey. This is regarded as a major success by all concerned with the well being of the peregrine. **The future is indeed promising.**

For further information, contact:

Washington
U.S. Fish and Wildlife Service
Department of the Interior
Office of Endangered Species
Washington, D.C. 20240

LOOK AT SHOREBIRDS!

Marvels of adaption, shorebirds open our minds to nature's mysterious arrangements.

I almost stepped on it! There, in the corn row at my feet, lay a fuzzy, newly-hatched killdeer. The brown and black markings had made it hard to spot against the rich topsoil. Overhead, three killdeer scolded and dived at me as I watched the chick walking on toothpick legs. One killdeer landed nearby and tried to lead me away with its famous broken-wing act.

That delightful experience happened when I was ten years old, but the memory of my introduction to the killdeer is still vivid. In succeeding years, killdeer were constant companions who always brightened the monotony of field work on our family farm.

The killdeer is just one of many shorebirds — including plovers, godwits, curlews, phalaropes, avocets, and sandpipers — which occur in Minnesota. Their natural history, extraordinary migrations, unique behavior, ease of observation, and tenuous status in the face of intensified land use make them important to an ecologically concernd public as well as to birdwatchers and ornithologists.

Plovers

The killdeer is probably the best known shorebird in Minnesota. Its plaintive call and adaptability to a variety of nesting sites make it a favorite of even the most casual birdwatcher. These hardy birds are among the first migrants to return in the spring. In addition to the killdeer are several other plovers: piping plover, semi-palmated plover, American golden plover, black-bellied plover, and ruddy turnstone.

Piping plovers resemble pale, miniature killdeer. They prefer to nest on undisturbed sandy beaches in the northern half of the state. As with killdeer, piping plovers use the broken-wing act to protect their nest and young from predators. By fluttering their secondary wing feathers and pushing themselves along the ground, they resemble crippled young plovers. Suckered by this act, many a hungry fox has been led away from a plover's nest.

One Sunday afternoon in the spring of 1976, I sat with my family and some friends on a large buffalo rubbing rock on the Lac qui Parle Wildlife Management Area south of Appleton. The prairie setting around us was highlighted by flocks of golden plovers appearing in the southern skies. Some passed right over our heads as they followed the valley of the ancient River Warren to their nesting grounds in the Artic. For thousands of years, plovers have passed over this rock each spring. Imagine what the scene was like when the rock was surrounded by buffalo instead of people!

Golden plovers winter in the grasslands of Brazil and Argentina. Their entire round trip covers 15,000-20,000 miles. These birds can best be seen from May 5 to 15 in central and western Minnesota as they search for insects in pastures and on burned grassland.

More uncommon than the golden plover, the black-bellied plover is usually encountered on mud flats and lakeshores. Arthur Bent described it in his book, **Life Histories of North American Shorebirds:**

"The black-bellied plover is an aristocrat among shorebirds, the largest and strongest of the plovers, a leader of its tribe. It is a distinguished-looking bird in its handsome spring livery of black and white; and its attitude, as it stands like a sentinel on the crest of a sand dune or on some distant mud flat, is always dignified and imposing."

Another beach bird is the semi-palmated plover which is similar to a miniature killdeer. Small flocks spread out and run in various directions — heads up, pausing, pecking at food. In contrast, sandpipers usually keep together, heads down, scampering back and forth at the edge of the waves in search of food.

Harlequin markings of the ruddy turnstone distinguish from other plovers. The name "turnstone" is appropriate. It comes from the bird's habit of overturning stones, shells, and dirt clods as it searches for the small creates which comprise its diet. By pushing with their breasts, ruddy turnstones overturn large stones and clods.

Larger Shorebirds

Although many shorebirds are small and non-distinctive, several are large and impressive. The whimbrel, for instance, has a large brown body, striped head, and long downcurved bill. The Hudsonian godwit has a slightly upturned bill. Birdwatchers identify it by the broad white band at the base of its black tail. Frequently it feeds in deeper water, probing with its head submerged. The migration route of the Hudsonian godwit is roughly similar to that of the golden plover.

Avocets normally prefer the alkaline wetlands of the central United States, but they occasionally nest in western Minnesota. The Salt Lake Wildlife Management Area in Lac qui Parle County is a likely place to see them. They are characterized by bluish legs, a black and white body, and a slender upturned bill.

The unusual upturned bill is an adapatation for the avocet's particular style of feeding. While wading in shallow water, the bill is held below the surface. The lower curve of the bill nearly touches bottom. Then the bird slowly runs, sweeping its bill from side to side to capture its aquatic prey.

Among the earliest migrant shorebirds to appear in Minnesota each spring are the yellow-legs — greater and lesser. Both species have bright yellow legs and white rumps and tails. The greater yellowlegs is the size of a willet and has a distinctive three-note call. The more abundant lesser yellowlegs is the size of a killdeer and has a two-note call.

While large shorebirds may capture your attention first during a wetland visit, the phalaropes will first capture your imagination with their beauty.

Dr. Edward Nelson's description of the phalaropes is as appropriate today as when it was written in 1887:

"Amid the loud tongued multitude (of ducks and geese) suddenly appears the graceful fairylike form of the northern phalarope, and there, riding on the water, scarcely making a ripple, floats this charming and elegant bird. It glides hither and thither on the water, apparently drifted by its fancy, and skims about the pool like an autumn leaf. The delicate tints and slender fragile form, combining grace of color and outline with a peculiarly dainty elegance of motion, render this the most lovely and attractive among its handsome congeners."

Wilson's phalaropes nest in western Minnesota. They look like feathered wind-up toys as they spin around and around in the shallow water. Their pattering feet stir up tiny aquatic morsels.

Also interesting are the reversed domestic roles of male and female phalaropes. The females are brightly colored, and the males are drab. After laying eggs, the female remains near the nest, but it's the male that incubates them.

Grassland Shorebirds

While phalaropes steal the show in the wetlands, several shorebirds dominate the scene on the surrounding grassy upland: long-billed curlew, willet, buff-breasted sandpiper, upland sandpiper, and marbled godwit. The upland sandpipers and marbled godwits remain to nest in Minnesota while the others pass on to more northerly nesting areas.

Marbled godwits must have a lasting impression on the first settlers in western Minnesota. In **The Birds of Minnesota,** T. S. Roberts described a trip he took in 1879 to Grant and Traverse Counties:

"The great marbled godwit was so abundant, and so insistent in its attention to the traveler on the prairie, and so noisy that at times it became an actual nuisance. They were continually hovering about the team, perfectly fearless and nearly deafening us with their loud, harsh cries — go-wit. Now and then the birds would all disappear and peace would enuse for a brief period, but they had only retired to muster their forces anew, for shortly a great company would bear down upon us, flying low over the prairie, and spread out in a wide array, all the birds silent, until, when almost upon us, they swerved suddenly upward over our heads, and broke out in a wild discordant clamor."

(Perhaps that is why a friend once referred to these birds as "garbled halfwits.")

The upland sandpiper is more widespread and numerous than the marbled godwit. Grassland hikers in central and western Minnesota can hear its warbling call and occasional wolf whistle.

The long-billed curlew's seven-inch bill is an unmistable characteristic of this striking, but uncommon species. It no longer nests in the state, but still finds suitable nesting habitat on the vast prairies of the Great Plains and central Canada.

The Eskimo curlew once migrated in great flocks through Minnesota, but it has been rare since the turn of the century and is now near extinction.

The large gray willet feeds either on the prairies or in wetlands. A broad white stripe on its black wings makes it easy to identify. This bird used to nest in Minnesota but it, too, has virtually vanished.

The willet could never regain its former numbers, but it is conceivable that wetland acquisition and management activities of the Department of Natural Resources, Fish and Wildlife Service, and The Nature Conservancy have helped restore some nesting habitat for these birds. Willets have been observed several times in recent summers.

Buff-breasted sandpipers resemble miniature upland sandpipers, and may sometimes be seen feeding with golden plovers. They migrate theough the state in May and again in late August.

Since 1918, grassland shorebirds have been protected by federal law, but the grassland itself has not been protected. As a result, the plow has probably eliminated more grassland shorebirds from the state than market shooters did in the 19th century. The future of our prairie fauna and flora lies with the preservation and management of remaining grasslands.

Marshes host an overwhelming variety of wetland sandpipers. Only one, the spotted sandpiper, commonly nests in Minnesota. The bird has consipicuous dark spots on its breast. A nickname, "teeter snipe," has been used for the spotted sandpiper because it constantly bobs its rump as it runs along a beach. Even newlyhatched chicks are rump bobbers.

The smallest sandpipers, collectively referred to as "peeps," include the least, semi-palmated, western, and Baird's. While it is difficult to distinguish these tiny species, that should not prevent someone from enjoying the birds as they busily glean aquatic life from a beach.

Least and semi-palmated sandpipers are the most abundant peeps, with the latter being more common. While least sandpipers gather in five to 20-bird flocks, semi-palmated sandpipers cluster in 20 to 100-bird flocks. Impressive in flight, they twist and turn in unison, alternately displaying their white breasts and gray backs.

Several medium-sized migrant sandpipers are no less interesting.

The solitary sandpiper nests in Canada in abandoned tree nests of robins and grackles. All other American shorebirds nest on the ground.

The slender pectoral sandpiper follows a migrantion route from Alaska and the Northwest Territories through Mexico and middle America to southern South America. Not to be outdone, the white-rumped sandpiper uses an Atlantic-coast route similar to that of the golden plover and migrates all the way to Tierra del Fuego, 1,000 miles beyond the destination of the pectoral sandpiper.

Sometimes referred to as the "timberdoodle," the American woodcock is probably best known for its remarkable spring courtship rites. At sunset, the male woodcock can be heard making a loud frequent buzzing note and sometimes it can be heard at close range making a soft guttural note that resembles the croaking of a frog. He may be strutting around in a small opening with his tail spread and erect, and with his bill pointing downward and resting on his breast. Suddenly the bird will take flight. Rising high in a spiraling flight, the woodcock will reach a height of 200 to 300 feet while making a muscial twittering call. The remarkable flight is climaxed by an earthward dive, highlighted by three repetitions of a call that sounds like "chickaree." The male returns to the starting point and again starts the buzzing call.

Woodcocks are highly regarded game birds that hold well to a pointing dog and offer excellent sport in some of Minnesota's forested bottomlands.

Top: Greater Yellowlegs.
Center right: Wilson's Phalarope.
Bottom: Semipalmated Plover.

Another marshland bird, the snipe, is also characterized by an impressive spring courtship display. Taking flight in mornings and evenings, the snipe may rise to a height of 150 feet or more above a marsh or meadow, circling around in an ever widening orbit. At intervals it swoops downward and produces a fascinating "winnowinng" sound. The sound is produced by air rushing through the two outer tail feathers on each side of the outspread tail. The snipe is also a very challenging game bird which has humbled many hunters with its elusive, erratic flight.

The long bill of the snipe is used for probing for food in shallow water and muck, while the woodcock's bill is used to probe for earthworms in soft soil.

In conclusion, consider the sanderling. Peter Matthiessen, in his book **The Wind Birds**, aptly captured in words the fascination and mystique of the sanderlings — and all other shorebirds:

"The sanderling is the white sandpiper or 'peep' of summer beaches, the tireless toy bird that runs before the surf. Because of the bold role it plays in its immense surroundings, it is the one sandpiper that most people have noticed. Yet how few ever asked them why it is there or where it might be going.

"We stand there heedless of an extraordinary accomplishment: the diminutive creature making way for us along the beaches of July may be returning from an annual spring voyage which took it from Central Chile to nesting grounds in northeast Greenland, a distance of eight thousand miles. One has only to consider the life force packed tight into that puff of feathers to lay the mind wide open to the mysteries — the order of things, the why and the beginning. As we contemplate that sanderling, there by the shinning sea, one question leads inevitably to another, and all questions come full circle to the questioner, paused momentarily in his own journey under the sun and sky."

From "Ducks At A Distance"
United States Department of Interior
Fish and Wildlife Service.

From "Ducks At A Distance"
United States Department of Interior
Fish and Wildlife Service.

DUCK FOOD PLANTS AND MANAGEMENT

Soil Conservation Service—U.S.D.A.

Sago Pondweed
A choice duck food*

General: Commercial sources of the aquatic plants which you need to improve your pond or wetland for ducks may be found in the back of any popular hunting and fishing magazine. However, before embarking on any expensive purchases, follow these suggestions:

1. New areas - aquatic plants generally start growing in a new area within two years. Unless you are in a big hurry, give native plants a chance to start on their own. It may save you considerable expense.

Cattails are good muskrat food

2. Old areas - areas that have been flooded for several years may suddenly stop attracting waterfowl. A drought occurs, the bottom is exposed for a growing season and the area is again used by waterfowl when the water is replenished. The same results may sometimes be obtained if an area is drained for a season and then reflooded. Water level control structures may be used to advantage. See your Soil Conservation Service technician.

3. Know your plants. Make sure you are not purchasing plants already present in your marsh.

4. Before purchasing plants make sure the plants are not available in nearby potholes or marshes. A permit must be obtained from the Commissioner of Conservation, St. Paul, Minnesota, before you may gather or transplant aquatic plants in public waters.

See other side for aquatic planting guide.

Smartweed is choice duck food

Wild millet is a choice duck food

UNITED STATES DEPARTMENT OF AGRICULTURE
SOIL CONSERVATION SERVICE

THE RING-NECKED PHEASANT

Pheasants are rugged, colorful birds whose lives are a constant struggle for survival. How does this magnificent gamebird thrive in Minnesota? There are those who say it can use some help.

Originally imported from Asia, the ring-necked pheasant is today as all-American as football and the flag. Long pursued by gamebird hunters, the colorful ring-neck is popular with Minnesotans in cities and rural communities throughout the southern one-third of the state. But despite its popularity, the pheasant is steadily disappearing on the very farmlands where it was first introduced.

Fat and Fit Birds

Pheasants were first brought to the U.S. in George Washington's time, but their successful introduction did not occur until the early 1880s.

It all started when Judge Owen W. Denny, U.S. Consulate General in Shanghai, began raising pheasants until they were "fat and fit for my table." Denny decided to experiment with raising the bird in America. He shipped several dozen to his brother in Oregon, who released the birds in the Willamette Valley. Spectacular increases in the bird's numbers followed and son many other states were raising and releasing the Chinese import.

In 1905, conservations brought 70 pairs to Minnesota from nearby Wisconsin and Illinois, but these birds produced less than 100 young. Ten years later, the Minnesota Division of Game and Fish established a game farm for raising pheasants. By 1922, the rugged ringneck became established and was flourishing in Minnesota just as it had in Oregon.

During Minnesota's first pheasant season in 1924, hunters took 300 birds. Six years later, 184,000 hunters harvested 531,000 ring-necks. The following year, over a million birds were taken. In the last ten years, about 100,000 small game hunters have been harvesting 200,000 birds annually. Fall populations (both hens and roosters) have ranged from over five million in the early 1940s to 700,000 in the past few years.

A Dashing Oriental

Known scientifically as *Phasianus colchicus*, this Oriental member of the Order Galliformes (which includes other chicken-like birds — turkey, grouse, quail, etc.) is primarily a mixture of Chinese, Korean, and Mongolian subspecies, though it is predominantly Chinese.

The typical rooster pheasant weighs just under three pounds and is a beautiful blend of russet and copper with flecks of green, white and black. Crimson patches frame his eyes and violets and greens gleam in his black headfeathers. Usually, but not always, he has a white neck ring.

Much of his body is rich brown broken by bars and scales of black and white that merge into a saddle of delicate, powder-blue feathers. His magnificent tail, sometimes over two feet long, is cross-barred with black and brown. The juvenile cock pheasant's short, blunt, light-colored spurs become hard, dark, and pointed in his second autumn.

In contrast, the somewhat smaller hen has a short tail and is well camouflaged. The soft-toned brown and black markings fade into light buff and cream color on her underparts. A faint violet sheen colors her upper sides and neck.

When chased, a rooster often breaks into the air with a hoarse "Ca-ak! Ca-ak!". Once airborne, the pheasant beats its small wings more than three times per second. One rooster pheasant was radar-timed at 38 miles per hour entering a glide, and 48 m.p.h. top speed. Pheasants may fly a mile or more and, though normally leveling off at 25 feet, have been known to fly at 100 feet for two miles.

A Rugged Game Bird

The pheasant is well equipped to withstand Minnesota's weather extremes if the keys to its survival — adequate cover and food — are available. However, the state-wide habitat picture is variable and changing.

Found in both rural and suburban areas, pheasants are usually most abundant in farm country which contains a "mix" of row crops (corn, soybeans), small grains (oats, wheat), hay, and pasture. Unfortunately, modern intensive farming has greatly altered this ideal mixture. Today, vast expanses of corn and soybeans cover the land during growing season. Come winter, the land becomes a barren sea of plowed soil.

The ring-neck is able to resist shortages in its staple diet — corn, wheat, oats, soybeans, and wild plant seeds — and can survive a week without food, even in severe weather. Food, however, quickly helps it regain weight, strength and cockiness. In addition, the pheasant is probably the least susceptible of gamebirds to disease and parasites.

Despite their basic ruggedness and resistance to disease, most pheasants survive less than one year. Winter storms, especially sleet followed by high winds and plunging temperatures have in some years slashed the pheasant population in half during a two-day period.

Pheasants are more accident prone than most birds. Highspeed haymowers take a deadly toll of pheasants in alfalfa fields and other nesting places. Some birds die in collissions with cars, especially in early morning or late after-

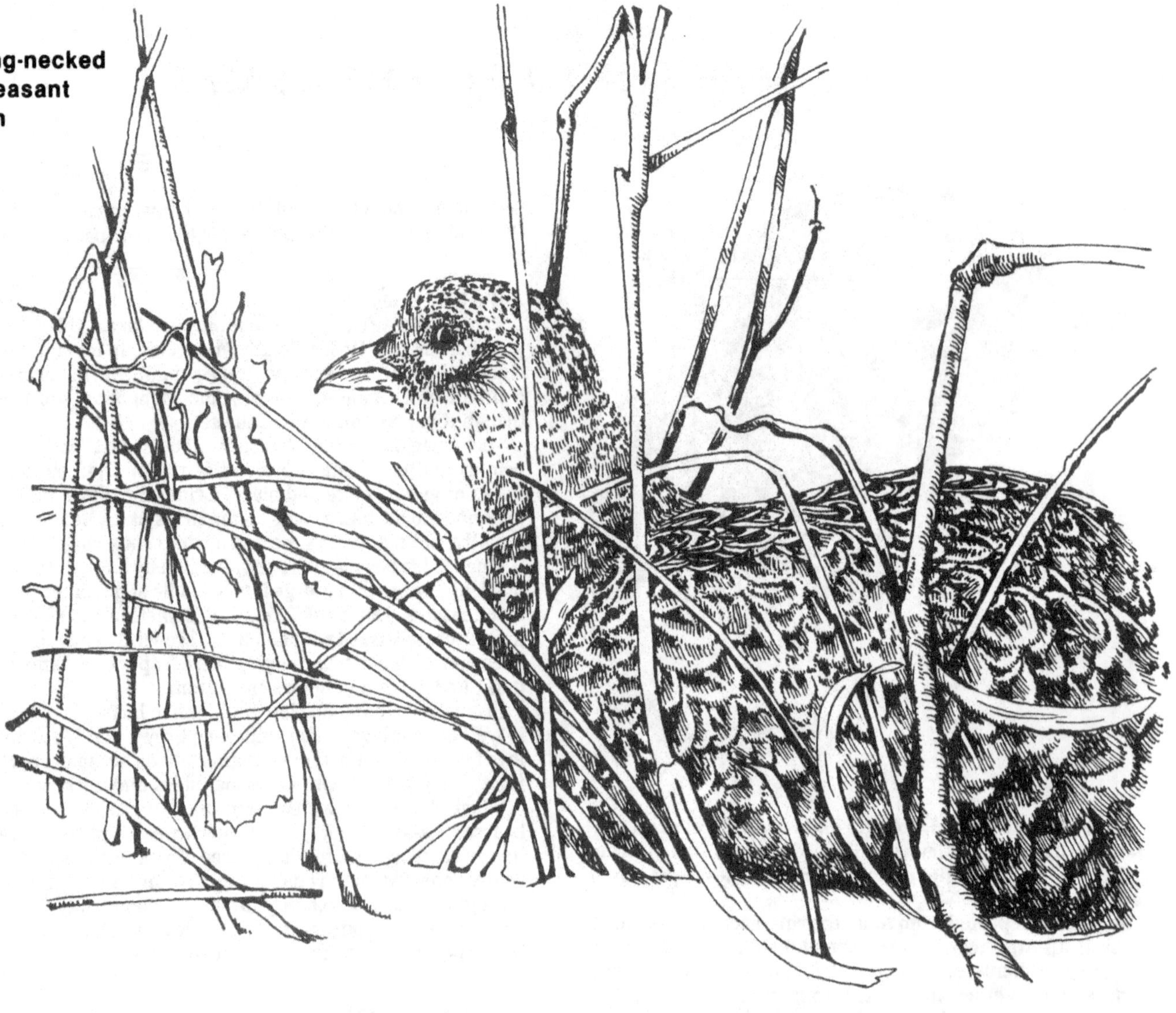

Ring-necked pheasant hen

noon. Predation by owls, foxes, and skunks are not a major threat to established pheasant populations.

Pheasant Numbers

Though plain in appearance compared to the rooster, the hen pheasant is the key to a stable or expanding population of ringnecks. A continuing supply of young is assured only when a plentiful supply of hens survives the winter in safe nesting cover. The number of roosters that survive winter is much less important. One rooster for every 15 hens is ample for reproduction.

To insure a supply of hens, Minnesota has maintained a closed season on hens for many years. Even so, the pheasant population has declined over 90 percent because of a comparable 90 percent loss of safe nesting cover.

Pheasants usually establish their nests during late April or early May. From two to 18 eggs (average 12) are laid over a period of days. A clutch of eggs requires 22-24 days incubation (warming by body heat). If the hen is forced to abandon the nest, or if the nest is destroyed, she will re-nest (lay another clutch) up to three times. Renesting can sometimes produce 50 percent of the year's chicks.

With favorable weather and abundant nesting cover, about 40 to 70 percent of the hends eventually raise a brood. The remaining hens are killed by haymowers, accidents, or predators, or may just be unsuccessful nesters.

Necessary Cover

Pheasant cover requirements differ throughout the year. Most important are safe nesting cover and winter cover.

Dense nesting cover is rapidly disappearing. This poses a serious threat to the future of the pheasant and to other ground-nesting wildlife such as rabbits, Hungarian partridge, and numerous songbirds.

Legumes or grasses provide cover for nesting, brooding, loafing, and escaping enemies. Early nesting pheasants usually establish nests in dry grass and sedges carried over from the previous fall. Later in spring, as soon as new plant growth is six to eight inches high, most hens begin nesting in alfalfa fields, grain fields, pastures, roadsides, and other grassy-type cover.

Today, because so much land has been converted to crops, the most important source of safe pheasant nesting cover in many areas of Minnesota is the unmowed roadside. Studies show that unmowed roadsides contain more than twice as many nests per acre as other kinds of nesting cover. It is important that nesting cover remain undisturbed for 35 to 40 days and not be mowed until after July 31st, if at all.

Without good cover, pheasants seldom survive a severe winter storm. During most winters, cattail marshes provide good cover. Other cover sources, in order of importance, in-

Pheasant Food

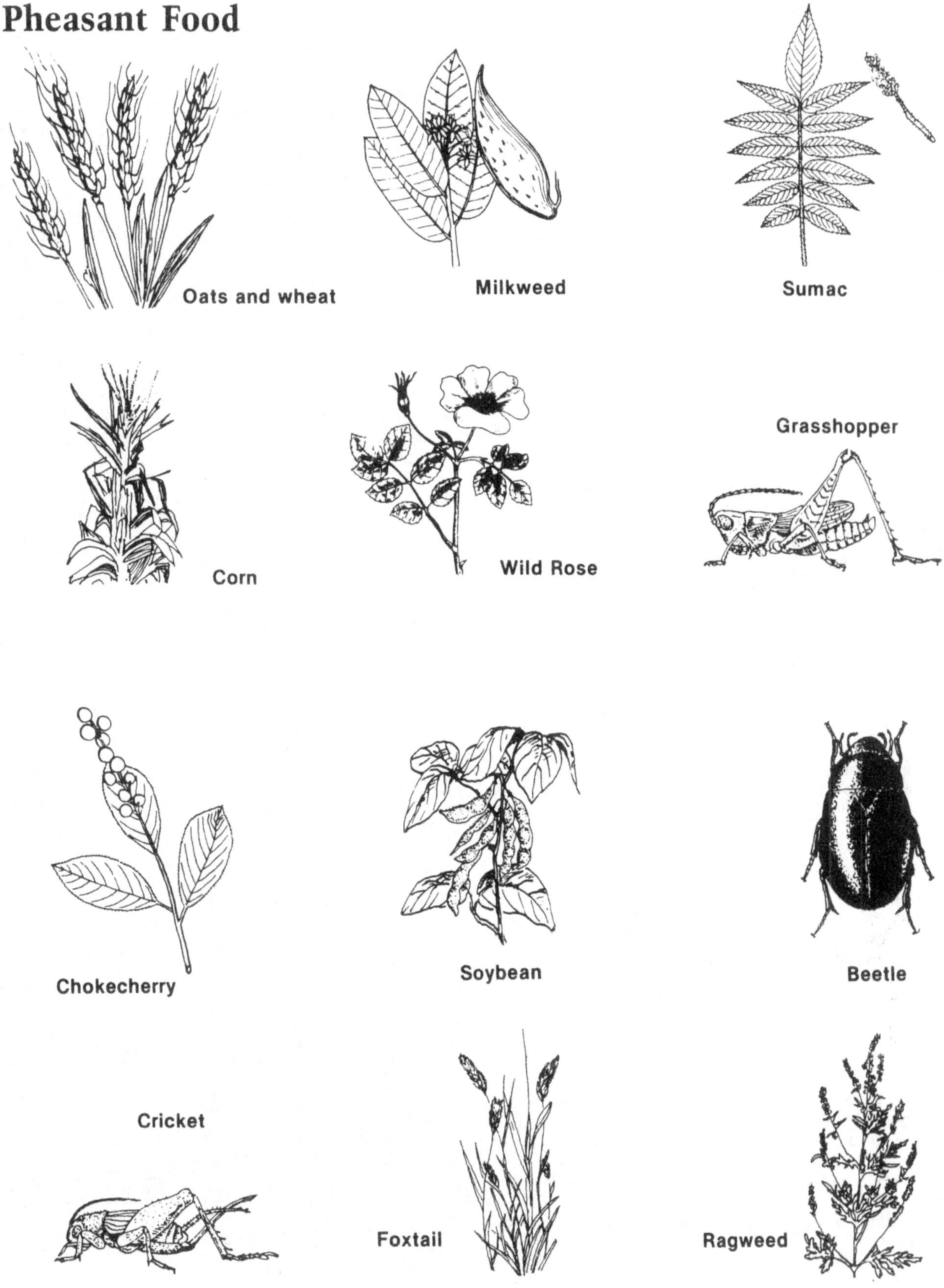

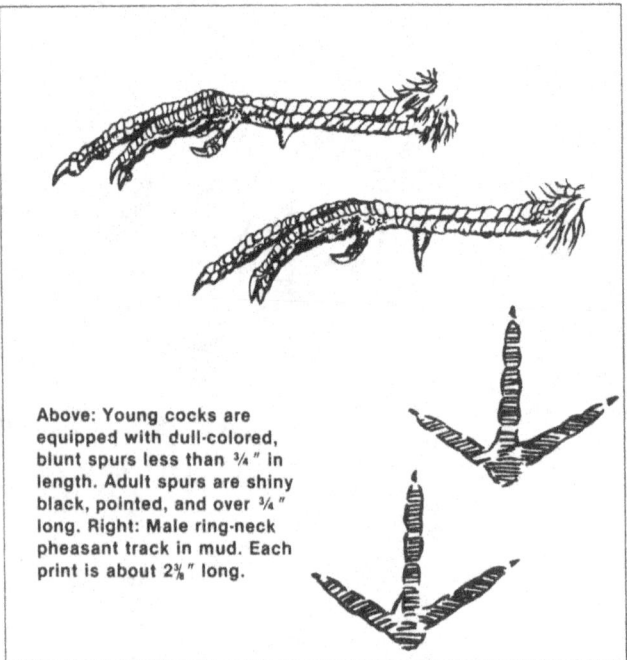

Above: Young cocks are equipped with dull-colored, blunt spurs less than ¾" in length. Adult spurs are shiny black, pointed, and over ¾" long. Right: Male ring-neck pheasant track in mud. Each print is about 2⅞" long.

clude brushy bottomlands, coniferous farm shelterbelts, small deciduous woodlots, and other areas of dense cover.

Many large cattail marshes have been drained or filled. So, pheasants and other farmland wildlife have been forced to depend more on farmstead windbreaks — rows of trees and shrubs and other woody cover for winter protection.

Well-designed windbreaks consist of ten or more rows of trees and shrubs of which at least four inside rows are conifers. These plantings, when 10 to 15 years or older, are large enough to provide shelter from wind-blown snow. During severe storms, narrower shelterbelts can become a death trap for wildlife seeking shelter from blowing snow.

Lack of winter food can be a factor which limits pheasant populations during severe winters, especially where fall plowing is extensive. Row crops such as corn and soybeans provide a sure source of winter food when left standing next to good winter cover until April 1st of the year following planting.

Hunting Facts

A rooster-only pheasant season has never been known to reduce the number of young born the following year. This fact is true regardless of the size of the pheasant population. About 90 percent of the fall rooster population is unnecessary for reproduction the following spring, but Minnesota hunters rarely take over 60 percent. Results: an underharvest of cocks, even during years of low pheasant population.

Fall rooster hunting is good management as far as the birds are concerned. However, we must also consider the hunter and the farmer. Part of the wildlife manager's goal is to distribute the birds fairly while providing maximum recreation. Several basic facts should be considered in this regard.

- Most corn is normally not harvested until the last week in October.
- Before this time, farmers are busy and most do not have time to hunt.
- Seventy-five to 85 percent of the season's kill and nearly an equal amount of hunting pressure occurs during the first nine days of a 30 to 40-day season.

If birds are plentiful, a large bag limit will allow plenty for all. If they are scarce, a smaller limit will give some assurance that the easy-to-get birds won't all be shot on the first weekend, thus distributing them among hunters over a longer period.

Wildlife managers recommend:

- A season on roosters every fall.
- A season as long as possible, regardless of the number of pheasants.
- The bag limit should be adjusted from year to year, according to the pheasant population, to distribute available roosters among Minnesota hunters.

Hunting Tips

The basic requirements for pheasant hunting are a small game hunting license, a shotgun, and access to pheasant country. An "almost" requirement is a good dog for locating birds and making retrieves. Labradors, springer spaniels, Brittany spaniels, and German shorthairs are commonly used. A well-trained hunting dog can double the satisfaction of the hunt and cut in half the number of cripples lost.

It takes a solid punch to put down a three-pound ring-neck. Few hunters will dispute the supremacy of No. 6 shot in a 12-gauge high base shell or, its equivalent, the 20-guage, three-inch magnum. Full-choked guns are best for the longer shots, but modified or improved cylinder barrels usually offer the best all-around combination, particularly when hunting with good dogs. However, gauge, shot size, choke, brand, and type of action are largely a matter of personal choice.

Early in the season, the birds scatter as singles or in small flocks. On mild days, they spend most of their time in standing corn, if it's available, but roost in hayfields or grain stubble. Grassy swales, idle acres, and wetland edges are frequent roosting areas for pheasants and worth checking early or late in the day.

As the season progresses, the birds seek heavier cover. A quiet approach on the downwind side of a heavy marsh, willow thicket, or woodlot can sometimes put a hunter in the midst of a flock of pheasants.

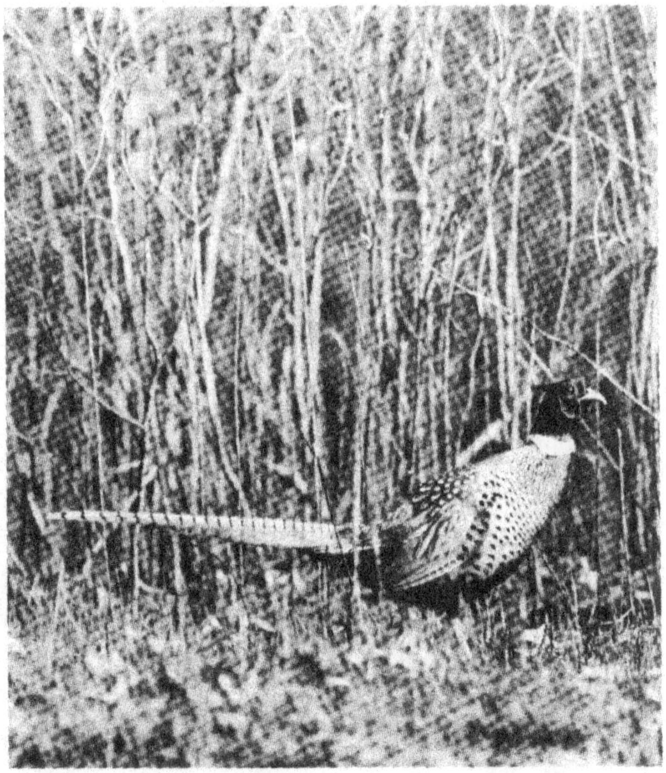

Pheasant hunting traditions vary considerably. Many people prefer to hunt large blocks of cover in large parties; others choose to hunt small patches of cover with only a dog.

Cooking Tips

Pheasant flesh is lean and thus dries quickly while being cooked. To reduce the loss of moisture from skinned birds, base with cooking oil or cover with strips of bacon before baking at 350° for two hours, or until the meat is brown and tender.

Another method is to cut a skinned bird into serving-size pieces and season with salt and pepper. Brown in butter or cooking oil and then allow to simmer in a covered pan or cook in a crock pot containing cream of mushroom soup. The quickest method is to de-bone the pheasant, cut into thin strips, and fry lightly.

Copy for this pamphlet was prepared by the Minnesota DNR Division of Fish and Wildlife.

Recommendations for Management of Ruffed Grouse Habitat in Northern Minnesota

Gordon W. Gullion

The ruffed grouse, or "partridge", has long been a resident of Minnesota's forests and woodlands. They were certainly present in the forests of this area throughout the 12,000 years since the melting of the last glacial ice-sheets, and probably were here during the inter-glacial periods for perhaps the last one-half million years or longer. So they are not newcomers in this region, as a species—and they surely long predate man's appearance in this area. Until the last three quarters of a century, they were dependent upon natural ecological agents such as fire and windstorm to maintain the forest disturbance that produces the early successional forests dominated by shade intolerant shrubs and trees.

Being typical grouse, they are primarily browsers, subsisting on the buds, twigs, leaves and fruits of various forest herbs, shrubs, and trees. They are non-migratory and tend to spend their entire life within a small area. It is probably a rare individual wich dies as far as 5 miles from where it was hatched, and research has shown that more than half of them live out their life-span less than one mile from where they were hatched. The male grouse spends its life in a drumming activity center which may encompass no more than 10 acres. All of the requirements for his livelihood must be met within this area. Hens are somewhat more mobile, but still their basic living requirements must be provided within a comparatively small area of less than 40 acres.

Ruffed grouse are essentially solitary in their habits; hens mate with any convenient male near his drumming log, and then may never see him again. The hen may select for her nest a site as far as one-half mile from the log of her mate, and she will incubate and rear her brood without any assistance from the male grouse.

These birds are short-lived under natural conditions. Normal over-winter losses take about 55 percent of the population each year. Among 1,000 15 week-old grouse alive in mid-September about 450 will be alive to breed that next spring. By the second nesting season there will only be about 185; about 78 left by the third nesting season; 41 by the fourth season; 13 by the fifth; and all will have died by the seventh year.

It appears that four or five males and accompanying hens per 40 acres is about the highest density of breeding grouse that we can expect on the best managed, most productive habitats. A spring breeding density such as this could produce hunting season populations of about five to six hundred grouse per square mile under optimum conditions. About one-half of the birds present each fall are surplus and could be harvested annually as a product of the forest. But it is seldom that hunters take more than 20 to 25 percent of the population.

Throughout the primary ruffed grouse range in Minnesota today, aspen or "popple" appears to be the most important plant contributing to their year-long welfare. During most winters and springs the flower buds of the **male** aspen are the most important single source of food. Most male grouse select drumming logs within sight of a group of male aspen, and hens, after being mated, evidently seek a suitable group of male aspen as a site for their nests (the female aspen is no more important than any other hardwood).

Although hazel, birch and ironwood catkins are sometimes heavily used for winter food, it seems to be essential to preserve an adequate supply of mature aspen throughout much of this species' range in Minnesota. The abundance of grouse in each covert is quite closely related to the quantity and quality of food available to them in that covert. Territorial behavior among these birds is such that there may be an excess of food in one small area, but the resident birds will not allow other grouse to utilize it—so an equitable distribution of satisfactory food resources throughout the habitat is preferable.

In addition to its value as a food resource, younger stands of "popple" provide these grouse with their best security from predators. Young, "dog-hair" stands of aspen suckers provide a high security habitat for broods through the summer, as well as a choice feeding site for older birds in the fall. Later, as these sucker stands develop into sapling and young pole stands 25 to 30 feet tall, they become the favored cover when there is not enough snow for burrow-roosting and as habitat for the drummers in spring and fall.

At every stage of its development, an aspen stand provides the highest quality of habitat resources for these birds at some time during their annual life-cycle. No other forest plant or tree fills this role in Minnesota's forests.

The most effective management and most abundant ruffed grouse populations occur where the cover provided by young aspen stands is close to aspen stands 30 or more years old. The older aspen provides the winter food supply. Where all needed resources are provided in a 6- to 10-acre area, we consistently find breeding ruffed grouse most abundant.

Two figures which follow show several alternative schemes which can be used to provide acceptable dispersion of habitat resources for these grouse. Figure 1 shows four variations in the treatment of a 40-acre forested tract. Variation A could be a commercial harvest and represent one 40 acre plot in a larger tract being managed for wood fiber and wildlife. The other three variations would seldom be commercially feasible, but they could be accomplished as a fuelwood cutting pattern.

Figure 2 illustrates how a larger forested area could be treated by a series of commercially harvested 10-acre strips, each 5 chains (330 feet) wide by 20 chains (1320 feet) long. The tract shown here would involve 520 acres.

In every case a nearly complete clear-cut of the aspen and all competing trees and shrubs is essential to assure an adequate density of aspen sucker regeneration. Ruffed grouse are quite selective in the quality of the habitats they will occupy, and, if aspen growth is too open, it will not be utilized. Initially the aspen stand should average one sucker for every two square feet, so that 10 years later, when the stand thins to the density acceptable to wintering and breeding adults, there should be at least one 25- to 30-foot aspen on each five square feet.

The third figure illustrates how a tree can be felled to make a drumming log. It should be 10 to 12 inches diameter at the butt end, and felled into heavy brush cover (not shown in this figure). The surrounding terrain should be free of other logs, brush piles, or other concealment for foxes or other 4-legged predators for a radius of at least 60 feet, and there should be flower-producing male aspens in the forest canopy overhead.

Hardwood stands lacking a good distribution of aspen, and forested sections dominated by pines, spruce, and fir must all be regarded as marginal or low density ruffed grouse habitats. Thus, prescribed treatments for these lands are difficult to recommend, other than doing whatever is necessary to preserve or encourage an adequate abundance and distribution of aspen on them.

Since the protective coloration of these grouse and their normal alertness are their best protection against predation, a major effort must be made to reduce the opportunities for grouse to be ambushed by their predators. Maintaining a forest free of brush-piles, slashing, and other accumulations of litter on the ground appears to be the best means of minimizing predator losses. Prescribed burning of newly cut-over areas appears to be vitally important as a method of maintaining the best quality ruffed grouse habitat.

The planting of conifers in grouse habitats for the purpose of providing cover is **not** recommended. Pines are especially undesirable since in about 20 years they will provide preferred cover for hawks and owls which are the primary predators of these grouse. Balsam fir and spruce seem to be less hazardous, but even the presence of these trees appear to depreciate the value of good hardwood habitats for ruffed grouse in central and northern Minnesota.

Perhaps the most important herbaceous food plant for these birds is the wild strawberry. The seeding of roadways and trails with white clover can be effectively used as a means of drawing the birds out to where they are more easily hunted in the fall (this practice does not appear to pose any threat to the over-shooting of local populations). Although ruffed grouse eat the fruits of many wild plants, these plants do not appear to be essential food items, and the planting of berry-producing shrubs seems to have little value other than as another method of concentrating grouse in the fall in situations where they are more easily hunted.

(Copyrighted Figures 1, 2 and 3 from "Improving Your Forested Lands for Ruffed Grouse", published by The Ruffed Grouse Society, are reproduced by permission of the author).

Prepared by: Gordon W. Gullion, Cloquet Forestry Center, University of Minnesota.
November, 1977.

FIGURE 2.

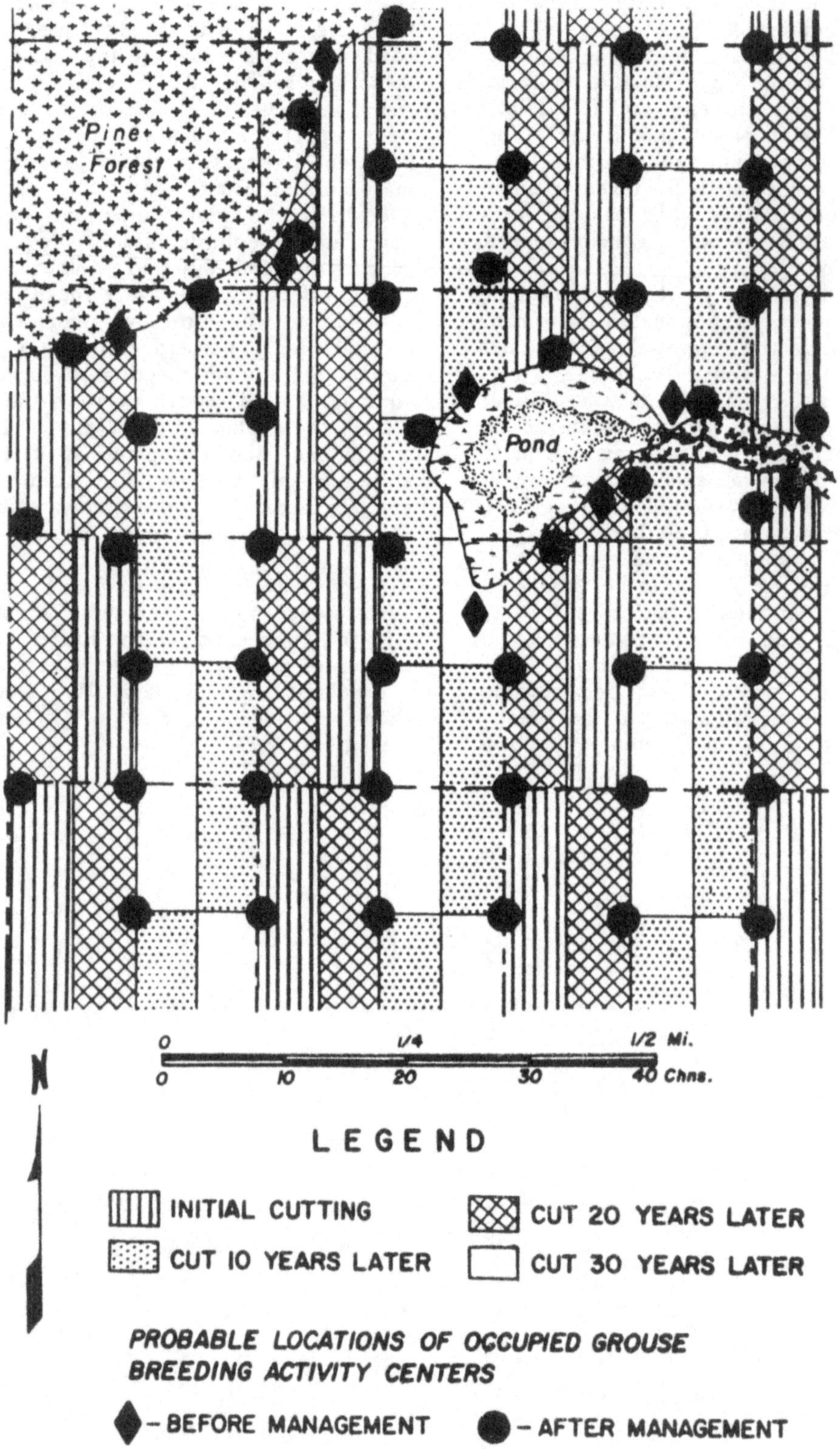

This figure suggests how fairly extensive commercial timber harvesting could be done in aspen or hardwood forests in a manner which should substantially benefit Ruffed Grouse and other forest wildlife species. Each of these rectangular strips consists of 10 acres, and 160 acres could be cut from a square mile in each operation, spaced at 10-year intervals. On the other hand, cuttings could be at 5-year intervals, cutting half as much each time.

FIGURE 1.

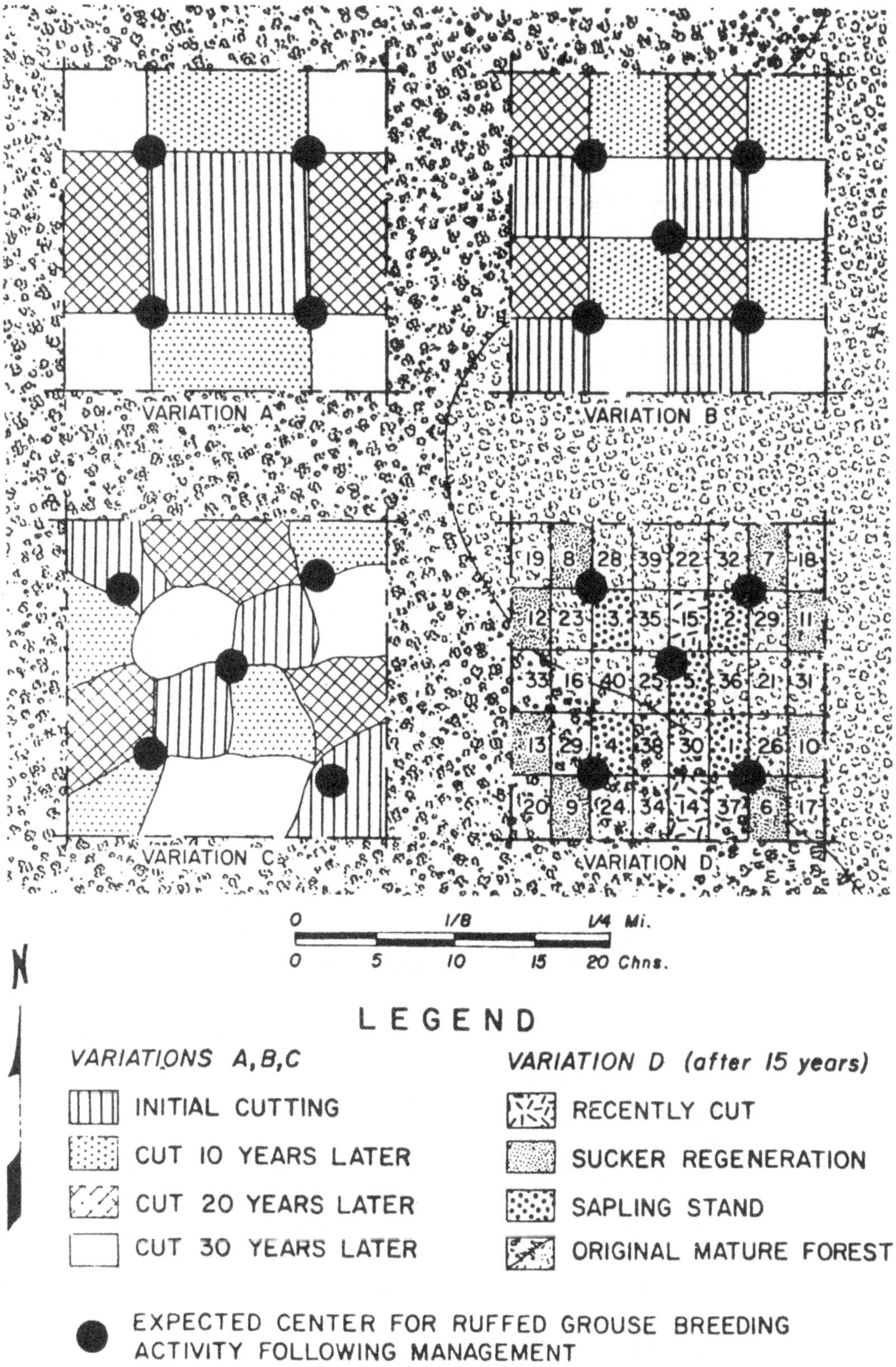

A management prescription for a 40-year forest tract, showing four variations in treating the same sized area. In Variation A the blocks include a central one of 10 acres, and four each of 2½ and 5 acres; in Variation B all the blocks are 2½ acres in size; and those in Variation C are about 2½ acres each, except across the bottom. Each block in Variation D is 1 acre, with the numbers indicating the year in which that piece would be cut. Forest condition for this variation is shown as it would appear 15 years after management commenced.

INVITE BIRDS TO YOUR HOME

By Wade H. Hamor, biologist, SCS

Birds are good neighbors. Their songs, colors, and lively activities add much to the joys of suburban or near-suburban living. They also help us control insects that attack flowers, lawns, gardens, and people.

You know about manmade bird attractors — feeders, birdbaths, birdhouses, nesting shelves, and the like. Let's consider some natural attractors — trees, shrubs, and vines — you can grow to invite birds to your home and garden.

Plantings can beautify your land as well as attract birds. Birds often feed on berries, and many of their favorites are bright-hued and decorative. Hedges and other dense shrubbery provide shelter and fit handsomely in the background scene. Trees for nesting and singing offer shade and beauty to the householder. Sunflowers and other colorful annuals are seed producers. A small wildlife pool is an attractive addition.

Blue Jay
acorns, cherry, sunflower, wild plum, cultivated grains

Meet the Birds' Needs

Birds have simple needs — they look for places to feed, sing, court, nest, rest, and hide. As you do your landscaping, you can provide these places with mixtures of trees, shrubs, vines, and other plants. Even in small yards, the right choice of plantings can meet many of these needs.

Mix and Blend Plants

Birds like variety — remember this when deciding what plants to use in your conservation-planned landscaping. Create a varied pattern with an intermingling of species, sizes, and shapes. Give birds a choice of places for their activities — from the crowns of tall trees down to low-growing flowers and grasses. Give them a choice of food sources — seeds, nuts, fruits, berries, and flower nectar. Many songbirds combine these plant foods with animal foods like insects, worms, and spiders.

Choose Plants of Wildlife Value

By knowing the wildlife value of plants you now have, you can plan additional plantings that will bring a diversity of plant forms, food producers, and shelter plants that would otherwise be missing. Many common shade trees and landscape shrubs, for example, yield little food for birds. Autumn-olive, cherry, or good fruit-bearing shrubs are helpful additions. Yards and grounds that have only deciduous trees and shrubs can be improved by adding junipers, cedars, yews, and other evergreens that provide winter shelter for birds.

Plants in this article are adapted and useful in the Midwest from Michigan and Indiana west to the Dakotas and Kansas.

Create a Landscape Design

You have endless choices of combinations to consider in creating a conservation landscape design; hardwoods and conifers; vines, shrubs, and trees; grasses; flowers; even weeds. If your yard is small, you may be limited to single specimens of different plants. With much larger grounds, you can use hedges, clumps, food plots, and other massed plantings. If you have a wooded area, a small clearing within it can create more edges for birds and lend variety to the landscape.

The upper left part of this design suggests plantings and their arrangement for a moderately large suburban house and lot. The rest of the design shows features that could be added for a larger tract. Your choice in species and their arrangement and placing will depend on your space.

Be sure your landscaping allows you to see the birds. Put the plants where they can be seen from a window, patio, or terrace. For best results, choose the kinds of plants reported to have high bird use. Careful attention to periods of bloom and the availability of choice foods makes it possible to have a succession of floral displays and bird foods throughout the year.

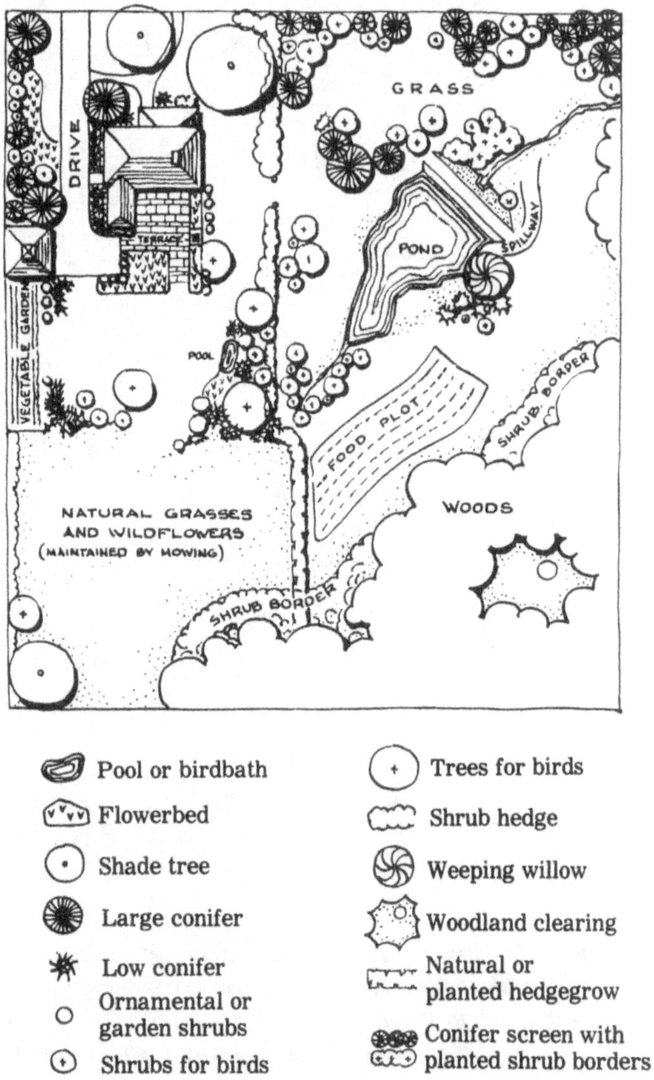

WAYS OF ATTRACTING BIRDS

Water Areas

Most birds need open water of some kind. A small pool with stones in the shallow edges draws birds to drink and bathe. They use the dry tops of the rocks for preening sites after bathing. A pond may attract some water birds at all seasons but most often during spring and fall migrations. Landscape the pond area with conifers, clumps of shrubs, and hedges for resting, nesting, and feeding. Encourage some aquatic growth along part of the shoreline but control its spread.

American Goldfinch
mulberry, sunflowers, seeds of garden flowers, weeds, and conifers

Living Screens

Hedges and rows of trees screen off unpleasant views and reduce noise from highways. In crowded neighborhoods they offer privacy for your backyard activities. And they attract birds to your place year after year. Conifers, autumn-olive, dogwood, cotoneaster, or a combination of these make good living screens.

Open Areas

Meadowlarks, bobolinks, and several kinds of sparrows favor open stretches of lawn and fields with few if any trees or shrubs. In seldom-mowed open areas, try planting a variety of native grasses. They protect the soil, and birds and mammals like the variety of foods.

Winter Protection

A windbreak of cedars, spruce, or pines with a crabapple tree tucked in on the sheltered side gives birds a warm, safe place to rest when the snow is deep. A food plot or feeder nearby helps keep the birds with you through the snowy months.

Living Fences

Hedges or rows of honeysuckle, dogwood, or autumn-olive can replace a wire fence between fields, can divide up fields, and can protect the house area. Cardinals, brown thrashers, and mockingbirds find living fences ideal.

Eastern Bluebird
dogwood, honeysuckle, mountain-ash, redcedar, various berries

Food Plots

"Seedeaters," such as goldfinches, cardinals, juncos, and sparrows, come to food plots of millets, grain sorghum, corn, or sunflowers; all are good bird foods to plant. Wild bristlegrasses and ragweeds also attract many birds. Plots can be small, perhaps several short rows, or large if you have the space.

THE BASIC RESOURCES—SOIL AND WATER

When planting for birds, you need to consider such things as soil, slope, drainage, exposure aspects, and climate as well as your personal wishes. Added benefits occur where plantings help provide shade, stabilize soil, and control potentially damaging water runoff.

In general, trees and shrubs that attract birds grow satisfactorily on well-drained, fairly fertile, somewhat loamy soils not particularly suited for vegetables and flowers. The ideal soil has a loose, loamy upper layer 18 inches or more deep. A neutral or slightly acid reaction is preferred.

When your house, street, sidewalk, and driveway were built, the normal pattern of water flow was changed and likely the runoff rate increased. Careful landscape planning can handle the extra water and check soil erosion. Plants described in this article that attract birds also help control soil erosion and water runoff.

Cardinal
autumn-olive, corn, dogwood, sunflower, various berries

HARMONIZE WITH MANMADE STRUCTURES

Try to blend your plantings with your house, walkways, and rock walls into a total landscape design. Put the permanent trees and shrubs in places where their roots will not interfere with masonry structures. Use annual species to provide more kinds of bird feeds, to balance landscape spacing, and to fill in along manmade structures. Planting on the contour adds natural beauty in pleasing patterns and helps control soil erosion and water runoff.

Bird houses, baths, and feeders can supplement the plantings to make your yard even more inviting. Select the kinds that blend in with your landscaping; the more simple and natural they look, the better they are. Feeders with fruits and grains are food sources in late winter after fruits from your plantings have been depleted. Certain kinds of nesting houses and shelves attract certain kinds of birds. The houses to select depend on what species nest nearby. But equally important is how urbanized your area is — some birds like city life but others do not.

FOR HELP

You can get further information on plants, soils, and conservation methods at your local soil and water conservation district office. The Soil Conservation Service, as part of its assistance to conservation districts, help people apply many conservation practices that increase wildlife. County agents, commercial nurserymen, landscape architects, and bird societies can also help.

U.S. Department of Agriculture pamphlets that may be helpful are:

- Autumn Olive (L-458)
- Gardening on the Contour (H&G-179)
- Making Land Produce Useful Wildlife (FB-2035)
- More Wildlife Through Soil and Water Conservation (AIB-175)
- Ponds and Marshes for Wild Ducks on Farms and Ranches in the Northern Plains (FB-2234)
- Soil Conservation at Home (AIB-244)
- Windbreaks for Conservation (AIB-339)

PA-982 December 1971

For sale by the Superintendent of Documents, U.S. Government Printing Office, Washington, D.C. 20402 — Price 20 cents. Stock Number 0100-1450.

Robin
cherry, cotoneaster, dogwood, Russian-olive, various berries

PLANTS FOR BIRDS, BEAUTY, PROTECTION

Cherry (Prunus spp.)
Bird use: 49 species
Ornamental values: Variable forms: shrubs, small to large trees; small fine-toothed leaves, yellow in fall; showy white flower clusters or drooping spikes; small bright-red to black fruits. Height: shrub, 5-15 ft.; tree, 20-75 ft.
Adaptions: Moist to dry soil; sun to light shade.
In bloom: April-June.
In fruit: Variable with species, June-November.
Sources: Commercial nurseries, wilding transplants.

Wild Plum (Prunus americana)
Bird use: 16 species
Ornamental values: Large shrub to small tree; suited to large yards or fields; spreads by suckers to form clumps; fragrant pink and white flowers; hardy red or yellow fruits.
Adaptations: Moist to well-drained loamy soil; sun.
In bloom: April-May
In fruit: July-October
Height: 10-30 ft.
Sources: Commercial nurseries, wilding transplants.

Ruby-Throated Hummingbird
nectar of columbine, delphinium, petunia, trumpetcreeper

Cotoneaster (Cotoneaster spp.)
Bird use: 6 species
Ornamental values: Generally medium-size shrubs; usually planted as a hedge but also as ground cover; dark-green leaves turning red-gold in fall; small pink or white flowers; showy red, orange, or black fruits.
Adaptations: Moist to well-drained soil; sun.
In bloom: May-June
In fruit: September-November
Height: 2-10 ft.
Sources: Commercial nurseries.

Baltimore Oriole
apple, cherry, elderberry, mountain-ash, various berries

Tatarian Honeysuckle (Lonicera tatarica)
Bird use: 18 species
Ornamental values: Large shrub; pink to yellow-white blooms; yellow to red fruits.
Adaptations: Well-drained to dry soil; sunlight to light shade.
In bloom: May-June
In fruit: July-September
Height: 5-15 ft.
Sources: Commercial nurseries.

Redcedar (Juniperus virginiana)
Bird use: 25 species
Ornamental values: Medium-size coniferous tree (many varieties); dense, green to blue-green needles; small dusty-blue, berrylike cones.
Adaptations: Moist to dry soil; sun to light shade.
In bloom: April-May
In fruit: September-May
Height: 15-40 ft.
Sources: Commercial nurseries, some state nurseries, wilding transplants.

Slate-Colored Junco
sunflowers, wheat, and seeds of grasses, weeds, and conifers

Bittersweet (Celastrus scandens)
Bird use: 12 species
Ornamental values: Twining vine; pale-green flowers; bright-red berries in yellow or orange husks.
Adaptations: Well-drained to dry soil; light shade.

Mockingbird
autumn-olive, elderberry, firethorn, highbush cranberry, holly

In bloom: May-June
In fruit: September-December
Height: climbs to 25 ft.
Sources: Commercial nurseries, some state nurseries, cuttings.

Holly (Ilex spp.)
Bird use: 20 species
Ornamental values: Variable forms: upright rounded shrubs, small to medium-size trees; many varieties; dark-green foliage, evergreen or deciduous; small whitish blooms; bright-red, black, or yellow fruits (very persistent).
Adaptations: Moist to well-drained soil; sun to shade.
In bloom: April-June
In fruit: September-May
Height: shrub, 5-15 ft.
Sources: Commercial nurseries.

Hawthorne (Crataegus spp.)
Bird use: 19 species
Ornamental values: Small trees; pale-green toothed leaves, abundant, clustered, white flowers; orange to red fruits (very persistent).
Adaptations: Deep, moist to dry soil; sun to shade.
In bloom: May-June
In fruit: October-March
Height: 15-30 ft.
Sources: Commercial nurseries.

DNR Reports

BIRDHOUSES IN MINNESOTA

A Guide for Building and Placing Houses for Birds.

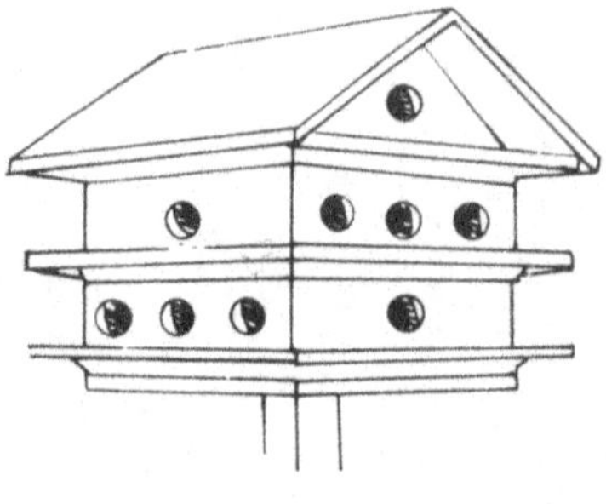

PROVIDING houses for wrens, bluebirds, and purple martins is a popular hobby for Minnesotans who enjoy wildlife around their homes. Houses built according to proper specifications, placed in proper habitat, and maintained regularly benefit birds. If they are not built and maintained properly, they can become either a "sparrow slum," remain unused, or actually become a death trap to nesting birds.

Although most people think of house wrens, bluebirds, purple martins, and wood ducks as the primary occupants of birdhouses, many bird species will occupy them. Birdhouses are built to simulate a natural cavity in a tree. There are actually 41 species of birds in Minnesota which will nest in tree holes – see the list on these pages. However, many birds will adapt to man-made structures if they have the opportunity. Man-made birdhouses are useful in areas that lack large, old, hollow trees that birds use for nesting.

Many types of birds need different kinds of cavities in different habitats. Following the general instructions below are accounts of how you can build and place birdhouses for several of the most popular occupants of bird homes in Minnesota.

THERE are several general instructions which apply to all birdhouse plans on these pages:

1. Do not make a box for "birds." Build it for a specific kind of bird. Different species have different house-size and entrance-hole requirements.
2. Provide a hinged door or roof so the house can be easily checked and cleaned out each winter.
3. Three or four ¼-inch drain holes should be drilled in the bottom of every birdhouse.

4. Although wren houses can be suspended from an anchor point under an eave or tree limb, all other birdhouses should be firmly attached to a support post, building, or tree. When you attach a nest box to a tree, use a lag bolt. Then, as years pass, gradually unscrew the bolts to allow for the growth of the tree.

5. Don't put perches on any birdhouses. Only English sparrows and starlings prefer perches.

6. The top-front edge of a birdhouse should overhang about one or two inches to help protect the entrance hole from wind-driven rain.

7. Two ¼-inch holes should be drilled near the top of the right and left sides of all birdhouses — except duck boxes — to provide ventilation.

8. Use galvanized nails to build all birdhouses.

9. The sides of a birdhouse should enclose the floorboard — don't nail them to the top of the floorboard. This keeps rain from seeping into the crack between sides and floor and then into the nest.

10. Do not use tin cans or metal for nests. Metals heat up in direct sun, kill the young, and overheat the eggs. However, commercial martin houses made from aluminum appear acceptable. Commercial plastic wood duck houses are also acceptable if they are placed in shady locations.

11. Wood is the best all-around material for birdhouses. Three-quarter-inch thick boards are the easiest to work with. Soft wood such as pine is fine for smaller nests, but preservative-treated cedar or cypress should be used for duck boxes.

12. Purple martins, goldeneyes, mergansers, and wood ducks do not defend territories around their nests. Therefore, martins should be provided with "apartment-type" houses. Duck and merganser boxes can be clustered in groups of two or four. Birdhouses for other bird species though should be spaced perhaps 50 feet apart to reduce territorial conflicts between males.

13. Small animals may take up residence in birdhouses, including mice, squirrels, bees, wasps, and raccoons. If not acceptable, remove mammal, starling, or English sparrow nests. Otherwise, you will probably need to put up additional houses to accommodate both the unexpected tenants and the desired bird species.

14. People like brightly colored birdhouses, but most birds prefer either unpainted houses, or houses painted in earth tones. Purple martin houses are an exception. They should be painted white to help reflect heat in hot, sunny weather.

15. When a birdhouse is inspected prior to the nesting season, it should be either sprayed with lysol or washed out with warm soapy water and treated with creolin. Creolin is available at most drugstores.

16. Any birdhouse entrance hole 1-3/8" in diameter or larger will admit English sparrows and any entrance hole larger than 1½" in diameter will admit starlings. These birds are persistent pests. The only way to discourage them is to remove their eggs and nests repeatedly during the nesting season.

This pamphlet was written by Carrol Henderson, Nongame Supervisor, DNR Section of Wildlife. Drawings of birdhouses by Betty Kennedy.

Cavity-Nesting Birds of Minnesota

- • Wood duck
- • Common goldeneye
- Bufflehead
- • Hooded merganser
- Common merganser
- Turkey vulture
- Peregrine falcon
- Merlin
- • American kestrel
- Barn owl
- • Screech owl
- Hawk owl
- Barred owl
- Boreal owl
- Chimney swift
- • Sawwhet owl
- • Common flicker
- Pileated woodpecker
- Red-bellied woodpecker
- Yellow-bellied sapsucker
- • Downy woodpecker
- Black-backed three-toed woodpecker
- Northern three-toed woodpecker
- • Great-crested flycatcher
- • Tree swallow
- • Purple martin
- • Black-capped chickadee
- Boreal chickadee
- Tufted titmouse
- • White-breasted nuthatch
- • Red-breasted nuthatch
- Brown creeper
- • House wren
- Winter wren
- Bewick's wren
- • Eastern bluebird
- • Starling
- Prothonotary warbler
- • English sparrow
- • Red-headed woodpecker
- • Hairy woodpecker

• These species will use nesting boxes.

HOUSE WREN

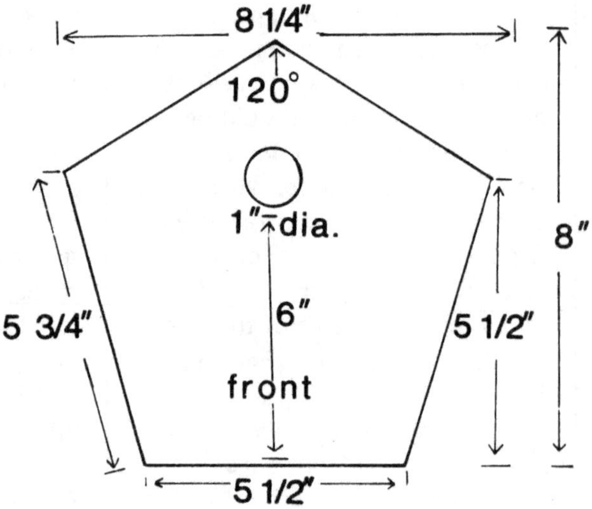

HOUSE wren structures should be 4 inches long, 4 or 6 inches wide, and 8 inches high. The hole should be 1 inch in diameter and centered 6 inches above the floor. Place the house 5 to 10 feet high under an eave or in a tree with an exposure of up to 60 percent sun. This is the easiest of all birds to attract to your home. House wren entrance holes are too small for sparrows or starlings to be a problem.

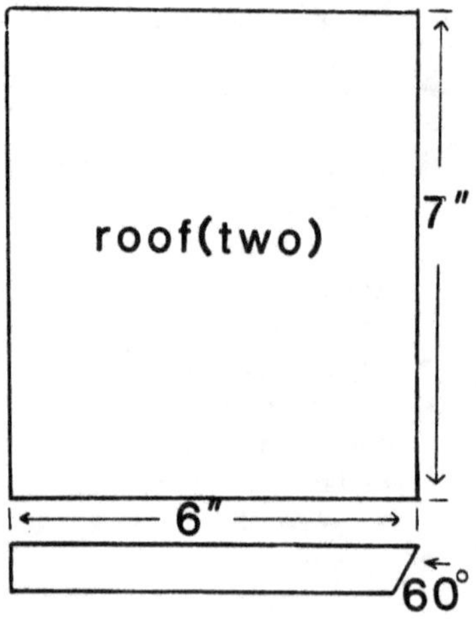

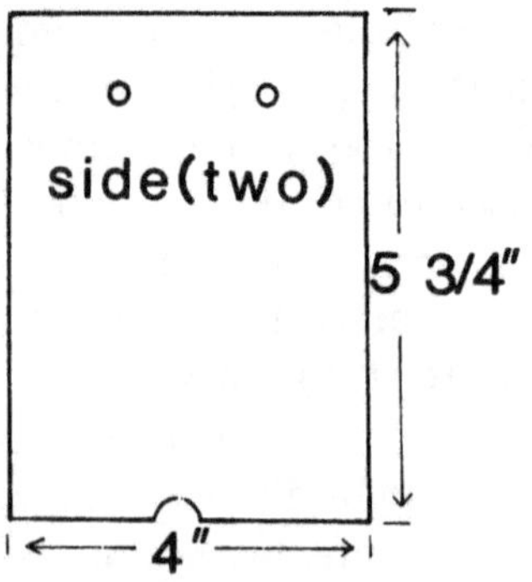

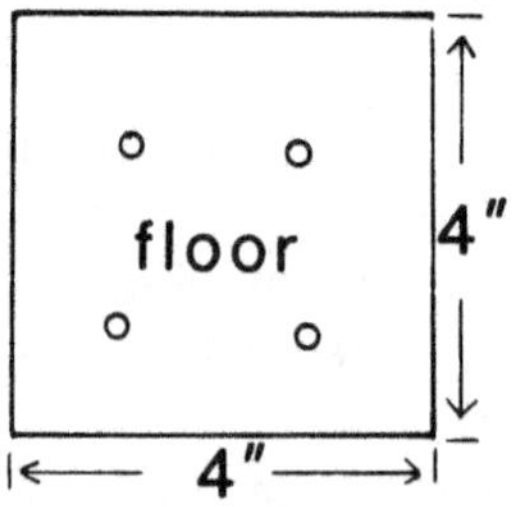

MOURNING DOVE

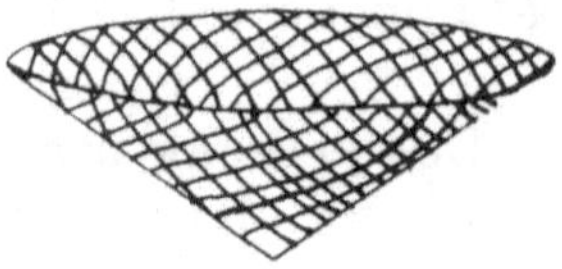

MOURNING doves do not nest in tree cavities or nest boxes, but in a very simple shallow cone of ¼-inch mesh or 3/8-inch mesh hardware cloth that can be placed in the crotches of tree limbs to provide instant nest sites for this very adaptable bird.

First cut out a 12-inch square piece of hardware cloth with tinsnips. Then trim the square to form a circle 12 inches in diameter.

Next, cut out a pie-shaped wedge – about 30 degrees of the 360 degree circle. After removing the pie-shaped piece, pull two cut edges together and overlap them about 2 or 3 inches. Wire the edges in this position. Place the cone in moderate shade in the crotch of a tree limb from 6 to 16 feet above the ground. There should be an open area around the nest site for easy access by the doves. Use roofing nails or staples to fasten the nest in place. Round the cut edges of the cone by bending them down after the nest is nailed to the branch.

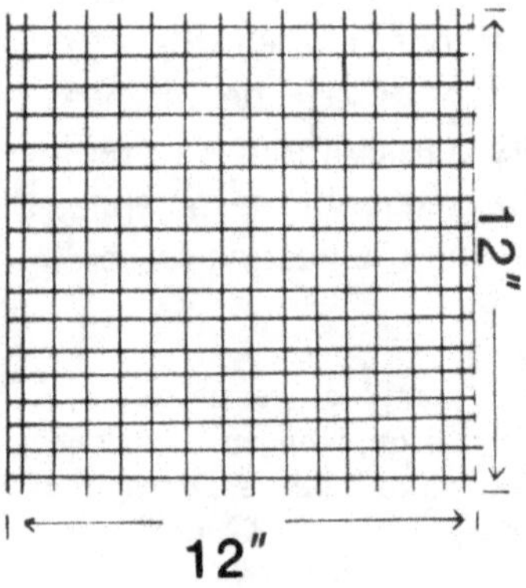

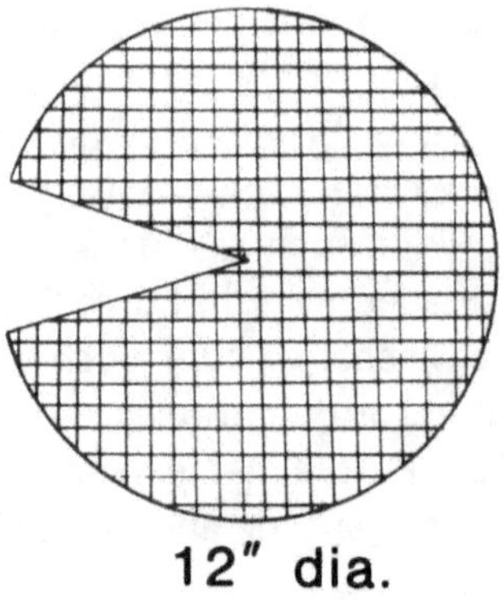

WOOD DUCK, COMMON GOLDENEYE, and HOODED MERGANSER

WOOD DUCKS have made a remarkable recovery in Minnesota during the past 20 years. They continue to amaze people by nesting in urban areas that certainly don't resemble typical "duck habitat." An additional consideration in northern Minnesota is that the traditional wood duck box will also be used by the beautiful hooded merganser and the common goldeneye.

The hooded merganser and common goldeneye nest primarily in the forested region of Minnesota which includes Cass, Beltrami, Itasca, Koochiching, Lake of the Woods, St. Louis, Lake, and Cook Counties. They nest along the larger lakes and rivers of this region. The wood duck is a state-wide resident, but is scarce in the Red River Valley and in the northeast.

A nesting box for these birds should be 12 inches wide, 12 inches long, and 24 inches high. The entrance hole should be an oval 3 inches high and 4 inches wide for the wood duck and hooded merganser. For the larger common goldeneye, the hole should be 3½ inches high and 4½ inches wide.

The hole should be centered 20 inches above the floor. An 18-inch by 3-inch strip of ¼-inch mesh hardware cloth should be cut out and the cut edges folded back. This should be attached inside the box under the entrance to function as a ladder for the newly hatched ducklings. At least 3 inches of mixed sawdust and wood shavings should be placed in the box to serve as nest material. The roof or a side of the box should be designed so the box can be opened for maintenance.

The house should be constructed of cedar, cypress, or other weather-resistant lumber. It should not be painted, stained, or creosoted. Ten feet of 1-inch by 12-inch rough lumber are required for one duck box.

The best post for a duck box is a 16-foot-long, 4-inch by 4-inch, cypress, cedar, or preservative-treated wood. Boxes placed in trees invite predation by raccoons unless the trees can be made predator proof. Attach the box to the tree with a lag bolt and a toe-nailed spike. Wood duck boxes in trees will be more successful if placed at a height of 20 to 25 feet.

Boxes over water should be about 6 to 8 feet above the water surface. If the water level of the site varies, the predator shield should be adjusted to remain about 3 feet above high water level.

One type of effective predator guard for the 4-inch by 4-inch post is a used, offset aluminum printing plate from a newspaper office. Just nail the aluminum sheet lengthwise around the post a couple feet under the nest. A 3-foot diameter conical predator guard of 26-gauge sheet metal can also be used under a duck box.

Another pest that needs to be considered is the starling. To keep starlings from pre-empting wood duck boxes, attach a bluebird type house with a 2-inch diameter entrance hole to the side of the wood duck box. The starlings apparently prefer the smaller box and, therefore, will not interfere with the use of the larger box by wood ducks.

Many woodies will use boxes far from water, but since the hen must lead the ducklings to water after hatching, houses should be placed so there are no significant obstacles like roads, rain gutters, or fences to pre-

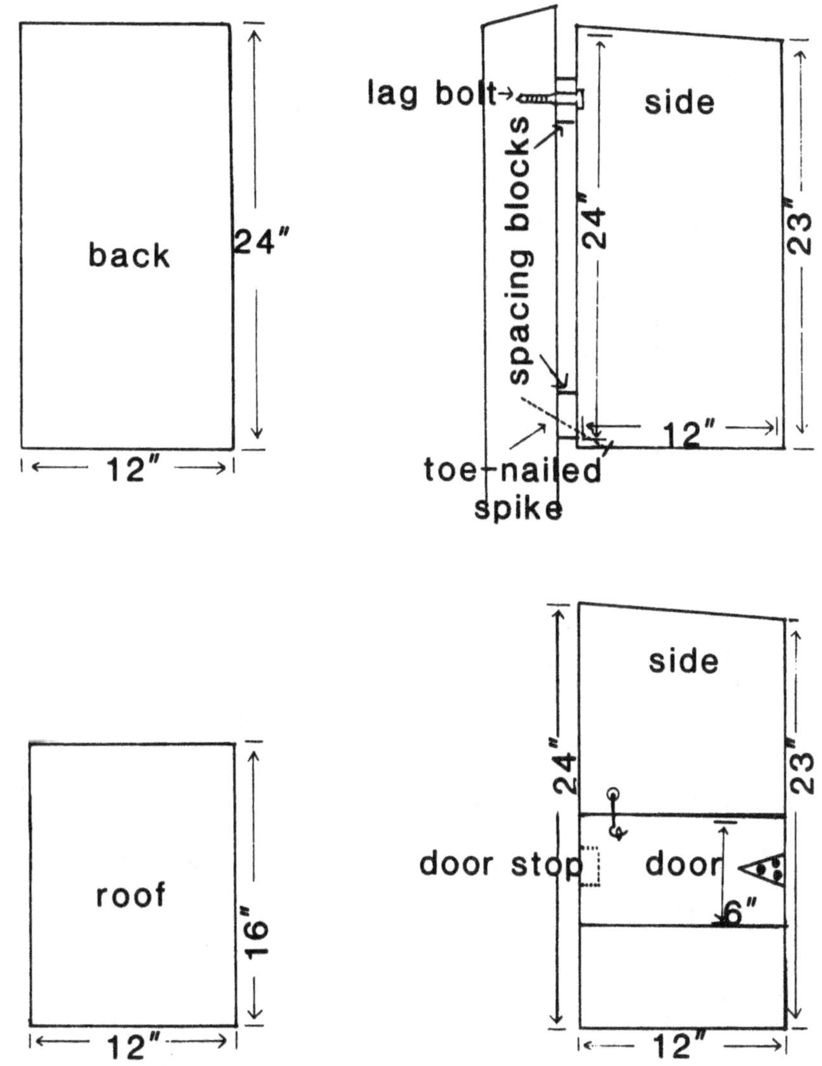

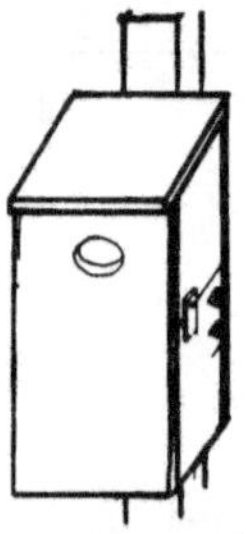

vent their movement.

Two duck boxes can be placed on the same post because they do not compete for territories like robins and some other birds.

All annual maintenance on wood duck boxes should be completed by March 1. Boxes should be opened, inspected, and more sawdust added to keep it 3 inches deep. The inside of the box should be sprayed with lysol or other disinfectant to discourage bees and wasps.

To facilitate annual maintenance of many boxes, it is often desirable to place the boxes on land where they are accessible by vehicle.

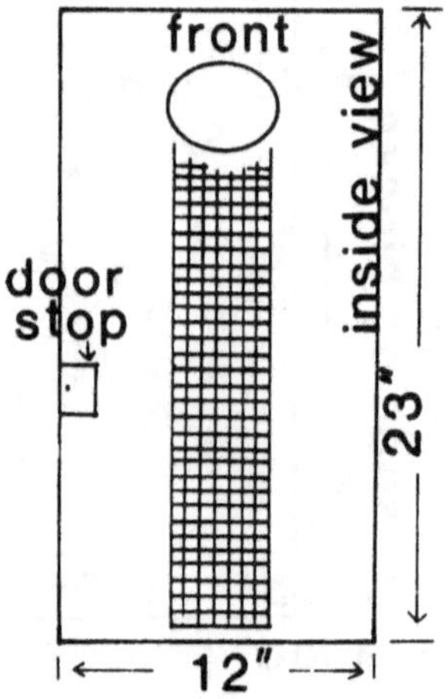

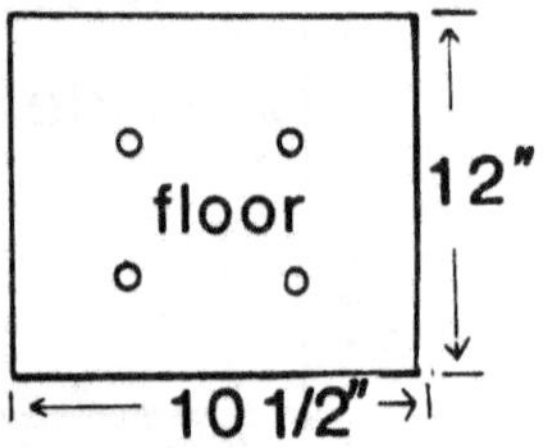

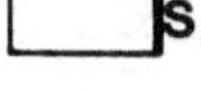

ROBIN, BARN SWALLOW, and PHOEBE

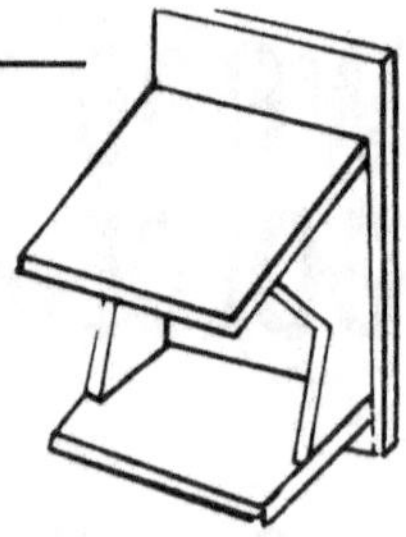

ROBINS, barn swallows, and phoebes do not nest in tree cavities or nest boxes, but they will nest on nest "platforms" which are open on three sides.

The platform nest is popular because it can be easily fastened under the eaves of houses and outbuildings. It can be placed near windows so that it is possible to observe the nesting season behavior of these three familiar species. This type of nest should be 7 inches wide, 7 inches long, and at least 8 inches high.

These nests should be 8 feet to 12 feet from the ground.

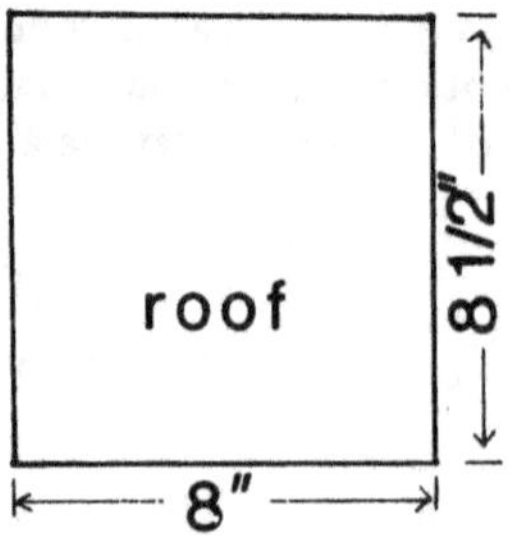

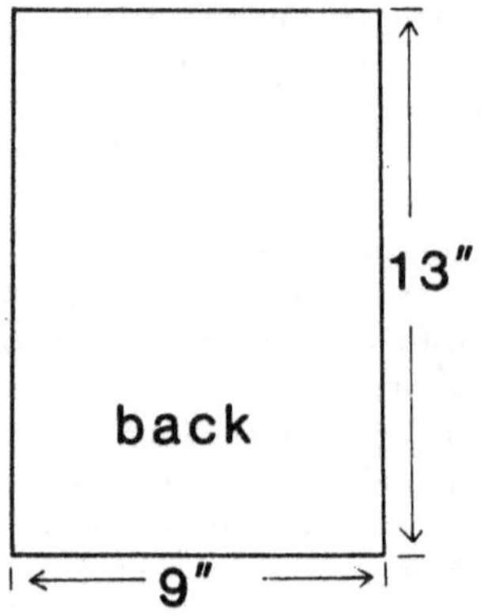

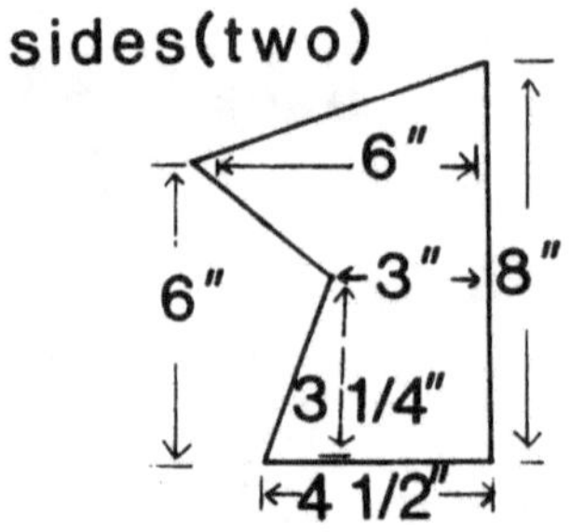

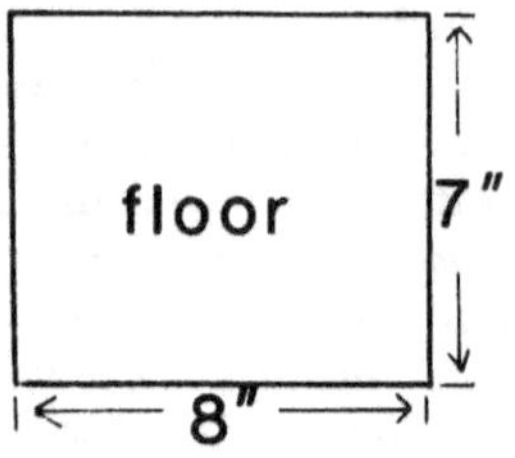

BLACK-CAPPED CHICKADEE

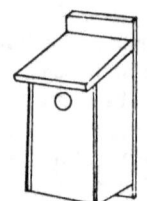

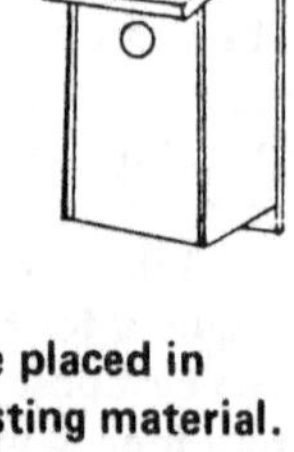

CHICKADEE houses should be 4 or 5 inches long, 4 or 5 inches wide, and 8 inches high. The hole should be 1-1/8 inches in diameter and centered 6 inches above the floor. The house should be mounted 5 to 15 feet high with 40 to 60 percent sunlight. There should be some large trees in the area.

Some sawdust should be placed in the house to provide nesting material. Chickadee entrance holes are too small for sparrows or starlings to be a problem. (Use the general design for the bluebird house — page 8 — but substitute the above dimensions. Make entrance hole 1-1/8" round, not oval.)

KESTREL AND SCREECH OWL

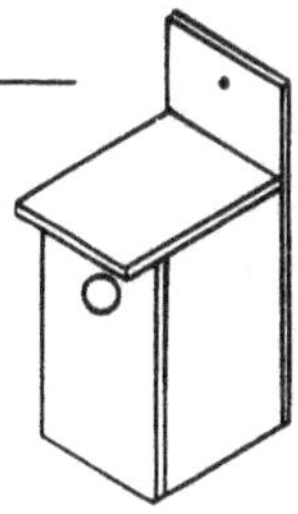

THE KESTREL, or sparrow hawk, is the smallest falcon which nests in Minnesota. It can be attracted to a house in an urban or rural area in a site facing an open field or pasture. The box should be 10 inches wide, 10 inches long, and 24 inches high. The hole should be 3 inches in diameter and 20 inches above the floor. The house can be 10 to 30 feet high and the approach to the entrance should be open — not obscured by branches.

Since kestrels inhabit relatively open country, houses can also be placed in isolated trees as well as in trees along woodlot edges. Some bird specialists recommend having the entrance facing west or south.

Since kestrels don't add nesting material to a cavity, about 3 inches of wood chips and sawdust should be placed in the bottom of the box. Starling control will be necessary to enable kestrels to use the box.

Screech owls require the same specifications as a kestrel, except that the box should be placed either in trees of open woods or on the edge of woodlots.

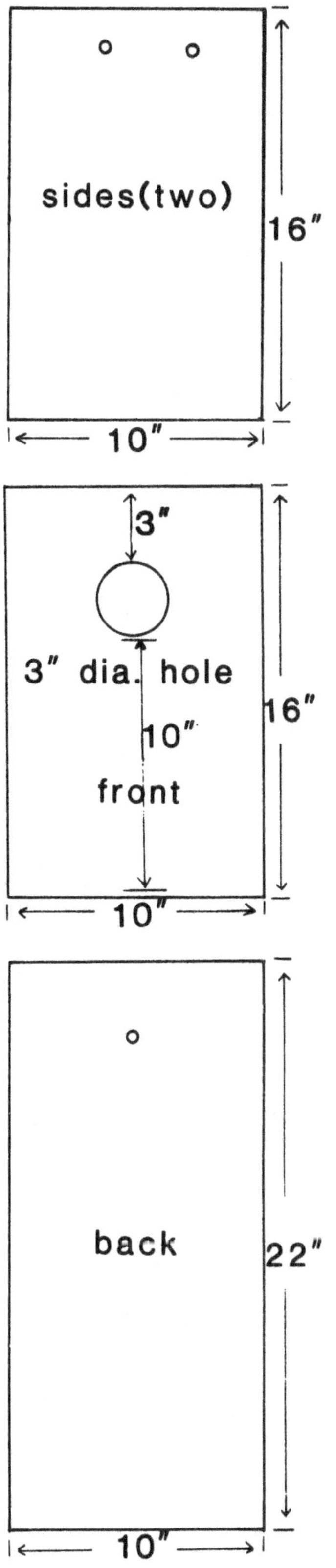

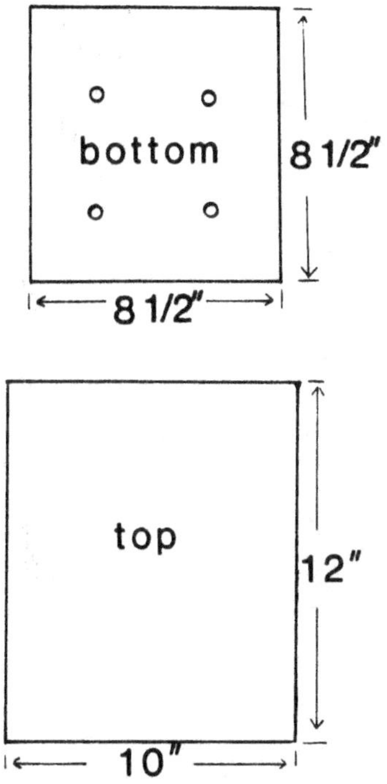

TREE SWALLOW

HOMES for tree swallows should be 5 inches wide, 5 inches long, and 6 inches high. The diameter of the hole should be 1½ inches. The hole should be centered 4 inches above the floor. The house can be mounted anywhere from 3 feet to 15 feet high.

Chances of attracting tree swallows are best in rural areas which are within two miles of a lake or pond. The nest site should be a relatively open place which receives 50 percent to 100 percent sun. Regular checks must be made to remove English sparrow nests and eggs.

To reduce sparrow competition, it may be necessary to cover tree swallow entrance holes in the winter and open them when the swallows arrive in spring.

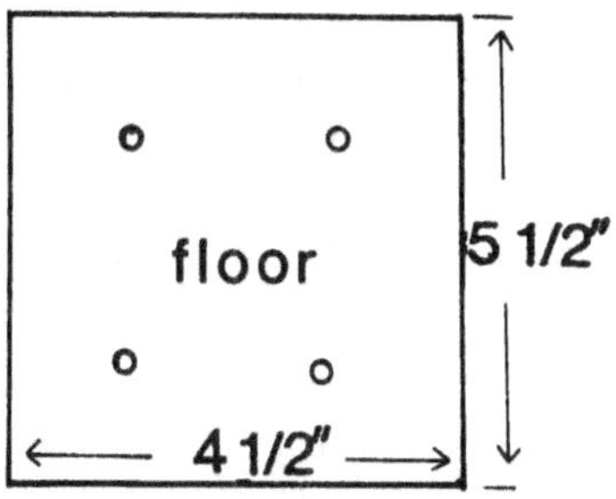

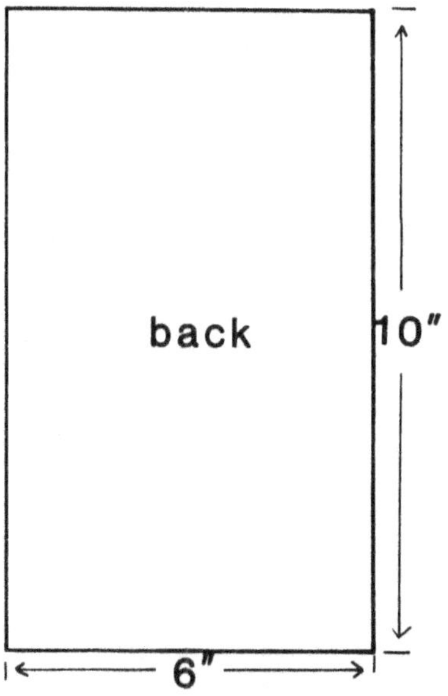

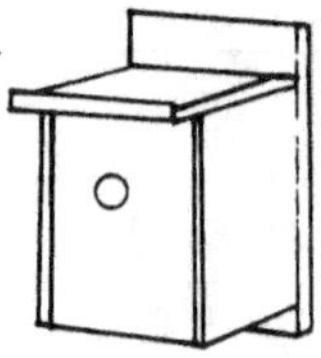

PURPLE MARTIN

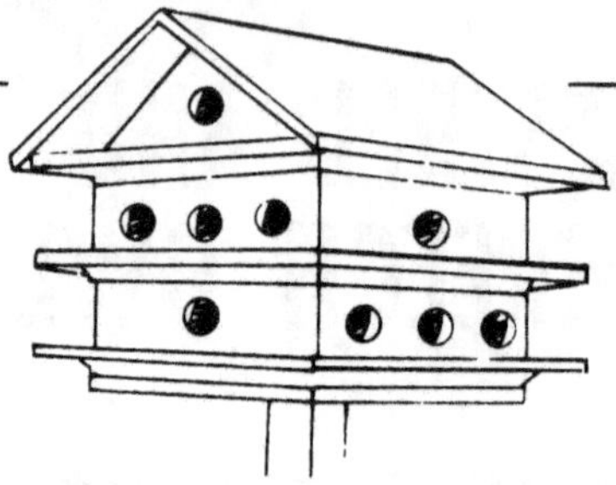

PURPLE martin houses are available commercially or they can be built at home. The multiple compartments should each be 6 inches wide, 6 inches long, and 6 inches high. Entrance holes should be 2½ inches in diameter with the base of the hole 2¼ inches above the floor. The house should be placed in an open site and set on a pole or post that is 15 to 20 feet high. An open yard with no tall trees is optimum for purple martins.

White is the preferred color for a martin house. To facilitate nest cleaning and maintenance, hinge the martin house pole so it can be tilted over and cleaned after the nesting season. Or keep a ladder handy to check the house regularly to remove English sparrow and starling nests and eggs.

To reduce competition by English sparrows, cover the entrance holes with small boards after the nesting season for martins. The holes should remain covered until you see the first martins arrive in spring. Then go out and uncover the holes.

Martins may arrive from late March through late May, but most arrive from mid-April through early May. A good general date for opening the martin house would be April 15.

Chances of attracting purple martins are best if a body of water is relatively near.

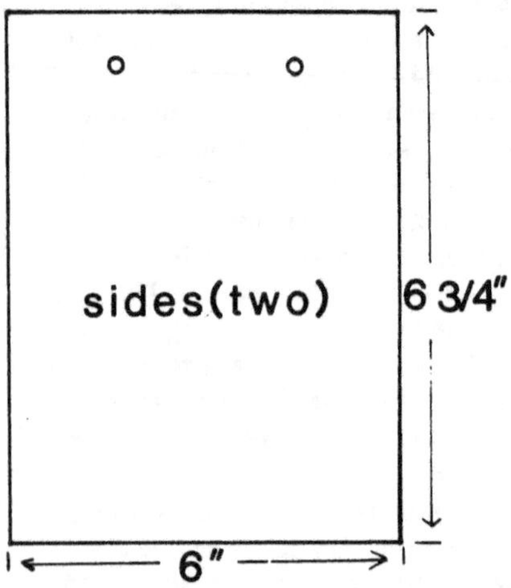

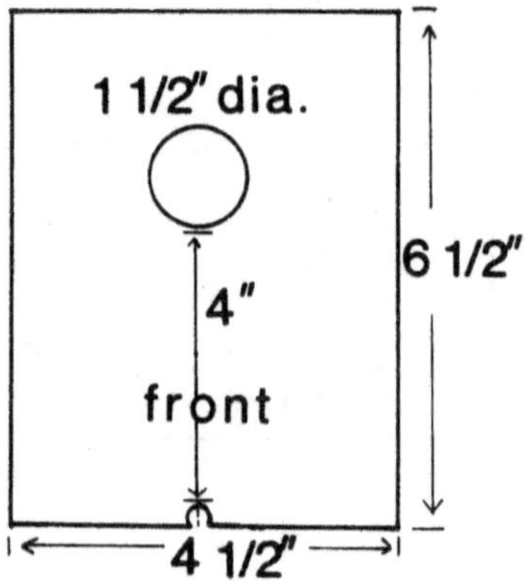

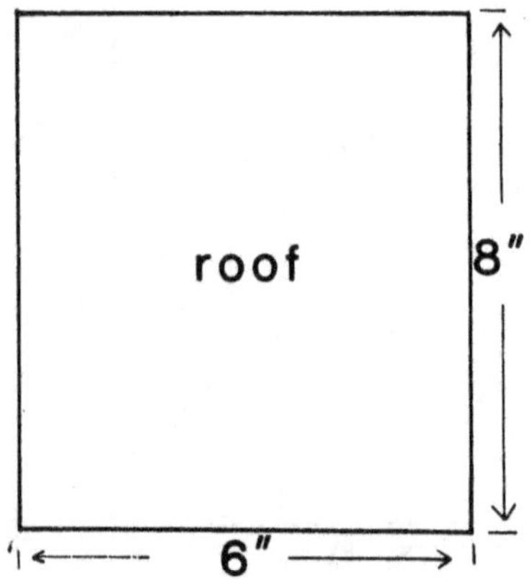

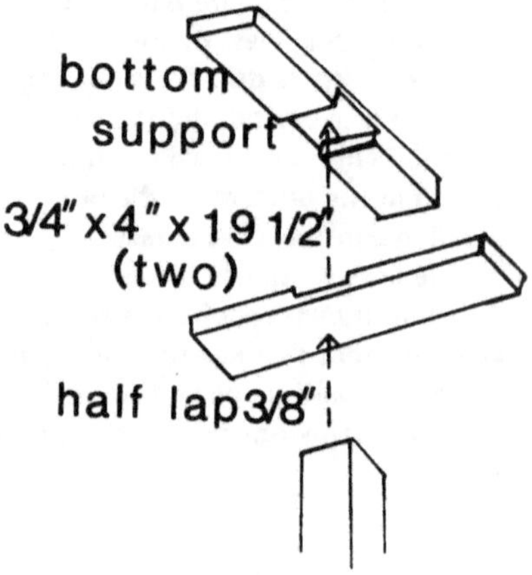

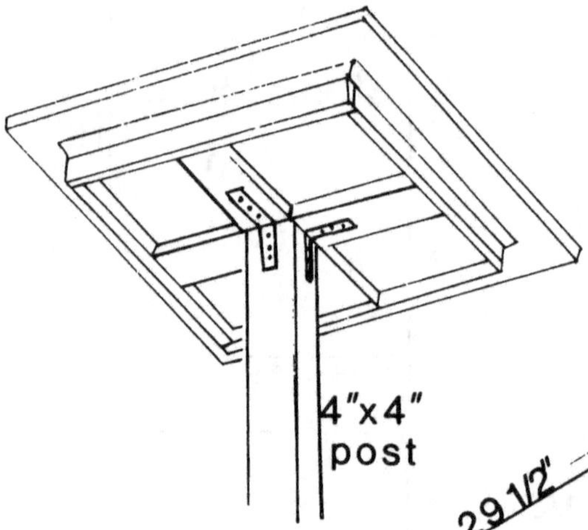

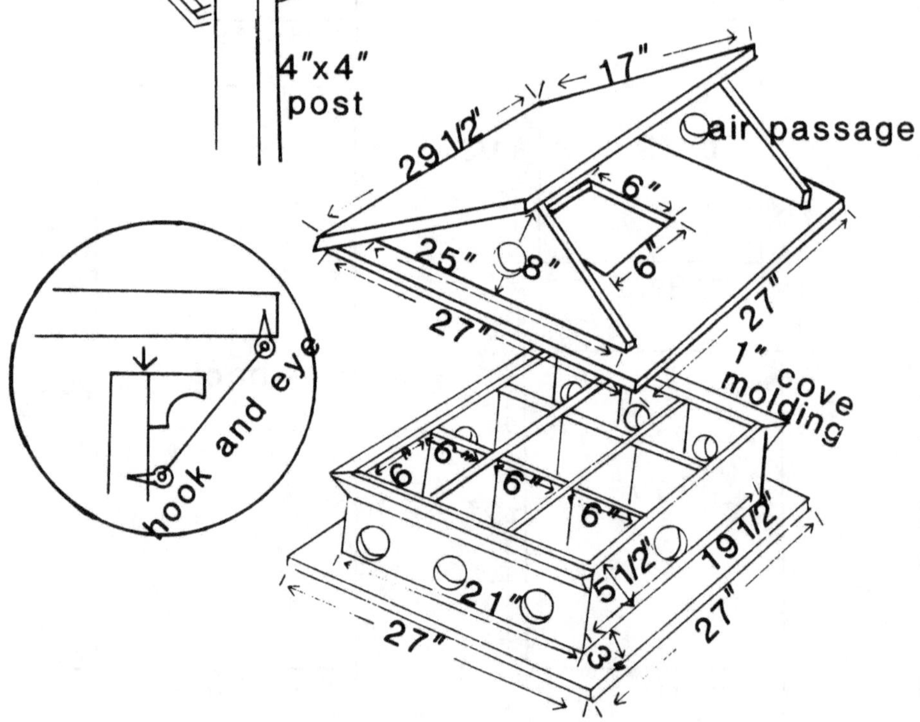

BLUEBIRD

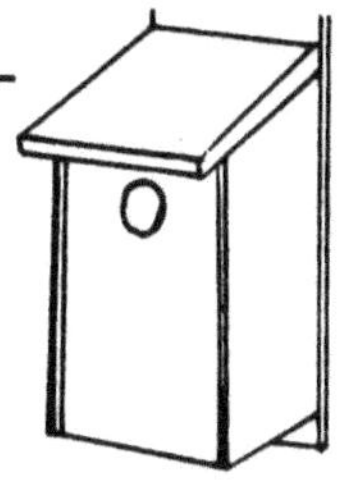

BLUEBIRDS need intensive birdhouse management because the starling, an introduced species, now nests in many of the natural cavities and in birdhouses that would otherwise be used by bluebirds.

A bluebird house should be 5 inches wide, 5 inches long, and 9 inches high. The oval hole should be 2 inches high and 1-3/8 inches wide. The hole should be centered 6 inches above the floor.

Earth tones are appropriate colors for the house.

Locate the house in an open, sunny place on a post 3 to 10 feet high. "Bluebird trails" can be created by placing houses on fenceposts along roads in rural areas. This makes them easy to check and maintain during the nesting season. Clean out the house as soon as the young bluebirds leave the nest. Then another pair of bluebirds can use the nest the same summer. Bluebird houses should be placed or cleaned out by April 12 each year, so they are ready in time for the spring migration.

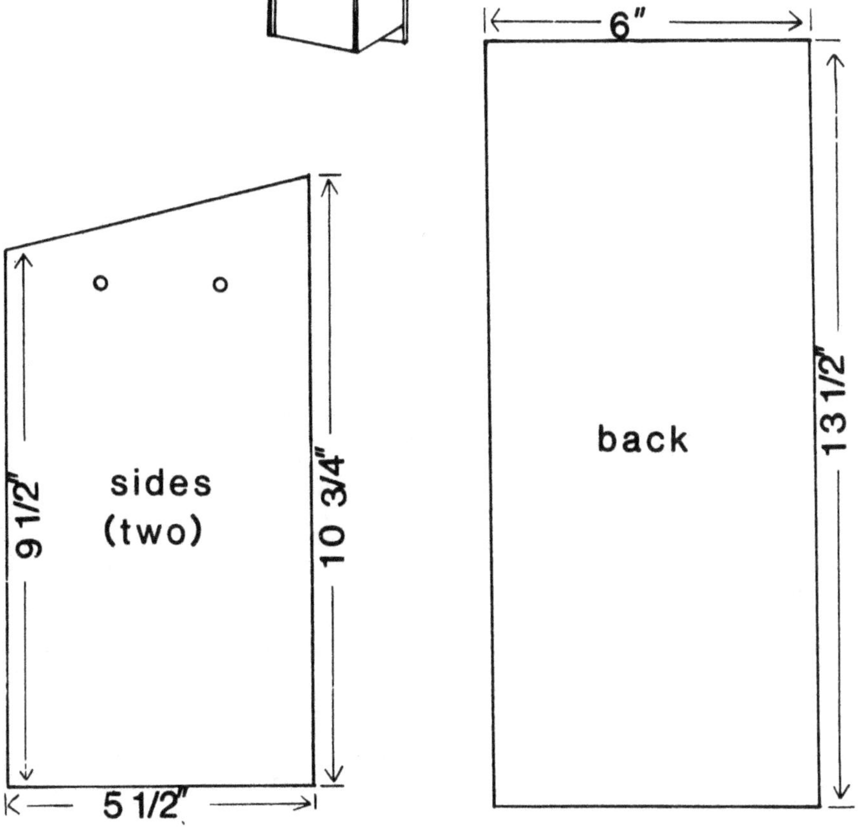

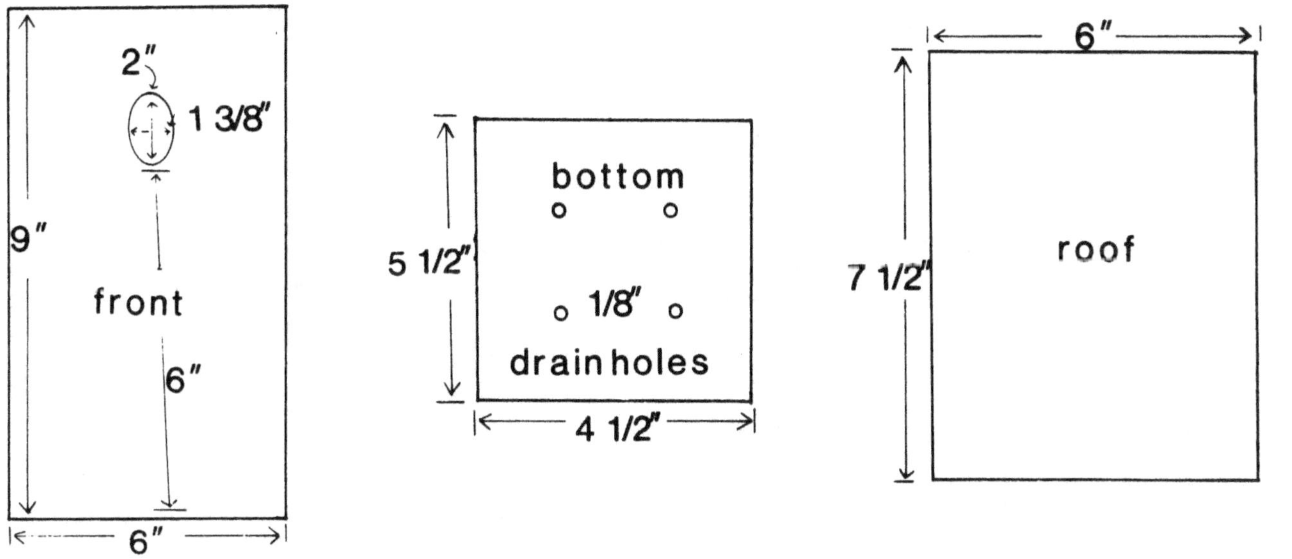

DNR Reports is a series of free publications on Minnesota's natural resources published and distributed by the Minnesota Department of Natural Resources, Bureau of Information and Education, 350 Centennial Office Building, St. Paul, MN 55155. Please write for a list of titles available in this series. This is title number:

77

5/79

GIVING THE BLUEBIRD A HOUSING HAND

by Mary Durant

They once migrated in flocks of hundreds and were among the most common of North American birds, a symbol of spring in the northern states. Nowadays a single pair of eastern bluebirds, newly arrived from their wintering grounds, constitutes an event — the two huddled together on a snowy branch in February or March like an illustration on a Victorian valentine, their rose-russet and blue plumage fluffed out against the cold.

The decline in the eastern bluebird population over the past forty years is estimated at an alarming ninety percent. Their natural nesting holes grow scarce as wooden fenceposts vanish and old trees are felled under the easy onslaught of chainsaws. Their habitat of fields and orchards is lost as farmlands diminish, and their summer diet of insects is curtailed by pesticdes, which presumably poison the birds as well. Severe winters also take their toll: The bluebird is not a hardy species.

These problems are minimal, however, compared to the overwhelming competition from starlings and house sparrows, both introduced from Europe in the last century, and both cavity-nesters, prodigious breeders, and bullies, They wage a fierce rivalry over nesting sites, driving out the bluebirds with murderous assaults, breaking eggs and killing young; and now that starlings and house sparrows have crossed the Rockies and blanketed the continent from coast to coast, the western and mountain bluebirds are also showing a downward trend.

Thoughtfully designed nesting boxes and solicitous attention are the most immediate, practical assistance that can be given the bluebird. In the design shown here, which was developed by Lawrence Zeleny, founder of the North American Bluebird Society, the 4- by 4-inch floor discourages the house sparrow, whose broods are larger than the bluebird's. The 1½-inch entrance hole effectively excludes the starling, and you can prevent woodpeckers from enlarging it by ringing the entrance with a metal band.

Mount the boxes 3 to 5 feet high — a further discouragement to house sparrows, which prefer loftier nesting sites. If they move in anyway, be ruthless. Toss out their nests daily. They eventually will give up and move on. And never put a perch on the front of the box. It offers starlings and house sparrows a handy spot from which to harass the resident bluebirds. As a guard against climbing predators (raccoons, cats, snakes) attach a cone-shaped metal skirt to the underside of the box. Or try a box mounted on a metal pole slathered with axle grease.

Since bluebirds are highly territorial, boxes should be 100 yards apart. Avoid shrubby overgrown areas, the habitat, attract to house wrens — also determined cavity-nesters and destroyers of bluebird eggs. Boxes should face open terrain with scattered trees, ideally with a tree, shrub, or fence within 25 to 100 feet and in direct view of the box. This gives the fledglings a safe haven to which they can head on their initial flight.

Clean and ready the boxes for occupancy in early spring, Clean them out again between broods (the bluebird will raise two to three broods per season). If there is evidence of bird lice or mites, dust with one percent rotenone powder. Be on the lookout for the most serious of bluebird parasites, a species of blowfly whose eggs are laid in the nesting material and whose larvae feed on the blood of the young birds, generally at night when the larvae cannot be seen and eaten by the adult birds. The larvae, which pupate deep in the nest, resemble tiny, dark brown, oblong eggs, and as the young birds fledge, the pupae hatch into adult blowflies and leave their hideaway to repeat the cycle in another nest.

If infestation is heavy (more than 150 larvae per nest), you might dust the nest with one percent rotenone powder or with powdered pyrethrum. But never do this after the eggs have hatched. In extreme cases, the young birds can be momentarily removed and their direct contact with the insecticide limited if you work it down in the nesting material. An infested nest also can be destroyed and quickly replaced with a "new" nest of clean dry grass, tightly packed. However, unless you're sure the problem is serious, do nothing. The young bluebirds usually survive and are free of larvae when they leave the nest.

For more information, send a self-addressed, stamped business envelope and 25 cents to the North American Bluebird Society, Box 6295, Silver Spring, Maryland 20906.

Mary Durant and her husband, Michael Harwood, are the authors of **On the Road with John James Audubon**, which was recently awarded the Burroughs Medal as the best book on natural history for the year.

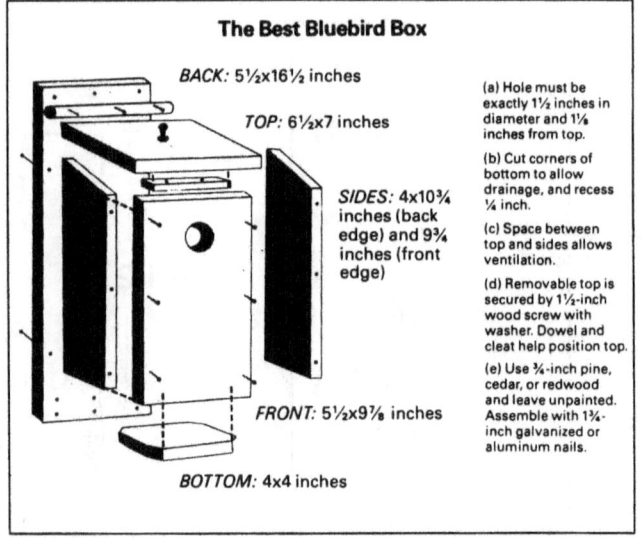

The Best Bluebird Box

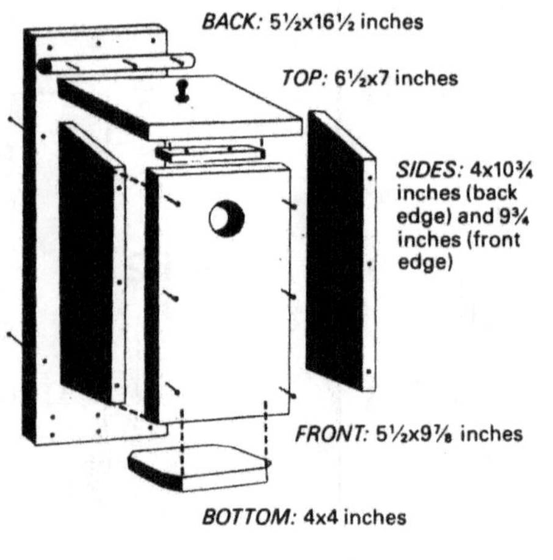

(a) Hole must be exactly 1½ inches in diameter and 1⅛ inches from top.

(b) Cut corners of bottom to allow drainage, and recess ¼ inch.

(c) Space between top and sides allows ventilation.

(d) Removable top is secured by 1½-inch wood screw with washer. Dowel and cleat help position top.

(e) Use ¾-inch pine, cedar, or redwood and leave unpainted. Assemble with 1¾-inch galvanized or aluminum nails.

SNAGS

Introduction

Minnesota has a wide variety of plants and animals (flora and fauna). This provides the resources which attract all ages to participate in outdoor recreation activities. Bird watching (birding) is one of the most popular activities. Young and old enjoy identifying birds or watching them build a nest and raise their young.

In Minnesota, 43 species of birds (table 1) and at least 25 species of mammals (table 2) use cavities in trees as nest or den sites. Of the birds, 21 species prefer to make cavities themselves (primary excavators) or have such specific requirements that they only nest in natural cavities. Removal of dead and dying trees, called snags, means loss of these nesting sites. Secondary excavators sublet: are willing to use homes in snags abandoned by another animal.

Continued high demand for forest products such as lumber and firewood may prove a serious threat to the survival of cavity nesting birds and mammals unless their requirements are considered a part of forest management plans. For those 21 species of birds, snags represent the sole source for present and future nest sites. There is no way humans can replace the lost snag with a manufactured substitute.

Table 1. Snag nesting species in Minnesota

Species	Excavators	
	Primary	Secondary
Wood duck		X
Goldeneye		X
Bufflehead		X
Hooded merganser		X
Common merganser		X
Red-breasted merganser		X
Turkey vulture		X
Merlin (pigeon hawk)		X
Am. kestrel (sparrow hawk)		X
Peregrine falcon		X
Barn owl		X
Hawk owl		X
Screech owl		X
Barred owl		X
Long-eared owl		X
Saw-whet owl		X
Chimney swift		X
Common flicker	X	
Pileated woodpecker	X	
Red-bellied woodpecker	X	
Red-headed woodpecker	X	
Yellow-bellied sapsucker	X	
Hairy woodpecker	X	
Downy woodpecker	X	
Black-backed three-toed woodpecker	X	
Northern three toed woodpecker	X	
Great creasted flycatcher		X
Tree swallow		X
Purple martin		X
Black-capped chickadee	X	
Boreal chickadee	X	
Tufted titmouse		X
White-breasted nuthatch		X
Red-breasted nuthatch	X	
Brown creeper	(occasionally)	X
House wren		X
Winter wren		X
Bewick's wren		X
Carolina wren		X
Easter bluebird		X
Starling		X
Prothonotory warbler		X
House sparrow		X
	12	31
Total species		43

Table 2. Minnesota mammals which use snags

Animals	Importance of snags to animals		
	High	Medium	Low
Squirrels			
Fox	X		
Gray	X		
Red	X		
Northern flying	X		
Southern flying	X		
Bats			
Little brown		X	
Big brown		X	
Pipistrelle		X	
Silver-haired	X		
Red	X		
Hoary	X		
Others			
Porcupine	X		
Raccoon	X		
Opossum		X	
Long-tailed weasel		X	
Ermine		X	
Short-tailed shrew			X
Masked shrew			X
Artic shrew			X
Least shrew			X
Deer mouse		X	
White-footed mouse		X	
Pine martin	X		
Fisher	X		
Least chipmunk	X		
Eastern chipmunk	X		

Snag Types

In this publication, a snag is any dead, partially dead or dying tree with a minimum diameter at breast height (d.b.h.) of 4 inches and a minimum height of 6 feet. This is considered the smallest snag birds use for nesting.

Will any snag do? No, each wildlife species has specific, characteristic habitat requirements. Sometimes these are quite flexible, at other times the requirements are very rigid.

Snags are classified as either hard or soft. Hard snags have some value as marketable wood. Soft snags are in advanced stages of decay and have no commercial value. Both classes of snags are valuable to wildlife. Some birds, such as chickadees, bluebirds, and nuthatches excavate nest cavities in soft snags. Others, primarily the woodpeckers, excavate cavities only in hard snags.

Forest Insect Control

It has not been conclusively established that insectiverous (insect-eating) birds are a major factor in the *suppression* of insect epidemics. There is, however, ample evidence establishing birds as major elements in *preventing* forest insect epidemics. This role is emphasized in northern latitudes during winter as the birds feed on overwintering adult and larval insects.

Birds should not be considered complete answers to insect outbreaks, but they are important components in a natural regulation system. In some European forests, insectiverous birds are considered so important that forest managers have placed as many as 400,000 nest boxes in an area of 345,947 acres.

Wildlife Use of Snags

Three conditions influence snag use by wildlife: the class of the snag—hard or soft, the surrounding plant community, and the size of the snag.

The types of snags change year by year. Hard snags eventually become soft. The speed of this process depends on such factors as, tree species, cause of death, tree condition at death, location, site conditions, and the types of decay organisms present. However, every stage in the process has value to some species of wildlife and a variety of dead and dying trees is needed to maintain a variety of wildlife.

Stages of the surrounding plant community influence snag use. Snags in open areas will be used by species of wildlife characteristic of open areas. Bluebirds and house wrens would use cavities in soft snags located in meadows or along woodlot edges; while chickadees and nuthatches would use soft snags in heavily wooded areas. Snags should be considered necessary in all stages of forest land management.

Each species of cavity-dependent wildlife has distinct requirements for the diameter and height of snags in which to build a nest or den. Physical size of the bird or animal determines the minimum snag diameter needed. For birds, snag height requirements are often related to feeding or perching habits.

Meeting Snag Requirements

Woodlot and forest managers might wish to remember: the same snag can be used by several species, birds of the same species will not usually excavate twice in the same snag, and large snags can substitute for small snags but not the other way around. There are a number of techniques which may be used to produce snags.

Long rotation periods for stands or individual trees provide snags naturally. Selective killing of trees by girdling or silvicides, rather than tree removal during stand thinnings, is an intensive practice for controlling the type and size of snags produced. Keeping existing snags during timber harvesting or thinnings is also a simple effective way of maintaining this resource.

This dead tree has excavations made by the pileated woodpecker.

Presence of soft snags is critical for the majority of snag-dependent wildlife. In intensively managed forests these snags are rare because they develop from hard snags which seldom are left standing long enough to soften.

All soft snags should be retained unless they present distinct fire or safety hazards. Normally, they are not fire hazards since the tops are broken off and do not attract lightning. They have no commercial value and rarely serve any forestry purpose such as wood production, beauty, wind or sun shelter, sound barrier, etc. If hard snag requirements for woodpeckers are met and all soft snags are retained, the requirements for all snag-dependent wildlife will be met.

It is really just guessing to suggest the minimum number of snags required per unit area; rather, this must be considered on a community basis. In many instances, there is insufficient data on key wildlife species to calculate these figures. Tables 3 and 4 contain hard snag requirements for several woodpeckers in various plant communities; these should be used as guides until data specific to Minnesota plant communities are available.

Most woodpeckers prefer to make their own excavations.

Table 3. Hard snag requirements for selected woodpecker species (adapted from Appendix 22 Agriculture Handbook No. 553,USFS, 1979)

Cavity dependent species	minimum snag d.b.h. (inches)	Snags required per 100 acres at selected population management levels				
		100%	70%	50%	30%	10%
Mixed conifer						
Pileated woodpecker	⩾ 20	14	9	7	4	1
Common flicker	⩾ 12	38	26	19	11	4
*Black-backed, three-toed woodpecker	⩾ 12	59	41	29	18	6
Hairy woodpecker	⩾ 10	180	126	90	54	18
Quaking aspen						
Common flicker	⩾ 12	38	26	19	11	4
Yellow-bellied sapsucker	⩾ 10	150	105	75	45	15
Downy woodpecker	⩾ 6	300	210	150	90	30

*Northern three-toed woodpecker has identical requirements for snags at these management levels.

Table 4. Generalized hard snag requirement for woodpeckers in mixed conifer and quaking aspen plant communities

Minimum snag d.b.h. (inches)	Snag required per 100 acres at selected population				
	Management levels				
	100%	70%	50%	30%	10%
Mixed conifer					
⩾ 20	14	9	7	4	1
⩾ 12	45	32	22	14	5
⩾ 10	121	85	61	36	12
Total	180	126	90	54	18
Quaking aspen					
⩾ 12	38	26	19	11	4
⩾ 10	112	79	56	34	11
⩾ 6	150	105	75	45	15
Total	300	210	150	90	30

Management Recommendations

Although the specific requirement for each species of cavity-dependent wildlife is not known, a useful list of general management practices follows. It is reproduced from a report prepared by Evans and Conner (1979).

- Manage for maximum feasible rotation age.
- Consider old growth a high priority. Select stands for deferred cutting as early as possible—20 years is optimum.
- Leave a 0.1 hectare (ha), about ¼ acre, clump permanently uncut in each 2 ha of regeneration cut.
- Discontinue removal of dead, dying, and decayed trees for use as materials or firewood in areas where nest cavity sites are limited.
- Consider management techniques such as providing artificial nest boxes and boring holes in suitably sized trees when cavity availability is limited.
- Consider leaving permanent uncut buffer strips on both sides of streams.
- Consider leaving shelterbelts or reestablishing them where they have been removed. This is especially needed in agricultural areas where woodlots (shelterbelts) are widely separated.

Additional References

Evans, K.E. and Conner, R.N. 1979. *Snag management,* in *Management of North Central and Northeastern forests for nongame birds.* USFS Gen. Tech. Rpt. NC. 51. North Central Forest Experiment Station, St. Paul, MN. 262 pp.

Minnesota Department of Natural Resources. 1979. *Bird houses in Minnesota: a guide for building and placing houses for birds.* DNR Report No. 77 Mn. DNR, St. Paul, MN.

Thomas, J.W. (editor). 1979. *Wildlife habitats in managed forests: the Blue Mountains of Oregon and Washington.* Agri. Handbook No. 553, USFS, Wash., D.C. 512 pp.

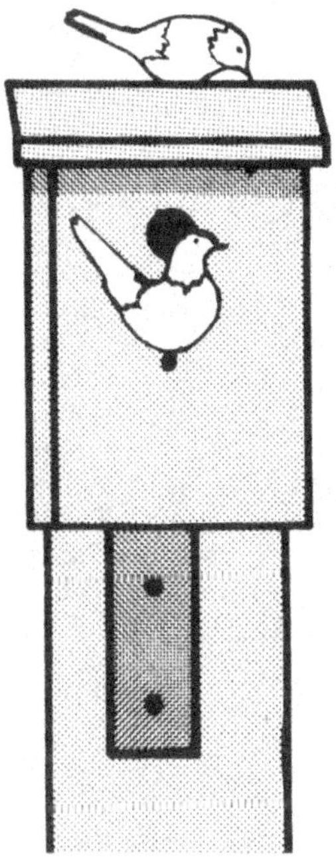

Nest boxes can be a replacement for snags. These house wrens like them.

Issued in furtherance of cooperative extension work in agriculture and home economics, acts of May 8 and June 30, 1914, in cooperation with the U.S. Department of Agriculture. Norman A. Brown, Director of Agricultural Extension Service, University of Minnesota, St. Paul, Minnesota 55108. The University of Minnesota, including the Agricultural Extension Service, is committed to the policy that all persons shall have equal access to its programs, facilities, and employment without regard to race, creed, color, sex, national origin, or handicap. **10 cents**

editor . . . . Mary K. O'Hearn
designer/illustrator . . . . Kelly Conner

OKAY, SO HOW FAST DOES A CHIPMUNK RUN?

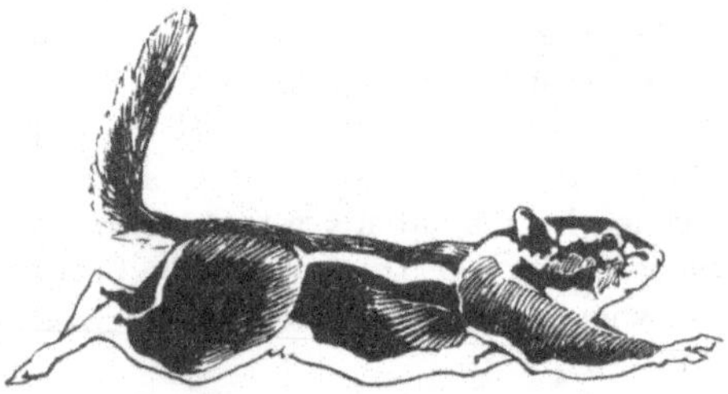

By Robert Kraske

A spunky chipmunk dashes across your lawn at 7½ mph. A squirrel, though, scampers up a tree as fast as it runs on level ground. How come? An inquiry into the speed of animals.

A white-tail deer bounding through the trees. A hooked pike streaking toward green lake depths, A ruby-throated hummingbird darting between flowers. Mammals, fish, and birds. How fast do they run, swim fly?

Wildlife watchers the world over ask these questions, but soon learn, once they delve into the matter, that accurate figures are hard to come by. The reason? Animals simply don't cooperate when it comes to speed trials over a measured course.

However, one fairly reliable way to time mammals and birds is to follow in a car while you watch the speedometer.

Roy Chapman Andrews, the famous naturalist, once trailed a herd of Mongolian gazelles in the Gobi Desert. "They ran so fast," he said, "that we could not see their legs any more than you can see the blades of an electric fan. We found they could leg it at sixty miles an hour for about half a mile and then slow down to forty or fifty."

Measuring two to three feet at the shoulder, a male black bear can flat-foot its 225-275 pound bulk to 25 mph.

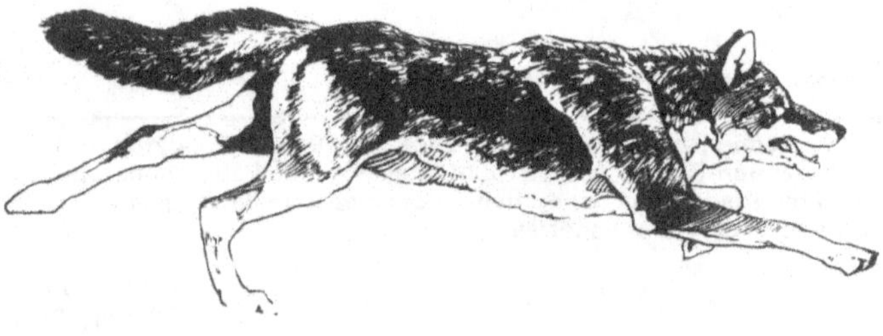

Wild relative of the domestic dog, symbol of wilderness, the timber wolf bounds after fleeing prey in short bursts of 40 mph.

White tail signaling alarm, a white-tailed deer bounds through open woodland at 35 mph.

Animals charging fleeing cars provide other figures. Lion, 50 mph; elephant, 25 mph; grizzly bear, 30 mph. The horn of one black rhinoceros prodded the tailgate of a truck traveling 35 mph.

The fastest mammal in North America? Naturalists agree on the pronghorn antelope. When paced by a car, one pronghorn ran 60 mph for two miles, then slacked off to 50 mph, and finally slowed to 40 mph.

Sixty miles-per-hour for one-half to two miles seems to be the highest sustained speed of any mammal, but it isn't the fastest. That honor belongs to the cheetah.

The London Times once reported on speed trials for these fleet cats. One cheetah accelerated to 45 mph in *two* seconds from a standing start. A few strides later, it was sprinting 103 feet per second, or 70-plus mph. No wonder rajahs in India once used these speedsters to hunt game.

Other animals have been timed on closed courses — the race horse and greyhound are two — and this is where stopwatch times come.

Man o' War, perhaps the greatest American racehorse, winner of 20 of 21 races, once ran a quarter-mile at 43 mph. For the longer mile, though, most racehorses run 35 mph. Champion quarter horses — a breed especially raised to sprinting 440 yards — hit 47.5 mph.

Fast animals like a horse or deer get an assist from long, lower-leg bones which provide a limber snapkick to propel these animals in swift, leaping strides.

Winning greyhounds at West Flagler Dog Track in Miami chase a mechanical rabbit at speeds up to 40 mph. Aided in full stride by a flexing spine, these fastest of all dogs run off the ground half the time, twice during each complete stride.

Speedy Fish/Birds

Calibrating the speed of fish is difficult at best. However, one fisherman found a unique way to measure the speed of a bluefin tuna. To his fishing rod he affixed a motorcycle speedometer and ran the line from the reel over a drive pully. Taking the lure, a 60-pound bluefin pulled 200 yards of 36-thread linen line through the water at a top speed of 44 mph!

An engineer, the fisherman later figured that the tuna, in pulling this length and weight of linen line at 44 mph, was towing the equivalent of 78 deadweight pounds. By extension, the tuna's feat is equivalent to a 150-pound man or woman running flat out while hauling a 200-pound sack of cement.

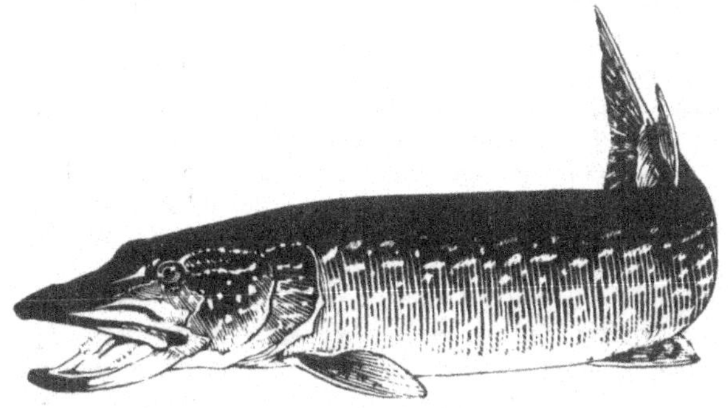

Thirty pounds and four feet of muscular challenge for the freshwater angler, the northern pike carves the water at 3½ mph.

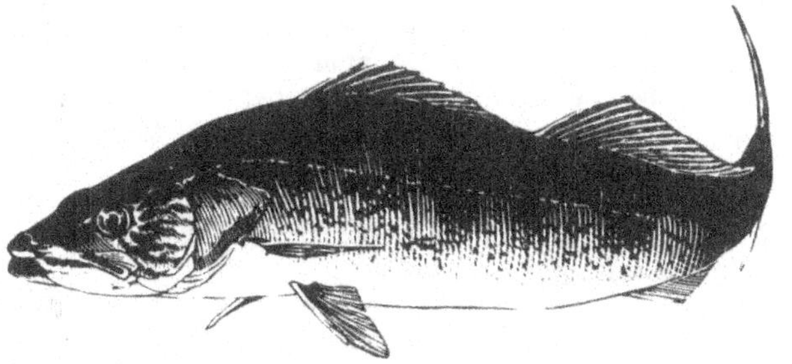

The night-feeding walleye favors yellow perch, minnows, and other small fish. Its top darting speed? About 3 mph.

One reason for the tuna's high speed is its tail which beats an incredible 10 to 20 times per second, as efficient, fisheries biologists say, as a rotating propeller on a boat. Another reason is special muscles under the tuna's skin that automatically adjust to slight changes in water pressure. These continuous muscle alterations create a smooth, non-turbulent flow of water along the tuna's flanks. In contrast, a turbulent flow robs fish — and boats — of speed.

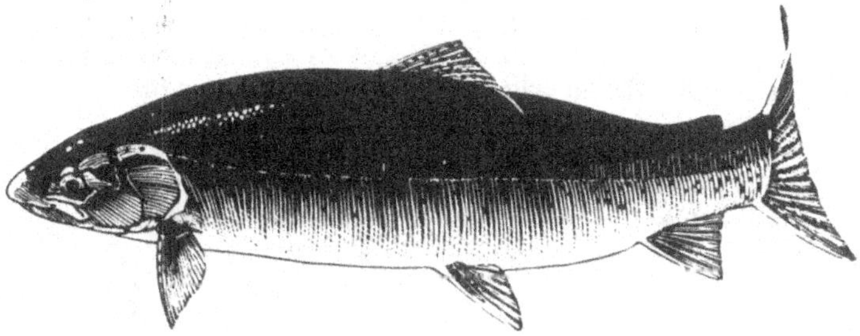

Wariest of trout, an immigrant (1884) from England, brown trout feed at night, spawn in fall and winter, and arrow through the water at 8½ mph.

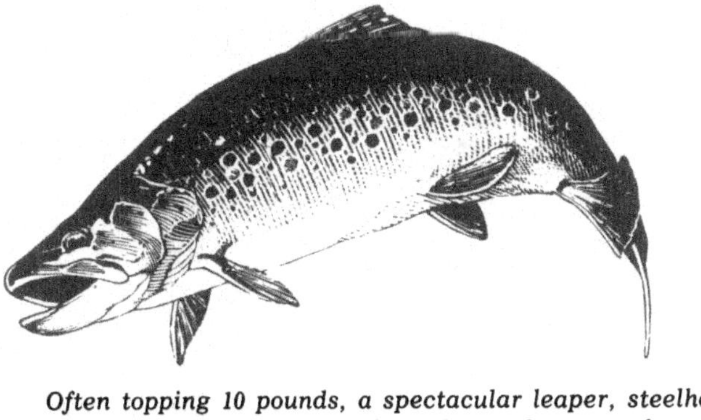

Often topping 10 pounds, a spectacular leaper, steelhead (rainbow) trout whirl a line through a reel at 18 mph.

The speediest animals of all are birds. The reasons are many: lightweight, hollow bones — the skeleton of a pigeon, for example, makes up only 4.4 percent of its body weight — well-developed lungs and circulatory systems; feathers shaping the body into streamlined contours; legs retracting in flight. All these features combine to produce smooth, swift passage through the air.

Among the fastest fliers are homing pigeons which have been clocked at 94.3 mph. Over an 80-mile route, pigeons have averaged 90 mph. One covered 182 miles at 73 mph.

Topping the pigeon though is the red-breasted merganser. One merganser was reported ahead of an airplane flying 80 mph into a 20 mph headwind — a ground speed calculated at 100 mph.

Before freeze-up, canvasbacks leave prairie nesting grounds in central Canada and, in wedge-shaped flocks, wing south at speeds up to 70 mph.

Biggest of game birds, the shy, elusive wild turkey sails a mile or more — with only an occasional wingbeat — at 55 mph.

Bib blazing red, the ruby-throated hummingbird darts from flower to feeder in spurts of 55-60 mph.

Most observers claim the fastest bird is the peregrin falcon. In level flight, its speed has been estimated at 66-75 mph; mourning doves, 55 mph; starlings, 45-50 mph; wading birds (sandpiper, curlew, etc.) 45-50 mph; crows, 40-45 mph; English sparrow, 35-40 mph; blue jay, 25-30 mph. A special radar once timed a quail at 44.5 mph and a pheasant at 38.2 mph.

Its wings partly closed, a peregrin falcon — perhaps nature's speediest creature — dives after prey. Her stooping speed: 180-200 mph or three miles per minute.

Food-thief, untidy nest builder, tough scrapper, the English sparrow cheers bleak winter days and wings from farmer's field to city lawn at 35-40 mph.

Fastest Man.

Among his fellow creatures on this planet, where does man fit as a speedster?

The fastest speed for competition runners has been recorded for 100 yards. In June 1963, Bob Hayes — "The World's Fastest Human" — set a world record of 9.1 seconds, a speed of 22.48 mph from a standing start. However, his top speed came between 60 and 75 yards. He covered those 15 yards in 1.1 seconds, 27.89 mph.

This speed capability puts man slightly behind wart hogs, emperor penguins, and domestic cats (30 mph), just ahead of galloping moose (25 mph), road runners (22 mph), and mountain goats (20 mph), but far in advance of running quail (15.5 mph), pigs (11 mph), and chickens (9 mph).

For sheer running ability among mammals, though, the nod has to go to the common squirrel. This furry speedster can run *up* a tree as fast as it runs on level ground — 12 mph.

Scientists who have studied this phenomenon claim that, to provide the spurt of energy necessary for its vertical dash, the squirrel increases its oxygen consumption. However, since the squirrel lifts only a pound or so, its oxygen increase is modest. In contrast, for a horse to run its 1,000-1,400 pound frame up a tree would require an oxygen increase of 630 percent.

Speed Ain't Everything.

Far down the list of animal speedsters is the giant tortoise who, in an all-out effort, can achieve 0.17 mph — back and forth along the length of a football field three times in one hour. The garden snail — 0.03 mph — would make it only to the 50-yard line.

Question is: If the tortoise and snail could go faster, would they? Speed, after all, isn't everything. Take the three-toed sloth, for example.

As a speedster — 0.15 mph — the sloth might challenge the tortoise. But it's not likely. Sloths avoid speed; they don't even like to move. A resident of South American forests, the sloth lives in trees and sleeps by hanging from a branch with one hook-like claw. So little does the sloth move that algae grows on its long, coarse fur.

Yet the sloth, for all its avoidance of speed, loves to feed on grasshoppers. How does this slowest of mammals capture the fleet grasshopper, an insect capable in a twinkling of leaping 20 times its own length?

Other animals that prey on grasshoppers rely on speed and dash, right? But not the sloth, Oh, no. The sloth uses what its admirers, and they are legion, claim is pure talent.

When a sloth spies a grasshopper on a branch, it cups its paws and ever so slowly places them on either side of its prey. Slow as a shifting shadow the paws come together. If the grasshopper notices anything, it is only a gradual dimming of light, like an eclipse at high noon. Then— *plink!* — off goes the light and grasshopper finds itself enfolded.

No rush. No fuss. Yet the sloth never starves. Like I said, it *(yawn)* takes talent.

Robert Kraske is the Associate editor of the Minnesota Volunteer. Drawings by Dan Metz.

MANAGING WOODLANDS FOR WHITETAILS

Jim Breyen, Minnesota D.N.R.

available. If a cedar stand or other dense conifer already exists within close proximity, planting conifer cover on your woodland will not be necessary. If cover is not available close by, a 40 acre tract that is comprised exclusively of aspen or other deciduous vegetation will likely be benefitted if a portion is planted to conifer. It doesn't take much. Three or four acres of evergreens planted in proper location should be adequate. Never establish more than 10% of your wood lot as winter cover, otherwise too much valuable summer and fall habitat will be sacrificed.

Although white cedar is considered the ideal winter cover species, balsam fir and jack pine run a close second. Cedar is more difficult to establish and is slower growing than balsam or jack pine, so you need to take that into consideration. In most situations, balsam has an edge over jack pine in preference by deer for winter cover. For a fast growing species such as balsam, the winter cover should begin providing protection in 12 to 15 years after being planted. Space the trees at a density of 300 to 500 trees per acre. In timber stands that already contain naturally growing balsam, make plans to preserve some of it as winter cover, up to 10% of the area.

In undulating terrain, establish the conifer (soils and hydrology permitting) in lowland areas or in valleys where the effects of wind are much less than on ridges or hill tops. On level terrain, it doesn't make much difference where winter cover is established except when a large open area adjoins the woodland on the west and/or north edges. Here you should establish the cover in the form of a dense windbreak on the north and west perimeter. Plant shrubs, such as honeysuckle, and small trees, such as plums, on the outside of the conifer.

Deer use conifer cover primarily for the protection it provides from harsh winter elements, but adequate browse on the edges is essential if the conifer is going to be attractive to wintering deer. Manage the edges of any winter cover area strictly for browse production so that deer do not have to travel far to meet their daily winter feeding requirements. Try to keep the browse strips less than 10 years old and at least 50 feet wide.

DUAL PURPOSE WOODLANDS

Dual purpose woodlands, those that are intended to benefit deer while at the same time grow conifers for income, limit the options for deer management, and thus, the carrying capacity for deer when compared to a well managed deciduous timber stand. But a conifer plantation that incorporates management practices designed to benefit deer will be more attractive than a plantation that is well stocked, fence row to fence row, with red pine. If you intend to establish a plantation on your property and want to maintain any semblence of deer habitat, you will have to sacrifice some evergreens.

A typical location that many landowners select for planting trees is an abandoned field that is located between blocks or strips of standing timber. The percentage of any open field that should be left unplanted is up to you, but the more conifer you plant the less desirable your land will be for deer. The long term effects of densely stocked pine and spruce plantations result in a shading out of understory vegetation that is preferred by deer. A thirty year old, well stocked plantation is essentially void of deer food, and it remains that way until the trees are harvested.

If you are managing for deer and trees, a general recommendation is to not plant open fields that are less than 10 acres in size. For fields larger than 10 acres, make plantings in blocks of 3 or 4 acres, leaving travel lanes of at least 100 feet around the blocks. And cut back on the stocking rate to 400 or 500 trees per acre. The travel lanes can be managed as wildlife openings and seeded to clover.

These recommendations are intended to provide general guidelines when developing a deer management plan for your woodlands. I suggest you contact a wildlife manager to obtain recommedations that pertain to your specific woodland tract. The extent to which you pursue recommended management practices depends, for the most part, on your available resources and size of your ownership. You do not have to accomplish everything overnight. Twenty-five percent of an even age aspen stand can be converted to the 1 to 10 year age class by cutting 10 acres in the first year, or by cutting on acre each year for 10 years. An opening can be put in when it is most convenient, and it doesn't have to be seeded to clover immediately. But a serious effort to incorporate the management practices I have discussed should bring more deer to your woodlands for you to enjoy.

is to coordinate deer management with adjacent landowners so that a larger area is affected. This, of course, is dependent on your neighbor having similar objectives for increasing deer numbers.

Aspen-birch stands that are older than 60 or 70 years of age need to be approached in a slightly different manner. Such stands are not likely to be around longer. The trees have reached the age where most are beginning to die and many have blown down. In this situation, instead of reducing the age of the aspen stand with a 40 year rotation, do it over 10 to 20 years. If you wait too long, there is a good possibility that not enough live aspen will be left to get sufficient suckering from the roots to rejuvenate and maintain the aspen type.

Oak and other northern hardwoods are a little more complicated to manage than is the aspen-birch type. But a management strategy similar to that used for aspen should make the stand more attractive to deer. The object is to cut some of the trees to allow more sunlight to reach the ground. Oak and red maple readily sprout from the stump, and the young saplings provide considerably more cover for deer than does a mature stand. The open canopy allows for a profusion of forbs and grasses for deer to graze on.

Oak stands should be cut in scattered blocks of 2 to 5 acres, but instead of cutting all the oak in a given block, leave 6 to 12 mature trees per acre as a source for acorns. Acorns are highly nutritious and deer will seek them out in years when acorns are produced. Since northern hardwoods are longer lived than aspen, rotation age should be longer. Strive for 10 to 15 percent of your wood lot in the 1 to 10 year age class.

In situations where oak or other northern hardwoods are growing on undulating terrain, invariably a fringe of aspen, dogwood and willow will skirt the hardwood fringe habitat, at least a good portion of it, less than 10 years old. Coupled with small cuts in the hardwood stand itself, the woodland will become more attractive to deer.

When small stand of oak (1/2 acre to 10 acres) are found within a much larger vegetative community dominated by other tree species such as aspen, efforts should be made to maintain the oak type as a mature, acorn producing stand. Some oaks eventually should be cut, but do so on a selective basis. Make sure the oak sprouts are given every opportunity to establish themselves as mature trees. Reduce competition from other trees and shrubs by cutting these away from the immediate vicinity of young oak sprouts.

WILDLIFE OPENINGS

Establishment of permanent grassy wildlife openings (grassy areas 1/2 acre to 2 acres in size that are without trees or shrubs) will complement a woodlands attractiveness to deer, especially in areas that are totally forested. Openings serve as an important source of highly nutritious food for deer, particularly in the spring and fall, and they appear to offer a habitat component that is beneficial to the social interactions of whitetails.

Whether or not you should establish openings depends on the size of your woodland and the proximity of open fields (hay fields) on adjacent land. Generally, if there is at least 20 acres in upland timber, an opening of 1 to 2 acres in the center will make it more attractive to deer. Try to place the opening on a gentle, south facing slope so that snow will melt sooner in the spring which make for earlier green-up of grasses and forbs. A rule of thumb is to have about 10% of your timber stand in wildlife openings. It is important that shrubs and trees be eliminated to realize the full value of a grassy opening.

If you have the time and want to maximize deer use of an opening, double-disc or otherwise disturb the soil and seed it to white dutch clover. Repeat this process every four or five years. Use by deer will be phenomenal, especially in the fall after frosts have turned all other vegetation brown.

Although deer are generally referred to as browsers, that classification is only partially accurate. For most of the spring, summer and early fall, deer rely almost exclusively on grasses and forbs for their food supply, meaning they are grazers for much of the year. Their reliance on browse (woody vegetation) comes about mainly out of necessity. In the winter time, green grasses and lush forbs are obviously not available, so deer are forced to rely on the less nutritious twigs of shrub species such as mountain maple, hazel, and red-osier dogwood. If it were not for the tremendous fat reserves that deer build up in late summer and early fall by feeding on high protein grasses and forbs, the poor food value of twigs would be insufficient to get deer through a normal winter. The primary importance of wildlife openings then is to provide an abundance of quality food that deer need in order to put on adequate fat reserves for the winter.

WINTER COVER

In northern forests, deer need conifer cover for surviving harsh winters, and some should be made

Wildlife managers are frequently requested to assist private individuals with the development of habitat improvement plans that can be used to make their woodlands more attractive to wildlife. The species that generates the most interest is the white-tailed deer. In fact, studies show a high percentage of private woodlands are owned specifically for recreational purpose, primarily for deer hunting. As a result, there is a strong interest on the part of the landowner to increase deer numbers on his land.

Although, there are general guidelines that can start the average woodland owner well on his way to more deer, each tract of woodland has its unique characteristics that dictate the specific habitat improvement practice that can be initiated to make it better deer habitat. Such things as size of the woodland, the percentage of upland, topography, and quality of adjacent habitat all come into play when evaluating each parcel and developing a habitat improvement prescription.

Generally, an over mature aspen-birch stand in northern Minnesota that is 60 to 70 years old and has not received treatment of any kind can be expected to provide habitat for about 8 to 12 deer per square mile, or about one deer for 80 acres. The same square mile, if managed intensively for deer, should harbor 30 to 40 deer once most management activities have been completed. Again, much depends on such things as the amount of upland, the amount of conifer, the severity of the previous winter, and the quality of adjacent habitat.

Deer populations in northern forested areas have a lower reproductive potential than deer populations found in agricultural areas or those found in the forested southeaster corner of the state. This is mainly due to a much lower percentage of female fawns in the north being bred in their first year. Similarly, winter survival rates are generally lower in the northern forests, because winter is consistently more severe. These factors make a difference on how rapidly deer respond to habitat improvement practices. It will take longer to build a deer population in the northern forests than it will in other parts of the state. So, if your woodland is in the north, don't compare the immediate results of habitat improvement to those of a friend who lives in the southeast. Give it time. The long-term results will be worthwhile.

Most private woodlands are comprised of relatively mature timber, and many are of the aspen-birch type. If there is any one recommendation that can be considered universal when discussing deer habitat improvement for private aspen-birch woodlands, it has to be: cut some of the trees. Deer populations are highest in habitats comprised of the early stages of forest succession.

Cutting prescriptions that benefit deer do not call for large cuts, and cuts can be spread out over a number of years to minimize disturbance. It is amazing how rapidly an aspen-birch cut-over rejuvenates, with only three or four years required for saplings to reach 10 feet or more in height on most sites. The dense new growth provides an abundant source of quality deer feed, and is also ideal cover for fawns. The increase in the amount of food means that more deer are able to obtain their required needs on a particular tract of land. Although leaves and sprouts of young aspen are eaten, it is the associated lush growth of grasses and forbs found in the young aspen stands that supply most of the feed preferred by deer. This lush and varied growth allows the deer to be more selective in their summer diets and to obtain the most nutritious intake of foods.

Once an aspen stand reaches twenty years of age it begins to slowly deteriorate as quality deer habitat. It your desire is to manage your woodland strictly for deer, then the majority of trees should be kept in the one to twenty year age class. But recycling aspen at 20 years of age can be an expensive proposition. A more common procedure is to manage the timber on a 40 year rotation where the stand is cut in 10 year increments, keeping 25% in the 1-10 year age class, 25% in the 11-20 year age class, and so on. Allowing the trees to reach 40 years of age makes it easier to attract loggers to do your cutting.

The size of a woodland dictates how detailed you will have to be when making timber cuttings. If you have placed a 40 acre timber stand on a 40 years rotation, then four - 2-1/2 acre cuts or two - 5 acre cut are all that is required, initially, to obtain 25% of the stand in the 1 to 10 year age class. Cuts should be scattered throughout the woodland, and not concentrated. When the initial cuts reach 10 years of age, added cuts should be made to keep 25% of the stand in the 1 to 10 year age class. For woodlands greater than 40 acres in size, cuts can be correspondingly larger, up to 10 acres, but maintaining 25% of the stand in age class 1 to 10 should remain the primary goal. One thing you may want to consider, if your ownership is limited,

HOW VARIOUS SPECIES OF TIMBER AFFECT WILDLIFE

North Central Forest Experiment Station
Forest Service U.S.D.A.—St. Paul
Excerpts From Several Publications

Aspen

Game Species

Ruffed grouse utilize aspen stands of all ages. Juvenile sucker stands at age 2 (12,000 to 14,000 stems per acre) up to about age 10 (6,000 to 8,000 per acre) are important brood habitats for grouse. Sapling and pole stands aged 10 to 25 are preferred overwintering and breeding cover. Aspen stands older than age 25 (when stem densities usually fall below 2,000 per acre) are devoid of breeding grouse but serve as nesting cover and as a very important food source. A primary year round food of grouse is aspen leaves and buds, best provided by stands nearing maturity. The staminate (male) flower buds of aspen are the most important nutritive source for grouse. Some male clones preferred by grouse may be 30 percent richer in proteins than male clones that are not eaten. Finally, snow accumulates earlier and deeper in aspen stands than in conifer stands. This provides burrowing cover, which is very important during most winters.

To increase ruffed grouse, aspen should be clearcut on a 40 to 50 year rotation, in patches no larger than 10 acres in each 40 acres, and at 10 year intervals. Male clones that grouse prefer should be favored during intermediate thinnings and during regeneration cuts (see "Expand Plus Clones"). Intermediate thinngs or short rotations are not recommended for grouse management.

White-tailed deer rely heavily on the aspen type, especially for spring and fall range, and for winter range within ½ mile of winter cover types. Herbaceous and shrubby growth associated with aspen is usually more abundant because the intolerant aspen admits more sunlight to the forest floor than do the more tolerant hardwoods and conifers. The quality and availability of herbaceous vegetation in the spring and fall greatly affect the vigor of northern deer herds. Conifer cover during winter to minimize body heat loss is even more important. Pure aspen or hardwood stands offer poor insulation and protection from wind compared to dense stands of lowland conifers (especially northern white cedar), balsam fir, or pines.

Deer populations can be increased by limiting aspen clearcuts to 40 acres (preferably 20 acres) and by short rotation management (25 to 30 years on 5 to 10 year intervals) of aspen stands within ½ mile of winter deer yards. (The cutting schedule recommended for grouse could also be applied for deer, with somewhat reduced benefits.) Ideal deer range should be 15 to 20 percent scattered conifer stands, and 5 to 10 percent sodded and brushy openings (which are important spring and fall feeding areas). Hardwood forest types should contain 25 to 35 percent of their area in aspen stands, and 25 percent of the aspen should be 1 to 10 years old.

Moose are dependent upon the aspen community to provide a large amount of browse. For moose management, clearcuts can be up to 100 acres; the stand composition should be similar to that for deer.

Rare and Endangered Species

Three rare or endangered species using the aspen type are the bald eagle, osprey, and eastern timber wolf. All are protected by Federal and State laws. The following tabulation lists restrictions on management activities for osprey and eagle nest trees:

Distance from Nest
Up to 350 ft.
350 to 650 ft.
700 and beyond
700 ft. to ¼ mile

Osprey nests
No activity anytime
No activity March to July
Normal activities OK

Eagle nests
No activity anytime
Thinning and pruning OK (no clearcutting) October to mid-February; no activity rest of year.
Normal activities October to mid-February only; no activities rest of year.

If areas more than ¼ mile away are visible from the eagle nest, the outer zone can be extended to ½ mile in that direction. Roads and trails within ¼ mile of eagle nests should be closed where possible. Scattered old growth trees, particularly the pines, should be reserved as much as possible for future nest trees.

Black Spruce

The black spruce type is utilized to some extent by many wildlife species. New harvest areas and young stands certainly producé different or more abundant browse and other wildlife food than mature black spruce stands. Therefore, shrubs and hardwoods should not be killed back with herbicide spraying until black spruce reproduction definitely needs release. And even then, all dogwood, willow, quaking aspen and other hardwoods should not be killed because some mixture of these shrubs and trees with black spruce probably enhances wildlife habitat.

The spruce grouse is of special interest because it depends on the black spruce type for most of its habitat needs. Spruce grouse apparently use noncommercial black spruce sites more than commercial sites, so their habitat may be only moderately affected by timber management. However, spruce grouse habitat can probably be maintained or enhanced on commercial sites by having the following kids of black spruce stands within compartments of 160 acres or less:

1. Mature stands with high basal areas (more than 150 square feet per acre) are important as display habitat for male grouse in late spring. These stands are characterized by little or no undergrowth and a ground cover of feather moss.
2. Young stands of black spruce 10 to 15 feet tall with dense shrub and herbaceous layers are used by female grouse for cover and feeding before and after nesting. Labrador-tea and leather-leaf are usually the dominant species of the undergrowth and sphagnum moss of the ground cover.
3. Mature spruce stands with moderate basal areas (80 to 100 square feet per acre) and little or no undergrowth, as well as young spruce stands with a dense shrub cover, are used for nesting.
4. Black spruce stands with trees 20 to 60 feet tall and about 150 square feet of basal area per acre are important habitat for both sexes in fall and winter.

Therefore, to provide the overall hábitat needs of spruce grouse, harvesting of black spruce should be planned carefully so that each compartment will have the kinds of stands just described. This can be done by clearcutting in strips or patches that are well distributed in the compartment and overtime. Further, the objective should be to break up extensive, pure stands of black spruce because transition zones with other forest types and some mixture of tamarack seem to benefit spruce grouse. Recording the kind and density of undergrowth in spruce stands during forest inventory would aid the manager interested in coordinating timber and grouse management.

Oaks

Oak forests provide habitat for numerous wildlife species. The principal game species include white tailed deer, turkey, fox and gray squirrels, and in some areas, ruffed grouse. Other important species include raccoon, opossum, red fox, bobcat, skunk, and a host of birds.

Creating and maintaining diverse vegetation is the key to providing a large variety of wildlife with suitable habitat. Regeneration clearcuts should be planned for a whole rotation and dispersed throughout a compartment or group of compartments to provide a well regulated range of age classes. This will result in several vegetation stages ranging from open, recently clearcut regeneration areas, through areas of saplings, poles, immature sawtimber, and mature sawtimber. Each of these stanges contributes to the habitat requirements of different groups of wildlife species.

Even so, special measures to enhance the habitat for species with specific requirements may sometimes be needed. Many of these measures will result in lower timber yields, but to what extent is unknown. During thinning operations, a few defective trees can be left to provide cavities or potential cavities for hole nesting wildlife species. About 30 cavity nesting bird species inhabit oak forests. In addition to cavities, some of these birds need an open overstory and a well developed mid story — conditions often found in old growth or overmature stands. Retaining selected stands beyond the normal rotation age will provide these conditions.

Leave dead snags standing in clearcut areas and kill unmerchantable trees during thinning operations instead of cutting them, so as to provide sites for hole nesting species. This should have little or no effect on timber production.

One of the most important contributions to wildlife from oak stands is mast, or acorns. Acorns are, of course, vital for regenerating oak, and any measures that will increase or optimize acorn production in a particular stand will benefit both wildlife and timber production. In general, thinnings that increase tree growth will also stimulate acorn production. Maintaining 40 to 60 percent of the area of each compartment in stands of mast bearing age will be optimum for wildlife. In compartments that contain a relatively high proportion of poor sites that would produce more timber if converted to pine, this may mean foregoing conversion and the higher yields it would bring.

High populations of acorn consuming wildlife can be detrimental to establishing oak reproduction. Even in good acorn years, essentially the entire crop may be lost in such areas. Acorns not damaged by insects will be eaten by squirrels, turkey, deer, or other acorn consumers, and thus not be available for reproducing the oaks.

Northern White Cedar

Deeryard management in the northern white cedar type is affected by the interaction of several factors such as yard size and condition, deer density, winter severity, and availability of browse in and around the yard. These factors very substantially in different parts of the Lake States and are often difficult or impossible to control. Further, considerable knowledge on managing deeryards in the white cedar type is based on experience rather than research, because the latter has been done mainly in areas with large yards and deep snow in upper Michigan. So, for these reasons, the present recommendations tend to be general and often cannot be applied directly to the various yarding situations found in the white cedar type. This means that to obtain desired results the manager must use good judgment and modify the recommendations to fit local conditions.

Whenever possible, deeryard management should be concentrated on areas that have: (1) special importance as traditional or potential yards; (2) deer densities that do not exceed the carrying capacity; (3) a site index of 30 (medium) or higher for northern white cedar; and (4) enough timber for commercial cutting.

In large deeryards (200 acres or more), the long range objective should be to organize compartments that contain five age classes each, with 15 to 20 years between classes, in patches of 40 to 160 acres. The size and distribution of these patches should be planned carefully so that adequate deer shelter and browse will always be available on separate patches within each compartment. Cutting of any kind should be done in as many compartments as possible to distribute the deer herd more and to rehabilitate vital yards faster. Of course, northern white cedar and hardwoods should be cut only during winter to provide deer with browse. Annual cutting is necessary to adequately feed deer where deep snow normally keeps them yarded most of the winter. Such cutting is also desirable where deer can move about more because it tends to attract them away from patches of young white cedar, which is vulnerable to overbrowsing.

The main objective in small, isolated yards (less than 200 acres) should be to obtain and maintain a closed evergreen canopy for optimum deer shelter. Specific practices for achieving this objective are prescribed and discussed under "Intermediate Treatment". Browse is usually scarce in small yards, so removal of hardwoods or northern white cedar should be done only by cutting during severe winters. This means most browse will have to come from young stands and winter cutting of other forest types in the surrounding area.

In areas where small yards are vital for deer shelter, the rotation can be extended well beyond the longest ones because northern white cedar is long lived. Eventually, however, these yards will need to be rehabilitated. Since they are probably too small to handle the series of age classes recommended for large yards, the whole yard should be reproduced as a single even aged stand by strip cutting. The time when a small yard is cut should be planned carefully so that, if possible, adquate deer shelter will be available elsewhere in the vicinity.

The white cedar type is utilized to some extent by many wildlife species besides white tailed deer. New openings and young stands certainly produce different or more abundant wildlife food than mature white cedar stands. For example, young stands should support substantial populations of snowshoe hare and their accompanying predators as soon as the tree crowns begin to close. Therefore, shrubs and hardwoods should not be killed back with herbicide spraying unless the growth or proportion of white cedar reproduction definitely needs to be increased. And even then, all stems should not be killed because a mixture of shrubs and hardwoods with white cedar and other conifers probably enhances wildlife habitat in general.

Some trout streams have their source in areas occupied by the northern white cedar type or they pass through such areas. So, to keep the water cool, areas cleared for new stands should probably: (1) have an uncut border between them and the streams; (2) not exceed 40 acres each; and (3) total only a small proportion of the surrounding watershed.

Red Pine

Red pine stands are generally considered poor habitat for game birds and animals but they provide cover and nesting sites for many species of wildlife. Large old growth trees are used by the American bald eagele as well as many songbirds.

Although red pine stands offer good shelter for wildlife, many of the favored food plants are not found in the understory. Managing stands near the minimum recommended stocking will favor a greater variety of understory plants. Prescribed burning may also be effective in developing a more favorable understory for wildlife food. Carefully planned landings can serve as wildlife openings providing some of the food plants needed. Landings should be at least ½ acre for an effective wildlife openings.

Northern Hardwoods

Northern Hardwood types harbor numerous kinds of birds and animals, both game and nongame. In northern hardwoods as in other forest types, diversity of species and size class leads to wide representation of animals and birds. Normal forest management can produce sufficient diversity to suit many wildlife species. In some cases, however, a single tree species must be favored because of its special benefit to wildlife.

In areas of deep snow and intense cold, coniferous species provide essential cover for white tailed deer. In some portions of the Lake States, for example, eastern hemlock provides a significant amount of cover and browse so that the regeneration and maintenance of this species is an important part of management. Hemlock is a long lived, very tolerant species that requires mineral soil seedbeds for regeneration. The same general procedures used for regeneration of yellow birch can be used for hemlock, except that hemlock should be seeded. It is necessary to seed this species in the fall or stratify seed artificially for 90 days before spring seeding. Because of the long life of hemlock, long rotations and small annual cutting areas are possible.

Where tolerant northern hardwoods border conifer swamp deer yarding areas, the amount of browse can be increased by increasing the frequency of cut and reducing residual stocking levels. Where even age management is used, a three cut shelterwood can be employed to increase the length of time that browse is available.

Some rules of thumb for management of browse for wildlife are: selection cutting procedures small amounts of browse per acre over a short period of time and relatively many plant species, but the requirements for producing good timber are frequently not met; shelterwood produces as much browse as clearcutting and meets timber production requirements but encourages fewer plant species.

A forest manager need not always modify practices in northern hardwoods to increase wildlife browse or shelter if other forest types that border or are interspersed with the northern hardwoods can provide these wildlife requirements.

Extensive heavy cutting in the past has resulted in large, unbroken areas of pole-size northern hardwoods whose understories are barren and poor habitat for birds and animals. Although small openings usually revegetate quickly and lose value for wildlife, their life can be extended to 20 or 30 years by: (1) cutting them in stands 50 years old or less where there is little or no advanced reproduction; (2) cutting them on either excessively poorly drained soils, or on soils that are shallow or in frost pockets. Openings should be at least an acre in size, not more than 10, and irregularly shaped.

Northern hardwoods frequently border streams and lakes where they may influence water quality, water temperatures, and fish populations. Silvicultural practices to enhance or preserve these features probably do not vary from those used in other forest types.

Jack Pine

Jack pine is generally considered a medium preference deer food, the same as aspen. Young trees may be heavily browsed where deer populations are high. Dense sapling and pole stands offer some wind protection and winter shelter but generally jack pine stands do not provide as good winter shelter as most other conifers. Because older stands of jack pine are usually less dense than other conifers, the understory shrubs and herbaceous plants have better growth and thus provide a better food supply.

Some wildlife species benefit from special stand conditions such as the excellent cover that dense young stands of jack pine provide for hares. Clumpy stands of young trees with branches reach the ground provide nesting sites for the endangered Kirtland's Warbler in the Lower Peninsula of Michigan. Most wildlife species that find food or shelter in jack pine forests will benefit from management efforts to provide a good distribution of age classes.

Chapter 20
Integrated Pest Management
Insect and Disease Management Guidelines

The use of cultural control methods in forest stands can no longer be ignored in stand and forest management decisions. Direct control of insect and disease agents in established stands can be a controversial, difficult, and/or expensive process. The proper implementation of silvicultural guidelines can reduce the severity of insect and disease outbreaks. Integrated pest management (IPM) is an approach to insect and disease control that utilizes a combination of cultural, biological, chemical, and/or mechanical techniques to achieve economical control in an environmentally sound manner. Integrated pest management can reduce the occurrence, severity, and spread of insect and disease problems and thereby lessen the problems associated with direct control.

The difficulty in instituting an integrated pest management program is that each stand must be considered both individually and as a part of the whole forest. Stand planning and regeneration is an extremely important aspect of cultural control. All stand manipulations affect the occurrence and severity of insects and disease. Once a stand is established many of the parameters governing insect and disease control are unalterable. In developing silvicultural guidelines species selection, regeneration methods, site characteristics, potential and historical pest problems, future stand manipulations, desired products, economics, rotations, and harvesting techniques must be considered. Prevention or suppression of future pest problems by integrated pest management techniques must be a major goal in forest management.

Specific pest management guidelines must be developed for individual stands. However, some **general** integrated pest management guidelines exist:

1. Consult with regional insect and disease specialists in future stand management planning, or when problems occur in established stands.
2. Check all stands routinely for insects and disease. Promptly remove, destroy, or chemically treat infested material where economically feasible.
3. Utilize regeneration methods that reduce or eliminate insect and disease potential for attack of young seedlings. See specific recommendations by type.
4. Favor insect and/or disease resistant species, or varieties.
5. Avoid replanting susceptible species in areas of historical insect and/or disease problems without an evaluation of cause.
6. Do not mix species which have common insect and disease problems.
7. Match tree species to the planting site.
8. All management practices must be aimed at promoting stand vigor.
   a. use proper planting techniques
   b. manipulate rotation age
   c. maintain proper stand density
9. Avoid planting frost pockets.
10. Avoid wounding growing stock during thinning and/or harvesting operations.
11. Diversify species whenever possible.

ASPEN TYPE
Insect and Disease Management Guidelines

Two defoliators and wood boring beetles are the major insects of Aspen. The Forest Tent Caterpillar, *Malacosoma disstria,* and the large aspen tortrix, *Choristoneura conflictana,* occasionally defoliate areas of several thousand square miles. Severe defoliation reduces growth but rarely causes mortality unless coupled with other stress. Wood boring beetles of the genus *Saperda* cause increased wind breakage and lumber and veneer degrade. As much as 64% of all mature aspen may be attacked. Larval tunnels serve as infection points for canker causing and wood rotting fungi. See Forest Insect and Disease Leaflet, #9 U.S.D.A. Forest Service.

The major diseases of Aspen are Hypoxylon canker *Hypoxylon mammatum* and white rot *Phellinus igniarius*. Hypoxylon causes annual losses which approach the net annual aspen growth. White rot decay can reduce gross merchantable yields by up to 10% with 50% of the trees infected. See How to Identify Hypoxylon Canker of Aspen and How to Identify and Minimize White Trunk Rot of Aspen. U.S.D.A. Forest Service North Central Forest Experiment. Management recommendations are as follows:

1. Check all aspen stands routinely for insects and disease.
2. **Rotation:** As stand age increases volume losses due to insects and disease increase.
   a. If 15 to 25 percent of the trees are infected with Hypoxylon, harvest the stand early and treat the site to encourage good aspen reproduction.
   b. If more than 25 percent of the trees are hypoxylon infected, harvest immediately and convert to other species.
   c. Lightly infected stands can be managed on rotations longer than 40 years.
   d. If surveys indicate heavy white rot infection over 30% of basal area/ acre of trees (with rot) a pathological rotation of 35-40 years is required to minimize losses.
   e. To estimate the total amount of white rot in an aspen stand, first determine the basal area of trunks with visible conks and then add 90% of this area to account for hidden decay.
   f. In stands that have sustained 2 to 3 years of successive defoliation, rotation age can be adjusted with years added if maximum fiber production is desired in the absence of severe Hypoxylon or white rot infection.
3. **Stocking:** If 30 to 40 thousand new stems per acre are produced after cutting or burning and densely stocked, uniform stands of pure aspen are maintained throughout the rotation, insect and disease losses will be reduced.
   a. Conduct harvesting operations during the winter to regenerate dense stands.
   b. Remove all residual species to regenerate pure and dense stands of sprouts.
4. **Site Index:** As low site index aspen (50) is more susceptible to insect and disease loss than high site aspen (70), good sites for aspen production should be favored.
5. **Species and Clonal Varieties:** The extent of losses due to insects and diseases varies within aspen species and between clones.
   a. Since bigtooth aspen is five times as resistant to Hypoxylon as quaking aspen, and balsam poplar is rarely infected, these species should be promoted in heavily infected areas.
   b. Selection of superior clones for their expansion in stands should include rating of their susceptibility to white rot and Hypoxylon.

BLACK SPRUCE - UPLAND AND LOWLAND
Insect and Disease Management Guidelines

The major problems on black spruce are dwarf mistletoe, *Arceuthobium pusillum,* and root and butt rots. Black spruce is attacked and killed in all stages of its development by dwarf mistletoe. Root and butt rots caused by *Armillariella, mellea* and *Inonotus tomentosus* are present in all stands over 30 years of age. Losses from root and butt rots may range up to 40% of the merchantable volume of the stand. Rots are the major contribution factor to wind damage.

(see: How to identify Easter Dwarf Mistletoe on Black Spruce by NCFES)

Dwarf Mistletoe is best controlled at the time of harvest. Specific harvest principles to consider include:

1. Kill all black spruce stems over 5 feet tall. Require this on all timber sales since it can be difficult to determine if an

individual tree is infected or not. Control is achieved by killing infected trees because misteletoe survives only on living trees.

2. Adjust timber sale boundaries to include the mistletoe pockets plus a two chain buffer strip on non-infected (no witches brooms) black spruce sites. This is to remove latent infections that cannot be seen.
3. Locate landings in mistletoe pockets if possible.
4. Mistletoe control should be a factor in preparing planned cutting lists to control mistletoe pockets while they are still small.
5. Establish buffer strips between infected and uninfected stands primarily between recent cutover sites and infected unmerchantable residuals.

Timber harvests with the 5 foot cutting rule included may not achieve adequate control of misteltoe in the following situations:

1. The stand is too heavily infected with mistletoe to get a logger to purchase the sale if the 5 foot cutting rule is included.
2. The stand to be cut is bordered by unproductive black spruce site infected with mistletoe.
3. The stand is immature but so badly infected with mistletoe it will never be merchantable and a new stand must be regenerated.

In these type of situations, other methods must be used to kill all the residual black spruce on the site, in an infected edge, or where entire immature stands must be removed. Methods which may be used are listed below:

1. If the acreage or the number of trees involved is small, a hand crew can be utilized.
2. With large projects and immature stands, it may be more economical to use a caterpillar with a KG blade to shear the standing timber.
3. A skidder can be used to shear and push over infected trees. If the site is being harvested, the most economical method would probably be to hire the logger who is harvesting the timber. A prescribed burn may have to be used in combination with this method to insure that small trees are killed and not just pushed over.
4. Prescribed burning is a proven method of eradicating dwarf mistletoe. Whenever possible, slash disposal regulations should stipulate that the slash will be lopped and evenly scattered on infected sites. Slash free alleys with a minimum width of 16 feet should be required.
5. In immature stands on better sites where the acreage or number of trees involved is small, so the entire stands do not have to be destroyed, direct control of affected areas (killing of ihfected trees) may be used.

Root and butt rots become critical in upland spruce sites, on mineral soils, which range between 70 and 100 years of age, and in swamp stands, on organic soils, when stands reach 100 years of age. Once rots become established, clearcut the stand to avoid losses. Treatment of just the pocket of rot is not sufficient since many nearby trees showing no symptoms are also infected.

Root and butt rots caused by *Inonotus tomentosus* seem to be most prevalent under the following conditions:

1. acidic soils with pH's ranging between 4 and 5;
2. soils low in nutrients;
3. soils low in moisture holding capacity;
4. and in situations in which root depth is limited by shallow soils or a hard pan.

If an upland stand of black spruce is growing under these conditions and rot is evident, clearcut the stand and convert to pine, balsam fir or hardwoods.

CENTRAL HARDWOODS

Insect and Disease Management Guidelines

The greatest volume losses in Central Hardwoods are the result of disease organisms, which discolor, decay, or deform standing timber. Silvicultural recommendations for these problems are discussed under the Northern Hardwood guidelines.

Dutch elm disease, Butternut decline, and terminal shoot losses in Black Walnut are serious problems in Central Hardwood Management. Dutch elm disease management guidelines are discussed under lowland hardwoods. Butternut canker, caused by the fungus *Sirococcos clavigigneti,* is resulting in severe losses throughout Minnesota. This disease has eliminated Butternut as a viable forest management species in Minnesota.

Black Walnut regeneration occurring throughout southern Minnesota is stunted and deformed. Deformed and reduced growth in Black Walnut regeneration due to loss of the terminal shoot is widespread in Southern Minnesota. The most significant cause of shoot mortality is late spring frost. Terminal bud loss is also caused by the Walnut tip moth *Acrobasis dometella*.

A canker disease apparently linked to summer pruning is currently under investigation.

See: How to Identify Butternut Canker, Northeastern Area State and Private Forestry, USDA Forest Service.

Management Recommendations are listed below:

1. All pruning in hardwood stands should be conducted during the dormant season.
2. Butternut regeneration should be discriminated against in all management practices.
3. Walnut Site Selection:
   a. avoid known frost pockets
   b. avoid planting in soils where the clay composition exhibits red, yellow, or grey mottling within the first three (3) feet of the surface.
   c. Restrict walnut planting to well-drained loams and clay loam soils at least three (3) feet in depth.

JACK PINE

Insect and Management Guidelines

The major insects attacking jack pine are the jack pine budworm, *Choristoneura pinus*; the pine tussock moth, *Dasychira pinicola;* white pine weevil, Pissodes strobi; and bark beetes, *Ips* spp. Insect damage includes deformed stems from weevil attacks, top-kill from budworm defoliation, or tree mortality resulting from bark beetle attacks or heavy defoliation by budworm and tussock moth.

The major diseases of jack pine include heart rots, particularly *Phellinus pini;* Armillaria root rot, *Armillariella mellea*. Armillaria root rot and the North American strain of Scleroderris canker are primarily confined to young trees. The European strain of Scleroderris, not yet present in Minnesota, has the potential of causing widespread mortality in all ages of jack pine. *P. pini* causes about 90% of the heart rot in all ages of jack pine, and this decay becomes more prevalent as the trees get older.

Control strategies for the white pine weevil can be found under the white pine type guidelines. A discussion of bark beetles, Armillaria root rot, can be found under the red pine guidelines.

Control strategies for the budworm and the tussock moth are as follows:

1. Use direct control strategies for budworm and tussock moth control in stands which are not salvageable or stands in which populations have built up high enough to cause damage. Consult the Regional Pest Specialist for specific control recommendations.
2. Large-crowned, open grown trees (wolf trees) and trees in the suppreseed and intermediate crown classes contribute to a budworm build-up due to the high production of staminate cones. Therefore, maintain an optimumly stocked stand, between 70 and 100 square feet of basal area to reduce the numbers of wolf trees and suppressed trees.
3. Stands with a site index of at least 55 should be favored in order to maintain vigorously growing stands of jack pine.
4. On good sites, 60+, and where economically feasible, thin jack pine from below to reduce the proportion of suppressed

and intermediate trees. Thin to 80 square feet of basal area, and thinning should be done as soon as a commercial sale is possible.

5. Where economically feasible, cut out the wolf trees in the stand.
6. Do not store mature jack pine on the stump. Rotation ages, based on budworm considerations, should be between 45 and 50 years. On only the very best sites, 70+, should jack pine be held longer than 50 years. The poorer the site, the earlier the harvest should be.
7. Break up large, extensive stands of jack pine with 2 to 5 chain wide buffer strips, hardwood stands, white spruce, or larch plantings. These will help to disrupt the dispersal of the budworm. Even-aged stands of jack pine should not exceed 20 to 40 acres. Do not break up the jack pine stands with other species of pine. During a budworm outbreak, other species of pine will be damaged by budworm feeding and may even aid in carrying over a budworm population during the low point of its population cycle.

Heart rot control strategies are as follows:

1. The major entry points for heart rotting fungi include mechanical wounds in the hole, rust cankers, dead branches or dead and broken tops. Therefore, when intermediate cultural activities take place in jack pine stands, care should be taken to avoid wounding the residual trees.
2. Priority for harvesting should be based on stand condition and stand age as follows:
   a. Stands with bole wounds or with broken and damaged tops possibly from adverse weather conditions should be given a high priority for harvesting, and
   b. Stands at or beyond the recommended rotation age should have a high priority for harvesting.

LOWLAND HARDWOODS
Insect and Disease Management Guidelines

The greatest timber losses in lowland hardwood forests are the result of disease organisms which deform, decay, or discolor standing timber. The two main causes of loss are canker diseases and wood decay in the form of heart, top, or butt rot. A discussion of these problems and general control guidelines can be found within the discussion of northern hardwood management.

Dutch elm disease, *Ceratocystis ulmi*, is one of the few diseases which causes widespread mortality in lowland hardwoods. This disease has essentially eliminated native elms as a viable management option. Dutch elm disease is presently spreading throughout the State of Minnesota. High value elm should be scheduled for removal to avoid loss of volume and/or value. Elm regeneration should be discriminated against in all management practices.

NORTHERN HARDWOODS
Insect and Disease Management Guidelines

Northern hardwoods suffer from very few insect or disease problems that result in tree mortality. The majority of losses in northern hardwood stands result from diseases which deform, discolor, or decrease the growth of standing trees. Approximately three quarters of all northern hardwood volume losses in Minnesota are due to decay and discoloration of standing timber. Wood decay caused by a variety of fungi can greatly affect management alternatives. Canker diseases r.g. Nectria *Nectria galligena* and *Eutypella parasitica* cause significant losses by reducing tree growth and weakening structural support. Canker also provide entrance for decay organisms.

Wood decay in the form of heart rot, butt rot, and top rot can be the result of actions by various fungi. The initial infection of a tree occurs through wounding. The three main factors initiating wood decay are logging, fire, and copice regeneration. The impact of wood decay can be reduced by improved silvicultural practices.

Canker diseases are also introduced through wounding. These diseases can lead to tree death, wind breakage, log degrade, heart rot, or significant growth reduction in hardwoods.

See: 1. How to Identify Nectria Canker on Hardwoods publication of the Northeastern Area State and Private Forestry, USDA Forest Service.
2. How to Identify and Minimize Damage Caused by Eutypella Canker of Maple. NA-FR-IO, USDA Forest Service, Northeastern Area State and Private Forestry.

Silvicultural practices should be maintained which promote or protect high quality hardwoods. Suggested practices are as follows:

1. Management should be based on a crop tree concept.
2. Sale specifications and guidelines are excellent tools for preventing future losses to wood decay by reducing damage to trees during logging operations. Pre-sale planning to locate skid trails, landings, and marked trees can greatly reduce injury to growing stock.
3. Promote elm harvest and discriminate against elm regeneration in all management practices.
4. Reduce or eliminate the impact of wildfires.
5. Avoid high grading in hardwood stands.
6. During thinning and harvesting operations remove cankered, or decayed trees (remembering wildlife considerations). Fuelwood demand should promote removal of infested material from the stand.
7. Favor sprouts originating on the root collar in regeneration and release practices.

NORTHERN WHITE CEDAR
Insect and Disease Management Guidelines

Compared to most other species, Northern White Cedar is a relatively insect and disease free species. There are no insects which cause significant damage. The black carpenter has caused some damage by attacking the heartwood of living trees, but this is always in trees which already have been extensively attacked by wood rotters. Wood rotting fungi *Pori subacida*. causing white stringy butt rot and *Polyporus spp*. causing brown cubical rot; however, can be a serious problem, mainly in overmature timber.

These are butt rotters, the rot column of which seldom extends more than 16" into the trunk of the tree. This has resulted in a volume loss of approximately 25-30% in cedar managed for sawbolts. This is because much of this type has been high-graded in the past and the remaining timber is overmature.

Specific management recommendations are listed below:

1. Northern White Cedar grows best on shallow, well-drained, slightly alkaline, well-decomposed organic soil and on moist, though well- drained slightly alkaline mineral soil. Sites other than these should be considered for management for other species such as tamarack, balsam fir, and black spruce.
2. To avoid wood rotting problems, grow stands to recommended rotation ages of approximately 80 years on good sites and 90 years on medium sites. Better site stands which show indications of much rot should be cut as soon as practical considering the need for the establishment of advance regeneration. Much of the time it is very difficult to determine the presence of rot, however, possible indicators of internal rot are the presence of basal fire scars, frost cracks, wind cracks, pine knots on the lower trunk, injuries to exposed roots, and woodpecker holes.
3. To maintain vigor and to provide for advance regeneration, thin stands to 130 square feet of basal area at middle-age and to 90 square feet at 10 year intervals until rotation age.

PAPER BIRCH
Betula papyrifers Marsh
Insect and Disease Management Guidelines

Paper birch has experienced a general decline through much of Minnesota. The major contributing factors to this decline include lack of soil moisture, increases in soil temperatures, restricted root growth, and invasion by the bronze birch borer, *Agrilus anxius* Gory.

(See Forest Pest Leaflet III, February, 1968, titled **The Bronze Birch Borer**).

Management recommendations to minimize birch decline are as follows:

1. Manage birch only on soils characterized by being acidic, relatively deep, moist, and moderately well drained to reduce susceptibility to decline.
2. Soil temperature increases as little as 4°F can cause root death leading to tree mortality. Therefore, avoid regenerating birch on south and west slopes, and attempt to maintain a closed canopy in existing stands.
3. Do not hold birch longer than 60 years on the poorer sites, site index 40, and 80 years on the good sites, site index 80. Prolonging the rotation ages will dramatically decrease the growth rate and accelerate stand deterioration through birch decline.
4. In old-growth birch stands, clearcut in order to stimulate dense birch regeneration. All residual stems larger than 1.5 inches DBH should be cut or killed to aid in stimulating regeneration. The dense regeneration will help shade the soil and prevent excessive soil temperatures.
5. Stands that are declining should be harvested within 2 years to prevent loss of the site to invading brush species.

RED PINE TYPE

Insect and Disease Management Guidelines

The major insect pest of red pine are Saratoga Spittlebug, Aphrophora saratogensis and several species of bark beetles. The Saratoga spittlebug causes branch and seedling mortality agent in overstocked, drought stressed, wind thrown, burned or poorly harvested plantations. See Forest Insect and Disease leaflet 3 and How to Identify and Control Pine Engraver Beetle Damage NCFES.

Red pine has two major diseases, Scleroderris canker, *Gremmeniellia abietina* and Armillaria root rot, Armillariella mellea. Two strains of the fungus causing Scleroderris canker occur in North America. The North American strain is present in Northeastern Minnesota where it kills branches of larger trees and can cause mortality to trees under six feet tall. The European strain is presently found only in the northeastern states and Canada where it has caused extensive mortality to a limited number of stands of red and Scotch pine of all ages. Armillaria root rot can be an important mortality factor in young stands on former hardwood sites. See Forest Insect and Disease leaflet 130 and the Root Rot Pest Alert.

Management recommendations are as follows:

1. Remove all overstory pine from a site scheduled for red pine plantings as they are potential insect and disease carriers.
2. New plantings should not be established adjacent to older pine plantations with existing insect or disease outbreaks without treatment of those problems.
3. Planting site preparation should include:
   a. removal of pine slash to reduce bark beetle brood material.
   b. weed control to eliminate or reduce the abundance of alternate hosts of Saratoga spittlebug.
   c. breakup of old hardwood stumps where possible to reduce Armillaria buildup.
4. To reduce the potential for Scleroderris occurrence and spread:
   a. do not plant red pine in frost pockets.
   b. do not plant red pine within one half mile of an existing Scleroderris infection.
   c. do not replant old infection sites to red pine.
5. Large stands of red pine are conducive to severe insect and disease outbreaks, plans to plant areas greater than 40 to 60 acres should include the potential use of alternate species (not pines) in strips of two chains or more to split up the red pine blocks.
6. To reduce bark beetle and wood borer damage:
   a. avoid cutting operations between April 1 and September
   b. when logging or thinning is done between these dates, remove all material larger than two inches in diameter within three weeks after cutting
   c. try not to wound trees during harvest
   d. leave branches with the green needles on to aid in rapid drying of unmerchantable stem wood and expose this slash to full sunlight
   e. trees stagnating due to wet soil, nutrient deficiencies, drought, defoliation or disease, and pockets of blowdown should be harvested as soon as possible.

SPRUCE - FIR

Insect and Disease Management Guidelines

The spruce budworm and various decays commonly attack, injure and kill all age classes of spruce-fir. The spruce budworm, *Choristoneura fumiferana* (Clemens) is the most destructive insect in Minnesota forests, causing the greatest volume loss. The budworm prefers balsam fir but can also be a problem on white spruce. Decays also cause a large volume loss. Root and butt rots caused by, *Armillariella mellea* (vahl. ex. Fr.) Karst and/or *Inonotus tomentosus* (Fr.) Gilbertson are present in most stands of spruce-fir older than 30 years. *Stereum sanguinolentum* (Alb. and Schw. ex. Fr.) Fr., a heart rot, enters the trees through broken tops, branches, and other injuries and causes the majority of the rot found in living fir trees.

Specific management recommendations are listed below:

1. Manage the spruce-fir on a rotation of approximately 40-45 years, since mature and overmature trees suffer the most severe damage.
2. Prevent stands from reaching maturity simultaneously over large areas by improving the distribution of age classes through planned cuttings.
3. Harvest spruce-fir by clearcutting. Fell all white spruce and balsam fir when harvesting stands where the next crop will be provided by balsam fir advanced regeneration. The residual trees may act as sources of bud worm infestation to the regeneration.
4. Avoid regenerating spruce-fir on coarse, acidic, or shallow soils as these sites favor the development of root and butt rots. Root and butt rots increase tree mortality when these stands are defoliated by spruce budworm.
5. Maintain or promote mixed species composition by encouraging white spruce and hardwoods and discouraging balsam fir to decrease the component of fir in the stands. Highest risk stands are pure balsam fir. White spruce suffers much less damage from budworm.
6. Break up susceptible spruce-fir types into small areas separated by other species.

If the budworm population become epidemic, the following steps should be taken:

1. Identify mature, overmature, and high volume stands; accelerate cutting on such stands.
2. Plant to salvage spruce-fir stands expected to suffer heavy losses.
3. Aerial spraying may be necessary to prevent heavy losses in valuable stands. Chemical control is aimed at foliage protection to keep trees alive rather than budworm control. A second buildup of budworm usually occurs after spraying and may actually prolong the duration of budworm out breaks.

TAMARACK

Insect and Disease Management Guidelines

The most serious insect pest of tamarack is the larch sawfly. This insect was responsible for widespread defoliation resulting in growth loss and mortality in the past. An outbreak during the years 1947-1969 resulted in 300,000 acres of timber being moderately to heavily defoliated. Recent introduction and disposal of exotic parasites have occurred. Since then losses have not been significant except in isolated circumstances.

Economic losses due to wood rotters, *Phellinus pini, Phaeolus schweinitzii*, and *Armillariella mellea*, are important in off site situations, overmature stands and where injury has occurred. Probably the most important injuries are due to abnormal water levels, both flooding or drought.

Specific management recommendations are listed below:

1. Tamarack grows best on the same type of soils as cedar except it will grow better on more acidic soils where cedar is best on slightly alkaline soils. The management decision here is to choose between tamarack with faster growth rate and poor regeneration possibilities and black spruce with usually slower growth and more favorable regeneration.
2. Where trees have been predisposed to some damaging agent such as drought, sawfly attack, overmaturity, etc., they may be attacked by the bark beetle *Dendroctonus simplex*. For specific bark beetle recommendations, see Red Pine.
3. Tamarack is one of those species for which many of the management recommendations may not be too reliable at present as much research is yet needed to determine optimum rotation ages, regeneration techniques, etc.

WHITE PINE

Insect and Disease Management Guidelines

White Pine Blister Rust *Cronartium ribicola* and White Pine Weevil *Pissodes strobi* are the major insect and disease problems of White Pine. These problems have restricted new plantings and greatly reduced the existing commercial range of the species. Introduced Pine Sawfly *Diprion similis* may at times be responsible for local severe defoliation with some top kill occurring. Management guidelines are as follows:

A. Recommendations for establishment of White Pine in the high and medium Blister Rust hazard zones of northern Minnesota are as follows:

1. Establishment must be attempted only in understory situations where it is possible to control the overstory through gradual removals. This will change understory humidity and temperature conditions so that Blister Rust spore infections will be reduced. It will also make weevil attack less likely or severe. Spacing in openings (space between crowns) should not exceed one-fourth the height of the surrounding trees. Once a stand is established, thinning must be carried out in a manner which allows maximum height growth to be maintained.
2. Plan to prune lower branches to 50% of live crown at age 5-7 and continuing every two years until there are no branches within 9 feet of ground level.
3. Manage for White Pine only in medium and high site index areas: site index 60 to 80.
4. Final release should be carried out when trees reach 35 feet.
5. Consider use of resistant stock when it become available.
6. For further recommendations, contact Forest Insect and Disease Specialists.

B. In central and southern Minnesota (hazzard zones 1 and 2), neither the Blister Rust or the White Pine Weevil are as severe a problem as in the north.

1. Only high and medium White Pine sites (S.I. 60-80) should be considered for thinning, pruning, or release work.
2. Thinning and pruning should be carried out as above.
3. Managed areas should not be located in areas where cold air collects at night or in or on the edges of forest openings.

In all cases, future plans for tending an intensively managed stand should be laid out prior to initiation of management and clearly adhered to.

The second generation of the Introduced Pine Sawfly can usually be controlled by the application of insecticide should it become a problem. In this case, contact the Regional Forest Insect and Disease Specialist.

WHITE SPRUCE

Insect and Disease Management Guidelines

The major insect and disease problems on white spruce are yellowheaded spruce sawfly, *Pikonema alaskensis*, spruce budworm, *Choristoneura fumiferana*, white pine weevil, *Pissodes strobi* and armillaria root rot, *Armillariella mellea*. Armillaria root rot (see Red Pine) can be an important killing agent especially when hardwood sites are converted to white spruce. White pine weevil (see White Pine) kills the leader and causes forked or crooked stems. It is primarily a problem when white spruce is grown close to or with white pine. Spruce budworm (see Spruce-fir) outbreaks can cause heavy mortality to white spruce growing within extensive acreages of the spruce-fir type. High levels of yellowheaded spruce sawfly are very destructive to spruce. The sawfly prefers open grown trees in grassy areas and seldom builds up to damaging levels on shaded trees. Completely defoliated trees may die by midsummer. Trees suffering 3 or 4 consecutive years of moderate to heavy defoliation will also be killed. Top kill may result from heavy defoliation.

Management recommendations for the yellow headed spruce sawfly are as follows:

1. Do not completely release white spruce plantations until the spruce are 10-12 feet in height. Twelve foot tall trees generally have enough foliate to withstand attack by high populations of the sawfly. White spruce is fairly tolerant and puts on good growth even when partially shaded. The sawfly prefers open-grown trees and seldom builds up to damaging levels on shaded trees.
2. Inspect open growing plantations with trees less than 10-12 feet tall in early June for sawfly larvae. If sawfly larval populations are high, especially if defoliation occurred last year and tree mortality or top kill are likely, direct control should be taken. Consult with the Region !nsect and Disease Specialist when deciding if direct control is necessary. Sawfly damage generally occurs in pockets in plantations. In that case, only the pockets of heavy damage should be sprayed rather than the entire plantation. If no tree mortality is likely to occur, the trees should not be sprayed to give the parasites and predators a chance to build up.

Chapter 21 — Periodicals and Publications

Periodicals With Forestry Topics
Alphabetical Listing

American Christmas Tree Journal - Coverage: Nationwide. Topics on the production and marketing of Christmas trees. Four issues per year, sent free to members of the National Christmas Tree Association. Published by: National Christmas Tree Association, Inc., 611 East Wells Street, Milwaukee, WI 53202.

American Forestry Association, 1319 18th Street NW, Washington, D.C. 20036.

American Forests - Coverage: Nationwide. Topics primarily on conservation and forests management. Issued monthly, $15.00 per year. Published by: The American Forestry Association, 1319 18th Street NW, Washington, D.C. 20036.

American Logger and Lumberman - Coverage: Nationwide. Topics primarily on logging, some on timber processing and forestry. Nine issues per year, $20.00 per year. Published by: Business Publications, Inc., 101 West Street, Hillsdale, NJ 07642.

Christmas Trees - Coverage: Primarily midwestern and north-central United States. Topics on management of Christmas tree plantations. Issued quarterly, $2.50 per year. Published by: Tree Publishers, Inc., Box 107, Lecompton, KS 66050.

Christmas Tree Growers News - Coverage: Upper Mid-West. Topics on Christmas tree culture, and marketing. Membership required. Issued quarterly. Published by: Minnesota Christmas Tree Growers Association, P.O. Box 124, Osseo, Minnesota 55369.

Forest Farmer - Coverage: Primarily southern United States. Topics on forest management and forest products marketing. Published monthly, offered only with association membership. Dues $20.00 per year. Published by: Forest Farmers Association, P.O. Box 95385, 4 Executive Park East, NE, Atlanta, GA 30347.

Forest Industries - Coverage: Nationwide, however, geared mainly to Pacific northwest forest industries. Topics on logging, timber processing and forestry. Issued monthly, free to management personnel of forest industries firms. Subscription rate, $25.00 per year to nonindustry personnel. Published by: Miller-Freeman Publication, 500 Howard Street, San Francisco, CA 94105.

Forest Products Journal - Coverage: Nationwide. Topics on forest products utilization research. Issued monthly, $45.00 per year. Published by: Forest Products Research Society, 2801 Marshall Court, Madison, WI 53705.

Journal of Forestry - Coverage: Nationwide. Topics on forest management policies and research. Issued monthly, $24.00 per year. Published by: Society of American Foresters, 5400 Grosvernor Lane, Washington, D.C. 20014.

Logging Management - Coverage: Nationwide. Topics primarily on logging, some on timber processing and forestry. Six issues annually, $15.00 per year. Published by: Vance Publishing Corporation, 300 West Adams, Chicago, IL 60606.

Michigan Forestry Association, 2791 East M-68, Indian River, Michigan 49749.

Michigan Forests - Coverage: Michigan. Topics on conservation, forestry, wildlife. Issued quarterly. $15.00 membership. Published by: Michigan Forest Association, Route 6, 802 So. 9 Mile Road, Midland, MI 48640.

Minnesota Forests - Coverage: Minnesota. Topics on forest management, legislation and policies. Issued quarterly, membership in Minnesota Forest Association required. Membership $10.00 per year. Published by: Minnesota Forestry Association, 220 First Avenue N.W., Room 210 Grand Rapids, MN 55744.

Minnesota Forest Products Marketing Bulletin - Coverage: Minnesota. Topics mainly of interest to forest industries, including classified marketing service and price reports. Issued quarterly, free. Published by: Agricultural Extension Service, University of Minnesota. Contact: Tom Milton, Area Extension Agent, 221-1/2 Jefferson North, Wadena, MN 56482.

National Woodland Owners Association, 374 Maple Avenue East, Suite 204, Vienna, VA 22180.

The American Tree Farmer - Free to tree farmers. Four issues per year. Published by: American Forest Industries, 1619 Massachusetts Ave. NW, Washington, D.C. 20036.

The Journal of Forest History - Coverage: United States. Topics on forest history. Published quarterly. $12.00 per year. Published by: Forest History Society, 109 Coral Street, Santa Cruz, CA 95060.

The Michigan Natural Resources Magazine - Coverage: Michigan. Topics same as for the Minnesota Volunteer. Six issues. $7.97 per year. Box 30034, Lansing, MI 48909.

The Minnesota Volunteer - Coverage: Minnesota. Topics on conservation, wildlife, outdoor recreation, and forestry. Six issues annually, free to year-round Minnesota residents. Published by: Minnesota Department of Natural Resources, Box 46, Centennial Building, St. Paul, MN 55155.

The National Woodlands Magazine - Coverage: Primarily the Lake States. Topics primarily on forest management, some on utilization. Six yearly. $9.50 per year. Published by: National Woodlands Publishing Co., 2228 Traverse City, MI 49684.

The Northern Logger and Timber Processor - Coverage: Primarily northeastern United States and Lake States. Topics primarily on harvesting, some on timber processing and forestry. Issued monthly, $5.00 per year. Published by: Northeastern Loggers Association, Old Forge, NY 13420.

Timber Bulletin - Coverage: Minnesota. Topics primarily on logging. Issued bimonthly. Published by: Minnesota Timber Producers Association, 208 Phoenix Building, Duluth, MN 55802.

Tree Farm News - The American Tree Farmer - Coverage: Nationwide. Topics on tree farm management. Issued quarterly, free to members of the American Tree Farm System. Published by: The American Forest Council, 1250 Connecticut Ave. NW, Washington, D.C. 20036.

Water, Woods and Wildlife - Coverage: Minnesota. Topics on conservation, recreation, wildlife and forestry. Six issues annually, $10.00 per year. Published by: Water, Woods, and Wildlife, Route 1, Warroad, MN 56763.

Wisconsin Forest Products Price Review - Coverage: Wisconsin and adjacent states. Topics: price and market information about stumpage, delivered timber, pulpwood, lumber and miscellaneous forest products. Eight issues per year, free. Published by: Cooperative Extension Service, Ted Peterson, Extension Forester, Department of Forestry, 1630 Linden Drive, University of Wisconsin, Madison, WI 53706.

Wisconsin Natural Resources - Coverage: Wisconsin. Topics are the same as for the Minnesota Volunteer. Six issues per year. Published by: Wisconsin Natural Resources, Box 7191, Madison, WI 53707.

Wisconsin Woodland Owners Association, P.O. Box 285, Stevens Point, WI 54481. 915-341-4798.

FORESTRY PUBLICATIONS
AVAILABLE FROM THE COOPERATIVE EXTENSION SERVICES OF MICHIGAN (MI), MINNESOTA (MN), AND WISCONSIN (WI)

List Provided by the Lake States Forestry Alliance

FOREST MANAGEMENT AND SILVILCULTURE . . . .

Growing Black Walnut in Minnesota, CD-BU-0505, (MN), $1.50

Information for Timber Management and Planning, AD-SB-2083, (MN), Free

Pruning Forest Trees, CD-FS-1064, (MN), $.20

Thinning Plantations and Natural Stands of Conifers, CD-FS-1065, (MN), $.20

Guidelines for Improving Northern Hardwood Timber, E0722, (MI), $.10

Woodlot Management for Fuelwood, E1486, (MI), $.20

Why Manage Your Woodlot?, E1492, (MI), $.20

Effects of Red Pine Thinning Regimes on Trunk Diameter (Tech. Info), RR423, (MI), $.35

Using Competition Quotient to Thin Red Pine (Tech. Info), RR434, (MI), $.25

Biological and Financial Rotation Age of Establishing Aspen Stands, RR478,(MI), $.50

Proceedings of National Northern White Cedar Conference, E0958, (MI), $.50

Pine Plantation Management, G1983, (WI), $.20

Aspen Management, G3162, (WI), $.25

Managing Northern Hardwood Stands, G3229, (WI), $.15

Thinning Planted Red Pine in Michigan, RR461, (MI), $.65

Intermediate Cuttings in Forest Management, G3398, (WI), $.25

Harvesting and Forest Management, G3424, (WI), $.30

Frequency of Suitable Prescribed Burning Weather, AD-TB-2045, (MN), Free

INVENTORY AND MEASUREMENTS . . . . . .

Sampling and Measuring Timber in the Private Woodland, CD-FO-3025, (MN), $.50

Point Sampling - Aid to Professional Forestry Practice, E1757, (MI), $1.00

Measuring Trees and Estimating Volume, G3332, (WI), $.30

Estimating and Interpreting Site Index, G3361, (WI), $.20

Estimating Stocking Conditions, G3362, (WI), $.30

CHRISTMAS TREES . . . .

Growing Christmas Trees in Minnesota, CD-BU-2022, (MN), $3.00

Shearing and Shaping Christmas Trees, E0719, (MI), $.05

Recommended Species for Christmas Tree Planting, E1155, (MI), $.35

Growing Christmas Trees in Michigan, E1172, (MI), $.15

Christmas Tree Shearing, G3268, (WI), $.20

HERBICIDES AND WEED CONTROL . . . . . .

Control of Weeds and Woody Plants on Rangelands, AD-BU-2344, (MN), $.75

Herbicides and Use for Forest Management in Minnesota, CD-FO-2499, (MN), $.50

Systemic Herbicides for Weed Control: Phenoxy Herbicides, AD-BU-2281, (MN), $.75

Chemical Weed Control in Shelterbelts & Forest Plantations, CD-FS-1069, (MN), $.20

Forest Plantings - Sprayer Calibration for Herbicide Application, E0916, (MI), $.10

Effective Herbicide Use in Christmas Tree Plantations, NCR251, (MI), $.70

Influences of Tillage and Herbicides Simazine and Diuron, RR426, (MI), $.30

PLANTING AND REFORESTATION . . . .

Growing Trees From Seed, CD-MI-0575, (MN), $2.00

How to Plant a Tree, CD-FS-1400, (MN), $.20

Planting Trees in Minnesota, CD-FO-0481, (MN), $1.50

Selecting Coniferous Planting Stock for Michigan Soils, E0721, (MI), $.10

Growing Evergreen Trees for Seed, E0723, (MI), $.05

Tree Planting in Michigan, E0771, (MI), $.20

Review of Jack Pine Regeneration in Lake States, RR443, (MI), $.70

Recommended Tree Seed Sources for Use in Michigan, RR445, (MI), $.30

Statistical Survey of the Michigan Tree Seedling Industry, RR445, (MI), $.50

Establishing Fuelwood Plantations in Michigan, E1572, (MI), $.35

A Guide to Forest Tree Collections of Known Source or Parentage, R3142, (WI), $2.00
Collecting & Planting Seeds of Cone-Bearing Trees, G1649, (WI), $.05

WINDBREAKS AND WIND EROSION........

Effect of Field Windbreak Design on Snow Distribution, AD-TB-2047, (MN), Free
Field Windbreaks, CD-F0-0824, (MN), $.50
Planting Trees for Farmstead Shelter, CD-BU-0468, (MN), $2.00
Studying Soil Erosion and Its Control, CD-FS-1043, (MN), $.20
Wind Erosion Control Methods, CD-FS-1216, (MN), $.20
Wind Erosion: Its Control in Minnesota, AG-F0-0774, (MN), $1.00
Wind Protection for Farmsteads, A2810, (WI), $.15

FOREST SOILS.............

Soil Survey Reports - Woodland Management, E1590, (MI), $.20
Studying Soil Texture and Its Influence, CD-FS-1044, (MN), $.20
Cation Exchange Capacity of Soils, AG-FS-1203, (MN), $.20
Cooperative Soil Survey Shows Soil Types, CD-F0-0877, (MN), $.50
Determining Soil Texture, CD-FS-1038, (MN), $.20
Exploring the Soil, CD-FS-1041, (MN), $.20
Fertility Status of Minnesota Soils as Shown by Soil Tests, AD-MR-2128, (MN), Free
Grassed Waterways Construction and Maintenance, CD-F0-0765, (MN), $.50
Measuring the Steepness of Land, CD-FS-1037, (MN), $.20

ECOLOGY AND PHYSIOLOGY........

Trees and Our Environment, CD-MI-0577, (MN), $1.50
Water Relations of the Aspens, AD-TB-2055, (MN), Free
Ecology of Upland Forest Communities and Implications, AD-TB-2029, (MN), Free
How Forest Trees Grow, G3277, (WI), $.25

LANDOWNER ASSISTANCE AND GUIDANCE.......

Sources of Information and Guidance for Woodland Owners, CD-F0-2328, (MN), $.50
Forestry Terms for the Landowner, E1238, (MI), $.25
Education and Assistance Programs for Michigan Forest Landowners, E1989, (MI), $1.30
Sources of Info & Guidance for Woodland Owners, G2096, (WI), $.15
Forestry Terms, G3018, (WI), $.20
1986 Michigan Consulting Foresters (Directory), NRM22, (MI), $.80
Consulting Forestry Services in Wisconsin, Unnumb., (WI), Free

COMPUTERS AND COMPUTER PROGRAMS.......

Forestry Programs for Programmable Calculators, E1601, (MI), $1.10
A Forest Inventory Processor for Microcomputers, NRM14, (MI), $1.00
Forest Inventory Processor Data Entry/Editor, NRM15, (MI), $1.00
Selecting Microcomputer Hardware for Forestry Applications, FF-17, (WI), Free
Selecting Microcomputer Software for Forestry Applications, FF-18, (WI), Free
SURVEY: For Computing the Area of a Closed Traverse, FF-19, (WI), Free
SITE: A Computer Program for Estimating Site Index, FF-21, (WI) Free
STOCK: A Computer Program for Estimating Stocking Conditions, FF-22. (WI), Free
QUE: A Computer Program to Estimate the Q-Ratio for Uneven-aged Stands, FF-25, (WI), Free
WORTH: An Investment Analysis Program For Forestry Applications, FF-23, (WI), Free
LEDGER: A Computerized Financial Record Book For Timber Grower, FF-24, (WI), Free
PRICE: A Computer Program to Track & Display Forest Products Price Information, FF-36, (WI), Free
Forestry Computer Quizzes, FF-40, (WI), Free
General Computer Program for Discounted Cash Flow Analysis, AD-TB-2046, (MN), Free
ACES: Allowable Cut Evaluation Simulator, (MN), $50
CASH: Investment Analysis Evaluation, (MN), $30
DPRG: Optimal Forest Stand Thinning Decisions, (MN), $30
RUNGROW: Implements GROW Subprogram of USDA Forest Service, (MN), $30
TARCV: Compiles Forest Inventory/Appraisal Data, (MN), $30
TRAVERSE: Surveying/Traverse Computations and Map, (MN), $25
UNEVEN – A Computer Model of Uneven-aged Forest Management, R3285, (WI), $2.35

MISCELLANEOUS FOREST MANAGEMENT INFORMATION.........

Minnesota's Forest Trees, CD-BU-0486, (MN), $2.00
Familiar Trees of Michigan, E0616, (MI), $.40
Using LANDSAT Imagery to Identify Coniferous Forest Types, RR448, (MI), $.40
Interpreting Michigan Forest Types through Color Infrared Aerial Photos, RR452, (MI), $.70
Planning Forest Use in Michigan, RR466, (MI), $.30
Potential of Exotic Species Imported Technologies to Michigan Forestry, RR476, (MI), $.55
Uhrenholdt Timber Harvest Demonstration Forest, FF-26, (WI), Free
Chilsen Timber Harvest Demonstration Forest, FF-27, (WI), Free
Dundee Timber Harvest Demonstration Forest, FF-33, (WI), Free
Carlin Timber Harvest Demonstration Forest, FF-37, (WI), Free
Forestry Matters Newsletter, published periodically, (WI), Free
Update...Forestry Newsletter, published periodically, (WI), Free

SAWLOGS AND PULPWOOD.......

Basic Specifications for Elm Sawlogs, CD-FS-1405, (MN), $.20
Quality of Aspen Bolts Delivered to Two Concentration Yards, AD-TB-2054, (MN), Free
Rate of Weight Loss in Piled Pulpwood, AD-SB-3036, (MN), $1.50
How much Lumber in that Tree?, E0461, (MI), $.10
Is it Red Oak or White Oak? Color Test Will Tell, G7FSRWO, (WI), $.20

LUMBER AND BOARD PRODUCTS......

Drying Eastern Hardwood Lumber, CD-BU-2821, (MN), $1.00
Drying Elm Lumber, CD-FS-1403, (MN), $.20
Finishing and Maintaining Wood Floors, CD-F0-1443, (MN), $.50
Finishing Exterior Plywood, Hardboard, and Particleboard, CD-F0-1439, (MN), $.50
Selecting Preservative Treated Wood, CD-F0-0897, (MN), $.50
Selection and Application of Exterior Finishes for Wood, CD-F0-1442, (MN), $.50
Useful Information About Wood for the Homeowner, CD-FS-1070, (MN), $.20
Identifying Wood - A Guide and a Key, E0746, (MI), $.35
Minimizing Warping of Solid Wood Panels, E0764, (MI), $.20
Production of High Quality Wood from Hard Maple Bolts, E0805, (MI), $.25
How Durable is Northern White Cedar, E0929, (MI), $.30
Wood and Decay, E1247, (MI), $.25
Chemical Preservatives, E1248, (MI), $.25
Treatment Methods, E1249, (MI), $.25
Using Treated Wood, E1250, (MI), $.25
Drying Eastern Hardwood Lumber, AAH528, (WI), $.60

OTHER FOREST PRODUCTS.......

Heating the Home with Wood, CD-BU-0531, (MN), $3.00
Homemade Maple Syrup, CD-FS-1067, (MN), $.20
Homemade Maple Syrup, E0703, (MI), $.10
Maple Syrup Making for Beginners, G3029, (WI), $.25
Selecting and Treating Minnesota Woods for Fence Posts, CD-F0-0633, (MN), $.50
How to Buy a Christmas Tree, HG189, (MI), Free
Using Elm for Firewood, A3185, (WI), $.15
Wood for Home Heating: the Creosote Problem & How to Reduce it, G2984, (WI), $.10
Wood for Home Heating: Measuring Firewood to Get Your Money's Worth, G2950 (WI), $.05
Wood for Home Heating: Safety & Wood Heating Systems--How to Install a Wood Burning System, G2936, (WI), $.15
Wood for Home Heating: Locating, Cutting & Gathering Wood, G2873, (WI), $.15

MISCELLANEOUS FOREST PRODUCT INFORMATION.......

Are They Really Termites?, AG-FS-0994, (MN), $.20
Carpenter Ants, AG-FS-1015, (MN), $.20
Minnesota Forest Products Directory, CD-BU-1390, (MN), $7.50
Minnesota Forest Products Marketing Bulletin, newsletter published 6 times annually, (MN), Free
Wisconsin Forest Products Price Review, published 6 times annually, (WI), Free
New Diversified Entrants Among U.S. Wood-Based Companies, AD-SB-2078, (MN), Free
The Effects of Pacific Rim Housing Markets and Log Supply, AD-TB-2295, (MN), Free
Use of Financial Indicators for Wood Products Firms, E0770, (MI), $.20

Internal Cost of Producing Sawmill Residues in Michigan, RR451, (MI), $.35

Wisconsin's Forest Resources: Present & Potential Uses, R2844, (WI), $.85

LOGGING.......

Employer Obligations in the Logging Industry, CD-BU-1394, (MN), $2.25

Felling and Bucking Hardwoods--How to Improve Your Profit, CD-MI-3098, (MN), $5.00

Logging Accidents--Reducing the Odds, AG-F0-0833, (MN), $.50

Logging Hazard Survey, AG-F0-0834, (MN), $.20

Safe Chainsaw Operation, AG-F0-2487, (MN), $.50

Safe Landing Layout and Use, AG-F0-2198, (MN), $.50

Safe Operation of Logging Equipment, AG-F0-0879, (MN), $1.00

Timber Harvesting News, newsletter published periodically, (MN), Free.

Protective Clothing for Chain Saw Operators, G3177, (WI), $.10

Safe Tree Harvesting, G3205, (WI), $.25

Buying a Safe Chain Saw, G3206, (WI), $.15

FOREST AND SHADE TREE PESTS........

Boxelder Bugs, AG-FS-0998, (MN), $.20

Bronze Birch Borer, The, AG-FS-1417, (MN), $.20

Cankerworms, AF-F0-0876, (MN), $.50

Cedar-Apple Rust, AG-FS-1150, (MN), $.20

Control of Scale Insects on Trees and Shrubs, AG-FS-1019, (MN), $.20

Damping-off of Seedlings, AG-FS-1167, (MN), $.20

Dial U Insect and Plant Information Request, AG-MI-2664, (MN), Free

Dutch Elm Disease, The, AG-BU-0518, (MN), $2.00

Dutch Elm Disease--Cause and Prevention, AG-MI-3265, (MN), $.50

Fire Blight, AG-FS-1159, (MN), $.20

Gypsy Moth in Minnesota: The Early Years, AG-F0-2363, (MN), $1.50

Insect Ecology: Papers Given in A.C. Hodson Ecology Series, AD-TB-2033, (MN), $4.00

Insect Pests of Evergreens, AG-F0-0865, (MN), $2.00

Integrated Approach to Dutch Elm Disease Management, AG-BU-0540, (MN), $2.00

Plant Galls, AG-FS-1009, (MN), $.20

Leaf Spot Disease of Deciduous Trees, AG-F0-0766, (MN), Free

Minnesota Fence, Boundary and Animal Control Law, AG-F0-0825, (MN), $.50

Minnesota Ticks and Their Control, AG-FS-1013, (MN), $.20

Native Elm Bark Beetle Control, AG-FS-1420, (MN), $.20

Noninfectious Diseases of Trees, AG-FS-1415, (MN), $.20

Oak Wilt, AG-F0-0612, (MN), $1.50

Protecting Trees From Animal Damage, CD-FS-1066, (MN), $.20

Root Graft Spread of Dutch Elm Disease, AG-FS-1401, (MN), $.20

Verticillium Wilt of Trees and Shrubs, AG-FS-1164, (MN), $.20

Weed Seedling Identification, AG-F0-0776, (MN), $.50

How to Keep Your Trees Healthy, E1076, (MI), $.05

Cytospora Canker of Spruce, E1078, (MI), $1.00

Habits and Control of the Eastern Pineshoot Borer, E1244, (MI), $.10

Spruce Budworm - Biology and Control, E1245, (MI), $.35

Terminal Bud Damage on Black Walnut, E1321, (MI), $.15

Eastern Tent Caterpillar Biology and Control, E1436, (MI), $.15

Jack Pine Budworm: How to Evaluate Damage, E1642, (MI), $.40

Gypsy Moth, E1678, (MI), $.45

European Pine Sawfly: Biology and Control, E1685, (MI), $.40

Gypsy Moth in Michigan - Guide for Home and Small Woodlot Owners, E1983, (MI), $.50

1987 Disease Control Guide for Conifers and Christmas Trees, E2023, (MI), $1.70.

Pine Needle Midge, E2061, (MI), $.30

Walnut and Butternut Toxicity, A3182, (WI), $.25

Pines (Pinus) Disorder: European Pine Sawfly, A3180, (WI), $.30

Deciduous Tree Disorder: Cankerworms, A3178, (WI), $.25

Ash (Fraxinus) Disorder: Ash Plant Bug, A3126, (WI), $.10

Maple (Acer) and Other Trees Disorder: Cottony Maple Scale, A3123, (WI), $.10

Deciduous Tree & Shrub Disorder: Skeletonizing by Rose Chafer, A3122, (WI), $.10

Dutch Elm Disease: a Lesson in Urban Forestry, A3071, (WI), $.05

Oak (Quercus) Disorder: Twolined Chestnut Borer, A2902, (WI), $.25

Homeowners Guide to Controlling Dutch Elm Disease with Systemic Fungicides, A2842, (WI), $.10

Maple (Acer) Disorder: Maple Petiole Borer, A2699, (WI), $.25

Maple (Acer) Disorder: Maple Gall Mites, A2691, (WI), $.20
Hickory (Carya) and Walnut (Juglans) Disorder: Downy Leaf Spot (White Mold), A2667, (WI), $.05
Colorado Blue Spruce (Picea pungens) and Other Species Disorder: Rhizosphaera Needle Blight, A2640, (WI), $.10
Spruce (Picea), Especially Blue Spruce, Disorder: Spruce (Cytospora) Canker, A2639, (WI), $.35
Oak (Quercus) and Other Trees Disorder: Chlorosis, A2638, (WI), $.35
Austrian Pine (Pinus nigra) and Certain Other Species of Pine Disorder: Dothistroma Needle Blight, A2620, (WI), $.35
Evergreens Disorder: Seasonal Needle Drop, A2614, (WI), $.10
Scotch, Red and Certain Other Pines (Pinus spp) Disorder: Needle Diseases - Brown Spot & Lophodermium, A2608, (WI), $.10
Mountain Ash (Sorbus) Disorder: Scab, A2594, (WI), $.20
Maple (Acer) Disorder: (Summer) Scorch, A2553, (WI), $.10
Oak (Quercus) and Numerous Other Species Disorder: Armillaria Root Rot, A2542, (WI), $.30
Maple (Acer) and Other Trees and Shrubs Disorder: Verticillium Wilt, A2537, (WI), $.25
Maple (Acer) Decline: Collar Rot and Basal Canker Complex, A2532, (WI), $.25
Controlling Dutch Elm Disease with Systemic Fungicides -- Questions and Answers, A2516, (WI), $.10
Shade Trees Disorder: Decline, Dieback or Early Senescene, A2510, (WI), $.15
Maple (Acer) and Other Trees Disorder: Decline Due to Accumulated Salt Injury, A2507, (WI), $.25
Dutch Elm Disease in Wisconsin, A2392, (WI), $.20
Birch (Betula) Disorder: Birch Leaf Miner, A2117, (WI), $.25
Recognizing Common Shade Tree Insects, A2079, (WI), $.25
Oak (Quercus) Disorder: Oak Wilt, G1693, (WI), $.35

PESTICIDES........

How to Inject Elms with Systemic Fungicides, AG-F0-0781, (MN), $.50
Insecticide Suggestions to Control Tree and Shrub Insects, AG-F0-0704, (MN), $.50
Insecticides, AG-BU-0499, (MN), $2.00
Pesticide Storage and Formulation Shed, AG-FS-0916, (MN), $.20
Pesticides and Pesticide Container Disposal, AG-FS-0925, (MN), $.20

FOREST ECONOMICS, MARKETING, AND POLICY.......

Economic Evaluation of the 1979 Forestry Incentives Program, AD-SB-2086, (MN), Free
Forestry Incentive Payment Recipients Ten Years Later, AD-SB-2529, (MN), $2.00
Forestry Incentive Program Investments in 1974, AD-SB-2087, (MN), Free
Minnesota Timber Policy Options: A Classification, AD-SB-2080, (MN), Free
State Forest Practice Laws and Regulations: A Review, AD-SB-2073, (MN), Free
Minnesota's County Forests: A Delphi Study of Options, AD-SB-2194, (MN), Free
Marketing Timber from the Private Woodland, CD-F0-2723, (MN), $.50
Michigan's Commercial Forest Act - Is it for You?, E1758, (MI), $.60
Costs of Forest Management Practices in the Lake States, RR457, (MI), $.30
Economics of Michigan Forest Industry Survey - Based Input/Output Survey, RR472, (MI), $1.00
The Woodland Tax Law and The Forest Crop Law, G1549, (WI), $.15
The Managed Forest Law Program, G3413, (WI), $.25
Marketing Timber, G3297, (WI), $.25
Income Tax Considerations for Forestland Owners, G3298, (WI), $.20
Casualties, Disasters & the Income Tax: Some Tips for Woodland Owners, FF-16, (WI), Free
The Wisconsin Pulpwood Trade - 1982, FF-20, (WI), Free
Financial Record Workbook - companion document for the LEDGER program, (WI), Free
Wisconsin's Forests are Gaining on Us!, FF-28, (WI), Free
Wisconsin's People and Trees -- Mirror Images on the State, FF-29, (WI), Free
Wisconsin's Forests--Who Owns Them?, FF-30, (WI), Free
Who Supplies the Wood From Wisconsin's Forests?, FF-31, (WI), Free
Wisconsin's Forest Bank Account: The Growth and Cut Situation, FF-32, (WI), Free
Wisconsin Hardwoods: Where & How Much, FF-34, (WI), Free
Wisconsin Softwoods: Where & How Much, FF-35, (WI), Free
What Will A Forest Tree Earn?, FF-38, (WI), Free
What's a Forest Tree Worth?, FF-39, (WI), Free

Pulpwood Demand & Supply in the Wisconsin Pulp & Paper Industry, 1950-82, R3375, (WI), $.25

Development Potential of Wisconsin's Red Pine Resource, R3373, (WI), $.25

Forest Resource Analyses -- a Profile of Wisconsin's Paper Industry, R3313, (WI), $.30

An Economic Perspective on the Wisconsin Hardwood Veneer & Plywood Industry, R3270, (WI), $.35

Investment-Efficient Stocking Guides for All-Aged Northern Hardwood Forests, R3129, (WI), $.40

Long-Term Forecasting of Timber Harvest, Growth & Inventory in Wisconsin, R3150, (WI), $1.75

PUBLICATIONS ARE AVAILABLE FROM THE FOLLOWING SOURCES............

(MI) Michigan.....

Bulletin Office
ANR Information Services
Michigan State University
10 Agriculture Hall
East Lansing, MI 48824
(517) 355-0240

(MN) Minnesota.....

Minnesota Extension Service
Distribution Center
University of Minnesota
3 Coffey Hall
1420 Eckles Avenue
St. Paul, MN 55108
(612) 625-8173

(WI) Wisconsin.....

Agriculture Bulletin Office
University of Wisconsin - Madison
Room 245
30 North Murray St.
Madison, WI 53715
(608) 262-3346

PARTIAL LISTING OF PUBLICATIONS AVAILABLE
FROM THE MINNESOTA STATE DOCUMENTS CENTER

Minnesota State Documents Center
Publication List (Partial)
117 University Ave.
St. Paul, MN 55155
Phone: (612) 297-3000 or 1-800-652-9747

Brainerd-Mille Lacs
Fishing Guide
Diversified Design, Inc.
$4.95

Birds of the Superior
National Forest
Superior National Forest
U.S. Dept. of Agriculture

Great Lake Trout & Salmon
Fishing
Paul L. Dorweiler
$7.95

Fish & Wildlife in
Minnesota
Dept. of Natural
$3.50

Maple Syrup Producer's Manual
U.S. Dept. of Agriculture
$8.00

Heating the Home with Wood
Agriculture Ext. Division
U of M
$3.75

Key to the Common Aquatic
Plants of Minnesota
Dept. of Natural Resources
Div. of Game & Fish
Section of Technical
$4.75

Billions of Year in Minnesota
The Geological Story of
the State
MN Historical Society
$5.50

Poster/Prints
Minnesota Loon
$2.50

Minnesota Guide
Dorn Books
$8.95

Prehistoric Peoples of
Minnesota
MN Historical Society
$2.50

Pelican Rapids-Itasca West
Fishing Guide
Lake Maps, Inc.
$4.95

Fishes of the Minnesota
Region
$12.95

Fifty Birds of Town & City
U.S. Dept. of the Interiors
$9.50

Minnesota Fishing Guide
Ron Schara, Mpls. Tribune
$7.95

Forestry in Minnesota
Schools
MN Dept. of Education
$3.50

Paddler's Guide to Boundary
Waters Canoe Area
Michael E. Duncanson
$4.95

Prairies, Woods & Islands A
Guide to the MN Preserves of
the Nature Conservancy,
MN Chapter
$6.75

Fishes of the Great Lakes
Fed. Dept. of Fish & Wildlife
$7.00

Minnesota Environmental
Organizations
$5.00

Park Rapids, Leech Lake,
Bemidji Fishing Guide
Diversified Design, Inc.
$4.95

Cavity Nesting Birds of
North America Forests
U.S. Dept. of Agriculture
$6.88

Homes for Birds
U.S. Dept. of the
Interior
$3.00

A Gathering of Waters
Dept. of Natural
Resources, Div. of Parks
and Recreation Rivers
$9.00

Twin Cities Fishing Guide
Sybil Smith
$8.95

Trees of Minnesota
Dept. of Natural
Resources
$3.00

Northland Wild Flowers
John & Evelyn Moyle,
University of Minnesota
$12.95

Streams & Rivers of
Minnesota
Thomas F. Waters, U of M
$7.95

Bird Portraits in Color
U of M Press
$12.95

Minnesota Guidebook to
State Agency Services
MN State Documents Center
$12.50

DNR REPORTS

AVAILABLE FROM: Department of Natural Resources, Bureau of Information & Education
Box 46, 500 Lafayette Road, St. Paul, MN 55146

1. Wildlife Rice
2. Before You Buy That Lakeshore Lot
3. Lost in the Woods!
4. Skunks
5. Minnesota's Colorful Snake Family
6. Minnesota's Mighty Moose
7. Heat Your Home With Wood
8. Muskrats Thrive in Minnesota
9. The Fisher: Strong, Swift, Elusive
10. Laws
11. Birds of Prey - Part 1
12. Birds of Prey - Part 2
13. Big Tree Hunt
14. Largemouth Bass
15. Rainbow Trout
16. Lake Trout
17. The Sauger
18. Northern Pike
19. Bullhead
20. Raccoons
21. Hungarian Partridge
22. Mink
23. Badger and Wolverine
24. The Coyote
25. Squirrels
26. Ruffed Grouse
27. Wild Turkey: Minnesota's Largest Game Bird
28. American Coot: The Reluctant Flyer
29. Mergansers
30. The Industrious Beaver
31. Smallmouth Bass
32. Crappie
33. Minnesota Catfish
34. Brown Trout
35. Brook Trout
36. Shoveler
37. Mallard
38. Canvasback
39. Ring-Necked Duck
40. Bufflehead
41. Gadwall
42. The Grassland Sea
43. This Treeless Solitude
44. Minnesota's Prairie Wildlife
45. Prescribed Burns: Managing Our Prairie Grasslands
46. Cooking Game Fish and Panfish
47. Look at Shorebirds
48. Tree Disease in Firewood
49. Grow Trees for Pleasure and Profit
50. Wild Rice Recipes
51. Peat in Your Future?
52. Bait Leeches
53. WMA
54. The Porcupine: Nature's Living Pincushion
55. The River Otter
56. Eutrophication: How Lakes Age and Die
57. Rabbits and Hares
58. Falconry Today
59. The Opossum
60. The Wood Duck
61. Redhead
62. Common Goldeneye
63. Old Squaw
64. White-Winged Scoter
65. American Wigeon
66. Lesser Scaup
67. Greater Scaup
68. Blue-Winged Teal
69. Green-Winged Teal
70. Pintail
71. Black Duck
72. Ruddy Duck
73. Wildcat!
74. The Lynx Link
75. Lakeshore Water Plants Are Not Weeds
76. White-Tailed Deer
77. Birdhouses in Minnesota
78. Prairie Plants of Lac Qui Parle and Vicinity
79. Minnesota's Foxes
80. The White Sucker
81. Do-It-Yourself Timber Harvesting
82. Introducing the Morning Dove
83. Water-Rich Minnesota
84. Chain Saw Fumbles
85. What Shoreland Owners Should Know About Waste Disposal
86. Danger! Thin Ice!
87. Get Started in Bird Watching!
88. New Ways to Reduce Flood Damage
89. Meet the Black Bear
90. Wildflowers in Minnesota's State Parks
91. Muskellunge
92. Attract Wildlife To Your Back Yard
93. Fishing The Elusive Walleye
94. Trapping Furbearers in Minnesota
95. Minnesota's Past Historic Sites
96. Native Trees For Ornamental Plantings
97. The Sandhill Crane
98. Grow Wildflowers In Your Back Yard
99. Great Horned Owl: Silent Predator
100. NO LONGER AVAILABLE
101. That Remarkable Night-Flyer, The Bat
102. They Used to Call Us Game Wardens
103. Stalking Our Woodland Mushrooms
104. Outdoor Hobby: Wildlife Photography
105. Ground Water
109. Bluegills: Fishing Techniques

MAPS

LAKE MAPS
Contact the Minnesota State Documents Center by calling 1-800-652-9797 or 297-3000 if you are in the Twin Cities and request a lake map catalog.

BIKE TRAIL MAPS
For a free catalog write The Minnesota Department of Transportation, Room B-20, Transportation Building, St. Paul, MN 55155 or call (612) 296-2216.

PUBLIC WATER ACCESS MAPS
Contact the DNR Trails and Waterways Unit, Water Access Section, Box 40, 500 Lafayette Rd., St. Paul, MN 55146 or call (612) 296-6699.

STATE FOREST TRAIL MAPS
Contact your local DNR or the DNR Trails and Waterways Unit, Information Center, Box 40, 500 Lafayette Rd., St. Paul, MN 55146 or call (612) 296-6699.

SNOWMOBILE TRAIL MAPS
Call (612)296-5029 in the Twin Cities, 1-800-652-9747 in Minnesota, and 1-800-328-1461 outside of Minnesota.

COUNTY MAPS
County maps can be obtained at the local courthouse, Department of Transportation office, County Engineers Office or contact MNDOT, Room B-20, St. Paul, MN 55155 or call (612) 296-2216.

COUNTY PLAT BOOKS
These books break down a county into townships and show the land ownership, as well as roads and lakes. Contact the County Extension Agent to locate a vendor for these books.

U.S.G.S. TOPOGRAPHIC MAPS
University of Minnesota, Minnesota Geological Survey, 2542 University Ave., St. Paul, MN 55114-1057.

WHEN ALL ELSE FAILS
Contact: The Map Store, 348 North Roberts At. 4th, St. Paul, MN 55101 or call (612) 227-6277. They have all kinds of maps!.

AERIAL PHOTOS
Contact the Agricultural Conservation and Stabilization Service in the county where the land is located for application blanks. Be sure you know the section, township and range numbers for the area of interest.

FORESTRY EQUIPMENT CATALOGS

Ben Meadows Company, 3589 Broad St., P.O. Box 80549, Atlanta, GA 30366

Forestry Suppliers Inc., 205 West Rankin St., P.O. Box 8397, Jackson, MS 39204-9987

North Star Evergreens Inc., Eastwood Plaza Bldg., Box 253, Park Rapids, MN 56470

TSI Company, P.O. Box 9239, Bridgeport, CT 06601

WANT TO PLANT TREES?

Consult This Check List For Government Cost Assistance

- ☐ **Step 1.** Contact your local forester or A.S.C.S. office in the county where the trees are to be planted. Check to see if your project is eligible for government cost sharing.
- ☐ **Step 2.** If your project is eligible for cost sharing fill out a "Request for Cost Sharing" (Form ACP 245) at the County A.S.C.S. office. In some counties the DNR Forester can help you fill it out in his office, in your home or through the mail.
- ☐ **Step 3.** Don't do anything until project approval by the County A.S.C.S. Committee. Save the approval (form 245). You're going to need it after the project is completed.
- ☐ **Step 4.** Be sure to get your copy of the "brief management plan" from the DNR Forester in charge of field checking of your project. This plan will specify species to plant, tree spacing requirements and suggested planting methods, and site preparation requirements.
- ☐ **Step 5.** Order your trees with the DNR tree order form or order your trees from a private nursery. If doing your own planting skip Steps 6 & 7. Trees must be ordered from the DNR between September 1 and March 15.
- ☐ **Step 6.** Send out bid forms to vendors who might plant your trees. Ask your Forester for tree planting bid forms.
- ☐ **Step 7.** Decide on a vendor and make up a tree planting contract. Ask a DNR Forester for contract forms.
- ☐ **Step 8.** Do site preparation - if needed.
- ☐ **Step 9.** Wait for notification that your trees are ready to pick up or wait for tree delivery. Notify your DNR Forester when you begin your tree planting.
- ☐ **Step 10.** Plant trees.
- ☐ **Step 11.** Notify forester that your planting is finished.
- ☐ **Step 12.** Pay planting bills **after being notified by forester that planting was satisfactory.**
- ☐ **Step 13.** Fill out, sign and return form "245" to the county A.S.C.S. office along with copies of planting related bills (trees, site preparation, equipment and labor).

Chapter 22

Forestry and Wildlife Glossary

ABIOTIC. The non-living components of the environment, such as air, rocks, soil, water, coal, peat, plant litter, etc.
ACID SOIL. A soil having a preponderance of hydrogen ions over hydroxyl ions in the soil solution; thus a soil giving an acid reaction (precisely below pH 7.0; practically, below pH 6.6)
ACRE. An area of land containing 43,560 square feet, roughly the size of a football field, or a square that is 208 feet on a side.
ADVENTITIOUS BUD. A bud which develops at the base of a needle cluster, or on woody tissue on a branch or leader, when the end of the branch or leader is injured or cut off.
AFFORESTATION. Establishing a forest on an area which has not previously had trees growing on it.
ALKALINE SOIL. Any soil that is alkaline in reaction (precisely, above pH 7.0; practically, above pH 7.3).
ALL-AGE FOREST. A forest stand in which trees of all ages and usually all sizes are present. This is in contrast to an "even-aged" forest.
ALTIMETER. An instrument used to determine the height of a tree.
ANNUAL RING. The growth layer of one year, as viewed on the cross section of a stem, branch, or root.
ANNUALS. Plants that live less than 12 months.
ANTHRACNOSE. A disease usually characterized by ulcer-like leaf or fruit spots and caused by fungi that produce asexual spores in the type of fruiting body called an acervulus.
ASPECT. The compass direction towards which a slope faces.
ASSOCIATION. An assemblage of plants having ecologically similar requirements and including one or more dominant species from which it derives a definite character.
AZIMUTH. Direction from a point, measured in degrees clockwise from true north.
BACKFILL. Excavated material used to build up a road higher than the original level.
BACKFIRE. 1) Fire set along the inner edge of a fire control line to stop a spreading wildfire by reducing the fuel or changing the direction of force of the fire's convection column. The term applies best where skill techniques are required for successful execution. Using such fire to consume unburned fuel inside the fireline to speed up line holding and mopup is usually distinguished as "burning out" or "clean burning." 2) A prescribed fire set to burn against the wind. Also called back-burn. 3) To set a backfire.
BASAL AREA. Of a tree- the cross-sectional area (in square feet) of the trunk at breast height (4-1/2 feet above ground). For example, the basal area of a tree 14 inches DBH is approximately 1 square foot. Of an acre of forest-the sum of basal areas of the individual trees on the acre. For example, a well stocked northern hardwood stand might contain 80-100 square feet of basal area.
BIENNIALS. Plants that live for two growing seasons.
BILTMORE STICK. A tool resembling a yardstick, calibrated to measure the diameter of a tree at breast height. Sticks are calibrated with different scales depending upon the reach (arm length) of the person using it.
BIOLOGICAL CONTROL. The use of organisms or viruses to control parasites, weeds, or other pests.
BLAZE. To mark a tree, usually by painting and/or cutting the bark. Boundaries of forest properties frequently are delineated by blazing trees along the boundary line.
BLOWDOWN. A tree pushed over by the wind, also called windthrow.
BOARD FOOT. A unit measuring wood volumes equalling 144 cubic inches which is commonly used to measure and express the amount of wood in a tree, sawlog, veneer log or individual piece of lumber. For example, a piece of wood 1 foot x 1 foot x 1 inch or one measuring 1 foot x 3 inches x 4 inches both contain 1 board foot of wood.
BOLE. The main trunk of a tree.
BOLT. A short log or a square timber cut from a log, commonly 8 feet long.
BREAST HEIGHT. The standard height, 4-1/2 feet above average ground level, at which the diameter of a standing tree is measured. Abbr. D.B.H.
BROADCAST BURNING. Burning over a considerable area and permitting fire to spread freely with or without the use of firebreaks.
BROWSE. Portions of woody plants including twigs, shoots, and leaves used as food by such animals as deer.
BROWSE LINE. The uppermost limit on trees and tall shrubs to which livestock and big game animals browse. Syn. Grazing line.
BUCK. To cut trees into shorter lengths, such as logs or cordwood.
BURN. An area over which fire recently has run.
BUTT LOG. This is the first log above the stump and is generally the most valuable log in a tree.
CACHE, FIRE-TOOL. A supply of fire tools and equipment assembled in planned quantities or standard units at a strategic point for exclusive use in fire suppression.
CALIPER (or calipers), TREE. An instrument to measure diameters of trees or logs.
CAMBIUM. The layer of cells between the inner bark and wood of a tree. This is where growth takes place.
CANDLE. The new bright green and tender growth all conifers grow in the spring.
CANOPY. The upper level of a forest, consisting of branches and leaves of taller trees.
CARNIVORE. A flesh eating animal.
CARRYING CAPACITY. The maximum number of animals possible in an area without inducing damage to vegetation or related resources; may vary from year to year because of fluctuating forage production.
CELL. The basic structural unit of all living organisms. An organism may be composed of a single cell (e.g. bacteria) or many cells working together (all "higher" organisms, including man).
CHAIN. A distance of 66 feet. Five chains makes a tally.
CHLOROPHYLL. The green photosynthetic substance in plants which allows them to capture solar energy.
CHOKER. A length of wire rope or chain with a loop or noose at one end used to secure trees or sections of trees for skidding.

Clearcut

CLEARCUT. A harvesting technique which removes all the trees (regardless of size) on an area in one operation. Clearcutting is most often used with species which require full sunlight to reproduce and grow well. Produces an even-aged forest stand.

CLINOMETER. An instrument used to determine the height of a tree.

CLONE. A plant group derived from a single individual through vegetative reproduction. Example: A clone of many aspen trees may sprout from the roots of a single aspen tree, after it is cut.

CO-DOMINANT. A tree receiving full light from above, but comparatively little from the sides. Such trees usually have medium sized crowns.

COMMERCIAL FORESTLAND. Any forested area capable of producing 20 cubic feet of timber per acre per year which has not been withdrawn from such use by law or statute.

COMMERCIAL TREATMENTS. Forestry operations, such as thinning or other TSI work, which generate income from sale of the trees that are removed.

COMMUNITY. A collection of living organisms functioning together in an organized system through which energy, nutrients, and water cycle.

CONIFER. A tree belonging to the order coniferales which is usually evergreen, cone-bearing and with needle, awl or scalelike leaves such as pine, spruce, fir and cedar; often referred to as a "softwood".

CONSERVATION. The protection, improvement, and use of natural resources according to principles that will assure their highest economic and social service.

CONSULTING FORESTER. A self-employed professional forester.

CONTACT HERBICIDE. An herbicide that kills primarily by contact with plant tissue rather than as a result of transloca-tion; only the portions of the plant which actually come in contact with the chemical are affected.

CONTROLLED BURNING. The planned application of fire with intent to confine it to a predetermined area.

COOPERAGE. Containers consisting of two round heads and a body composed of staves held together with hoops.

COPPICE FOREST. A forest consisting wholly or mainly of sprouts.

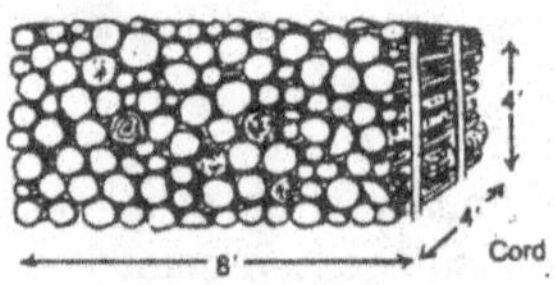

Cord

CORD. A pile of wood 4 feet high, 4 feet wide, and 8 feet long, measuring 128 cubic feet. Actual volume of solid wood in a cord will vary from 60 to 100 cubic feet, depending on size of individual pieces and orderliness of stacking. In the Lake States, pulpwood cords are usually 4' x 4' x 100" and contain 133 cubic feet.

CORDWOOD. Small diameter and/or low quality wood suit-able for firewood, pulp, or chips, but not for sawlogs.

COVER TYPE. Classification of lands according to pre-dominating vegetative cover.

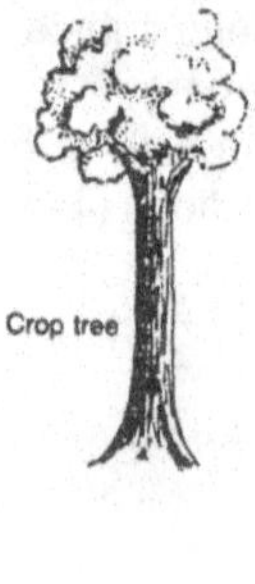
Crop tree

CROOK. A defect of a tree characterized by a sharp bend in the main stem.

CROP TREE. A tree identified to be grown to maturity and which is not removed from the forest before the final harvest cut. Usually selected on the basis of its location with respect to other trees and its quality.

CROWN. The branches and foliage of a tree; the upper portion of a tree.

CROWN CLASSIFICATION. Individual trees in a stand may be classified according to the relative size and height of their crowns compared to other trees in the stand. In descending order of crown height and size the classes are: dominant, co-dominant, intermediate, suppressed.

CROWN COVER. The canopy of green leaves and branches formed by the crowns of all trees in a forest. Syn. Leaf Canopy.

CROWN FIRE. A fire which runs through the tops of living trees, brush or chaparral.

CROWN RATIO OR LIVE-CROWN RATIO. The ratio of the portion of a tree height with leaves to the total tree height.

CRUISE. A survey of forest land to locate timber and estimate its quantity by species, products, size, quality, or other characteristics. Also refers to an estimate derived from such a survey.

CRUISER. One who cruises timber. Syn. Estimator; Land looker; Valuer.

CUBIC FOOT. A wood volume measurement containing 1,728 cubic inches, such as a piece of wood measuring 1 foot on a side. A cubic foot of wood contains approximately 6 to 10 usable board feet of wood.

CULL. 1) A tree or log of merchantable size rendered un-merchantable because of poor form, limbyness, rot or other defect. 2) The deduction from gross volume made to adjust for defect. 3) To cut a small portion of a stand by selecting one or a few of the best trees. 4) To reject a tree, log, or board in scaling or grading. 5) Any item of production which does not meet specifications.

CUNIT. A unit of measure in cube scaling equal to 100 cubic feet of wood.

CUTTING CYCLE. The planned time interval between major harvesting operations in the same stand. The term is usually applied to uneven-aged stands. For example, a cutting cycle of 10 years means that every 10 years a harvest would be carried out in the stand.

DAMPING-OFF. The killing of young seedlings by certain fungi that cause decay of the stem or roots.

DIAMETER, BREAST HIGH (DBH). The diameter of a tree at 4.5 feet above average ground level, except that in National Forest practice it is measured from the highest ground level.

DEBARK. The action of removing bark from trees or sec-tions of trees. Debark generally denotes mechanical means as opposed to manual peeling. Syn. Bark; Barking.

DECIDUOUS TREE. A tree which loses all of its leaves during the winter season.

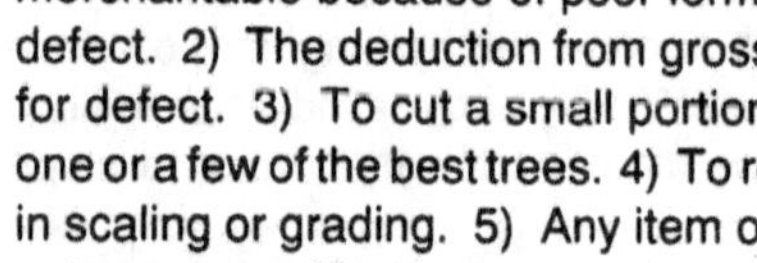

DEFECT. Any irregular-ity or imperfection in a tree, log, piece product, or lumber that reduces the volume of sound wood or lowers its durability, strength, or utility value. Defects in lumber may result from such factors as insect or fungus attack, growth conditions and abnormali-ties, manufacturing or seasoning practices, etc.

DEFOLIATION. The loss of leaves or foliage on a plant or tree.

DENDROLOGY. The study of the identification, habits and

distribution of trees.

DEN TREE. A hollow tree used as a home by a mammal.

DIAMETER. Tree diameter is usually measured 4-1/2 feet above ground level (see DBH).

DIAMETER-LIMIT SALE. A timber sale in which all trees over a specified dbh may be cut. Diameter-limit sales often result in high grading.

DIAMETER TAPE. A tape measure, calibrated to determine the diameter of a tree by measuring its circumference.

DIB (d.i.b.). Diameter inside bark, usually measured at the small end of a log.

DIEBACK. The progressive dying, from the tip downward, of twigs, branches or tops.

DIMENSION LUMBER. Hardwood dimension lumber processed so it can be used virtually in the sizes provided, in the manufacture of furniture or other products. Softwood dimension lumber consists of boards more than 2 inches thick but less than 5 inches thick. Such wood is used in construction and is sold by units such as 2 x 4s, 4 x 8s, or 2 x 10s.

DIURNAL. Active during the day.

DOMINANT. The tallest, fastest growing trees in a plantation or natural stand.

DOMINANT TREES. Those trees within a forest stand which extend their crowns above surrounding trees and capture sunlight from above and around the crown.

DOYLE RULE. One of several log rules designed to estimate the board-foot volume of lumber which can be sawn from logs of a given length and diameter. See log rule.

DUFF. Forest litter and other organic debris in various stages of decomposition, on top of the mineral soil, typical of conifer forests in cool climates where rate of decomposition is slow, and where litter accumulation exceeds decay.

ECOLOGICAL NICHE. The role a particular organism plays in the environment.

ECOLOGY. The study of interactions between living organisms and their environment.

ECOSYSTEM. An interacting system of living organisms (plants and/or animals), soil and climatic factors. Foresters consider a forest an ecosystem.

ECOTONE. A transition between two distinct communities.

EDGE. The boundary between open land and woodland or two other ecological communities. This transaction area between environments provides valuable wildlife habitat. Consideration of edge can reduce visual impact of a timber harvest.

EDGE EFFECT. The increased richness of plants and animals resulting from the mixing of two communities where they join.

ENDANGERED SPECIES. A species designated as being in danger of becoming extinct.

ENDEMIC. Native or confined to a certain area.

ENTOMOLOGY, FOREST. The science that deals with insects in their relation to forests and forest products.

ENVIRONMENT. The prevailing conditions which reflect the combined influence of climate, soil, topography and biology (other plants and animals) factors present in an area.

Epicormic branching

EPICORMIC BRANCHING. Branches which grow out of the main stem of a tree, arising from buds under the bark. Severe epicormic branching increases knottiness, thereby reducing the quality of lumber sawn.

EPIDEMIC. A temporary widespread outbreak of disease.

Even-aged stand

EVEN-AGED FOREST. A forest in which all of the trees present are essentially the same age (within 10 to 20 years).

EVERGREEN. Trees which retain green foliage throughout the year. Not all conifers are evergreens. An example is tamarack.

EXTINCT. Being no longer found anywhere in the world- e.g.; passenger pigeon.

FACE CORD. Stove length wood with a "face" of 32 square feet (a stack four feet high and eight feet wide). The volume of a face cord depends on its length.

FAUNA. Animals.

FELL. The process of severing a tree from the stump so that it drops to the ground.

FINAL CUT. In even-aged management, the amount or volume of material removed in harvesting the main crop at the end of the rotation.

FIREBREAK. A natural or constructed barrier utilized to stop or check fires that may occur or to provide a control line from which to work. Sometimes called a fire lane.

FIRE PLOW. A heavy duty, usually specialized machine, either of the share or disk type, designed solely for abusive work in the woods and used with either horses or tractors to construct firebreaks and fire lines.

FLORA. Plants.

FOAMING AGENT. A material, designed to reduce drift, which causes a pesticide mixture to form a thick foam.

FOLIAGE. Growth of leaves on a tree or other plant.

FOOD CHAIN. A group of plants, animals, and/or microorganisms linked together as sources and consumers of food.

FOREST. A plant community in which the dominant vegetation is trees and other woody plants.

FORESTATION. The establishment of forest naturally or artificially upon areas where it is at present absent or insufficient. See Afforestation; Reforestation.

FORESTED WETLAND. An area characterized by woody vegetation over 20 feet tall where soil is at least periodically saturated with or covered by water.

FOREST FIRE. Any fire on forest land which is not being used as a tool in forest protection or management in accordance with an authorized plan.

FOREST FLOOR. All dead vegetable matter on the mineral soil surface in the forest, including litter and unincorporated humus.

FOREST MANAGEMENT. Giving the forest the proper care so that it remains healthy and vigorous and provides the products and amenities the landowner desires. (Technical Definition: The application of technical forestry principles and practices and business techniques (such as accounting, benefit-cost analysis, etc.) to the management of a forest.)

FORESTRY. The scientific management of forests for the continuous production of goods and services.

FOREST SANITATION. The destruction, removal, or treatment of infected or infested material for the purpose of reducing disease and insect incidence in the forest. See also Cutting, methods of.

FOREST SURVEY. An inventory of forest land to determine area, condition, timber volume, and species for specific purposes such as timber purchase, forest management or as a basis for forest policies and programs.

FOREST TYPE. A group of tree species which, because of their environmental requirements and tolerance for shade and moisture, are repeatedly found growing together. Examples are the jack pine type and the aspen-paper birch type.
FORK. A defect characterized by division of the main stem or bole of a tree into two or more stems.

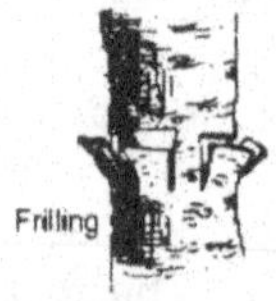

FRILLING. A method of killing trees by inflicting a series of cuts around the bole (stem) and applying an herbicide to the wounds. Frilling or girdling of trees may be used to reduce the density of a stand or to kill individual undesirable trees.
FROST CRACK. A vertical split in the wood of a tree, generally near the base of the bole, from internal stresses and low temperatures.
FULLY STOCKED STAND. A forest stand in which all growing space is effectively occupied but having ample room for development of the crop trees. Syn. Normal stand.
GALL. A pronounced swelling or outgrowth on a plant.

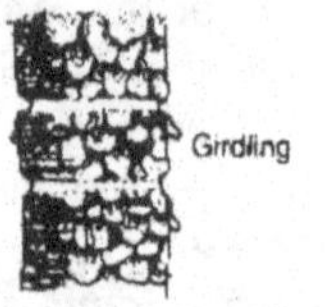

GIRDLE. To encircle the stem of a living tree with cuts that completely sever bark and cambium and often are carried well into the outer sapwood, for the purpose of killing the tree by preventing passage of nutrients or by introducing toxic materials. Besides girdling proper, or removal of bark and cambium in a band of appreciable width girdling may take several forms, viz: 1) Hacking or frilling — A single line of overlapping downward axe cuts, leaving a frill into which toxic materials may be poured. 2) Double hacking — Girdling by means of a double frill cut around the tree and the removal of the chips between them. 3) Notching — Ringing the tree with notches cut well into the sapwood. 4) Stripping — Peeling off a band of bark completely around the tree.
GRADE. (see slope) The slope of a surface such as a roadway. Also, the elevation of a real or planned surface or structure.
GRADING. Evaluating and sorting trees or logs according to quality.
GROUP SELECTION. A process of harvesting patches of selected trees to create openings in the forest canopy and to encourage reproduction of uneven-aged stands.
GROWTH RATE. With reference to wood, the rate at which the wood substance has been added to the tree at any particular point; usually expressed in terms of number of rings per inch. Growth rate bears an inverse relationship to number of rings per inch. Also applies to volume, value, or other types of increase in trees or stands.

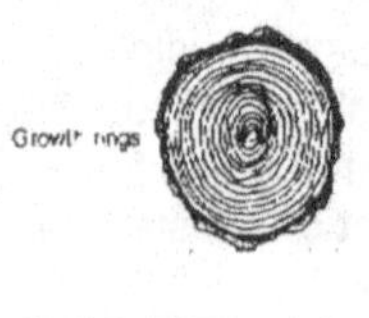

GROWTH RINGS. The layers of wood laid down each growing season, also called annual rings. These rings frequently are visible when a tree is cut and may be used to estimate the age of the tree, as well as to determine the rate of its growth.
GUYLINE. A line used to stay or support spar trees, booms, etc.
HABITAT. The local environment in which a plant or animal lives.
HARDWOOD. A term used to describe broadleaf, usually deciduous, trees such as oaks, maples, ashes, elms, etc. It does not necessarily refer to the hardness of the wood.
HARVEST. A general term for the removal of trees.
HEARTWOOD. The inner core of a woody stem, wholly composed of nonliving cells and usually differentiated from the outer enveloping layer (sapwood) by its darker color.
HEEL-IN. To store young trees prior to planting by placing them in a trench and covering the roots or rooting portions with soil.
HEIGHT, MERCHANTABLE. The height of a tree (or length of its trunk) up to which a particular product may be obtained. For example, if the minimum usable diameter of pulpwood sticks is 4 inches, the merchantable height of a straight pine tree would be its height up to a trunk diameter of 4 inches. Note, one must know the product being cut to estimate merchantable height.
HEIGHT, TOTAL. The height of a tree from the ground level to the top of its crown.
HERBACEOUS VEGETATION. The low-growing, non-woody plants in a forest understory, including wildflowers and ferns.
HERBICIDE. A chemical which kills herbaceous (non-woody) plants. In common usage, however, often used interchangeably with the words phytocide (plant killer) and silvicide (tree killer).
HERBIVORE. A plant eating animal.
HIBERNATION. A condition where an animal's metabolism is purposely slowed to endure prolonged periods of adverse environmental conditions, normally several months at a time.
HIGH GRADING. Removing the mature, high quality trees from a stand and leaving inferior species and defective trees. "Take the best and leave the rest." Generally regarded as a poor forestry practice.
HOME RANGE. The area which an animal uses during its normal activities, not to be confused with territory.
HORIZON, SOIL. A layer of soil approximately parallel to the land surface with more or less well-defined characteristics that have been produced through the operation of soilbuilding processes. 1) A-horizon — The upper horizon of the mineral soil, from which material has been removed by percolating waters. The horizon of eluviation. Commonly divided into a dark colored A1 horizon containing a relatively high content of organic matter, and light-colored A2 horizon of maximum leaching. 2) B-Horizon — The horizon of deposition, to which materials have been added by percolating waters, the horizon of illuviation. 3) C-Horizon — The weathered parent material.
HUMUS LAYER. The top portion of the soil which owes its characteristic features to its content of humus. The humus may be incorporated or unincorporated in the mineral soil.
HYPSOMETER. Any of several tools or instruments designed to measure the height of trees. The altimeter is such a tool.
IMPRINTING. A short-term rapid learning process early in life which is generally irreversible. More prevalent in precocial young.
IMPROVEMENT CUT. A cutting made in a stand past the sapling stage for the purpose of improving its composition and character, by removing trees of less desirable species, form and condition in the main crown canopy.

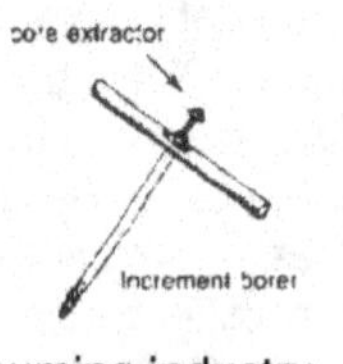

INCREMENT BORER. An auger-like instrument with a hollow bit, used to extract cores from trees for growth and age determination.
INDUSTRY FORESTER. A professional forester working for a wood consuming industry.
INSECTICIDE. Any chemical used to destroy insects and

other small invertebrates.

INSECTIVORE. An animal that eats insects.

INSTAR. A stage in the development of an insect between two successive molts.

INTEGRATED PEST MANAGEMENT (IPM). An ecological approach to pest management in which all available necessary techniques are consolidated into a unified program so that pest populations can be managed in such a manner that economic damaged is avoided and adverse side effects are minimized.

INTENSIVE FORESTRY. The practice of forestry with the objective of obtaining the maximum in volume and quality of products per unit of area through the application of the best techniques of silviculture and management.

INTERMEDIATE CROWN CLASS. Trees with crowns extending into the canopy with dominant and codominant trees. These trees receive little direct sunlight from above and none from the sides. Crowns generally are small and crowded on all sides.

INTERMEDIATE CUT. The removal of immature trees from the forest sometime between establishment and major harvest with the primary objective of improving the quality of the remaining forest stand. Contrast with harvest cut. An intermediate cut may generate income (commercial cutting) or, in some cases, may actually cost the forest landowner (a non-commercial cutting).

INTERMEDIATE CUTTINGS. Cuttings made in a stand between the time of its formation and its major harvest. Included are cleaning, liberation, weeding, release, thinning, improvement, salvage, and sanitation cuttings.

INTERNATIONAL RULE. One of several log rules designed to estimate the volume of lumber which may be sawn from a given log. See log rule.

INTERPLANT. To set young trees among existing forest growth of similar age and/or size, planted or natural, to bring the stand to a fully stocked condition.

INTOLERANCE. The characteristic of certain tree species which does not permit them to survive in the shade of other trees. Example: Aspen and Norway pine are intolerant; sugar maple and balsam fir are tolerant.

LANDING. An area where wood is concentrated in a harvest operation prior to hauling to the mill. Yard is synonymous.

LAYERING. Process of regenerating a tree by covering a lower branch with soil after which the branch develops roots and can stand alone as a new tree.

LEACHING. Downward movement of a pesticide or other soluble material through the soil as a result of water movement.

LEADER. A terminal leader is the uppermost branch or vertical tip of the tree. It eventually becomes the tree stem or trunk.

LITTER. The uppermost layer of the organic debris, composed of freshly fallen or slightly decomposed organic materials.

LODGED TREE. A tree that has not fallen to the ground after being partly or wholly separated from its slump or otherwise displaced from its natural position.

LOG. A piece of the woody stem of a tree. The trunk portion of a tree. A unit of measurement of a tree stem section exactly 16 feet in length.

LOGGER. An individual whose profession is cutting timber.

LOG RULE. A device, usually presented in tabular form, which expresses log volume content based on log diameter (inside bark of the small end) and length. A log rule expresses the volume of cut logs. A tree rule expresses the volume of standing trees.

LOP. 1) To chop branches, tops, or small trees after felling so that the slash will lie close to the ground. 2) To cut the limbs from a felled tree. Syn. Toplop; Limb.

LUMP-SUM SALE. A timber sale in which payment is based on the APPRAISED value of the tract, distinguished from a sale in which payment is based on the volume HARVESTED and SCALED.

MACHINE, PLANTING. Mechanical equipment which opens a hole or furrow and closes it again and firms the soil about a tree seedling which is usually inserted by hand.

MAIN STEM. The portion of a tree between ground level and the division into major branches, usually referred to as the bole.

MANAGEMENT PLAN. A written plan for the operation of a forest property using forestry principles. It usually records data and prescribes measures designed to provide for optimum use of all forest resources.

MARKING TIMBER. The process of indicating what trees are to be cut or otherwise treated. Prior to timber sales it is advisable to mark with paint each tree to be harvested. One spot of paint at eye level and one on the stump portion will help determine whether unmarked trees have been cut.

MAST. Nutlike fruits of trees, such as acorns, beech, and chestnuts. Mast is valuable as a source of food for many wildlife species.

MATURE TREE. A tree that has reached the desired size or age for its intended use. Size or age will vary considerably depending on the species and intended use.

MENSURATION, FOREST. A science dealing with the measurement of volume, growth and development of individual trees and stands and the determination of various products obtainable from them.

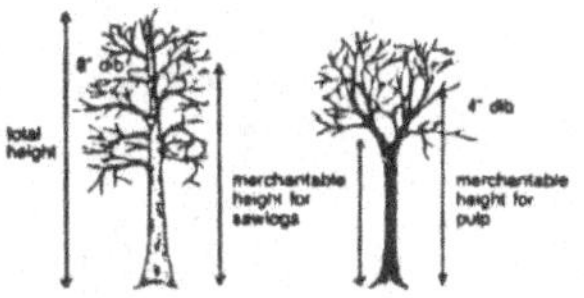

MERCHANTABLE HEIGHT. The point on a tree stem at which diameter limit requirements for a certain product are not met. Limits are: the point at which a sawlog tree is less than 8 inches in diameter, measured inside the bark (dib); a pulpwood tree less than 4 inches dib; or the point on any tree at which a defect is found that cannot be processed out.

MERCHANTABLE TIMBER. A tree or stand of trees which may be disposed of at a profit through conversion to salable products.

MILACRE. A sample plot of 1/1000 acre (usually 1/10 chain square) used in reproduction or vegetation surveys.

MIXED STAND. A stand in which less than 80 percent of the trees in the main crown canopy are of a single species.

MOLT. To shed the hair, outer skin, or feathers at certain intervals, to be soon replaced by a new growth.

MORTALITY. Death or destruction of forest trees as a result of competition, disease, insect damage, drought, wind, fire, and other factors.

MULTIPLE USE. Using and managing a forested area to provide more than one benefit simultaneously. Common uses may include, wildlife, timber, recreation, and water.

NON-COMMERCIAL CUTTING. A cutting which does not yield a net income, usually because the trees cut are too small, poor quality or not marketable.

NONINDUSTRIAL PRIVATE FORESTLAND (NIPF). Forest land owned by a private individual, group, or corporation not involved in wood processing.

NONSELECTIVE HERBICIDE. An herbicide which will kill or harm all or most plant species.

NOXIOUS WEED. A plant defined by law as being especially undesirable, troublesome, and difficult to control.

OMNIVORE. An animal that eats both plants and animals.

ORNITHOLOGY. The study of birds.

OVERBROWSING. Excessive use of browse usually found where there is an over-population of game. Similar to overgrazing, except that overgrazing refers to grasses and forbs, while overbrowsing refers to shrubs and trees.

OVERMATURE FOREST. A forest in which, as the result of age, growth has almost entirely ceased, and decay and deterioration has accelerated.

OVERSTOCKED. The situation in which trees are so closely spaced that they are competing for resources, resulting in less than full-growth potential for individual trees.

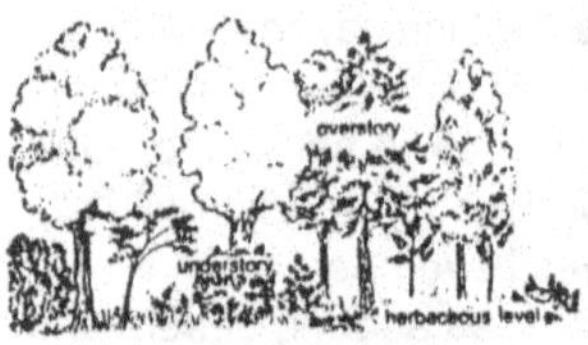

OVERSTORY. The canopy in a stand of trees. In contrast to the understory which is low growing woody or herbaceous vegetation forming a layer beneath the overstory.

OVERTOPPED CROWN CLASS. Trees with crowns entirely below the general level of the crown cover receiving no direct light either from above or from the sides. Syn. Suppressed.

PAIR BOND. The attachment that either of the mated pair has for the other.

PARTIAL CUT. A cutting by which only a part of the stand is removed. It usually implies a series of such cuttings.

PATHOGEN. A living organism capable of causing disease in a particular species or range of species.

PATHOLOGY, FOREST. The science that deals with diseases of forest trees or stands, and to the deterioration of forest products by organisms.

PHLOEM. The tissue in higher plants which transports organic nutrients manufactured in the leaves to other portions of the plant.

PHYTOTOXICITY. Injury to plants due to exposure to a chemical.

PICKAROON. A device with a head similar to an axe but with a point rather than a blade mounted on the end of a handle which is used to assist in the lifting and placement of bolts of wood.

PLANTATION. An artificially reforested area established by planting or direct seeding. Contrast with a natural forest stand which is established naturally.

PLANTING BAR. A hand tool used in making a slit-hole in which trees are planted.

PLANT PATHOLOGY. The science that deals with the nature and causes of plant disease.

PLOT. An area of land usually less than one acre on which trees and sometimes other vegetation are measured during a cruise (or inventory).

POLE. A young tree 4 inches or more in diameter breast high. The maximum size of poles is usually though not invariably taken to be some diameter breast high between 8 and 12 inches.

PRECOCIAL. Young born with eyes open, down or fur covered, and are quite mobile in the first day or two, e.g. ruffed grouse young.

PRECOMMERCIAL OPERATIONS. Cutting conducted in forest stands which removes wood of a size too small to be marketed. Such operations usually are designed to improve species composition and increase quality, growth, and vigor of the remaining trees.

PREDATOR. Any animal that kills and feeds on other animals.

PRESCRIBED BURNING. Skillful application of fire to natural fuels that will allow confinement of the fire to a predetermined area and at the same time will produce certain planned benefits.

PRESUPPRESSION, FIRE. Activities in advance of fire occurrence to insure effective suppression action. Includes recruiting and training, planning the organization, maintaining fire equipment and fire control improvements, and procuring equipment and supplies.

PREVENTION, FIRE. Activities directed at reducing the number of fires that start, including public education, law enforcement, personal contact and reduction of fuel hazards.

PRUNING. The removal of live or dead branches from standing trees. With forest trees, pruning is generally done along the trunk to remove the side branches (which cause knots in the wood) to produce a higher quality wood (knot-free).

PULPWOOD. Wood cut or prepared primarily for manufacture into wood pulp, for subsequent manufacture into paper, fiber board, or other products, depending largely on the species cut and the pulping process. Generally trees 5" to 9" DBH.

RANGE. The geographic area in which a tree species grows. Natural range is the entire geographic area where a species is known to occur under natural conditions; commercial range is the geographic area in which a species is harvested for commercial purposes.

RAPTOR. The birds of prey — including falcons, hawks, owls, eagles, and ospreys.

REFORESTATION. The natural or artificial restocking of an area with forest trees; most commonly used in reference to the latter.

REGENERATION CUT. A timber harvest designed to promote and enhance natural establishment of trees. Even-aged stands are perpetuated by seed tree, shelterwood, and clearcuts. Uneven-aged stands are perpetuated by selection of individual or small groups of trees.

RELEASE. To free trees from competition by cutting or otherwise removing or killing nearby vegetation and branches. Usually applied to young stands.

REPRODUCTION. The process by which the forest is replaced or renewed. This may be: Artificial Reproduction, by means of seeding or planting. Natural Reproduction, from natural seeding or sprouting.

RESIDUAL STAND. Trees remaining uncut following any cutting operation.

RESTRICTED-USE PESTICIDE. A pesticide which is designated as such by the Environmental Protection Agency because it is felt that it may generally cause, without additional regulatory restrictions, unreasonable adverse effects on the environment, including injury to the applicator. A "restricted-use" pesticide may be used only by, or under the direct supervision of, a certified applicator.

RICK. One-third of a standard cord. 37 cubic feet unsplit, 40 cubic feet split.

ROOTS. That portion of the tree which is generally underground and which functions in nutrient absorption, anchorage and storage of food and waste products.

ROT. A defect characterized by decay of wood in a standing tree or log.

ROTATION. The planned time interval between regeneration cuts in a forest stand.

ROTATION AGE. The age at which the stand is considered ready for harvesting under the adopted plan of management.

SALVAGE CUT. A harvest made to remove trees killed or damaged by fire, insects, fungi, or other harmful agents, to utilize available wood fiber before further deterioration occurs.

SANITATION CUT. A cutting made to remove trees killed or injured by fire, insects, fungi, or other harmful agencies (and sometimes trees susceptible to such injuries), for the purpose of preventing the spread of insects or disease.

SAP. The moisture in unseasoned wood and all that it holds in solution.

SAPLING. A tree at least 4.5 feet tall and up to 4 inches diameter.

SAWLOG. A log large enough to produce lumber or other products that can be sawed. Its size and quality vary with the utilization practices of the region.

SAWMILL. A plant at which logs are sawed into salable products. It includes all the machinery and buildings necessary for the operation of the plant.

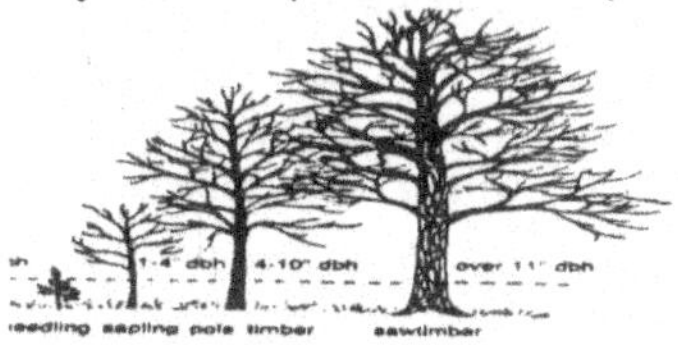

SAWTIMBER. Trees that yield logs suitable in size and quality for the production of lumber.

SCALING. Process of measuring wood products, usually pulpwood and sawlogs, after the trees are felled.

SCALE STICK. A flat stick, similar to a yardstick, which is calibrated so log volumes can be read directly when the stick is placed on the small end of the log of known length.

SCALPING. Removing a patch or strip of sod in preparation for planting trees.

SCARIFY. 1) To break up the forest floor and top soil preparatory to natural regeneration or direct seeding. 2) As applied to seed, to wear down by abrasion or by acid treatment an outer more or less impervious seed coat to facilitate or hasten germination.

SECOND GROWTH. Forests that originate naturally after removal of a previous stand as by cutting, fire, or other cause. A loosely used term for young stands.

SEEDBED. 1) In natural regeneration, the soil or forest floor on which seed falls. 2) In nursery practice a prepared area in which seed is sown.

SEEDING. A method of establishing a forest artificially by sowing seed. In broadcast seeding seed is sown over the entire area. Partial seeding may be done in strips, furrows or trenches, plots or spots.

Seed trees

SEED TREE. Any tree which bears seed; specifically, a tree left to provide the seed for natural reproduction. Syn. Mother tree.

Selection cut

SELECTION CUT. A regeneration cut designed to create and perpetuate an uneven-aged stand. Trees may be removed singly or in small groups. A well designed selection cut removes trees of lesser quality and trees in all diameter classes along with merchantable and mature high-quality sawlog trees.

SELECTIVE HERBICIDE. An herbicide which is effective only against certain species and is able to control unwanted plants without serious injury to desirable species.

SHADE TOLERANCE. Relative ability of a tree species to reproduce and grow under shade. Tree species are usually classified in descending order of shade tolerance as: very tolerant, tolerant, intermediate, intolerant, and very intolerant.

SHELTERBELT. A wind barrier of living trees and shrubs maintained for the purpose of protecting farm fields. As applied to individual farmsteads, termed "Windbreak". Syn. Belt.

Shelterwood

SHELTERWOOD HARVEST CUTTING. A harvest cutting in which trees on the harvest area are removed in a series of two or more cuttings to allow the establishment and early growth of new seedlings under partial shade and protection of older trees. Produces an even-aged forest.

SHRUB. A low-growing perennial plant with a persistent woody stem and low branching habit.

SILVICULTURE. The art of producing and tending a forest; the application of the knowledge of silvics in the treatment of a forest; the theory and practice of controlling forest establishment, composition, and growth.

SITE. An area evaluated as to its capacity to produce a particular forest or other vegetation based on the combination of biological, climatic and soil factors present.

SITE INDEX. An expression of forest site quality based on the expected height of dominant trees at a specified age (usually 50 years in the eastern United States).

SITE PREPARATION. Treatment of a site with mechanical clearing, burning, or herbicides, to prepare a site for planting.

SKIDDING. The act of moving trees from the site of felling to a loading area or landing. Skidding may be accomplished by tractors, horses, or specialized logging equipment. The method of skidding can greatly affect the impact of logging on soil and the residual stand.

SKID ROAD. A road or trail leading from the stump to the skidway or landing.

SLASH. Debris left after logging, pruning, thinning, or brush cutting; also, large accumulation of debris after wind or fire. It includes logs, chunks, bark, branches, stumps and broken understory trees or brush.

SLASH DISPOSAL. Treatment of slash to reduce the fire hazard or for other purposes.

SLOPE. (See grade). A term of measurement in percent and means the increase in height over the distance measured. An increase of 1 foot over a distance of 5 feet is expressed as a 20 percent slope.

SNAG. A standing dead tree used by many species of birds and mammals for feeding and nesting.

SOFTWOOD. Generally, one of the botanical groups of trees that in most cases have needle or scale-like leaves; the conifers; also, the wood produced by such trees.

SOIL. The top layer of the earth's surface, composed of finely divided disintegrated rock containing more or less organic material, which is penetrated by the roots of plants. It includes the surface soil (horizon A), the subsoil (horizon B) and the upper portion of the substratum (horizon C) to the extent that it is penetrated by plant and tree roots. The average soil is composed of 45% mineral, 25% air, 25% water and 5% vegetation.

SOIL REACTION. The degree of acidity or alkalinity of the soil mass expressed in pH values or in words as follows: extremely acid, below 4.5; very strongly acid, 4.5-5.0; strongly acid, 5.1-5.5; medium acid, 5.6-6.0; slightly acid, 6.1-6.5; neutral, 6.6-.7.3 (strictly 7.0): mildly alkaline, 7.4-8.0; strongly alkaline, 8.1-9.0; very strongly alkaline, over 9.1.

SOIL TEXTURE. The feel or composition of a soil based on the proportion of sand, silt, and clay in the soil.

SOLD AS APPRAISED SALES. Wood is sold "on the stump" and the sale price is based on the appraised volume determined by the forester. This volume is only an estimate.

SPECIES COMPOSITION. The mix of tree species occurring together in the same stand.

SPOT FIRE. Fire set outside the perimeter of the main fire by flying sparks or embers.

SPROUT. A tree that grows from the stump or sucker root of a parent tree: it is not of seed origin. Basswood is frequently of sprout origin.

STAND. A group of trees occupying a given area and sufficiently uniform in species composition, age and condition so as to be distinguishable from the forest on adjoining areas. A forest stand is said to be "pure" if 80 percent or more of the trees present are of the same species. If less than 80 percent of all trees present are of the same species, the stand is said to be "mixed".

STAND. A part of the forest that, due to it's age, species composition, and other conditions, is identifiably different from its surroundings. A forest is comprised of many stands. Example: A "pure stand" of aspen may be comprised of 80% aspen, where all aspen trees were about 40 years old. This stand would be called "even-aged". A "mixed stand" of northern hardwoods may be comprised of sugar maple, basswood, and other species of various ages. This stand would be called "uneven-aged". A stand is sometimes called a "type".

STAND DENSITY. The quantity of trees per unit area. Density usually is evaluated in terms of basal area or percent-crown cover. See basal area, crown cover, stocking.

STEM. The portion of a tree that supports the branches; also called the bole.

STICK, BILTMORE. A rule graduated in such a way that the diameter of standing tree may be estimated when the stick is held tangent to the surface at right angles to the main axis of the tree, and at a distance from the eye for which the stick is graduated.

STICK, SCALE. A graduated stick for measuring the diameters and contents of logs; both measures are stamped on the stick.

STOCKING. An indication of the number of trees in a stand as compared to the desirable number for best growth and management, such as well-stocked, overstocked, partially stocked.

STOMATA. Minute openings on the surfaces of leaves and stems through which gases (e.g. oxygen, carbon dioxide, water vapor) and some dissolved materials pass into and out of plants.

STUMPAGE. The value of a tree of a stand as it stands in the woods uncut (on the stump).

STUMPAGE. Uncut trees standing in the forest. Sometimes used to mean the commercial value of standing trees.

SUCCESSION. The gradual replacement of one plant community by another.

SUCKER. Synonymous with sprout.

SUGAR BUSH. A stand mostly of sugar maple which is currently used for gathering sap for the production of maple syrup.

SUPPRESSED. The condition of a tree characterized by low growth rate and low vigor due to competition with overtopping trees. See overtopped.

SUPPRESSION, FIRE. All the work of extinguishing or confining a fire beginning with its discovery.

SUSTAINED YIELD. An ideal forest management objective at which point the volume of wood removed is equal to growth within the total forest.

SWEEP. Tree defect resulting from a gradual curve in the main stem of the tree.

TALLY. A system of recording trees counted during a timber cruise. The commonly used dot-tally system is illustrated.

TAPROOT. The main root of a tree which strikes downward with or without heavy branching until it either reaches an impenetrable layer or one so lacking in oxygen or moisture that further downward growth is impossible.

TERRITORY. The area which an animal defends, usually during breeding season, against intruders of its own species. Territories are smaller and are normally located within the animal's home range.

THINNING. Removal of trees in an overstocked stand to give the remaining trees adequate room for growth.

THREATENED SPECIES. Species that could become endangered in the forseeable future.

TIMBER. Standing trees, usually of commercial size.

TIMBER INVENTORY. A collection of information about a timber stand made by measuring tree and stand characteristics such as tree volume and grade and stand density.

TSI (TIMBER STAND IMPROVEMENT). A practice in which the quality of a residual forest stand is improved by removing less desirable trees, vines and, occasionally, large shrubs to achieve the desired stocking of the best quality trees.

TOLERANCE. The capacity of a tree to develop and grow in the shade of and in competition with other trees. Trees able to grow in full or partial shade are considered "tolerant". Trees requiring full sunlight for survival are considered "intolerant".

TRANSPLANT. A tree which has been removed from its original seedbed and replanted one or more times in a nursery.

TREE. A woody plant having a well-defined stem, more or less definitely formed crown and usually attaining a height of at least 10 feet.

TREE FARM. A privately owned forest (woodland) dedicated to the production of timber crops. Additionally, it may be recognized as a "Tree Farm" by the Tree Farm Program, an organization sponsored by the American Forest Industries.

TREE INJECTOR. Equipment specially designed to inject chemicals, usually phytocides, into the trunk of a tree.

TRIM ALLOWANCE. Excess length of a log to allow for square trimming the lumber to an exact length. Syn. Overlength.

TRUNK. Main stem or bole of a tree.

UNDERCUT. 1) In logging, the notch cut in a tree to govern the direction in which the tree is to fall and to prevent splitting. 2) In forest management, the harvesting of a quantity of timber less than the budgeted cut.

UNDERPLANT. To set out young trees or sow seed under an existing stand.

UNDERSTOCKED. A stand of trees so widely spaced that, even with full growth potential realized, crown closure will not occur. Understocking indicates a waste of resources, as the site is not fully occupied.

UNDERSTORY. The lesser vegetation (shrubs, seedlings, saplings, small trees) within a forest stand which forms a layer between the overstory and the herbaceous plants of the forest floor.

UNEVEN-AGED STAND. A group of trees of a variety of ages and sizes growing together on a uniform site.

VENEER. Thin sheets of wood (usually less than 1/4" thick) produced by slicing or peeling a log.

VENEER LOG. A log of high quality and desirable species suitable for conversion to veneer. Logs must be large, straight, of minimum taper, and free from defects.

VIRGIN FOREST. A wooded area with old-growth trees which never has been harvested or altered by humans.

VOLUME. The amount of wood in a tree or stand according to some unit of measurement, (board feet, cubic feet, etc.) or some standard of use (pulpwood, sawtimber, etc.)

VOLUME TABLE. A table of figures used to estimate the volume of wood contained in a standing tree, based on dbh and merchantable height.

WATERSHED. A region or area defined by patterns of stream drainage. A watershed includes all the land from which a particular stream or river is supplied.

WATER TABLE. The highest point in a soil profile where water saturates the soil on a seasonal or permanent basis.

WEED. An unwanted plant.

WELL STOCKED. The situation in which a forest stand contains trees spaced widely enough to prevent competition yet closely enough to utilize the entire site.

WHORL. Two to 10 or more branches growing in a ring at a node, surrounding the central leader or stem.

WILDFIRE. 1) An unplanned fire requiring suppression action, as contrasted with a prescribed fire burning within prepared lines enclosing a designated area, under prescribed conditions. 2) A free-burning fire unaffected by fire suppression measures.

WILDLIFE HABITAT. The native environment of an animal, ideally providing all elements required for life and growth: food, water, cover, and space.

WINDBREAK. A wind barrier of living trees and shrubs maintained for the purpose of protecting the farm home, other buildings, garden, orchard or feedlots. See Shelterbelt. Syn. Farmstead windbreak; Farmstead plant.

WINDTHROW. A tree pushed over by wind. Windthrows (blowdowns) are more common among shallow-rooted species and in areas where cutting has reduced the density of a stand so that individual trees remain unprotected from the force of the wind.

WOLF TREE. A tree which occupies more space in the forest than its value justifies. Usually a tree which is older, larger or more branchy than other trees in the stand.

WOODY PLANTS. Plants which live longer than two years and have a thick, tough stem or trunk covered with a layer of cork.

WOOD PULP. Mechanically ground or chemically digested wood (composed primarily of wood fiber) which is used in the manufacture of paper, fiberboard, etc.

XYLEM. The tissue in higher plants which transports water, dissolved salts, and other materials (e.g. pesticides) from the roots to aerial portions of the plant.

ZOOLOGY. The study and classification of animals and animal life.

Minnesota's Forest Treasures

Funding and Technical Support:

Minnesota Department of Natural Resources
Division of Forestry

Minnesota Department of Education
Environmental Education Unit

Minnesota Forest Industries

Minnesota Forestry Association

Illustrations by Amy Beyer

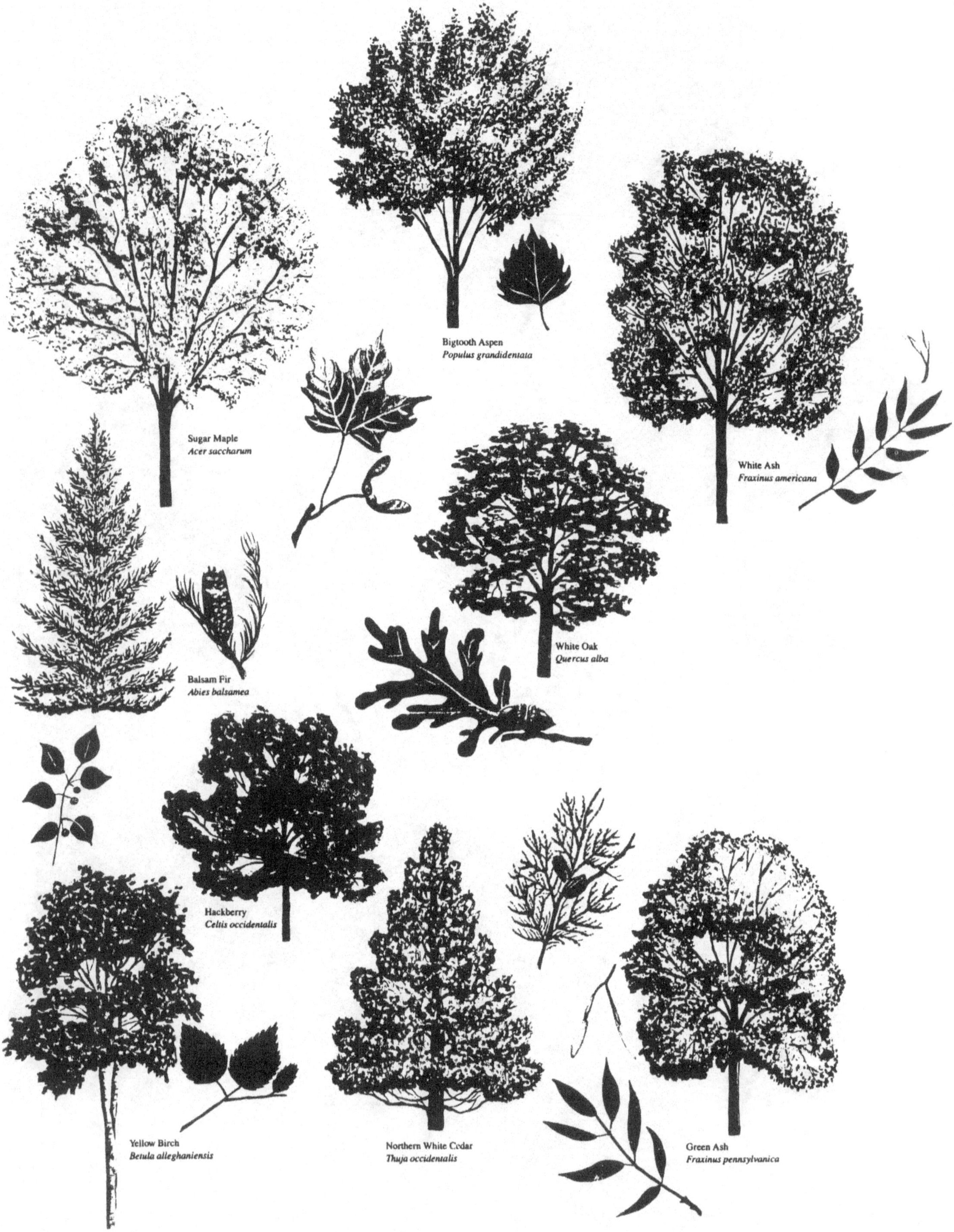

Bigtooth Aspen
Populus grandidentata
Sugar Maple
Acer saccharum
White Ash
Fraxinus americana
White Oak
Quercus alba
Balsam Fir
Abies balsamea
Hackberry
Celtis occidentalis
Yellow Birch
Betula alleghaniensis
Northern White Cedar
Thuja occidentalis
Green Ash
Fraxinus pennsylvanica

Eastern White Pine
Pinus strobus
Black Ash
Fraxinus nigra
White Spruce
Picea glauca
American Elm
Ulmus americana
Silver Maple
Acer saccharinum
Eastern Larch (Tamarack)
Larix laricina
Basswood (American Linden)
Tilia americana
Butternut
Juglans cinerea

INDEX

D

E

F

G

H

I

J

K

L

M

N

O

P

Q

R

S

T

U

V

W

Z

www.ingramcontent.com/pod-product-compliance
Lightning Source LLC
Chambersburg PA
CBHW060504300426
44112CB00017B/2546
9780913163368